Environmental Contaminants

Environmental Contaminants: Assessment and Control

Daniel A. Vallero, Ph.D.

ELSEVIER
ACADEMIC
PRESS

AMSTERDAM • BOSTON • HEIDEIBERG • LONDON
NEW YORK • OXFORD • PARIS • SAN DIEGO
SAN FRANCISCO • SINGAPORE • SYDNEY • TOKYO

Elsevier Academic Press
200 Wheeler Road, 6th Floor, Burlington, MA 01803, USA
525 B Street, Suite 1900, San Diego, California 92101-4495, USA
84 Theobald's Road, London WC1X 8RR, UK

This book is printed on acid-free paper. ∞

Library of Congress Cataloging-in-Publication Data
Application submitted.

British Library Cataloguing in Publication Data
A catalogue record for this book is available from the British Library

ISBN: 0-12-710057-1

For all information on all Academic Press publications visit our Web site at www.academicpress.com

Printed in the United States of America
04 05 06 07 08 09 9 8 7 6 5 4 3 2 1

To our children's children's children. . . . With thanks to the Moody Blues.

Contents

Preface

Why a Book on Environmental Contaminants

My principal objective in writing this book is to help the environmental professional, professor, student, and citizen to apply the science, engineering, and technology for assessing environmental risks and cleaning up environmental problems in air, water, soil, sediment, and living systems. I do so by introducing a key topic related to environmental risk assessment or methods to control or reduce risks and, when appropriate, follow it with examples of problems and solutions. Each solution includes a discussion of the basic and applied sciences as well as other considerations, such as when the equations and applied principles may not work, where uncertainties may exist, and how these applications may or may not work in the "real world."

Straightforward examples and quantitative demonstrations are important tools for explaining environmental phenomena. I recently conducted an unscientific study of introductory environmental texts, such as those used to teach courses like Introduction to Environmental Studies, Problems in Environmental Biology, or even Environmental Science, and found many to be almost devoid of problems. At the other end, texts on environmental hydrogeology and geophysics, biotransformation, and environmental general or organic chemistry were laden with theory and seemingly endless derivations of equations and formulae, making them excellent textbooks for a course mentored by a seasoned professor, but rendering them less than completely useful to the practicing environmental professional. I fear (and recall as a student, I must confess) that much of the information in these courses is lost within weeks—or even hours—after the final exam. Of course there are exceptions, such as the fine and very useful text, *Chemical Fate and Transport in the Environment* by Harold F. Hemond and Elizabeth J. Fechner-Levy (Second Edition, 2000, Academic Press, San Diego, Calif.), and the standard of many environmental engineering and science courses, *Environmental Chemistry* by Stanley E. Manahan (Sixth Edition, 1994,

Lewis Publishers, Boca Raton, Fla.). There are also excellent manuals and handbooks of science and engineering, which are quite strong on reference material, equations, and quantitative methods but that are not designed to describe the environmental systems in detail. Such references, which are published by both the private and public sectors, are important sources for risk assessment, environmental sampling and analysis, exposure assessment, transport and fate calculations, and engineering. However, they were never intended to be a "good read"! That said, I am striving to strike a balance between the description of environmental systems and a rigorous scientific and engineering approach. I believe that a text should be useful beyond the classroom, providing feasible approaches to deal with contaminants.

My discussions of feasibility vary, depending on the topic. When appropriate, I discuss technical limitations, such as the differences between the controlled environment in the laboratory and the heterogeneous conditions in the industrial setting or ambient environment. Practical solutions will also include how the scientific and engineering fundamentals need to be tempered with reality, such as some of the lessons learned in how to communicate risks and how environmental managers must evaluate the scientific soundness within the comprehensive decision-making processes dictated by policy, political, regulatory, economic, and social milieus.

Many readers may benefit from a book that bridges quantitative environmental science with methods for practical applications:

1. Faculty in risk assessment, environmental sciences, and environmental engineering departments
2. Faculty in chemistry, chemical engineering, environmental toxicology, soil sciences, ecology, and geosciences
3. Environmental professionals practicing in the fields of engineering, research, environmental audit, emergency response, community-based initiatives, and risk assessment
4. Graduate students enrolled in risk assessment, waste management, contaminant transport, environmental hydrogeology, and environmental engineering courses
5. Senior and upper-level undergraduates enrolled in environmental engineering and environmental chemistry, physics, and biology courses
6. Professionals and students preparing for professional examinations, such as Professional Engineer's (PE) and Future Engineer (FE) exams, especially the environmental engineering, chemistry, fluid dynamics, and engineering ethics sections of the exams
7. Interested members of the public who need to learn more about environmental sciences

All but the last of these groups need an understanding of the first principles of science as they apply to the environment, and they need to know how and when to apply these principles to make good environmental decisions. The groups vary only in the extent and type of applications of the sciences. Obviously, the faculty in the basic sciences are well aware of the principles but may be less certain about when it is appropriate to apply them in their teaching and research. The faculty in engineering and the applied sciences, conversely, may want a reference to remind them which principles underlie their practice. The professionals in the field are looking for ways to address problems that are at their doorstep. Their need may be more for deployment with understanding than enhancing knowledge for the sake of knowledge (although I know of many practitioners whose thirst for wisdom is every bit as strong as that of my academic colleagues). And, the students are moving from the basic sciences to an increasingly greater focus on their chosen area of environmental science and engineering.

The seventh group, the public audience, is the more eclectic readership. It includes individuals ranging from highly motivated and technically trained individuals (medical doctors, engineers from other disciplines, attorneys, etc.) who may have only recently become interested in the specifics of environmental science. The precipitating event may be the potential location of some environmentally threatening facility in their neighborhood (a landfill, a road, a power plant) or the peril of an important resource due to development (a coastline, a wetland, a historic treasure). The group also includes people who would never classify themselves as "techies" but who share the same risks as their more technical citizenry. They need to know why we environmental "experts" are making decisions that could threaten the things that they hold dear. This last group has gotten more attention recently, including the "environmental justice" communities, where it has been found that minority and lower socioeconomic status neighborhoods are more likely to have environmental degradation and more likely to have waste disposal landfills and other perceived and real environmental menaces sited in their communities.[1] This book, then, is also a resource for neighborhood groups and individuals who may want to "do the math" themselves on whether the specific environmental decisions that will affect them are sound and fair.

Structure and Emphasis

Since the major motivation of this book is scientific and quantitative, I do not spend a large amount of time dealing with history and the more general social context of environmental programs. The book introduces the major policy and legislative history, but more importantly, it points out policy implications in the discussions of specific problems, even those that seem to be purely scientific. For example, after I present the problem of how to line a landfill, I point out potential implementation and policy issues, such as the rationale for and the adequacy of the "mismanagement" scenario assumptions of the federally mandated Toxicity Characteristic Leaching Procedure (TCLP) for pollutants. Herein lies an example of a very well-defined technical method that incorporates environmental management to achieve results.

The book has four parts. The first gives the policy context for environmental risk assessment and introduces the reader to how environmental science and engineering are used in decision making. The second introduces the reader to the fundamental principles underlying environmental assessment and response actions. The third introduces risk assessment and environmental toxicology, with guidance on how hazards and dose responses are determined, how exposures can be estimated, and techniques for calculating risks under a number of realistic scenarios. This part closely follows the companion book, *Engineering the Risks of Hazardous Wastes* (Elsevier, 2003), which addressed only hazardous wastes, principally targeting engineers and engineering students.

The book's fourth part shows how the practitioner can put these fundamental principles to work to clean up environmental problems, and introduces environmental management considerations, and the expectations of environmental professionals. The discussions are heavily annotated with example problems, sidebars, and case study discussion boxes. Numerous illustrations and case studies address a number of the myriad ethical, environmental management, and professional issues that are confronting the environmental community today.

Notation and terms are introduced and defined within the contexts of the discussion of contaminant behavior. These and other environmental terms are defined in the Glossary.

This text subscribes to the axiom that "everything matters in environmental science." It frequently holds that the only correct answer to almost any question involving the environment is, "it depends." However, this is not an invitation for arbitrary and unscientific solutions, nor does it obviate the need for sound solutions based upon strong quantitative methods in environmental science and engineering. In fact, it increases the need for them. Complex problems with social import depend on the applications of the best and the right science. I hope this book enhances readers' appreciation for the intricate nature of environmental problems and increases their confidence that the solutions adhere to strong scientific principles.

Strengths and Weaknesses, Realities and Perceptions, Myths and History

One of the major challenges of environmental science and engineering is that in our professions we must always be mindful of the quantifiable and the subjective elements needed to make environmentally sound decisions.

The "environment" is simultaneously the object of rigorous scientific inquiry and a matter of great social import. Although we must be ever mindful that our science needs to be "relevant," this duality has, on occasion, led some to perceive the environmental fields as "soft" when compared to our sister scientific disciplines in chemistry and physics, and to other engineering disciplines such as structural and chemical engineering.

There is validity in these arguments and elements of truth to these perceptions. Our fields have been co-opted by so-called "environmentalists" who may lack the ability or will to require scientific rigor underlying their conclusions (see Discussion Box "Issues in Environmental Science: Real versus Junk Science"). Environmentalists and other advocacy groups are motivated to protect the environment, but they may or may not be motivated by technically strong and reproducible scientific evidence. "No wetlands, no seafood" bumper stickers express this sentiment, but they are not necessarily well-thought-out conclusions based upon the first principles of science and the application of the scientific method. That is, even with no wetlands we would have seafood, but the diversity would be greatly diminished and we humans (and other organisms) may not enjoy eating many of the remaining species. However, to get the attention of the public and politicians, environmentalists have felt the need to dramatize many issues "beyond the supporting data." Our wetland/seafood relationship is not necessarily "wrong" per se; it is simply an overextension of what is known. Unfortunately, a more correct and scientifically supportable slogan would lose its political punch and effectiveness as mass communication. You are unlikely to see T-shirts and bumpers emblazoned with "Reduced productivity and surface area of wetlands is leading to reduced diversity of aquatic species, including those with high economic value." Likewise, arguments for and against oil exploration in Alaska are seldom made purely from positions of science and often are made from ideological perspectives. Extreme positions, not reasoned arguments, often carry the day in the political realm.

Issues in Environmental Science: Real versus Junk Science

The Royal Society of London of the seventeenth century, led by none other than Robert Boyle, the English chemist, is often credited with codifying the "scientific method" that still exists today. Boyle argued that the three requirements for acceptable science were that objectivity (in the form of an experiment) was paramount, that a "literary technology" (publication of methods and results) must be maintained, and that the research had to be witnessed by objective scientists (peer review).

These elements still provide the metric of whether scientific methods and results are deemed credible by the scientific community.

Unfortunately, at times things get out of kilter. The peer review may be weak or lacking, or may be too late in the process (e.g., results are released to the public before sufficient review by the scientific community). The results, while being credible insofar as the research was designed, may be "overextended" by the researchers or others who have a particular agenda (e.g., a pesticide may be found to move from the soil to the air under certain conditions in a laboratory chamber, but the results are "overgeneralized" by an antipesticide group to conclude that people will be breathing greater amounts of the pesticide. They ignore the special conditions in the laboratory that showed the pesticide transport and the caveats recommended by the researchers).

There is a great debate between people who want to push agendas and those who want to be careful about releasing the findings. The first group is sometimes accused of being "sloppy" or, in the worst cases, "deceitful," while the second group is accused of being "overly careful," or "myopic," or at worst "fearful." Many (perhaps most) scientists and engineers avoid the policy arena for these reasons, but it is important to be aware that whatever information is given, even when carefully gathered and shared with the public, there is always the possibility that it will be misused or even abused.

Let us consider the "White Paper on Potential Developmental Effects of Atrazine on Amphibians"[ii] recently released by a United States government science advisory panel. Much concern has risen about amphibians in the popular and professional presses, with quite a few fingers being pointed at the widely used herbicide atrazine ($C_8H_{14}C_{15}N_5$). Atrazine has been found in surface water at concentrations of 20 to 40 parts per billion (ppb). Atrazine is used throughout the United States to control weeds on farms. It is moderately volatile and soluble in water and resists breakdown by microbes. Its physicochemical properties and widespread application as an herbicide have led to concerns about possible risks to aquatic organisms. Pictures and stories about two-headed and other deformed frogs have been prominent (see Figure P.1). However, the white paper suggested that none of the scientifically credible studies supports the "overall weight-of-evidence" regarding "whether or not atrazine exposure adversely affects amphibian development." In other words, the scientific community isn't sure, but what is currently known *does not* link atrazine consistently with the problems in amphibians. This is not the same as a "clean bill of health," but it is a cautionary report on the need to be careful about linking cause with effect in terms of pesticide exposure and developmental disorders in amphibians.

On the other side of the ledger, the white paper points out that even though "the weight-of-evidence does not show that atrazine produces a consistent, reproducible effect, both laboratory and field

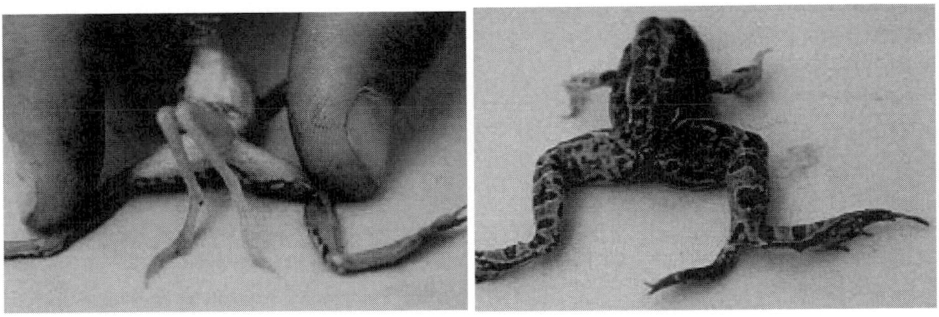

FIGURE P.1. Frog with extra legs. (Source: 1997, *Environmental Health Perspectives*, Vol. 105, No. 10, October.)

studies provide evidence that atrazine exposure may be associated with effects on gonadal development and secondary sexual characteristics" (page 75).[2] The reason that results are not reproducible across studies might be attributable "to an inconsistency in the methods used by the various research teams, and the absence of a dose-response at this point do not refute the hypothesis that atrazine exposure may result in gonadal developmental effects in amphibians" (page 75).[2] In other words, different approaches and different research objectives do not in themselves mean "inconsistency."

In short, a lot more work needs to be done. Unfortunately, groups on both sides of the argument have run with premature conclusions and selective use of findings to support their cause. This is known as "junk science." Even good science (well designed and adhering to the scientific method) can become junk science in the wrong hands. For example, a University of California study has stated that atrazine concentrations as low as 0.1 ppb caused either multiple or both male and female sex organs to develop in male frogs.

Two problems with these particular studies have been noted. The first is that more developmental disorders in frogs were observed at low doses than at high doses. This violates one of the tests for causality: according to the "biological gradient," the higher the dose the greater the effect. Although this does not always hold (e.g., at very high doses of a carcinogen, tumors may not form because the dose is so high that it kills the cell), it is very unlikely and points to the possibility that some other factors may be causing the problem. The second problem is that atrazine has been applied for decades but only recently has been linked to amphibian endocrine effects. This may be because people hadn't noticed the problems, or began to notice the deformities only recently and felt the need to report them due to the heightened awareness of the problem. It may also be due to the exposure of frogs to chemical transformation products or a synergy of atrazine with newly used products. The white paper gave little credence to these possibilities; only recently did various government agencies begin to postulate that certain biological mechanisms could trigger the deformities, and so they began to identify possible environmental causes and the need for research. The National Reporting Center for Amphibian Malformities (NARCAM) has been established at the Northern Prairie Wildlife Research Center in North Dakota to document observed deformities.

The National Institute for Environmental Health Sciences (NIEHS)[3] has evaluated possible explanations for the deformities, including chemical contamination, greater exposure to ultraviolet (UV) radiation as a result of ozone depletion, parasitic infestation, or even combinations of these and other unknown factors. Frogs spend most of their time in surface water, so they may be subjected to chemical contamination. A number of the places where deformed frogs are found are agricultural, so pesticides cannot be ruled out, especially when synergies between the chemical exposures and other factors are taken into account.

So you decide. Are the associations of amphibian deformities with pesticide exposures based on real or junk science?

Even practitioners in certain environmental fields avoid the quantitative aspects of our professions. I have recently reviewed some undergraduate texts in environmental science and found that most of them deal with concepts and content. However, the students are seldom challenged to apply the concepts and information with rigorous problem solving. Most of the material is presented for its importance in terms of policy and doing the right thing, but in the rare instances where equations are presented, they are very simplistic and are not applied to actual conditions. There is even a textbook that boasts of being math-free! This is difficult to understand, since mathematics is the "language" of science. How can one sufficiently address scientific subject matter without using its language?[4] For example, there may be a discussion of diversity and why it is important in biological systems, or of the field of human toxicology, but there is no mention of the Shannon Diversity Index or Lifetime Average Daily Dose, which are expressions of diversity in ecosystems and exposures in human populations, respectively (these terms are discussed at length in Chapter 9). And if such metrics are mentioned, the text lacks example problems or study questions on how to quantify diversity in ecosystems or risk and exposure in human populations. Hence, if our "best and brightest" are not exposed to rigorous quantitative environmental science during their undergraduate experiences, how can we expect them to apply these tools in professional life?

Environmental engineering, as a discipline within civil or chemical engineering in most universities, does exhort students to build a solid analytical and quantitative tool kit in preparation for professional life. Again, many aspects of other engineering fields are enviable compared to environmental engineering. Uncertainty is problematic for structural engineers and chemical engineers, but I would argue that the number of variables for building design and chemical reactors are far fewer than those confronting the environmental engineer. For one thing, we deal with living systems. Biology is much "messier" than chemistry or physics. One can even argue that biology is really a complicated set of functions *of* chemistry and physics; biology is a set of "second principles" of science. If so, environmental engineering must then be a set of "third principles" of science, since what we do is put the first and second principles to work. In this way, environmental engineering has more than its share of ill-posed and chaotic problems. Even if we observe the occasional linear relationship in the environment, there is always reason to think that it will change if we watch it long enough!

Prologue: The Challenge

Since stepping back and considering the environmental concepts from a fundamental standpoint is a major tool used in this text, let us consider an example to demonstrate what seems to be a simple, straightforward finding in the laboratory that, upon further investigation, quickly becomes very complicated in the real world. I asked three of the smartest and most well-respected environmental engineers in the world the following question:

RE: Technical Question

From: Dan Vallero
To: Environmental Expert

I need your advice on a technical question. I know you're (busy, retired, etc.) but whom better to ask?

I have an applied biodegradation question for you. It kind of demonstrates the difference between basic science and engineering.

Empirically, it has been found that microbes degrade benzene as:

$$C_6H_6 + 7.5O_2 \rightarrow 6H_2O + 6CO_2 + \text{microbial biomass} \qquad \text{Reaction P–1}$$

Looking only at stoichiometry, it appears that to destroy 1 g of benzene this way, all you need is a bit more than 3 g of oxygen.

This is obviously too good to be true in real life because otherwise all I'd have to do is pump 15 g of air (since it's about 20% O_2) for each mole of benzene into a contaminated aquifer and all my benzene is gone.

Here's my question: Is there a factor that takes into account the mass transfer, etc. of oxygen into a biosolid or film around a particle? For example, I know that molecular diffusion is not very important unless you have almost quiescent conditions. If there is such a factor, where can I find this? In other words, I can do the chemistry, but it assumes that all of the oxygen is available and being used by the bugs. What fraction is REALLY available and being used by the bugs?

This is complicated even further with more complex organic molecules and mixtures (e.g., solubility of the mixture and as the contaminant concentration increases, we may not be able to assume that water is the only solvent!).

Thanks,
Dan

As a reminder, the three people I sent this problem to are all in the same field (environmental engineering), and I know that they have worked together on research and applied biological treatment methods. I should also remind the reader that I have been in the field for nearly three decades, and I believe there should be a nice, neat answer, readily available, without consulting these mentors!

The range of responses is revealing:[5]

Expert Response A

Benzene can be easily metabolized in a completely mixed activated sludge system. The basic problem with benzene is its toxicity. It is essential that the bacteria metabolize the benzene before the benzene reacts with the bacteria. The difficulty with subsoil treatment of benzene lies in the relatively small number of bacteria. Subsoil metabolism of benzene occurs at the edge of the contamination and moves slowly inward. While oxygen is important, nitrogen, phosphorus, and trace metals are essential for metabolism. As the bacteria grow, they tend to fill the void spaces. Water is critical for bacteria metabolism. Without water, the bacteria cannot survive.

Another problem with subsurface metabolism is the lack of mixing of the bacteria, the substrate, nutrients, and trace metals. Without complete mixing of all materials, metabolism does not occur. I believe the lack of effective mixing in subsurface contaminants is the primary reason we do not see better metabolism. Adding soluble nutrients into water pumped into the ground results in the contaminants being pushed rather than mixed.

In activated sludge the oxygen concentration inside the bacteria is a function of the DO concentration in the water. Oxygen is poorly soluble in water. DO saturation at 20°C is only 9.1 mg L^{-1} with air or around 42 mg L^{-1} with pure oxygen gas.[6] In an excess of DO, the rate of metabolism is a function of the bacteria mass and the rate at which bacteria can process the nutrients. The maximum rate of metabolism with mixed bacteria populations at 20°C is about 2.8 hrs to double the cell mass. As long as the DO is about 0.1 to 0.2 mg L^{-1}, oxygen will not limit metabolism.

Expert Response B

I don't know the answer, but I can make some observations. The situation is similar to many others where you are asking certain molecules, chemicals, and/or organisms to react. In theory, you will get every single organism/chemical to react with every single organism/chemical you want them to react with, but the probability of the last two (unloved) organisms/molecules finding each other is essentially zero.

It is similar to the case of recycling—finding that last glass bottle so you can get 100% recycling. You will spend most of your energy and funds finding the last of the bottles. Likewise, you will spend enormous energy/effort/time matching up the last benzene molecule with the last oxygen molecule. So you have to have O_2 in excess in order to have the reaction take place in a reasonable time.

I don't think it is a problem of diffusion through membranes or any other step within the decomposition phase that might be rate limiting. I think it is the physical transfer of the oxygen to the benzene that is the problem.

I could be all wet. As someone once commented about me, "Often wrong, but never in doubt."

Expert Response C

Great question!

As always, "it depends" (. . . on a lot of factors and there is no answer to your question). Let's write a research proposal and search for some answers!

The experts' answers ranged from the logic of first principles, to the differences between theory and practice, to the quest for better science. Who is right and who is wrong? I contend that all three are right! Environmental science is more often a "yes, and" proposition than an "either, or" dichotomy. And the expert responses are emblematic of the multidisciplinary and ever-searching aspects of environmental science and engineering. I jokingly assert that the four of us all have "degrees in philosophy" (the last one each of us received). Philosophy directly translated means a "love of knowledge," so we are encouraged and driven to fill this knowledge gap.

From a chemical mass balance perspective, it would appear that the question would simply require stepping up laboratory results from mass transfer of oxygen into a microbial cell. However, the question becomes increasingly complicated by first principles (e.g., Does benzene chemically react to produce intermediate degradation products before generating carbon dioxide, and how does the gas diffuse into various cells?) and second principles (e.g., How does the biology of the system change as concentrations of benzene drop and new chemicals are formed? This is another way of saying that if the kinetics is changing we cannot assume a first-order decay rate). The environmental engineering (third principle) questions concern things like how much air can be mechanically pumped into this particular soil or aquifer without shearing the biofilm from the particles? This gives rise to an engineering economics question: Would it then be more cost-effective to diffuse pure oxygen rather than pump air to achieve oxygen mass transfer rates? We will return to the benzene question in the Epilogue (Chapter 14), after considering all of these and other principles.

The purpose of this text is to introduce the scientific and engineering fundamentals of environmental pollution through a series of real-life (and some theoretical) questions, problems, and exercises. These are followed by solutions and discussions of some of the meaningful aspects of environmental science and engineering.

The science, engineering, and technology needed to assess and to manage the risks posed by environmental contaminants are our major focus. Since environmental science is a cacophony of disciplines and scientific perspectives, the text introduces the reader to realistic scenarios that could be encountered in the practice of environmental professions.

Many of the cases and example problems deal with hazardous chemicals in the various components of the environment and the individual person so this text is, in effect, complementary to the discussions in my recent book, *Engineering the Risks of Hazardous Wastes* (Elsevier, 2003). Each of the two books can stand on its own, but I would like to think that, when consulted together to address the topical areas of risk and environmental engineering, they would provide a synergy of ideas and problem solving. I have strived to have the two be mutually supportive. For example, this text provides the instructor with ideas for homework and exercises to complement the lectures and discussions based on *Engineering the Risks of Hazardous Wastes*.

After a brief discussion on history and policy, we can get started with some questions and exercises about what is important in environmental risk assessment and engineering.

Acknowledgments

This book is only possible because of the wealth of knowledge and wisdom shared with me over the years by mentors and colleagues. I am particularly grateful to Jeffrey Peirce, with whom I have collaborated on a number of the scientific and engineering projects at Duke that have provided the real-life and laboratory lessons noted here. I also want to thank others who have selflessly shared their research results and insights, which I have incorporated either directly or indirectly in this book. I am particularly grateful to Ross McKinney, who has formally retired from the University of Kansas, but who will never retire from engineering and the pursuit of knowledge. My colleagues at the U.S. Environmental Protection Agency, including Charles Lewis, Leonard Stockburger, Robert Lewis, Mario Mangino, Robert Seila, Erick Swartz, Mack Wilkins, Jerry Blancato, Robert Stevens, and Gary Foley, have provided excellent examples. My ongoing discussions with Aarne Vesilind at Bucknell University have been quite helpful in several parts of this book. The thoughtful and groundbreaking discussions and examples regarding ultraviolet light were provided by Karl Linden and his research associate at Duke, Erik Rosenfeldt. Cynthia Yu of the EPA provided valuable information regarding risk communication. My colleagues in the Sound Management of Chemicals Program, part of the side agreement to the North American Free Trade Agreement, shared valuable information about metals and persistent chemicals. And, I would like to thank the many other authors of texts in environmental chemistry and engineering who have provided the framework for this text.

The engineering students at Duke and the biology students at North Carolina Central University have been a rich source of quality control for this book. In particular, I want to recognize the students in Duke's undergraduate course Control of Hazardous and Toxic Waste (Civil Engineering 249), in which I team teach the chemistry and risk module with Jeff Peirce.

The students identified needed corrections and clarifications in the explanations of dose-response curves, caveats on applying slope factors to risk calculations (e.g., the saturation effect), and consistency on units in Chapters 9 and 10. In fact, the section on food addition would not have appeared had a particularly attentive Duke engineering student not needed this calculation to conduct his risk assessment of ethylene oxide exposures in schools. The students in my Environmental Problems course (Biology 2700) at NCCU have embraced the "engineering" ethos although many will engage in fields other than environmental disciplines. For example, a future biology teacher pointed out some missing information in a previous draft of the acid rain discussions, that is, the assumed molar concentration (i.e. 350 ppm) of carbon dioxide needed to estimate the pH of normal rain. I have learned that study guides for students are an excellent way to find out if a book will be useful as a teaching device. I hope that the teachers and students will benefit from this "QA check" as I have.

I have thoroughly enjoyed my relationship with Elsevier/Academic Press. I appreciate very much the insights of my editor, Christine Minihane, for the inception of this project.

The copyediting process was impressive. Almost every question from Harbor Hodder and Kyle Sarofeen led to an improvement in clarity and content. I have particularly enjoyed another opportunity to work with Kyle, who was so helpful in my previous book, *Engineering the Risks of Hazardous Wastes* (Elsevier/Butterworth-Heinemann, Boston, 2003). In fact, our discussions following that project were key to my decision to write this book.

Notes and Commentary

1. For example, a landmark study (Commission for Racial Justice, United Church of Christ, 1987, *Toxic Wastes and Race in the United States*) found that the rates of landfill siting and the presence of hazardous waste sites in a community were disproportionately higher in African American communities. This report was instrumental in the recent requirements across federal governmental agencies to include environmental justice as a criterion in decisions. This is articulated in Presidential Executive Order 12898, 1994, "Federal Actions to Address Environmental Justice in Minority Populations and Low-Income Populations," February 11.
2. U.S. Environmental Protection Agency, Federal Insecticide, Fungicide, and Rodenticide Scientific Advisory Panel, 2003, "White Paper on Potential Developmental Effects of Atrazine on Amphibians" (submitted for review and comment in support of an interim reregistration eligibility decision on atrazine), Office of Prevention, Pesticides and Toxic Substances Office of Pesticide Programs Environmental Fate and Effects Division, Washington, D.C.
3. Frog Deformity Research Not Leaping to Conclusions, 1997, *Environmental Health Perspectives*, Vol. 105, No. 10, October.
4. Yes, we can study cultures other than our own in the liberal arts. But, if we follow the example of true humanities research, there comes a time when the student must learn the *lingua franca* of the culture. The social nuances are

simply too subtle to understand when one's entire perspective is based upon other people's translations. Many divinity students and Biblical scholars, for example, eventually come to a time when they must apply their own interpretations with respect to the original Hebrew, Aramaic, and Greek. Likewise, students of the philosophers must read the original texts of Socrates and his ilk in Greek, and the historians of science must be prepared to read and understand the Latin of the early natural philosophers up to the Renaissance.

5. The responses are slightly paraphrased from the actual responses (to protect the "innocent," but mainly to remove any personal comments not relevant to remediation. After all, they are friends as well as colleagues!).
6. Same as above.
7. By the way, we will revisit diffusion and mass transfer several times. We will also consider several other interesting points made by Expert A, including microbial toxicity of organic contaminants (e.g., benzene) and the composition of microbial biomass. Environmental science, by its very nature, is full of examples of not knowing where to start to explain something. I had this problem in deciding whether to talk about physics before chemistry and introducing organic molecules before the formal environmental organic chemistry discussion. This seems to be less of a problem for today's students, who seem to be able to pick up a discussion midstream. For us more "linear learners," however, it can be maddening. So please be patient.

Author's Note on Discussion Boxes, Equations, and Concentration Units

Author's Note Regarding Discussion Boxes

I have tried to design all discussion boxes to be free-standing. They appear at points in the text where the discussion is relevant, but the boxes can be understood outside of the specific context. This should enhance their usefulness as homework projects, meeting handouts, or attachments to correspondence.

Author's Note Regarding Equations and Concentration Units

For clarity and consistency with most recent style requirements of environmental publications, particularly environmental chemistry and engineering journals, this text uses exponents in equations, rather than fraction (i.e. the slash "/"), to represent one variable or factor with respect to another. For example, rather than mg/L, this text would represent this concentration as $\mathrm{mg\,L^{-1}}$ and rather than $\mathrm{mg/m^3}$, the text would give units in $\mathrm{mg\,m^{-3}}$. Also, scientific notation is used in equations (e.g., 5×10^{-7} or 5.00 E–7) and discussions, where appropriate. These uses of exponents help to avoid the representation of multiple fractions (i.e., numerators and denominators within numerators and denominators), often obviating the need for brackets and parentheses. For example $[(A/B)\cdot(X/Y)\cdot 0.000001]/Z$ is more clearly presented as:

$$\frac{AB^{-1}\cdot XY^{-1}\cdot 10^{-6}}{2}$$

This style also provides for easier conversion of units and presents the mathematics and arithmetic in a more straightforward manner.

I have strived for consistency in using concentration units throughout the text. However, environmental science and engineering applies various ways of expressing concentrations of contaminants and other substances in environmental media (e.g., water, air, soil, sediment, and biota), as the concentration variable in the equations employed to calculate dose, exposure, and risk (e.g., in inhaled air, ingested water, food, and soil, and dermal intake), and in expressions of the removal of contaminants and the presence of other substances in pollution control technologies (e.g., concentrations of target contaminants in stack gases, effluent from waste treatment plants, and finished water distributed from drinking water treatment facilities). The environmental disciplines have inherited much from chemistry and chemical engineering in how we express concentration. For example, the traditional synonym for concentration, at least to chemists, is molarity. Molarity is the amount of a substance in a solution expressed in moles of solute per liter of solution. This has value to environmental expressions of concentration but is limited, i.e., mass is only expressed in moles and molarity only applies to dissolved concentrations. In addition to solutions, environmental concentrations include suspensions, aerosols, and sorbed surfaces. Certainly, molar concentration is used throughout the text, particularly when discussing fundamental concepts and in discussing models. Molar concentration is also important in understanding pH, ionic strength, and other water quality factors. It is, however, a rather uncommon way to express contamination in most environmental situations.

Whenever possible, I express concentration as mass per unit volume concentrations for water (e.g., $mg\,L^{-1}$ or $\mu g\,L^{-1}$) and air (e.g., $mg\,m^{-3}$ or $\mu g\,m^{-3}$), and mass per unit mass for solid matrices (e.g., $mg\,kg^{-1}$), such as soil, sediment, sludge (biosolids), and food.[1] However, in both lay and technical literature, the concentration of substances is frequently expressed as the volume of substances dissolved or suspended within a specific volume of a liquid or gas, referred to as volume per unit volume. In almost every environmental situation, the liquid is water and the gas is air. Arguably, the most widely understood type of volume per volume (V : V) concentration is the percentage. For example, if we ignore any moisture, the air in the lowest level of the atmosphere (i.e., the troposphere) is comprised of about 78% nitrogen and about 21% oxygen by volume. Such V : V units are fine for high concentrations like these, but contaminants and other substances found in the environment exist at much lower concentrations. Commonly, scientists and engineers use parts per million (ppm) to express contaminant concentrations. A ppm is the volume of contaminant per million volumes of the water or air where the contaminant is found. Any volume unit can be used so long as the units for both numerator and the denominator are the same (e.g., liters of contaminant per million liters of air or water). So, if we think of percent as parts per hundred, then the conversion from percent to ppm is readily seen to be:

$$(\% \text{ by volume}) \times (10{,}000) = \text{ppm} \qquad \text{Equation AN–1}$$

For liquids, V:V concentration can be converted to mass per volume (M:V) concentrations, such as $\text{mg}\,\text{L}^{-1}$, if the density of the concentrated substance and the density of the liquid are known. Many environmental texts and models use shorthand terms of "solvent" and "solute," however, not all substances of concern are dissolved (e.g., some contaminants are suspended as particles or in emulsions in water). Since the liquid that we are usually most interested in is water, we want to express how much of the bad stuff (pollutants) or good stuff (e.g., dissolved oxygen) is in the water. The density of water under most environmental conditions is very nearly unity, so the V:V concentration can be converted to M:V concentration simply as:

$$\text{ppm} = \text{C} \times \rho \qquad \text{Equation AN–2}$$

Where, C = concentration of substance in water ($\text{mg}\,\text{L}^{-1}$) and ρ = density of the substance ($\text{g}\,\text{mL}^{-1}$).

Thus, if density of the contaminant is nearly $1\,\text{g}\,\text{cm}^{-1}$ (i.e., about the same as water at about 25°C temperature and 1 atmosphere (at)m pressure) and the M:V concentration is micrograms per liter ($\mu\text{g}\,\text{L}^{-1}$), the V:V concentration will be in parts per billion (ppb); and if the M:V concentration is nanograms per liter ($\text{ng}\,\text{L}^{-1}$) the V:V concentration will be in parts per trillion (ppt).

Water Concentration Example

The density of ethylene glycol at 25°C, the major ingredient of many radiator antifreeze products, is about $1.11\,\text{g}\,\text{cm}^{-1}$. If the creek near where you park your car has an ethylene glycol concentration of 10 ppm, what is the mass to volume concentration at 25°C and 1 atm?

Answer

If ppm = $C \times \rho$, then C = ppm/ρ, so:

$\text{C mg}\,\text{L}^{-1} = 10\,\text{ppm}/1.11\,\text{g}\,\text{mL}^{-1} = 9.0\,\text{mg}$ ethylene glycol per liter of water.

Concentrations in air can also be expressed as either M:V or V:V, with the two related as:

$$1\,\text{ppm} = \frac{1 \text{ volume of substance}}{10^6 \text{ total volumes}} \qquad \text{Equation AN–3}$$

Converting from V:V to M:V concentrations in air is a bit more complicated than that in water, because gas densities depend on the gas law, which states that the product of pressure (P) and the volume occupied by the gas (V) is equal to the product of the number of moles (n), the gas constant (R), and the absolute temperature (T) in degrees:

$$PV = nRT \qquad \text{Equation AN–4}$$

Where, P = absolute pressure of the gaseous substance (atm)
V = volume of the gaseous substance
n = number of moles of the gaseous substance
T = absolute temperature (°K), where °K is degrees Kelvin, which equals °C + 273°
R = Universal Gas Constant, $0.082056\,\text{L}\cdot\text{atm}\cdot{}^\circ\text{K}^{-1}\cdot\text{mol}^{-1}$

Common atmospheric M:V concentrations are $\mu\text{g}\,\text{m}^{-3}$ or $\text{mg}\,\text{m}^{-3}$. So, pressure, temperature and the molecular weight of the substance determine the relationship between M:V and V:V concentrations.

Air Concentration Example

The air near a water heater in the basement of an old home is measured for carbon monoxide (CO) and found to contain 0.1% by volume CO at 22°C and 1 atm. Express the CO concentration in the basement in $\mu\text{g}\,\text{m}^{-3}$.

Answer

Let us express the V:V in liters since these are the units in the gas law:

$$0.1\% = 10^{-3}\ \text{ppm} = \frac{10^{3}\ \text{L CO}}{10^{6}\ \text{L air}}$$

$T = 273 + 22 = 295°\text{K}$
$P = 1\,\text{atm}$
$R = 0.082056\,\text{L}\cdot\text{atm}\cdot{}^\circ\text{K}^{-1}\cdot\text{mol}^{-1}$
Molecular weight of CO = Atomic mass of C and O = 12 + 16
= $28\,\text{g}\,\text{mol}^{-1}$

Applying the ideal gas law and solving for the weight of CO gives:

$$\text{Weight of CO} = \frac{1\,\text{atm}\cdot 10^{-3}\,\text{L}\cdot 28\,\text{g}\,\text{mol}^{-1}}{0.082056\,\text{L}\,\text{atm}\,\text{mol}^{-1}\,\text{K}^{-1}\cdot 295°\text{K}}$$

$$\cong 1.2 \times 10^{-3}\,\text{g}$$

Thus, rearranging the ppm equation and solving for concentration gives:

$$= 1.2 \times 10^{-9}\,\text{g CO} = 1.2 \times 10^{-3}\,\mu\text{g}\,\text{L}^{-1}.$$

Since $1\,\text{m}^{-3}$ contains 1000 L, the CO concentration in the basement is $1.2\,\mu\text{g}\,\text{m}^{-3}$.

In soil, sediment, sludge, food or other solid or partially solid matrices, concentration is often given as mass per mass (M:M), such as $\text{mg}\,\text{kg}^{-1}$. Volume to volume concentrations are usually not used, because the matrices contain void spaces. Also, since all of these matrices contain fluids, especially water, in these void spaces, the concentrations are expressed as dry weight. So, a common soil concentration, for example, is X milligrams of contaminant per kilogram of soil (dry weight basis). This text assumes dry weight for all soil and sediment concentrations.

Note and Commentary

1 Biomarkers, i.e., indicators that signal the presence of a substance in the body used to identify exposure, also are expressions of concentrations. Concentration of a substance in blood is usually expressed as a mass per unit volume, most often as mass per decaliter (e.g., $\text{mg}\,\text{dL}^{-1}$), a common pharmacological and medical unit. Contaminant biomarker concentrations in hair and adipose (fat) are usually expressed as mass per mass concentrations (e.g., $\mu\text{g}\,\text{g}^{-1}$), just as in solid matrices.

Part I

An Environmental Policy Primer

CHAPTER 1

Scientific and Engineering Perspectives of Environmental Contaminants

The Evolution and Progress of Environmental Science and Engineering

Environmental science and engineering are young professions compared to many other disciplines in the physical and natural sciences and engineering. Understanding how organisms interact with their surroundings was at first a subdiscipline of the biological sciences.

Farmers, herders, and most people in ancient times knew, at least intuitively, of the interconnections and relationships among organisms and their environments. However, as scientists often do, explanations and systems of understanding needed to be applied to these practical understandings. Biologists and their subdisciplines thus began to specialize in what came to be known as the environmental sciences. Renaissance health scientists, like Paracelus[1] and William Harvey,[2] provided insights into how the human body interacts with and reacts to environmental stimuli. In fact, Paracelus' sixteenth century studies of metal contamination and exposure among miners may well be among the earliest examples of *environmental epidemiology.*

Environmental specialists, however, only appeared recently. The professions of ecologist, environmental scientist, and environmental engineer came into their own in the twentieth century. Ecology is the study of how organisms relate to one another and to their environments. Environmental science applies the fundamentals of chemistry, physics, and biology, and their derivative sciences such as geology and meteorology, to understand these abiotic[3] and biotic relationships. Expanding these observations to begin to control outcomes is the province of environmental engineering.

Not only are the environmental disciplines young, but many of the environmental problems faced today differ from those of most of the earth's history. The difference is in both kind and degree. For example, the synthesis of chemicals, especially organic compounds, has grown exponentially since the mid-1900s. Most organisms had no mechanisms to metabolize and eliminate these new compounds. Also, the stresses put on only small parts of ecosystems prior to the Industrial Revolution were small in the extent of their damage. For example, pollutants have been discharged into creeks and rivers throughout human history, but only recently have the discharges been so large and long-lasting that they have diminished the quality of entire ecosystems.

Assessment Example 1

Why has the term *environmental engineering* for the most part replaced *sanitary engineering* in the United States?

Discussion

There were many reasons for the name change. One certainly is the greater appreciation for the interconnections among abiotic and biotic systems in the protection of ecosystems and human health. Starting with the New Deal in the 1930s engineers engaged in "public works" projects, which in the second half of the twentieth century evolved to include sanitary engineering projects, especially wastewater treatment plants, water supplies, and sanitary landfills. The realization that there was much more beyond these "structures" has led engineers to comprehensive solutions to environmental problems. Certainly, structural solutions are still very important, but these are now seen as a part of an overall set of solutions. Thus, systems engineering, optimizations, and the application of more than physical principles (adding chemical and biological foundations) are better reflected in "environmental engineering" than in sanitary engineering. As mentioned by Vesilind and colleagues,[4] "everything seems to matter in environmental engineering."

Another possible reason for the name change is that "sanitary" implies human health, while "environmental" brings to mind ecological welfare as well as human health as the primary objectives of the profession. Sanitation is the province of industrial hygienists and public health professionals. The protection of the environment is a broader mandate for engineers.

Assessment Example 2

Why is environmental engineering often a field in the general area of civil engineering, and not chemical engineering?

Solution and Discussion

The historical "inertia" may help to explain why environmental engineering is a discipline of civil rather than chemical engineering. As mentioned in Assessment Example 1, environmental protection grew out of civil engineering projects of the New Deal and beyond. Chemical engineering is most concerned with the design and building of systems (e.g., "reactors") that convert raw materials into useful products. Chemical engineering is thus a kind of mirror image of environmental engineering, which often strives to return complex chemicals to simpler compounds (ultimately CO_2, CH_4, and H_2O). One could then view the two fields as a chemical equilibrium where the reactions in each direction are equal. Most importantly, both fields are crucial in hazardous waste management and contribute in unique ways.

What Is a Contaminant?

The term *contamination* is daunting. If you were told that your yard, your home, your water supply, or your air is contaminated, it is very likely that you would be greatly troubled. You would probably want to know the extent of the contamination, its source, what harm you may have already suffered from it, and what you can do to reduce it. Contamination is also a term that is applied differently by scientists and the general public, as well as among scientists from different disciplines.

What, then, is *contamination*? The dictionary[5] definition of the verb "*contaminate*" is "to corrupt by contact or association," or "to make inferior, impure, or unfit." These are fairly good descriptions of what environmental contaminants do. When they come into contact with people, ecosystems, crops, materials, or anything that society values, they cause harm. They make resources less valuable or less fit to perform their useful purposes. For example, when water pollution experts talk about a stream not meeting its "designated use," such as recreation or public water supply, because the stream contains certain chemicals, the experts are saying that this is contamination (see discussion in "Environmental Manager's Journal Entry: Beneficial Use in the Great Lakes"). Likewise, when the air we breathe contains substances that detract from healthy living, these substances are by definition *contaminants*.

Environmental Manager's Journal Entry: Beneficial Use in the Great Lakes

Great Lakes Areas of Concern (AOCs) are severely degraded geographic areas within the Great Lakes Basin. The AOCs are defined by the United States–Canada Great Lakes Water Quality Agreement (Annex 2 of the 1987 Protocol agreement between the two countries) as "geographic areas that fail to meet the general or specific objectives of the agreement where such failure has caused or is likely to cause impairment of beneficial use of the area's ability to support aquatic life." The two governments have identified 43 AOCs (See Figure 1.1); 26 in the United States and 17 in Canada (five are shared between the United States and Canada on connecting river systems). The two federal governments are collaborating with state and provincial governments to carry out Remedial Action Plans (RAPs) in each AOC. The RAPs are written to achieve and maintain 14 beneficial uses. An impaired beneficial use means a change in the chemical, physical, or biological integrity of surface waters. These include:

- restrictions on fish and wildlife consumption
- tainting of fish and wildlife flavor
- degradation of fish wildlife populations
- fish tumors or other deformities
- bird or animal deformities or reproduction problems
- degradation of benthos
- restrictions on dredging activities
- eutrophication or undesirable algae
- restrictions on drinking water consumption, or taste and odor problems
- beach closings
- degradation of aesthetics
- added costs to agriculture or industry
- degradation of phytoplankton and zooplankton populations
- loss of fish and wildlife habitat

An example of a remedial action plan to address impaired uses is that in the Cuyahoga River, located in northeast Ohio. The river runs for about a hundred miles from Geauga County, flowing south to Cuyahoga Falls where it turns sharply north until it empties into Lake Erie. The river drains 813 square miles of six counties.

The river's notoriety in the 1960s was widespread. In 1969, the river actually caught fire! This was one of the seminal events that led to the pollution abatement programs that were codified in the Clean

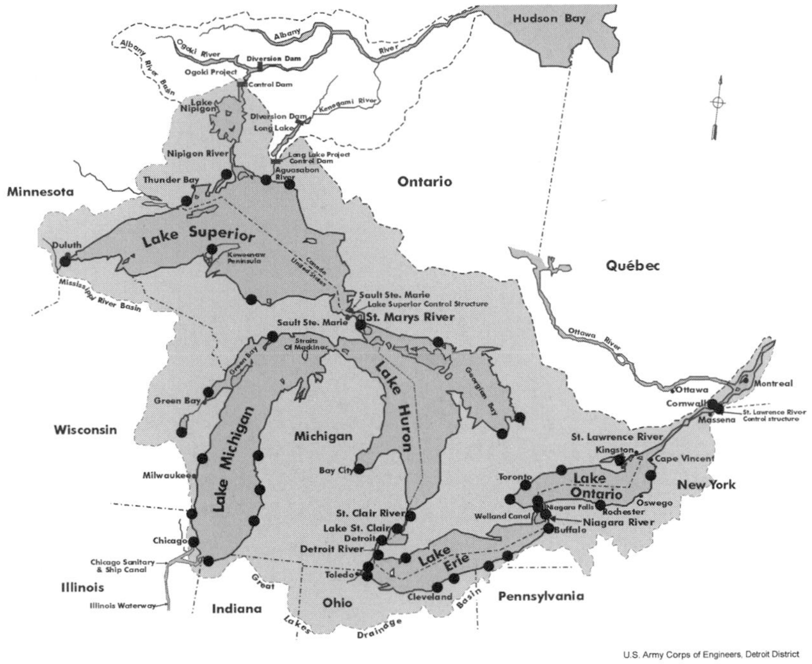

FIGURE 1.1. Areas of Concern in the Great Lakes. (Source of map: U.S. Army Corps of Engineers, Detroit District. Source of sites: U.S. Environmental Protection Agency.)

Water Act, Great Lakes Water Quality Agreement, and the spawning of the federal and state Environmental Protection Agencies.

The Cuyahoga AOC embodies the lower 45 miles of the river from the Ohio Edison Dam to the river's mouth, along with 10 miles of the Lake Erie shoreline. The AOC also includes 22 miles of urbanized stream between Akron and Cleveland.

Beneficial Use Impairments[6]

Ten of 14 use impairments have been identified in the Cuyahoga basin through the Remedial Action Plan (RAP) process. The environmental degradation resulted from nutrient loading, toxic substances (including polychlorinated biphenyls, or PCBs, and heavy metals), bacterial contamination, habitat change and loss, and sedimentation. Sources for these contaminants include municipal and industrial discharges, bank erosion, commercial/residential development, atmospheric deposition, hazardous waste disposal sites, urban stormwater runoff, combined sewer overflows (CSOs), and wastewater treatment plant bypasses.

Restrictions on Fish Consumption

In 1994, an advisory about eating fish was issued for Lake Erie and the Cuyahoga River AOC. The basis for the advisory was elevated PCB levels in fish tissue. The advisory restricted the consumption of white sucker, carp, brown bullhead, and yellow bullhead in the Cuyahoga River AOC, and walleye, freshwater drum, carp, steelhead trout, white perch, Coho salmon, Chinook salmon, small mouth bass, white bass, channel catfish, and lake trout in Lake Erie.

Degradation of Fish Populations

Beginning at the Ohio Edison Gorge and extending downstream to Lake Erie, measures of fish population conditions ranged from fair to very poor and were below the use criteria applicable to Ohio warmwater aquatic life habitats. Although fish communities have recovered significantly compared to the historically depleted segments of the Cuyahoga River, pollution-tolerant species continue to compose the dominant fish population.

Degradation of Wildlife Populations

Anecdotal information indicates some recovery of Great Blue Heron nesting in the Cuyahoga River watershed. Resident populations of black-crowned night herons have been noted in the navigation channel. The RAP is seeking partners to undertake research in this area in order that an evaluation may be made.

Fish Tumors or Other Deformities

Although deformities, eroded fins, lesions and external tumors (i.e., so-called DELT anomalies) have declined throughout the watershed, significant impairments continue to be found in the headwaters to the near shore areas of Lake Erie.

Bird or Animal Deformities or Reproductive Problems

No data have been found to suggest this is impaired in the Cuyahoga River AOC, but "no data" is not the same as "no problem."

Degradation of Benthos

Macroinvertebrate populations living at or near the bottom (i.e., benthic organisms) of the Cuyahoga River remain impaired at certain locations.

However, there are indications of substantial recovery, ranging from good to marginally good throughout most free-flowing sections of the river. Some fair and even poor designations are still seen, though.

Restrictions on Dredging Activities

The U.S. EPA restricts the disposal of dredged sediment in most of the Cuyahoga AOC due to high concentrations of heavy metals. Almost all of the sediment dredged from the river is classified by the U.S. EPA as "heavily polluted," meaning that more than 350,000 m^3 of sediment dredged each year must be disposed in a confined facility.

Eutrophication or Undesirable Algae

The Cuyahoga navigation channel seems to be impaired due to extreme oxygen depletion during summer months. The oxygen demand of the sediment is a factor.

Restrictions on Drinking Water Consumption, or Taste and Odor

The AOC contains no public drinking water sources, but contaminated aquifers and surface waters may still be threats to individual supplies and wells.

Beach Closings and Recreational Access

High bacterial counts after rain events episodically pollute two beaches in the AOC. Swimming advisories are issued after a storm, or if microbial counts exceed certain thresholds.

Degradation of Aesthetics

Aesthetics are impaired throughout the AOC due to soil erosion, surface water contamination from debris, improperly functioning septic systems, combined sewer overflows, and illegal dumping of pollutants.

Degradation of Phytoplankton and Zooplankton Populations

According to some studies, phytoplankton populations in the AOC are impaired. No standards exist for zooplankton communities as well.

Added Cost to Agriculture and Industry

No registered water withdrawals for agricultural purposes are taking place in the AOC. Industry does not appear to be adversely affected.

Loss of Fish and Wildlife Habitat

Channelization, nonexistent riparian cover, silt, bank reinforcement with concrete and sheet piling, alterations of littoral areas and shorelines, and dredging are contributors to the impairment of fish and wildlife in the AOC.

Planning and remediation are ongoing, but much work remains to be done.

In the environmental sciences, contamination is usually meant to be "chemical contamination" and this most often is within the context of human health. However, contaminants can also be biological (e.g., viruses released from treatment plants, landfills, or other facilities).

Contaminants may also be physical (such as the loss of wetlands from dredging, or the energy from ultraviolet light). A classic example of a physical contaminant is the effect of warm water effluents to rivers. For much of the twentieth century, power plants used river water to cool their turbines. Plant intakes diverted the water, which would run over the hot parts of the turbines, and then directly discharge the water back into the river. This aptly named "once-through cooling" process led to temperature increases in the river throughout the year. Aquatic biota can be quite sensitive and respond to minute changes in temperature. The whole structure of the river system is at risk of change. Some of these changes are directly related to the temperature, while others are the result of chemical changes induced by the increased temperature, especially drops in dissolved oxygen concentrations in the water. Game fish, like trout and salmon, require cool waters for optimum health. Some stages in an organism's life are more sensitive to temperature shifts. As water temperatures increase above certain levels, the fish become physically stressed and more susceptible to microbial infections. Many game fish die when water temperatures exceed a threshold (e.g., 25°C for salmon).

Although public health is usually the principal driver for assessing and controlling environmental contaminants, ecosystems are also important *receptors* of contaminants. Contaminants also impact structures and other engineered systems, including historically and culturally important monuments and icons, such as the contaminants in rainfall (e.g., nitrates and sulfates) that render it more corrosive than would normally be expected (i.e., *acid rain*).

Although we will for the most part consider chemical contamination in this text, these other "stressors" can wreak havoc in the environment.

Understanding Policy by Understanding Science

Too often, it seems, we are asked to manage or lead without an adequate understanding of what it is that we are managing or leading. To assess and address environmental issues and problems appropriately and thus make sound environmental decisions require at least a basic understanding of the underlying scientific principles affecting those issues, problems, and decisions. The military coda that one should lead by example, and that one should not ask others to do what one would not also do, holds true for environmental management. Although I will introduce and cover many of the scientific topics in much more detail later, it is helpful to consider how quantitative approaches underlie environmental policies. Let's begin by asking a few questions that demonstrate that even some of the most profound and apparently well-understood subjects of environmental issues, such as oxygen in water, are not as they may seem without careful examination. Such a lack or deficiency of understanding can be further propagated at the next levels of risk management and environmental rule-making.

Assessment Example 3

Why do concentrations of dissolved oxygen fall when water temperatures increase? Put another way, why can cold water hold more oxygen than warm water?

Solution and Discussion

Physicists, chemists, and meteorologists will tell you that gases dissolve in liquids, forming solutions. This involves an equilibrium process, which depends on an equilibrium constant. The equilibrium between molecular oxygen gas and dissolved oxygen in water[7] is:

$$O_2(aq) \leftrightarrow O_2(g) \qquad \text{Reaction 1–1}$$

The equilibrium constant for this relationship is:

$$K = p(O_2)/[O_2]. \qquad \text{Equation 1–1}$$

That is, the concentration of a solute gas in a solution is directly proportional to the partial pressure (p) of that gas above the solution. This is an expression of the so-called "leaving tendency" or the phase partitioning from liquid solution to the gas, as expressed by Henry's Law.[8] The mass of a gas that remains in solution is a function of the pressure-concentration ratio called Henry's law:

TABLE 1.1
Fugacity to the Gas Phase, as Expressed by Henry's Law Constants for Relatively Abundant Gases in the Earth's Troposphere

Gas	K_H *(Pa mol^{-1} L^{-1})*	K_H *(atm mol^{-1} L^{-1})*
Helium (He)	2.83×10^8	2865
Oxygen (O_2)	7.47×10^7	756.7
Nitrogen (N_2)	1.55×10^8	1600
Hydrogen (H_2)	1.21×10^8	1228
Carbon dioxide (CO_2)	2.94×10^6	29.76
Ammonia (NH_3)	5.69×10^6	56.9

$$p = K_H[c] \qquad \text{Equation 1–2}$$

Where, K_H = Henry's Law constant
p = Partial pressure of the gas
$[c]$ = Molar concentration of the gas

Henry's Law describes the behavior of gases dissolving in liquids at relatively low concentrations and partial pressures. This is known as *fugacity* or the leaving tendency from a solution to the atmosphere. It is consistent with the ideal gas law relationships. Gas-liquid solutions consistent with Henry's Law are considered to be ideal dilute solutions. Table 1.1 provides some selected values of Henry's Law constants for tropospheric[9] gases dissolved in water, expressed in two units that we will use to arrive at a solution to the problem.

Henry's Law constants are temperature dependent. Generally, K_H increases as the solvent temperature increases, meaning that the solubility of a gas generally decreases with increasing temperature. Heating water in a pan gives an example of this relationship. Bubbles of air appear on the sides of the pan long before the water actually begins to boil. The oxygen is being released from solution in the water that was air-saturated at lower temperatures, but at higher temperature, the maximum amount of oxygen in the water, or the molar solubility of the gas, decreases.

The decrease in solubility of gases with increasing temperature is an example of the operation of *Le Chatelier's Principle*, which states that a system in equilibrium responds to any stress by restoring the equilibrium. Since the change in heat (or enthalpy change) of the dissolution of most gases is negative (i.e., exothermic), then increasing the temperature leads to a greater release of a gas. In our example the enthalpy change, as expressed by Henry's law, means the concentration of dissolved oxygen will decrease.

Although we will cover this topic in greater detail later, let's bring some closure to our whole energy/matter discussion about oxygen in biological systems with another example.

Assessment Example 4

What other factors besides dissolved oxygen may contribute to the mortality or morbidity of certain fish species with elevated water temperatures?

Solution and Discussion

Microbial growth is greatly accelerated in warmer water. These microbiological changes include pathogenic bacteria and fungi. Also, algae will grow so rapidly that so-called "blooms" can be seen in the water (a thick covering). This growth leads to changes in the water's pH. The algal growth also uses more dissolved oxygen, so the oxygen available to the fish decreased even more than what would have resulted from the higher vapor pressures induced by increasing the water temperature.

These questions thus demonstrate the interconnectivity among the three major scientific disciplines applied in addressing environmental questions:

- Physics (e.g., the laws of motion, thermodynamics, and partitioning from one physical phase to another);
- Chemistry (e.g., the solubility and reactions of gases, especially oxygen, in various parts of the environment); and
- Biology (e.g., microbial degradation and interdependence on physical and chemical conditions).

We are now ready to consider these relationships in greater detail.

Connections and Interrelationships of Environmental Science

One of the advantages of working in an environmental profession is that it is so diverse. Many aspects of a problem have to be considered in any environmental decision. From a scientific perspective, this means consideration must be given to the characteristics of the pollutant *and* the characteristics of the place where the chemical is found. This place is known as the "environmental medium." The major environmental media are air, water, soil, sediment, and even biota.

The media can be further subdivided. Water, for example, is commonly divided between "surface water" and "groundwater." The former includes everything from puddles and rivulets, to large rivers and lakes, to the oceans.

The names that we give things, the ways we describe them, and how we classify them for better understanding is uniquely important to each discipline. Although the various environmental fields use some common language, they each have their own lexicons and systems of taxonomy. Sometimes the difference is subtle, such as different conventions in nomenclature and symbols.[10] This is more akin to different slang in the same language.

Sometimes the differences are profound, such as the use of the term "particle". In atmospheric dispersion modeling, a "particle" is the theoretical point that is followed in a fluid (see Figure 1.2). We will discuss this as part of the fluid properties discussion in Chapter 3, Applied Contaminant Physics: Fluid Properties. The point represents the path that the pollutant in the air stream is expected to take. Particle is also used interchangeably with the term *aerosol* in atmospheric sciences and exposure studies (see Figure 1.3). Particle is also commonly used to describe one part of an unconsolidated material, such as a soil or sediment particle (see Figure 1.4). We

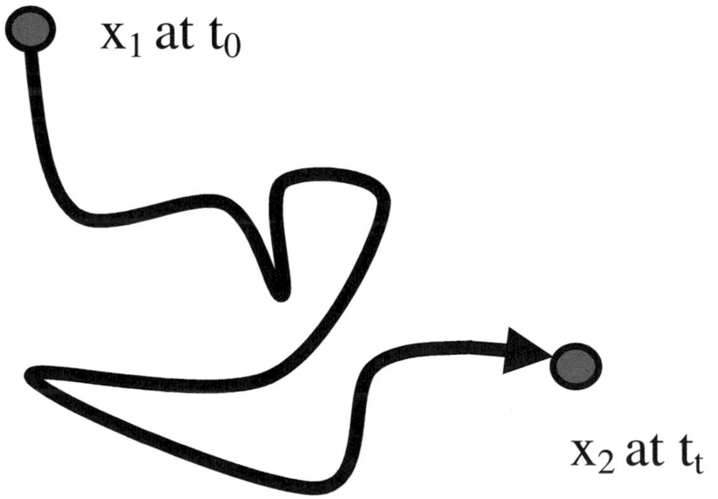

FIGURE 1.2. Atmospheric modeling definition of a particle; i.e., a hypothetical point that is moving in a random path during time interval ($t_0 - t_t$). This is the theoretical basis of a Lagrangian model.

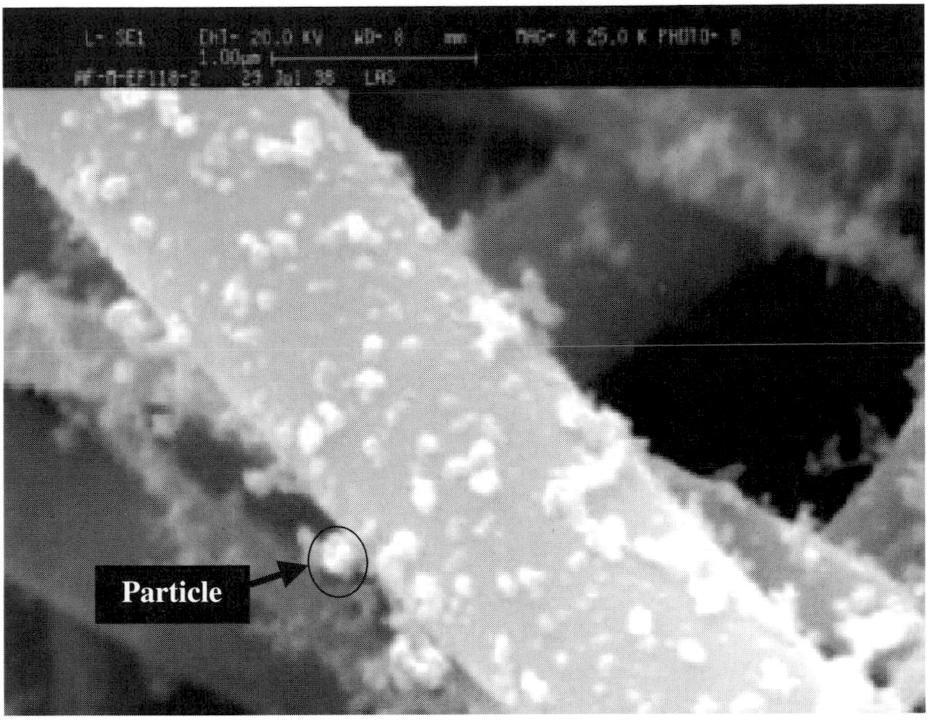

FIGURE 1.3. Electron micrograph (>45,000 X enlargement) showing an example of a particle type of air pollutant. These particles were collected from the exhaust of an F-118 aircraft under high throttle (military) conditions. The particles were collected with glass fiber filters (the 1 μm width tubular structures in the micrograph). Such particles are also referred to as particulate matter (PM) or aerosols. The size of the particle is important, since the small particles are able to infiltrate the lungs and penetrate more deeply into tissues, which increases the likelihood of pulmonary and cardiovascular health problems. (Source: L. Shumway, 2002, "Characterization of Jet Engine Exhaust Particulates for the F404, F118, T64, and T58 Aircraft Engines," U.S. Navy, Technical Report 1881, San Diego, Calif.)

will also consider the engineering mechanics' definition of particle as it applies to kinematics; that is, a body in motion that is not rotating is called a particle. At an even more basic level, particle is half of the particle-wave dichotomy of physics, so the quantum mechanical properties of a particle, such as a photon, are fundamental to detecting chemicals using chromatography. Different members of the environmental science community use all of these terms.

Let us consider a realistic example of the challenge of science communications related to the environment. There is concern that particles emitted from a power plant are increasing aerosol concentrations in your

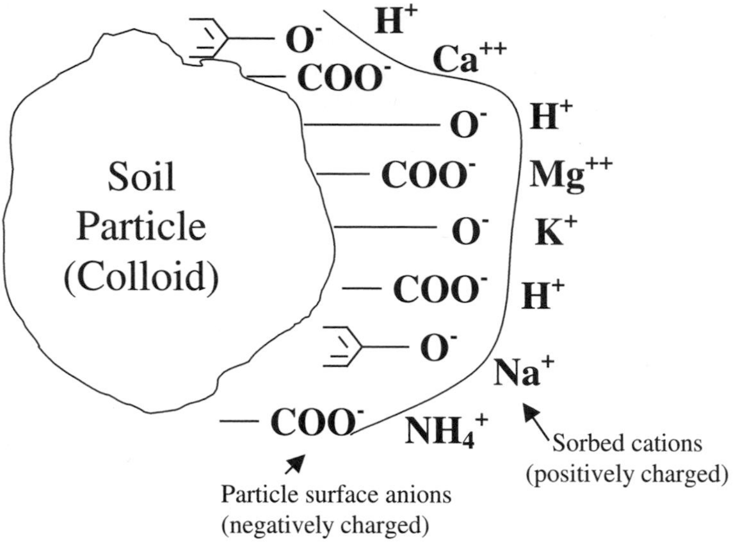

FIGURE 1.4. Particle of soil (or sediment) material; in this instance, humic matter with a negative surface that sorbs cations. The outer layer's extent depends on the size of the cations. For example, a layer of larger sodium (Na^+) cations will lead to a larger zone of influence than will a layer of smaller magnesium (Mg^{++}) cations.

town. To determine if this is the case, the state authorizes the use of a Lagrangian (particle) dispersion model to see if the particles are moving from the source to the town. The aerosols are carrying pollutants that are deposited onto soil particles, so the state asks for a soil analysis to be run. One of the steps in this study is to extract the pollutants from individual soil particles before analysis. The soil scientist turns his extract over to an analytical chemist who uses chromatography (which is based on quantum physics and particle-wave dichotomy) to analyze the sample. You invite the dispersion modeler, the soil scientist, the chromatographer, as well as an exposure scientist to explain the meaning of their findings. In the process of each explanation, the scientists keep referring to particles. They are all correct within their specific discipline, but considered together they leave you confused. This is akin to trying to understand the difference in homonyms (i.e., words with the same spelling but different meanings such as "bay" the body of water or "bay" the tree) or the use of different languages altogether.

Another example is the definitions of terms used to describe groundwater quality. Groundwater includes all water below the surface, but depending upon the profession, may be further differentiated from soil-bound water. Engineers commonly differentiate water in soil from groundwater because the soil water greatly affects the physical and mechanical properties of every soil. Environmental engineering publications frequently describe soil water according to the amount of void space filled, or the water filled pore space (WFPS), which is the percentage of void space containing water. The WFPS is another way of expressing the degree of saturation. Almost all environmental science professions classify the water below the soil layer based upon whether the unconsolidated material (e.g., gravel and sand) is completely saturated or unsaturated. The saturated zone lies under the unsaturated zone. Hydrogeologists refer to the unsaturated zone as the "vadose zone." It is also referred to as the zone of aeration. Another type of groundwater, albeit rare, is the Karst groundwater system, which is actually made up of underground lakes and streams that flow through fractured limestone and dolomite rock strata. Small cracks in the rock erode over time to allow rapidly flowing water to move at rates usually seen only on the earth's surface. Usually, groundwater flows quite slowly, but in these caves and caverns, water moves rapidly enough for its flow to be turbulent. We will cover all of these topics in detail when we discuss pollutant transport, drawing from several different scientific disciplines.

Soil is classified into various types. This is not a soil science text, but it is important to understand that for many decades, soil scientists have struggled with uniformity in the classification and taxonomy of soil. Much of the rich history and foundation of soil scientists has been associated with agricultural productivity. The very essence of a soil's "value" has been its capacity to support plant life, especially crops. Even forest soil knowledge owes much to the agricultural perspective, since much of the reason for investing in forests has been monetary. A stand of trees are seen by many to be a "standing crop." In the United States, for example, the National Forest Service is an agency of the U.S. Department of Agriculture. The engineers have been concerned about the statics and dynamics of soil systems, improving the understanding of soil mechanics so that they may support, literally and figuratively, the built environment. The agricultural and engineering perspectives have provided valuable information about soil that environmental professionals can put to use. The information is certainly necessary, but not completely sufficient, to understand how pollutants move through soils, how the soils themselves are affected by the pollutants (e.g., loss of productivity and diversity of soil microbes), and how the soils and contaminants interact chemically (e.g., changes in soil pH will change the chemical and biochemical transformation of organic compounds). At a minimum, environmental scientists must understand and classify soils according to their texture or grain size (see Table 1.2), ion exchange capacities, ionic strength, pH, microbial populations, and soil organic matter content.

Whereas air and water are fluids, sediment is a lot like soil in that it is a matrix made up of various components, including organic matter and unconsolidated material. The matrix also contains liquids ("substrate" to the chemist and engineer) within its interstices. Much of the substrate of

TABLE 1.2
Commonly Used Soil Texture Classifications[11]

Name	*Size Range (mm)*
Gravel	>2.0
Very coarse sand	1.0–1.999
Coarse sand	0.500–0.999
Medium sand	0.250–0.499
Fine sand	0.100–0.249
Very fine sand	0.050–0.099
Silt	0.002–0.049
Clay	<0.002

TABLE 1.3
Percent Composition of Two Environmentally Important Gases in Soil Air from a Study of Soils of Three Different Textures[13]

Depth from Surface (cm)	*Silty Clay*		*Silty Clay Loam*		*Sandy Loam*	
	O_2 (% Volume of Air)	*CO_2 (% Volume of Air)*	*O_2 (% Volume of Air)*	*CO_2 (% Volume of Air)*	*O_2 (% Volume of Air)*	*CO_2 (% Volume of Air)*
30	18.2	1.7	19.8	1.0	19.9	0.8
61	16.7	2.8	17.9	3.2	19.4	1.3
91	15.6	3.7	16.8	4.6	19.1	1.5
122	12.3	7.9	16.0	6.2	18.3	2.1
152	8.8	10.6	15.3	7.1	17.9	2.7
183	4.6	10.3	14.8	7.0	17.5	3.0

this matrix is water with varying amounts of solutes. At least for most environmental conditions, air and water are solutions of very dilute amounts of compounds. For example, air's highest-concentration solutes represent only small percentages (e.g., <3%) of the solution (e.g., water vapor), and the majority of solutes represent parts per million (a bit more than 300 ppm carbon dioxide). Most contaminants (i.e., the "bad solutes")[12] in air and water, thankfully, are found in the parts per billion range (ppb) range, if found at all. On the other hand, soil and sediment themselves are conglomerations of all states of matter. Soil is predominantly solid, but frequently has large fractions of liquid (soil water) and gas (soil air, methane, carbon dioxide) that make up the matrix. The composition of each fraction is highly variable. For example, soil gas concentrations are different from those in the atmosphere and change profoundly with their depth from the surface (Table 1.3 shows the inverse relationship between carbon dioxide and oxygen). Sediment is really an underwater soil. It is a collection of particles that have settled on the bottom of water bodies.

Ecosystems are combinations of these media. For example, a wetland system consists of plants that grow in soil, sediment, and water. The water flows through living and nonliving materials. Microbial populations live in the surface water, with aerobic species congregating near the water surface and anaerobic microbes increasing with depth due to the decrease in oxygen levels and to the reduced conditions. Air is not only important at the water and soil interfaces, but it is a vehicle for nutrients and contaminants delivered to the wetland. The groundwater is fed by the surface water during high water conditions, and feeds the wetland during low water.

Another way to think about these environmental media is that they are *compartments*, each with boundary conditions, kinetics, and partitioning relationships within a compartment or among other compartments. Chemicals, whether nutrients or contaminants, change as a result of the time spent in each compartment. The environmental professional's challenge is to describe, characterize, and predict the behaviors of various chemical species as they move through the media. When something is amiss, the cause and cure lie within the physics, chemistry, and biology of the system. It is up to the professional to properly apply the principles.

Connections Project 1

Group Project

How can one interpret the meaning of "Everything matters in environmental engineering?"

Make three lists with the following headings: (1) Physical Sciences, (2) Biological Sciences, and (3) Social Sciences and Humanities. List under each heading as many possible factors (and no less than five) that could affect decisions regarding hazardous wastes. For example, under Biological Sciences one could mention "microbes," and under Social Sciences and Humanities "fairness." Bring the lists to the next class or meeting and break into small groups for 10 minutes. Compare and contrast the items on everyone's lists. Compile a "master list" for each group, then report and compare it to those of the other groups. Did your group miss something important? Do you agree with the other groups? Were there some factors shared by all groups? Why are these common to everyone's list?

Suggestions

The objectives of this exercise are to introduce the complexity of hazardous waste engineering and to begin team building in the class or group. Team projects are an excellent way to deal with hazardous waste issues because there is seldom only one right answer. The students and other participants will also begin thinking from interdisciplinary perspectives.

Here are some of the factors that may be identified:

Physical Sciences

- Fluid mechanics
  - Air flow rates
  - Flow rate in groundwater
  - Flow rate in surface water
- Movement through soils
- Atmospheric plume development, characterization, and movement
- Sorption in soil, sediment, and biota
- Vapor pressure (rate of volatilization)

Chemistry

- Concentrations of contaminants in environmental media
- Rates of chemical reactions (including degradation or decay rates, lambdas)
- Solubility of contaminants
- Solvents (including water)
- Sampling protocols
- Analytical methods needed to measure pollutants

Biological Sciences

- Taxonomy of flora and fauna
- Taxonomy of microbes (bacteria, fungi, algae, viruses)
- Species diversity and sensitivity to contaminants
- Concentrations of contaminants in organisms
- Biochemistry of contaminants in organisms
- Biochemistry of contaminants in the environment
- Rates of uptake by organisms
- Diseases from exposures (toxicology)
- Population ecology
- Comparative animal studies as indications of human disease
- Epidemiology and human health studies

The discussions that ensue will be an indication of the students' knowledge of hazardous wastes, as well as a first step of matching skills and interests (core competencies) in teams.[14]

You may wish to save the lists and revisit them periodically over time as a "reality check" that the class or project team is covering all the major topics of interest.

Connections Project 2

Group Project

Write a one-page personal and/or family history, emphasizing the health or environmental risks that you or your family members have encountered. Be specific about the risks (what was the potential harm, the likelihood of the harm occurring, and the extent of the potential damage). The paper should explain any difference in how the risks were perceived when they were encountered versus how they appear after-the-fact and after some objective analysis.[15]

Suggestions

Follow the discussion with an exercise in which the class or project team uses scientific terminology to "label" these risk histories. For example, if someone's uncle immigrated to the United States and worked in a coal mine, describe this experience in terms of the hazards (pneumoconiosis, cave-ins), agents (coal dust, methane, structural failure of geologic strata), dose-response (more exposure, more disease, or adverse outcome), exposure scenarios (8 hours per day, 6 days per week, 30 years), and risk characterization (overall likelihood of adverse outcome given the dose, exposure, and characteristic hazards). This can bring a new realization to everyday risk versus quantitative risk calculations.

Connections Project 3

Group Project

How have concerns about homeland security and terrorism changed environmental engineering?

Suggestion

This is an intentionally open-ended question. I have found that there is a natural reluctance by some faculty to cover current topics, even when they are very relevant to the course (see D. Vallero, "Teachable moments and the tyranny of the syllabus: The September Eleventh Case," *Journal of Professional Issues in Engineering Education,* 129 (2), 2003, 100–105). Discuss the answers as a group. You may even want to break into small groups before the whole class or project team discusses it.

The bottom line is that current undergraduate students are living in a world different from the one that their professors and parents were exposed to during their educational experiences. The challenge of "intentional malevolence" goes beyond the problems of corruption, greed, carelessness, and myopia that have led to many of the environmental threats and problems that engineers are asked to prevent and remediate.

Environmental Assessment and Intervention

We have considered a fairly explicit relationship between direct pollution (heat), indirect pollution (lower concentrations of dissolved oxygen), effects (algal blooms and loss of game fish species), and how environmental professionals can intervene to prevent such pollution or at least ameliorate conditions. Let us now apply the advantage of hindsight to see how we may be able to avoid environmental disasters or lessen the effects of potential environmental problems.

Assessment Example 5

Consider the contamination of the Love Canal site in upstate New York. What aspects of this problem can be considered to be the result of honest mistakes in judgment? Which can be seen as breaches of public trust and malignant neglect? Prepare an event tree or critical path of the decisions and consequences. Explain how risks could have been avoided or ameliorated at various decision points.

Discussion

The Love Canal was not only the "watershed" case that many believe led to the passage of the Superfund, but it is also still considered to be one of the most complex. Scholars and practitioners alike have studied its convoluted history and the critical paths of its decisions. It is sometimes easier to label the involved parties as "good" and "bad" guys, but it would behoove students to find out more about the event trees that led to military, commercial, and civilian governmental decisions. An event tree shows the sequences of each decision and action that led to (or could lead to) a specific outcome or result. One particularly interesting event tree is that of the public school district's decisions to accept the donation of land and building the school on the property. Was there political coercion or malevolent "hand washing" that led to

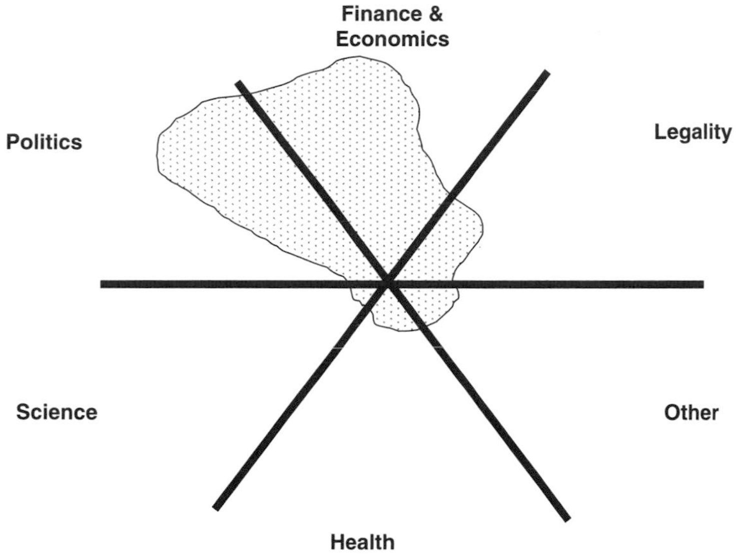

FIGURE 1.5. Theoretical decision force field for Love Canal.

this decision? Explore the factors that led to the decision. Draw up critical paths after identifying the major motivations of each party. This can get quite complex, so start with a decision force field, beginning with the factors that appear to have led to most of the decisions. The force field is a graphical depiction of the importance of key factors bearing on a decision. It may look something like the one in Figure 1.5.

Next, show how this changed over time. As depicted in Figure 1.6, for example, the science and health sectors grew as these factors were considered in the mid-1970s, and even more as the cleanup decisions were made. In addition, more laws were passed and new court decisions and legal precedents were established in the realm of toxic substances. Other factors may be added over time.

The event trees or critical paths can start with a Gantt[16] chart of milestones, enhanced by PERT[17] charts of critical activities, decisions, and outcomes.

Follow the decision analysis format of Edmond G. Seebauer and Robert L. Barry in *Fundamentals of Ethics for Scientists and Engineers*

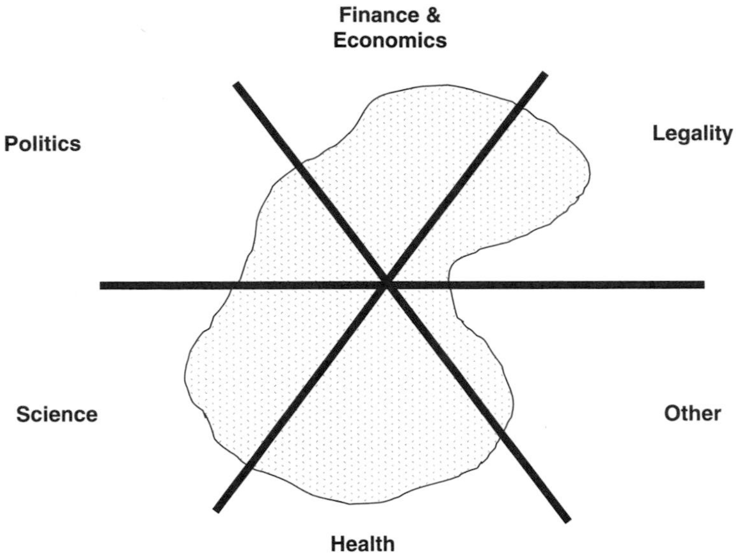

FIGURE 1.6. Theoretical decision force field for Love Canal, as available science and engineering information increases.

(New York: Oxford University Press, 2001), by first listing all of the characters and agencies involved in each key decision and each of their key interests, describing the circumstances under which the decision is made, and based on this information, drawing the event tree in the format shown in Figure 1.7.

Visit the EPA website for descriptions and discussions on each cleanup step. Each team can consider a different decision. Consider the steps needed to reach a final record of decision, with each contributing a decision analyzed with an event tree.

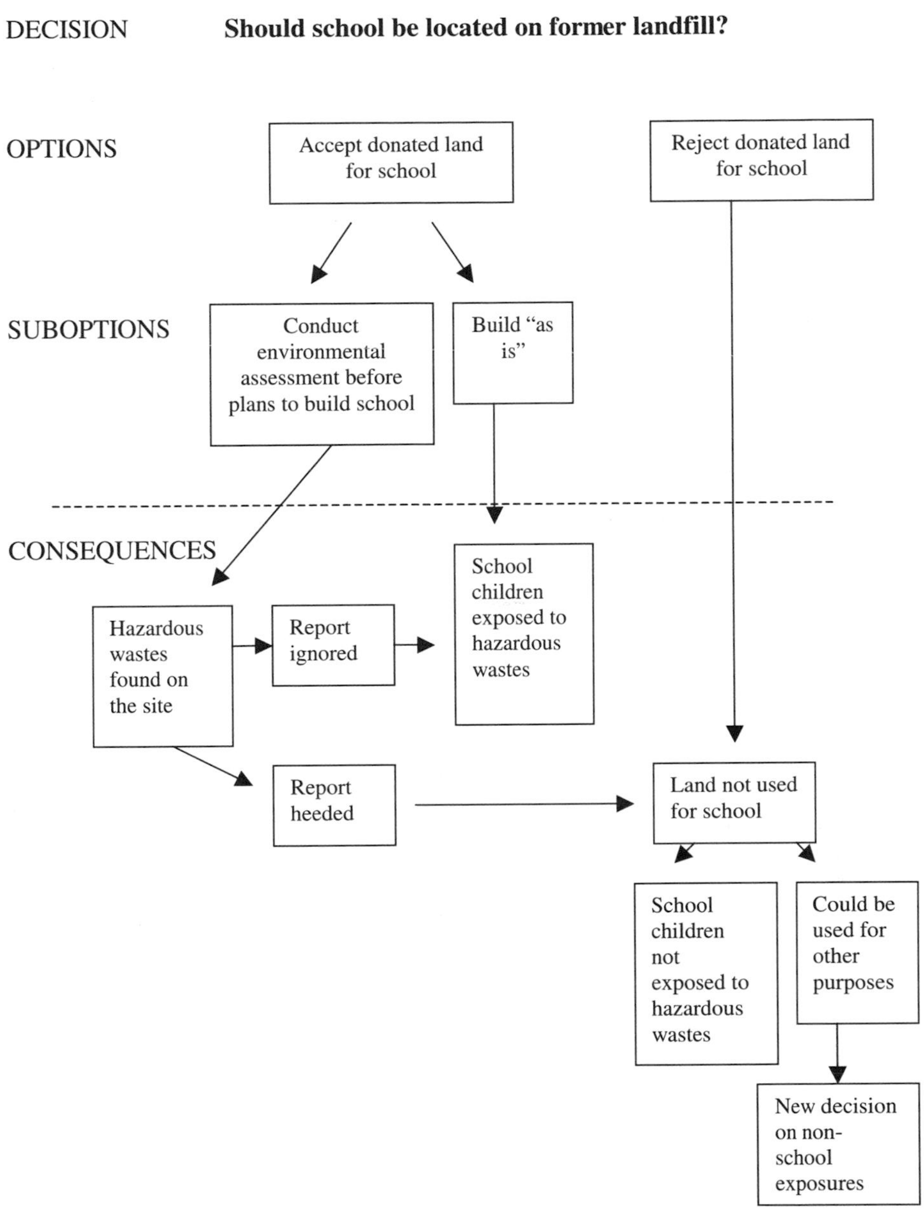

FIGURE 1.7. Theoretical ethical event tree for one key decision affecting Love Canal.

Assessment Example 6

Break into small groups to conduct a research project. Each group considers a famous (or infamous) hazardous waste site:

Times Beach, Missouri (http://www.epa.gov/history/topics/times/index.htm);
Valley of the Drums, Kentucky (http://www.epa.gov/history/topics/drums/index.htm);
Hanford Nuclear Site (http://www.hanford.gov/rl/index.asp); and
Warren County, North Carolina landfill site.

Evaluate each of these hazardous waste sites and the remediation steps taken. Design an optimal cleanup process that includes each step listed in the Discussion Box "Engineering Technical Note: Cleaning up a Hazardous Waste Site."

Discussion

These are historically important sites, but you may want to visit the EPA website's National Priority Listing for sites closer to you or have students or project team members select sites in or near their hometowns. The starting point is to become familiar with each cleanup step listed in the Love Canal case (for example, visit http://onlinesthics.org/environment/lcanal/timeline.html) and in the Discussion Box "Cleaning up a Hazardous Waste Site."

Engineering Technical Note: Cleaning up a Hazardous Waste Site

The U.S. EPA has established a set of steps to determine the potential for a release of contaminants from a hazardous waste site. These steps are known as the "Superfund Cleanup Process." The first step is a *Preliminary Assessment/Site Inspection* (PA/SI), from which the site is ranked in the Agency's *Hazard Ranking System* (HRS). The HRS is a process that screens the threats of each site to determine if the site should be listed on the *National Priority Listing* (NPL), the list of most serious sites identified for possible long-term cleanup, and what the rank of a listed site should be. Following the initial investigation, a formal

Remedial Investigation/Feasibility Study (RI/FS) is conducted to assess the nature and the extent of contamination. The next formal step is the *Record of Decision* (ROD), which describes the various possible alternatives for cleanup to be used at an NPL site. Next, a *Remedial Design/Remedial Action* (RD/RA) plan is prepared and implemented. RD/RA specifies which remedies will be undertaken at the site and lays out all plans for meeting cleanup standards for all environmental media. The Construction Completion step identifies the activities that were completed to achieve cleanup. After completion of all actions identified in the RD/RA, a program for Operation and Maintenance (O&M) is carried out to ensure that all actions are as effective as expected, and that the measures are operating properly and according to the plan. Finally, after cleanup and demonstrated success, the site may be deleted from the NPL.

Scientists and engineers are often called upon to be consultants to a company or to a government agency to lead remediation efforts and to advise on how cleanup should proceed at a hazardous waste site. Although all sites are unique, a number of steps must be taken for any hazardous waste facility. First, the location of the site and its boundaries should be clearly specified, including the formal address and geodetic coordinates. The history of the site, including all present and past owners and operators should be documented. The search for this background information should include both formal (e.g., public records) and informal documentation (e.g., newspapers and discussions with neighborhood groups).[18]

The main or most recent businesses that have operated on the site, as well as any ancillary or previous interests, should be documented and investigated. For example, in the famous Times Beach, Missouri incident, the operator's main business was an oiling operation to control dust and to pave roads. Unfortunately, the operator also ran an ancillary waste oil hauling and disposal business. The operator combined these two businesses, spraying waste oil that had been contaminated with dioxins, which led to the widespread problem and numerous Superfund sites in Missouri, including the relocation of the entire town of Times Beach.

The investigation at this point should include *all* past and present owners and operators. Any decisions regarding *de minimus* interests (i.e. parties that only contributed "insignificant" or "very low" levels of pollution) will be made at a later time (by the government agencies and attorneys). At this point, one should be searching for every potentially responsible party (PRP). A particularly important part of this review is to document all sales of the property, or any parts of the property. Also, all commercial, manufacturing, and transportation concerns should be known, as these may indicate the types of wastes that have been generated or handled at the site. Even an interest of short duration can be very important, if this interest produced highly persistent and toxic substances that may still be on-site, or that may have migrated off-site. The investigation should also determine either on-site or through manifest reports, whether any attempts were made to dispose of wastes from operations and whether any wastes were shipped off-site. A detailed account should be given of all waste reporting, including air emission and water discharge permits, and of voluntary audits that include tests like the *Toxicity Characteristic Leaching Procedure* (TCLP). These results should be compared to benchmark levels, especially to determine if any of the concentrations of contaminants exceed the U.S. EPA hazardous waste limit (40 CFR 261). For example, the TLCP limit for lead (Pb) is 5 mg L^{-1}. Any instances of exceeding this federal limit in the soil or sand on the site must be reported.

Initial monitoring and chemical testing should be conducted to target those contaminants that may have resulted from past or ongoing activities. A more general surveillance is also needed to identify a broader suite of contaminants. This is particularly important in soil and groundwater, since their rates of migration (Q) is quite slow compared to the rates usually found in air and surface water transport. Thus, the likelihood of finding remnant compounds is greater in soil and groundwater. Also, in addition to parent chemical compounds, chemical degradation products should also be targeted, since decades may have passed since the waste was buried, spilled, or released into the environment.

An important part of the preliminary investigation is the identification of possible exposures, both human and environmental. For example, the investigation should document the proximity of the site to schools, parks, water supplies, residential neighborhoods, shopping areas, and businesses.

One means of efficiently implementing a hazardous waste remedial plan is for the present owners (and past owners, for that matter) to work voluntarily with government health and environmental agencies. States often have voluntary action programs that can be an effective means of expediting the process, and that allow companies to participate in, and even lead, the *Remedial Investigation and Feasibility Study* (RI/FS) consistent with a state-approved work plan (which can be drafted by their consulting engineer).

The feasibility study (FS) delineates potential remedial alternatives, comparing the cost-effectiveness to assess each alternative approach's ability to mitigate potential risks associated with the contamination. The FS also includes a field study to retrieve and chemically analyze (at a state approved laboratory) water and soil samples from all environmental media on the site. Soil and vadose zone contamination will likely require that test pits be excavated to determine the type and extent of contamination. Samples from the pit are collected for laboratory analysis to determine general chemical composition (e.g., a so-called "total analyte list") and TCLP levels (that indicate leaching, i.e., the rate of movement of the contaminants).

An iterative approach may be appropriate as the data are derived. For example, if the results from the screening (e.g., total analytical tests) and the leaching tests indicate that the site's main problem is with one or just a few contaminants, then a more focused approach to cleanup may be in order. For instance, if preliminary investigation indicated that for most of the site's history a metal foundry was in operation, then the first focus should be on metals. If no other contaminants are identified in the subsequent investigation, a remedial action that best contains metals may be in order. If a clay layer is identified at the site from test pit activities and extends laterally beneath the foundry's more porous overburden material, the clay layer should be sampled to see if any screening levels have been exceeded. If groundwater contamination has not been found beneath the metal-laden material, an appropriate interim action removal may be appropriate, followed by a metal treatment process for any soil or environmental media laden with metal wastes. For example, metal-laden waste has recently been treated by applying a buffered phosphate and stabilizing chemicals to inhibit Pb leaching and migration.

During and after remediation, water and soil environmental performance standards must be met and confirmed by sampling and analysis. Poststabilization sampling and TCLP analytical methods to assess contaminant leaching (e.g., to ensure that Pb does not violate the federal standard of $5\,mg\,L^{-1}$). Confirmation samples must be analyzed to verify complete removal of contaminated soil and media in the lateral and vertical extent within the site.

The remediation steps should be clearly delineated in the final plan for remedial action, such as the total surface area of the site to be cleaned up and the total volume of waste to be decontaminated. At a minimum, a Remedial Action is evaluated on the basis of the current and proposed land use around the site, applicable local, state, and federal laws and regulations, and a risk assessment specifically addresses the hazards and possible exposures at or near the site. Any proposed plan should summarize the environmental assessment and the potential risks to public health and the environment posed by the site. The plan should clearly delineate all remedial alternatives that have been considered. It should also include data and information on the background and history of the property, the results of the previous investigations, and the objectives of the remedial actions. Since this is an official document, the State environmental agency must abide by federal and state requirements for public notice, as well as to provide a sufficient public comment period (about 20 days).

The final plan must address all comments. The Final Plan of remedial action must clearly designate the selected remedial action, which will include the target cleanup values for the contaminants, as well as all monitoring that will be undertaken during and after the remediation. It must include both quantitative (e.g., to mitigate risks posed by metal-laden material with total Pb > $1000\,mg\,kg^{-1}$ and the TCLP fraction of Pb ≥ $5.0\,mg\,L^{-1}$) and qualitative objectives (e.g., control measures and management to ensure limited exposures during cleanup). The plan should also include a discussion on planned and potential uses of the site following remediation (e.g., will it be zoned for industrial use or changed to another land use). The plan should distinguish between interim and final actions, as well as interim and final cleanup standards. The Proposed Plan and the Final Plan then constitute the "Remedial Decision Record."

The ultimate goal of the remediation is to ensure that all hazardous material on the site has either been removed or rendered nonhazardous through treatment and stabilization. The nonhazardous, stabilized material can then be properly disposed of, for example in a nonhazardous waste landfill.

Social Aspects of Environmental Science

Environmental quality is important to everyone. This text stresses the importance of sound science, but it should always be understood that when it comes to the environment and public health, the social sciences can never be ignored. Let us consider the importance of value systems in the following question.

Assessment Example 7

What are examples of competing values between a scientist or engineer and people living near a waste site?

Discussion

The major differences between environmental professionals and citizens living in the area are related to process and outcomes. The professionals are likely to adhere to strong science and data supporting their recommended actions. While the citizenry appreciates this process, their decisions are likely also to rely on emotions, perceptions, and history. You may want to look ahead to Chapter 13, "Environmental Decisions and Professionalism," for cues on how to bridge science to the myriad ways that people perceive its meaning and importance.

Develop a composite list of values. Group the values into "bins," and then choose the most important set of values for scientists and neighborhoods. One means of accomplishing this sorting process is "multivoting," where each group member has one vote for the most important value for the scientists and one vote for the most important citizens' value. Tally up the votes and see how close the class comes to a consensus.

You may also want to see which values are easiest to deal with (or most feasible) versus those that are most important. Give one green "sticky" for each student to vote for the most feasible, and one red "sticky" for the most important value.

This exercise reveals the complexity of peoples' value systems. Some of these values have been passed down for generations, while others have been recently adapted to changes. For example, if a person or neighborhood has just experienced a nasty battle with an industry or governmental organization regarding an environmental decision, such as the location of landfill or prohibition of a certain type of use of their own land, they may be reticent to trust the environmental professional who promises that it will be different this time. The second half of the twentieth century, particularly from the mid-1960s through the 1970s, ushered in a new environmental ethos, or at least memorialized the fact that most, if not all, people expect a certain quality of the environment. The laws and policies stemming from these years are still evolving.

Introduction to Environmental Policy

Environmental awareness grew in the second half of the twentieth century. With this awareness the public demand for environmental safeguards and remedies to environmental problems was an expectation of a greater role for government. A number of laws were on the books prior to the 1960s, such as early versions of federal legislation to address limited types of water and air pollution, and some solid waste issues, such as the need to eliminate open dumping. In fact, key legislation to protect waterways and riparian ecosystems was written at the end of the nineteenth century as the Rivers and Harbors Act (the law that set the stage for the U.S. Army Corps of Engineers to permit proper dredging operations, later enhanced by Section 404 of the Clean Water Act).

The real growth, however, followed the tumultuous decade of the 1960s. The environment had become a social cause, akin to the civil rights and antiwar movements. Major public demonstrations on the need to protect "spaceship earth" encouraged elected officials to address environmental problems, exemplified by air pollution "inversions" that capped polluted air in urban valleys, leading to acute diseases and increased mortality from inhalation hazards, the "death" of Erie Canal, and rivers that caught on fire in Ohio and Oregon.

The National Environmental Policy Act

The movement was institutionalized in the United States by a series of new laws and legislative amendments. The National Environmental Policy Act (NEPA) was in many ways symbolic of the new federal commitment to environmental stewardship. It was signed into law in 1970 after contentious hearings in the U.S. Congress. NEPA was not drafted to deal directly with any specific type of pollution. Rather, it did two main things. It created the Environmental Impact Statement (EIS) and established the Council on Environmental Quality (CEQ) in the Office of the President. Of the two, the EIS represented a sea change in how the federal government was to conduct business. Agencies were required to prepare EISs on any major action that they were considering that could "significantly" affect the quality of the environment. From the outset, the agencies had to reconcile often-competing values: their mission and the protection of the environment.

The CEQ was charged with developing guidance for all federal agencies on NEPA compliance, especially when and how to prepare an EIS. The EIS process combines scientific assessment with public review. The process is similar for most federal agencies. For example, the National Aeronautics and Space Administration (NASA) decision flowchart is shown in Figure 1.8. Local and state governments have also adopted similar requirements for their projects (e.g., the North Carolina process is shown in Table 1.4). Agencies often strive to receive a so-called "FONSI"[19] or a "finding of no significant impact," so that they may proceed unencumbered on a mission-oriented project.[20] The Federal Highway Administration's FONSI process (Figure 1.9) provides an example of the steps needed to obtain a FONSI for a project.

Whether a project either leads to a full EIS or a waiver through the FONSI process, it will have to undergo an evaluation. This step is referred to as an "environmental assessment." An incomplete or inadequate assessment will lead to delays and increases the chance of an unsuccessful project, so sound science is needed from the outset of the project design.

The final step is the Record of Decision (ROD). The ROD describes the alternatives and the rationale for the final selection of the best alter-

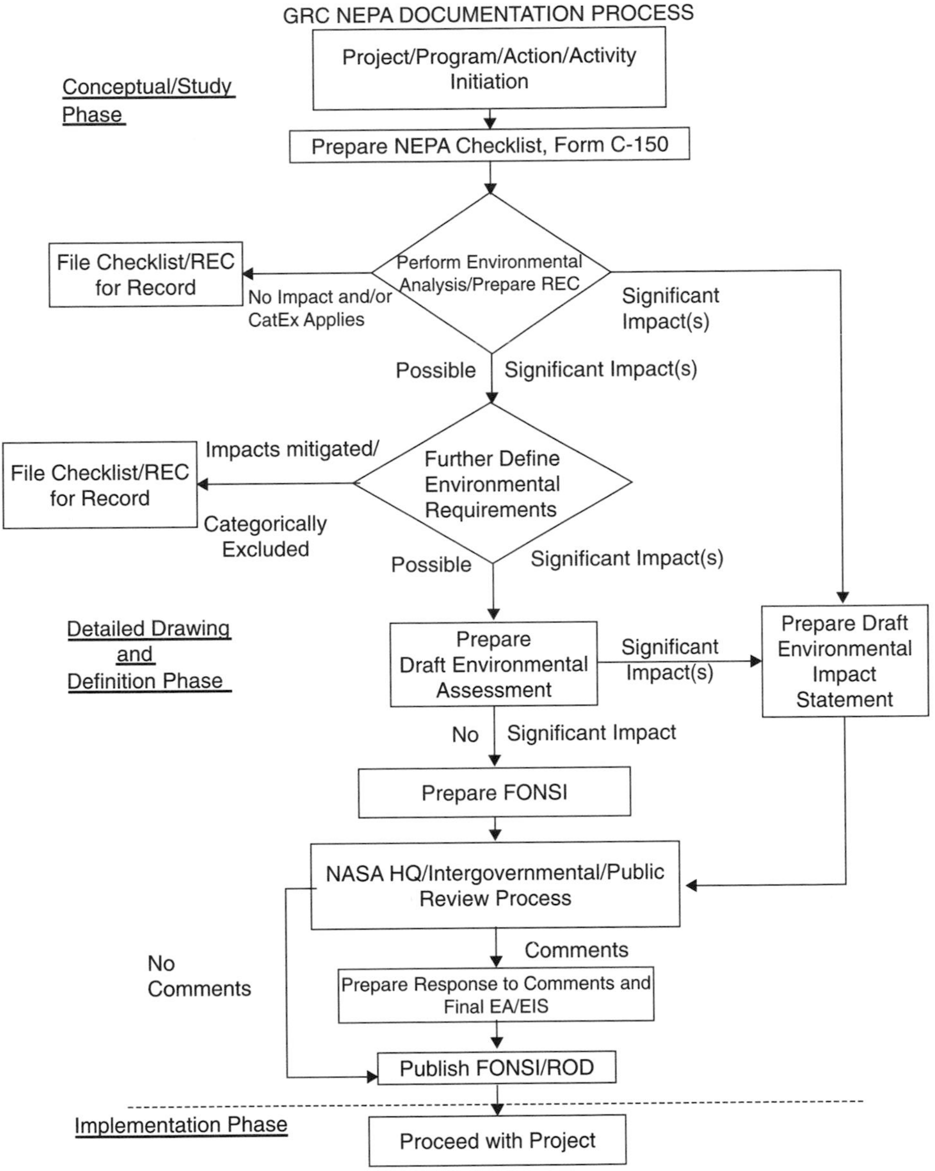

FIGURE 1.8. Decision flowchart for Environmental Impact Statements at the National Aeronautics and Space Administration.

TABLE 1.4
North Carolina's State Environmental Policy Act Review Process

Step I: Applicant consults/meets with Department of Environment and Natural Resources (DENR) about potential need for a State Environmental Policy Act (SEPA) document and to identify scope/issues of concern.

Step II: Applicant submits draft environmental document to DENR.

- Environmental document is either an environmental assessment (EA) or an environmental impact statement (EIS).

Step III: DENR-Lead Division reviews environmental document.

Step IV: DENR-Other Divisions review environmental document.

- 15–25 calendar days.
- DENR issues must be resolved prior to sending to the Department of Administration—State Clearinghouse (SCH) review.

Step V: DENR-Lead Division sends environmental document and FONSI to SCH.

Step VI: SCH publishes Notice of availability for environmental document in NC Environmental Bulletin. Copies of environmental document and FONSI are sent to appropriate state agencies and regional clearinghouse for comments.

- Interested parties have either 30 (EA) or 45 (EIS) calendar days from the Bulletin publication date to provide comments.

Step VII: SCH forwards copies of environmental document comments to DENR-Lead Division who ensures that applicant addresses comments.

- SCH reviews applicant's responses to comments and recommends whether environmental document is adequate to meet SEPA.
- Substantial comments may cause applicant to submit revised environmental document to DENR-Lead Division. This will result in repeating of Steps III–VI.

Step VIII: Applicant submits final environmental document to DENR-Lead Division.

Step IX: DENR-Lead Division sends final environmental document and FONSI (in case of EA and if not previously prepared) to SCH.

Environmental Assessment (EA)

Step X: SCH provides letter stating one of the following:

- Document needs supplemental information, or
- Document does not satisfy a FONSI, and an EIS should be prepared, or
- Document is adequate; SEPA is complete.

Environmental Impact Statement (EIS)

Step XI: After lead agency determines the FEIS is adequate, SCH publishes a Record of Decision (ROD) in the NC Environmental Bulletin.

Notes: PUBLIC HEARING(S) ARE RECOMMENDED (BUT NOT REQUIRED) DURING THE DRAFT STAGE OF DOCUMENT PREPARATION FOR BOTH EA AND EIS.

For an EA, if no significant environmental impacts are predicted, the lead agency (or sometimes the applicant) will submit both the EA and the Finding of No Significant Impact (FONSI) to SCH for review (either early or later in the process).

Finding of No Significant Impact (FONSI): Statement prepared by Lead Division that states proposed project will have only minimal impact on the environment.

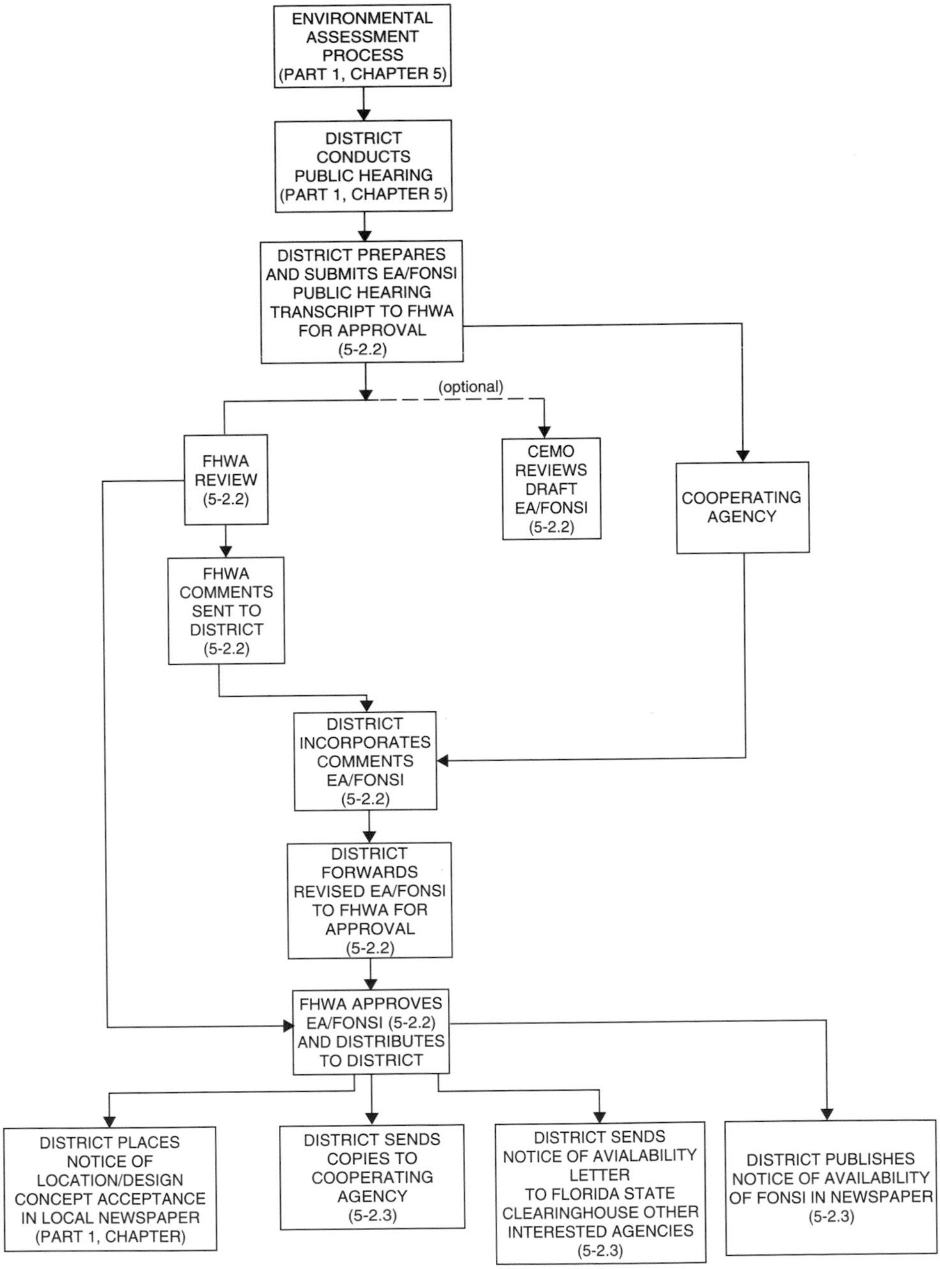

FIGURE 1.9. Decision flowchart for a Finding of No Significant Impact (FONSI) for Federal Highway Administration Projects. [Source: Federal Highway Administration, 1987, Technical Advisory T6640.8A. "Guidance for Preparing and Processing Environmental and Section 4(f) Documents."]

native. It also summarizes the comments received during the public reviews and how the comments were addressed. Many states have adopted similar requirements for their RODs.

The EIS documents are supposed to be a type of "full disclosure" of actual or possible problems if a federal project is carried out. This has been accomplished by looking at all of the potential impacts to the environment from any of the proposed alternatives, and comparing those outcomes to a "no action" alternative. At first, many federal agencies tried to demonstrate that their "business as usual" was in fact very environmentally sound. In other words, the environment would be better off with the project than without it (action is better than no action). Too often, however, an EIS was written to justify the agency's mission-oriented project. One of the key advocates for the need for a national environmental policy, Lynton Caldwell, is said to have referred to this as the federal agencies using EIS to "make an environmental silk purse from a mission-oriented sow's ear!"[21]

The courts adjudicated some very important laws along the way, requiring federal agencies to take NEPA seriously. Some aspects of the "give and take" and evolution of federal agencies' growing commitment to environmental protection was the acceptance of the need for sound science in assessing environmental conditions and possible impacts, and the very large role of the public in deciding on the environmental worth of a highway, airport, dam, waterworks, treatment plant, or any other major project sponsored by or regulated by the federal government. This has been a major impetus in the growth of the environmental disciplines since the 1970s. We needed experts who could not only "do the science" but who could communicate what their science means to the public.

Issues in Environmental Science: Confessions of an EIS Preparer

My first environmental job was as an EIS preparer. Although the mission of the agency where I was employed was actually to protect and to enhance the quality of the environment, even this mission ran up against the true intentions of NEPA. My first project as a 22-year-old, green, newly-minted environmental "professional" was to prepare EISs for a large coal-fired power plant that needed a sophisticated (at least by the standards of the mid-1970s) effluent discharge permit.[22] I believe that I have the dubious honor of having written the last EIS approved for a "once-through cooling" power generation facility in the United States. By the time I inherited the project, some major administrative decisions had been made about stationary point sources. The federal agencies had decided that for sources applying for both a "dredge and fill" permit from the Army Corps and an effluent discharge permit from the EPA, the lead agency for all "existing" sources, would be the Army and the lead agency for all "new" sources would be the EPA. Being the lead agency meant that that agency would have to take the lead and expend the resources needed to prepare the EIS, with support from other agencies.

I put quotation marks around the terms *new* and *existing*, because these sources were often more literal than actual. For example, for some time the power plant in question was considered to be an existing source because it was the newly designed fourth unit that was in need of its own permits. The power cooperative applying for the permits contended it was all part of the same power generating facility. However, after further investigation, the agencies decided that the fourth unit was really an independent and "new" facility in need of new permits. At that point, the EIS responsibility shifted to the EPA. This is an example of how governmental policy is iterative and changing up to and even after decision points on a critical path.

After inheriting the project, I was subjected to business logic and arguments about how the power plant was going to improve the lives of people on the power grid, and that it was without question an environmental benefit to all species. I began asking for additional information to support these contentions, such as an inventory of the species that would be potentially affected. I also asked that other alternatives be explored beyond taking water from the Missouri River, passing by turbines for heat exchange, and discharging it again to the river. I was advised by some cagey professionals in the wastewater permit program of the potential cumulative heating effect of all the power facilities along the Missouri River (as much as 3°C above baseline, if memory serves). I was also the beneficiary of lunchtime games of Euchre with three bright, young EIS reviewers who shared stories of the many games played and rationalizations given by agencies trying to get their EIS approved and their projects started. After all, they were my seniors by at least one year, so I had to listen to their stories!

I learned the meaning of mitigation. I first asked why other cooling approaches, especially cooling towers and lakes, were not being designed into this large, 600+ megawatt facility. After all of the risk assessment and management decisions were made, the power company had to add some features, such as sloughs and co-generation and sharing heated water with a nearby chemical company. Actually, in retrospect, the advice that I received from my "senior" colleagues was soon vindicated by the federal decision to eliminate all once-through cooling systems shortly after the acceptance of my EIS.

My second EIS was also somewhat controversial, but for different reasons. The EIS was again called for because of wastewater discharge issues, but this time it was a city's wastewater treatment program, funded under the so-called "Construction Grants" program pursuant to Section 201 of the Federal Water Pollution Control Act Amendments. The city also needed a discharge permit. I sometimes wonder to this day why this facility, of the hundreds being constructed in the mid-1970s with federal dollars (often 75% of the total project costs), rose to meet the two EIS metrics—that is, being a "major federal action" and one with a "significant environmental impact." It must have been the innovation of the land application of the wastewater. It was not your typical way of releasing wastes to the environment. The paradigm was, and still is in most instances, that if the waste came to the plant in the form of wastewater, it was to be released to a water body.

This was a classic case of pleasing the technologists and professionals while alienating the rest of the public. After all, the nutrients in the wastewater are really fertilizers, so the plants growing in the sprayed fields would benefit. The engineers were happy, the agricultural scientists were happy, and the city planners were happy. We were turning a waste into a resource, after all!

That was my take up to my first public hearing. What a surprise! The questions and indictments went something like this: "What about drift?" "This is sewage, and sewage is loaded with pathogens, can you guarantee that they will not find their way to the air that I breathe or the water that I drink?" "Don't things like viruses pass through systems untreated? So, aren't you just putting this waste out there to infect me?"

I cannot speak for my colleagues, but I was stunned. Why couldn't they see? As the EIS progressed and I learned more about land application, many if not all of the concerns were also being voiced around the world. And the scientific community was ill-equipped to allay these fears. I learned many valuable lessons about environmental assessments in that relatively small town on the High Plains. One was that risk communications should be ever on the mind of the environmental professional. And beyond communications, one should learn to listen to the community members. They live there and they will be affected by the "expert" decisions long after we leave.

Elmo Roper had it right back in 1942. The famous pollster said "many of us make two mistakes in our judgment of the common man. We overestimate the amount of information he has; and underestimate his intelligence." Roper seemed to be surprised that the general public frequently has too little information to decide on important matters. Roper was even more surprised that in spite of this lack of sufficient information, the common person's "native intelligence generally brings him to a sound conclusion." This is important to keep in mind in dealing with people who will potentially be affected by the recommendations of environmental experts.

All federal agencies must follow the CEQ regulations[23] to "adopt procedures to ensure that decisions are made in accordance with the policies and purposes of the Act." Agencies are required to identify the major decisions called for by their principal programs and to make certain that the NEPA process addresses them. This process must be set up in advance, early in the agency's planning stages. For example, if waste remediation or reclamation is a possible action, the NEPA process must be woven into the remedial action planning processes from the beginning, with the identification of the need for and possible kinds of actions being considered.

Noncompliance or inadequate compliance with NEPA rules regulations can lead to severe consequences, including lawsuits, increased project costs, delays, and the loss of the public's loss of trust and confidence, even if the project is designed to improve the environment, and even if the compliance problems seem to be only "procedural."

The U.S. EPA is responsible for reviewing the environmental effects of all federal agencies' actions. This authority was written into Section 309 of the Clean Air Act (CAA). The review must be followed with the EPA's public comments on the environmental impacts of any matter related to the duties, responsibilities, and authorities of the EPA's administrator, including EISs. The EPA's rating system (see Appendix 1) is designed to determine whether a proposed action by a federal agency is unsatisfactory from the standpoint of public health, environmental quality, or public welfare. This determination is published in the *Federal Register* (for significant projects) and referred to the CEQ.

Clean Air Legislation

A watershed[24] year in environmental awareness was 1970. The 1970 amendments to the Clean Air Act arguably ushered in the era of environmental legislation with enforceable rules. The 1970 version of the Clean Air Act was enacted to provide a comprehensive set of regulations to control air emissions from area, stationary, and mobile sources. This law authorized the EPA to establish National Ambient Air Quality Standards (NAAQS) to protect public health and the environment from "conventional" (as opposed to "toxic") pollutants: carbon monoxide; particulate matter; oxides of nitrogen; oxides of sulfur; and photochemical oxidant smog or ozone (see Discussion Box "Issues in Environmental Science: Evolution of Environmental Indicators for Air and Water"). The metal lead (Pb) was later added as the sixth NAAQS pollutant.

Issues in Environmental Science: Evolution of Environmental Indicators for Air and Water

The term *smog* is a shorthand combination of *smoke* and *fog*. However, it is really the code word for photochemical oxidant smog, the brown haze that can be seen when flying into Los Angeles, St. Louis, Denver, and other metropolitan areas around the world (see Figure 1.10). In fact, to make smog, at least three ingredients are needed: light, hydrocarbons, and radical sources, such the oxides of nitrogen. Therefore, smog is found most often in the warmer months of the year, not because of temperature but because these are the months with greater amounts of sunlight. More sunlight is available for two reasons, both attributed to the earth's tilt on its axis. In the summer, the earth is tilted toward the sun, so the angle of inclination of sunlight is greater than when the sun is tipped away from the earth, leading to more intensity of light per earth surface area. Also, the days are longer in the summer, so these two factors increase the light budget.

Hydrocarbons come from many sources, but the fact that internal combustion engines burn gasoline, diesel fuel, and other mixtures of hydrocarbons makes them a ready source. Complete combustion results in carbon dioxide and water, but anything short of complete combustion will be a source of hydrocarbons, including some of the original ones found in the fuels, as well as new ones formed during combustion. The compounds that become *free radicals*, like the oxides of nitrogen, are also readily available from internal combustion engines, since the ambient air is more than three-quarters molecular nitrogen (N_2). Although N_2 is, relatively speaking, not chemically reactive, it will react in conditions of high temperature and pressure, such as those in the internal combustion engine. Thus it will combine with the O_2 from the fuel/air mix and generate oxides that can provide electrons to the photochemical reactions.

The pollutant most closely associated with smog is ozone (O_3), which forms from the photochemical reactions mentioned above. In the early days of smog control efforts, O_3 was used more as a surrogate or *marker* for smog, since one could not really take a sample of smog. Later, O_3 became recognized as a pollutant in its own right, since it was increasingly linked to respiratory diseases.

This transition from an indicator to an actual pollutant in its own right is not uncommon in environmental protection. Witness fecal coliforms in water. These are bacteria that are naturally found enterically, that is, in the intestines of humans and animals. At first the fecal coliform count was mainly used to link potential sources contributing pollutants to a water body. Elaborate marker systems, especially fecal

FIGURE 1.10. Photo of smog episode in Los Angeles, California, taken in May of 1972. (Source: Documerica, U.S. Environmental Protection Agency's Photo Gallery; Photographer: Gene Daniels.)

coliform/fecal streptococcal ratios[25] were developed to ascertain whether the pollutant source was human (e.g., wastewater outfalls or septic tank leaks) or animal. The ratios were even subdivided into types of animals: a duck had one ratio, a pig another, and bears another (please forgive the imagery and the inevitable question about a bear in the woods!). The enteric bacterium *Escherichia coli* was used as a marker or indicator of fecal coliform. In the 1970s and 1980s environmental scientists and engineers were not particularly concerned about the *E. coli* bacterium itself, but, like the relationship of O_3 to smog, the presence of *E. coli* simply indicated that there was a source of human wastes, such as effluent from a treatment plant or septic tank leach fields. This attitude changed in the 1990s, when virulent strains of *E. coli* were found to present a major threat to public health.

The original goal was to set and achieve NAAQS in every state by 1975. These new standards were combined with charging the 50 states to develop state implementation plans (SIPs) to address industrial sources in the state. The ambient atmospheric concentrations are measured at over 4000 monitoring sites across the United States. The ambient levels have continuously decreased, as shown in Table 1.5.

The Clean Air Act Amendments of 1977 set new dates to achieve attainment of NAAQS (many areas of the country had not met the prescribed dates set in 1970). Other amendments were targeted at insufficiently addressed types of air pollution, including acidic deposition (so-called "acid rain"), tropospheric ozone pollution, depletion of the stratospheric ozone layer, and a new program for air toxics, the National Emission Standards for Hazardous Air Pollutants (NESHAPS).

The 1990 Amendments to the Clean Air Act profoundly changed the law, by adding new initiatives and imposing dates to meet the law's new requirements. Here are some of the major provisions.

TABLE 1.5
Percentage Decrease in Ambient Concentrations of National Ambient Air Quality Standard Pollutants from 1985 through 1994 (Source: U.S. Environmental Protection Agency)

Pollutant	*Decrease in Concentration*
CO	28%
Lead	86%
NO_2	9%
Ozone	12%
PM_{10}	20%
SO_2	25%

Urban Air Pollution

Cities that failed to achieve human health standards as required by NAAQS were required to reach attainment within six years of passage, although Los Angeles was given 20 years, since it was dealing with major challenges in reducing ozone concentrations.

Almost 100 cities failed to achieve ozone standards, and were ranked from *marginal* to *extreme*. The more severe the pollution, the more rigorous the required controls, although additional time was given to those extreme cities to achieve the standard. Measures included new or enhanced inspection/maintenance (I/M) programs for autos; installation of vapor recovery systems at gas stations and other controls of hydrocarbon emissions from small sources; and new transportation controls to offset increases in the number of miles traveled by vehicles. Major stationary sources of nitrogen oxides would have to reduce emissions.

The 41 cities failing to meet carbon monoxide standards were ranked *moderate* or *serious*; states may have to initiate or upgrade inspection and maintenance programs and adopt transportation controls. The 72 urban areas that did not meet particulate matter (PM_{10}) standards were ranked *moderate*; states will have to implement Reasonably Available Control Technology (RACT); use of wood stoves and fireplaces may have to be curtailed.

The standards promulgated from the Clean Air Act Amendments are provided in Table 1.6. Note that the new particulate standard addresses smaller particles, or particles with diameters ≤ 2.5 microns ($PM_{2.5}$). Research has shown that exposure to these smaller particles is more likely to lead to health problems than exposure to larger particles. Smaller particles are able to penetrate further into the lungs and likely are more *bioavailable* than the larger PM_{10}.

Mobile Sources

Vehicular tailpipe emissions of hydrocarbons, carbon monoxide, and oxides of nitrogen were to be reduced with the 1994 models. Standards now have to be maintained over a longer vehicle life. Evaporative emission controls were mentioned as a means for reducing hydrocarbons. Beginning in 1992, "oxyfuel" gasolines blended with alcohol began to be sold during winter months in cities with severe carbon monoxide problems. In 1995, reformulated gasolines with aromatic compounds were introduced in the nine cities with the worst ozone problems; but other cities were allowed to participate. Later, a pilot program introduced 150,000 low-emitting vehicles to California that meet tighter emission limits through a combination of vehicle technology and substitutes for gasoline or blends of substitutes with gasoline. Other states are also participating in this initiative.

TABLE 1.6
National Ambient Air Quality Standards

Pollutant	*Averaging Period*[20]	*Standard*	*Primary Standards*[21]	*Secondary Standards*[22]
Ozone	1-hr	Cannot be at or above this level on more than three days over three years.	125 ppb	125 ppb
	8-hr	The average of the annual fourth highest daily eight-hour maximum over a three-year period cannot be at or above this level.	85 ppb	85 ppb
Carbon Monoxide	1-hr	Cannot be at or above this level more than once per calendar year.	35.5 ppm	35.5 ppm
	8-hr	Cannot be at or above this level more than once per calendar year.	9.5 ppm	9.5 ppm
Sulfur Dioxide	3-hr	Cannot be at or above this level more than once per calendar year.	—	550 ppb
	24-hr	Cannot be at or above this level more than once per calendar year.	145 ppb	—
	Annual	Cannot be at or above this level.	35 ppb	—

Nitrogen Dioxide	Annual	Cannot be at or above this level.	54 ppb	54 ppb
Respirable Particulate Matter (aerodynamic diameter ≤ 10 microns = PM_{10})	24-hr	The three-year average of the annual 99th percentile for each monitor within an area cannot be at or above this level.	155 $\mu g\,m^{-3}$	155 $\mu g\,m^{-3}$
	Annual	The three-year average of annual arithmetic mean concentrations at each monitor within an area cannot be at or above this level.	51 $\mu g\,m^{-3}$	51 $\mu g\,m^{-3}$
Respirable Particulate Matter (aerodynamic diameter ≤ 2.5 microns = $PM_{2.5}$)	24-hr	The three-year average of the annual 98th percentile for each population-oriented monitor within an area cannot be at or above this level.	66 $\mu g\,m^{-3}$	66 $\mu g\,m^{-3}$
	Annual	The three-year average of annual arithmetic mean concentrations from single or multiple community-oriented monitors cannot be at or above this level.	15.1 $\mu g\,m^{-3}$	15.1 $\mu g\,m^{-3}$
Lead	Quarter	Cannot be at or above this level.	1.55 $\mu g\,m^{-3}$	1.55 $\mu g\,m^{-3}$

Toxic Air Pollutants

The number of toxic air pollutants covered by the Clean Air Act was increased to 189 compounds in 1990 (see Appendix 3). Most of these are carcinogenic, mutagenic, or toxic to neurological, endocrine, reproductive, and developmental systems. All 189 compound emissions were to be reduced within 10 years. The EPA published a list of source categories and issued Maximum Achievable Control (MACT) standards for each category over a specified timetable.

The next step beyond MACT standards is to begin to address chronic health risks that would still be expected if the sources meet these standards. This is known as residual risk reduction. The first step was to assess the health risks from air toxics emitted by stationary sources that emit air toxics after technology-based (MACT) standards are in place. The residual risk provision sets additional standards if MACT does not protect public health with an "ample margin of safety," as well as additional standards if they are needed to prevent adverse environmental effects.

What an "ample margin of safety" means is still up for debate, but one proposal for airborne carcinogens is shown in Figure 1.11. That is, if a source can demonstrate that it will not contribute to greater than 10^{-6} cancer risk, then it meets the ample margin of safety requirements for air toxics. The ample margin needed to protect populations from noncancer toxins, such as neurotoxins, is being debated, but it will involve the application of the hazard quotient (HQ). The HQ is the ratio of the potential exposure to the substance and the level at which no adverse effects are expected. An HQ < 1 means that the exposure levels to a chemical should not lead to adverse health effects. Conversely, an HQ > 1 means that adverse health effects are possible. Due to uncertainties and the feedback that is still coming from the business and scientific communities, the ample margin of safety threshold is presently ranging from HQ = 0.2 to 1.0. Therefore, if a source can demonstrate that it will not contribute to greater than the threshold (whether it is 0.2, 1.0, or

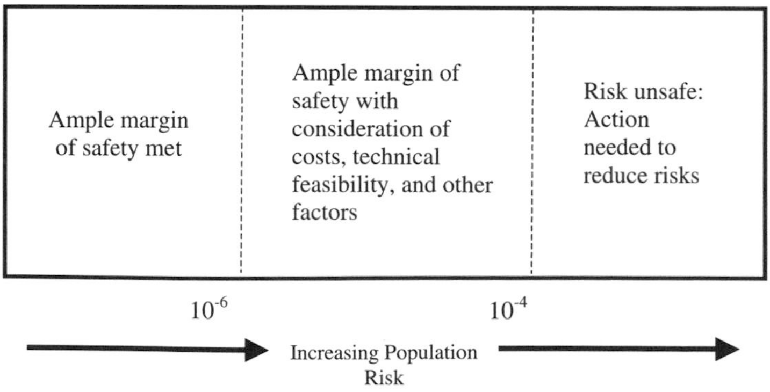

FIGURE 1.11. Ample margin of safety based on airborne contaminant's cancer risk.

some other level established by the federal government) for noncancer risk, it meets the ample margin of safety requirements for air toxics.

Acid Deposition

The introduction of acidic substances to flora, soil, and surface waters has been collectively called "acid rain" or "acid deposition." Acid rain is generally limited to the so-called "wet" deposition (low pH precipitation), but acidic materials can also reach the earth's surface by dry deposition (acid aerosols) and acid fog (airborne droplets of water that contains sulfuric acid or nitric acid). Generally, acid deposition is not simply materials of $pH < 7$, but usually of $pH < 5.7$, since "normal" rainfall has a pH of about 5.7, due to the ionization of absorbed carbon dioxide (see Discussion Box "Environmental Field Log Entry: Normal Rain *Is* 'Acid Rain'"). Thus it is the contribution of the human-generated acidic materials, especially the oxides of sulfur and the oxides of nitrogen, that are considered to be the sources of "acid rain."

Environmental Field Log Entry

Discussion: Normal Rain *Is* "Acid Rain"

There has been much concern about acid deposition because aquatic biota can be significantly harmed by only slight changes in pH. The problem of acidified soils and surface waters is a function of both the increase in acidity of the precipitation and the ability of the receiving waters and soil to resist the change in soil pH (the "buffering capacity" of the soil is discussed in detail in the soil discussion in later chapters).

We hear the term acid rain frequently, but what is it, really, since much of the rain falling in North America has been acidic, even before industrialization? The main source of natural acidity is carbon dioxide gas that is dissolved into water droplets in the atmosphere and the resulting carbonic acid is ionized, lowering the water's pH.

Given the mean pressure of CO_2 in the air is 3.0×10^{-4} atm, let us calculate the pH of water in equilibrium with the air at 25°C, and the concentration of all species present in this solution. We can also assume that the mean concentration of CO_2 in the troposphere is 350 ppm, but this concentration is rising by some estimates at a rate of 1 ppm per year.

The concentration of the water droplet's CO_2 in water in equilibrium with air is obtained from the partial pressure of Henry's law constant:[26]

$$p_{CO_2} = K_H[CO_2]_{aq} \qquad \text{Equation 1–3}$$

The change from carbon dioxide in the atmosphere to carbonate ions in water droplets follows a sequence of equilibrium reactions:

$$CO_{2(g)} \xleftrightarrow{K_H} CO_{2(aq)} \xleftrightarrow{Kr} H_2CO_{3(aq)} \xleftrightarrow{K_{a1}} HCO^-_{3(aq)} \xleftrightarrow{K_{a2}} CO^{2-}_{3(aq)}$$

Reaction 1–2

The Henry's Law constant is a function of a substance's solubility and vapor pressure. The concentration of carbon dioxide [CO_2] is constant, since the CO_2 in solution is in equilibrium with the air that has a constant partial pressure of CO_2. And the two reactions and ionization constants for carbonic acid are:

$$H_2CO_3 + H_2O \leftrightarrow HCO_3^- + H_3O^+ \qquad K_{a1} = 4.3 \times 10^{-7} \qquad \text{Reaction 1–3}$$

$$HCO_3^- + H_2O \leftrightarrow CO_3^{-2} + H_3O^+ \qquad K_{a2} = 4.7 \times 10^{-11} \qquad \text{Reaction 1–4}$$

K_{a1} is four orders of magnitude greater than K_{a2}, so the second reaction can be ignored for environmental acid rain considerations. The solubility of gases in liquids can be described quantitatively by Henry's Law, so for CO_2 in the atmosphere at 25 degrees °C, we can apply the Henry's Law constant and the partial pressure to find the equilibrium. The K_H for $CO_2 = 3.4 \times 10^{-2}\,\text{mol}\,L^{-1}\,\text{atm}^{-1}$. We can find the partial pressure of CO_2 by calculating the fraction of CO_2 in the atmosphere. Since the mean concentration of CO_2 in the earth's troposphere is 350 ppm by volume in the atmosphere, the fraction of CO_2 must be 350 divided by 1,000,000 or 0.000350 atm.

Thus, the carbon dioxide and carbonic acid molar concentration can now be found:

$$[CO_2] = [H_2CO_3] = 3.4 \times 10^{-2}\,\text{mol}\,L^{-1}\ \text{atm}^{-1} \times 0.000350\ \text{atm} = 1.2 \times 10^{-5}\,M$$

Thus, the equilibrium is $[H_3O^+] = [HCO^-]$.

Taking this and our carbon dioxide molar concentration, gives us:

$$K_{a1} = 4.3 \times 10^{-7} = \frac{[HCO_3^-][H_3O^+]}{CO_2} = \frac{[H_3O^+]^2}{1.2 \times 10^{-5}} \qquad \text{Equation 1–4}$$

$$[H_3O^+]^2 = 5.2 \times 10^{-12} \qquad \text{Equation 1–5}$$

$$[H_3O^+]^2 = 2.1 \times 10^{-6} M \qquad \text{Equation 1–6}$$

Or, the droplet pH is about 5.6.

If the concentration of CO_2 in the atmosphere increases to 400 ppm, what will happen to the pH of "natural rain"? The new molar concentration would be $3.4 \times 10^{-2}\,\mathrm{mol\,L^{-1}\,atm^{-1}} \times 0.000400\,\mathrm{atm} = 1.4 \times 10^{-5} M$, so $4.3 \times 10^{-7} = \frac{[H_3O^+]^2}{1.4 \times 10^{-5}}$ and $[H_3O^+]^2 = 6.0 \times 10^{-12}$ and $[H_3O^+] = 3.0 \times 10^{-6} M$.

The droplet pH would be about 5.5. So the incremental increase in atmospheric carbon dioxide may also contribute to greater acidity in natural rainfall.

Other processes have led to even further increases in droplet acidity, especially sulfuric and nitric acid formation from the oxides of sulfur ionizations further decrease this pH level to those harmful to aquatic organisms. Acid precipitation can result from the emissions of plumes of strong acids, such as sulfuric acid (H_2SO_4) or hydrochloric acid (HCl) in the forms of acid mists, but most commonly the pH drop is the result of secondary reactions of the acid gases SO_2 and NO_2:

$$SO_2 + \frac{1}{2}O_2 + H_2O \rightarrow \text{several intermediate reactions} \rightarrow \{2H^+ + SO_4^{2-}\}_{aq}$$

Reaction 1–5

$$2NO_2 + \frac{1}{2}O_2 + H_2O \rightarrow \text{several intermediate reactions} \rightarrow \{2H^+ + NO_3^-\}_{aq}$$

Reaction 1–6

The 1990 amendments introduced a two-phase, market-based system to reduce sulfur dioxide emissions from power plants by more than half. Total annual emissions were to be capped at 8.9 million tons, a reduction of 10 million tons from the 1980 baseline levels. Power facilities were issued allowances based on fixed emission rates set in the law, as well as their previous fossil-fuel use. Penalties were issued for exceedences, although the allowances could be banked or traded within the fuel burning industry. In Phase I, large high-emission plants in Eastern and Midwestern United States were required to reduce emissions by 1995. Phase II began in 2000 to set emission limits on smaller, cleaner plants and further tighten the Phase I plants' emissions. All sources were required to install continuous emission monitors to assure compliance. Reductions in the oxides of nitrogen (NO_x) were also to be reduced; however, the approach differed from the oxides of sulfur (SO_x), using EPA performance standards, instead of the two-phase system.

Protecting the Ozone Layer[27]

The ozone layer filters out significant amounts of ultraviolet radiation from the sun. This UV radiation can cause skin damage and lead to certain forms of cancer, including the most fatal form, melanoma. Therefore, the international scientific and policy communities have been concerned about the release of chemicals that find their way to the stratosphere and accelerate the breakdown of the ozone (see Discussion Box "Environmental Field Log Entry: Ozone—Location, Location, and Location").

Environmental Field Log Entry: Ozone—Location, Location, Location

Like the three rules of real estate, the location of ozone (O_3) determines whether it is essential or harmful. As shown in Figure 1.12, O_3 concentrations are small, but increase in the stratosphere (about 90% of the atmosphere's O_3 lies in the layer between 10 and 17 kilometers above the earth's surface up to an altitude of about 50 kilometers). This is commonly known as the ozone layer. Most of the remaining ozone is in the lower part of the atmosphere, or the troposphere. The stratospheric O_3 concentrations must be protected, while the tropospheric O_3 concentrations must be reduced.

Stratospheric ozone (the "good ozone") absorbs most of the biologically damaging ultraviolet sunlight (UV-B), allowing only a small amount to reach the earth's surface. The absorption of ultraviolet radiation by ozone generates heat, which is why Figure 1.12 shows an increase in temperature in the stratosphere. Without the filtering action of the ozone layer, greater amounts of the Sun's UV-B radiation can penetrate the atmosphere and reach the earth's surface. Many experimental studies of plants and animals as well as clinical studies of humans have shown the harmful effects of excessive exposure to UV-B radiation.

In the troposphere, O_3 exposure is destructive ("bad ozone"), because it is highly reactive with tissues, leading to ecological and welfare effects, such as forest damage and reduced crop production, and human health effects, especially cardiopulmonary diseases.

Chlorofluorocarbons (CFCs), along with other chlorine- and bromine-containing compounds, can accelerate the depletion of stratospheric O_3 layer. CFCs were first developed in the early 1930s for many industrial, commercial, and household products. They are generally compressible, nonflammable, and nonreactive, which led to

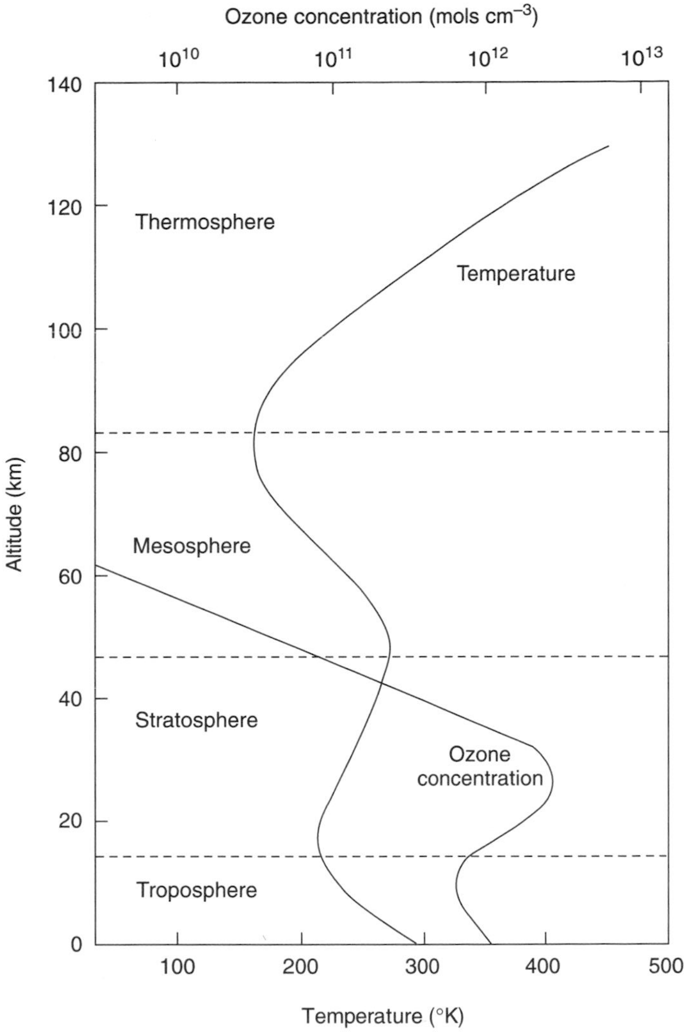

FIGURE 1.12. Ozone concentrations and temperature profile of the earth's atmosphere. (Source: R. T. Watson, M. A. Geller, R. S. Stolarski, and R. F. Hampson, 1986, *Present State of Knowledge of the Upper Atmosphere: An Assessment Report*, NASA Reference Publication.)

many CFC uses, including as coolants for commercial and home refrigeration units and aerosol propellants. In 1973, chlorine was found to catalyze ozone destruction. Catalytic destruction of ozone removes the odd-numbered oxygen species (atomic oxygen [O] and ozone [O_3]), but leaves the chlorine unaffected. A complex scenario involving atmospheric dynamics, solar radiation, and chemical reactions accounts for spring thinning of the ozone layer at the earth's poles. Global monitoring of ozone levels from space using NASA's Total Ozone Mapping Spectrometer (TOMS) instrument has shown significant downward trends in ozone concentrations at all of the earth's latitudes, except the tropics. Even with the international bans and actions, stratospheric ozone levels are expected to be lower than predepletion levels for many years because CFCs are persistent in the troposphere, from where they are transported to the stratosphere. In the high-energy stratosphere, the compounds undergo hundreds of catalytic cycles involving O_3 before the CFCs are scavenged by other chemicals.

A search for more environmentally benign substances has been underway for some time. One set of potential substitutes is the hydrochlorofluorocarbons (HCFCs). Obviously, the HCFC molecules contain Cl atoms, but the hydrogen increases the reactivity of the HCFCs with other tropospheric chemical species. These low-altitude reactions decrease the probability of a Cl atom finding its way to the stratosphere. Hydrofluorocarbons (HFCs), which lack chlorine, are potential substitutes for CFCs.

The new Clean Air Act Amendments of 1990 built upon the Montreal Protocol on Substances that Deplete the Ozone Layer, the international treaty in which nations agreed to reduce or eliminate ozone-destroying gas production and uses of chemicals that pose a threat to the ozone layer. The amendments further restricted the use, emissions, and disposal of these chemicals, including the phasing out of the production of CFCs, as well as other chemicals that lead to ozone-attacking halogens, such as tetrachloromethane (commonly called carbon tetrachloride) and methyl bromide by the year 2000, and methyl chloroform by 2002. The new Clean Air Act Amendments will also will freeze the production of CFCs in 2015 and require that CFCs be phased out completely by 2030. Companies servicing air conditioning for vehicles are now required to purchase certified recycling equipment and train employees to properly use and dispose of ozone-depletive chemicals. EPA regulations require reduced emissions from all other refrigeration sectors to the lowest achievable levels. "Nonessential" CFC applications are prohibited. The Clean Air Act increases the labeling requirements of the Toxic Substances Control Act by mandating the placement of warning labels on all containers and products (such as cooling equipment, refrigerators, and insulation) that contain CFCs and other ozone-depleting chemicals.

Water Quality Legislation

Environmental water laws come in two forms: those aimed at providing "clean water" to drink and use, and those cleaning up "dirty water." [Section 101(a)(2)] The Safe Drinking Water Act is the principal federal law designed to provide the U.S. population with potable water. The Clean Water Act addresses the many aspects of water pollution.

Drinking Water

The Safe Drinking Water Act (SDWA) was passed in 1974 to protect public drinking water supplies from harmful contaminants, assuring that the concentrations of these contaminants in drinking water stay below Maximum Contaminant Levels (MCLs). The Act, as amended, authorizes a set of regulatory programs to establish standards and treatment requirements for drinking water, as well as the control of underground injection of wastes that may contaminate water supplies and the protection of groundwater resources.

The SDWA was extensively amended in 1986 to strengthen the standard-setting procedures, increase enforcement authority, and provide for additional groundwater protection programs. The U.S. EPA was mandated to issue drinking water regulations for 83 specified contaminants by 1989 and for 25 additional contaminants every three years thereafter (see Appendix 2). The 1986 Amendments also required that all public water systems using surface water disinfect, and possibly filter, water supplies. Thus far, the EPA has regulated 84 contaminants. The law covers all public water systems with piped water to be used for human consumption with at least 15 service connections or a system that regularly serves at least 25 people.

Most serious violations of drinking water regulations have occurred in small water systems serving populations of less than 3300. This is often the result of limited financial and technical resources that can be devoted to water monitoring and treatment. Because 87% of all community water systems are small, amendments to the SDWA have focused on increasing the abilities of these communities to meet regulatory requirements, known as "capacity building."

New emphases have been on how to balance risks and costs in setting standards, how and whether to discourage the formation of new drinking water systems that are unlikely to comply, and the appropriate state and federal roles in providing high-quality water supplies.

Water Pollution Abatement

The Clean Water Act (CWA), passed in 1972, represents the myriad programs aimed at surface water quality protection in the United States, employing a variety of regulatory and nonregulatory approaches needed to reduce direct pollutant discharges into U.S. waterways. This has been accomplished through the issuance of effluent discharge permits, designation of water quality protection levels for water bodies, financing municipal wastewater treatment facilities, and managing nonpoint sources to allow polluted runoff to enter surface waters. The goal of these actions is to restore and to maintain the chemical, physical, and biological integrity of the nation's waters so that they can support, in the words of the Act, "the protection and propagation of fish, shellfish, and wildlife and recreation in and on the water." [Section 101(a)(2)]

For years after the CWA became law, the major focus of the federal government, states, and Native American tribes was on the chemical aspects of the "integrity" goal. More recently, greater attention has been paid to other provisions of the Act to maintain physical and biological integrity. The implementation of the law has also moved from almost completely focusing on the regulation of pollutant discharges from "point source" facilities, like municipal wastewater treatment plants and industrial facilities, to now giving greater attention to pollution from mining operations, roads, construction sites, farms, and other "nonpoint sources."

In many ways, nonpoint problems are more intractable than point sources, because they more heavily depend on management and comprehensive planning programs. Some successful nonpoint programs have included voluntary programs, like cost sharing with landowners. So-called "wet weather point sources" such as urban storm sewer systems and construction sites, require regulatory actions.

The traditional "command and control" programs of enforcement and compliance programs have been evolving into a greater number of programs that consist of comprehensive watershed-based strategies, where more equity exists between the protection of healthy waters and the restoration of impaired surface waters. Such holistic approaches depend heavily on public involvement and coalition building to achieve water quality objectives that are both technically sound and publicly acceptable.

Solid and Hazardous Wastes Laws

The two principal U.S. laws governing solid wastes are the Resource Conservation and Recovery Act (RCRA) and the Superfund. The RCRA law covers both hazardous and solid wastes, while Superfund and its amendments generally address abandoned hazardous waste sites. The RCRA addresses active hazardous waste sites.

Management of Active Hazardous Waste Facilities

With the RCRA, the U.S. EPA received the authority to control hazardous waste throughout the waste's entire life cycle, known as "cradle-to-grave."

This means that manifests must be prepared to keep track of the waste, including its generation, transportation, treatment, storage, and disposal. The RCRA also set forth a framework for the management of nonhazardous wastes in Subtitle D.

The Federal Hazardous and Solid Waste Amendments (HSWA) to the RCRA required the phase out of the land disposal of hazardous waste. HSWA also increased the federal enforcement authority related to hazardous waste actions, set more stringent hazardous waste management standards, and provided for a comprehensive underground storage tank program.

The 1986 amendments to the RCRA allowed the federal government to address potential environmental problems from underground storage tanks (USTs) for petroleum and other hazardous substances.

Addressing Abandoned Hazardous Wastes

The Comprehensive Environmental Response, Compensation, and Liability Act (CERCLA) is commonly known as Superfund. Congress enacted it in 1980 to create a tax on the chemical and petroleum industries and to provide extensive federal authority for responding directly to releases or threatened releases of hazardous substances that may endanger public health or the environment.

The Superfund law established prohibitions and requirements concerning closed and abandoned hazardous waste sites; established provisions for the liability of persons responsible for releases of hazardous waste at these sites; and established a trust fund to provide for cleanup when no responsible party could be identified.

The CERCLA response actions include:

- Short-term removals, where actions may be taken to address releases or threatened releases requiring prompt response. This is intended to eliminate or reduce exposures to possible contaminants.
- Long-term remedial response actions to reduce or eliminate the hazards and risks associated with releases or threats of releases of hazardous substances that are serious, but not immediately life threatening. These actions can be conducted only at sites listed on the EPA's National Priorities List (NPL).

Superfund also revised the National Contingency Plan (NCP), which sets guidelines and procedures required when responding to releases and threatened releases of hazardous substances.

CERCLA was amended by the Superfund Amendments and Reauthorization Act (SARA) in 1986. These amendments stressed the importance of permanent remedies and innovative treatment technologies in cleaning up hazardous waste sites. SARA required that Superfund actions consider the standards and requirements found in other State and Federal environmental laws and regulations and provided revised enforcement authorities and new settlement tools. The amendments also increased State involvement in every aspect of the Superfund program, increased the focus on human health problems posed by hazardous waste sites, encouraged more extensive citizen participation in site cleanup decisions, and increased the size of the Superfund trust fund.

SARA also mandated that the Hazard Ranking System (HRS) be revised to make sure of the adequacy of the assessment of the relative degree of risk to human health and the environment posed by uncontrolled hazardous waste sites that may be placed on the National Priorities List (NPL).

Environmental Product and Consumer Protection Laws

Although most of the authorizing legislation targeted at protecting and improving the environment is based on actions needed in specific media, that is, air, water, soil, and sediment, some laws have been written in an attempt to prevent environmental and public health problems while products are being developed and before their usage.

The predominant product laws designed to protect the environment are the Federal Food, Drug, and Cosmetics Act (FFDCA); the Federal Insecticide, Fungicide, and Rodenticide Act (FIFRA); and the Toxic Substances Control Act (TSCA). These three laws look at products in terms of potential risks for yet-to-be-released products and estimated risks for products already in use. If the risks are unacceptable, new products may not be released as formulated or the uses will be strictly limited to applications that meet minimum risk standards. For products already in the marketplace, the risks are periodically reviewed. For example, pesticides have to be periodically reregistered with the government. This reregistration process consists of reviews of new research and information regarding health and environmental impacts discovered since the product's last registration.

FIFRA's major mandate is to control the distribution, sale, and applications of pesticides. This not only includes studying the health and environmental consequences of pesticide usage, but also to require that those applying the pesticides register when they purchase the products. Commercial applicators must be certified by successfully passing exams on the safe use of pesticides. FIFRA requires that the EPA license any pesticide used in the United States. The licensing and registration makes sure that pesticide is properly labeled and will not cause unreasonable environmental harm.

An important, recent product production law is the Food Protection Act (FQPA), including new provisions to protect children and limit their risks from carcinogens and other toxic substances. The law is actually an amendment to the FIFRA and FFDCA that includes new requirements for safety standard–reasonable certainty of no harm that must be applied to all pesticides used on foods. FQPA mandates a single, health-based standard for all pesticides in all foods; gives special protections for infants and children; expedites approval of pesticides likely to be safer than those in use; provides incentives for effective crop protection tools for farmers; and requires regular reevaluation of pesticide registrations and tolerances so that the scientific data supporting pesticide registrations includes current findings.

Another product-related development in recent years is the screening program for endocrine disrupting substances. Research suggests a link between exposure to certain chemicals and damage to the endocrine system in humans and wildlife. Because of the potentially serious consequences of human exposure to endocrine disrupting chemicals, Congress added specific language on endocrine disruption in the FQPA and recent amendments to the SDWA. The FQPA mandated that the EPA develop an endocrine disruptor screening program, and the SDWA authorizes the EPA to screen endocrine disruptors found in drinking water systems. The newly developed Endocrine Disruptor Screening Program focuses on methods and procedures to detect and to characterize the endocrine activity of pesticides and other chemicals (see Figure 1.13). The scientific data needed for the estimated 87,000 chemicals in commerce does not exist to conduct adequate assessments of potential risks. The screening program is being used by the EPA to collect this information for endocrine disruptors and to decide on appropriate regulatory action by first assigning each chemical to an endocrine disruption category.

Chemicals will undergo sorting into four categories according to the available existing, scientifically relevant information:

- Category 1 chemicals have sufficient, scientifically relevant information to determine that they are not likely to interact with the estrogen, androgen, or thyroid systems. This category includes some polymers and certain exempted chemicals.
- Category 2 chemicals have insufficient information to determine whether they are likely to interact with the estrogen, androgen, or thyroid systems, thus they will need screening data.
- Category 3 chemicals have sufficient screening data to indicate endocrine activity, but data to characterize actual effects are inadequate and will need testing.
- Category 4 chemicals already have sufficient data for the EPA to perform a hazard assessment.

The TSCA gives the EPA the authority to track 75,000 industrial chemicals currently produced or imported into the United States. This is accomplished through screening of the chemicals and requiring that reporting and testing be done for any substance that presents a hazard to human health or the environment. If a chemical poses a potential or actual risk that is unreasonable, the EPA may ban the manufacture and import of that chemical.

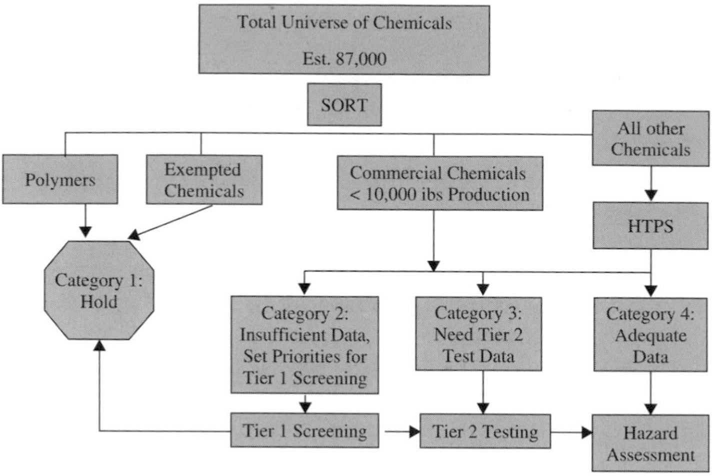

FIGURE 1.13. Endocrine Disruptor Screening Program of the U.S. Environmental Protection Agency. Note: HTPS = High Throughput Prescreening Program. (Source: U.S. EPA, 2000, "Report to Congress: Endocrine Disruptor Screening Program.")

The EPA has tracked thousands of new chemicals being developed by industries each year, if those chemicals have either unknown or dangerous characteristics. This information is used to determine the type of control that these chemicals would need to protect human health and the environment. Manufacturers and importers of chemical substances first submitted information about chemical substances already on the market during an initial inventory. Since the initial inventory was published, commercial manufacturers or importers of substances not on the inventory have been required to submit notices to the EPA, which has developed guidance about how to identify chemical substances to assign a unique and unambiguous description of each substance for the inventory. The categories include:

- polymeric substances;
- certain chemical substances containing varying carbon chains;
- products containing two or more substances, formulated and statutory mixtures; and chemical substances of unknown or variable composition, complex reaction products, and biological materials (UVCB substance).

Now that we have seen the importance of environmental policy and health assessments, let us turn our attention to the fundamental science underlying sound decision making. We are particularly interested in the principles that lead to a contaminant's movement and changes, and the interfaces of the contaminant with the abiotic and biotic components of the environment.

Notes and Commentary

1. Born in 1493 as Theophratus Phillipus Bombastas von Hohenheim and considered by many to be the founder of modern epidemiology and toxicology.
2. Born in 1578, and published in 1628 *An Anatomical Study of the Motion of the Heart and of the Blood of Animals* considered to be the seminal work in explaining the circulatory system.
3. The term *abiotic* includes all elements of the environment that are nonliving. What is living and nonliving may appear to be a straightforward dichotomy, but so much of what we call "ecosystems" is a mixture. For example, some soils are completely abiotic (e.g., clean sands), but others are rich in biotic components, such as soil microbes. Vegetation, such as roots and rhizomes, are part of the soil column (especially in the "A" horizon or topsoil). Formerly living substances, such as detritus, exist as lignin and cellulose in the soil organic matter (SOM). In fact, one of the problems with toxic chemicals is that the biocidal properties kill living organisms, reducing or eliminating the soil's productivity.
4. This quote comes from P. Aarne Vesilind, J. Jeffrey Peirce, and Ruth F. Weiner, 2003, *Environmental Engineering*, 4th Edition, Butterworth-Heinemann, Boston, Mass., p. xiii. The text is an excellent introduction to the field of environmental engineering and one of the sources of inspiration for this book.
5. *Webster's Ninth New Collegiate Dictionary*, 1990, Merriam-Webster, Springfield, Mass.
6. The source for this section is the U.S. Environmental Protection Agency's website: http://www.epa.gov/glnpo/aoc/index.html.
7. Common chemical nomenclature is to show the phases in parentheses: (aq) for aqueous, (g) for gas, (s) for solid. A downward pointing arrow (↓) depicts a precipitate and an upward pointing arrow (↑) means the release of a gas.
8. This gas relationship was articulated by J. W. Henry in 1800, long before the accepted concepts of chemical equilibria were established. Chemistry has had more than its share of such amazing intuition or creative thinking.
9. The troposphere is the part of the atmosphere where we live, the layer of air from the surface to an elevation of between 9 and 16 km.
10. This text intentionally uses different conventions in explaining concepts and providing examples. One reason is that is how the information is presented. Environmental information comes in many forms. A telling example is the convention the use of K. In hydrogeology, this means hydraulic conductivity, in chemistry it is an equilibrium constant, and in engineering it can be a number of coefficients. Likewise, units other than metric will be used on occasion, because that is the convention in some areas, and because it demonstrates the need to apply proper dimensional analysis and conversions. Many mistakes have been made in these two areas!
11. T. Loxnachar, K. Brown, T. Cooper, and M. Milford, 1999, *Sustaining Our Soils and Society,* American Geological Institute, Soil Science Society of America, USDA Natural Resource Conservation Service, Washington, D.C.

12. "Good" or "bad" solutes can either be those that when present in any concentration are harmful (e.g., mercury or dioxin), or those that are beneficial, but that become harmful at higher concentrations (e.g., $CO_2 > 350$ ppm).
13. V. P. Evangelou, 1998, *Environmental Soil and Water Chemistry: Principles and Applications*, John Wiley & Sons, New York, N.Y.
14. Team projects are very useful in engineering. The fields of hazardous waste engineering and science are so diverse and mutually dependent that it is always beneficial to have varied perspectives, including students with an aptitude for health, engineering, and law. This does not preclude the instructor or team leader from ensuring that each student or team member gains an understanding of all the aspects involved (e.g., the member interested in health issues should not be allowed to go on "cruise control" while a fellow team member deals with engineering). However, much can be gained by having the students teach one another from their own experiences and perspectives.
15. *Warning:* If this is done as a seminar, homework, workshop or classroom project, be clear that only information that may be shared with the whole class should be included. The confidentially and privacy of participants must be held paramount.
16. In 1917, the engineer Henry Gantt developed a horizontal bar chart to keep track of production. The charts are widely used today for quality control and project tracking. They can be produced using various spreadsheet and project management software packages.
17. Program Evaluation and Review Technique (PERT) is an event tracking system for steps beginning with planning through project completion.
18. Many community resources are available, from formal public meetings held by governmental authorities to informal groups, such as homeowner association meetings and neighborhood "watch" and crime prevention group meetings. Any research related activities should adhere to federal and other governmental regulations regarding privacy, intrusion, and human subjects considerations. Privacy rules have been written according to the Privacy Act and the Paperwork Reduction Act (e.g., the Office of Management and Budget limits the type and amount of information that U.S. agencies may collect in what is referred to as an Information Collection Budget).

    Any research that affects human subjects, at a minimum, should have prior approval for informed consent of participants and thoughtful consideration of the need for an institutional review board (IRB) approval. See Figure 13.2 and Table 13.2 for some highlights of the federal decision-making process for projects involving human subject approvals.
19. Pronounced "Fonzy," as in the nickname for the character Arthur Fonzerelli portrayed by Henry Winkler in the television show, *Happy Days*.
20. This is understandable if the agency is in the business of something not directly related to environmental work, but even the natural resources and environmental agencies have asserted that there is no significant impact to their projects. It causes the cynic to ask, then, why are they engaged in any project that

has no significant impact? The answer is that the term "significant impact" is really understood to mean "significant adverse impact" to the human environment.

21. This quote was passed on to me by Timothy Kubiak, one of Professor Caldwell's former graduate students in Indiana University's Environmental Policy Program. Kubiak has since gone on to become a successful environmental policymaker in his own right, first at EPA and then at the U.S. Fish and Wildlife Service. By the way, Kubiak was one of my "senior Euchre partners!"
22. The National Pollutant Discharge Elimination System (NPDES) Permit, under Section 401 of the Federal Water Pollution Control Act Amendments of 1972, was designed to control the type and amount of effluent going into the waters of the United States.
23. 40 CFR 1507.3
24. Pun intended! (Or should it be an "air shed" year?)
25. Traditionally, the sources of fecal matter have been identified by counting the colonies of bacteria that ferment glucose at 44.5°C. Two major subdivisions of bacteria have been coliform and streptococci (spherically shaped). The ratios were considered somewhat unique for different animal genera, especially the ratios for humans (e.g., in wastewater effluent) and the ratios for farm animals. For example, it has been a general rule that a fecal strep/fecal coliform (FS/FC) > 4 indicates a human source, while FS/FC < 0.7 is likely to be a farm animal source, particularly poultry or livestock. However, the current version (20th Edition) of *Standard Methods for Examination of Water and Wastewater*, 1998, L.S. Clesceri, A.D. Eaton, and A.E. Greenberg (editors), American Public Health Association, Washington, D.C., does not recommend using these ratios because the streptococci counts have been found to vary in the environment. Some species survive, while others die off quickly. Also, the various streptococcus species' sensitivity to disinfectants is highly variable.

    Current research is attempting to improve the ratio. This includes the use of antibiotic resistance in coliforms and streptococci as a means to identify sources (e.g., agriculture, treatment plants, and septic tanks). Another approach being explored is DNA identification techniques.
26. For a complete explanation of the Henry's Law constant, including how it is calculated and example problems, see Chapter 5, "Movement of Contaminants in the Environment".
27. Sources for this discussion are "Frequently Asked Questions" of the World Meteorological Organization/United Nations Environment Programme report, Scientific Assessment of Ozone Depletion: 1998, WMO Global Ozone Research and Monitoring Project-Report No. 44, Geneva, 1999; and the Center for International Earth Science Information Network (CIESIN) website: http://www.ciesin.org/index.html.

Part II

Fundamentals of Environmental Science and Engineering

Introduction to Part II

Environmental risk assessment is a multifaceted and complex mix of science, engineering, and technology. Risk assessment is a process distinct from risk management, but the two are deeply interrelated and require continuous feedback with one another. What really sets risk assessment apart from the actual management and policy decisions is that it must follow the prototypical rigors of scientific investigation and interpretation. All of the science upon which assessments and decisions are made must adhere to tenets of science, beginning with the *scientific method*. In the absence of this sound scientific foundation, there is no reason to think that the proper environmental decisions will be made and, to the contrary, very dangerous outcomes can result.

Two types of dangers can occur in the absence of sound science. The first is the *false negative*, or reporting that there is no problem when one in fact exists. This may be at the forefront of the positions taken by environmental and public health agencies and advocacy groups. They ask questions like:

- What if this substance really does cause cancer but the tests are unreliable?
- What if people are in fact being exposed, but we don't know enough about how the contaminant is moving though the environment?
- What if there is a relationship that is different from the laboratory when this substance is released into the "real world," such as the difference between how a chemical behaves in the human body by itself as opposed to when other chemicals are present?

The second type of concern is the *false positive*. This can be a big problem for public health agencies charged with protecting populations from exposures to hazardous substances. For example, what if previous evidence shows that an agency had listed a compound as a potential endocrine disruptor, only to find that a wealth of new information is now showing that it has no effect? Perhaps the conclusions were based upon faulty models, or models that only work well for lower organisms, but later models that were developed took into consideration the physical, chemical, and biological complexities of higher-level organisms, including humans. The problems associated with false positives is that they may mean we are spending time and resources to deal with so-called "nonproblems," that we have inappropriately sent out alarms over potentially useful products, and that the scientific and policy community has lost credibility with the people we have been asked to protect.

Both types of problems are rooted in science. Therefore, professionalism demands that those of us engaged in environmental risk assessment have a working knowledge of the basic and applied sciences upon which risk is determined.

In Part II we will focus on the important scientific principles and concepts that one must apply to assess risks. Only after the risks have been adequately identified and, usually, quantified with an appropriate degree of certainly can the relevant remedies and actions be put into place to respond to the risks, reduce them, and, we hope, eliminate them.

Therefore, we will spend ample time addressing the science needed. Each compartment of the risk assessment paradigm, shown in Figure II.1 is

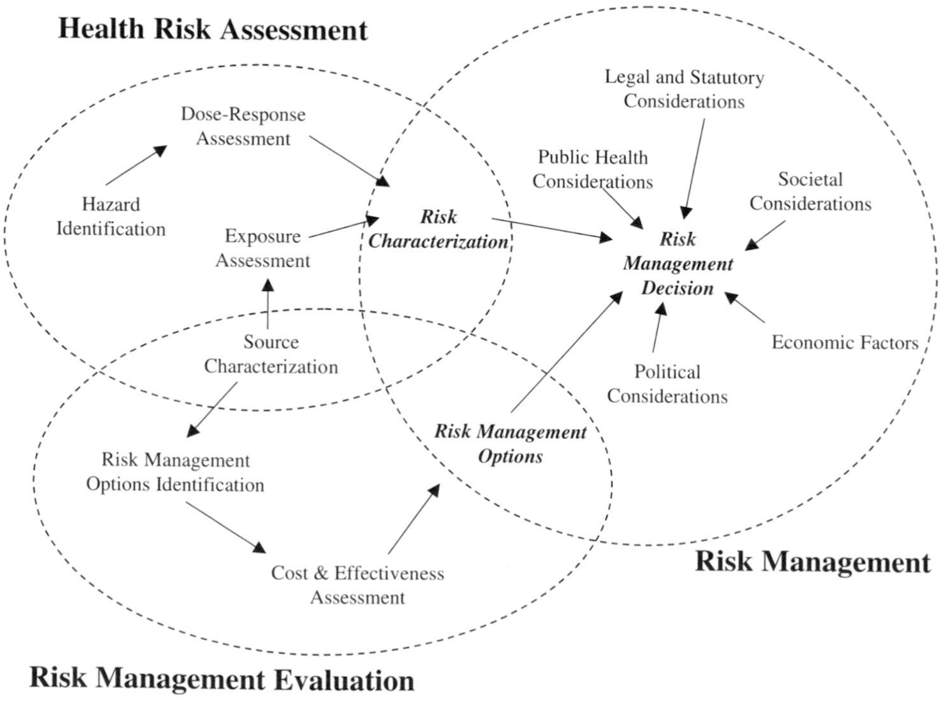

FIGURE II.1. The risk assessment paradigm. (Source: U.S. Environmental Protection Agency.)

underpinned by science. To use engineering language, the risk assessment process is a "critical path" in which any unacceptable error or uncertainty along the way will lead to an inadequate risk assessment and, quite likely, a bad environmental decision.

Importance of Physics in Environmental Contamination and Risk

From the perspective of the engineer, everything begins with physics. It is how we understand matter and energy. As such, it provides the basis for chemistry, which in turn establishes biological principles. How things move and how efficiently energy is transferred among compartments lie at the heart of understanding pollution. Every step in risk assessment requires an understanding of physics. First, the determination of whether a substance is *hazardous* is a physical concept.

Human health effects and exposure assessments require that the movement of and changes to contaminants be understood. This requires a solid grounding in the physical, chemical, and biological principles covered in Part II. Likewise, ecological risk assessments require an understanding of physicochemical, hydrological, and geological concepts to appreciate how chemical nutrients and contaminants cycle through the environment, how physical changes may impact ecosystems, and the many possible ways that ecological resources are put at risk. In a very basic way, we must consider first principles to make rigorous environmental assessments and to give reliable information to the engineers and decision makers who will respond to health and ecological risks. Good policy stands on the shoulders of quality science.

Let us note a few areas of environmental risk that are heavily dependent upon physics. First, energy is often described as a system's capacity to do work, so getting things done in the environment is really an expression of how efficiently energy is transformed from one form to another. Energy and matter relationships determine how things move in the environment. That is why we begin our discussion of applied physics with environmental transport. The physical movement of contaminants among environmental *compartments* is central to risk assessment. After a contaminant is released, physical processes will go to work on transporting the contaminant and allow for *receptors* (like people and ecosystems) to be exposed. Transport is one of two processes (the other is *transformation*) that determine a contaminant's *fate* in the environment. Figure II.2 shows the major steps needed to study a contaminant as it moves through the environment.

Much of what scientists and engineers consider in environmental contamination deals with fluids. The fluids that are important at all scales, from molecular to global, are water and air. However, fluid properties, statics, and dynamics are involved at every step in the risk assessment process. To identify a hazard and dose-response associated with the

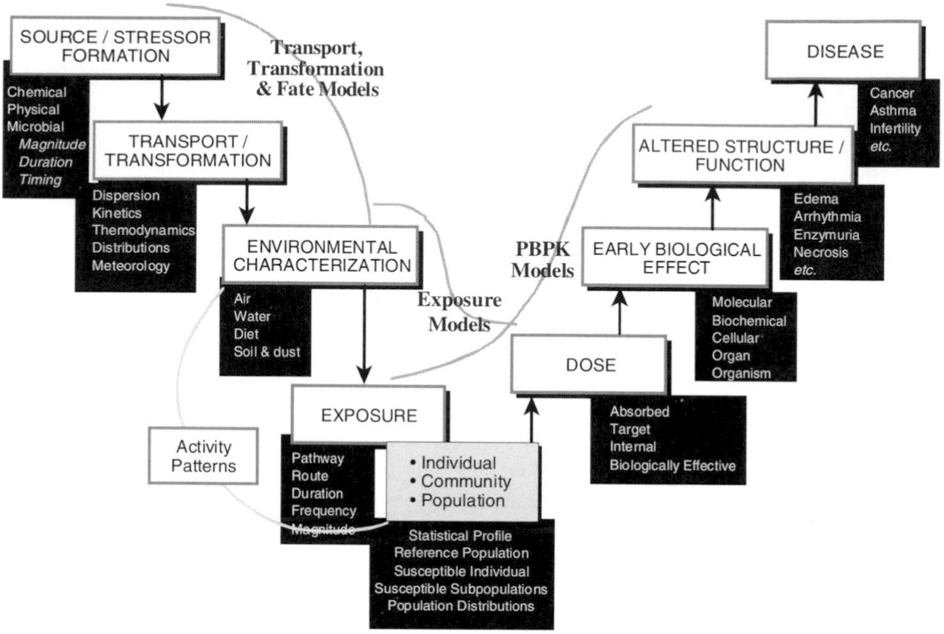

FIGURE II.2. Components of exposure science: From characterizing the source of a pollutant to its transport, transformation, and fate after release, to its ultimate contact with human beings and ecosystems, an exposure assessment must be based upon scientific information of the highest possible quality. (Source: U.S. Environmental Protection Agency.) Note: PBPK = Physiologically based, phramacokinetic.

chemical, the fluid properties must be understood. For example, if a contaminant's fluid properties make it insoluble in water and blood, then the target tissues may be more likely to be lipids. If a chemical is easily absorbed, the hazard may be higher. However, if it does not change phases under certain cellular conditions, it could be more or less toxic, depending on the organ.

In determining exposures, in addition to the transport phenomena already mentioned, the fluid properties of a chemical or biological agent may determine how and where the contaminant is likely to be found in the environment (e.g., in the air as a vapor, sorbed to a particle, dissolved in water, or taken up by biota).

Another perspective on the importance of physicochemical transport and fate to exposure assessment is the observation that physical models often bridge the sources to the effects, whether they are ecological or health. Figure II.3 shows that even if one has high-quality emission or release data for a contaminant, some means of reducing, digesting, and interpreting

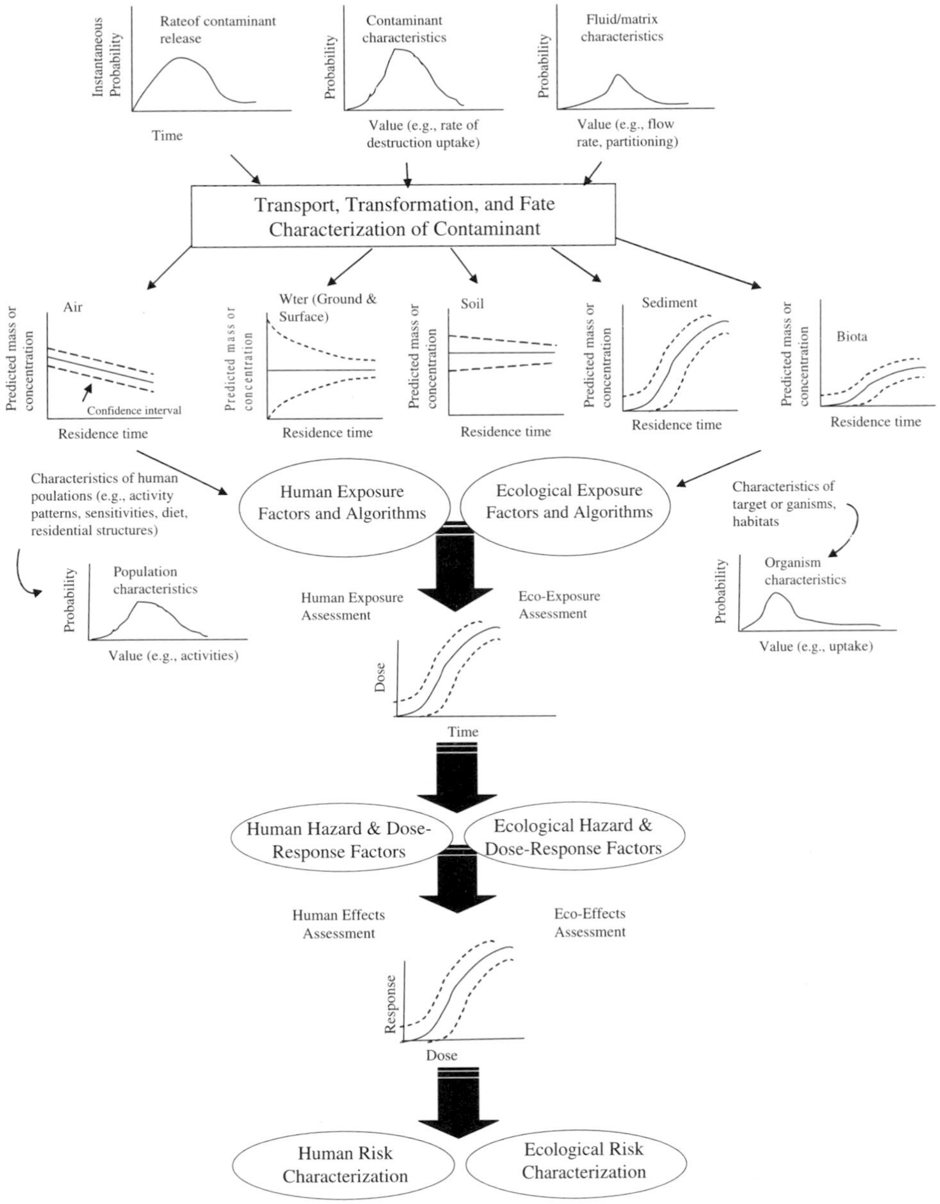

FIGURE II.3. Importance of transport, transformation, and fate processes to human and ecological exposure and risk assessments. (Source: Top three layers (ecological components) adapted from: G. Suter, 1993, "Predictive Risk Assessment of Chemicals," in *Ecological Risk Assessment*, edited by G. Suter, Lewis Publishers, Chelsea, Mich.)

those data and linking them to environmental compartments is crucial to exposure assessments and, ultimately, risk characterizations.

Energy is stored and used in organisms. The thermodynamics involved as the energy is transformed and released in the environment will determine the efficiency and health of microbes as they help us treat contaminants, how ecosystems transform the sun's energy throughout the food chain, and which chemical reactions will break down contaminants, metabolize substances, and ultimately participate in the toxicological response of organisms, including humans. Therefore, physical mechanisms will be revisited in discussions of bioremediation in Chapter 8, "Biological Principles of Environmental Contamination" and in discussions of biological treatment in Chapter 12, "Intervention: Managing the Risks of Environmental Contamination."

Importance of Chemistry in Environmental Contamination and Risk

Chemistry may be the first scientific discipline that one thinks of when addressing risks. The literature is ripe with lists of chemical compounds that have been associated with diseases. Much of risk assessment follows the toxicological paradigm (which is really an enhancement of the pharmacological paradigm), wherein chemicals are considered from a dose-response perspective (see Figure II.3). Indeed, exposure assessments rely on analytical chemistry in determining the presence and quantity of a chemical contaminant in the environment. Also, toxicology begins with an understanding of the chemical characteristics of an agent, followed by investigations of how the agent changes after release and the steps leading up to the adverse effect. Whether the chemistry is inorganic or organic; whether it is induced, mediated, or complemented by biological processes or is simply abiotic chemistry; and whether it is applied at the molecular or global scale, every risk assessment, every intervention, and every engineering activity is a chemical expression.

The reader is here introduced to environmental chemistry by building on the introductory physics and contaminant transport discussions. Several of the topics considered in transport provide a transition from physics to environmental chemistry, especially phase partitioning and fluid properties. In fact, no environmental chemistry discussion is complete without discussions of mass balance and partitioning. Solubility, sorption, and volatilization could just as easily have been included under chemistry as physics. And the biological processes are indeed extensions and elaborations of the physical and chemical mechanisms.

The discussion begins with environmental inorganic chemistry, since many of the principles and concepts also apply to organic and organometallic chemistry. The abiotic chemical processes are generally inclusive of the biochemical processes. For example, metabolism, respiration, and growth combine numerous processes that are introduced in the inorganic and organic chemistry discussions, such as redox, hydrolysis, and ionization.

Importance of Biology in Environmental Contamination and Risk

Risk assessment is in essence biological. The burgeoning field of risk assessment can be considered to be a discipline of the life sciences that asks how living things are affected by exposures to environmental contaminants. Also, biological processes are involved in the transformation of contaminants after release (see Figures II.2 and II.3). Microbial biology, biophysics, toxicology, and, ultimately, risk characterization are all part of the life cycle of an environmental agent. Biology becomes an integrator of physics and chemistry. Important biological processes like photosynthesis, metabolism, and respiration are essentially energy transfer and cell synthesis systems. Thus, whether in a unicellular organism, like a bacterium, or a complex mammalian system, the way that energy and matter change and move will determine the health of a person or an ecosystem, whether a treatment process is effective, and whether the deoxyribonucleic acid (DNA) strands will signal a healthy cell or cancer.

Another aspect of biology has to do with extrapolation. What can we learn from studies of cells in petri dishes about a chemical's potential effects? If a chemical causes a mutation in a bacterium, will it cause this or some other mutation, including cancer, in higher organisms? Biological principles can guide us as we model expectations in humans from these simple studies, as well as from comparative studies in animals. Likewise, biology is a foundation for *epidemiology*, which is often employed by risk assessors. If a population or a sector of the population is exposed to a contaminant, what adverse effects can be expected? Are the effects seen to be more severe in certain subpopulations? What is the "expected" or background levels of exposures, and what happens to people who have been exposed? In the emerging area of eco-epidemiology, the effects of contaminants on ecosystem status and condition are considered. Biological indicators are needed to determine the health of either a human population or an ecosystem.

Beyond Basic Science

Part II also introduces some of the "derived" sciences, such as geology, hydrology, hydrogeology, and meteorology, insofar as they relate to environmental risk, especially with regard to contaminant transport. These are extensions of the physical and chemical sciences. These sciences will further apply the foundations covered in the first chapters of Part II. They are commonly used in risk assessments and need to be covered before we discuss remedial responses, treatment technologies, and risk management decisions.

CHAPTER 2

Fundamentals of Environmental Physics

Physics is the science that concerns itself with matter, energy, motion, and force. Some would argue that all other sciences are simply branches of physics. Even chemistry, which is the science that deals with the composition, properties, transformations, and forms of matter, is merely a discipline within physics. We will not enter into this argument, but it is a challenge when describing environmental phenomena to draw "bright lines" between physics and chemistry. When necessary, then, some of the concepts introduced here will be addressed again in Chapter 6, Fundamentals of "Environmental Chemistry," Chapter 7, "Chemical Reactions in the Environment," and Chapter 8, "Biological Principles of Environmental Contamination," as well as in the applications and remediation engineering discussions in Part III, "Contaminant Risk."

Often, in this book and elsewhere, the dichotomy is avoided by using the term *physicochemical* to address properties and characteristics that are included in both physics and chemistry. Force, velocity, flow rates, discharge, and friction are clearly terms of physics. Likewise, redox, acidity/alkalinity, stoichiometry, and chirality are terms of chemistry. However, kinetics, sorption, solubility, vapor pressure, and fugacity are "physiochemical" terms.

This chapter provides an introduction to the branches of physics that are important to environmental science and engineering. We will cover those areas that directly apply to natural resources and public health, identifying the fundamentals of energy and mass as they apply to environmental solids and fluids. The next chapter will address the application and extension of these fundamentals to the physical transport of pollutants.

Principles and Concepts of Energy and Matter Important to the Environment

Every crucial environmental issue or problem can be represented, explained, and resolved using energy and matter fundamentals. How contaminants are formed, how they change and move through the environment, the diseases and problems they cause, and the types of treatment technologies needed to eliminate them or reduce the exposures of people and ecosystems can be seen through the prisms of energy and matter.

The relationship between energy and matter has only recently been characterized scientifically. Most simply, energy is the ability to do work, and work involves motion. *Kinetic energy* is the energy that is due to motion. The kinetic energy of a mass (m) moving with the velocity v is:

$$E_{\text{kinetic}} = \frac{1}{2}mv^2 \qquad \text{Equation 2–1}$$

We will begin our discussions of energy with a discussion of environmental mechanics.

Energy is also defined as the ability to cause change, so this aspect of energy will be considered in this chapter, as well as in the following chapters on environmental chemistry and environmental biology. Energy has a positional aspect to it. That is, *potential energy* is the energy resulting from one body with respect to another body. The potential energy of a mass m that is raised through a distance h is:

$$E_{\text{potential}} = mgh \qquad \text{Equation 2–2}$$

Where g = acceleration due to gravity.

Matter is anything that has both mass and volume. Matter is found in three basic *phases*: solids, liquids, and gases. The phases are very important for environmental science and engineering. The same substance in one phase may be relatively "safe," but in another phase very hazardous.[1] For example, in the solid and liquid phase, a highly toxic compound may be much more manageable than it is in the gas, particularly if the most dangerous route of exposure is via inhalation. Within the same phase, solid and liquid aerosols are more of a problem when they are very small than when they are large, because larger particles settle out earlier than do lighter particles, and small particles may penetrate airways more efficiently than coarse particles.

Mass and Work

Mass was introduced in the discussions of potential and kinetic energy, but we have yet to formally define it. Mass is the property of matter that is an expression of matter's inertia. We will consider the elements, forms, composition, and properties of matter in the Chapter 6, "Fundamentals of Environmental Chemistry."

The capacity of a mass to do work is known as the energy of the mass. This energy may be stored or it may be released. The energy may be mechanical, electrical, thermal, nuclear, or magnetic. The first four types have obvious importance to environmental applications. The movement of fluids as they carry pollutants is an example of mechanical energy. Electrical energy is applied in many treatment technologies, such as electrostatic precipitation (ESP), which changes the charge of particles in stack gases so that they may be collected rather than being released to the atmosphere. Thermal energy is important for incineration and sludge treatment processes. Nuclear energy is converted to heat that is used to form steam and turn a turbine by which mechanical energy is converted to electrical energy. The environmental problems and challenges associated with these energy conversions include heat transfer, release of radiation, and long half-lives of certain isotopes that are formed from fission. Even the fifth form, magnetic energy, has importance to environmental measurements in its application to gauges and meters.

Energy is a scalar quantity; that is, it is quantified by a single magnitude. This contrasts with a vector quantity, which has both magnitude and direction, and which we will discuss in some detail shortly. Although energy is a positive scalar quantity, a change in energy may be either positive or negative. A body's total energy can be ascertained from its mass, m, and its *specific energy*, or the amount of energy per unit mass. The *law of conservation of energy* states that energy cannot be created nor destroyed, but it may be converted among its different forms. In the environment, then, we often see the conversion of mechanical energy into electrical energy (e.g., a turbine), which in turn is converted to heat (hence the need for cooling before make up water[2] is added to replace the water that is lost due to evaporation cooling, as well as water lost as steam from the turbine, purging of boilers, stack washing, wastewater treatment, and water supply for plant employees). The key of the law is that the sum of all forms of energy remains constant:

$$\sum E = \text{constant} \qquad \text{Equation 2–3}$$

We have been using the term *work* but still have yet to define it completely. Work (W) is the act of changing the energy of a system or a body. An external force does external work, while internal work is done by an internal force. Work is positive when the force is acting in the direction of a motion, helping to move the body from one location to another, and work is negative when the force acts in the opposing direction; for example, friction can only do negative work in a system.

Returning to our brief discussion on potential energy and kinetic energy, potential energy is lost when the elevation of a body is decreased. The lost potential energy is usually converted to kinetic energy. If friction and other nonconservative forces are absent, the change in the potential energy of a body is equal to the work needed to change the elevation of the body:

$$W = \Delta E_{\text{potential}} \qquad \text{Equation 2–4}$$

The *work-energy principle* states that, in keeping with the conservation law, external work that is performed on a system will go into changing the system's total energy:

$$W = \Delta E = E_2 - E_1 \qquad \text{Equation 2–5}$$

This principle is generally limited to mechanical energy relationships.

Mass and Work Example 1

Calculate the work done by 4 million kg of effluent pumped from a sluice gate into a holding pond if the water starts from rest, accelerates uniformly to a constant stream velocity of $1\,\text{m}\,\text{sec}^{-1}$, then decelerates uniformly to stop 2 meters higher than the initial position in the sluice. Neglect friction and other losses.

Solution

Applying the work-energy principle, the work done on the effluent is equal to the change in the effluent's energy. Since the initial and final kinetic energy is zero (i.e., the effluent starts at rest and stops again), the only change in mechanical energy is the change in potential energy. Using the initial elevation of the effluent as the reference height, or $h_1 = 0$, then:

$$W = E_{2\,potential} - E_{1\,potential} = mg(h_2 - h_1)$$
$$= (4\times10^6\ \text{kg})(9.81\,\text{m}\,\text{sec}^{-2})(2\ \text{m}) = 7.85\times10^7\ \text{kg}\,\text{m}^{-1}\,\text{sec}^{-2} = \underline{7.85\times10^7\,\text{J}}$$

Converting one energy form to another is in keeping with the conservation law. Most conversions are actually special cases of the work-energy principle. If a falling body is acted on by gravity, for example, the conversion of potential energy into kinetic energy is really just a way of equating the work done by the gravitational force (constant) to the change in kinetic energy. *Joule's Law* states that one energy form can be converted to another energy form without loss. Regarding thermodynamic applications, Joule's law says that in internal energy of an ideal[3] gas is a function of the temperature change and not the change in volume.

Mass and Work Example 2

An aerosol weighing 2 μg is emitted from a stack straight up into the atmosphere with an initial velocity of $5\,\text{m}\,\text{sec}^{-1}$. Calculate the kinetic energy immediately following the stack emission. Ignore air friction and external forces, such as winds.

Solution

From Equation 2–1, we can calculate the kinetic energy:

$$E_{kinetic} = \frac{1}{2}\text{mv}^2 = \frac{1}{2}(2\times10^{-9}\,\text{kg})(5\,\text{m}\,\text{sec}^{-1})^2 = \underline{5\times10^{-7}\,\text{kg}\,\text{m}^2\,\text{sec}^{-2}}$$

Mass and Work Example 3

Calculate the kinetic energy and the potential energy at the maximum height for the problem in Mass and Work Example 2.

Solution

Wherever we find the point of maximum height, by definition the velocity is zero, so a close look at Equation 2–1 shows that the kinetic energy must also be zero. By definition, at the maximum height, all the kinetic energy has been converted to potential energy. So the value found earlier for the kinetic energy immediately after the emission is now the value for potential energy of the system; that is:

$$5\times10^{-7}\,\text{kg}\,\text{m}^2\,\text{sec}^{-2}$$

Mass and Work Example 4

What is the total energy in Mass and Work Example 3 at the elevation where the particle velocity has fallen to $0.5\,\text{m}\,\text{sec}^{-1}$?

Solution

Even though some (even most) of the kinetic energy has been converted to potential energy, the total energy of the system remains at $\underline{5\times10^{-7}\,\text{kg}\,\text{m}^2\,\text{sec}^{-2}}$.

Mass and Work Example 5

What is the maximum height reached by the aerosol?

Solution

All of the kinetic energy is converted to potential energy at the maximum height (no more energy is available to lift the particle). So we can use Equation 2–2:

$$E_{\text{potential}} = mgh$$

$$5 \times 10^{-7}\,\text{kg}\,\text{m}^2\,\text{sec}^{-2} = (2 \times 10^{-9}\,\text{kg}) \times (9.81\,\text{m}\,\text{sec}^{-2})h$$

$$h = \frac{(5 \times 10^{-7}\,\text{kg}\,\text{m}^2\,\text{sec}^{-2})}{(2 \times 10^{-9}\,\text{kg})(9.81\,\text{m}\,\text{sec}^{-2})}$$

$$h \cong 25.5\,\text{m}$$

Although particle matter can reach the winds aloft, it is not due to the ejection energy. That is, it is not as if the particle were a rocket launched with a great deal of energy and that much of the lift is owed to the initial launch. The particle's trek is determined by other energy sources along the way. Since we ignored friction and winds, we cannot accurately predict the height to be reached by a released aerosol. In fact, friction will play a large role as an opposing force, and winds will redirect the particle. However, this problem does show how a source will theoretically eject a pollutant. It is up to the environmental scientist and engineer to identify contravening forces and account for them in models.

Power and Efficiency

The amount of work done per unit time is *power* (P). Like energy, power is a scalar quantity:

$$P = W/\Delta t \qquad \text{Equation 2–6}$$

Power can also be expressed as a function of force and velocity:

$$P = Fv \qquad \text{Equation 2–7}$$

Mass and Work Example 6

The emission of oxides of nitrogen (NO_x) from an older car's exhaust is 100 mg per kilometer traveled. If this increases by 10 mg per kilometer for each additional horsepower (hp) expended, how much additional NO_x would be released if the car traveling $100\,\mathrm{km\,h^{-1}}$ supplies a constant horizontal force of 50 newtons (N) to carry a trailer?

Note: $N = \mathrm{kg\,m\,s^{-2}}$
Watt (W) $= \mathrm{N\,m\,s^{-1}}$
1 hp = 0.7457 kW

Solution:

First, we must calculate the tractive power (hp) required to tow the trailer using Equation 2–7:

$$
\begin{aligned}
P &= Fv \\
&= \frac{(50\mathrm{N})(100\,\mathrm{km\,h^{-1}})(1000\,\mathrm{m\,km^{-1}})}{(60\,\mathrm{sec\,min^{-1}})(60\,\mathrm{min\,h^{-1}})(1000\,\mathrm{W\,kW^{-1}}} \\
&= [(50\mathrm{N})(100\,\mathrm{km\,h^{-1}})(1000\,\mathrm{m\,km^{-1}})] \\
&\quad \times [(60\,\mathrm{sec\,min^{-1}})(60\,\mathrm{min\,h^{-1}})(1000\,\mathrm{W\,kW^{-1}}]^{-1} \\
&= 1.389\,\mathrm{kW}
\end{aligned}
$$

$$1\,\mathrm{hp} = 0.7457\,\mathrm{kW}, \text{ so } P = 1.86\,\mathrm{hp}$$

Therefore, towing the trailer at this speed adds 1.86 × 10 mg, or *18.6 mg NO_x* to the atmosphere for each km traveled. This means that at $100\,\mathrm{km\,h^{-1}}$, the old car is releasing 118.6 mg (i.e. 100 mg + the additional 18.6 mg from towing the trailer) of NO_x for each km it travels.

Mass and Work Example 7

How much will the NO_x be reduced if the old car above produces 90 mg NO_x for each mile traveled $50\,\mathrm{km\,h^{-1}}$ and the NO_x increase from towing falls to 5 mg per kilometer for each hp expended?

Solution:

Once again, we use Equation 2–7:

$$
\begin{aligned}
P &= Fv \\
&= \frac{(50\,\text{N})(50\,\text{km h}^{-1})(1000\,\text{m km}^{-1})}{(60\,\text{sec min}^{-1})(60\,\text{min h}^{-1})(1000\,\text{W kW}^{-1})} \\
&= 0.695\,\text{kW} \\
&= 0.93\,\text{hp}
\end{aligned}
$$

Therefore, towing the trailer at this speed adds 0.93 × 5 mg, or *4.7 mg NO_x* to the atmosphere for each km traveled. This means that at 50 km h^{-1}, the old car is releasing 90 + 4.7 = 94.7 mg NO_x for each km it travels. So, by slowing down, the NO_x emissions drop 23.9 mg for each km traveled.

Let us next deal with mass and its relationships with energy, beginning with mass and motion as they apply to environmental concepts.

Environmental Mechanics

Mechanics is the field of physics concerned with the motion and the equilibrium of bodies within particular frames of reference. Environmental sciences make use of the mechanical principles in practically every aspect of pollution, from the movement of fluids that carry contaminants, to the forces within substances that affect their properties, to the relationships between matter and energy within organisms and ecosystems. Engineering mechanics is important to environmental science because it includes statics and dynamics. Fluid mechanics and soil mechanics are two branches of mechanics particularly important to the environment.

Statics is the branch of mechanics that is concerned with bodies at rest with relation to some frame of reference, with the forces between the bodies, and with the equilibrium of the system. It addresses rigid bodies that are at rest or moving with constant velocity. Hydrostatics is a branch of statics that is essential to environmental science and engineering in that it is concerned with the equilibrium of fluids (liquids and gases) and their stationary interactions with solid bodies, such as pressure. While many fluids are considered by environmental assessments, the principal fluids are water and air.

Dynamics is the branch of mechanics that deals with forces that change or move bodies. It is concerned with accelerated motion of bodies. It is an especially important science and engineering discipline because it is fundamental to understanding the movement of contaminants through the environment. Dynamics is sometimes used synonymously with kinetics. However, we will use the engineering approach and treat kinetics as

TABLE 2.1
Contrasts between Plumes in Ground Water and Atmosphere

	Ground Water Plume	*Air Mass Plume*
General Flow Type	Laminar	Turbulent
Compressibility	Incompressible	Compressible
Viscosity	Low viscosity (1×10^{-3} kg m^{-1} s^{-1} @ 288°K)	Very low viscosity (1.781×10^{-5} kg m^{-1} s^{-1} @ 288°K)

one of the two branches of dynamics, with the other being kinematics. Dynamics combines the properties of the fluid and the means by which it moves. This means that the continuum fluid mechanics vary according to whether the fluid is viscous or inviscid, compressible or incompressible, and whether flow is laminar or turbulent. For example, the properties of the two principal environmental fluids, water in an aquifer and an air mass in the troposphere, are shown in Table 2.1.

Let us begin with environmental statics. When the forces acting on a body balance one another, the body is at rest. We will briefly consider the static equilibrium of particles and rigid bodies, and discuss other concepts of statics including moments of inertia and friction, which are fundamental to environmental fluids.

Environmental Determinate Statics: A Review of the Basic Physics

For a rigid body to be stationary, it must be in "static equilibrium," which means that no unbalanced forces are acting on it.[4] One of the key concepts in statics important to environmental science and engineering is *force*. A push or pull by one body on another body is known as a *force* (**F**). A force is any action that has a tendency to alter a body's state of rest or uniform motion along a straight line. (We will discuss Newton's Laws regarding these concepts shortly when we address dynamics and kinetics.) Forces come in two major types: *external forces* and *internal forces*.

External forces on a rigid body result from other bodies. The external force may result from physical contact with another body, known as "pushing," or due to the body being in close proximity, but not touching, the other body, such as gravitational and electrical forces. When the forces are unbalanced, the body will be put into motion. Internal forces are those that keep the rigid body in one piece. As such, these are compressive and tensile forces within the body that can be found by multiplying the stress and area of a part of the body. Internal forces never cause motion, but can lead to deformation. Since force has both magnitude and direction, it is a vector quantity, so let us briefly discuss vectors as they apply to determinate statics.

A scalar is a quantity with a magnitude only, but no direction. A vector has both magnitude and direction. A vector is a directed line segment in space that represents a force, as well as a velocity or a displacement. *Basic vectors* (always denoted with bold-faced letters, **i**, **j**, and **k**), are directed line segments in a three-dimensional, rectangular coordinate system from the origin (0,0,0) to points (1,0,0), (0,1,0), and (0,0,1), respectively.

At a point in space, the *position vector* is denoted as:

$$\overrightarrow{OP} = x\mathbf{i} + y\mathbf{j} + z\mathbf{k} \qquad \text{Equation 2–8}$$

The position vector is a directed line segment from the origin (0,0,0) to a point at coordinate P (x, y, z), as shown in Figure 2.1. The vector **A** moving from one point to another, that is from $P_1 = (x_1, y1, z1)$ to $P_2 = (x_2, y_2, z_2)$, is expressed as:

$$\overrightarrow{P_1P_2} = \mathbf{A} = a_x\mathbf{i} + a_y\mathbf{j} + a_z\mathbf{k} \qquad \text{Equation 2–9}$$

The "a" terms denote displacement, where $a_x = x_2 - x_1$, $a_y = y_2 - y_1$, and $a_z = z_2 - z_1$. This notation is quite common in environmental modeling, mapping, plume characterization, and contaminant transport. The length of the vector depicts its magnitude. For vector **A**, the magnitude is:

$$|\mathbf{A}| = \sqrt{(a_x)^2 + (a_y)^2 + (a_z)^2} \qquad \text{Equation 2–10}$$

Any vector with a magnitude of 1 is a unit vector (i.e., the vector's magnitude is unity). So, for $\mathbf{A} = a_x\mathbf{i} + a_y\mathbf{j} + a_z\mathbf{k}$, the unit vector in direction of **A** is $\mathbf{A}/|\mathbf{A}|$. Mathematical operations between vectors and scalars are used in numerous environmental applications. These include:

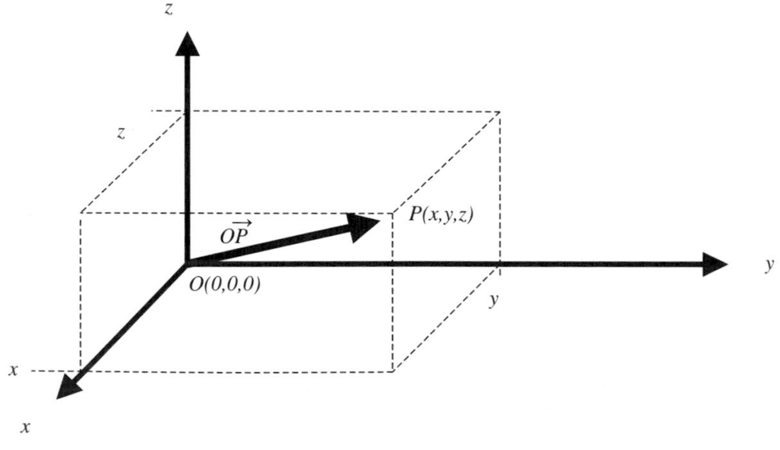

FIGURE 2.1. Position vector.

1. Addition and subtractions of vectors:

$$\mathbf{A} \pm \mathbf{B} = (a_x \pm b_x)\mathbf{i} + (a_y \pm b_y)\mathbf{j} + (a_z \pm b_z)\mathbf{k} \qquad \text{Equation 2–11}$$

2. Scalar multiplication: If c is a scalar (real number), then:

$$c\mathbf{A} = ca_x\mathbf{i} + ca_y\mathbf{j} + ca_z\mathbf{k} \qquad \text{Equation 2–12}$$

3. Dot product operation:

$$\mathbf{A} \cdot \mathbf{B} = a_x b_x + a_y b_y + a_z b_z = |\mathbf{A}||\mathbf{B}|\cos\theta, \qquad \text{Equation 2–13}$$

Where θ = angle between vectors **A** and **B**. **A** and **B** are perpendicular when $\mathbf{A} \cdot \mathbf{B} = 0$, when **A** nor **B** are nonzero vectors.

4. Cross-product operation (using matrix algebra):

$$\mathbf{A} \times \mathbf{B} = \begin{vmatrix} \mathbf{i} & \mathbf{j} & \mathbf{k} \\ a_x & a_y & a_z \\ b_x & b_y & b_z \end{vmatrix} = |\mathbf{A}||\mathbf{B}|\mathbf{n}\sin\theta, \qquad \text{Equation 2–14}$$

Where **n** = unit vector perpendicular to the plane that is formed by **A** and **B**. This follows the "right-hand rule" as shown in Figure 2.2. Thus, **n** points

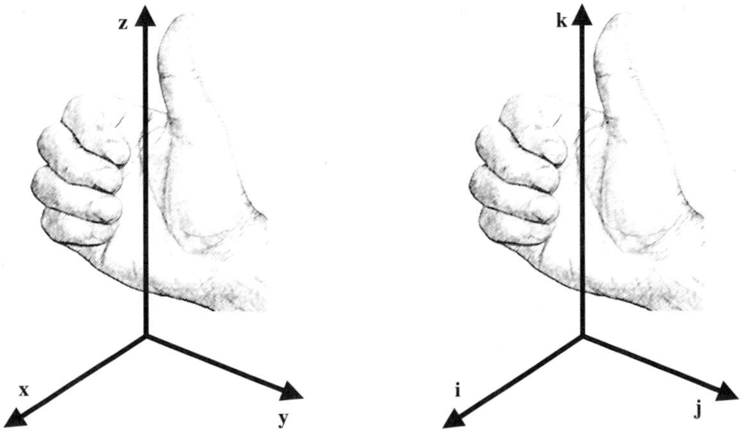

FIGURE 2.2. The right-hand rule for vector cross-products. If you follow the direction of the fingers to go from the x-axis to the y-axis, then the thumb points in the direction of the z-axis. The unit vectors **i**, **j**, and **k** point in these same directions. Using the right-hand rule for cross-products, cross (**i**, **j**) will equal **k**, as illustrated in the drawing on the right. The length of the cross-product is also the area of the parallelogram determined by the two vectors.

in the direction that your right thumb points if your fingers curl to the angle θ from **A** to **B**. For nonzero vectors **A** and **B** to be parallel to each other, $\mathbf{A} \times \mathbf{B} = 0$. The magnitude of $\mathbf{A} \times \mathbf{B}$ is the area of the parallelogram determined by **A** and **B**. This geometric aspect of vectors is important in particle and other surfaces, especially when a surface resembles a plane when observed at small scales.

Statics Example 1

What is the magnitude of A, if A = 6**i** – 12**j** + 4k?

Solution

$$|A| = \sqrt{(6)^2 + (-12)^2 + (4)^2} = \sqrt{196} = \underline{14}$$

Statics Example 2

What is the unit vector in the direction of A?

Solution

$$\frac{\mathbf{A}}{|\mathrm{A}|} = \frac{9\mathbf{i} - 6\mathbf{j} + 2\mathbf{k}}{14} = \frac{9}{14}\mathbf{i} - \frac{6}{14}\mathbf{j} + \frac{4}{14}\mathbf{k} = \frac{9}{14}\mathbf{i} - \frac{3}{7}\mathbf{j} + \frac{2}{7}\mathbf{k}$$

Let us now return to forces. A point force (or concentrated force) is a vector with magnitude, direction, and location. The force's line of action is the line in the direction of force that is extended forward and backward. So a slightly modified representation of Figure 2.1 shown in Figure 2.3 and an adaptation of Equation 2–2 shows the three-dimensional force given by:

$$\mathbf{F} = F_x\mathbf{i} + F_y\mathbf{j} + F_z\mathbf{k} \qquad \text{Equation 2–15}$$

From the discussion on vectors, if **u** is a unit vector in the direction of the force represented in Figure 2.3, then the force is represented as $\mathbf{F} = F\mathbf{u}$ and the components are then found using the direction cosines of the angels made by the force vector with the three axes (the force **F** and its unit vector **u** are found along the line of action):

$$F_x = F\cos\theta_x \qquad \text{Equation 2–16}$$

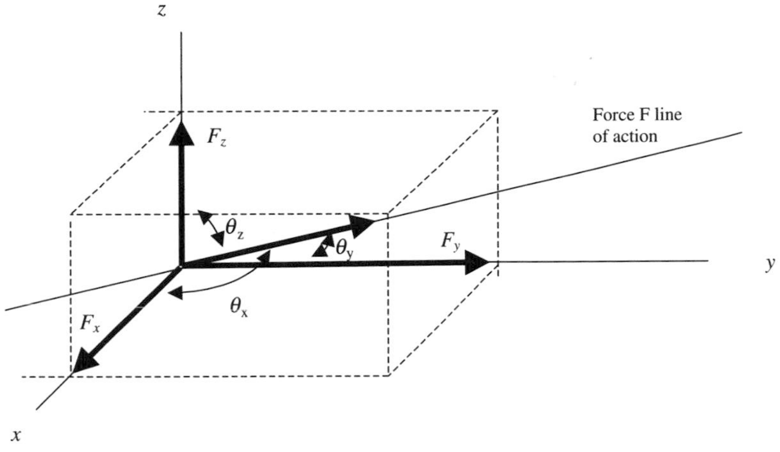

FIGURE 2.3. Components and direction of angles of a force.

$$F_y = F\cos\theta_y \qquad \text{Equation 2–17}$$

$$F_z = F\cos\theta_z \qquad \text{Equation 2–18}$$

So,

$$F = \sqrt{(F_x)^2 + (F_y)^2 + (F_z)^2} \qquad \text{Equation 2–19}$$

Another aspect of force is the *moment*, which is the tendency of a force to rotate, twist, or turn a rigid body around a pivot. In other words, when a body is acted upon by a moment, the body will rotate. But even if the body does not actually rotate because it is being restrained, the moment still exists. So the units of a moment are length × force (e.g., Newton-meters). The moment is zero when the line of action of the force passes through the center of rotation (pivot). The moment is a vector that is the cross-product of the force and the vector from the pivot point. The line of action is found using the right-hand rule. The moment has two parts; the moment of a single force and the moment of a couple. We will define both.

A moment important to environmental engineering is the moment of force about a line. Pumps, for example, have a fixed rotational axis, which means that it turns around a line not about a pivot. Unlike the moment rotating about a pivot ($\mathbf{M}_O$), the moment around a line ($\mathbf{M}_{OL}$) is a scalar (dot product):

$$\mathbf{M}_{OL} = \mathbf{a}\cdot\mathbf{M}_O = \mathbf{a}\cdot(\mathbf{r}\times\mathbf{F}) = \begin{vmatrix} a_x & a_y & a_z \\ x_P - x_O & y_P - y_O & z_P - z_O \\ F_x & F_y & F_z \end{vmatrix}, \qquad \text{Equation 2–20}$$

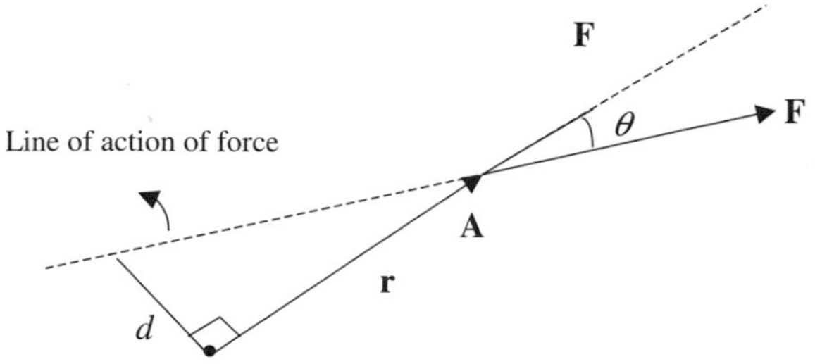

FIGURE 2.4. Moment about a single force.

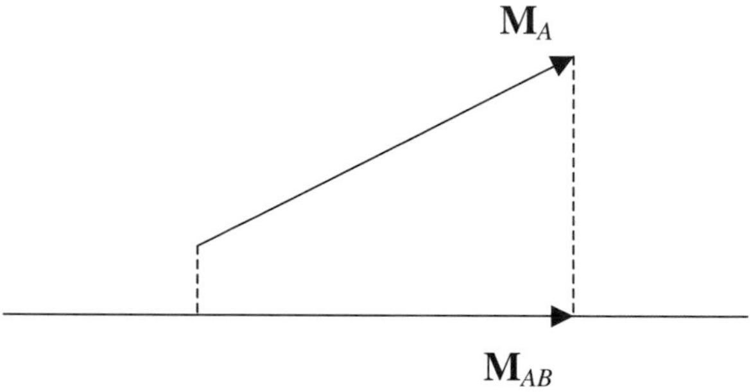

FIGURE 2.5. Determining the moment of a couple.

Making point O the origin, Equation 2–13 becomes:

$$\mathbf{M}_{OL} = \begin{vmatrix} a_x & a_y & a_z \\ x & y & z \\ F_x & F_y & F_z \end{vmatrix}, \qquad \text{Equation 2–21}$$

The moment about a single force is shown in Figure 2.4. The moment M of a force F about point A in the figure is the product of the force and the perpendicular distance (d) from that point to the line of action for the force. So the magnitude of this moment is:

$$M_A = Fd \qquad \text{Equation 2–22}$$

The moment may also be determined by vector analysis:

$$\mathbf{M}_A = \mathbf{r} \times \mathbf{F} \qquad \text{Equation 2–23}$$

Where, r = position vector from point A to any point of the line of action of force.

A couple is formed when two equal and parallel forces do not share lines of action and are opposite in sense. The sense of a moment in two dimensions is the direction of the moment clockwise or counterclockwise. In three dimensions the sense is found by using the right hand rule described in this chapter. The moment of a couple is determined from the product of the force and the minimum distance between the two forces (see Figure 2.5). Like the moment in a point of space, the equation for the moment of the couple is $M = Fd$.

Let us now resolve forces. First, a force can be resolved in two dimensions, that is, along any two axes. The standard set of axes is the rectangular coordinate system.

Statics Example 3

Resolve the forces in Figure 2.6 into their respective x and y coordinate components.

Solution

We can use Equations 2–10 and 2–11 to solve for $\mathbf{F}_1$ and $\mathbf{F}_3$ since we are given their angles:

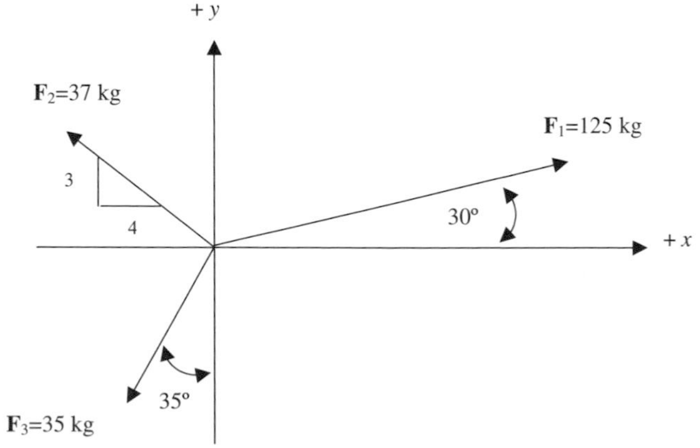

FIGURE 2.6. Resolving forces in two dimensions.

$$\mathbf{F}_{1x} = 125\cos 30 = 125 \times 0.87 = 100.75\,\text{kg}$$

$$\mathbf{F}_{1y} = 125\sin 30 = 125 \times 0.5 = 62.5\,\text{kg}$$

$$\mathbf{F}_{3x} = -35\cos 35 = -35 \times 0.82 = -28.7\,\text{kg}$$

$$\mathbf{F}_{3x} = -35\sin 35 = -35 \times 0.57 = -20.0\,\text{kg}$$

Since we know the slope for $\mathbf{F}_2$, we can use the Pythagorean Theorem to resolve this force. That is, the segment of the line of action is the hypotenuse for the right triangle formed by sides (rise = 3 and run = 4). So the sum of the squares of the sides is 25, and this is equal to the square of the hypotenuse. Thus, the hypotenuse is 5:

$$\mathbf{F}_{2x} = -37\left(\frac{4}{5}\right) = -21.6\,\text{kg}$$

$$\mathbf{F}_{2y} = 37\left(\frac{3}{5}\right) = 22.2\,\text{kg}$$

Note that the signs depend on the direction of the force away from the origin with respect to each axis.

Statics Example 4

Calculate the concurrent forces in Figure 2.7.

Solution

The resultant force **R** is found by resolving two rectangular components R_x and R_y:

$$\mathbf{R} = \Sigma\mathbf{F}$$

$$R_x = \Sigma F_x, \qquad R_x = \Sigma F_x,$$

$$R_x = \sqrt{R_x^2 + R_y^2}$$

$$\theta = \tan^{-1}\frac{R_y}{R_x}$$

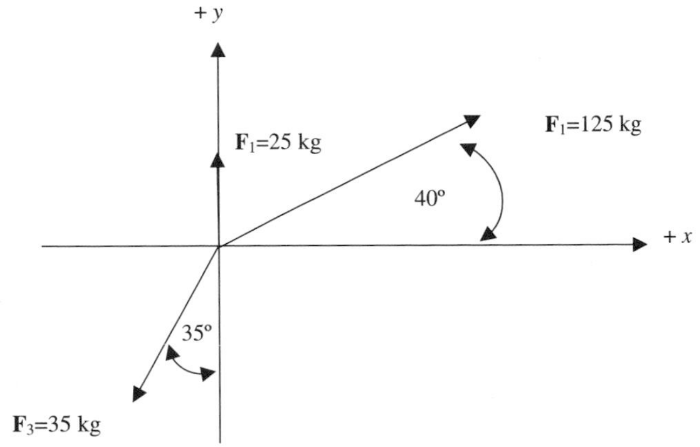

FIGURE 2.7. Resolving concurrent forces in two dimensions.

Where,

$$R_x = F_{1x} + F_{2x} + F_{3x} = 125\cos 40 - 35\cos 35 = 95.8 - 28.7 = 67.1\text{kg}$$

$$R_y = F_{1y} + F_{2y} + F_{3y} = 125\sin 40 + 25 - 35\sin 35 = 80.3 + 25 - 20.1 = 125.5\text{kg}$$

Forces can also be resolved in a similar manner in three dimensions, using the vector determination approaches described earlier; that is, adding the *z*-axis.

Environmental Dynamics

Dynamics is the general area of physics concerned with moving objects. It includes kinematics and kinetics. *Kinematics* is concerned with the study of a body in motion independent of forces acting on the body. That is, kinematics is the branch of mechanics concerned with the motion of bodies with reference to force or mass. This is accomplished by studying the geometry of motion irrespective of what is causing the motion. Therefore, kinematics relates position, velocity, acceleration, and time.

Hydrodynamics is the important branch of environmental mechanics that is concerned with deformable bodies. It is concerned with the motion of fluids. Therefore, it is an important underlying aspect of contaminant transport and movements of fluids, and considers fluid properties such as compressibility and viscosity. These are key to understanding water distribution and treatment systems, flows in pipes, and design of pumps and fluid exchange systems.

Kinetics is the study of motion and the forces that cause motion. This includes analyzing force and mass as they relate to translational motion. Kinetics also considers the relationship between torque and moment of inertia for rotational motion.

A key concept for environmental dynamics is that of *linear momentum,* which is the product of mass and velocity. A body's momentum is conserved unless an external force acts upon a body. Kinetics is based on Newton's *first law of motion,* which states that a body will remain in a state of rest or will continue to move with constant velocity unless an unbalanced external force acts on it. Stated as the *law of conservation of momentum,* linear momentum is unchanged if no unbalanced forces act on a body. Or, if the resultant external force acting on a body is zero, the linear momentum of the body is constant.

Kinetics is also based on Newton's *second law of motion,* which states that the acceleration of a body is directly proportional to the force acting on that body, and inversely proportional to the body's mass. The direction of acceleration is the same as the force of direction. The equation for the second law is:

$$\mathbf{F} = \frac{d\mathbf{p}}{dt} \qquad \text{Equation 2–24}$$

Where, $\mathbf{p}$ = momentum.

Newton's *third law of motion* states that for every acting force between two bodies, there is an equal but opposite reacting force on the same line of action, or:

$$\mathbf{F}_{reacting} = -\mathbf{F}_{acting} \qquad \text{Equation 2–25}$$

Another force that is important to environmental systems is friction, which is a force that always resists motion or an impending motion. Friction acts parallel to the contacting surfaces. When bodies come into contact with one another, friction acts in the direction opposite to what is bringing the objects into contact.

We can now apply these physical principles to environmental contamination.

Notes and Commentary

1. We will address the concepts of hazard, risk, and safety in Part III, "Contaminant Risk."
2. The term "make up water" is simply the amount of water that must be continuously added to a power generation facility to ensure the efficient and safe

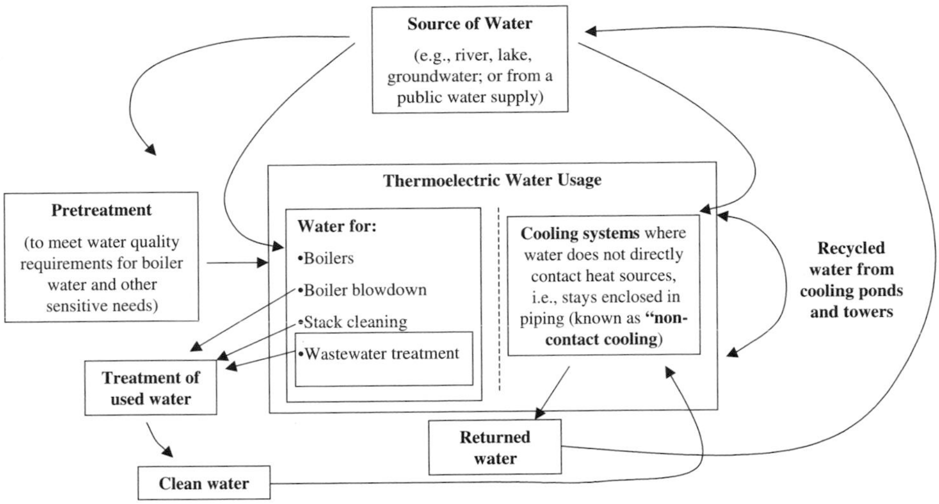

FIGURE 2.8. Thermoelectric water use at a prototypical power generating facility.

production of electrical energy. Figure 2.8 shows that water is important to every operational aspect of a power plant. Thus, any water losses must be replaced immediately by process make up water. The water usage is either for heat dissipation (thermal) or energy production (electric), so the plant operators are concerned with "thermoelectric" water usage.

3. An ideal gas is one that conforms to Boyle's Law and that has zero heat of free expansion (i.e., conforms to Charles' Law). Boyle's Law states that for a given mass, at constant temperature, the product of pressure and volume is constant:

$$pV = C \qquad \text{Equation 2–26}$$

4. Please pardon the double negative, but this is one of the few occasions where stating something positively loses some of its meaning. "A rigid body having balanced forces acting on it" is not the same as "A rigid body having no unbalanced forces acting on it."

CHAPTER 3

Applied Contaminant Physics: Fluid Properties

In the last chapter we discussed the basic principles and concepts of physics. We can now use this knowledge to understand how and why pollutants can move in the environment. The movement may be within one environmental compartment, such as a dissolved contaminant moving within a lake. Pollutants, however, often move among numerous compartments, such as when a contaminant moves from water to soil, then to the atmosphere, until it is deposited again to the soil and surface waters, where it is taken up by plants and eaten by animals.[1]

The general behavior of contaminants after they are released is shown in Figure 3.1. The movement of pollutants is known as *transport*. This is half of the often–cited duo of environmental "fate and transport." *Fate* is an expression of what a contaminant becomes after all the physical, chemical, and biological processes of the environment have acted. It is the ultimate site of a pollutant after its release. The pollutant will undergo numerous changes in location and form before reaching its fate. Throughout the contaminant's journey it will be physically transported and undergo coincidental chemical processes, known as *transformations*, such as photochemical and biochemical reactions.[2]

Physical transport must obviously deal with the kinematics and mechanics of fluids, but it must also consider how and when these processes reach equilibrium, such as when a chemical is sequestered and stored. Fate is often described according to environmental media or compartments.

Physical Properties of Environmental Fluids

To understand transport, we must first consider the characteristics of environmental fluids. A *fluid* is a collective term that includes all liquids and

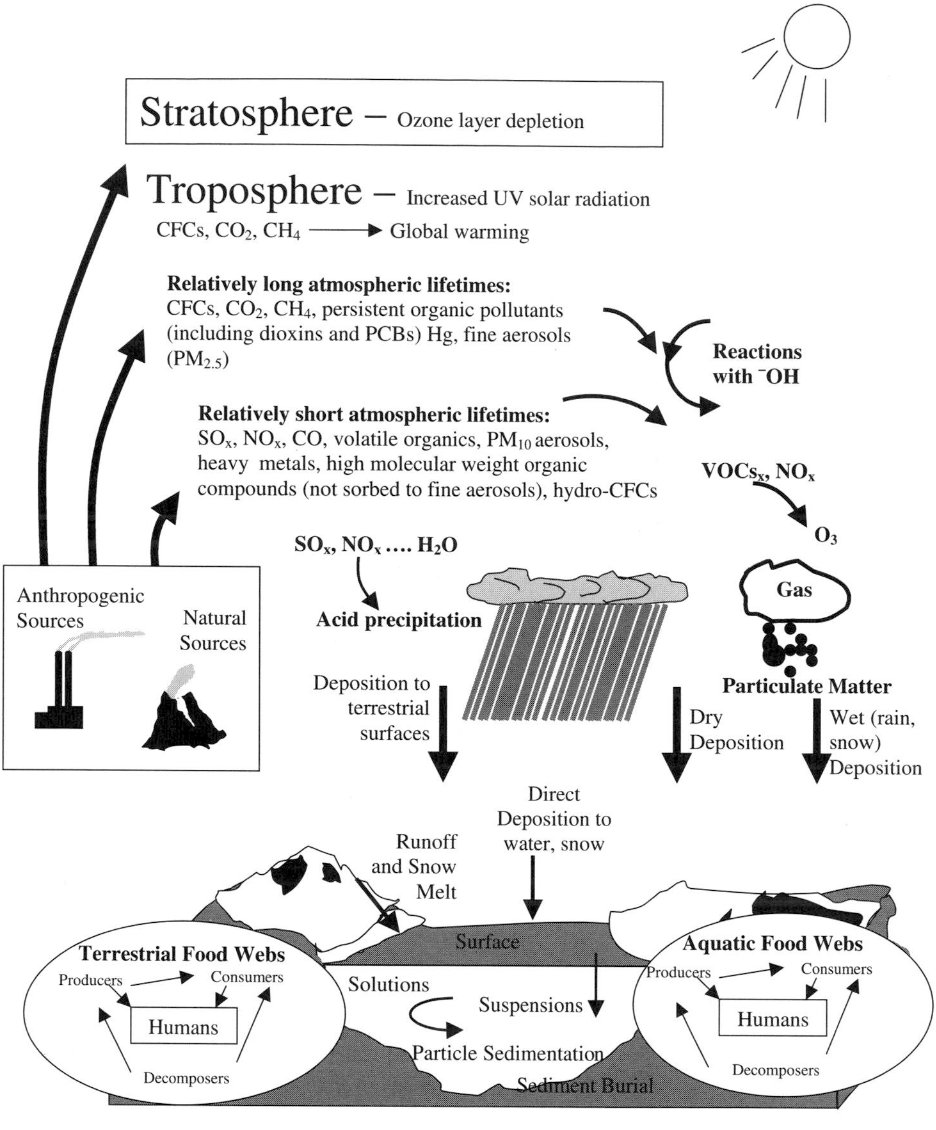

FIGURE 3.1. The physical movement, transformation, and accumulation of global contaminants after release. (Source: Adapted from the Commission for Environmental Cooperation of North America, 2002, "The Sound Management of Chemicals [SMOC] Initiative of the Commission for Environmental Cooperation of North America: Overview and Update," Montreal, Canada.)

gases. A liquid is matter that is composed of molecules that move freely among themselves without separating from each other. A gas is matter composed of molecules that move freely and are infinitely able to occupy the space with which they are contained at a constant temperature. Engineers define a fluid as a substance that will deform continuously upon the application of a shear stress; that is, a stress in which the material on one side of a surface pushes on the material on the other side of the surface with a force parallel to the surface.

Fluids are generally divided into two types: *ideal* and *real.* The former has zero viscosity and, thus, no resistance to shear (explained below). An ideal fluid is incompressible and flows with uniform velocity distributions. It also has no friction between moving layers and no turbulence (i.e., eddy currents).

On the contrary, a real fluid has finite viscosity, has nonuniform velocity distributions, is compressible, and experiences friction and turbulence. Real fluids are further subdivided according to their viscosities. A *Newtonian fluid* is one that has a constant viscosity at all shear rates at a constant temperature and pressure. Water and most solvents are Newtonian fluids. However, environmental engineers are confronted with non-Newtonian fluids, or those with viscosities not constant at all shear rates. Sites contaminated with drilling fluids and oils have large quantities of non-Newtonian fluids on-site, for example.

At this point, let us consider three engineering concepts that must be understood before considering fluid properties. Physicists use the term *particle* to mean a theoretical point that has a rest-mass and location but no geometric extension. We can observe this particle as it moves within the fluid as a representation of where that portion of the fluid is going and at what velocity it is moving. Another important concept is that of the *control volume,* which is an arbitrary region in space that is defined by boundaries. The boundaries may be either stationary or moving. The control volume is used to determine how much material and at what rate the material is moving through the air, water, or soil. The third concept, which is included in the definition of a fluid, is *stress.* As we saw when we discussed forces, the forces acting on a fluid may be *body forces* or *surface forces.* The former are forces that act on every particle within the fluid, occurring without actually making physical contact, such as gravitational force. The latter are forces that are applied directly to the fluid's surface by physical contact.

Stress represents the total force per unit area acting on a fluid at any point within the fluid volume. Stress at any point P is thus:

$$\sigma(P) = \lim_{\delta A \to 0} \frac{\delta F}{\delta A} \qquad \text{Equation 3–1}$$

Where, $\sigma(P)$ = vector stress at point P

δA = infinitesimal area at point P

δF = force acting on δA

Fluid properties are characteristics of the fluid that are used to predict how the fluid will react when subjected to applied forces. We will discuss the chemical characteristics in the Chapter 6, "Fundamental of Environmental Chemistry." We can start by considering some, but certainly not all, fluid properties that are important to environmental systems.

If a fluid is considered to be infinitely divisible, that is, it is made up of many molecules that are constantly in motion and colliding with one another, this fluid is in *continuum*. Such a fluid acts as though it has no holes or voids, meaning its properties are continuous (i.e., temperature, volume, and pressure fields are continuous). If we make the assumption that a fluid is a continuum we can consider the fluid's properties to be functions of position and time. We can then represent the fluid properties as two fields. The density field is:

$$\rho = \rho(x, y, z, t) \qquad \text{Equation 3–2}$$

Where, ρ = density of the fluid
x, y, z = coordinates in space
t = time

The other fluid field is the velocity field:

$$\vec{v} = \vec{v}(x, y, z, t) \qquad \text{Equation 3–3}$$

Thus, if the fluid properties and the flow characteristics at each position do not vary with time, the fluid is said to be at *steady flow*:

$$\rho = \rho(x, y, z) \quad \text{or} \quad \frac{\partial \rho}{\partial t} = 0 \qquad \text{Equation 3–4}$$

and

$$\vec{v} = \vec{v}(x, y, z) \quad \text{or} \quad \frac{\partial \vec{v}}{\partial t} = 0 \qquad \text{Equation 3–5}$$

Conversely, a *time-dependent flow* is considered to be an *unsteady flow*. Any flow with unchanging magnitude and direction of the velocity vector $\vec{v}$ is considered to be a *uniform flow*.

Fluids, then, can be classified according to observable physical characteristics of flow fields. A continuum fluid mechanics classification is shown in Figure 3.2. *Laminar flow* is in layers, while *turbulent flow* has random movements of fluid particles in all directions. In *incompressible flow*, the variations in density are assumed to be constant, while the *compressible flow* has density variations, which must be included in flow calculations. *Viscous flows* must account for viscosity, while *inviscid flows* assume that viscosity is zero.

The velocity field is very important in environmental modeling, especially in modeling plumes in the atmosphere and in groundwater, since the

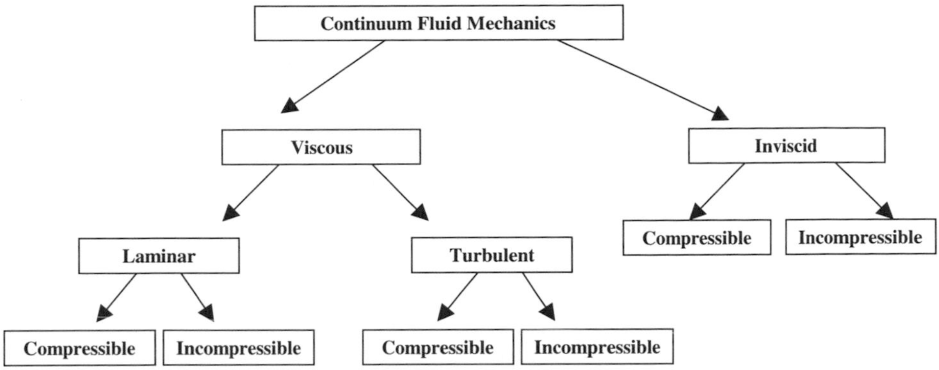

FIGURE 3.2. Classification of fluids based on continuum fluid mechanics. (Source: Research and Education Association, 1987, *The Essentials of Fluid Mechanics and Dynamics I*, REA, Piscataway, N.J.)

velocity field is a way to characterize the motion of fluid particles and provides the means for computing these motions. The velocity field may be described mathematically using Equation 3–3. This is known as the *Eularian* viewpoint.

Another way to characterize the fluid movement, or flow, is to follow the particle as it moves, using time functions that correspond to each particle, as shown in Figure 1.2. This is the *Lagrangian* viewpoint, which is expressed mathematically as:

$$\vec{v} = [x(t), y(t), z(t)] \qquad \text{Equation 3–6}$$

Most environmental transport models are either Eularian or Lagrangian. Let us now consider the specific fluid properties crucial to assessing environmental contaminants.

Velocity

The time rate of change a fluid particle's position in space is the fluid velocity (V). This is a vector field quantity. Speed (V) is the magnitude of the vector velocity V at some given point in the fluid, and average speed ($\overline{V}$) is the mean fluid speed through a control volume's surface. Therefore, velocity is a vector quantity (magnitude and direction), while speed is a scalar quantity (magnitude only). The standard units of velocity and speed are meter per second ($m\,sec^{-1}$).

Obviously, velocity is important to determine pollution, such as mixing rates after an effluent is discharged to a stream, how rapidly an aquifer will become contaminated, and the ability of liners to slow the movement of leachate from a landfill toward the groundwater. The distinction between velocity and speed is seldom made in nontechnical publications, but scientists and engineers should be clear about which they are using to describe fluid movement.

To measure surface water velocity, a velocity index can be related to mean velocity, which is then multiplied by cross-sectional area. The index velocity may be measured at either a point or along a line, using hydrological monitoring equipment (see Figure 3.3). The measured velocity along a

FIGURE 3.3. Monitoring station that provides hydrodynamic data, including mean velocity, of surface waters. (Source: U.S. Geological Survey, 2003, *Hydrodynamics of the Southwest Coast Estuaries*, E. Patino and V. Levesque, Principal Investigators, http://sofia.usgs.gov/publications/posters/sw_hydro/.)

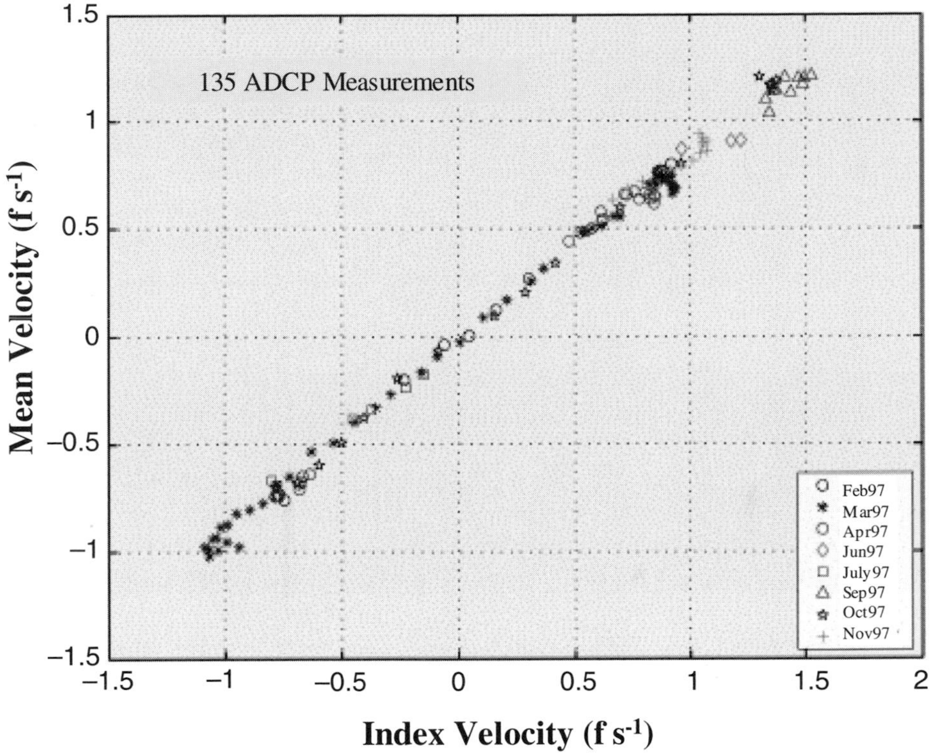

FIGURE 3.4. Mean velocity versus index velocity from a monitoring station in southwestern Florida estuary. *Note*: Data were collected using acoustic doppler current profiler (ADCP) on fixed platforms to measure the detailed velocity profiles across the river. (Source: U.S. Geological Survey, 2003, *Hydrodynamics of the Southwest Coast Estuaries*, E. Patino and V. Levesque, Principal Investigators, http://sofia.usgs.gov/publications/posters/sw_hydro/.)

line usually is preferred to the point velocity, because the former relates better to average stream velocity and is thus a better index of mean velocity than is point velocity (see Figure 3.4). The line velocity can be measured without regard to flow direction by using an acoustic velocity meter, such as the acoustic doppler current profiler (ADCP).

Discharge and Flow

Now that we have an understanding of velocity, let us consider the same concept in three dimensions. The amount of water flowing in a stream (streamflow) is a common measurement around the world. In the United

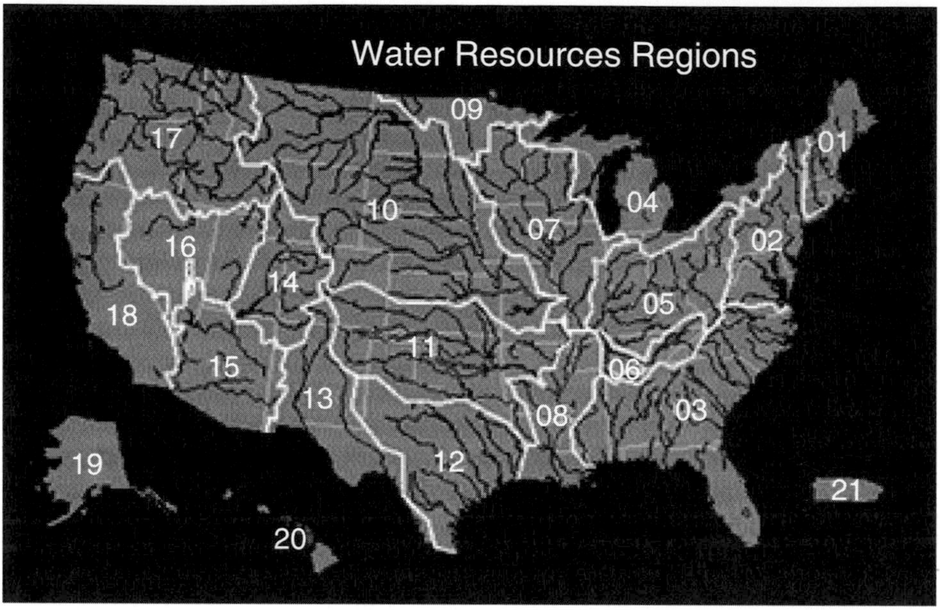

FIGURE 3.5. The 21 hydrologic regions of the United States. (Source: U.S. Geological Survey, 2003, http://waterdata.usgs.gov/.)

States., the U.S. Geological Survey maintains sites in hydrological regions, which are part of the standardized watershed classification system developed by the U.S.G.S. in the mid-1970s. The system consists of 21 hydrologic units for drainage areas in the United States. The hydrologic units are watershed boundaries organized in a nested hierarchy by sizes; the largest units are regions (see Figure 3.5), down to local watersheds.

Surface water flow is known as *stream discharge*, Q, with units of volume per time. Although the appropriate units are $m^3 sec^{-1}$, most stream discharge data in the United States is reported as number of cubic feet of water flowing past a point each second (cfs). Discharge is derived by measuring a stream's velocity at numerous points across the stream. Since heights (and volume of water) in a stream change with meteorological and other conditions, stream-stage/stream-discharge relationships are found by measuring stream discharge during different stream stages. The flow of a stream is estimated based on many measurements. The mean of the flow measurements at all stage heights is reported as the estimated discharge.

Flow measurements taken from a subsection of a large stream can be used to estimate the discharge in the stream's cross-section, as shown in Figure 3.6. The calculation of discharge of the stream of width w_s is the sum of the products of mean depth, mean width, and mean velocity:[3]

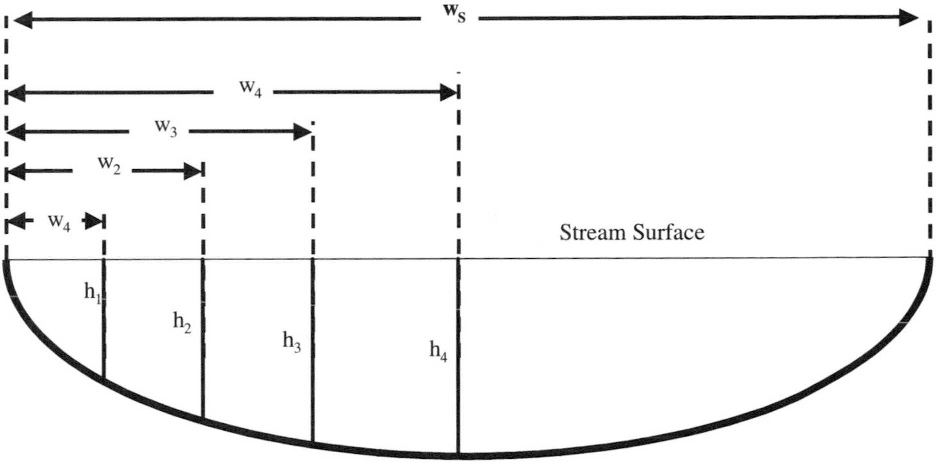

FIGURE 3.6. Stream discharge measurement approach based on stream flow measurements from subsections at various stream heights (h_n) and the associated stream widths (h_n). (Adapted from: C. Lee and S. Lin, eds., 1999, *Handbook of Environmental Engineering Calculations*, McGraw-Hill, New York, N.Y.)

$$Q = \sum_{n=1}^{n} \frac{1}{2}(h_n + h_{n-1})(w_n + w_{n-1}) \times \frac{1}{2}(v_n + v_{n-1})\frac{1}{2}(h_n + h_{n-1})$$

Equation 3–7

Where, Q = Discharge ($m^3 sec^1$)
w_n = nth distance from baseline or initial point of measurement (m)
h_n = nth water depth (m)
v_n = nth velocity ($m sec^1$) from velocity meter

Stream Flow Example

If measurements are made at Dukeheel Creek, as shown in Figure 3.7, what is the average discharge of this stream?

Solution

The U.S. Geological Survey calls for measuring the flow in a stream according to its profile (width and depth). First, you would venture out into the creek to take these measurements. Figure 3.7 shows a cross-

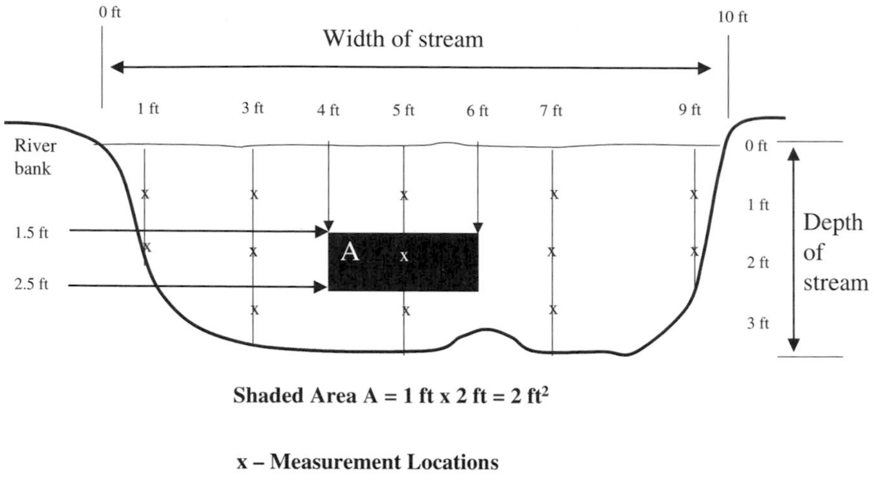

FIGURE 3.7. Profile of Dukeheel Creek. (Adapted from: U.S. Geological Survey.)

section of Dukeheel Creek, which is 10 feet wide. The stream-measurement procedure is to take measurements traversing the stream at defined intervals and to measure the total depth and the velocity of the water at selected depths at each interval across the stream.

Velocity measurements would be taken at each point (x) in the profile. Each area of the measured interval is determined (Box A). The lines in the figure demonstrate that water depth and velocity measurements are obtained horizontally across the stream at 1, 3, 5, 7, and 9 feet. Let us use the water depth/velocity measurement obtained at a point 5 feet from the edge of the stream. The total depth is 3 feet and velocity readings are obtained at depts of 1.5 ft and 2.5 ft in Box A, which represents an area that is midway between this measurement point and the measurement points on either side of the creek. The area is 2 ft across and 1 ft high, or $2\,ft^2$. The measured velocity at the top of Box A is $2\,ft\,sec^{-1}$, and $1\,ft\,sec^{-1}$ at the bottom of Box A, so the Box A average velocity is $1.5\,ft\,sec^{-1}$. To find the volume of water flowing in Box A each second, find the product of Box A and the velocity of the water:

- 2 feet wide × 1 foot high = $2\,ft^2$
- $\therefore\ Q = 2\,ft^2 \times 1.5\,ft\,sec^{-1} = 3\,ft^3\,sec^{-1}$.

A hydrologist computes the total stream discharge using control cross-sectional areas like Box A between all of the measurements and applying the average velocity of the water in every box. The total stream discharge is the sum of all the boxed areas.

This example is a simplification in that when taking actual stream measurements, many more measurement points are averaged into the stream-flow estimate. When a real measurement is made, the hydrologist really takes measurements at about 20 points across the stream. The goal is to have no one vertical cross-section contain more than 5% of the total stream discharge.

The U.S. Geological Survey maintains sites across the nation (see Figure 3.8). Data are recorded as frequently as daily and reported over varying lengths of time. Figure 3.9 shows flow data for the past 30 years at a monitoring station on the Potomac River in Maryland.

Pressure

Pressure is a very important environmental fluid property because it influences where a contaminant is going to move and even determines the state of matter, i.e., solid, liquid, or gas, of the fluid carrying the contaminant and of the contaminant itself. Pressure (p) is a force per unit area:

$$p = \frac{F}{A} \qquad \text{Equation 3–8}$$

So p is a type of stress that is exerted uniformly in all directions. It is common to use pressure instead of force to describe the factors that influence the behavior of fluids. The standard unit of p is the Pascal (P), which is equal to $1\,\mathrm{N\,m^{-2}}$. Therefore, pressure will vary when the area varies, as shown in Figure 3.10. In this example, the same weight (force) over different areas leads to different pressures, and much higher pressure when the same force is distributed over a smaller area.

For a liquid at rest, the medium is considered to be a continuous distribution of matter. However, when considering p for a gas, the pressure is an average of the forces against the vessel walls, that is, the gas pressure. Fluid pressure is a measure of energy per unit volume per the *Bernoulli Equation*, which states that the static pressure in the flow plus one half of the density times the velocity squared is equal to a constant throughout the flow, referred to as the total pressure of the flow:

FIGURE 3.8. U.S. Geological Survey's Station Number 01638500, Potomac River at Points of Rocks, Maryland. (Source: U.S. Geological Survey, 2003, http://waterdata.usgs.gov/nwis/dv/.)

$$\rho + \frac{1}{2}\rho V^2 + \rho g h = \text{constant} \qquad \text{Equation 3–9}$$

Where, P = pressure
V = fluid velocity

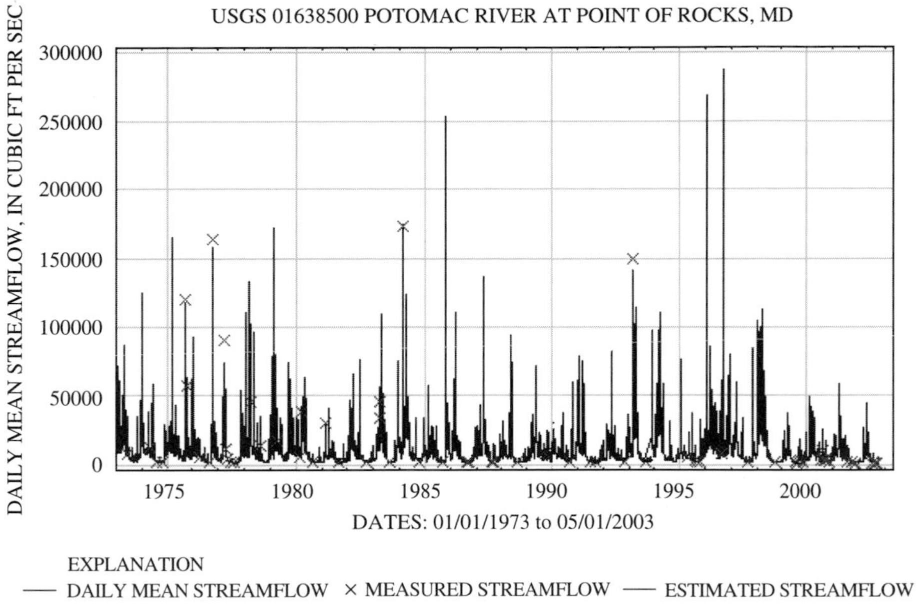

FIGURE 3.9. Daily mean streamflow data at Potomac River at Points of Rocks, Frederick County, Maryland (Latitude 39°16′24.9″, Longitude 77°32′35.2″ NAD83). (Source: U.S. Geological Survey, 2003, http://waterdata.usgs.gov/nwis/dv/.)

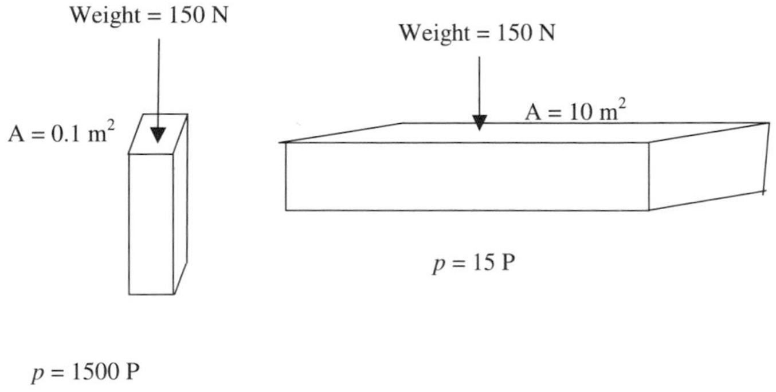

FIGURE 3.10. Difference in pressure with same weight over different areas.

h = elevation
g = gravitational acceleration

This also means that, in keeping with the conservation of energy principle, a flowing fluid will maintain the energy, but velocity and pressure can change. In fact, velocity and pressure will compensate for each other to adhere to the conservation principle, as stated in the Bernoulli Equation:

$$\rho_1 + \frac{1}{2}\rho V_1^2 + \rho g h_1 = \rho_2 + \frac{1}{2}\rho V_2^2 + \rho g h_2 \qquad \text{Equation 3–10}$$

This is shown graphically in Figure 3.11. The so-called "Bernoulli effect" occurs when increased fluid speed leads to decreased internal pressure.

In environmental applications, fluid pressure is measured against two references: *zero pressure* and *atmospheric pressure*. *Absolute pressure* is compared to true zero pressure and *gauge pressure* is reported in reference to atmospheric pressure. To be able to tell which type of pressure is reported, the letter "a" and the letter "g" are added to units to designate whether the pressure is absolute or gauge, respectively. Thus it is common to see pounds per square inch designated as "psia" or inches of water as "in wg." If no letter is designated, the pressure can be assumed to be absolute pressure.

When a gauge measurement is taken, and the actual atmospheric pressure is known, absolute and gauge pressure are related:

$$p_{absolute} = p_{gauge} + p_{atmospheric} \qquad \text{Equation 3–11}$$

Barometric and atmospheric pressure are synonymous. A negative gauge pressure implies a *vacuum* measurement. A reported vacuum quantity is to be subtracted from the atmospheric pressure. When a piece of equipment is operating with 20 kilopascals (kP) vacuum, the absolute pressure is 101.3 kP – 20 kP = 81.3 kP. (Note: The standard atmospheric pressure = 101.3 kPa = 1.013 bars.) Thus, the relationship between vacuums, which are always given as positive numbers, and absolute pressure is:

$$p_{absolute} = p_{atmospheric} - p_{vacuum} \qquad \text{Equation 3–12}$$

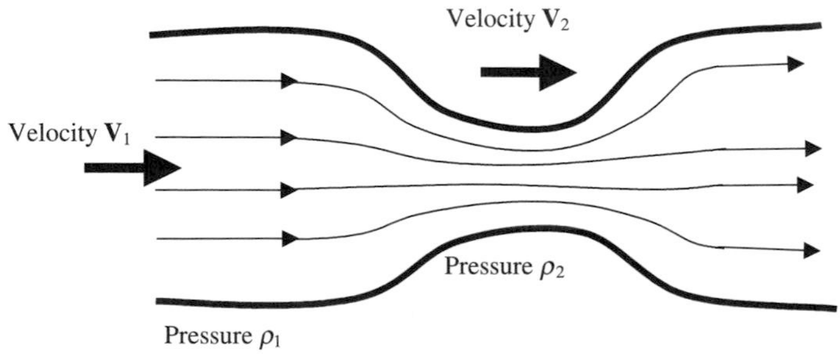

FIGURE 3.11. Bernoulli principle and the effect of relationship between pressure energy, area, and velocity. As the cross-sectional area of flow decreases, the velocity increases and the pressure decreases.

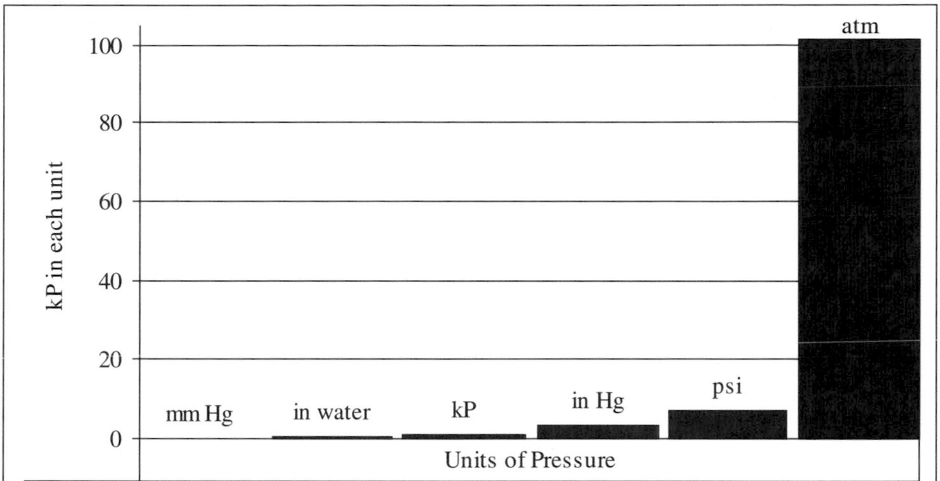

■ mm Hg	0.133
■ in water	0.249
■ kP	1.000
■ in Hg	3.374
■ psi	6.895
■ atm	101.300

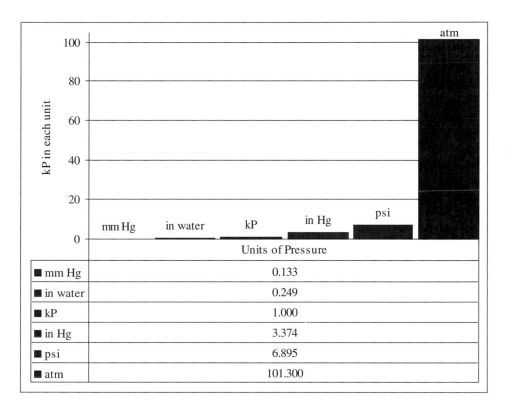

FIGURE 3.12. Comparison of the size of pressure units.

Pressure is used throughout this text, as well as in any discussion of physics, chemistry, and biology. Numerous units are used. The preferred unit in this book is the kilopascal (kP), since the standard metric unit of pressure is the Pascal, which is quite small. See Figure 3.12 for a comparison of relative size of pressure units commonly used in environmental assessments, research studies, and textbooks.

Acceleration

When discussing forces and the properties of fluids, we often include acceleration in the equations. For example, any discussion of potential and kinetic energies includes acceleration due to gravity. In many ways, it seems that acceleration was a major reason for Isaac Newton's need to develop "the calculus."[4] Mathematics needed a way to deal with this concept that was understood, possibly intuitively, by such great minds as Galileo, Kepler, and others, but it also needed the structure brought by the calculus. Calculus (the definite article is commonly dropped nowadays) was known as the mathematics of change, which is what acceleration is all about. Newton needed a way to express mathematically his new law of motion.

Acceleration is the time rate of change in the velocity of a fluid particle. In terms of calculus, it is a second derivative; that is, it is the derivative of the velocity function. And a derivative of a function is itself a function, giving its rate of change. This explains why the second derivative must be a function showing the rate of change of the rate of change. This is obvious when one looks at the units of acceleration: length per time per time ($m\,sec^{-2}$).

Fluid Acceleration Example

If a fluid is moving at the constant velocity of $4\,m\,sec^{-1}$, what is the rate of change of the velocity? What is the second derivative of the fluid's movement?

The function s = f(t) shows the distance the fluid has moved (s) after t seconds. If the fluid is traveling at $4\,m\,sec^{-1}$, then it must travel 4 meters for each second, or 4 t meters after t seconds. The rate of change of distance (how fast the distance is changing) is the speed. We know that this is $4\,m\,sec^{-1}$. So:

$$s = f(t) = 4t \quad \text{and,}$$
$$ds/dt = f'(t) = 4 \qquad \text{Equation 3–13}$$

In acceleration, we are interested in the rate of change of the rate of change. This is the rate of change of our fluid velocity. Since the fluid is moving at constant velocity, it is not accelerating.

So acceleration = 0.

This is another way of saying that when we differentiate for a second time (called the *second derivative*), we find it is zero.

Displacement, Velocity, and Acceleration

We can now combine these three concepts to describe fluid movement. If we are given the function *f(t)* as the displacement of a particle in the fluid at time *t*, the derivative of this function *f′(t)* represents the velocity. The second derivative *f″(t)* represents the acceleration of the particle at time *t*:

$$s = f(t) \qquad \text{Equation 3–14}$$

$$v = \frac{ds}{dt} = f'(t) \qquad \text{Equation 3–15}$$

$$a = \frac{d^2s}{dt^2} = f''(t) \qquad \text{Equation 3–16}$$

Engineer's Notebook Entry: Stationary Points in a Fluid

The derivative of a function can be described graphically (see Figure 3.13). If the derivative is zero, the function is flat and must therefore reside where the graph is turning. We are able to identify the turning points of a function by differentiating

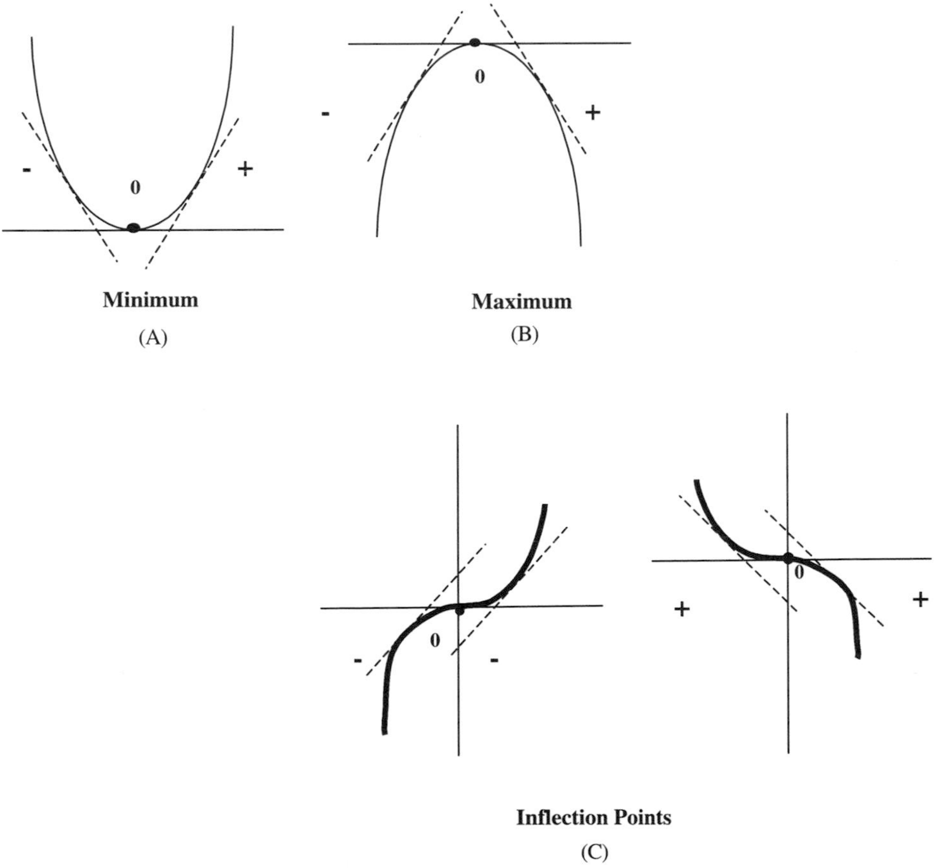

FIGURE 3.13. Stationary points important to displacement, velocity, and acceleration of particles in a fluid.

and setting the derivative equal to zero. Turning points may be of three types: minima (Figure 3.13A), maxima (Figure 3.13B), and points of inflexion (Figure 3.13C). The graph shows how the derivatives are changing around each of these stationary points:

Near the point where the derivative is changing from negative to positive, it is increasing. In other words, the rate of change in velocity is positive. The derivative of the derivative, or the second derivative, then, must be positive. When the second derivative is positive at a given turning point, this is the minimum point. Likewise, at the maximum, negative to positive means that the derivative is decreasing, that is, that the rate of change is negative. This means that when the second derivative is negative at a given turning point, this must be a maximum point.

At the inflection points, the rate of change is neither positive nor negative; the rate of change is zero. Keep in mind that zero is also a possible value for the second derivative at a maximum or minimum.

Density

The relationship between mass and volume is important in both environmental physics and chemistry, and is a fundamental property of fluids. The density (ρ) of a fluid is defined as its mass per unit volume. Its metric units are kg m^{-3}. The density of an ideal gas is found using the specific gas constant and applying the ideal gas law:

$$\rho = p(RT)^{-1} \qquad \text{Equation 3–17}$$

Where, p = gas pressure
R = specific gas constant
T = absolute temperature.

The specific gas constant must be known to calculate gas density. For example, the R for air is $287\,\text{J kg}^{-1}\,\text{K}^{-1}$. The specific gas constant for methane (R_{CH4}) is $518\,\text{J kg}^{-1}\,\text{K}^{-1}$.

Density is a very important fluid property for environmental situations. For example, a first responder[5] must know the density of substances in an emergency situation. If a substance is burning, whether it is of greater or lesser density than water will be one of the factors on how to extinguish the fire. If the substance is less dense than water, the water will likely settle below the layer of water, making water a poor choice for fighting the fire. Any flammable substance with a density less than water (see Table 3.1), such as benzene or acetone, will require fire-extinguishing substances other than water. For substances heavier than water, like carbon disulfide, water may be a good choice.

Another important comparison in Table 3.1 is that of pure water and seawater. The density difference between these two water types is important for marine and estuarine ecosystems. Salt water contains a significantly greater mass of ions than does freshwater (see Table 3.2). The denser saline water can wedge beneath freshwaters and pollute surface waters and groundwater (see Figure 3.14). This phenomenon, known as "saltwater

TABLE 3.1
Densities of Some Important Environmental Fluids

Fluid	*Density ($kg\,m^{-3}$) at 20°C unless otherwise noted*
Air at standard temperature and pressure (STP) = 0°C and $101.3\,N\,m^{-2}$	1.29
Air at 21°C	1.20
Ammonia	602
Diethyl ether	740
Ethanol	790
Acetone	791
Gasoline	700
Kerosene	820
Turpentine	870
Benzene	879
Pure water	1000
Seawater	1025
Carbon disulfide	1274
Chloroform	1489
Tetrachloromethane (carbon tetrachloride)	1595
Lead (Pb)	11340
Mercury (Hg)	13600

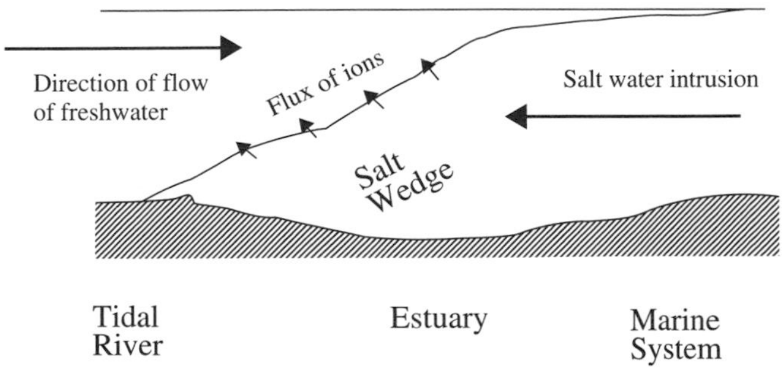

FIGURE 3.14. Saltwater intrusion into a freshwater system. This denser saltwater submerges under the lighter freshwater system. The same phenomenon can occur in coastal aquifers.

TABLE 3.2
Composition of Freshwaters (River) and Marine Waters for Some Important Ions

Composition	*River Water*	*Salt Water*
pH	6–9	8
Ca^{2+}	4×10^{-5} M	1×10^{-2} M
Cl^-	2×10^{-4} M	6×10^{-1} M
HCO_3^-	1×10^{-4} M	2×10^{-3} M
K^+	6×10^{-5} M	1×10^{-2} M
Mg^{2+}	2×10^{-4} M	5×10^{-2} M
Na^+	4×10^{-4} M	5×10^{-1} M
SO_4^{2-}	1×10^{-4} M	3×10^{-2} M

Sources: K. A. Hunter, J. P. Kim, and M. R. Reid, 1999, Factors influencing the inorganic speciation of trace metal cations in fresh waters, *Marine Freshwater Research*, vol. 50, pp. 367–372; and R. R. Schwarzenbach, P. M. Gschwend, and D. M. Imboden, 1993, *Environmental Organic Chemistry*, Wiley Interscience New York, N.Y.

intrusion," can significantly alter an ecosystem's structure and function, and threaten freshwater organisms. It can also pose a huge challenge to coastal communities who depend on aquifers for their water supply. Part of the problem and the solution to the problem can be found in dealing with the density differentials between fresh and saline waters.

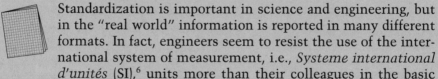

Engineer's Notebook Entry: Units in Handbooks and Reference Manuals

Standardization is important in science and engineering, but in the "real world" information is reported in many different formats. In fact, engineers seem to resist the use of the international system of measurement, i.e., *Systeme international d'unités* (SI),[6] units more than their colleagues in the basic sciences. This may, at least in part, be due to the historic inertia of engineering, where many equations were derived from English units. When an equation is based on one set of units and is only reported in those units, it can take much effort to convert them to SI units. Exponents in many water quality, water supply, and sludge equations have been empirically derived from studies that applied English units [e.g., pounds (lbs), inches (in), gallons (g)].

Some equations may use either English or SI units, such as the commonly used Hazen-Williams formula for mean velocity flow (v) in pressure pipes:

$$v = 1.318C \cdot r^{0.63} s^{0.54} \quad \text{Equation 3–18}$$

Where, r is the hydraulic radius in feet or meters, s is slope of the hydraulic grade line (head divided by length), and C is the friction coefficient (a function of pipe roughness). The exponents apply without regard to units. Other formulae, however, require that a specific set of units be used. An example is the fundamental equation for kinetic energy. Two different equations are needed when using either the SI system or the English system, which requires the gravitation conversion constant (g_c) in the denominator. These are, respectively:

$$E_{\text{kinetic}} = \frac{mv^2}{2} \quad \text{Equation 3–19}$$

$$E_{\text{kinetic}} = \frac{mv^2}{2g_c} \quad \text{(in ft-lbf)} \quad \text{Equation 3–20}$$

Two other important physical equations, potential energy and pressure, require the insertion of the gravitation conversion constant into their denominators:

$$E_{\text{potential}} = \frac{mgz}{g_c} \quad \text{(in ft-lbf)} \quad \text{Equation 3–21}$$

$$p = \frac{\rho g h}{g_c} \quad (\text{in ft-lbf ft}^{-2}) \quad \text{Equation 3–22}$$

Where g is the gravitational acceleration, ρ is density, and h is the height.

With this in mind, it is sometimes better to simply apply the formulae using English units and convert to metric or SI units following the calculation. In other words, rather than try to change the exponent to address the difference in feet and meters, just use the units called for in the empirically derived equation. After completing the calculation, convert the answer to the correct units. This may seem contrary to the need to standardize units, but it may save time and effort in the long run. Either way, it is mathematically acceptable dimension analysis.

Another variation in units is how coefficients and constants are reported. For example, the octanol-water coefficient seems to be reported more often as $\log K_{ow}$ than simply as K_{ow}. This is usually because the ranges of K_{ow} values can be so large. One compound may have a coefficient of 0.001, while another has one of 1000. Thus, it may be more manageable to report the $\log K_{ow}$ values as –3 and 3, respectively.

Further, chemists and engineers are comfortable with the "p" notation as representative of the negative log. This could be because pH and pOH are common parameters. Thus one may see the negative logarithm used with units in handbooks. For example, vapor pressure is sometimes reported as a negative log.

Therefore, examples and problems in this text make use of several different units as they are encountered in the environmental literature.

Engineering Technical Note: Density as a Factor in Emergency Response

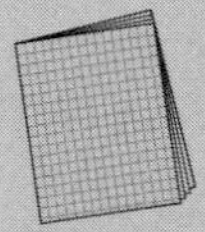

The density of the important environmental fluids—air, freshwater, and seawater—is a key property that must be included in emergency response protocols, such as part of the U.S. Coast Guard's engineering calculations laid out in its *Hazard Assessment Handbook*.[7] For example, the density of water will influence the rates at which pollutants will be transported and the size of the contaminant plume. Likewise, in the air, if a gaseous contaminant has a density near that of air, slight changes in temperatures can influence the distance and route that a plume will travel. The values in Table 3.3 for freshwater are those measured in pure water. Although the values for the water of lakes and streams are different from those of pure water, no generally recognized "standard" freshwater is accepted throughout the engineering and scientific community. Thus, the values for pure water are generally adopted.

The U.S. Coast Guard's "standard" seawater is water that contains 35 grams of salts per kilogram of solution. The values for the water of tidal systems and estuaries vary a bit from those of "standard" seawater because these are generally zones of dilution and ion flux (see Figure 3.14), making the salinity levels between those of fresh and seawaters.

The U.S. Coast Guard value for the density of air has been derived from the ideal gas law, which is applied to air that is dry and at 1 atmosphere (atm) pressure.[8] Water content and pressure will both affect air density.

Other emergency response agencies, such as the Centers for Disease Control and Prevention (CDC) and the U.S. EPA, include density along with other important fluid properties of contaminants in their spill and release contingency planning. They particularly make

TABLE 3.3
Density of Freshwater, Sea Water, and Air

DENSITY OF FRESHWATER		*DENSITY OF SEA WATER*		*DENSITY OF ICE*		*DENSITY OF DRY AIR (at 1 atm.)*	
Temperature (°F)	*Pounds per cubic foot*	*Temperature (°F)*	*Pounds per cubic foot*	*Temperature (°F)*	*Pounds per cubic foot*	*Temperature (°F)*	*Pounds per cubic foot*
32	62.410	30	64.250	–50	57.670	–10	0.088
40	62.418	40	64.200	–40	57.625	0	0.086
50	62.401	50	64.170	–30	57.600	10	0.085
60	62.358	60	64.100	–20	57.582	20	0.083
70	62.293	70	64.020	–10	57.541	30	0.081
80	62.208	80	63.950	0	57.105	40	0.079
90	62.105	90	63.800	10	57.490	50	0.078
100	61.986	100	63.700	20	57.455	60	0.076
110	61.852			30	57.410	70	0.075
120	61.704					80	0.074
						90	0.072
						100	0.071
						110	0.070
						120	0.068

Source: U.S. Coast Guard, 2003, Chemical Hazards Response Information System [CHRIS], http://www.chrismanual.com/Intro/prop.htm.

use of the Agency for Toxic Substances and Disease Registry's toxicological profiles.[9] These documents, which are available on the Internet,[10] are succinct characterizations of the toxicological and adverse health effects information about numerous hazardous substances. Each profile identifies and reviews the key literature that describes a hazardous substance's toxicological properties. Table 3.4 provides an example of the section of the ethylene glycol toxicological profile dealing with the physical and chemical properties of this compound, and the related compound propylene glycol.[11]

For an underground plume that is approximately the same density as water, the flow will be similar to that shown in Figure 3.15. However, density differentials among fluids also commonly occur when organic compounds contaminate groundwater. Those organics more dense than water (so-called dense nonaqueous phase liquids, or DNAPLs) will penetrate more

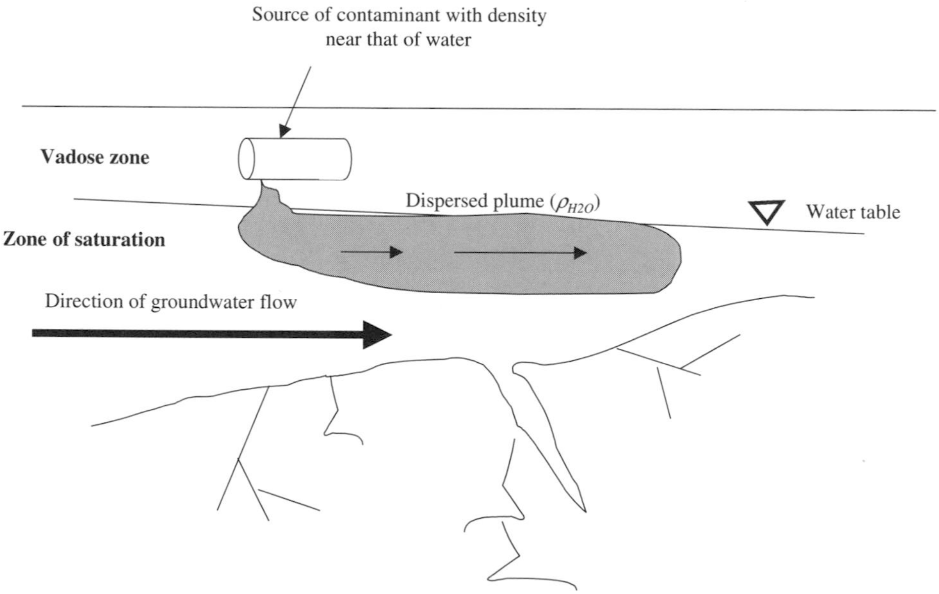

FIGURE 3.15. Importance of density of fluids in groundwater contamination. The plume of a contaminant having a density equal to that of water will disperse and move in the direction of the general groundwater flow system. (Adapted from: M. N. Sara, 1991, "Groundwater Monitoring System Design," in *Practical Handbook of Ground-Water Monitoring*, edited by D. M. Nielsen, Lewis Publishers, Chelsea, Mich.)

TABLE 3.4
Physicochemical Properties of Ethylene Glycol and Propylene Glycol as Provided in the Toxicological Profile

Property	*Ethylene Glycol*[a]	*Propylene Glycol*[b]
Molecular weight	62.07[c,d]	76.11[c]
Color	Clear, colorless[f]	Colorless[e]
Physical state	Liquid[c]	Liquid[c]
Melting point	–11.5°C[d]	–60°C[f,c] (forms glass)
Boiling point	198°C[d]	187.6°C; 188.2°C[c]
Density:		
at 20°C (g/cm^3)	1.1135[c]	1.0361[d]
at 30°C (g/cm^3)	1.1065[c]	No data
Odor	Odorless	Odorless
Odor threshold	No data	No data
Solubility: water at 20°C	Miscible with water	Miscible with water
Organic solvent(s)	Soluble in lower aliphatic alcohols, glycerol, acetic acid, acetone;[c] slightly soluble in ether; practically insoluble in benzene, chlorinated hydrocarbons, petroleum ether, oils	Soluble in alcohol, ether, benzene; soluble in acetone, chloroform[c]
Partition coefficients:		
Log K_{ow}	–1.36	–0.92[g,h]
Log K_{oc}	0.592[f]	0.88[f], 0.76[h]
Vapor pressure at 20°C	0.06 mm Hg	0.07 mm Hg[also a]
Henry's Law constant:		
at 25°C	2.34×10^{-10} atm-m^3/mole	1.2×10^{-8} atm-m^3/mole 1.7×10^{-8} atm-m^3/mole[h]
Autoignition temperature	412.93°C[i] 398°C[j]	421.26°C[i] 371°C[j]
Flashpoint	111.26°C[i,j]	99.04°C[i,j]
Flammability limits	3.2–21.6%[i,j]	2.6–12.5%[i,j]
Conversion factors	1 ppm = 2.54 mg/m^3[k] 1 mg/L = 365.0 ppm[k]	1 ppm = 3.11 mg/m^3[k] 1 mg/L = 321.6 ppm[k]
Explosive limits	No data	No data

[a] Unless otherwise noted all references for ethylene glycol are HSDB 1995a.
[b] Unless otherwise noted all references for propylene glycol are HSDB 1995b.
[c] Merck 1989.
[d] Weast 1988.
[e] Lewis 1993.
[f] Daubert and Danner 1980.
[g] EPA 1987a.
[h] ASTER 1995.
[i] Daubert and Danner 1989.
[j] NFPA 1994.
[k] Rowe and Wolf 1982.
Source: Agency for Toxic Substances and Disease Registry, 2003, http://www.atsdr.cdc.gov/toxprofiles/tp96-c3.pdf.
See the ATSDR website for full citations of these references.

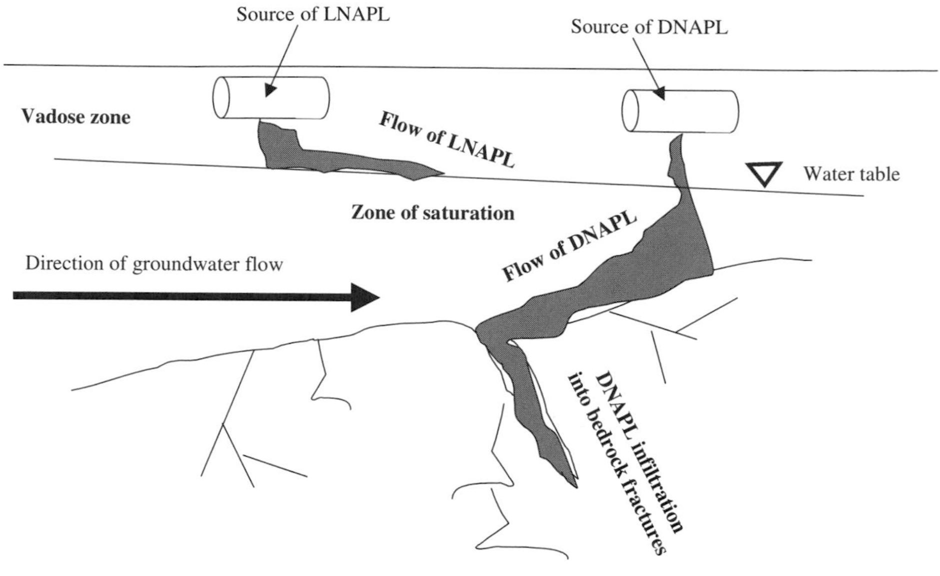

FIGURE 3.16. Importance of density of fluids in groundwater contamination scenarios. The dense nonaqueous phase liquids (DNAPLs) can penetrate more deeply into the aquifer than do the light nonaqueous phase liquids (LNAPLs). The density reference for whether a compound is a DNAPL or an LNAPL is whether it is denser or lighter than water, respectively. The DNAPL movement may even be against the general flow of the groundwater. (Adapted from: H. Hemond and E. Fechner-Levy, 2000, *Chemical Fate and Transport in the Environment*, Academic Press, San Diego, Calif.)

deeply, while the lighter organics (light nonaqueous phase liquids, or LNAPLs) will float near the top of the zone of saturation (see Figure 3.16). We will cover this topic in greater detail in Chapter 4, "Environmental Equilibrium Partitioning and Balances," and when discussing groundwater contamination and remediation in Chapter 12, "Intervention: Managing the Risks of Environmental Contamination."

Specific Gravity

The ratio of a substance's density to a standard reference density is known as specific gravity (SG). Since this is a ratio of densities, specific gravity is dimensionless. The standard reference for most liquids and solids is pure water, but there is some variability in this reference, since various applications apply the density of water at different temperatures. Hazardous materials and first responders have used water at 4°C as a reference, but others have applied 21°C or even 16°C. If given a value for SG for a substance, it

TABLE 3.5
Commonly Reported Values for Standard Temperature and Pressure (STP)

System Reporting STP	*Pressure*	*Temperature*
SI (International System)	101.325 kP	273.15°K
Scientific	760 mm Hg	0.0°C
Engineering (U.S.)	14.696 psia	32°F
Natural Gas Industry (U.S.)	14.65, 14.73, 15.025 psia.	60°F
Natural Gas Industry (Canada)	14.696 psia.	60°F

is helpful to know which reference has been applied (e.g., in chemical engineering and synthesis operations). For most environmental situations, however, the variability of water's density between 4°C and 21°C is not important, since the densities are within three significant figures of one another. The liquid SG is:

$$SG_{liquid} = \rho_{liquid}(\rho_{water})^{-1} \qquad \text{Equation 3–23}$$

The reference density for gases is air. Thus, the SG for a gas is:

$$SG_{gas} = \rho_{gas}(\rho_{air})^{-1} \qquad \text{Equation 3–24}$$

Gas density is more sensitive to temperature and pressure differences than are liquids and solids, so it is more important to know these variables for air. Usually, the air reference density is specified to be at standard temperature and pressure (STP). Table 3.5 provides some of the common conversions for STP. Note the differences in units, but more importantly, the variability even within the same industry (i.e., natural gas).

Engineer's Notebook Entry: Fluid Properties

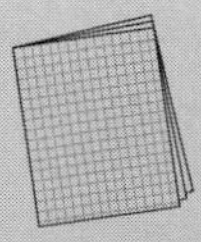

Our discussion of fluids has addressed density and specific gravity; we will consider fluid velocity separately. However, engineers use numerous fluid properties in their characterizations and calculations. Here are a few:

Specific Volume

The reciprocal of a substance's density is known as its specific volume (ν). This is the volume occupied by a unit mass of a fluid. The units of ν are reciprocal density units ($m^3 kg^{-1}$). Stated mathematically, this is:

$$\nu = \rho^{-1} \qquad \text{Equation 3–25}$$

Specific Weight

The weight of a fluid per its volume is known as *specific weight* (γ). Civil engineers sometimes use the term interchangeably with *density*. Geoscientists frequently refer to a substance's specific weight. A substance's γ is not an absolute fluid property because in depends on the fluid itself and the local gravitational force:

$$\gamma = gp \qquad \text{Equation 3–26}$$

The units are the same as those for density; e.g., $kg m^{-3}$.

Mole Fraction

In a composition of a fluid made up of two or more substances (A, B, C, . . .), the mole fraction (x_A, x_B, x_C, . . .) is the number of moles of each substance divided by the total number of moles for the whole fluid:

$$x_A = \frac{n_A}{n_A + n_B + n_c + \ldots} \qquad \text{Equation 3–27}$$

The mole fraction value is always between 0 and 1. The mole fraction may be converted to a mole percent as:

$$x_{A\%} = x_A \times 100 \qquad \text{Equation 3–28}$$

For gases, the mole fraction is the same as the volumetric fraction of each gas in a mixture of more than one gas.

Mole Fraction Example

112 g of $MgCl_2$ are dissolved in 1 L of water. The density of this solution is 1.089 $g cm^{-3}$. What is the mole fraction of $MgCl_2$ in the solution at standard temperature and pressure?

Solution

The number of moles of $MgCl_2$ is determined from its molecular weight:

$$\frac{112\,\text{g}}{95.22\,\text{g}} = 1.18\,\text{mole}$$

Next, we calculate the number of moles of water:

$$\text{Mass of water} = 1.00\,\text{L} \times (1000\,\text{cm}^{-3}\,\text{L}^{-1}) \times (1.00\,\text{g}\,\text{cm}^{-3}) = 1000\,\text{g water}$$

and,

$$\text{moles of water} = \frac{1000\,\text{g}}{18.02\,\text{g mol}^{-1}} = 55.49\,\text{mol}$$

Thus, $x_{MgCl_2} = \dfrac{1.18\,\text{mol}}{55.49 + 1.18} = 0.021$.

The mole percent of $MgCl_2$ is 2.1%.

Compressibility

The fractional change in a fluid's volume per unit change in pressure at constant temperature is the fluid's coefficient of compressibility. Any fluid can be compressed in response to the application of pressure (p). For example, water's compressibility at 1 atm is $4.9 \times 10^{-5}\,\text{atm}^{-1}$. This compares to the lesser compressibility of mercury ($3.9 \times 10^{-6}\,\text{atm}^{-1}$) and the greater compressibility of hydrogen ($1.6 \times 10^{-3}\,\text{atm}^{-1}$).

A fluid's bulk modulus, E (analogous to the modulus of elasticity in solids), is a function of stress and strain on the fluid (see Figure 3.17), is a description of its compressibility, and is defined according to the fluid volume (V):

$$E = \frac{\text{stress}}{\text{strain}} = -\frac{dp}{dV/V_1} \qquad \text{Equation 3–29}$$

E is expressed in units of pressure (e.g., kP). Water's $E = 2.2 \times 10^6\,\text{kP}$ at 20°C.

Surface Tension and Capillarity

Surface tension effects occur at liquid surfaces (interfaces of liquid-liquid, liquid-gas, liquid-solid). Surface tension, σ, is the force in the liquid surface normal to a line of unit length drawn in the surface. Surface tension decreases with temperature and depends on the contact fluid. Surface tension is involved in capillary rise and drop. Water has a very high σ value (approximately $0.07\,\text{N}\,\text{m}^{-2}$ at 20°C). Of the environmental fluids, only mercury has a higher σ (see Table 3.6).

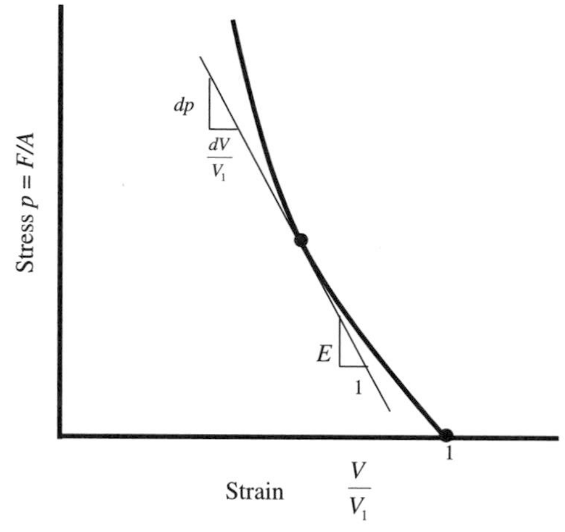

FIGURE 3.17. Stress and strain on a fluid, and the bulk modulus of fluids.

TABLE 3.6
Surface Tension (Contact with Air) of Selected Environmental Fluids

Fluid	*Surface Tension, σ (Nm^{-1} at 20°C)*
Acetone	0.0236
Benzene	0.0289
Ethanol	0.0236
Glycerin	0.0631
Kerosene	0.0260
Mercury	0.519
n-Octane	0.0270
Tetrachloromethane	0.0236
Toluene	0.0285
Water	0.0728

The high surface tension creates a type of skin on a free surface, which is how an object more dense than water (e.g., a steel needle) can "float" on a still water surface. It is the reason insects can sit comfortably on water surfaces. Surface tension is somewhat dependent upon the gas that is contacting the free surface. If not indicated, it is usually safe to assume that the gas is the air in the troposphere.

Capillarity is a particularly important fluid property of groundwater flow and the movement of contaminants above the water table. In fact, the zone immediately above the water table is called the *cap-*

illary fringe. Regardless of how densely soil particles are arranged, void spaces (i.e., pore spaces) will exist between the particles. By definition, the pore spaces below the water table are filled exclusively with water. However, above the water table, the spaces are filled with a mixture of air and water. As shown in Figure 3.18, the spaces between unconsolidated material (e.g., gravel, sand, or clay) are interconnected and behave like small conduits or pipes in their ability to distribute water. Depending on the grain size and density of packing, the conduits will vary in diameter, ranging from large pores (macropores), to medium pore sizes (mesopores), to extremely small pores (micropores).

Fluid pressures above the water table are negative with respect to atmospheric pressure, creating tension. Water rises for two reasons: its adhesion to a surface, plus the cohesion of water molecules to one another. Higher relative surface tension causes a fluid to rise in a tube

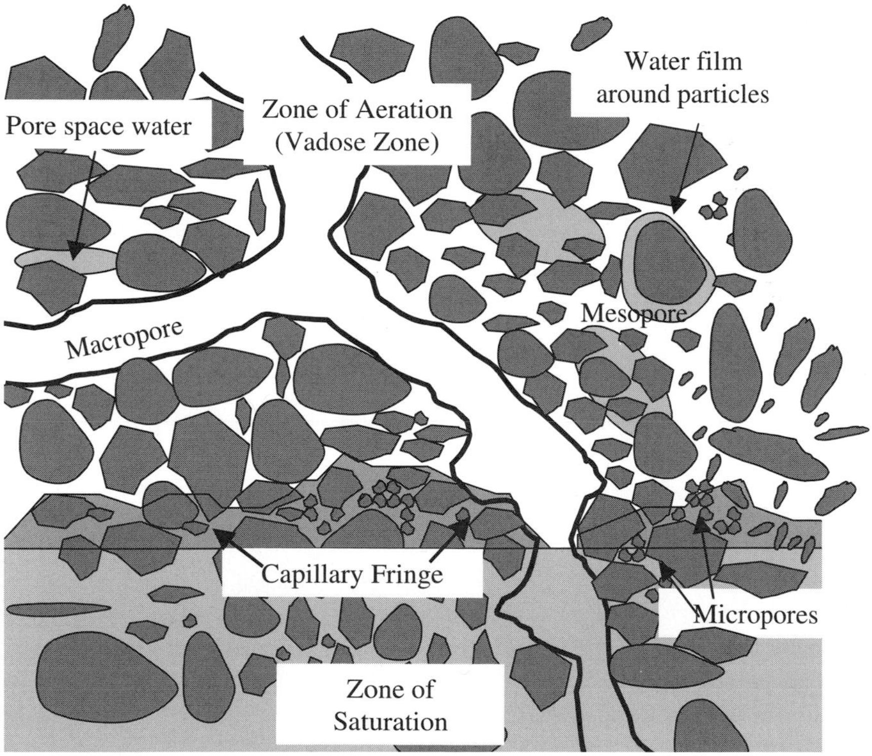

FIGURE 3.18. Capillary fringe above the water table.

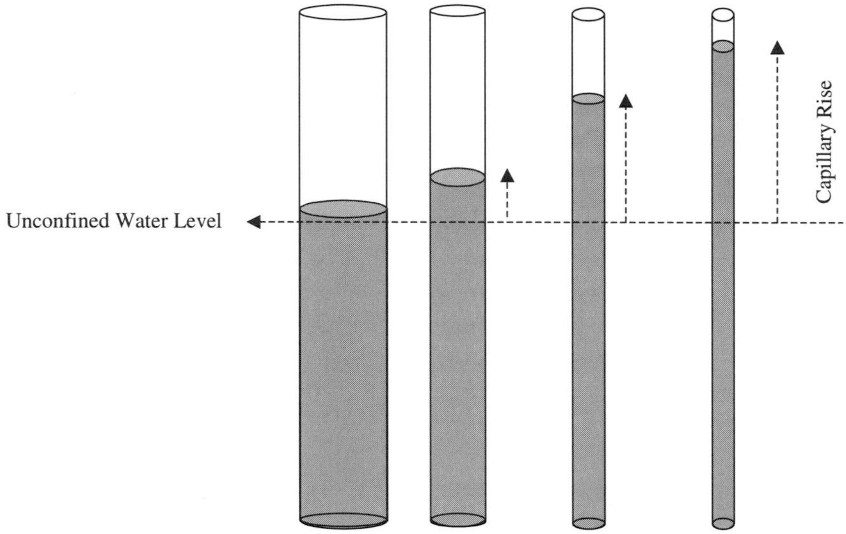

FIGURE 3.19. Capillary rise of water with respect to diameter of conduit.

(or a pore) and is indirectly proportional to the diameter of the tube. In other words, capillarity is greater the smaller the inside diameter of the tube (see Figure 3.19). The rise is limited by the weight of the fluid in the tube. The rise ($h_{capillary}$) of the fluid in a capillary is expressed as (Figure 3.20 shows the variables):

$$h_{capillary} = \frac{2\sigma \cos \lambda}{\rho_w g R} \qquad \text{Equation 3–30}$$

Where, σ = fluid surface tension ($g\,s^{-2}$)
λ = angle of meniscus (concavity of fluid) in capillary (degrees)
ρ_w = fluid density ($g\,cm^{-3}$)
g = gravitational acceleration ($cm\,sec^{-1}$)
R = radius of capillary (cm)

The contact angle indicates whether cohesive or adhesive forces are dominant in the capillarity. When λ values are greater than 90°, cohesive forces are dominant; when $\lambda < 90°$, adhesive forces dominate. Thus, λ is dependent upon both the type of fluid and the surface to which it comes into contact. For example, water-glass $\lambda = 0°$; ethanol-glass $\lambda = 0°$; glycerin-glass $\lambda = 19°$; kerosene-glass $\lambda = 26°$; water-paraffin $\lambda = 107°$; and mercury-glass $\lambda = 140°$.

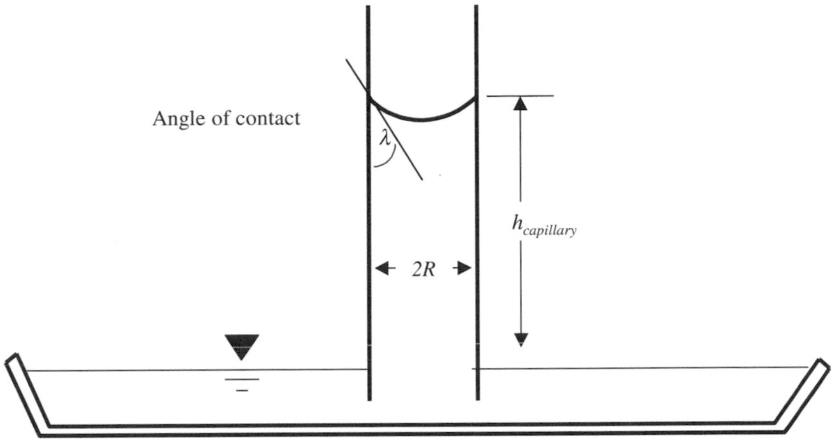

FIGURE 3.20. Rise of a fluid in a capillary. In this example, adhesive forces within the fluid are dominant, so the meniscus is concave (i.e., a valley). This is the case for most fluids. However, if cohesive forces dominate, such as the extremely cohesive liquid mercury, the meniscus will be convex (i.e., a hill).

In the lowest level of the capillary fringe, the soil is saturated without regard to pore size. In the vadose zone, however, the capillary rise of water will be highest in the micropores, where relative surface tension and the effects of water cohesion are greatest.

Capillarity Example

What is the rise of contaminated water (i.e., a solution of water and soluble and insoluble contaminants) in a capillary fringe with an average pore space diameter of 0.1 cm, at 18°C and a density of 0.999 g cm^{-3}, under surface tension of 50 g sec^{-1} if the angle of contact of the meniscus is 30°?

What would happen if the average pore space were 0.01 cm, with all other variables remaining as stated?

Solution

$$h_{capillary} = \frac{2\sigma \cos\lambda}{\rho_w g R} = \frac{2 \times 80 \times \cos 30}{0.999 \times 980 \times 0.05}\,\text{cm} \cong 0.25\,\text{cm}$$

If the pore space were 0.01 cm in diameter, the rise would be 2.5 cm. However, it is likely that the angle of contact would have also decreased since the angle is influenced by the diameter of the column (approaching zero with decreasing diameter).

Also note that since the solution is not 100% water, the curvature of the meniscus will be different, so the contact angle λ will likely be greater (i.e., less curvature) than the meniscus of water alone. The lower surface tension of the mixture also means that the capillary rise will be less.

Engineer's Notebook Entry: Viscosity

How much a fluid resists flow when it is acted on by an external force, especially a pressure differential or gravity, is the fluid's *viscosity*. This is a crucial fluid property used in numerous environmental applications, but particularly in air pollution plume characterization, sludge management, and wastewater and drinking water treatment and distribution systems.

Recall from Bernoulli's Equation and Figure 3.11 that if a fluid is flowing in a long, horizontal pipe with a constant cross-sectional area, the pressure along the pipe must be constant. But why, if we measure the pressure as the fluid moves in the pipe, will there be a pressure drop? A pressure difference is needed to push the fluid through the pipe to overcome the drag force exerted by the pipe walls on the layer of fluid that is making contact with the walls. Since there is a drag force exerted by each successive layer of the fluid on each adjacent layer that is moving at its own velocity, a pressure difference is needed (see Figure 3.21). The drag forces are known as *viscous forces*. Thus, the fluid velocity is not constant across the pipe's diameter, owing to

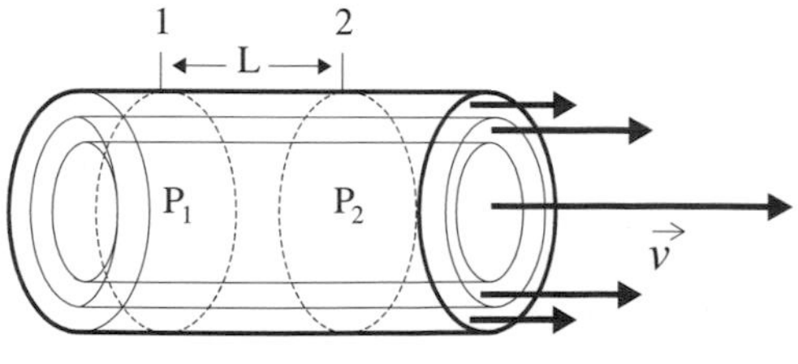

FIGURE 3.21. Viscous flow through a horizontal pipe. The highest velocity is at the center of the pipe. As the fluid approaches the pipe wall, the velocity approaches zero.

the viscous forces. The greatest velocity is at the center (furthest away from the walls), and the lowest velocity is found at the walls. In fact, at the point of contact with the walls, the fluid velocity is zero.

So, if P_1 is the pressure at point 1, and P_2 is the pressure at point 2, with the two points separated by distance L, the pressure drop (ΔP) is proportional to the flow rate:

$$\Delta P = P_1 - P_2 \qquad \text{Equation 3–31}$$

and,

$$\Delta P = P_1 - P_2 = I_v R \qquad \text{Equation 3–32}$$

where, I_v is volume flow rate and R is the proportionality constant representing the resistance to the flow. R depends on the length (L) of pipe section, the pipe's radius, and the fluid's viscosity.

Viscosity Example 1

Workers have been exposed to a chemical that is known to decrease blood pressure in the capillaries, small arteries, and major arteries and veins after the blood is pumped from the aorta. If high-dose studies show an acute drop in the gauge pressure of the circulatory system from 100 torr to zero torr at a volume flow of $0.7\,\text{L}\,\text{sec}^{-1}$, give the total resistance of the circulatory system.

Solution

Solving for R from Equation 3–32, and converting to SI units gives us:

$$\begin{aligned} R &= \Delta P(I_v)^{-1} \\ &= (100\,\text{torr})(0.7\,\text{L}\,\text{sec}^{-1})^{-1}(133.3\,\text{Pa})(1\,\text{torr})^{-1} \\ &\quad \times (1\,\text{L})(10^3\,\text{cm}^{-3})^{-1}(1\,\text{cm}^{-3})(10^{-6}\,\text{m}^{-3}) \\ &= 1.45 \times 10^7\,\text{Pa} \cdot \text{s} \cdot \text{m}^{-3} \\ &= 1.45 \times 10^7\,\text{N} \cdot \text{s} \cdot \text{m}^{-5} \end{aligned}$$

We will now consider the two types of viscosity: absolute viscosity and kinematic viscosity.

Absolute Viscosity

Physicists define the fluid's coefficient of viscosity by assuming that the fluid is confined between two parallel, rigid plates with equal area.

The absolute viscosity of a fluid can be measured a number of ways, but engineers commonly use the *sliding plate viscometer test*. The test applies two plates separated by the fluid to be measured (see Figure 3.22).

For Newtonian fluids, the force applied in the viscometer test has been found to be in direct proportion to the velocity of the moving plate and inversely proportional to the length of separation of the two plates:

$$\frac{F}{A} \propto \frac{dv}{dy} \qquad \text{Equation 3–33}$$

Making this proportionality into an equality requires a constant:

$$\frac{F}{A} = \mu \frac{dv}{dy} \qquad \text{Equation 3–34}$$

This equation is known at *Newton's Law of Viscosity*. Fluids that conform to this law are referred to as Newtonian fluids.[12] The constant, μ, is the fluid's *absolute viscosity*. The μ is also known as the *coefficient of viscosity*, but environmental texts often refer to μ as *dynamic viscosity*. The term *fluidity* is the reciprocal of dynamic viscosity.

The inverse relationship between viscosity and fluidity makes sense if you think about it! Since the definition of viscosity is the resistance to flow when an external force is applied, then it stands to

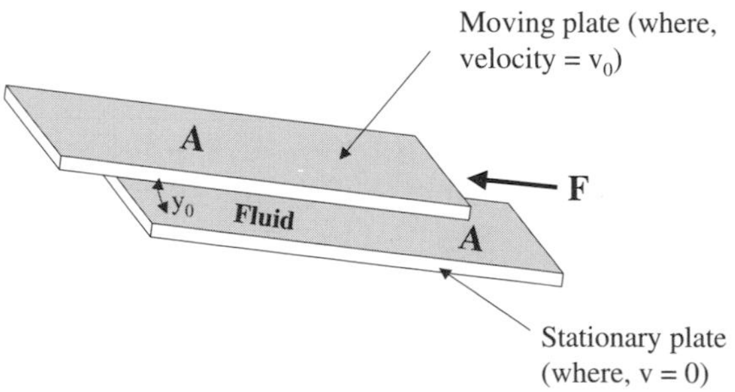

FIGURE 3.22. The sliding plate viscometer. A fluid of thickness y_0 is placed between two plates of area A. The top plate moves at the constant velocity v_0 by the exertion of force F. (Adapted from: M. Lindeberg, 2001, *Civil Engineering Reference Manual for the PE Exam*, 8th Edition, Professional Publications, Belmont, Calif.)

reason that if it doesn't do a good job resisting the flow, the substance has a lot of fluidity. An electrical analogy might be that of conductivity and resistance. If copper wire has much less resistance to electrical flow than does latex rubber, we say that copper must be a good conductor. Likewise, if water at 35°C is less effective at resisting flow downhill (i.e., gravity is applying our force) than is motor oil at the same temperature, we say that the water has less dynamic viscosity than the motor oil. We also say that the water has more fluidity. Before the modern blends of multi-viscosity motor oils, the temperature-viscosity relationship was part of the seasonal rituals of the oil change. You had to put in less viscous motor oil (say a 10W) in your car's engine to prepare for the lower temperatures in winter, so that the starter could "turn over" the engine (less viscous oil = less resistance to the force of the starter moving the pistons). Conversely, in preparing for summer, you needed a more viscous motor oil (commonly 40W)[13] because the high temperatures in the engine would allow the oil to "blow out" through the piston rings or elsewhere (because the oil wasn't doing a good job of resisting the force applied by the pistons and shot out of the engine!). The newer oil formulations (e.g., 10W40) maintain a smaller range of viscosities, so automobile owners worry less about the viscosity.

The $\frac{F}{A}$ term is known as the *shear stress*, τ, of the fluid. The $\frac{dv}{dy}$ term is known as the *velocity gradient* or the *rate of shear formation.*[14] So, the shear stress is linear; that is, it can be expressed as a straight line (in the form $y = mx + b$):

$$\tau = \mu \frac{dv}{dy} \qquad \text{Equation 3–35}$$

The relationship between the two sides of this equality determines the types of fluids, as shown in Figure 3.23. Most fluids encountered in environmental studies are Newtonian, including water, all gases, alcohols, and most solvents. Most solutions also behave as Newtonian fluids. Slurries, muds, motor grease and oils, and many polymers behave as *pseudoplastic fluids*; that is, viscosities decrease with increasing velocity gradient. They are easily pumped, since higher pumping rates lead to a less viscous fluid. Some slurries behave as *Bingham fluids* (e.g., they behave like toothpaste or bread dough), where the shear formation is resisted up to a point. The rare *dilatant fluids* are sometimes encountered in environmental engineering applications, such as clay slurries used as landfill liners and when starches and certain paints and coatings are spilled. These can be difficult fluids

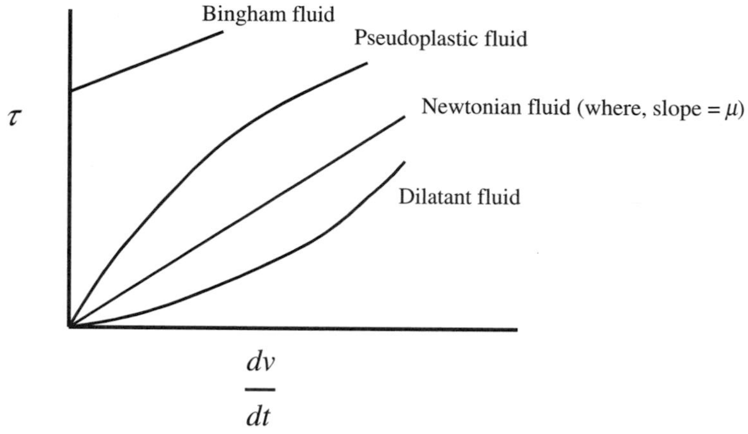

FIGURE 3.23. Hypothetical fluid types according to shear stress (τ) behavior relative to velocity gradient. (Adapted from: M. Lindeberg, 2001, *Civil Engineering Reference Manual for the PE Exam*, 8th Edition, Professional Publications, Belmont, Calif.)

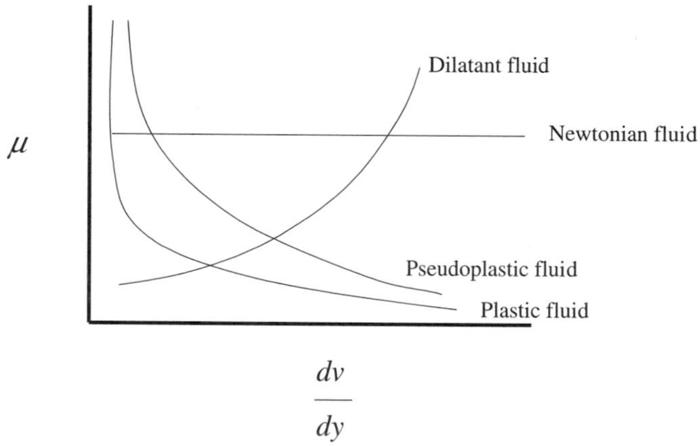

FIGURE 3.24. Hypothetical fluid types according to viscosity (μ) and shear rate (velocity). (Adapted from: M. Lindeberg, 2001, *Civil Engineering Reference Manual for the PE Exam*, 8th Edition, Professional Publications, Belmont, Calif.)

to remove and clean up, since their viscosities increase with increasing velocity gradient, so pumping these fluids at higher rates can lead to their becoming almost solid with a sufficiently high shear rate. *Plastic fluids* (see Figure 3.24) require the application of a finite force before any fluid movement.

Categorizing and characterizing fluids according to their behavior under shear stress and velocity gradient is not absolute. For example, a Bingham fluid can resist shear stresses indefinitely so long as they are small, but these fluids will become pseudoplastic at higher stresses. Even when all conditions remain constant, viscosity can also change with time. A *rheopectic fluid* is one where viscosity increases with time, and a *thixotropic fluid* is one that has decreasing viscosity with time. Those fluids that do not change with time are referred to as *time-independent fluids.* Colloidal materials, like certain components of sludges, sediments, and soils, act like thixotropic fluids; that is, they experience a decrease in viscosity when the shear is increased. However, there is no hysteresis; the viscosity does not return to the original state with the ceasing of the agitation.

There is a seeming paradox between viscosity and temperature. As a general rule, temperature is inversely proportional to viscosity of liquids, but temperature is directly proportional to the viscosity of gases. Viscosity of liquids is predominantly caused by molecular cohesion. These cohesive forces decrease with increasing temperature, which is why viscosity decreases with increasing temperature. Gas viscosity is mainly kinetic-molecular in its origin, so increasing temperature means that more collisions will occur between molecules. The more the gas is agitated, the greater the viscosity, so gas velocity increases with increasing temperatures.

There is only a very slight increase in the viscosity of liquids with increasing pressure. Under environmental conditions, absolute viscosity can be considered to be independent of pressure.

Absolute viscosity units are mass per length per time (e.g., $g\,cm^{-1}\,sec^{-1}$). The coefficients for some common fluids are provided in Table 3.7. Note the importance of temperature in a substance's absolute viscosity; for example, the several orders of magnitude decrease with only a 20°C increase in glycerin.

Viscosity Example 2

A liquid with the absolute viscosity of $3 \times 10^{-5}\,g\,sec\,cm^{-1}$ flows through a rectangular tube in a wastewater treatment plant. The velocity gradient is $0.5\,m\,sec^{-1}\,cm^{-1}$. What is the shear stress in the fluid at this velocity gradient?

TABLE 3.7
Absolute Viscosity of Fluids Important to Health and Environmental Studies

Fluid	*Temperature (°C)*	*Absolute Viscosity, μ (Pa·s)*
Water	0	1.8×10^{-3}
	20	1×10^{-3}
	60	6.5×10^{-2}
Whole human blood	37	4×10^{-3}
SAE 10 motor oil	30	2×10^{-1}
Glycerin	0	10
	20	1.4
	60	8.1×10^{-2}
Air	20	1.8×10^{-5}

Source: P. Tipler, 1999, *Physics for Scientists and Engineers*, Volume 1, W.H. Freeman and Co., New York, N.Y.

Solution

$$\tau = \mu \frac{dv}{dy}$$
$$= (3 \times 10^{-5}\,\mathrm{g\,sec\,cm^{-1}})(0.5\,\mathrm{m\,sec^{-1}\,cm^{-1}})(100\,\mathrm{cm\,m^{-1}})$$
$$= 1.5 \times 10^{-3}\,\mathrm{g\,cm^{-2}}$$

Kinematic Viscosity

In environmental engineering, the ratio of absolute viscosity to mass density is known as *kinematic viscosity* (ν):

$$\nu = \mu\rho^{-1} \qquad \text{Equation 3–36}$$

The units of ν are area per sec (e.g., $\mathrm{cm^2\,sec^{-1}}$ = stoke). Because kinematic viscosity is inversely proportional to a fluid's density, ν is highly dependent on temperature and pressure. Recall that absolute viscosity is only slightly affected by pressure. Since viscosity is reported in so many different units, use Table 3.8 to convert most of these reported units.

Laminar versus Turbulent Flow: The Reynolds Number

At a sufficiently high velocity, a fluid's flow ceases to be laminar and becomes turbulent. Engineers make use of the dimensionless Reynolds number (N_R) to differentiate types of flow. The N_R is expressed as the ratio of inertial to viscous forces in a fluid:

TABLE 3.8
Viscosity Units and Conversions

Multiply:	*By:*	*To Obtain:*
	Absolute Viscosity (μ)	
centipoise (cP)	1.0197×10^{-4}	kgf·s m^{-2}
cP	2.0885×10^{-5}	lbf-s ft^{-2}
cP	1×10^{-3}	Pa·s
Pa·s	2.0885×10^{-3}	lbf-sec ft^{-2}
Pa·s	1000	cP
dyne·s cm^{-2}	0.10	Pa·s
lbf-s ft^{-2}	478.8	poise (P)
slug ft^{-1} sec^{-1}	47.88	Pa·s
	Kinematic Viscosity (v)	
ft^2 sec^{-1}	9.2903×10^{4}	centistoke (cSt)
ft^2 sec^{-1}	9.2903×10^{-2}	m^2 s^{-1}
m^2 s^{-1}	10.7639	ft^2 sec^{-1}
m^2 s^{-1}	1×10^{6}	cSt
cSt	1×10^{-6}	m^2 s^{-1}
cSt	1.0764×10^{-5}	ft^2 sec^{-1}
	μ to v	
cP	$1/\rho$ (g cm^{-3})	cSt
cP	$6.7195 \times 10^{-4}/\rho$ in lbm ft^{-3}	cSt
lbf-sec ft^{-2}	$32.174/\rho$ in lbm ft^{-3}	ft^2 sec^{-1}
kgf·s m^{-2}	$9.807/\rho$ in kg m^{-3}	m^2 s^{-1}
Pa·s	$1000/\rho$ in g cm^{-3}	cSt
	v to μ	
cSt	ρ in g cm^{-3}	cP
cSt	1.6×10^{-5}	Pa·s
m^2 s^{-1}	$0.10197 \times \rho$ in kg m^{-3}	kgf·s m^{-2}
m^2 s^{-1}	$1000 \times \rho$ in g cm^{-3}	Pa·s
ft^2 sec^{-1}	$3.1081 \times 10^{-2} \times \rho$ in lbm ft^{-3}	lbf-sec ft^{-2}
ft^2 sec^{-1}	$1.4882 \times 10^{3} \times \rho$ in lbm ft^{-3}	cP

$$N_R = \frac{\text{Inertial Forces}}{\text{Viscous Forces}} \qquad \text{Equation 3–37}$$

The inertial forces are proportional to the velocity and density of the fluid, as well as to the diameter of the conduit in which the fluid is moving. An increase in any of these factors will lead to a proportional increase in the momentum of the flowing fluid. We know from our previous discussion that the coefficient of viscosity or absolute viscosity (μ) represents the total viscous force of the fluid, so, N_R can be calculated as:

$$N_R = \frac{D_e v \rho}{\mu} \qquad \text{Equation 3–38}$$

Where, D_e is the conduit's equivalent diameter, which is a so-called "characteristic dimension"[15] that evaluates the fluid flow as a physical length. It is actually the inside diameter (i.d.) of the conduit, vent, or pipe. Recall that $\mu\ \rho^{-1}$ is the kinematic viscosity ν, so the Reynolds number can be stated as the relationship between the size of the channel or pipe, the average fluid velocity v, and ν:

$$N_R = \frac{D_e v}{\upsilon} \qquad \text{Equation 3–39}$$

When fluids move at very low velocities, the bulk material moves in discrete layers parallel to one another. The only movement across the fluid layers is molecular motion, which creates viscosity. Such a flow is *laminar* (see Figure 3.25). Laminar flow is common in most groundwater systems.

With increasing fluid velocity, the bulk movement changes, forming eddy currents that create three-dimensional mixing across the flow stream. This is known as *turbulent* flow. Most pollution control equipment and atmospheric plumes are subjected to turbulent flow. (See Figure 3.26.)

Flows in closed conduits with Reynolds numbers under 2,100 are usually laminar.[16] Due to the relatively low velocities associated with this type of flow, they are mainly encountered with liquids such as

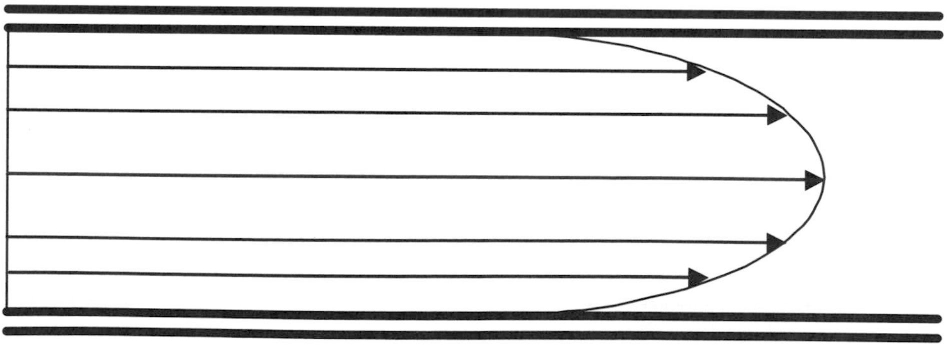

FIGURE 3.25. Laminar flow in closed conduit.

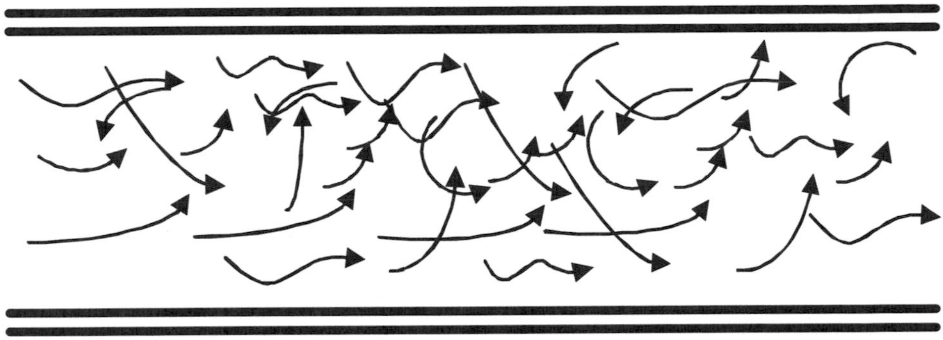

FIGURE 3.26. Turbulent flow in a closed conduit.

water moving through underground strata and blood flowing in arteries. In open atmospheric conditions, such as a plume of an air pollutant, laminar flow is quite rare. Flows with Reynolds numbers greater than 4000 are usually turbulent. The range of N_R values between these thresholds are considered "critical flows" or "transitional flows," that show properties of both laminar and turbulent flow in the flow streams. Usually, if the flow is in the transition region, engineers will design equipment as if the flow were turbulent, as this is the most conservative design assumption.

Under laminar conditions, the fluid particles adhere to the wall conduit. The closer to the wall that a particle gets, the more likely it will adhere to the wall. Laminar flow is, therefore, parabolic, and its velocity at the conduit wall is zero (see the left-hand diagram of Figure 3.27). Laminar flow velocity is greatest at the pipe's center (v_{max} in the figure), and is twice the value of the average velocity, $v_{average}$:

$$v_{average} = \frac{\dot{V}}{A} = \frac{v_{max}}{2} \quad \text{[laminar]} \qquad \text{Equation 3–40}$$

Where, $\dot{V}$ is the volumetric fluid velocity and A is the cross-sectional area of the pipe.

Turbulent flow velocity, on the other hand, has no relationship with the proximity to the wall due to the mixing (see the right-hand diagram of Figure 3.27). So, all fluid particles in a turbulent system are assumed to share the same velocity, known as the average velocity or bulk velocity:

$$v_{average} = \frac{\dot{V}}{A} \qquad \text{Equation 3–41}$$

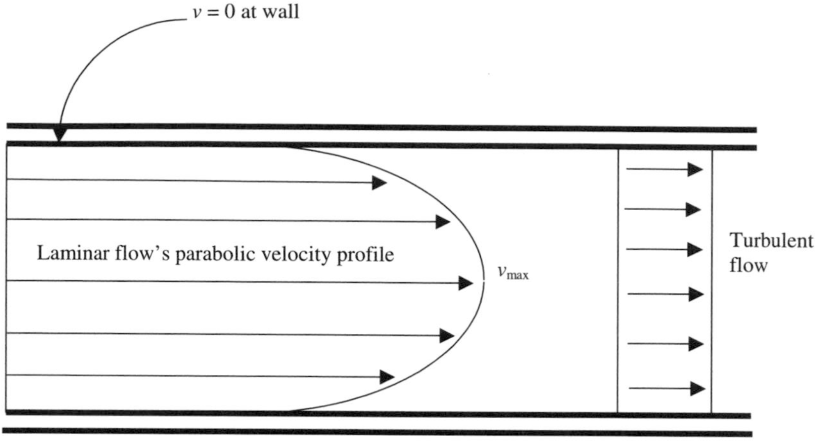

FIGURE 3.27. Velocity distributions of laminar and turbulent flows.

There is a thin layer of turbulent flow near the wall of the conduit where the velocity increases from zero to $v_{average}$, known as the *boundary layer*. In fact, no flow is entirely turbulent and there is some difference between the centerline velocity and $v_{average}$. However, for most environmental applications, the assumption of consistently mixed flow is acceptable.

Reynolds Number Example 1

Find the Reynold's number of water flowing in a 0.2 m (i.d.) pipe at $0.1\,\mathrm{m\,sec^{-1}}$. Assume that the water's coefficient of viscosity is $8 \times 10^{-3}\,\mathrm{N{\cdot}s\,m^{-3}}$ and the density is $1000\,\mathrm{kg\,m^{-3}}$.

Solution

Use Equation 3–38.

$$
\begin{aligned}
N_R &= \frac{D_e v \rho}{\mu} \\
&= \frac{(1000\,\mathrm{kg \cdot m^3})(0.1\,\mathrm{m \cdot s^{-1}})(0.2\,\mathrm{m})}{8 \times 10^{-3}\,\mathrm{N \cdot s \cdot m^{-2}}} \\
&= 2500
\end{aligned}
$$

Reynolds Number Example 2

How is this flow characterized? Assuming this flow is representative of a discharge from an industrial process, what kind of flow should be assumed in selecting water pollution control equipment to treat wastes moving from this pipe?

Solution

Since the N_R is greater than 2100, but less than 4000, the flow is considered transitional or critical. Therefore, the conservative design of pollution equipment calls for an assumption that the flow is turbulent.

Notes and Commentary

1. Although this book strives to compartmentalize the science discussions among physics, chemistry, and biology, complex topics like transport require that all three of the sciences be considered. Thus, though the focus of this chapter is predominantly on physical transport, I must interject it with chemical and biological topics to explain the concepts properly. In the next chapters, we will cover in greater detail some of the chemical and biological topics simply introduced here.
2. Fate may also include some remediation reactions, such as thermal and microbial treatment, but in discussions of fate and transport, the reactions are usually those that occur in the ambient environment. The treatment and remediation processes usually fall under the category of environmental engineering.
3. From C. Lee and S. Lin, editors, 1999, *Handbook of Environmental Engineering Calculations*, McGraw-Hill, New York, N.Y.
4. Newton actually co-invented the calculus with Willhelm Leibniz in the seventeenth century. Both are credited with devising the symbolism and the system of rules for computing derivatives and integrals, but their notation and emphases differed. A debate rages on who did what first, but both of these giants had good reason to revise the language of science, or mathematics, to explain motion.
5. As the name implies, the "first responder" is a person or group who first arrives or is asked to respond to an emergency. This includes firefighters, police, HAZMAT teams, and emergency response teams from the federal Office of Homeland Security (including the Federal Emergency Management Administration), U.S. Coast Guard, National Guard, and the U.S. Environmental Protection Agency.

6. The International System of Units (SI) is the modern system of measurement that is based entirely on the metric system. See Publication 811 (SP811), *Guide for the Use of the International System of Units (SI)*, by Barry N. Taylor, 1995, National Institute of Standards and Technology (NIST). Publication SP811 provides important information about the policies for using SI, including the classes (i.e., base units, derived units, and supplementary units) and prefixes (See Table 3.9). Note for example that many environmental rules and regulations written by local, state, and federal agencies predominantly use the SI system.
7. U.S. Coast Guard, U.S. Department of Transportation, 1985, *Hazard Assessment Handbook*, Commandant Instruction, Report Number M.16465.12A, Washington, D.C.
8. An atm = 1.01325×10^5 Pascals (P) or 1.01325×10^2 kilopascals (kP). The typically given pressure at sea level is approximately 1 atm. The variability is mainly attributed to ambient temperature.
9. The U.S. Congress has mandated the Agency for Toxic Substances and Disease Registry (ATSDR) to prepare toxicological profiles for hazardous substances found at National Priorities List (NPL) sites, the "worst" of the Superfund sites. The substances are ranked according to their frequency of occurrence at NPL sites, toxicity, and potential for human exposure. The ATSDR also prepares toxicological profiles for the Department of Defense (DoD) and the Department of Energy (DOE) on substances related to federal sites. Toxicological profiles are developed in two stages. The toxicological profiles are first produced as drafts. ATSDR announces in the Federal Register the release of these draft profiles for a 90-day public comment period. After the 90-day comment period,

TABLE 3.9
Prefixes Used in the International System of Units (SI). For example 10^3 grams = a kilogram or 1 kg, 10^6 seconds = a megasecond or 1 Ms, and 10^{-12} meter = a picometer or 1 pm. Source: B.N. Taylor, 1995, National Institute of Standards and Technology, Publication 811, *Guide for the Use of the International System of Units (SI)*, Washington, DC

Factor	*Prefix*	*Symbol*	*Factor*	*Prefix*	*Symbol*
$10^{24} = (10^3)^8$	yotta	Y	10^{-1}	deci	d
$10^{21} = (10^3)^7$	zetta	Z	10^{-2}	centi	c
$10^{18} = (10^3)^6$	exa	E	$10^{-3} = (10^3)^{-1}$	milli	m
$10^{15} = (10^3)^5$	peta	P	$10^{-6} = (10^3)^{-2}$	micro	μ
$10^{12} = (10^3)^4$	tera	T	$10^{-9} = (10^3)^{-3}$	nano	n
$10^9 = (10^3)^3$	giga	G	$10^{-12} = (10^3)^{-4}$	pico	p
$10^6 = (10^3)^2$	mega	M	$10^{-15} = (10^3)^{-5}$	femto	f
$10^3 = (10^3)^1$	kilo	k	$10^{-18} = (10^3)^{-6}$	atto	a
10^2	hecto	h	$10^{-21} = (10^3)^{-7}$	zepto	z
10^1	deka	da	$10^{-24} = (10^3)^{-8}$	yocto	y

ATSDR considers incorporating all comments into the documents. ATSDR finalizes the profiles and the National Technical Information Service (NTIS) distributes them.

10. See http://www.atsdr.cdc.gov/toxpro2.html. As of January 2004, the website provides toxicological profiles for 275 substances, 244 of which have been published as "final."
11. Interestingly, the toxicity of ethylene glycol and propylene glycol are quite different. Visit the ATSDR website http://www.atsdr.cdc.gov/toxprofiles/ to review the ethylene glycol profile and to query the complete listing.
12. See discussion of Newtonian and non-Newtonian fluids under the "Engineer's Notebook Entry: Viscosity" in this chapter.
13. However, some of us "motorheads" used 50 W or even higher viscosity racing formulas even if we never really allowed our cars to ever reach racing temperatures! We often used the same logic for slicks, glass packs, four-barrel carburetors, and other racing equipment that was really never needed, but looked and sounded awesome!
14. The $\frac{dv}{dy}$ term is also known as the *rate of strain* and the *shear rate.*
15. Other equivalent diameters for fully flowing conduits are the annulus, square, and rectangle. Equivalent diameters for partial flows in conduits are the half-filled circle, rectangle, wide and shallow stream, and trapezoid. For calculations of these diameters, see M. Lindeberg, 2001, *Civil Engineering Reference Manual for the PE Exam*, 8th Edition, Professional Publications, Belmont, Calif.
16. The literature is not consistent on the exact Reynolds numbers as thresholds for laminar versus turbulent flow. Another value used by engineers is 2300.

CHAPTER 4

Environmental Equilibrium, Partitioning, and Balances

The understanding of the properties of fluids can be extended to environmental systems by considering physical and chemical equilibria, mass and energy balances, and their effects on the movement and change of contaminants from one environmental medium to another.

Fundamentals of Environmental Equilibria

Partitioning among phases is an equilibrium concept, and partitioning coefficients are equilibrium constructs. Equilibrium is both a physical and chemical concept. It is the state of a system where the energy and mass of that system are distributed in a statistically most probable manner, obeying the laws of conservation of mass, conservation of energy (first law of thermodynamics), and efficiency (second law of thermodynamics). Therefore, if the reactants and products in a given reaction are in a constant ratio—that is, the forward reaction and the reverse reactions occur at the same rate—then that system is in equilibrium. Up to the point where the reactions are yet to reach equilibrium, the process is *kinetic*, or the rates of particular reactions are considered.

In environmental situations, we are mainly concerned with thermodynamic and chemical equilibria. That is, we must ascertain whether a system's influences and reactions are in balance. We know from the conservation laws that everything is balanced eventually, but since we only observe systems within finite time frames and confined spatial frameworks, we may only be seeing some of the steps in reaching equilibrium. Thus, for example, it is not uncommon in the environmental literature to see nonequilibrium constants (i.e., kinetic coefficients).[1]

To understand the concepts of environmental equilibria, let us begin with some fundamental chemical concepts. The first is that chemical reactions depend on "colligative" (collective) relationships between reactants and products. Colligative properties are expressions of the number of solute particles available for a chemical reaction. In a liquid solvent like water, then, the number of solute particles determines the property of the solution. This means that the concentration of solute determines the colligative properties of a chemical solution. These solute particle concentrations for pollutants are expressed as either mass-per-mass (e.g., $mg\,kg^{-1}$) or, most commonly, as mass-per-volume (e.g., $mg\,L^{-1}$) concentrations. In gas solutions, the concentrations are expressed as mass-per-volume ($mg\,m^{-3}$). Colligative properties may also be expressed as mole fractions, where the sum of all mole fractions in any solution equals 1.

Equilibrium Example

What is the equilibrium involved in dissolving 1 g sucrose in 9 g water. The total mass of this solution would be 10 g. The given sugar solution contains 240 g sucrose per 1000 g water.

Solution

Sucrose is our *solute*. Water is our *solvent*. The gram molecular weight of sucrose ($C_6H_{12}O_6$) is 180 g, so we would have 240/180 = 1.3 moles sucrose in 1000 g water.

Since the molecular weight of H_2O is 18, the mole-fraction of our sugar solvent =

$$\frac{moles(solute)}{moles(solute) + moles(solvent)} = \frac{240/180}{240/180 + 1000/18} = 1.3/56.9 = 0.02.$$

And, the mole fraction of water is $\frac{1000/18}{1000/18 + 240/18} = 0.98.$

Thus, the mole fraction (expressed as mole-percent) of our solute is approximately 2% and the mole-percent of our solvent is about 98%. The sum of all mole-percentages is 100% because the sum of all mole fractions is 1.

Colligative properties depend directly on concentration. One important property is vapor pressure, which is decreased with increased temperature. This is why water will require higher temperatures to boil when a solvent is present. For example, pure water will boil at 100°C and one atmosphere (760 mm Hg) of pressure, because under these conditions the water escapes as water vapor. In the case of pure water, all of the molecules are water (100% mole fraction). By adding solute to the pure water, we change the mole fraction. For example, if we heat the example (2% sucrose) solution, our vapor pressure is lowered by 2%, so that rather than 760 mm Hg, our vapor pressure = (0.98 water mole fraction) (760 mm Hg) = 745 mm Hg. Thus, the vapor pressure of the solvent (P) in any solution is found by:

$$P = X_A P^0 \qquad \text{Equation 4–1}$$

Where, X_A = Mole fraction of solvent
P^0 = Vapor pressure of 100% solvent

Solution Equilibria

A body is considered to be in thermal equilibrium if there is no heat exchange within the body and between that body and its environment. Analogously, a system is said to be in chemical equilibrium when the forward and reverse reactions proceed at equal rates. Again, since we are looking at finite space and time, such as a spill or an emission, or movement through the environment, reactions within that time and space may be either nonequilibrium ($xA + yB \rightarrow zC + wD$) or equilibrium ($xA + yB \Leftrightarrow zC + wD$) chemical reactions. The x, y, z, and w terms are the *stoichiometric* coefficients, which represent the relative number of molecules of each reactant (A and B) and each product (C and D) involved in the reaction. To have chemical equilibrium, the reaction must be reversible, so that the concentrations of the reactants and the concentrations of the products are constant with time.

The Law of Concentration Effects states that the concentration of each reactant in a chemical reaction dictates the rate of the reaction. Using our equilibrium reaction ($xA + yB \Leftrightarrow zC + wD$), we see that the rate of the forward reaction, or the rate that the reaction moves to the right, is most often dictated by the concentrations of A and B. Thus we can express the forward reaction as:

$$r_1 = k_1[A]^x[B]^y \qquad \text{Equation 4–2}$$

The brackets indicate molar concentrations of each chemical species (i.e., all products and reactants). Further, the rate of the reverse reaction can be expressed as:

$$r_2 = k_2[C]^z[D]^w \qquad \text{Equation 4–3}$$

Since at equilibrium, $r_1 = r_2$ and $k_1[A]^x[B]^y = k_2[C]^z[D]^w$ we can rearrange the terms to find the equilibrium constant K_{eq} for the reversible reaction:

$$\frac{k_1}{k_2} = \frac{[C]^z[D]^w}{[A]^x[B]^y} = K_{eq} \qquad \text{Equation 4–4}$$

The equilibrium constant for a chemical reaction depends on the environmental conditions, especially temperature and ionic strength of the solution. An example of a thermodynamic equilibrium reaction is a chemical precipitation water treatment process.[2] This is a heterogeneous reaction in that it involves more than one physical state. For an equilibrium reaction to occur between solid and liquid phases the solution must be saturated, and undissolved solids must be present. At a high hydroxyl ion concentration (pH = 10), the solid phase calcium carbonate ($CaCO_3$) in the water reaches equilibrium with divalent calcium (Ca^{2+}) cations and divalent carbonate (CO_3^{2-}) anions in solution. When a saturated solution of $CaCO_3$ then contacts solid $CaCO_3$, the equilibrium is:

$$CaCO_3(s) \Leftrightarrow Ca^{2+}(aq) + CO_3^{2-}(aq) \qquad \text{Equation 4–5}$$

The (s) and (aq) designate that chemical species are in solid and aqueous phases, respectively.

Thus, applying the equilibrium constant relationship in Equation 4–3, the dissolution (precipitation) of calcium carbonate is:

$$K_{eq} = \frac{[Ca^{2+}] + [CO_3^{2-}]}{[CaCO_3]} \qquad \text{Equation 4–6}$$

The solid phase concentration is considered to be a constant K_s. In this instance, the solid $CaCO_3$ is represented by K_s, so:

$$K_{eq}K_s = [Ca^{2+}] + [CO_3^{2-}] = K_{sp} \qquad \text{Equation 4–7}$$

K_{sp} is known as the solubility product constant. These K_{sp} constants for inorganic compounds are published in engineering handbooks (e.g., in Part 1, Appendix C of the *Handbook of Environmental Engineering Calculations*). Other equilibrium constants, such as the Freundlich Constant (K_d) discussed in the sorption section of Chapter 5, "Movement of Contaminants in the Environment," are also published for organic compounds (e.g., in Part 1, Appendix D of the *Handbook of Environmental Engineering Calculations*).

Gas Equilibria

For gases, the thermodynamic "equation of state" expresses the relationships of pressure (p), volume (V), and thermodynamic temperature (T) in a defined quantity (n) of a substance. For gases, this relationship is defined most simply in the ideal gas law:

$$pV = nRT \qquad \text{Equation 4–8}$$

Where R = the universal gas constant or molar gas constant = $8.31434\,\mathrm{J\,mol^{-1}\,{}^{\circ}K^{-1}}$

It should be noted that the ideal gas law only applies to ideal gases, or those that are made up of molecules taking up negligible space, with negligible spaces between the gas molecules. For real gases, the equilibrium relationship is:

$$(p + k)(V - nb) = nRT \qquad \text{Equation 4–9}$$

Where,

k = factor for the decreased pressure on the walls of the container due to gas particle attractions
nb = volume occupied by gas particles at infinitely high pressure.

Further, the van der Waals equation of state is:

$$k = \frac{n^2 a}{V^2} \qquad \text{Equation 4–10}$$

Where, a is a constant.

The van der Waals equation generally reflects the equilibria of real gases. It was developed in the early twentieth century and has been updated, but these newer equations can be quite complicated.

Gas reactions, therefore, depend on partial pressures. The gas equilibrium K_p is a quotient of the partial pressures of the products and reactants, expressed as:

$$K_p = \frac{p_C^z p_D^w}{p_A^x p_B^y} \qquad \text{Equation 4–11}$$

And from Equations 4–1, 4–5, 4–6, and 4–7, K_p can also be expressed as:

$$K_p = K_{eq}(RT)^{\Delta v} \qquad \text{Equation 4–12}$$

Where Δv is defined as the difference in stoichiometric coefficients.

Free Energy

Equilibrium constants can be ascertained thermodynamically by employing the Gibbs free energy (G) change for the complete reaction. Free energy is the measure of a system's ability to do work, in this case to drive the chemical reactions. This is expressed as:

$$G = H - TS \qquad \text{Equation 4–13}$$

Where G is the energy liberated or absorbed in the equilibrium by the reaction at constant T. H is the system's enthalpy and S is its entropy.

Enthalpy is the thermodynamic property expressed as:

$$H = U + pV \qquad \text{Equation 4–14}$$

Where U is the system's internal energy.

Entropy is a measure of a system's energy that is unavailable to do work. Numerous handbooks[3] explain the relationship between Gibbs free energy and chemical equilibria. The relationship between a change in free energy and equilibria can be expressed by:

$$\Delta G^{*} = \Delta G_f^{*0} + RT \ln K_{eq} \qquad \text{Equation 4–15}$$

Where,

ΔG_f^{*0}= Free energy of formation at steady state (kJ gmol^{-1}).

Importance of Free Energy in Microbial Metabolism

Metabolism is the cellular process that derives energy from a cell's surroundings. Energy to do chemical work is exemplified by cellular processes. Microbes, like bacteria and fungi, are essentially tiny, efficient chemical factories that mediate reactions at various rates (kinetics) until they reach equilibrium. These "simple" organisms (and complex organisms alike) need to transfer energy from one site to another to power their machinery needed to stay alive and reproduce. Microbes play a large role in degrading pollutants, whether in natural attenuation, where the available microbial populations adapt to the hazardous wastes as an energy source, or in engineered systems that do the same in a more highly concentrated substrate (see Table 4.1). Free energy is an important factor in microbial metabolism.

The reactant and product concentrations and pH of the substrate affect the observed ΔG^{*} values. If a reaction's ΔG^{*} is a negative value, the free energy is released, the reaction will occur spontaneously, and

TABLE 4.1
Genera of Microbes Able to Degrade a Persistent Organic Contaminant: Crude Oil

Bacteria	*Fungi*
Achromobbacter	*Allescheria*
Acinetobacter	*Aspergillus*
Actinomyces	*Aureobasidium*
Aeromonas	*Botrytis*
Alcaligenes	*Candida*
Arthrobacter	*Cephaiosporium*
Bacillus	*Cladosporium*
Beneckea	*Cunninghamella*
Brevebacterium	*Debaromyces*
Coryneforms	*Fusarium*
Erwinia	*Gonytrichum*
Flavobacterium	*Hansenula*
Klebsiella	*Helminthosporium*
Lactobacillus	*Mucor*
Leucothrix	*Oidiodendrum*
Moraxella	*Paecylomyces*
Nocardia	*Penicillium*
Peptococcus	*Phialophora*
Pseudomonas	*Rhodosporidium*
Sarcina	*Rhodotorula*
Spherotilus	*Saccharomyces*
Spirillum	*Saccharomycopisis*
Streptomyces	*Scopulariopsis*
Vibrio	*Sporobolomyces*
Xanthomyces	*Torulopsis*
	Trichoderma
	Trichosporon

Source: U.S. Congress, Office of Technology Assessment, 1991, "Bioremediation for Marine Oil Spills," Background Paper, OTA-RP-O-70, U.S. Government Printing Office. Washington, D.C.

the reaction is *exergonic*. If a reaction's ΔG^{*} is positive, the reaction will not occur spontaneously. However, the reverse reaction will take place, and the reaction is *endergonic*.

Time and energy are limiting factors that determine whether a microbe can efficiently mediate a chemical reaction, so catalytic processes are usually needed. Since an *enzyme* is a biological *catalyst*, these compounds (proteins) speed up the chemical reactions of degradation without themselves being used up (See Table 4.1 for species able

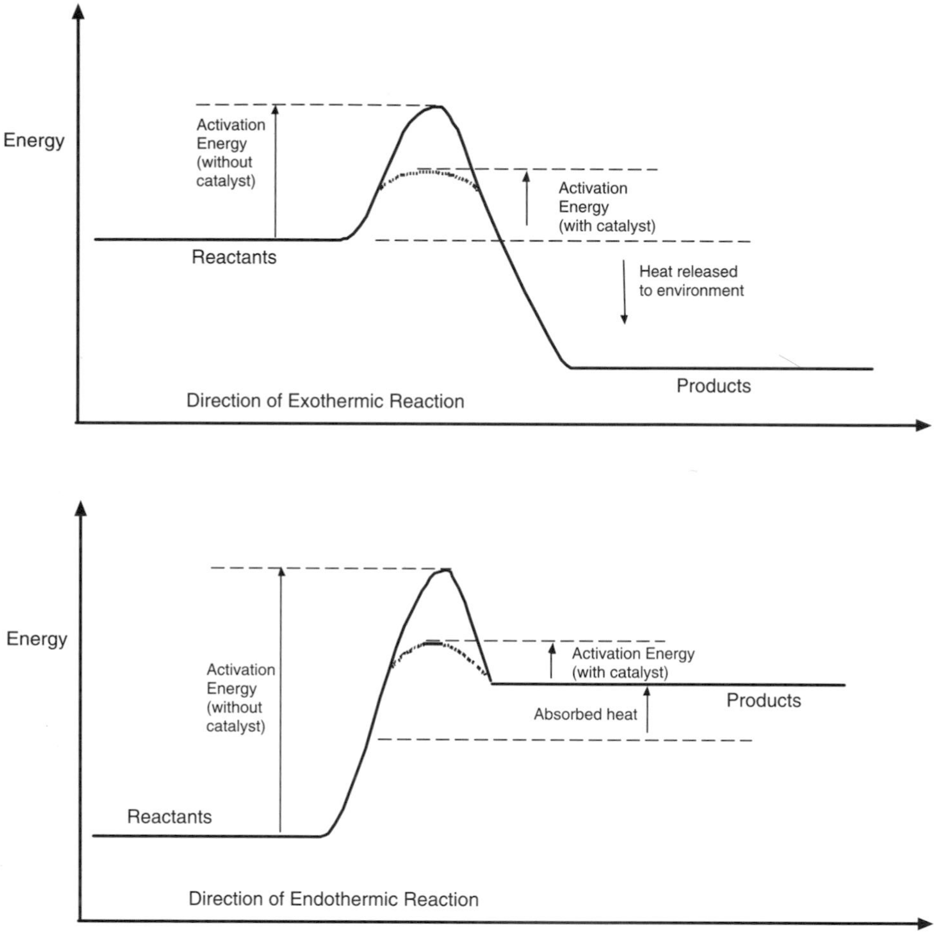

FIGURE 4.1. Effect of a catalyst on an exothermic reaction (top) and on an endothermic reaction (bottom).

to catalyze reactions). They do so by helping to break chemical bonds in the reactant molecules (see Figure 4.1). Enzymes play a very large part in microbial metabolism. They reduce the reaction's *activation energy*, which is the minimum free energy required for a molecule to undergo a specific reaction. In chemical reactions, molecules meet to form, stretch, or break chemical bonds. During this process, the energy in the system is maximized, and is then decreased to the energy level of the products. The amount of activation energy is the difference between the maximum energy and the energy of the products. This difference represents the energy barrier that must be overcome for a chemical reaction to take place. Catalysts (in this case, microbial enzymes) speed up and increase the likelihood of a reaction by reducing the amount of energy, or the activation energy, needed for the reaction.

The most common microbial coupling of exergonic and endergonic reactions by means of high-energy molecules to yield a net negative free energy is that of the nucleotide, adenosine triphosphate (ATP) with $\Delta G^* = -12$ to $-15\,\text{kcal mole}^{-1}$. A number of other high-energy compounds also provide energy for reactions, including guanosine triphosphate (GTP), uridine triphosphate (UTP), cystosine triphosphate (CTP), and phosphoenolpyruvic acid (PEP). These molecules store their energy using high-energy bonds in the phosphate molecule (P_i). An example of free energy in microbial degradation is the possible first step in acetate metabolism by bacteria:

$$\text{Acetate} + \text{ATP} \rightarrow \text{acetyl-coenzyme A} + \text{ADP} + P_i \qquad \text{Equation 4–16}$$

In this case, the P_i represents a release of energy available to the cell. Conversely, to add the phosphate to the two-P_i structure ADP to form the three-P_i ATP requires energy (i.e., it is an *endothermic process*). Thus, the microbe stores energy for later use when it adds the P_i to the ATP.

Solubility as a Physical and Chemical Phenomenon

The measure of the amount of chemical that can dissolve in a liquid is called *solubility*. It is usually expressed in units of mass of solute (that which is dissolved) in the volume of solvent (that which dissolves). Usually, when scientists use the term solubility without any other attributes, they mean the measure of the amount of the solute in water, or the aqueous solubility. Otherwise, the solubility will be listed along with the solvent, such as solubility in benzene, solubility in methanol, or solubility in hexane. Solubility may also be expressed in mass per mass or volume per volume, represented as parts per million (ppm), parts per billion (ppb), or parts per trillion (ppt). Occasionally, solubility is expressed as a percent or in parts per thousand, however, this is uncommon for contaminants and is usually reserved for nutrients and essential gases (e.g., percent carbon dioxide in water or ppt water vapor in the air).

The solubility of a compound is very important to environmental transport. The diversity of solubilities in various solvents is a strong indication of where one is likely to find the compound. For example, the various solubilities of the most toxic form of dioxin, tetrachlorodibenzo-*para*-dioxin

TABLE 4.2
Solubility of Tetrachlorodibenzo-*para*-dioxin in Water and Organic Solvents

Solvent	*Solubility (mg L⁻¹)*	*Reference*
Water	1.93×10^{-5}	Podoll, et al., 1986, *Environmental Science and Technology* 20: 490–492.
Water	6.90×10^{-4} (25°C)	Fiedler, et al., 1990, *Chemosphere* (20): 1597–1602.
Methanol	10	International Agency for Research on Cancer[4] (IARC)
Lard oil	40	IARC
n-Octanol	50	IARC
Acetone	110	IARC
Chloroform	370	IARC
Benzene	570	IARC
Chlorobenzene	720	IARC
Orthochlorobenzene	1400	IARC

(TCDD) are provided in Table 4.2. From these solubilities, one would expect TCDD to have a much greater affinity for sediment, organic particles, and the organic fraction of soils. The low water solubilities indicate that dissolved TCDD in the water column should be at only extremely low concentrations.

Polarity

A number of physicochemical characteristics of a substance come into play in determining its solubility. One is a substance's *polarity*. The polarity of a molecule is its unevenness in charge. The water molecule's oxygen and two hydrogen atoms are aligned so that there is a slightly negative charge at the oxygen end and a slightly positive charge at the hydrogen ends. Since "like dissolves like," polar substances have an affinity to become dissolved in water, and nonpolar substances resist being dissolved in water.

Consider the very polar water molecule (see Figure 4.2). The hydrogen atoms form an angle of 105° with the oxygen atom. The asymmetry of the water molecule leads to a dipole moment (see the discussion in the next section) in the symmetry plane pointed toward the more positive hydrogen atoms. Since the water molecule is highly polar, it will more readily dissolve other polar compounds than nonpolar compounds.

An element's ability to attract electrons toward itself is known as *electronegativity*. It is a measure of an atom's ability to attract shared electrons toward itself. The values for electronegativity range from 0 to 4, with fluorine (electronegativity = 4) being the most electronegative (see Table 4.3).

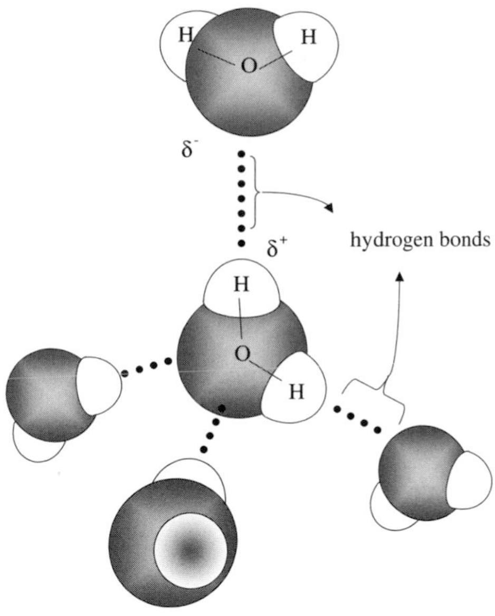

FIGURE 4.2. Configuration of the water molecule, showing the electronegativity (δ) at each end. The hydrogen atoms form an angle of 105° with the oxygen atom.

Each atom is uniquely able to attract electrons to varying degrees owing to its size, the charge of its nucleus, and the number of core (i.e., nonvalent) electrons. Values vary with the element's position in the periodic table, with electronegativity increasing from left to right across a row and decreasing downwardly within each group. This is due to the fact that smaller atoms allow electrons to get closer to the positively charged nucleus. Thus the higher the net charge of the combined nucleus plus the electrons of the filled, inner shells (collectively referred to as the *kernel*), the greater the electronegativity and the tendency of the atom to attract electrons.

The strength of a chemical bond in molecules is determined by the energy needed to hold the like and unlike atoms together with a covalent bond (i.e., a bond where electrons are shared between two or more atoms). The bond energy is expressed by the bond dissociation *enthalpy* (ΔH_{AB}). For a two-atom or *diatomic* molecule, the ΔH_{AB} is the heat change of the gas phase reaction. That is, at constant temperature and pressure, ΔH_{AB} is:

$$\text{A-B} \rightarrow \text{A}\bullet + \bullet\text{B} \qquad \text{Equation 4–17}$$

Where, A-B is the *educt* and A• and •B are the products of the reaction. The enthalpies and bond lengths for some of the bonds important in environmental engineering and science are given in Table 4.4.

TABLE 4.3
Electronegativity of the Elements

	IA	*IIA*	*IIIB*	*IVB*	*VB*	*VIB*	*VIIB*	*VIIIB*			*IB*	*IIB*	*IIIA*	*IVA*	*VA*	*VIA*	*VIIA*	*VIII*
1	H																He	
	2.1																	
2		4											B	C	N	O	F	Ne
	Li	Be											2.0	2.5	3.0	3.5	4.0	
	1.0	1.5																
3	Na	Mg											Al	Si	P	S	Cl	Ar
	0.9	1.2											1.5	1.8	2.1	2.5	3.0	
4	K	Ca	Sc	Ti	V	Cr	Mn	Fe	Co	Ni	Cu	Zn	Ga	Ge	As	Se	Br	Kr
	0.8	1.0	1.3	1.5	1.6	1.6	1.5	1.8	1.8	1.8	1.9	1.6	1.6	1.8	2.0	2.4	2.8	
5	Rb	Sr	Y	Zr	Nb	Mo	Tc	Ru	Rh	Pd	Ag	Cd	In	Sn	Sb	Te	I	Xe
	0.8	1.0	1.3	1.4	1.6	1.8	1.9	2.2	2.2	2.2	1.9	1.7	1.7	1.8	1.9	2.1	2.5	
6	Cs	Ba	La	Hf	Ta	W	Re	Os	Ir	Pt	Au	Hg	Tl	Pb	Bi	Po	At	Rn
	0.7	0.9	1.1	1.3	1.5	1.7	1.9	2.2	2.2	2.2	2.4	1.9	1.8	1.6	1.9	2.0	2.2	

TABLE 4.4
Bond Lengths and Enthalpies for Bonds in Molecules Important in Environmental Studies

Bond	*Bond Length (angstroms)*	*Enthalpy, ΔH_{AB}, (kJmol^{-1})*	*Notes*
		Diatomic Molecules	
H—H	0.74	436	
H—F	0.92	566	
H—Cl	1.27	432	
H—Br	1.41	367	
H—I	1.60	298	
F—F	1.42	155	
Cl—Cl	1.99	243	
Br—Br	2.28	193	
I—I	2.67	152	
O=O	1.21	4.98	
N≡N	1.10	9.46	
		Organic Compounds[5]	
H—C	1.11	415	
H—N	1.00	390	
H—O	0.96	465	
H—S	1.33	348	
C—C	1.54	348	
C—N	1.47	306	
C—O	1.41	360	
C—S	1.81	275	
C—F	1.38	486	
C—Cl	1.78	339	
C—Br	1.94	281	
C—I	2.14	216	
C=C	1.34	612	
C=N	1.28	608	
C=S	1.56	536	In carbon disulfide
C=O	1.20	737	In aldehydes
C=O	1.20	750	In ketones
C=O	1.16	804	In carbon dioxide
C≡C	1.20	838	
C≡N	1.16	888	

Source: R. Schwarzenbach, P. Gschwend, and D. Imboden, 1993, *Environmental Organic Chemistry*, John Wiley & Sons, New York, N.Y.

To grasp the concept of electronegativity, imagine an "electron cloud" between the nuclei of the two atoms of a diatomic molecule. The cloud is the average positions of the electrons that are bent toward the atom that is most attractive to the electrons. Or, as we have stated, the cloud is distorted in the direction of the more electronegative atom. Thus, when atoms bind, such as the O—H in water, the absolute value of the difference in electronegativity determines the electronegativity of the bonded atoms. If the difference is less than 1, the bond is considered to be nonpolar. Oxygen is very electronegative, so it attracts electrons to increase stability. Hydrogen is far less electronegative than oxygen, so each hydrogen gives part of its electron density to the oxygen, leaving the hydrogen with a partial positive charge (denoted by δ^+ in Figure 4.2), and oxygen with a partial negative charge (δ^-). A molecule with an uneven charge is considered to be polar, so water is one of the most polar molecules encountered in environmental situations.

Solubility Example

What is the electronegativity of the O—H bond of water? How does this compare with the C—H bond in methane?

Solution

The absolute value of the difference of electronegativity between atoms is the bond's electronegativity. Since, according to Table 4.3, hydrogen's electronegativity value is 2.1 and oxygen's electronegativity value is 3.5, the H—O bond has the electronegativity of $3.5 - 2.1 = \underline{1.4}$.

The electronegativity of carbon is 2.5, so the C—H bond's electronegativity is $2.5 - 2.1 = \underline{0.4}$.

Therefore, the O—H bond is polar, and the C—H bond is nonpolar, since the former is greater than 1 and the latter is less than 1.

In multiple bond molecules, each bond exerts its own polarity, with the sum of all of the molecule's bonds providing the polarity for the whole molecule.

Intramolecular Bonds, Intermolecular Forces, and Molecular Dipole Moments

The kinetic molecular theory tells us that gas particles are constantly moving randomly and colliding. The theory also states that the diameters of the particles are quite small in comparison to the distance between the particles. Solid-state matter holds its shape in a matrix of solid geometry. Liquids share some properties with gases and some with solids. They, like gases, conform to their container's shape. Like solids, they are not able to expand to fill their container. Liquids are very difficult to compress.

To explain molecular motion, we should distinguish between intramolecular bonds and intermolecular forces. The water molecule demonstrates the difference (see Figure 4.2). The covalent bonds between the hydrogen and oxygen atom within the water molecule are examples of intramolecular bonds, while the attraction between two water molecule neighbors is an example of intermolecular forces. The intramolecular bonds keeping each water molecule together are much stronger than the intermolecular forces between water molecules. For example, 463 kJ of energy is needed to break a H—O bond, but only 50 kJ can detach the intermolecular forces between H_2O molecules.

Increasing temperature, or increased kinetic energy, in a system increases the velocity of the molecules so that intermolecular forces are weakened. With increasing temperature, the molecular velocity becomes sufficiently large so as to overcome all intermolecular forces, so that the liquid boils (vaporizes). Intermolecular forces may be relatively weak or strong. The weak forces in liquids and gases are often called van der Waals forces.

Although the total charge of a molecule must be zero, the nature of chemical bonds is such that the positive and negative charges do not completely overlap in most molecules. Such molecules are said to be polar because they have a permanent dipole moment. To understand this concept, let us revisit the polar water molecule. The ionic nature of polar covalent bonds is determined by the *charge separation* in bonds of atoms having different electronegativities. How much a molecule takes on a *partial ionic character*[6] via these polar covalent bonds is one factor in predicting the behavior and how reactive the compounds formed by these molecules will be in the environment. The greater the difference in electronegativity, generally, the more ionic will be the bond between the two atoms sharing the electrons. This partial charge separation causes each bond between dissimilar atoms to become a *dipole*. The sum of all bond dipoles in a molecule's structure yields what is known as the *dipole moment* of the molecule. The dipole moment is a measurable property of the molecule and is very useful in fate and transport studies. Most important are the dipole moments of each bond regarding the interactions of a compound with its neighboring molecules. The solvent acetone is an example of dipole-dipole forces (see Figure 4.3). The acetone dipole-dipole forces are weak, requiring only a small amount of energy to pull the individual molecules apart. Covalent bonds (e.g., C—C, C—H, and C═O) within the acetone molecule are much stronger than the dipole-dipole forces. The actual strength of a dipole-dipole force is dependent upon the proximity of molecules and the magnitude of the dipole moment.

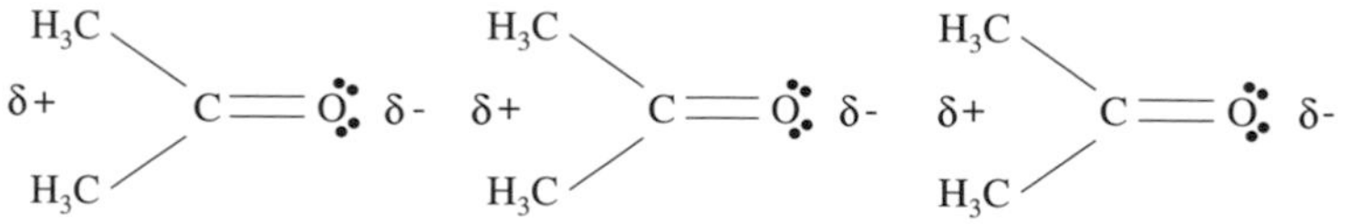

FIGURE 4.3. Acetone molecules showing dipole-dipole ($\delta-$ $\delta+$) force between each molecule. The dipole moment is generated by the C=O bond in each acetone molecule.

Molecules with mirror symmetry, like oxygen, nitrogen, carbon dioxide, and carbon tetrachloride, lack dipole moments. Even without a permanent dipole moment, however, a dipole moment may be induced by applying an electric field. This is called polarization, and the magnitude of the dipole moment induced is a measure of the polarizability of the molecular species. Dipole-induced dipole forces can occur when solvents are mixed. Let us consider a molecule with no dipole moment, e.g. tetrachloromethane, or CCl_4 (also commonly known as carbon tetrachloride). The individual bonds in CCl_4 (C—Cl) are polar because of the electronegativity between C and Cl. The molecular shape is a tetrahedron (four Cl atoms around one C atom), so the polarities of the bonds cancel out one another. However, the electrons are in constant motion, so there is an opportunity that they can be "caught" by intermolecular forces. Thus, if we mix CCl_4 with acetone, which we know has a dipole moment, when the CCl_4 molecule gets sufficiently close to an acetone molecule the electrons of the molecule will shift positions to generate a very small dipole moment (see Figure 4.3.). A polar molecule can distort the electron cloud of a molecular neighbor, making for a dipole-induced dipole intermolecular force.

A word of caution: The published electronegativities, such as those in Table 4-3, are a very rough scale of the relative electron attractiveness of elements. The type of substitutions of the atoms that form the compound is also a factor in determining the extent and direction of polarity of the compound.

Fluid Solubility/Density Relationships

The discussion on density was our first entrée into the fundamentals underlying the transport and fate of chemical contaminants. We will cover the movement and change of contaminants within and among environmental compartments (air, water, soil, sediment, and biota) in greater detail in Chapter 5, "Movement of Contaminants in the Environment." However, the relationship between the density of a fluid and its solubility in other fluids should be introduced at this point.

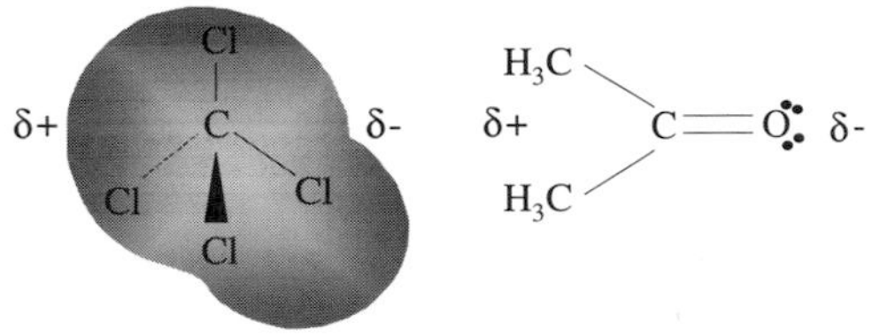

FIGURE 4.4. Dipole-induced dipole forces. The electron cloud of the tetrachloromethane molecule becomes distorted as the polar acetone molecule gets close. The edge of the electron cloud (the darker region of higher electron density) is attracted to the partial positive charge of the acetone molecule. The tetrachloromethane's small dipole moment is induced by the charge difference.

We are embarking on the subject of *phase partitioning,* which is also sometimes called "phase distribution." It is a principal subject matter of equilibrium physics and chemistry (and contrasted with *kinetic* physics and chemistry). Revisiting Figure 3-16, the LNAPL and DNAPL plumes are shown as if there is little or no solubility of the contaminating fluid in fresh water. Hence, the "NAP" (i.e., non-aqueous phase) in LNAPL and DNAPL is quite telling.

Environmental investigations use shorthand ways of describing a substance's "solubility." First, one must know how soluble a substance is in water. If it is quite soluble, or easily dissolved in water under normal environmental conditions of temperature and pressure, it is known to be *hydrophilic.* If, conversely, a substance is not easily dissolved in water under these conditions, it is said to be *hydrophobic.* Since many contaminants are organic (i.e., consist of molecules containing covalent carbon-to-carbon bonds and/or carbon-to-hydrogen bonds), the solubility can be further differentiated as to whether under normal environmental conditions of temperature and pressure, the substance is easily dissolved in organic solvents. If so, the substance is said to be *lipophilic* (i.e., readily dissolved in lipids). If, conversely, a substance is not easily dissolved in organic solvents under these conditions, it is said to be *lipophobic.*

It turns out that for most organic compounds, hydrophilic compounds are usually lipophobic, and lipophilic compounds are usually hydrophobic. We thus have a somewhat mutually exclusive relationship between these two types of substances.[7] If something dissolves readily in water, it is very likely not to dissolve easily in organic solvents. We will investigate this in detail in the dissolution section of Chapter 5, but this relationship allows for an important environmental partitioning; that is, the octanol-water partition coefficient (K_{ow}). The K_{ow} is the ratio of a substance's concentration in octanol ($C_7H_{13}CH_2OH$) to the substance's concentration in water at equilibrium (i.e., the reactions have all reached their final expected chemical composition in a control volume of the fluid).

TABLE 4.5
Solubility, Octanol-Water Partitioning Coefficient, and Density Values for Some Environmental Pollutants

Chemical	*Water Solubility (mg L⁻¹)*	K_{ow}	*Density (kg m⁻³)*
Atrazine	33	724	
Benzene	1780	135	879
Chlorobenzene	472	832	1110
Cyclohexane	60	2754	780
1,1-Dichloroethane	4960	62	1180
1,2-Dichloroethane	8426	30	1240
Ethanol	Completely miscible	0.49	790
Toluene	515	490	870
Vinyl chloride	2790	4	910
Tetrachlorodibenzo-*para*-dioxin (TCDD)	1.9×10^{-4}	6.3×10^{6}	

Source: H. F. Hemond and E. J. Fechner-Levy, 2000, *Chemical Fate and Transport in the Environment*, Academic Press, San Diego, Calif.; TCDD data from the NTP Chemical Repository, 2003, National Environmental Health Sciences Institute; and U.S. Environmental Protection Agency, 2003, Technical Fact Sheet on Dioxin [2,3,7,8-TCDD].

Octanol, an eight carbon (C_8) alcohol, has been chosen by scientists, including environmental engineers and scientists, as a surrogate for the organic phase. In general, like other C_1 to C_{10} alcohols, octanol can dissolve both water soluble (hydrophilic) compounds and fat soluble (lipophilic compounds), that is, octanol is *amphiphilic*. In a way, this mimics some of the uptake and metabolic processes in organic tissue, i.e. when a lipophilic organic compound is taken up, the metabolic processes often work to make it more hydrophilic, e.g. by adding an–OH group (in a sense making it an organic alcohol). The K_{ow} reflects the "NAP" part of our LNAPL and DNAPL classification. If a substance is aqueous, by definition it is not an LNAPL or DNAPL. Since, as mentioned, the ratio forming the K_{ow} is $[C_7H_{13}CH_2OH]$: $[H_2O]$, then the larger the K_{ow} value, the more lipophilic the substance. Values for solubility in water and K_{ow} values of some important environmental compounds, along with their densities, are shown in Table 4.5.

Solubility/Density Example 1

Review the data in Table 4.5. Which of the compounds leaked from a surface source are likely to reach the zone of aeration of an aquifer first? Which are most likely to move with the groundwater flow?

Solution

The compounds with densities much greater than that of water are more likely to settle first. Those with densities greater than water by at least 10% are chlorobenzene, 1,1-dichloroethane, and 1,2-dichloroethane.

The lighter substances that have low water solubility are likely to stay near the water table at the interface of the zone of aeration and the zone of saturation compared to those less dense than water; that is, the only one of these listed is cyclohexane. Benzene, ethanol, toluene, and vinyl chloride have sufficiently high water solubility that they may well become diffuse in the zone of aeration and move in a much larger plume with the groundwater flow.

Solubility/Density Example 2

Which of the compounds in Table 4.5 are likely to be classified as LNAPLs? Which will be DNAPLs?

Solution

The compounds with densities less than that of water which are "non-aqueous" are LNAPLs. Our measure of the organic versus aqueous partitioning is reflected in the K_{ow} values for the compounds. Every compound listed, except atrazine and ethanol, is a "nonaqueous phase liquid" (NAPL). Atrazine is not an NAPL because it is in the solid phase under environmental conditions. Ethanol is not a NAPL because its K_{ow} value is less than 1. The LNAPLs are thus those that meet both criteria for density and phase partitioning: benzene, cyclohexane, toluene, and vinyl chloride. Likewise, the DNAPLs are those that meet both criteria for density greater than that of water and phase partitioning: chlorobenzene; 1,1-dichloroethane; and 1,2-dichloroethane.

Another quick glance at Table 4.5 tells us a bit more about solubility and organic/aqueous phase distribution. Water solubility is somewhat inversely related to K_{ow}, but the relationship is uneven. This results from the fact that various organic compounds are likely to have affinities for neither, either, or both the organic and the aqueous phases. Most compounds are not completely associated with either phase; they have some amount of "amphiphilicity."

Also, what seem to be minor structural changes to a molecule can make quite a difference in phase partitioning and in density. Even the isomers (i.e., same chemical composition with a different arrangement) vary in their K_{ow} values and densities. (Note that the "1,1" versus "1,2" arrangements of chlorine atoms on 1,1-dichloroethane, and 1,2-dichloroethane, causes the former to have a slightly decreased density but twice the K_{ow} value than the latter!) The location of the chlorine atoms alone accounts for a significant difference in water solubility in the two compounds.

Taking into account the relationship between density and organic/aqueous phase partitioning, let us now revisit our contaminated groundwater example in Figure 3-16. The transport of the NAPLs through the vadose zone assumes that the NAPLs have extremely high K_{ow} values and extremely low water solubility. What would happen if, as is common in polluted systems, some of the NAPLs have relatively low K_{ow} values and high water solubilities? If a portion of the leaking waste has a density near that of water (i.e., is neither dense nor light) and is hydrophilic (highly miscible in water), while the remaining portion is a DNAPL, the flow may be altered to look more like Figure 4.5.

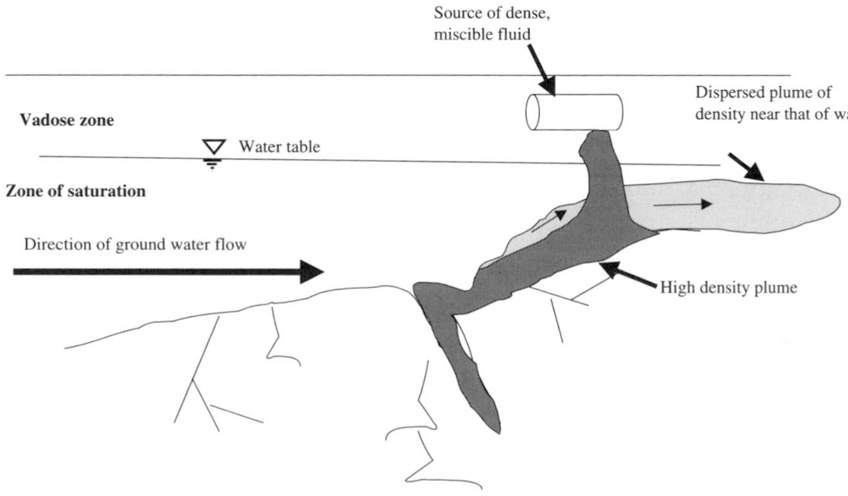

FIGURE 4.5. Hypothetical plume of dense, highly hydrophilic fluid. (Adapted from: M.N. Sara, 1991, "Groundwater Monitoring System Design," in *Practical Handbook of Ground-Water Monitoring*, edited by D.M. Nielsen, Lewis Publishers, Chelsea, Mich.)

When a dense, miscible fluid seeps into the zone of saturation, the dense contaminants move downward. When these contaminants reach the bottom of the aquifer, the shape dictates their continued movement and slope of the underlying bedrock or other relatively impervious layer, which may well be in a direction other than the flow of the groundwater in the aquifer. Solution and dispersion near the boundaries of the plume will have a secondary plume that will generally follow the overall direction of groundwater flow. The physics of this system points out that deciding where the plume is heading will entail more than the fluid densities, including solubility and phase partitioning. Even when one knows that the source was only a DNAPL, monitoring wells will need to be installed upstream and downstream from the source to account for partitioning.

If a source consists entirely of a light, hydrophilic fluid, the plume may be characterized as shown in Figure 4.6. Low-density organic fluids, however, often are highly volatile; their vapor pressures are sufficiently high to change phases from liquid to gas. This means that we must consider another physicochemical property of environmental fluids, i.e. vapor pressure, along with density and solubility.

Another important process in plume migration is that of *co-solvation*, the process where a substance is first dissolved in one solvent and then the new solution is mixed with another solvent. This can be an important con-

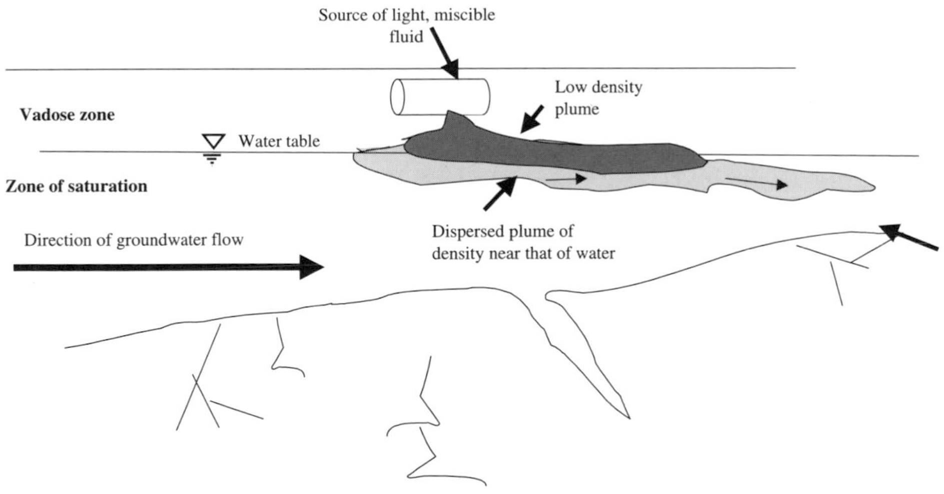

FIGURE 4.6. Hypothetical plume of light, highly hydrophilic fluid. (Adapted from: M.N. Sara, 1991, "Groundwater Monitoring System Design," in *Practical Handbook of Ground-Water Monitoring,* edited by D.M. Nielsen, Lewis Publishers, Chelsea, Mich.)

taminant transport phenomenon for NAPLs in groundwater. For example, a *hydrophobic* compound like DDT will have very low concentrations in pure water, but can migrate into and within groundwater if it is first dissolved in an NAPL that serves as an organic solvent. A DNAPL (e.g., chlorobenzene or one of the dichloroethane isomers), then, will move downward because its density is less than that of water and is transported in the DNAPL, which has undergone co-solvation with the water. Likewise, the pesticide dichlorodiphenyl trichloroethane (DDT) can be transported in the vadose zone or upper part of the zone of saturation when the DDT undergoes co-solvation with an LNAPL (e.g., toluene) and water.

Environmental Thermodynamics

Describing an environmental compartment and the contaminants that may exist in that compartment is a description of the properties of matter. Such descriptions are the subject matter of thermodynamics, the study of the relationships among properties of matter and the changes that occur to these properties. The changes may result spontaneously or from interactions with other materials. Thermodynamics concerns itself with the thermal systems, including ecosystems and organic systems in humans, especially how these systems work.

First, although it may seem obvious, we should define what we mean when we use the term *system*. In thermodynamics, a system is simply a sector or region in space or some parcel that has at least one substance that is ordered into phases (the phase diagram depicted in Figure 4.7 is an example of the phases that are available in this ordering). The more general understanding of scientists and technicians is that a "system" is a method of organization, from smaller to larger aggregations. The "ecosystem" and the "organism" are examples of both types of systems. They consist of physical phases and order (e.g., producer-consumer-decomposer; predator-prey; individual-association-community; or cell-tissue-organ-system). They are also a means for understanding how matter and energy move and change within a parcel of matter.

Systems are classified into two major types: closed and open. Both exist and are important in the environment. A *closed system* does not allow material to enter or leave the system (engineers refer to a closed system as a "control mass"). The *open system* allows material to enter and leave the system (such a system is known as a *control volume*).

Another thermodynamic concept is that of the "property." As mentioned in the discussions of the fluid properties of contaminants, in Chapter 3, a property is some trait or attribute that can be used to describe a system and to differentiate that system from others. A property must be able to be stated at a specific time independently of its value at any other time and

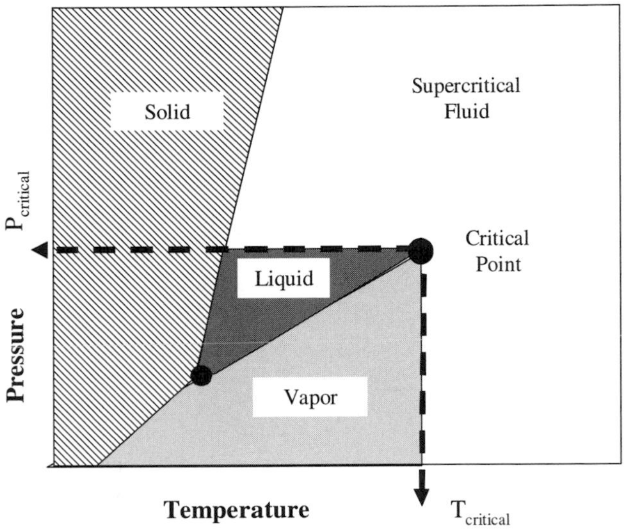

FIGURE 4.7. Phase diagram for a hypothetical substance. All of the polygons shown in the diagram will exist, but their shapes and slopes of the polygon sides and points of phase changes will differ. The vapor-liquid boundary line temperature is the boiling point, and the pressure is the vapor pressure. The line between the vapor and solid phases is known as the sublimation point (i.e., a phase change from solid to gas, or gas to solid, without first becoming a liquid). The liquid-solid boundary line temperature is the freezing and melting point. $P_{critical}$ and $T_{critical}$ are the pressure and temperature, respectively, of a fluid at its critical point, or the point at which a gas cannot be liquefied by an increase of pressure.

unconstrained by the process that induced the condition (state). An intensive property is independent of the system's mass (such as pressure and temperature). An extensive property is a proportionality to the mass of the system (such as density or volume). Dividing the value of an extensive property by the system's mass gives a "specific property," such as *specific heat*, *specific volume*, or *specific gravity*.

The thermodynamics term for the description of the change of a system from one state (e.g., equilibrium) to another is a *process*. Processes may be reversible or irreversible, and they may be adiabatic (no gain or loss of heat, so all energy transfers occur through work interactions). Other processes include isometric (constant volume), isothermal (constant temperature), isobaric (constant pressure), isentropic (constant entropy), and isenthalpic (constant enthalpy).

Fluid Volatility/Solubility/Density Relationships

Substances of low molecular weight and certain molecular structures have high enough vapor pressures that they can exist in either the liquid or gas phases under environmental conditions.

The vapor pressure (P^0) of a contaminant in the liquid or solid phase is the pressure that is exerted by its vapor when the liquid and vapor are in dynamic equilibrium (see Figure 4-8). This is really an expression of the *partial pressure* of a chemical substance in a gas phase that is in equilibrium with the nongaseous phases. The *ideal gas law* can be used to convert P^0 into moles of vapor per unit volume:

$$\frac{n}{V} = \frac{P^0}{RT} \qquad \text{Equation 4–18}$$

Where, V = volume of the container
n = number of moles of chemical
R = molar gas constant

The value $\frac{n}{V}$ is the gas phase concentration (moles L^{-1}) of the chemical.

The P^0 that is published in texts and handbooks is an expression of a chemical in its pure form; that is, P^0 is the force per unit area exerted by a vapor in an equilibrium state with its pure solid, liquid, or solution at a given temperature (see Table 4.6).

P^0 is a measure of a substance's propensity to evaporate; increasing exponentially with an increase in temperature (see Figure 4.8); a statement of P^0 must thus always be accompanied by a temperature for that P^0. For example, the P^0 of trichloroethene at 21.0°C is about 7.5 kP, but at 25.5°C rises to about 9.5 kP.[8] We will address P^0 in the environmental chemistry chapter, Chapter 6, as a relative measure of chemical volatility. As such, P^0 is a component of partitioning coefficients and volatilization rate constants.

Air pollution experts frequently categorize contaminants according to their vapor pressures. For example, volatile organic compounds (VOCs) have P^0 values greater than 10^{-2} kP; semivolatile organic compounds (SVOCs) have P^0 values between 10^{-5} and 10^{-2} kP; and the so-called "nonvolatile organic compounds" have P^0 values less t 10^{-5}. This is a general guideline, since various experts apply different ranges.

As noted earlier, P^0 is highly temperature dependent. The values in Table 4-6 are for P^0 values at various temperatures, but we can expect the temperature to drop down to 12°C at night. This means that the vapor pressure will be between the published P^0 at night, or between 170 and 355 kP. Certainly, it will not affect the answer to this specific question. However, for a narrow temperature range, the vapor pressure can be found using the *Antoine equation*:

TABLE 4.6
Vapor Pressures at 20°C for Some Environmental Pollutants

Chemical	*Vapor Pressure (kP) at 0°C*	*Vapor Pressure (kP) at 20°C*	*Vapor Pressure (kP) at 25°C*	*Vapor Pressure (kP) at 50°C*
Atrazine		4.0×10^{-8}		
Benzene	3.3		12.7	36.2
Chlorobenzene			1.6	
Cyclohexane			13.0	36.3
1,1-Dichloroethane	9.6		30.5	79.2
1,2-Dichloroethane	2.8	9.2	10.6	31.4
Ethanol	1.5		7.9	29.5
Toluene			3.8	
Vinyl chloride	170	344	355	
Tetrachlorodibenzo-*para*-dioxin (TCDD)		4.8×10^{-9}	5.6×10^{-3}	

Sources: Column 2: H. Hemond and E. Fechner-Levy, 2000, *Chemical Fate and Transport in the Environment*, Academic Press, San Diego, Calif.; Columns 3 and 4: D. Lide, ed., 1995, *CRC Handbook of Chemistry and Physics*, 76th Edition, CRC, Boca Raton, Fl. TCDD data from the NTP Chemical Repository, National Environmental Health Sciences Institute, 2003; and U.S. Environmental Protection Agency, 2003, Technical Fact sheet on Dioxin [2,3,7,8-TCDD].

Vapor Pressure Example

A train derailed and two of its cars emptied their contents onto a clay soil near the track. The spill occurred in autumn with the diurnal temperature ranging from 12 to 21°C. The first car contained the pesticide atrazine in a solid form, and the second car, a tanker, spilled vinyl chloride. Which car is most likely to result in air pollution?

Solution

Table 4.6 shows that the pesticide in its dry form has a vapor pressure that is 10 orders of magnitude less than vinyl chloride. In fact, vinyl chloride is so volatile, it is considered by some to be a VVOC, or a very volatile organic compound. At these vapor pressures, the vinyl chloride can be expected to partition to the gas phase readily, and should be considered a major a pollutant in the plume leaving the spill.

$$\ln P^0 = \frac{-B}{T+C} + A \qquad \text{Equation 4–19}$$

Where T is the absolute temperature, and A, B, and C are constants based upon a statistical regression fitted to the vapor pressure data taken at several temperatures.

Any substance, depending upon the temperature, can exist in any phase (see phase diagram in Figure 4.7). However, in many environmental contexts, a vapor refers to a substance that is in its gas phase, but under typical environmental conditions exists as a liquid or solid under a given set of conditions.

Although the pressure in the closed container in Figure 4.8 is constant, molecules of the vapor will continue to condense into the liquid phase, and molecules of the liquid will continue to evaporate into the vapor phase. However, the rate of these two processes is equal, meaning no net change in the amount of vapor or liquid. This is an example of dynamic equilibrium, or *equilibrium vapor pressure.*

At the boiling point temperature, a liquid's vapor pressure is equal to the external pressure. This is why in Denver (the "Mile High City"), the boiling point of water is less than the boiling point of water in

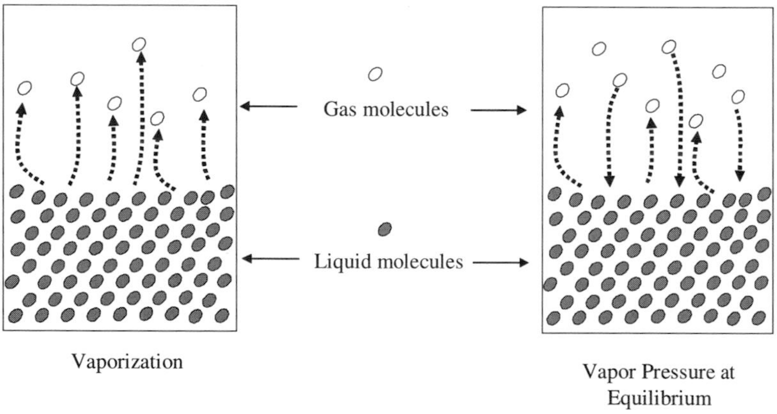

FIGURE 4.8. Vapor pressure of a fluid during vaporization and at equilibrium. A portion of a substance in an evacuated, closed container with limited headspace will vaporize. The pressure in the space above the liquid increases from zero and eventually stabilizes at a constant value. This value is what is known as the vapor pressure of that substance. Substances not in closed container (i.e., infinitely available headspace) will also vaporize, but will continue to vaporize until all of the substance has partitioned to the gas phase. Heating the system increases the activity, so vapor pressure increases (see Figure 4.9).

Wilmington, North Carolina (near sea level). The column of air above Denver is about a mile less than the air column above Wilmington (see Figure 4.10).

Generally, the higher the substance's vapor pressure at a given temperature, the lower the boiling point. Compounds with high vapor pressures are classified as "volatile," meaning they form higher concentrations of vapor above the liquid.[8] This means that they are potential air pollutants from storage tanks and other containers; it also means that they can present problems to first responders. For example, if a volatile compound is also flammable, there is a higher fire and explosion hazard than if the substance were less volatile.

We are now ready to take into account the effect of volatilization on phase partitioning. In groundwater, if a source of LNAPLs includes a relatively insoluble substance that distributes between liquid and gas phases (see Figure 4.11), the fluid will infiltrate and move along the water table along the top of the zone of saturation, just above the capillary fringe. (See Chapter 3.) However, some of the contaminant fluid lags behind the plume and slowly solubilizes in the pore spaces of the soil and unconsolidated material. These more water soluble forms of the fluid find their way to the zone of saturation and move with the general groundwater flow. The higher vapor pressures of portions of the plume will lead to upward movement of volatile compounds in the gas phase. Thus, this system has at least three plumes as a result of the solubility, density, and vapor pressure of the fluid components and environmental conditions.

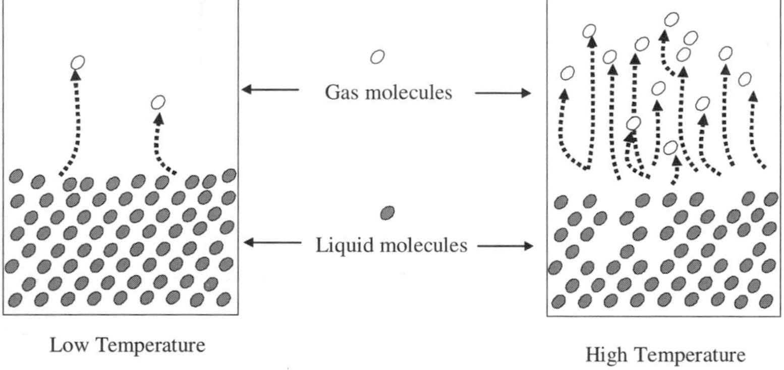

FIGURE 4.9. Exponential increase in vapor pressure with increasing temperature. Vapor pressure values must always be accompanied by the temperature at which each measured vapor pressure is occurring.

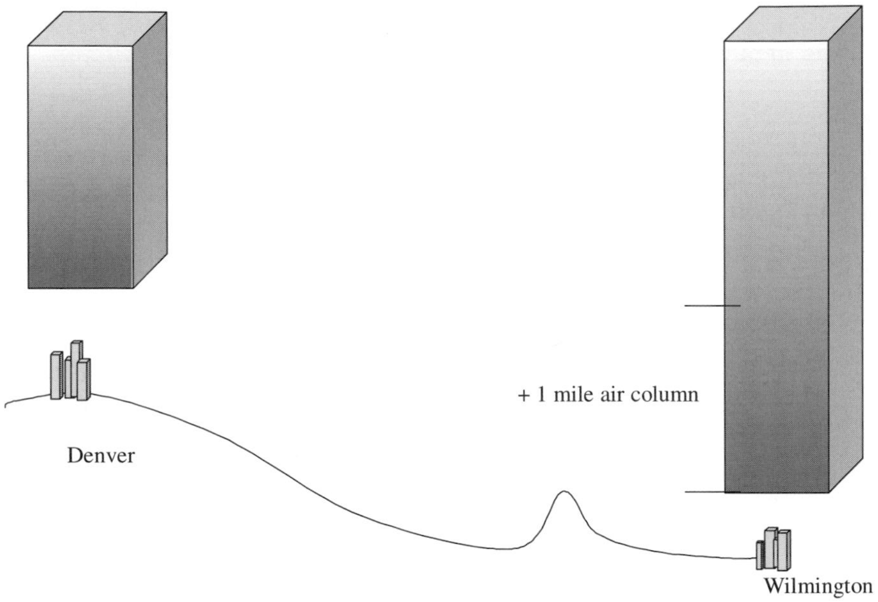

FIGURE 4.10. Increased atmospheric pressure due to larger column of air above Wilmington, N.C., than above Denver. This is the reason for the lower boiling point for Denver.

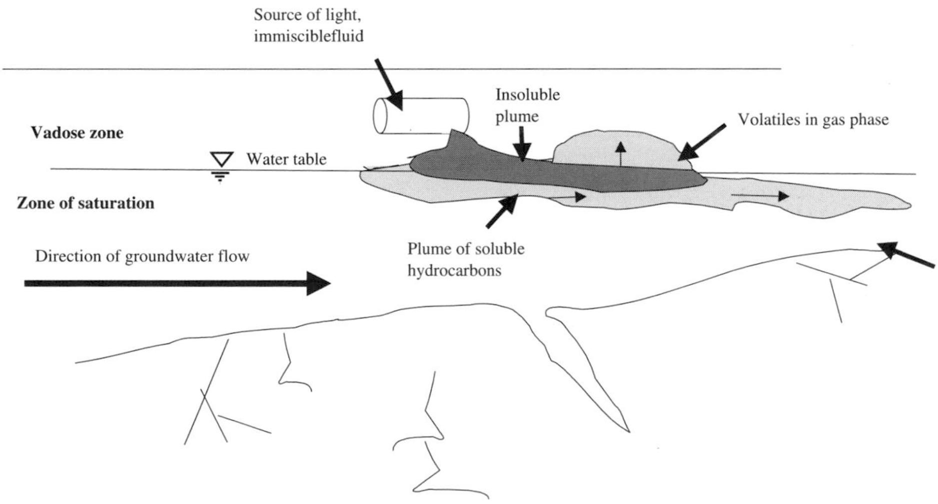

FIGURE 4.11. Hypothetical plume of hydrophobic fluid. (Adapted from: M.N. Sara, 1991, "Groundwater Monitoring System Design," in *Practical Handbook of Ground-Water Monitoring,* edited by D.M. Nielsen, Lewis Publishers, Chelsea, Mich.)

Lab Notebook Entry: Partitioning in the Laboratory

A major concept in environmental science and engineering is that of partitioning. Specifically, we are often concerned about how and why a substance changes from one phase or moves from one environmental compartment to another. Laboratory techniques apply the physical and chemical properties of substances. One example is what chemists refer to as *separation science.*

The analytical chemist must physically and spatially separate a compound from a medium before that compound may be detected. By far the most useful technique for such separations is chromatography. There are many forms of chromatography that apply to environmental analyses. The type of chromatography used is related to the phase distribution equilibrium, which is a function of the differing affinities for distinct physical phases of the compound. For example, a separation of sodium chloride from molecular iodine may take advantage of the two compounds' very different solubilities in tetrachloromethane versus water. At its most basic level, chromatography is achieved by transferring a compound between two distinct phases or states of matter, and by taking advantage of the compound's unique physicochemical properties.

Consider gel chromatography as an example of phase partitioning.[9] Gel chromatography, also known as size exclusion chromatography, is a procedure where solutes (i.e., the compounds that are dissolved in the solvent) are fractionated by molecular size. The process is actually a molecular "sieve" using gels made up of cross-linked neutral (i.e., without electrical charge) polymers that do not chemically react with the compounds being separated. The individual gel particles contain extremely small pore spaces through which the smaller molecules can be transferred. However, the larger molecules cannot pass through the gel, i.e., they are "size excluded."[11]

Figure 4.12 shows the separation of two different sized solute molecules flowing through a column of gel particles (beads). A band of the two different solutes to be separated (or *analytes*) is injected onto the top of the gel column. The molecules with diameters smaller than the pores in the gel particles travel through the void spaces between the beads. The smaller molecules diffuse in and out of the gel beads because their diameters are sufficiently small to enter the gel pores. The probability of diffusion increases as the solution moves down the column, so at a slow enough rate of travel, the molecules

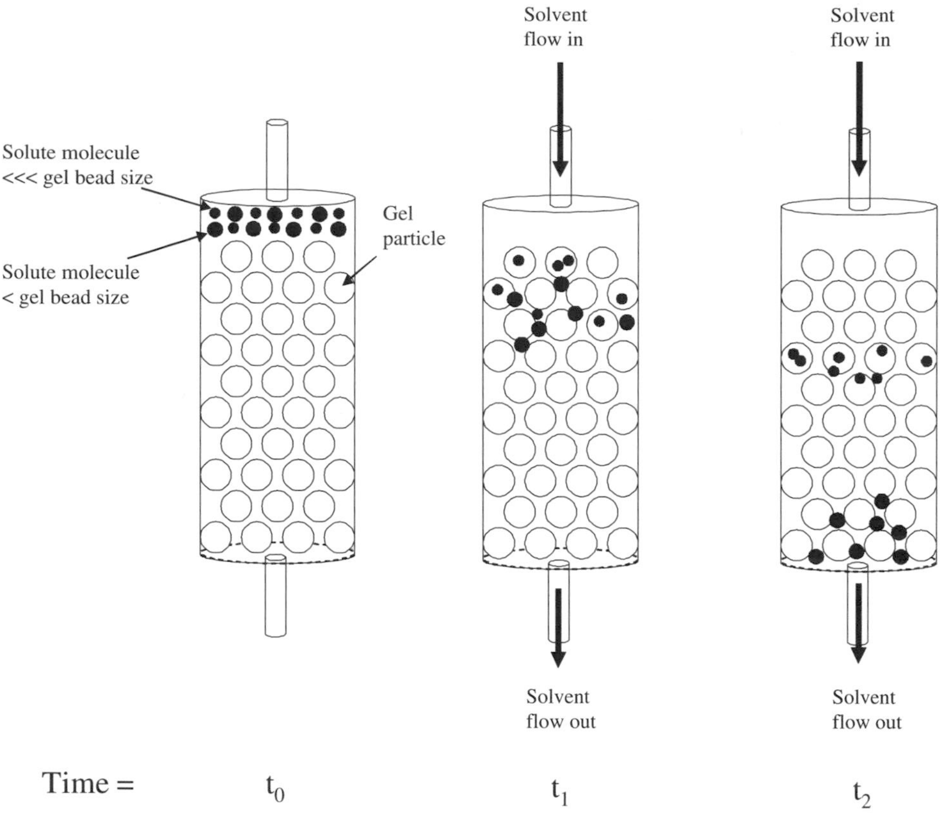

FIGURE 4.12. Flow of solution of two differently sized molecules through a gel bead column.[10]

are allowed to equilibrate with the gel particles at each tier. This means that if the shapes are the same, the molecules are eluted (i.e., the solution leaves the column) according to their molecular size.[10]

The larger molecules are eluted first, so they have the shortest *residence time* or *retention time* in the column, while the smaller molecules spend more time in the column. In other words, at t_0, immediately following their injection onto the surface of the column, all of the molecules are dispersed evenly throughout the solution. However, due to a physicochemical characteristic (in this instance, molecular size), the two types of molecules become distributed according to their respective sizes, as shown at t_1. Finally, at t_2, the two classes of compounds have completely "parted" or "partitioned." At some future time, the column will only contain the smaller molecules and, ultimately, all of the solute molecules will be eluted. Chemists refer to this as *separation* (hence "separation science").

Gel chromatography provides a straightforward, physical example of what is known as the *partition coefficient*, K_d. In this instance, the partitioning is between the gel and the solvent being fed into the column:[12]

$$K_d = \frac{C_{gel}}{C_{sol}} = \frac{m_{in}}{C_{gel} V_{in}} \qquad \text{Equation 4–20}$$

Where, C_{gel} = solute concentration in the gel
C_{gel} = solute concentration in solvent feed
m_{in} = mass of solute in the gel
V_{in} = volume of the gel particle's pores that is accessible to the solvent

By controlling the physicochemical variables discussed previously, such as temperature, solubility, viscosity, sorption, and vapor pressure, the environmental analytical chemist can separate numerous compounds so that they may be identified and quantified. Therefore, compounds will be characterized according to their affinities for phases, such as solid, liquid, and gas, as well as aqueous versus organic. They may also be characterized according to their affinities to air versus water, or among various solvents. Such characterizations are quite useful when estimating or modeling transport and fate. For example, if a compound has a very strong affinity for lipids, it is more likely to be found in sediment than in the water column (or at least higher concentrations would be expected in the former than the latter).

Environmental Balances

Although many possible outcomes can occur after a contaminant is released into the environment, the possibilities fall into three basic categories:

- The chemical may remain where it is released and retain its physicochemical characteristics (at least within a specified time);
- The substance may be transported to another location; or
- The substance may be changed chemically, known as the *transformation* of the chemical.

This is a restatement of the conservation laws mentioned earlier. If we focus on mass within a control volume, it is a statement of *mass balance*. Every bit of mass moving into and out of the control volume must be accounted for, as well as any chemical changes to the contaminant that take place within the control volume. A control volume may be a nice, neat geometric shape, such as a cube (Figure 4.13A) through which contaminant fluxes are calculated. However, a control volume can also be an organism (e.g., what it eats, metabolizes, and eliminates) or an ecosystem, such as the pond in Figure 4.13B.

Much of the work of environmental assessment is an accounting for the mass on both sides of the mass balance equation. The change in storage of a substance's mass is equal to the difference between the mass of the chemical transported into the system less the mass of the chemical transported out of the system. However, the actual chemical species transported in may be different from what is transported out, due to the chemical and biological reactions taking place within the control volume. The mass balance equation thus may be written as:

Accumulation or loss of contaminant A = Equation 4–21

Mass of A transported in – Mass of A transported out ± Reactions.

The reactions may be either those that generate chemical A (i.e., *sources*), or those that destroy chemical A (i.e., *sinks*).

The amount of mass transported in is the inflow to the system that includes pollutant discharges, transfer from other control volumes and other media (for example, if the control volume is soil, the water and air may contribute mass of chemical A), and formation of chemical A by abiotic chemistry and biological transformation. Conversely, the outflow is the mass transported out of the control volume, which includes uptake, by biota, transfer to other compartments (e.g., volatilization to the atmosphere), and abiotic and biological degradation of chemical A. This means the rate of change of mass in a control volume is equal to the rate of chemical A transported in, minus the rate of chemical A transported out, plus the rate of production from sources, and minus the rate of elimination

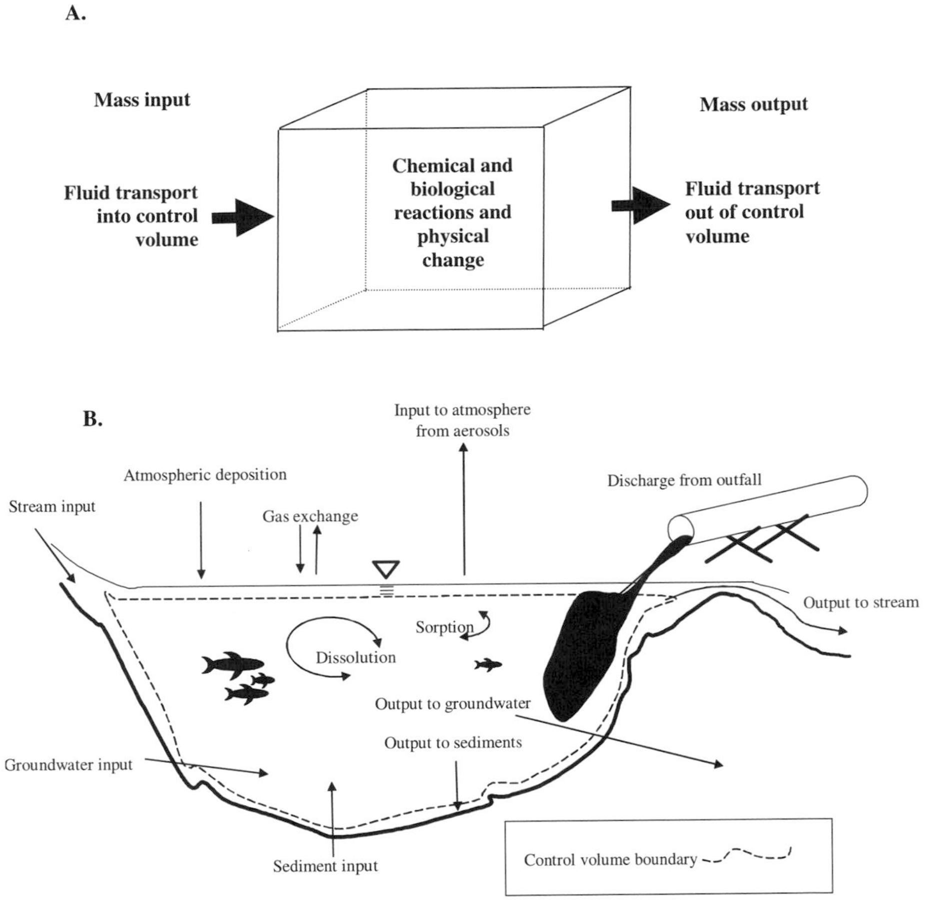

FIGURE 4.13. Two types of control volumes when considering environmental mass. **A.** Control volume of an environmental matrix (e.g., soil, sediment, or other unconsolidated material) or fluid (e.g., water, air, or blood). **B.** A pond. Both volumes have equal masses entering and exiting, with transformations and physical changes taking place within the control volume.

by sinks. Stated as a differential equation, the rate of change contaminant A is:

$$\frac{d[A]}{dt} = -v \cdot \frac{d[A]}{dx} + \frac{d}{dx}\left(D \cdot \frac{d[A]}{dx}\right) + r \qquad \text{Equation 4–22}$$

Where, v = fluid velocity

D = compartmental rate constant

$$\frac{d[A]}{dx} = \text{concentration gradient of chemical A}$$

$$r = \text{internal sinks and sources within the control volume}$$

We are now prepared to consider the transport mechanisms responsible for the movement of a contaminant within a system. Then we can move on and address the transformation term, r. In fact, reactions that lead to transformation contaminants are considered in detail in Chapter 7.

Technical Note: Control Volumes for Environmental Contamination

Two "control volumes" that are commonly considered in environmental exposure and risk assessments are the organism and a defined environmental volume around the organism. Thus, scientists commonly calculate mass balances for the classic control cube and adapt it to the environment (see Figure 4.13). Not surprisingly, the most studied control volume organism is the human. Human beings meet the same criteria as our cube and pond, in that we must fully account for the pollutant mass in and out, as well as the processes that occur within these control volumes.

Exposure assessments in human populations must account for the amount of a contaminant that a person contacts. However, just because a contaminant finds its way to one's mouth, nose, or skin, does not necessarily mean the contaminant will harm the individual, or that the contaminant will find its way to a susceptible cell. Organisms have numerous protective mechanisms in their metabolic processes. For example, many compounds are converted to harmless metabolites. Unfortunately, other compounds are *activated* during metabolism and can even become more toxic. For example, many researchers believe that the epoxide formed when trying to metabolize benzo(a)pyrene is a likely carcinogen. Conducting a mass balance in an organism is complicated.

It is sometimes advantageous to look at the human being as a control volume for the mass balance of a contaminant. *Body burden* is the total amount of the contaminant in the body at a given time of measurements. This is an indication of the behavior of the contaminant in the control volume (i.e., the person). Some contaminants accumulate in the body and are stored in fat or bone, or they simply are metabolized more slowly and tend to be retained for longer time

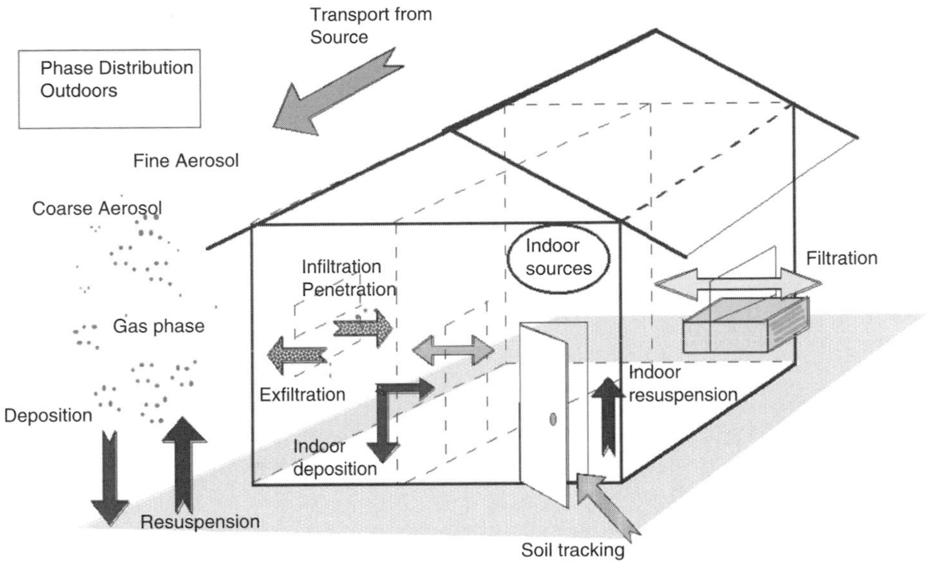

FIGURE 4.14. The home as a control volume. (Source: U.S. Department of Energy, Lawrence Berkeley Laboratory, 2003, http://eetd.lbl.gov/ied/ERA/CalEx/partmatter.html.)

periods. This concept is at the core of what are known as *physiologically based pharmacokinetic* (PBPK) models. These models attempt to describe what happens to a chemical after it enters the body, showing its points of entry, its distribution (i.e., where it goes after entry), how it is altered by the body, and how it ultimately is eliminated by the body. This is almost identical to the processes that take place in a stream, wetland, or other system.

Recently, engineers and scientists have applied mass balance approaches to the home. Unlike the well-defined boundary conditions of the small control volume, a home has numerous inflows and outflows, as well as sources, sinks, and transformation reactions. Some are shown in Figure 4.14. Modeling these dynamics is a growing priority for agencies charged with estimating the exposures of people to toxic substances.

The Environmental Mass Balance Reaction Term

Environmental reactions represented in Equations 4–21 and 4–22 seldom result from a simple and direct change of a reactant to product, i.e., a parent compound to the final degradation product.[13] Often, they include complex sequences. The final product is formed from the parent compound after a series of reactions. A consecutive reaction, therefore, can be depicted as the formation of product C from reactant A, only after A forms an intermediate product B:

$$A \xrightarrow{R_1} B \xrightarrow{R_2} C \qquad \text{Reaction 4–1}$$

As shown in Figure 4.15, each reaction step has its own reaction rate, i.e., the change in the concentration in the reactant to a product per unit time. Thus A degrades to B at rate R_1 and B degrades to C at rate R_2. The oxidation of the cyanide anion (CN^-) is an example.[14] An oxidizer, such as hydrogen peroxide (H_2O_2), is used to treat wastewaters that contain cyanide in a reactor. At pH between 9.5 and 10.5, cyanide is oxidized to cyanate while the H_2O_2 is reduced to water:

$$CN^- + H_2O_2 \rightarrow CNO^- + H_2O \qquad \text{Reaction 4–2}$$

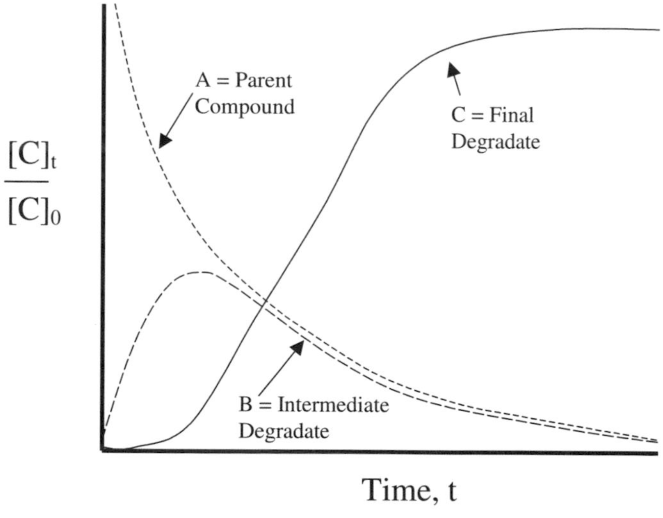

FIGURE 4.15. Distribution of chemical species for consecutive environmental reactions (first-order degradation of parent compound A). $[C]_t/[C]_0$ is the proportion of concentration of the compound at time t to the concentration at the time reaction begins. Source: Adapted from W.J. Weber, Jr. and F.A. DiGiano, 1996, *Process Dynamics in Environmental Systems*, John Wiley & Sons, New York, N.Y.

This first reaction is followed by a reaction of cyanate with water to form ammonia and carbon dioxide:

$$CNO^{-} + 2H_2O \rightarrow NH_3 + CO_2 \qquad \text{Reaction 4–3}$$

Like many environmental and waste treatment reactions, the reaction rate, in this instance the oxidation rate, depends on temperature and reactant concentration, i.e., the CN^{-} concentration, as well as excess H_2O_2. The rate is sped up with catalysts.

Since none of the concentration curves in Figure 4.15 are straight lines, i.e., they are not linear, the reaction rates are changing with time. In fact, the rate of degradation of the parent product in this instance is concentration dependent, so the rate is higher at the initial time t_0 than at any other part of curve A. Likewise, the rate at which B is formed is highest at time t_0 and slows down considerably after its peak. The concentration curve for the final product is sigmoidal, i.e., S-shaped, reflecting the rates of degradation of A and formation and subsequent degradation of B to C. The rates' concentration dependence reflects a first-order reaction sequence, that is, the reaction rate is proportional to the concentration of the single substance undergoing change:

$$\frac{dC_A}{dt} = kC_A \qquad \text{Reaction 4–4}$$

Where C_A is the concentration of reactant *A*, *t* is time, and *k* is the first-order reaction constant, i.e., the fraction of A degrading per unit of time.

Another important consideration of the mass balance conditions expressed in Equations 4–21 and 4–22 is that the reaction term, r, in these equations may occur in various combinations of physical phases and environmental compartments.[15] Reactions that take place in a single phase are known as *homogeneous reactions.* These are the most straightforward to determine. However, since partitioning among phases and compartments is taking place while the reactions occur, it is important to keep in mind that many are likely taking place in more than one phase, i.e., they are *heterogeneous reactions.* Figure 4.16 shows a simplified diagram of a heterogeneous reaction.

The movement and change of contaminants across phase boundaries and compartmental interfaces can be envisioned as the separation and joining of films. Perhaps the easiest system to contemplate phase distribution is the *two-film model* depicted in Figure 4.17, which demonstrates the relationship between the liquid and gas phases at a microscopic scale.[16] The model is designed under the assumption that the gas and liquid phases are in turbulent contact with each other, but segregated by an interface, where contaminants may cross in either direction. Each mass transfer zone, known as a film, comprises a small volume of the gas and liquid phases on either

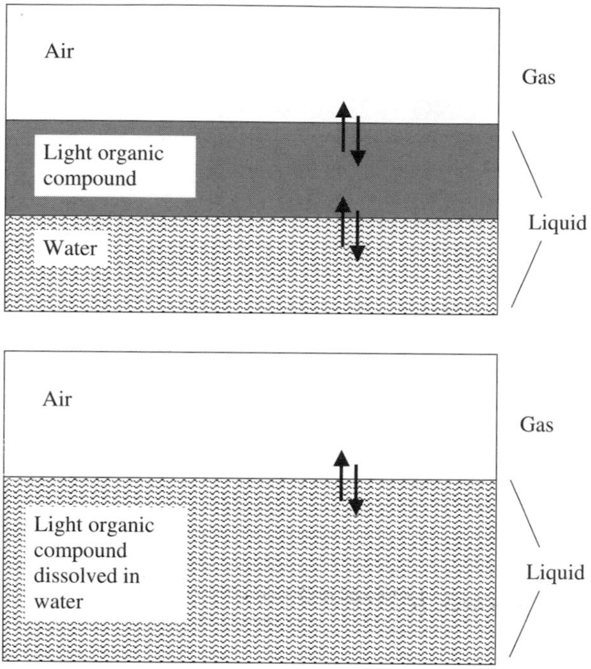

FIGURE 4.16. Top diagram: Transport and reactions occurring within and among two phases (liquid and gas) and three compartments (water, organic compound, and air) in an environmental system. Bottom diagram: Transport and reactions if the substances in the same phase are miscible (i.e., the organic compound is dissolved in the water). Source: Adapted from W.J. Weber, Jr. and F.A. DiGiano, 1996, *Process Dynamics in Environmental Systems*, John Wiley & Sons, New York, N.Y.

side of the interface. The flow in the mass transfer zones are assumed to be laminar and the flow in the bulk liquid and gas regions is generally turbulent (see discussions on the Reynolds number in Chapter 3). Assuming complete mixing in the bulk phases and equilibrium between the substance's molecule and the molecules of the interface, for a substance to move from the gas phase to the liquid phase, it must first migrate from the bulk gas phase into the gas film, then diffuse through the gas film (a function of the gas' partial pressure), before crossing the interface. After it crosses the interface, the substances must diffuse through the liquid film and mix into the bulk liquid (a function of the substance's concentration in the liquid). This means that all resistance to movement occurs as the substance's molecules diffuse through the films to reach the interface region. That is why the two-film theory is also referred to as the *double resistance theory.* In liquid-to-gas transport, the concentration in the liquid changes as it moves from the bulk liquid phase through the liquid film. The rate of mass transfer from one

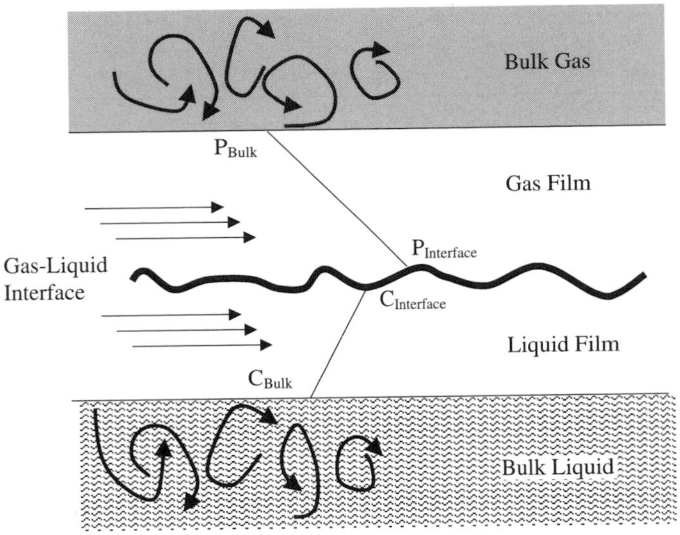

FIGURE 4.17. Two-film model. C_{Bulk} is the concentration of a substance in the bulk liquid and $C_{Interface}$ is the concentration of the substance at the gas-liquid interface. P_{Bulk} is the partial pressure of the substance in the bulk liquid and $P_{Interface}$ is the partial pressure of the substance at the gas-liquid interface. Flow lines indicate laminar flow in films and turbulent flow in bulk phases. Based on W.G. Whitman, 1923, The two-film theory of gas absorption, *Chemical and Metallurgical Engineering.* 29, p. 147.

phase to the other then equals the product of the amount of the substance being transferred times the resistance the substance receives when moving through the films. Thus, the difference in a contaminant's partial pressure in the bulk gas and at the interface is known as the *gas-side impedance,* while the difference in a contaminant's concentration in the bulk liquid and at the interface is known as the *liquid-side impedance.*[17]

Therefore, chemical reactions can occur in the bulk liquid, in the bulk solid, but also in the mass transfer zones. The reaction rates and characteristics may vary according to which phase or mass transfer zone the contaminant is located.

Chemical kinetics is also important for treatment technologies, such as the cyanide process discussed earlier, and for treating persistent organic compounds. For example, hexachlorobenzene can be dehalogenated in a manner similar to that shown in Figure 4.15, when the consecutive reaction steps occur in the presence of organometallic compounds, such as vitamin B_{12}. In this case, the theoretical curves of Figure 4.15 would represent hexachlorobenzene (parent compound A), pentachlorobenzene (intermediate degradate B), and tetrachlorobenzene (final product C). Obviously,

TABLE 4.7
Common Species of Tributyltin

Chemical Species	*Formula*
Tributyltin benzoate	$C_{19}H_{32}O_2Sn$
Tributyltin chloride	$C_{12}H_{27}ClSn$
Tributyltin fluoride	$C_{12}H_{27}FSn$
Tributyltin linoleate	$C_{30}H_{58}O_2Sn$
Tributyltin methacrylate	$C_{16}H_{32}O_2Sn$
Tributyltin oxide	$C_{24}H_{54}OSn_2$

the sequence can continue to produce unsubstituted benzene (i.e., no chlorine atoms).

The good news is that with each sequence of chlorine atom removal, the toxicity and persistence of each new reactant generally falls. However, reactions can also give products that are more toxic than the parent compound, such as certain metabolites of polycyclic aromatic hydrocarbons produced from catalytic (i.e., enzymatic) reactions in cells. Thus, a mass balance must consider not only transport mechanisms, but transformation processes as well.

Figure 4.15 also demonstrates speciation, i.e., the phenomenon where a compound changes into various chemical forms. For example, metals can exist in diverse molecular and ionic forms, each with unique physical, chemical, and biological characteristics. Some may be more toxic, some more mobile, and some more persistent in various environmental media. For example, the compound tributyltin (TBT), a potent biocide (a substance that kills biota), has recently been banned because it has been found to impair endocrine systems in certain aquatic fauna. However, TBT was once widely used in North America as an antifouling compound in marine paints. As a result, TBT contamination has occurred in harbors and shore environments. The TBT itself can exist in several forms (see Table 4.7). The TBT structure is shown in Figure 4.18.

Organic forms of tin (i.e., organotins) degrade in a step-wise manner:

$$R_4Sn \rightarrow R_3SnX \rightarrow R_2SnX_2 \rightarrow RS_nX_3 \rightarrow SnX_4 \qquad \text{Reaction 4–6}$$

Where R is an organic functional group [e.g., a butyl chain or a benzene ring (phenyl) group]; X is a halide group (e.g., Cl^-) or other negatively charged (anionic) group (e.g., OH^-). Thus, a TBT compound is configured as R_3SnX. This type of degradation, which progresses toward the inorganic species (SnX_4) is known as *mineralization.*

The TBT is degraded in the marine environment to dibutyltin (R_2SnX_2), monobutyltin ($RSnX_3$), and inorganic tin (Sn), in a manner similar to that

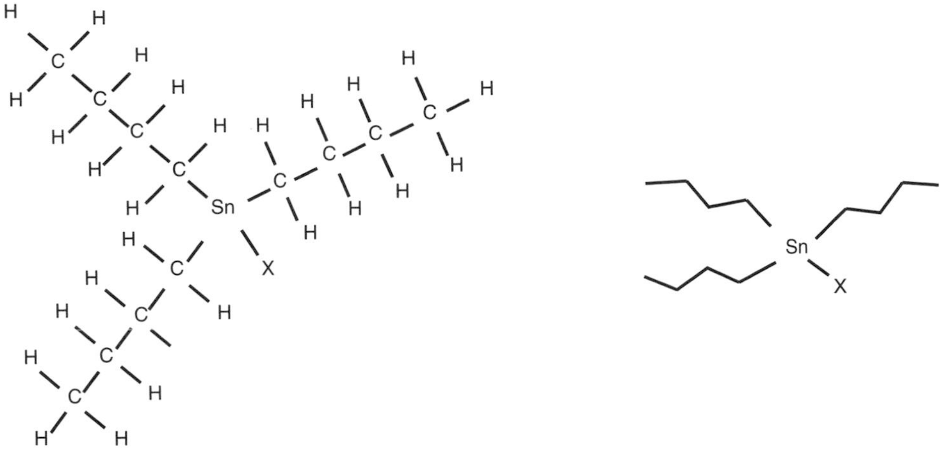

FIGURE 4.18. Structure of a tributyltin compound, showing the three butyl (C_4H_9) functional groups. The X designates an anionic group, which may be either inorganic, such as chlorine (Cl^-), or organic, such as benzoate ($C_7H_5O_2^-$). Shorthand structure shown on the right.

shown in Figure 4.15. The reaction rates, especially those associated with cleaving the Sn-C bonds are generally greater for TBT to dibutyltin versus dibutyltin to monobutyltin, or monobutyltin to inorganic Sn species. So, the chemical species dibutyltin may be the most persistent chemical species after TBT is released into the environment. Thus, the environmental scientists must look for a number of species, especially dibutyltin, even though the contaminant was TBT.

A question worth considering is whether it is the parent compound alone, i.e., TBT, that is responsible for the toxicity, or do the other more persistent forms, e.g., dibutyltin, also account for some of the observed aquatic effects? If the more persistent forms are toxic or if they lead to other toxic species, then the concept of *rate limitation* becomes paramount. That is, the slowest step in the overall reaction determines the rate of the overall persistence of the series of contaminants in the environment. So, a cleanup action or environmental condition that speeds up the degradation of this most persistent species will speed up the overall degradation of the contaminants in the environment. However, it may be that when these conditions occur they may or may not induce a faster breakdown of the next slowest chemical species.

If a reactant is part of more than one reaction in the environment, it may undergo parallel and unrelated reactions. For example, microbial and abiotic degradation of an organic compound may occur simultaneously. In this case, rather than the sequence in Reaction 4–1, we would see parallel reactions:

$$A \xrightarrow{R_1} B$$
$$A \xrightarrow{R_2} C \qquad \text{Reaction 4–5}$$

In this case, the overall degradation rate of A would be the sum of rates R_1 and R_2.

Thus, the environmental mass balance is determined in part by the behavior and fate of contaminants. Highly persistent compounds, i.e., those having very slow reaction rates, will remain in environmental media for years, or even decades and centuries. This means that monitoring programs must not only look for the parent compounds, but must carefully consider the reaction kinetics to determine other intermediate degradates that will have formed as the parent contaminants react.

Notes and Commentary

1. This is one of many counterintuitive and seemingly oxymoronic terms common in environmental sciences, such as "dynamic equilibrium." However, this is the nature of environmental science and engineering. The field is divided between theoretical and empirical concepts. We may know that the sorption or dissolution or volatilization is incomplete, but we have seen similar situations in the laboratory and field so often that we can prepare a "nonequilibrium constant" for a "pseudosteady state condition."
2. For the calculations and discussions of solubility equilibrium, including this example, see C.C. Lee and S.D. Lin, eds., 2000, *Handbook of Environmental Engineering Calculations*, McGraw-Hill, New York, N.Y., pp. 1.368–73.
3. See Michael LaGrega, Phillip Buckingham, and Jeffrey Evans, 2001, *Hazardous Waste Management*, 2nd Edition, Boston, Mass.: McGraw-Hill.
4. Reference for all of the organic solvents: International Agency for Research on Cancer, 1977, *Monographs on the Evaluation of the Carcinogenic Risk of Chemicals to Man: 1972–Present*, World Health Organization, Geneva, Switzerland.
5. The single bond lengths given are as if the partner atoms are not involved in double or triple bonds. If that were not the case, the bond lengths would be shorter.
6. See discussions in V. Evangelou, 1998, *Environmental Soil and Water Chemistry: Principles and Applications*, John Wiley & Sons, New York, N.Y.; and in R. Schwarzenbach, P. Gschwend, and D. Imboden, 1993, *Environmental Organic Chemistry*, John Wiley & Sons, New York, N.Y.
7. One prominent exception is the alkylphenols, which have both hydrophilic and lipophilic properties at either end of the molecule. That is why alkylphenols are used as surfactants and detergents.
8. T. Boublik, V. Fried, and L. Brown, 1984, *The Vapour Pressures of Pure Substances*, 2nd Edition, Elsevier, Amsterdam.

9. This is an important aspect of chromatography as well. The lower boiling point compounds, or the VOCs, usually come off the column first. That is, as the gas chromatograph's oven increases the temperature of the column, the more volatile compounds leave the column first and hit the detector first, so their peaks show up before the less volatile compounds. Other chemical factors such as halogenation and sorption affect this relationship, so this is not always the case. But as a general rule, a compound's boiling point is a good first indicator of residence time on a column.
10. The principal reference for this discussion is R. Probstein, 1994, *Physicochemical Hydrodynamics: An Introduction*, 2nd Edition, John Wiley & Sons, New York, N.Y.
11. Size exclusion can be visualized as a series of traps with increasing door sizes to catch smaller animals first. As animals pass by each successive trap, bigger animals enter as the door size increases. The biggest animals never get caught. The smaller animals are held until conditions change (the trapper opens the doors). Such "door openings" in chromatography are brought about by changes in pressure and temperature, until all the analytes (animals in our analogy) are successively released. So, larger molecules are size excluded in gel chromatography.
12. The symbols and nomenclature are inconsistent for partition coefficients specifically and equilibrium constants in general. Different scientific disciplines use various symbols for partition coefficients. Engineers and chemists frequently use K_d or P, while hydrogeologists may use the symbol sigma, δ, (perhaps because K is often used in groundwater work to represent hydraulic conductivity).
13. For a more detailed discussion of the process dynamics touched on here, see W.J. Weber, Jr. and F.A. DiGiano, 1996, *Process Dynamics in Environmental Systems*, John Wiley & Sons, New York, N.Y., particularly Chapters 4 and 5.
14. U.S. EPA, 2000, Capsule Report: Managing Cyanide in Metal Finishing, Report No. EPA 625/R-99/009
15. The environmental literature is not completely consistent in its use of the term "phase." The term may be interpreted to mean exactly what physical chemists have defined, i.e., phase is the distinct state of matter in a physical system. Such matter is identical in chemical composition and physical state and separated from other material by the phase boundary as represented in Figure 4.–7. However, phase may be more akin in environmental contexts to "fractional solubility," meaning that a substance can be found in varying amounts in different solvents. One of the most common examples of fractional solubility is demarcation between the concentration of a contaminant in the "organic phase" and "aqueous phase" as represented by the octanol-water partition coefficient (K_{ow}), discussed in this chapter. Another less commonly applied connotation of environmental phases is its synonymous usage with environmental "media" or "compartments." For example, one may hear environmental engineers discussing the transport of a contaminant from the water phase to the

air phase, or the soil phase to the water and air phases. Context, in fact, is important in many environmental discussions, and reading in context is often the only way to know what a publication means. As discussed in Chapter 1, since environmental science, engineering, and technology draws from so many fields, numerous terms (e.g., particle) have various meanings which can only be understood within the context of the specific discussion. Further, even within the environmental disciplines, uses of terms like phase will vary. For example, air experts may use the physical chemists' definition when discussing phase distributions within a stack gas, but may use the solubility fractionation definition when discussing the transformation and transport of a pollutant between a raindrop and the air (i.e., air-water partitioning). Likewise, soil and water scientists may apply the physicochemical definition of phases when in the laboratory (e.g., measuring the amount of a liquid or solid phase analyte volatilizing to the gas phase from the gas chromatograph's column and carried as a gas to the detector). However, in the field they may refer to phase distribution as the movements among soil, water, and air compartments.

16. W. G. Whitman, 1923, The two-film theory of gas absorption, *Chemical and Metallurgical Engineering.* 29, p. 147.
17. Several other theories predict interphase transfer, including other film models, as well as penetration, and surface renewal models. For an excellent discussion of these theories, see Chapter 4 of W.J. Weber, Jr. and F.A. DiGiano, 1996, *Process Dynamics in Environmental Systems*, John Wiley & Sons, New York, N.Y.

CHAPTER 5

Movement of Contaminants in the Environment

The scientific community is sometimes subdivided between groups most concerned with environmental measurements and groups whose foremost concern is environmental modeling. In fact, however, the two are highly complementary. Models are very useful in environmental assessments because measurements cannot be taken at all times and in all places with the intensity and focus needed to assess all environmental conditions. Models are thus needed to "fill in the gaps" as well as to extend the measurement data to characterize and evaluate an environmental problem or study. Likewise, models depend on high-quality measurements so that they may be "ground-truthed" and verified. Often, data will be collected to compare actual measurements to those predicted by a model.

The various physical and chemical conditions and factors we have considered so far will interact in complex ways. Scientists are continuously devising new models to predict the movement and change of compounds in the environment. Models provide a means for representing a real system in an understandable way.[1] They take many forms, beginning with "conceptual models" that explain the way a system works, such as a delineation of all the factors and parameters of how a particle moves in the atmosphere after its release from a power plant. Conceptual models help to identify the major influences on where a chemical is likely to be found in the environment, and as such, need to be developed to help target sources of data needed to assess an environmental problem.

Research scientists often develop "physical" or "dynamic" models to estimate the location where a contaminant would be expected to move under controlled conditions, only on a much smaller scale. For example, the U.S. Environmental Protection Agency's wind tunnel facility in Research Triangle Park, North Cardina is sometimes used when mathematical models have too many data gaps or when terrain and other conditions are so complex as to render the models practically useless. Recently, for example, the wind tunnel built a scaled model of the town of East Liverpool, Ohio, and its surrounding terrain to estimate the movement of the plume from an incinerator.[2] The plume could be observed under varying conditions, including wind direction and height of release. Only a few such facilities exist, however, so most hazardous waste and other sites will have to make use of more virtual tools, such as computer simulations and geographic information systems (GIS). Like all models, the dynamic models' accuracy is dictated by the degree to which the actual conditions can be simulated and the quality of the information that is used.

Transport and fate models can be statistical or "deterministic." Statistical models include the pollutant dispersion models, such as the Lagrangian models, which assume Gaussian distributions of pollutants from a point of release (see Figure 5.1). That is, the pollutant concentrations are normally distributed in both the vertical and horizontal directions from the source. The Lagrangian approach is common for atmospheric releases. "Stochastic" models are statistical models that assume that the events affecting the behavior of a chemical in the environment are random, so such models are based upon probabilities.

Deterministic models are used when the physical, chemical, and other processes are sufficiently understood so as to be incorporated to reflect the movement and fate of chemicals. These are very difficult models to develop because each process must be represented by a set of algorithms in the model. Also, the relationship between and among the systems, such as the kinetics and mass balances, must also be represented.

Thus, the modeler must "parameterize" every important event following a chemical's release to the environment. Often, hybrid models using both statistical and deterministic approaches are used, for example, when one part of a system tends to behave more randomly, while another has a very strong basis in physical principals.

Numerous models are available to address the movement of chemicals through a single environmental media, but increasingly, environmental scientists and engineers have begun to develop "multimedia models," such as compartmental models that help to predict the behavior and changes to chemicals as they move within and among the soil, water, air, sediment, and biota (see Figure 5.2).[3] Such models will likely see increased use in all environmental science and engineering.

Environmental Chemodymamics Models

Environmental chemodynamics is concerned with how chemicals move and change in the environment. Up to this point we have discussed some of the fluid properties that have a great bearing on the movement and dis-

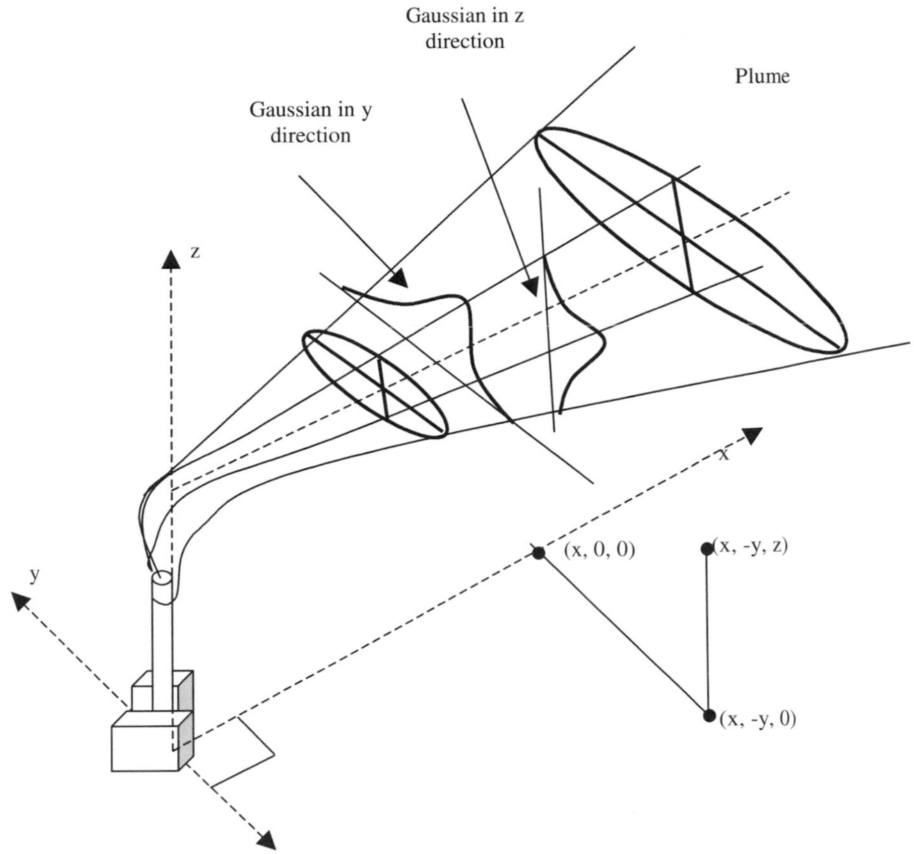

FIGURE 5.1. Atmospheric plume model based upon random distributions in the horizontal and vertical directions. (Adapted from: D. Turner, 1970, *Workbook of Atmospheric Dispersion Estimates*, Office of Air Programs Publication No. AP-26 [NTIS PB 191 482], U.S. Environmental Protection Agency, Research Triangle Park, N.C.)

tribution of contaminants. However, we are now ready to consider four specific partitioning relationships that control the "leaving" and "gaining" of pollutants among particle, soil, and sediment surfaces, the atmosphere, organic tissues, and water as they may be applied to estimating and modeling where a contaminant will go after it is released. These relationships are sorption, solubility, volatilization, and organic carbon-water partitioning, which are expressed, respectively, by coefficients of sorption (distribution coefficient, K_D, or solid-water partition coefficient, K_p); dissolution or solubility coefficients; air-water partitioning (and the Henry's Law [K_H] constant); and organic carbon-water (K_{oc}).

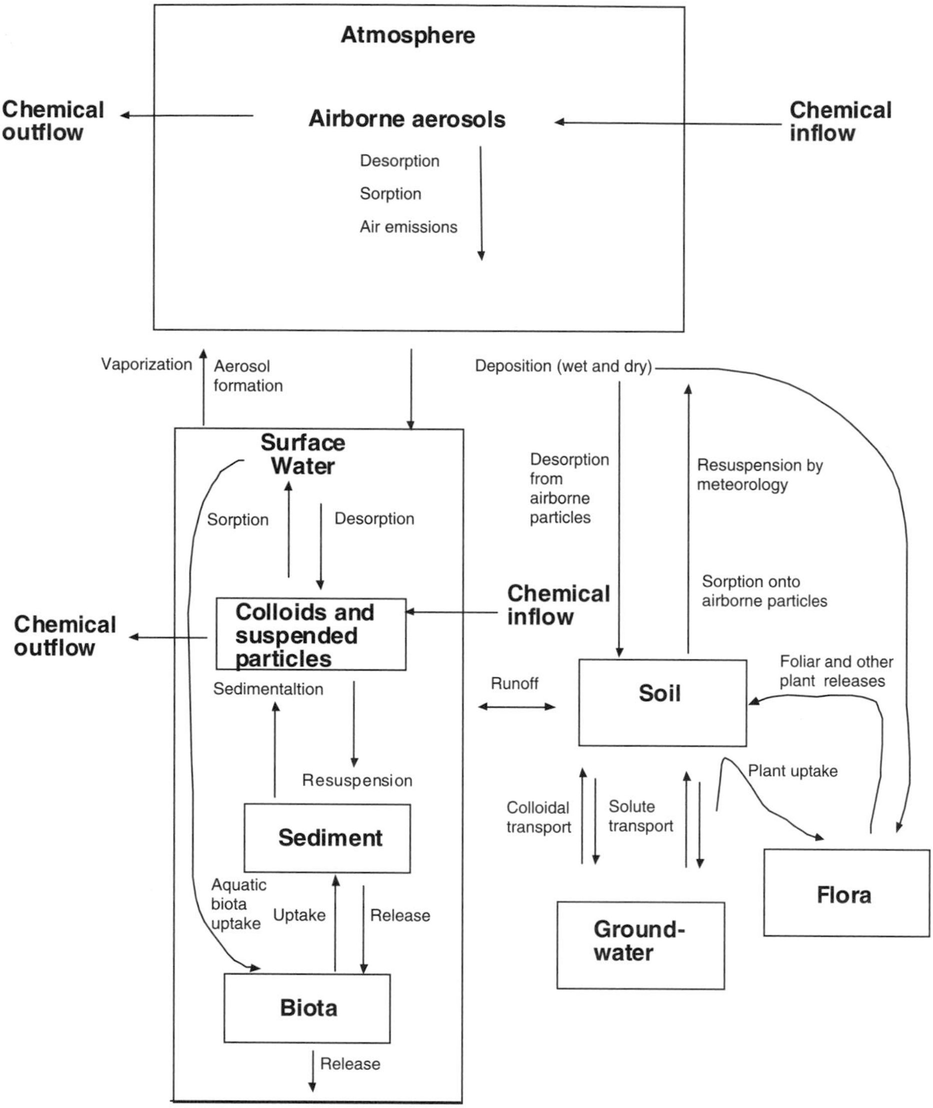

FIGURE 5.2. Example framework and flow of a multimedia, compartmental chemical transport and transformation model. Algorithms and quantities must be provided for each box. Equilibrium constants (e.g., partitioning coefficients) must be developed for each arrow. Steady-state conditions may not be assumed, so reaction rates and other chemical kinetics must be developed for each arrow and box.

In chemodynamics, the environment is subdivided into finite compartments. Recall from the discussions on mass balances and from Figure 4.13 that the mass of the contaminant entering the control volume and the mass leaving the control volume must be balanced by what remains within the control volume (a la the conservation laws). Likewise, within that control volume, each compartment may be a gainer or loser of the contaminant mass, but the overall mass must balance. The generally inclusive term for these compartmental changes is known as *fugacity* or the "fleeing potential" of a substance. It is the propensity of a chemical to escape from one type of environmental compartment to another. Combining the relationships between and among all of the partitioning terms is one means of modeling chemical transport in the environment.[4] This is accomplished by using thermodynamic principles and, hence, fugacity is a thermodynamic term.

The simplest chemodynamic approach addresses each compartment where a contaminant is found in discrete phases of air, water, soil, sediment, and biota. However, a complicating factor in environmental chemodynamics is that even within a single compartment, a contaminant may exist in various phases (e.g., dissolved in water and sorbed to a particle in the solid phase). This points to the importance of interphase reactions, or the physical interactions of the contaminant at the interface between each compartment. Within a compartment, a contaminant may remain unchanged (at least during the designated study period), or it may move physically, or it may be transformed chemically into another substance. Actually, in many cases all three mechanisms will take place. A mass fraction will remain unmoved and unchanged. Another fraction remains unchanged but is transported to a different compartment. Another fraction becomes chemically transformed with all remaining products staying in the compartment where they were generated. And a fraction of the original contaminant is transformed and then moved to another compartment. Thus, on release from a source, the contaminant moves as a result of thermodynamics. We were introduced to fugacity principles in our discussion of the fluid properties K_{ow} as well as vapor pressure and the partial pressure of gases.

Fugacity requires that at least two phases be in contact with the contaminant. For example, recall that the K_{ow} value is an indication of a compound's likelihood to exist in the organic versus aqueous phase. This means that if a substance is dissolved in water and the water comes into contact with another substance, such as octanol, the substance will have a tendency to move from the water to the octanol. Its octanol-water partitioning coefficient reflects just how much of the substance will move until the aqueous and organic solvents (phases) will reach equilibrium. For example, if a spill of equal amounts of the polychlorinated biphenyl, decachlorobiphenyl ($\log K_{ow}$ of 8.23), and the pesticide chlordane ($\log K_{ow}$ of 2.78), the PCB has much greater affinity for the organic phases than does the chlordane (more than five orders of magnitude). This does not mean than a great amount of either of the compounds is likely to stay in the water column, since they are both hydrophobic, but it does mean that they will vary in the time and mass of each contaminant moving between phases. The rate (kinetics) is different, so the time it takes for the PCB and chlordane to reach equilibrium will be different. This can be visualized by plotting the concentration of each compound with time (see Figure 5.3). When the concentrations plateau, the compounds are at equilibrium with their phase.

When phases contact one another, a contaminant will escape from one to another until the contaminant reaches equilibrium among the phases that are in contact with one another. Kinetics takes place until equilibrium is achieved.

We can now consider the key partitioning factors needed for a simple chemodynamic model.

Partitioning to Solids: Sorption

Some experts spend a lot of time differentiating the various ways that a contaminant will attach to or permeate into surfaces of solid phase particles. Others, including many environmental engineers, lump these processes together into a general phenomenon called *sorption*, which is the process in which a contaminant or other solute becomes associated, physically or chemically, with a solid sorbent.

Sorption is arguably the most important transfer process that determines how bioavailable or toxic a compound will be in surface waters. The

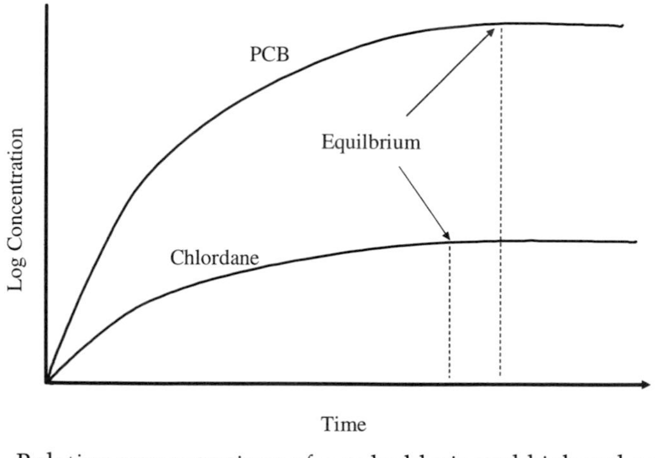

FIGURE 5.3. Relative concentrations of a polychlorinated biphenyl and chlordane in octanol with time.

physicochemical transfer[5] of a chemical, A, from liquid to solid phase is expressed as:

$$A_{(solution)} + \text{solid} = \text{A-solid} \qquad \text{Equation 5–1}$$

The interaction of the solute (i.e., the chemical being sorbed) with the surface of a solid surface can be complex and dependent upon the properties of the chemical and the water. Other fluids are often of such small concentrations that they do not determine the ultimate solid-liquid partitioning. Although it is often acceptable to consider "net" sorption, let us consider briefly the four basic types or mechanisms of sorption:

1. *Adsorption* is the process wherein the chemical in solution attaches to a solid surface, which is a common sorption process in clay and organic constituents in soils. This simple adsorption mechanism can occur on clay particles where little carbon is available, such as in groundwater.
2. *Absorption* is the process that often occurs in porous materials so that the solute can diffuse into the particle and be sorbed onto the inside surfaces of the particle. This commonly results from short-range electrostatic interactions between the surface and the contaminant.
3. *Chemisorption* is the process of integrating a chemical into porous materials via chemical reaction. In soil, this is usually the result of a covalent reaction between a mineral surface and the contaminant.
4. *Ion exchange* is the process by which positively charged ions (cations) are attracted to negatively charged particle surfaces, or negatively charged ions (anions) are attracted to positively charged particle surfaces, causing ions on the particle surfaces to be displaced. Particles undergoing ion exchange can include soils, sediment, airborne particulate matter, or even biota, such as pollen particles. Cation exchange has been characterized as being the second most important chemical process on earth, after photosynthesis. This is because the cation exchange capacity (CEC), and to a lesser degree anion exchange capacity (AEC) in tropical soils, is the means by which nutrients are made available to plant roots. Without this process, the atmospheric nutrients and the minerals in the soil would not come together to provide for the abundant plant life on planet earth.[6]

These four types of sorption are a mix of physics and chemistry. The first two are predominantly controlled by physical factors, and the second two are combinations of chemical reactions and physical processes. We will spend a bit more time covering these specific types of sorption when we consider the surface effects of soils. Generally, sorption reactions affect three processes[7] in aquatic systems:

1. The chemical contaminant's transport in water due to distributions between the aqueous phase and particles;
2. The aggregation and transport of the contaminant as a result of electrostatic properties of suspended solids; and,
3. Surface reactions such as dissociation, surface-catalysis, and precipitation of the chemical contaminant.

When a contaminant enters a soil, some of the chemical remains in soil solution and some is adsorbed onto the surfaces of the soil particles. Sometimes this sorption is strong due to cations adsorbing to the negatively charged soil particles. In other cases the attraction is weak. Sorption of chemicals on solid surfaces needs to be understood because they hold onto contaminants, not allowing them to move freely with the pore water or the soil solution. Therefore sorption slows the rate at which contaminants move downwardly through the soil profile.

Contaminants will eventually establish a balance between the mass on the solid surfaces and the mass that is in solution. Recall that molecules will migrate from one phase to another to maintain this balance (see Figure 5.4). The properties of both the contaminant and the soil (or other matrix) will determine how and at what rates the molecules partition into the solid and liquid phases. These physicochemical relationships, known as *sorption isotherms*, are found experimentally. Figure 5.5 gives three isotherms for pyrene from experiments using different soils and sediments.

The x-axis shows the concentration of pyrene dissolved in water, and the y-axis shows the concentration in the solid phase. Each line represents the relationship between these concentrations for a single soil or sediment. A straight-line segment through the origin represents the data well for the range of concentrations shown. Not all portions of an isotherm are linear, particularly at high concentrations of the contaminant. Linear chemical partitioning can be expressed as:

$$S = K_D C_W \qquad \text{Equation 5–2}$$

Where S = concentration of contaminant in the solid phase (mass of solute per mass of soil or sediment)

C_W = concentration of contaminant in the liquid phase (mass of solute per volume of pore water)

K_D = partition coefficient (volume of pore water per mass of soil or sediment) for this contaminant in this soil or sediment.

For many soils and chemicals, the partition coefficient can be estimated using:

$$K_D = K_{OC} OC \qquad \text{Equation 5–3}$$

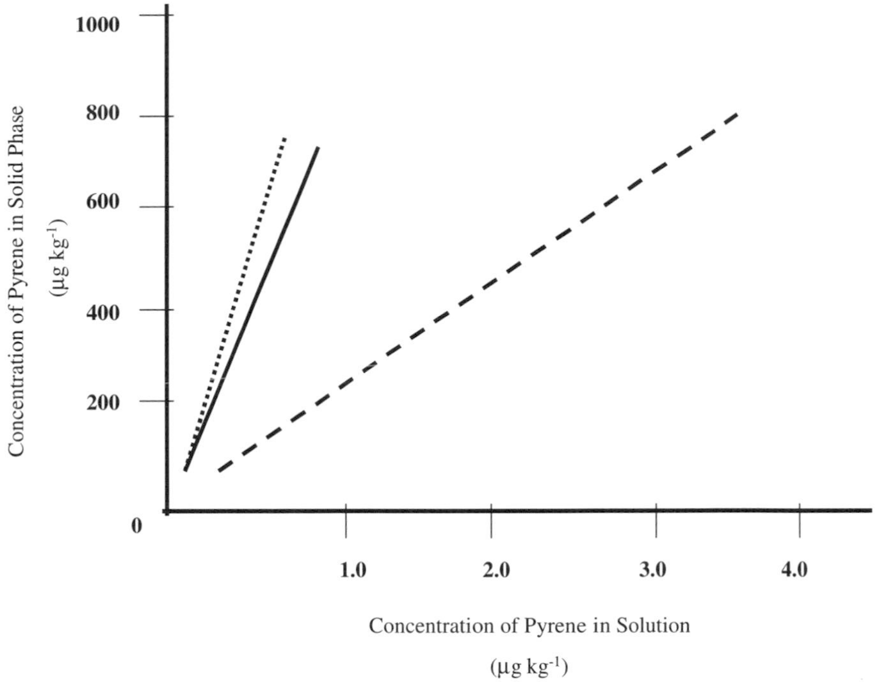

FIGURE 5.4. Three experimentally determined sorption isotherms for the polycyclic aromatic hydrocarbon, pyrene. (Source: J. Hassett and W. Banwart, 1989, "The Sorption of Nonpolar Organics by Soils and Sediments," in *Reactions and Movement of Organic Chemicals in Soils*, edited by B. Sawhney and K. Brown, Soil Science Society of America Special Publication 22, p. 35.)

Where K_{OC} = organic carbon partition coefficient (volume of pore water per mass of organic carbon)
OC = soil organic matter (mass of organic carbon per mass of soil)

This relationship is a very useful tool for estimating K_D from the known K_{OC} of the contaminant and the organic carbon content of the soil horizon of interest. The actual derivation of K_D is:

$$K_D = C_S(C_W)^{-1} \qquad \text{Equation 5–4}$$

Where C_S is the equilibrium concentration of the solute in the solid phase and C_W is the equilibrium concentration of the solute in the water.

Therefore, K_D is a direct expression of the partitioning between the aqueous and solid (soil or sediment) phases. A strongly sorbed chemi-

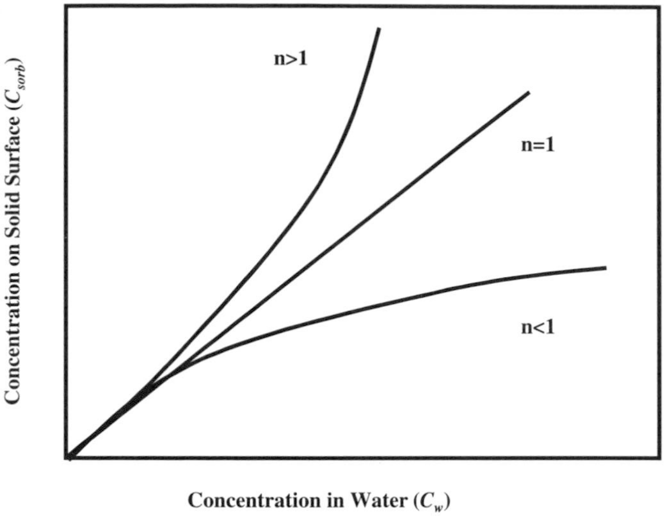

FIGURE 5.5. Hypothetical Freundlich isotherms with exponents (n) less than, equal to, and greater than 1, as applied to the equation $C_{sorb} = K_F C^n$. (Sources: R. Schwarzenbach, P. Gschwend, and D. Imboden, 1993, *Environmental Organic Chemistry*, John Wiley & Sons, New York, N.Y.; and H. F. Hemond and E. J. Fechner-Levy, 2000, *Chemical Fate and Transport in the Environment*, Academic Press, San Diego, Calif.)

cal like a dioxin or the banned pesticide DDT can have a K_D value exceeding 10^6. Conversely, a highly hydrophilic, miscible substance like ethanol, acetone, or vinyl chloride, will have K_D values less than 1. This relationship between the two phases demonstrated by Equation 5–4 and Figure 5.5 is roughly what environmental scientists call the *Freundlich Sorption Isotherm*:

$$C_{sorb} = K_F C^n \qquad \text{Equation 5–5}$$

Where C_{sorb} is the concentration of the sorbed contaminant, or the mass sorbed at equilibrium per mass of sorbent, and K_F is the Freundlich isotherm constant. The exponent determines the linearity or order of the reaction. Thus, if n = 1, then the isotherm is linear; meaning the more of the contaminant in solution, the more would be expected to be sorbed to surfaces. For values of n < 1, the amount of sorption is in smaller proportion to the amount of solution and, conversely, for values of n > 1, a greater proportion of sorption occurs with less contaminant in solution. These three isotherms are shown in Figure 5.5. Also note that if n = 1, then Equation 5–4 and the Freundlich sorption isotherm are identical.

Research has shown that when organic matter content is elevated in soil and sediment, the amount of a contaminant that is sorbed is directly proportional to the soil/sediment organic matter content. This allows us to convert the K_D values from those that depend on specific soil or sediment conditions to those that are soil/sediment independent sorption constants, K_{OC}:

$$K_{OC} = K_D(f_{OC})^{-1} \qquad \text{Equation 5–6}$$

Where f_{OC} is the dimensionless weight fraction of organic carbon in the soil or sediment. The K_{OC} and K_D have units of mass per volume. Table 5.1 provides the $\log K_{OC}$ values that are calculated from chemical structure and those measured empirically for several organic compounds, and compares them to the respective K_{ow} values.

Partitioning to the Liquid Phase: Dissolution

We have already devoted significant attention to the partitioning to the liquid phase in our earlier discussion of solubility, electronegativity, and polarity, as well as in the previous section dealing with sorption. One thing to keep in mind, however, is that the "default" in solubility discussions is that the solvent is water. Technical handbooks and manuals should designate whether the dissolution processes they describe or "aqueous solubility" or solubility in some other solvent. Unless otherwise stated, one can assume that when a compound is described as insoluble that such a statement means that the compound is *hydrophobic*. However, in this business, it is never really acceptable to "assume anything." A good resource for contaminant solubilities for water, dimethylsulfoxide (DMSO), ethanol, acetone, methanol, and toluene is the National Toxicology Program's *Chemical Solubility Compendium*[8] and the program's Health and Safety reports.[9] The latter are updated frequently and provide useful data on the properties of toxic chemicals.

Most characterizations of contaminants will describe solubility in water and provide values for aqueous solubility, as well as a substance's solubility in other organic solvents, such as methanol or acetone. This is important and valuable information when considering the fugacity of a compound. First, if a compound is highly hydrophobic, one may be led to assume that it will not be found in the *water column* in an environmental study. This is a reasonable expectation theoretically, and is based on the expectation that the only solvent in water bodies is water. However, surface and groundwater is never completely devoid of other solvents.

The process of *co-solvation* is a very important mechanism for getting a highly lipophilic and hydrophobic compound into water. In other words, if a compound is hydrophobic and nonpolar, but is easily dissolved

TABLE 5.1
Calculated and Experimental Organic Carbon Coefficients (K_{oc}) for Selected Contaminants Found at Hazardous Waste Sites

Chemical	*log K_{ow}*	*Calculated*		*Measured*	
		log K_{oc}	*K_{oc}*	*log K_{oc}*	*K_{oc} (geomean)*
Benzene	2.13	1.77	59	1.79	61.7
Bromoform	2.35	1.94	87	2.10	126
Carbon tetrachloride*	2.73	2.24	174	2.18	152
Chlorobenzene	2.86	2.34	219	2.35	224
Chloroform	1.92	4.60	40	1.72	52.5
Dichlorobenzene, 1,2- *(o)*	3.43	2.79	617	2.58	379
Dichlorobenzene, 1,4- *(p)*	3.42	2.79	617	2.79	616
Dichlorethane, 1,1-	1.79	1.50	32	1.73	53.4
Dichlorethane, 1,2-	1.47	1.24	17	1.58	38.0
Dichloroethylene, 1,1-	2.13	1.77	59	1.81	65
Dichloroethylene, *trans* -1,2-	2.07	1.72	52	1.58	38
Dichloropropane, 1,2-	1.97	1.64	44	1.67	47.0
Dieldrin	5.37	4.33	21,380	4.41	25,546
Endosulfan	4.10	3.33	2,138	3.31	2,040
Endrin	5.06	4.09	12,303	4.03	10,811
Ethylbenzene	3.14	2.56	363	2.31	204
Hexachlorobenzene	5.89	4.74	54,954	4.90	80,000
Methyl bromide	1.19	1.02	10	0.95	9.0
Methyl chloride	0.91	0.80	6	0.78	6.0
Methylene chloride	1.25	1.07	12	1.00	10
Pentachlorobenzene	5.26	4.24	17,378	4.51	32,148
Tetrachloroethane, 1,1,2,2-	2.39	1.97	93	1.90	79.0
Tetrachloroethylene	2.67	2.19	155	2.42	265
Toluene	2.75	2.26	182	2.15	140
Trichlorobenzene, 1,2,4-	4.01	3.25	1,778	3.22	1,659
Trichloroethane, 1,1,1-	2.48	2.04	110	2.13	135
Trichloroethane, 1,1,2-	2.05	1.70	50	1.88	75.0
Trichloroethylene	2.71	2.22	166	1.97	94.3
Xylene, *o*-	3.13	2.56	363	2.38	241
Xylene, *m*-	3.20	2.61	407	2.29	196
Xylene, *p*-	3.17	2.59	389	2.49	311

*Tetrachloromethane.
Source: U.S. Environmental Protection Agency, 1996, Soil Screening Program.

in acetone or methanol, it may well end up in the water because these organic solvents are highly miscible in water. The organic solvent and water mix easily, and a hydrophobic compound will remain in the water column because it is dissolved in the organic solvent, which in turn has mixed with the water. Compounds like PCBs and dioxins may be transported as co-solutes in water by this means. The combination of hydrophobic compounds being sorbed to suspended materials and co-solvated in organic co-solvents that are miscible in water can mean that they are able to move in water bodies and that receptors can be exposed through the water pathways.

The rate of dissolution is dependent on the concentration of a contaminant being released to a water body (i.e., the volume of contaminant versus the volume of the receiving waters). However, concentrations of contaminants are usually at the ppm or lower level, so this is seldom a limiting factor in environmental situations. Other factors that influence dissolution are the turbulence of the water, temperature, *ionic strength*, dissolved organic matter present in the water body, the aqueous solubility of the contaminant, and the presence of co-solvents.[10]

Solubility is determined from saturation studies. In other words, in the laboratory at a certain temperature, as much of the solute is added to a solvent until the solvent can no longer dissolve the substance being added. If Compound A has a published solubility of $10\,\text{mg}\,\text{L}^{-1}$ in water at 20°C, this means that the one liter of water could only dissolve 10 mg of that substance. If, under identical conditions, Compound B has a published aqueous solubility of $20\,\text{mg}\,\text{L}^{-1}$, this means that one liter of water could dissolve 20 mg of Compound B, and that Compound B has twice the aqueous solubility of Compound A.

Actually, solutions are really in "dynamic equilibrium" because the solute is leaving and entering the solution at all times, but the average amount of solute in solution is the same. The functional groups on a molecule determine whether it will be more or less polar. Compounds with hydroxyl groups are therefore more likely to form H-bonds with water. Thus, methane is less soluble in water than methanol. Also, since water interacts strongly with ions, salts are usually quite hydrophilic. The less the charge of the ion, the greater the solubility in water.

Partitioning to the Gas Phase: Volatilization

The next form of partitioning to be introduced is the change of a contaminant's phase to a gas. In its simplest connotation, volatilization is a function of the concentration of a contaminant in solution and the contaminant's partial pressure (see the previous discussion in Chapter 4 on vapor pressure).

Henry's Law states that the concentration of a dissolved gas is directly proportional to the partial pressure of that gas above the solution:

$$p_a = K_{\text{H}}[c] \qquad \text{Equation 5–7}$$

Where, K_{H} = Henry's Law constant
p_{a} = Partial pressure of the gas
$[c]$ = Molar concentration of the gas

or,

$$p_A = K_H C_W \qquad \text{Equation 5–8}$$

Where C_W is the concentration of gas in water.

Thus, for any chemical contaminant, we can establish a proportionality between the solubility and vapor pressure. Henry's Law is an expression of this proportionality between the concentration of a dissolved contaminant and its partial pressure in the headspace (including the open atmosphere) at equilibrium. A dimensionless version of the partitioning is similar to that of sorption, except that instead of the partitioning between solid and water phases, it is between the air and water phases (K_{AW}):

$$K_{AW} = \frac{C_A}{C_W} \qquad \text{Equation 5–9}$$

Where C_A is the concentration of gas A in the air.

The relationship between the air/water partition coefficient and Henry's Law constant for a substance is:

$$K_{AW} = \frac{K_H}{RT} \qquad \text{Equation 5–10}$$

Where R is the gas constant ($8.21 \times 10^{-2}\,\mathrm{L\,atm\,mol^{-1}\,K^{-1}}$) and T is the temperature (°K).

Henry's Law relationships work well for most environmental conditions. It represents a limiting factor for systems where a substance's partial pressure is approaching zero. At very high partial pressures (e.g., 30 Pascals) or at very high contaminant concentrations (e.g., >1,000 ppm), Henry's Law assumptions cannot be met. Such vapor pressures and concentrations are seldom seen in ambient environmental situations, but may be seen in industrial and other source situations. Thus, in modeling and estimating the tendency for a substance's release in vapor form, Henry's Law is a good metric and is often used in compartmental transport models to indicate the fugacity from the water to the atmosphere.

Henry's Law Example

At 25°C, the log Henry's Law constant ($\log K_H$) for 1,2-dimethylbenzene (C_8H_{10}) is 0.71 $\mathrm{L\,atm\,mol^{-1}}$ and the log octanol-water coefficient ($\log K_{ow}$) is 3.12. The $\log K_H$ for the pesticide parathion ($C_{10}H_{14}NO_5PS$) is −3.42 $\mathrm{L\,atm\,mol^{-1}}$, but its $\log K_{ow}$ is 3.81. Explain how these substances can have similar values for octanol-water partitioning yet so different Henry's Law constants. What principle physicochemical properties account for much of this difference?

Answer

Dimethylbenzene and parathion both have an affinity for the organic phase compared to the aqueous phase. Since Henry's Law constants are a function of both vapor pressure and water solubility, and both compounds have similar octanol-water coefficients, the difference in the Henry's Law characteristics must be mainly attributable to the compounds' respective water solubilities, their vapor pressures, or both.

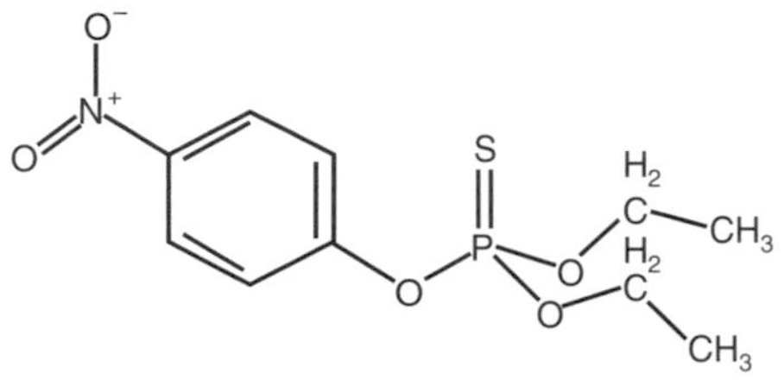

Parathion

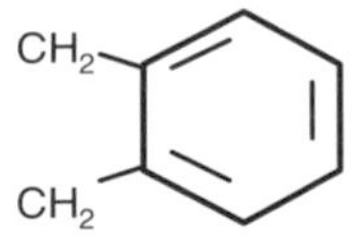

Ortho-xylene (1,2-dimethylbenzene)

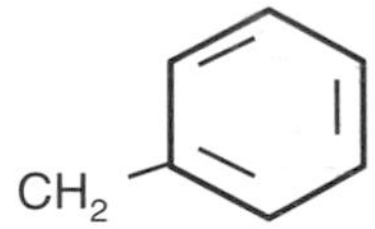

Toluene

Benzene

FIGURE 5.6. Molecular structure of the pesticide parathion and the solvents ortho-xylene, toluene, and benzene.

Parathion is considered "semivolatile" because its vapor pressure at 20°C is only 1.3×10^{-3} kPa. Parathion's solubility[11] in water is 12.4 mg L^{-1} at 25°C.

1,2-Dimethylbenzene is also known as ortho-xylene (*o*-xylene). The xylenes are simply benzenes with more two methyl groups. The xylenes have very high vapor pressures: 0.9 kPa, and water solubilities[12] of about 200 at mg L^{-1} at 25°C.

Thus, since the solubilities are both relatively low, it appears that the difference in vapor pressures is responsible for the large difference in the Henry' Law constants, or the much larger tendency of the xylene to leave the water and enter the atmosphere. Some of this tendency may result from the higher molecular weight of the parathion, but it is also attributable to the additional functional groups on the parathion benzene than the two methyl groups on the xylene (see Figure 5.6).

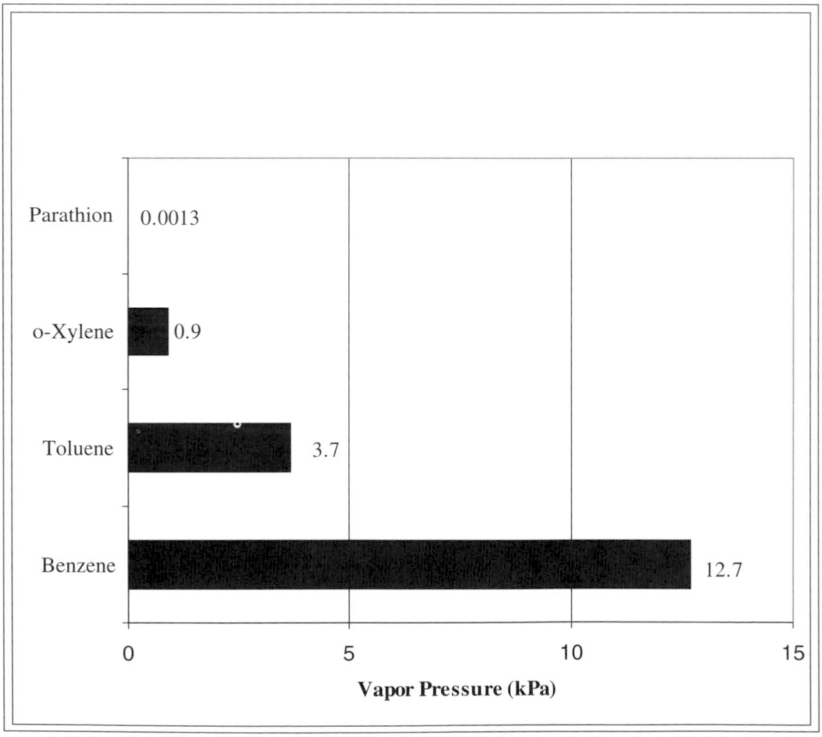

FIGURE 5.7. Effect of functional group substitutions on vapor pressure of four organic aromatic compounds at 20°C.

Another way to look at the chemical structures is to see them as the result of adding increasingly complex functional groups. In other words, moving from the unsubstituted benzene to the single methylated benzene (toluene) to *o*-xylene to parathion. The substitutions result in progressively decreasing vapor pressures:

$$\text{Benzene's } P^0 \text{ at } 20°\text{C} = 12.7\,\text{kPa}$$

$$\text{Toluene's } P^0 \text{ at } 20°\text{C} = 3.7\,\text{kPa}$$

$$o\text{-Xylene's } P^0 \text{ at } 20°\text{C} = 0.9\,\text{kPa}$$

$$\text{Parathion's } P^0 \text{ at } 20°\text{C} = 1.3 \times 10^{-3}\,\text{kPa}$$

The effect of these functional group additions on vapor pressure is even more obvious when seen graphically (Figure 5.7).

It is important to keep in mind that Henry's Law constants are highly dependent upon temperature, since both vapor pressure and solubility are also temperature dependent. When using published K_H values, one must compare them isothermically. Also, when combining different partitioning coefficients in a model or study, it is important either to use only values derived at the same temperature (e.g., sorption, solubility, and volatilization all at 20°C), or adjust them accordingly. A general adjustment is an increase of a factor of 2 in K_H for each 8°C temperature increase.

Also, any sorbed or otherwise bound fraction of the contaminant will not exert a partial pressure, so this fraction should not be included in calculations of partitioning from water to air. For example, it is important to differentiate between the mass of the contaminant in solution (available for the K_{AW} calculation) and that in the suspended solids (unavailable for K_{AW} calculation). This is crucial for many hydrophobic organic contaminants, where they are most likely not to be dissolved in the water column (except as co-solutes), with the largest mass fraction in the water column being sorbed to particles.

The relationship between K_H and K_{ow} is also important. It is often used to estimate the *environmental persistence*, as reflected in the chemical *half-life* ($T_{1/2}$) of a contaminant. However, many other variables determine the actual persistence of a compound after its release. Note in Table 5.1, for

TABLE 5.2
Properties of Chemicals Used in Atmospheric Compartmental Modeling

Compound	*Half-Life (days)*	*Log K_{ow}*	*Log K_H*
Benzene	7.7	2.1	–0.6
Chloroform	360	1.97	–0.7
DDT	50	6.5	–2.8
Ethyl benzene	1.4	3.14	0.37
Formaldehyde	1.6	0.35	–5.0
Hexachlorobenzene	708	5.5	–3.5
Methyl chloride	470	0.94	–0.44
Methylene chloride	150	1.26	–0.9
PCBs	40	6.4	–1.8
1,1,1 Trichloroethane	718	2.47	0.77

Source: D. Toro and F. Hellweger, 1999, "Long-Range Transport and Deposition: The Role of Henry's Law Constant," Final Report, International Council of Chemical Associations.

example, that benzene and chloroform have nearly identical values of K_{oc} and K_{ow}, and in Table 5.2 that K_{ow} and K_H are nearly the same, yet benzene is far less persistent in the environment. We will consider these other factors in Chapter 12, when we discuss abiotic chemical destruction and biodegradation.

With these caveats in mind, however, relative affinity for a substance to reside in air and water can be used to estimate the potential for the substance to partition not only between water and air, but more generally between the atmosphere and biosphere, especially when considering the long-range transport of contaminants (e.g., across continents and oceans).[13] Such long-range transport estimates make use of both atmospheric $T_{1/2}$ and K_H. Also, the relationship between octanol-water and air-water coefficients can be an important part of predicting a contaminant's transport. For example, Figure 5.8 provides some general classifications according to various substances' K_{AW} and K_{ow} relationships. In general, chemicals in the upper-left-hand group have a great affinity for the atmosphere, so unless there are contravening factors, this is where to look for them. Conversely, substances with relatively low K_{AW} and K_{ow} values are less likely to be transported long distances in the air.

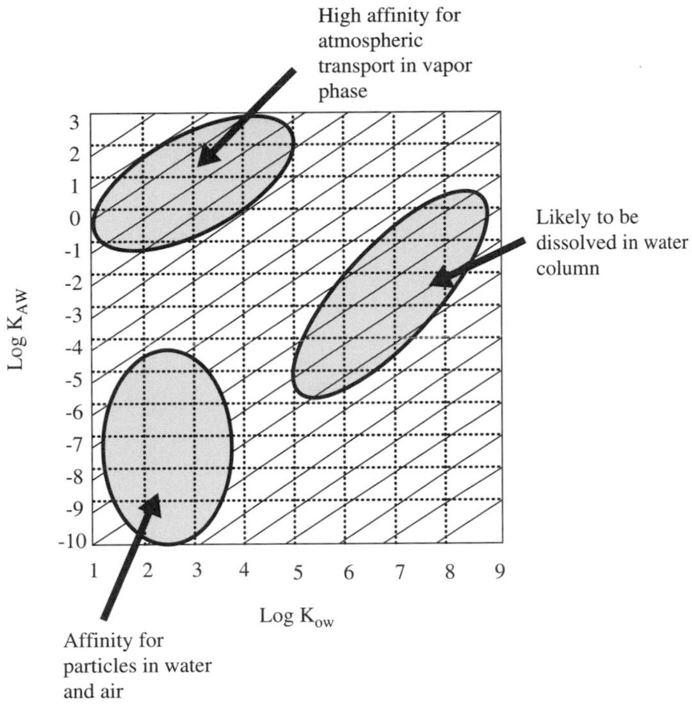

FIGURE 5.8. Relationship between air-water partitioning and octanol-water partitioning and affinity of classes of contaminants for certain environmental compartments. (Source: D. van de Meent, T. McKone, T. Parkerton, M. Matthies, M. Scheringer, F. Wania, R. Purdy, and D. Bennett, 1999, "Persistence and Transport Potential of Chemicals in a Multimedia Environment, " in *Proceedings of the SETAC Pellston Workshop on Criteria for Persistence and Long-Range Transport of Chemicals in the Environment, 14–19 July 1998, Fairmont Hot Springs, British Columbia, Canada,* Society of Environmental Toxicology and Chemistry, Pensacola, Fla.)

Engineering Technical Note: Phase Distributions within the Air Compartment—Atmospheric Measurements of the World Trade Center Plume

The September 11, 2001 attack on the World Trade Center (WTC) resulted in an intense fire (about 1000°C) and the subsequent, complete collapse of the two main structures and adjacent buildings, as well as significant damage to many surrounding buildings within and around the WTC complex. This 16-acre area has become known as Ground Zero. The collapse of the buildings and the fires created a large plume comprised of both particles and gases that were injected into the New York City air shed. The plume began at an elevation 80 to 90 stories above ground, with the initial combustion of the jet fuel and building materials. After the collapse of the buildings, aerosols were emitted from ground level, moving downwind and reaching many outdoor and indoor locations downwind. For the first 12 to 18 hours after the collapse, the winds transported the plume to the east and then to the southeast toward Brooklyn, New York.

Characterizing the chemical composition of an atmospheric pollutant plume is needed to assess people's exposures to contaminants. Following the World Trade Center attacks, people living nearby as well as emergency responders were exposed to gases and particles released directly from the site. Particles not only included those released from the fire that burned for many weeks, but also those that became resuspended by air turbulence and mechanical disturbance. The likely pathways of exposure were inhalation, ingestion of deposited particles, and dermal. Some contaminants that were found in the measurements at the WTC have been associated with human health effects that include carcinogenic compounds (e.g., benzo[a]pyrene and other *polycyclic aromatic hydrocarbons* [PAHs] from smoldering fires), endocrine disruptors (e.g., phthalates and styrene derivatives from plastics), and neurotoxins (such as dioxins from incomplete combustion).

The importance of airborne particles led to the taking of many measurements around Ground Zero, but the chemical composition of these particles can be very complex. Analyses[14] at WTC showed the particulate matter in settled dust contained pulverized building material, rendering it alkaline (pH > 9) with significant fractions of inorganic matter and metals (e.g., $>35\,\mu g\,g^{-1}$ calcium, $>110\,\mu g\,g^{-1}$ magnesium, $>1500\,\mu g\,g^{-1}$ titanium, and $>500\,\mu g\,g^{-1}$ aluminum). The cement/carbon ratio ranged between 37% and 50%, and the glass fiber content of the dust was 40%. This chemical composition was also observed in the fine fraction (<2.5 mm diameter), which frequently forms during combustion, thus usually containing larger percentages of carbon than typically found in coarse particles. Remnants of organic matter (e.g., wood, paper, wool, and cotton) were also found in the dust samples (e.g., cellulose content of 9% to 20%). The analyses also show elevated concentrations of products of incomplete combustion (e.g., $>200\,mg\,g^{-1}$ total polycyclic aromatic hydrocarbons and $100\,ng\,g^{-1}$ of dioxin total equivalents).

Organic compounds are highly diverse in their physical characteristics and chemical composition. Volatile organic compounds (VOCs), for example, exist in the ambient atmosphere almost entirely in the gas phase, since their vapor pressures in the environment are usually greater than 10^{-2} kilopascals, while semivolatile organic compounds (SVOCs), with vapor pressures between 10^{-2} and 10^{-5} kilopascals can exist in both the gas and particle phases in the ambient air. Nonvolatile organic compounds (NVOCs) with vapor pressures $<10^{-5}$ kilopascals, are predominantly found sorbed to particles, unless significant energy is added to increase their volatility. These values correspond to the classifications that are based upon observations of the compounds' behaviors during air sampling.[15]

The compounds' phase distribution between gas and particle depends on environmental conditions, such as relative humidity and temperature. Measurements of phase distribution can be difficult due

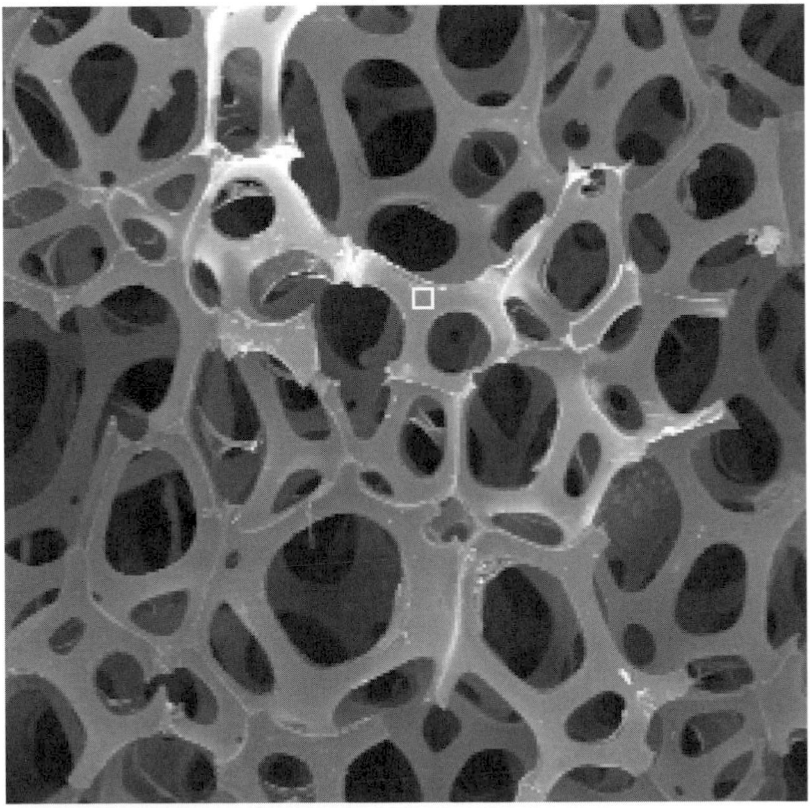

FIGURE 5.9. Scanning electron micrograph (40X magnification) showing the large amount of surface area in polyurethane foam (PUF) found in a semivolatile organic compound trap.

to the various inherent engineering features of the monitoring equipment, such as the need to maintain sampling flow rate, the pore size of filters, and the amount of time an organic compound resides on the filter or trap.[16] For example, a compound will sorb to a trapping material (see Figure 5.9), such as a polymer (e.g., XAD) or polyurethane foam (PUF).

Systems for collecting, extracting, and analyzing SVOCs have been recommended by the American Society for Testing and Materials[17] and by the U.S. Food and Drug Administration.[18] Gas phase air pollutants are collected as they diffuse into media, such as PUF and XAD sorbant traps and denuders.[19] Denuders are devices that contain glass surfaces on which compounds are collected. Most of a gas volume, including the gas phase of semivolatile organic compounds, immediately diffuses onto surfaces like those on the PUF; for example, SO_2 gas collection efficiencies[20] in a denuder have been found to be 99.2% and 99.6%. About 90% of the gas collected is in the first 10% of the denuder length.[21] The PUF is analogous to a "nest" of micro-denuders with its many crevices and large amount of surface area (Figure 5.9), so similar collection dynamics are assumed. Recently, similar findings have been found for PUF and the polymer collection media.[22]

The amount of time that the air is pumped through the trap is directly related to the amount of air. Thus, if the flow volume is known (e.g., for SVOC monitoring, the target flow rate of air is about $350\,\mathrm{L\,min^{-1}}$), the concentration can be calculated. If 20 mg of total dioxins are collected on the trap through which air has flowed for 24 hours at $350\,\mathrm{L\,min^{-1}}$, the integrated concentration of dioxins for that day would be:

$$\frac{20\,\mathrm{mg}}{24\,\mathrm{hour} \times \frac{60\,\mathrm{min}}{\mathrm{hour}} \cdot \frac{350\,\mathrm{L}}{\mathrm{min}}} = 4 \times 10^{-5}\,\mathrm{mg\,L^{-1}} \text{ or } 40\,\mathrm{ng\,L^{-1}}.$$

However, concentrations of contaminants are usually expressed in terms of cubic meters of air, so the total dioxin concentration on this day would be $40{,}000\,\mathrm{ng\,m^{-3}}$ or $40\,\mu\mathrm{g\,m^{-3}}$.

At the New York sites, a state-of-the-art high-capacity integrated organic gas and particle sampling system (see Figure 5.10) making use of these principles was deployed. Let us now look at some of the results from the sampling as they apply to phase partitioning in the air compartment.

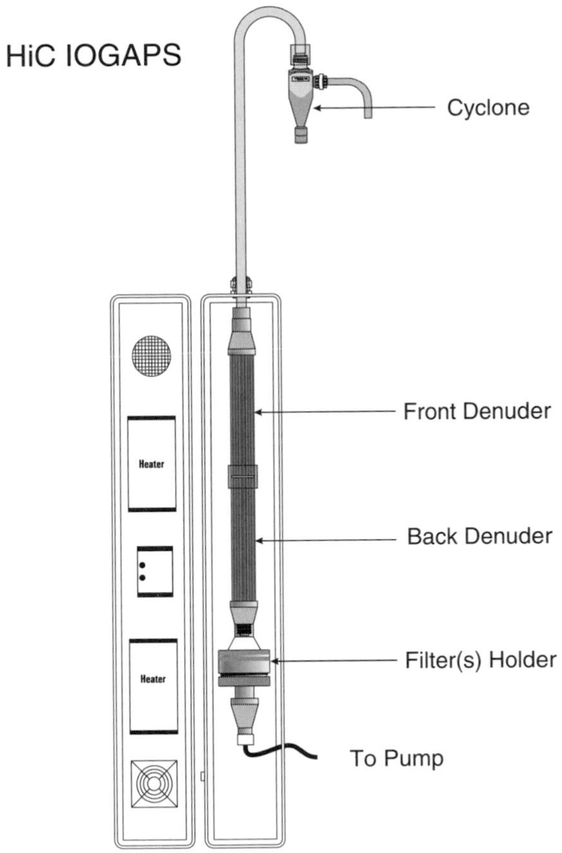

FIGURE 5.10. Schematic of the high-capacity Integrated Organic Gas and Particulate (HiC IOGAP) sampler with a 2.5 m cyclone inlet for particle discrimination that was used in Lower Manhattan, New York, to collect semivolatile gases and particles for speciation of organic compounds. The sampler utilizes two sorbant (XAD-4) coated eight-channel annular denuders (52 mm outer diameter, 285 mm length) to collect the gas phase species and a prebaked quartz filter followed by three XAD-4 impregnated quartz filters to collect the particle phase. The XAD-4 impregnated quartz filters were used to collect those compounds that desorb from the particles on the quartz filters and are not removed by the denuders. The volumetric flow rate of the HiC IOGAP sampler was set to $85\,L\,min^{-1}$ and temperature controlled at 4°C above ambient to prevent condensation of water. (Drawing used with permission from URG Corporation.)

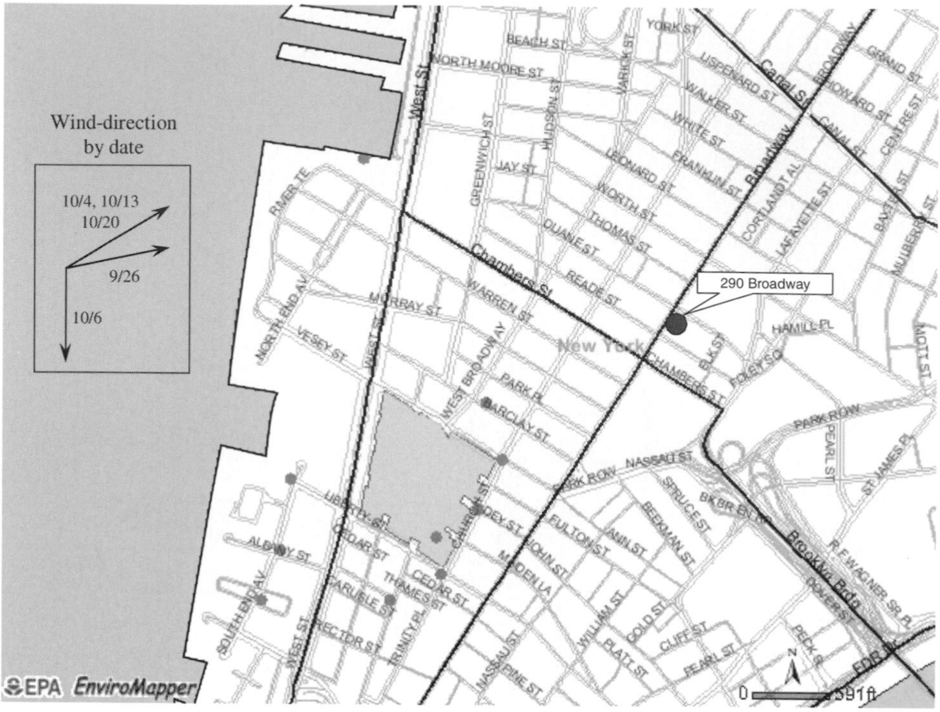

FIGURE 5.11. Sampling location and dominant wind direction and relative wind magnitude (length of vectors) for gas and particle phase pollutant measurements at lower Manhattan, following the collapse of the World Trade Center towers. (Source: E. Swartz, L. Stockburger, and D. Vallero, 2003, "Polyaromatic Hydrocarbons and Other Semi-Volatile Organic Compounds Collected in New York City in Response to the Events of 9/11," *Environmental Science and Technology*, 37(16), pp 3537–3546.)

Air samples were taken about 500 meters from the WTC at a height of about 50 meters (see Figure 5.11 for the map of the sampling site relative to the WTC). The reason for the location and height of sampling was that the sophisticated equipment that was needed to collect very low concentrations of semivolatile contaminants must run on alternating current (AC) electrical power (e.g., to accommodate the pumping of large amounts of the air necessary in low-concentration measurements). The sampling setup is shown in Figures 5.12 and 5.13. The volumetric flow rate of the HiC IOGAP sampler was set to $85\,L\,min^{-1}$ and temperature controlled at 4°C above ambient to prevent

FIGURE 5.12. Site of air monitoring equipment in lower Manhattan used to characterize the plume of semivolatile organic compounds leaving Ground Zero, facing the southeast. Erick Swartz, a U.S. EPA researcher is adjusting the equipment as part of the daily monitoring activities in October, 2001. The site is 16 floors above ground. The World Trade Center site is to the lower right of the photograph. (Source: Photo courtesy of Leonard Stockburger, U.S. EPA.)

condensation of water. The 24-hour integrated sampling typically began at 12:00 P.M. (local time).[23]

Polycyclic Aromatic Hydrocarbons

Table 5.3 provides the mean concentrations of the polycyclic aromatic hydrocarbons (PAHs) observed during the sampling periods. After the initial destruction of the WTC the remaining air plumes from the disaster site were, in part, comprised of pollutants typically associated with fossil fuel emissions and their combustion products. The likely

FIGURE 5.13. Storage and analytical hardware, pumps, and regulators associated with the high-volume air samplers used to measure semivolatile compounds in lower Manhattan, New York. (Source: Photo courtesy of Leonard Stockburger, U.S. EPA.)

sources at the WTC site are the large number of diesel generators and outsized vehicles used in the removal phases.

Figure 5.14 shows the concentration of a number of PAHs collected by the various samplers. This is a gross indication of phase distribution between aerosols and gases. The lighter compounds to the left of the figure are fairly well distributed between the vapor and particle phases. This is indicated by their being captured in the gas phase by denuders and sorbant materials (i.e., the polymer XAD-4), and in the particle phase by the quartz filters. The heavier compounds to the right of the figure are mainly captured on quartz filters, that is, in the particle phase.

Care must be taken, however, not to "overinterpret" such data. For example, since the sampling at 290 Broadway was approximately

TABLE 5.3
Total Concentrations ($ng\,m^{-3}$) of PAHs Measured at Lower Manhattan, New York

Dates	*09/26–09/27* *Total*	*10/04–10/05* *Total*	*10/06–10/07* *Total*	*10/12–10/13* *Total*	*10/20–10/21* *Total*
Naphthalene°	699 (*34*)	824 (*405*)	109	22	42
2-Methyl naphthalene°	267 (*6*)	323 (*37*)	54	37	165
1-Methyl naphthalene°	178 (*5*)	212 (*27*)	29	11	103
Biphenyl°	182 (*4*)	224 (*9*)	11	90	190
2-Ethyl naphthalene*	42 (*1*)	42 (*1*)	7	30	46
1-Ethyl naphthalene*	17 (*1*)	24 (*1*)	4	10	26
2,6-dimethyl naphthalene*	81 (*2*)	77 (*2*)	19	75	99
1,6-dimethyl naphthalene*	44 (*2*)	63 (*BD*)	BD	45	75
Acenaphthylene	14 (*BD*)	3 (*BD*)	BD	6	2
Acenaphthene	49 (*1*)	55 (*1*)	9	37	46
2,3,5-Trimethyl naphthalene*	29 (*1*)	36 (*1*)	6	48	37
Fluorene	36 (*1*)	52 (*1*)	8	57	47
1-Methyl-9H-fluorene*	3 (*BD*)	9 (*BD*)	3	5	4
Dibenzothiophene*	12 (*BD*)	10 (*BD*)	1	12	10
Phenanthrene	299 (*6*)	411 (*8*)	14	276	212
Anthracene	23 (*BD*)	13 (*BD*)	BD	14	8
Carbazole*	5 (*BD*)	5 (*BD*)	BD	2	1
Fluoranthene	111 (*7*)	179 (*6*)	4	61	52

TABLE 5.3 *(continued)*

Dates	*09/26–09/27* *Total*	*10/04–10/05* *Total*	*10/06–10/07* *Total*	*10/12–10/13* *Total*	*10/20–10/21* *Total*
Pyrene	59 (*5*)	93 (*4*)	2	32	27
Retene	9 (*1*)	10 (*1*)	BD	18	13
Benzo(a)anthracene	6 (*3*)	10 (*3*)	BD	2	2
Chrysene/triphenylene	15 (*10*)	30 (*11*)	BD	7	8
Benzo[b]fluoranthene	11 (*11*)	22 (*20*)	BD	4	6
Benzo[k[fluoranthene	4 (*4*)	8 (*7*)	BD	1	2
Benzo[a]pyrene	2 (*2*)	3 (*3*)	BD	1	1
Benzo[e]pyrene*	4 (*4*)	8 (*8*)	BD	1	2
Perylene*	BD (*BD*)	1 (*1*)	BD	BD	BD
Indeno[1,2,3-cd]pyrene	2 (*2*)	4 (*4*)	BD	BD	1
Dibenzo[a,h]anthracene	1 (*1*)	1 (*1*)	BD	BD	BD
Benzo[g,h,i]perylene	3 (*3*)	5 (*5*)	BD	1	1

* Estimated concentrations based on calibrations of similar compounds
° Possible breakthrough of the species through the system
BD = Below Detection
NR = Not Reported
Italicized data represents particle-only concentrations.

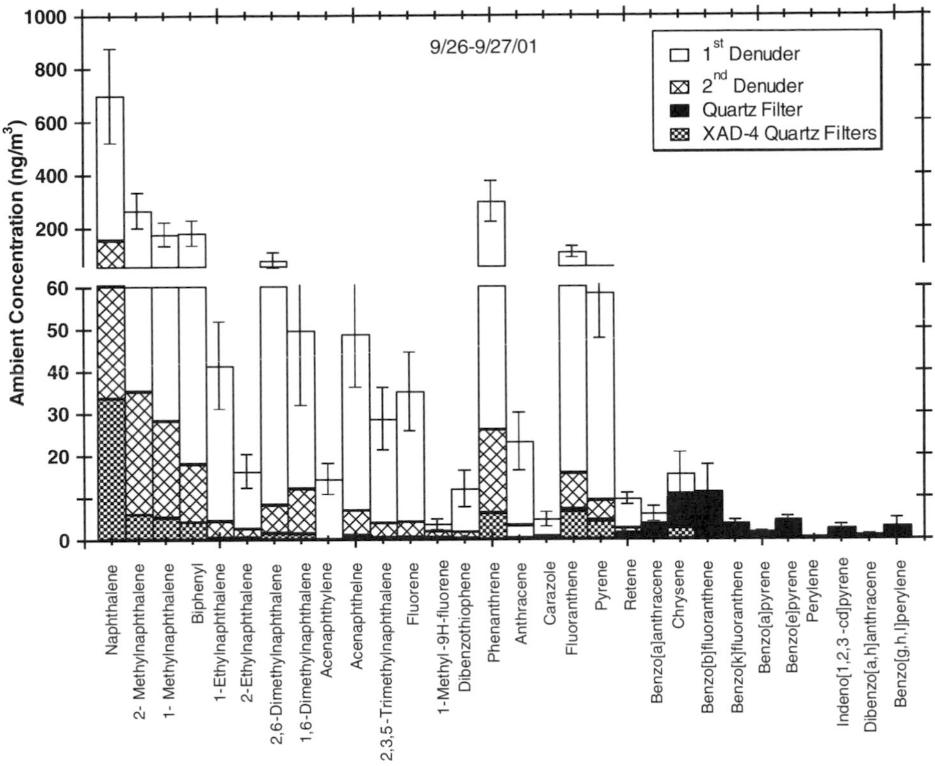

FIGURE 5.14. Average concentration of polycyclic aromatic hydrocarbons sampled at lower Manhattan, September 26 and 27, 2001. Note broken bar between 60 and 200 ng m^{-3} concentrations.

500 meters from Ground Zero and 50 meters above ground, only a general statement can be made about the concentrations directly at Ground Zero. From the mass data by filter measurements both at a ground site (north side) and at the 290 Broadway sampling site, the ground site was typically a factor of 4 to 8 higher, this includes data taken in the cleaner period (after October 7, 2001).[24] The difference in sampling height and dispersion may, at least in part, account for these observations.

Phase Partitioning Used to Identify Sources of Pollution

Particular ratios of PAHs have been used to identify possible sources of pollution.[25] Table 5.4 provides the ratios of methyl phenanthrenes/phenanthrene, MePh/Ph; fluoranthene/(fluoranthene + pyrene), Fl/(Fl+Py); benzo[a]anthracene/(benzo[a]anthracene + chrysene/triphenylene), BaA/(BaA+CT); benzo[e]pyrene/(benzo[e]pyrene + benzo[a]pyrene), BeP/(BeP+BaP), and indeno[1,2,3-cd]pyrene/(indeno [1,2,3-cd]pyrene + benzo[ghi]perylene, IP/(IP+BgP). The PAHs' ratios are similar to the data expected in an intense photochemical smog episode. In addition, the ratios from diesel and gasoline engines are provided in the table. The MePh/Ph ratio can be used to investigate the source. For example, a MePh/Ph > 1 for vehicular exhaust and <1 for a stationary combustion source. The values of Fl/(Pr+Fl) are similar to those reported in diesel engines. The values for BaA/(CT+BaA) are also similar to those found in oil combustion sources and heavy trucks. The IP/(BgP+IP) ratio observed are indicative of diesel-fueled equipment.

Shortly after the collapse of the WTC, the area was sealed off to all but rescue teams and security personnel, so few internal combustion sources were present. Later, during the removal phase, many diesel generators and outsized vehicles were permitted into the area, but general traffic was rerouted. As expected, the PAH ratios show significant contributions coming from diesel and other heavy combustion sources that found their way to the WTC plume during the

TABLE 5.4
Polycyclic Aromatic Hydrocarbon Ratios

	Ratios Found at WTC	*Heavy Diesel Engines Ratios*	*Catalytic Gasoline Engines Ratios*
MePh/Ph	<0.2	2.57	2.44
		NR	NR
Fl/(Pr + Fl)	0.66	0.37	0.44
		0.60–0.70	0.40
BaA/(CT + BaA)	0.24	0.27	0.33
		0.38–0.70	0.43
IP/(IP + BgP)	0.45	NR	0.09
		0.35–0.70	0.18
BeP/(BaP + BeP)	0.64	0.67	0.51
		0.29–0.40	0.6–0.8

Abbreviations and references discussed in text.

removal phase. Some research[26] has shown that BeP/(BaP+BeP) ratio components are emitted at almost equal concentrations, or ratio = 0.5, although others[27] found the emission ratio to be higher for heavy diesel engines (0.67). Since benzo(a)pyrene reacts in the atmosphere at a much faster rate than benzo(e)pyrene, this ratio can also be used to determine the age of the particle.

Other Semivolatile Organic Compounds

Table 5.5 gives the total concentrations of the various semivolatile organic compounds beyond PAHs measured in lower Manhattan. A noteworthy compound, 1,3-diphenyl propane [1′,1′-(1,3-propanediyl)bis-benzene], was observed in the ambient sampling. Table 5.5 shows that the species is primarily found in the gas phase (90% of the mass found on the front denuder). Although the source of the compound in this study is not known, it may have formed during the combustion of polystyrene or other polymers. 1,3-diphenyl propane has been found to co-occur with polystyrene plastics,[28] so another possibility is that the compound was already present and

TABLE 5.5
Total Concentrations (ng m^{-3}) of Other Semivolatile Organic Compounds Measured at 290 Broadway (16th floor) Lower Manhattan, New York

Dates	*09/26–09/27 Total*	*10/04–10/05 Total*	*10/06–10/07 Total*	*10/12–10/13 Total*	*10/20–10/21 Total*
Diphenyl ether*	7 (*BD*)	BD (*BD*)	BD	9	9
Dibenzofuran	99 (*1*)	138 (*3*)	9	106	97
Bibenzyl*	23 (*1*)	34 (*1*)	BD	42	43
1,3-Diphenyl propane*	190 (*5*)	594 (*5*)	5	693	541
Pristane	46 (*4*)	38 (*4*)	16	68	48
Phytane	39 (*4*)	31 (*3*)	13	55	40
1,4a-dimethyl-7-(methylethyl)-1,2,3,4,9,10,10a,4a-octahydrophenanthrene*	3 (*BD*)	3 (*BD*)	BD	8	7
1,4a-dimethyl-7-(methylethyl)-1,2,3,4,9,10,10a,4a-octahydrophenanthrene*	4 (*BD*)	4 (*BD*)	BD	8	7

* Estimated concentrations based on calibrations of similar compounds.
° Possible breakthrough of the species through the system
BD = Below Detection
NR = Not Reported
(Italicized data represents particle-only concentrations.)

encapsulated in large volumes of plastics in the buildings and was off-gassed during the pulverization process.

Along with the presence of 1,3-diphenyl propane, there is further evidence that the plume contained emissions of burning and remnant materials from the WTC site. The molecular markers for these emissions include retene and 1,4a-dimethyl-7-(methylethyl)-1,2,3,4,9,10,10a,4a-octahydrophenanthrene, which typically are associated with combustion of biogenic materials.[29] For example, retene is a known marker for wood smoke and was seen in all samples analyzed from the WTC.

Organic Chains

Figure 5.15 provides the gas-particle distributions of the n-alkanes on October 4, 2001. As expected from the decreasing volatility associated with increased molecular mass (i.e., number of carbons), the longer chains have a higher affinity for the particle phase than for the vapor phase.

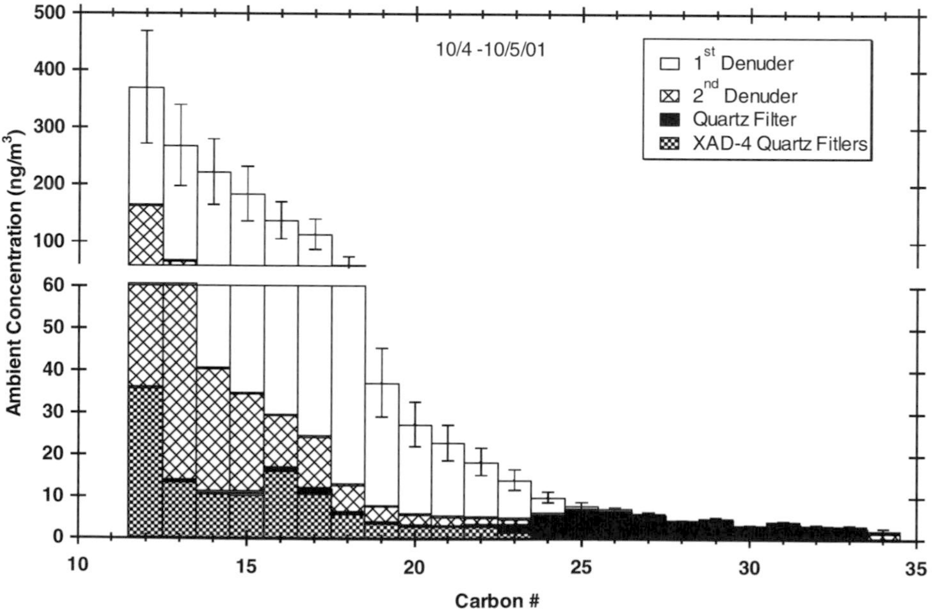

FIGURE 5.15. Average concentration of n-alkanes from sampling near the World Trade Center during 10/4–10/5/01 sampling. Note broken bar between 60 and 200 $ng\,m^{-3}$ concentrations.

Gas-Particle Partitioning

Analogous to solid-water partitioning coefficient (K_d), the gas-particle partitioning coefficient describes the distribution of semivolatile compounds between the gas and particle phase. The measured gas-particle partitioning coefficient $K_{gas\text{-}particle}$ is defined as:

$$K_{gas\text{-}particle} = \frac{C_F/TSP}{C_g} \qquad \text{Equation 5–11}$$

Where C_F is the mass of the contaminant found on the filters (e.g., all filters), C_g is the mass of the contaminant found in the gas phase (e.g., denuders), and TSP is the mass of the total suspended particulate matter collected. The units of $K_{gas\text{-}particle}$ are inverse concentrations (i.e., volume per mass), commonly $m^3 \mu g^{-1}$. Gas-particle partitioning has been found to be a function of the super-cooled liquid vapor pressure ($p_L°$), and often has a linear relationship during a sampling event and type of compound (see Figure 5.16). This relationship is expressed as:

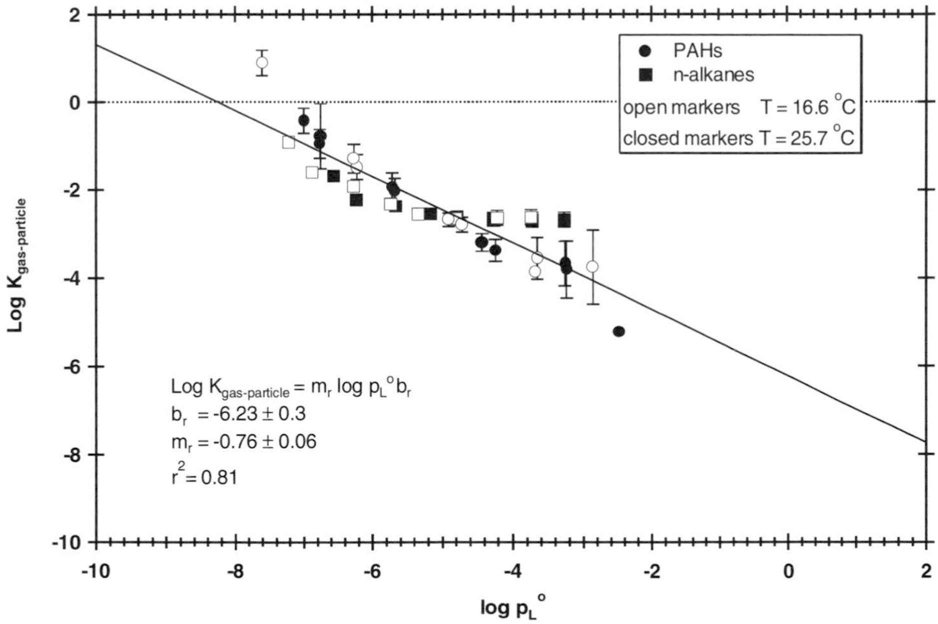

FIGURE 5.16. The gas-particle partitioning ($K_{gas\text{-}particle}$) coefficient as a function of super-cooled liquid vapor pressure ($p_L°$), based on World Trade Center measurements.

$$\log K_{gas-particle} = m_r \log p_L{}^{o} + b_r$$

Experimentally the slope m_r has been determined to be close to −1, where the intercept b_r is typically associated with the chemical compound class. There does not seem to be any observable temperature dependence on $K_{gas\text{-}particle}$, within the temperature ranges 16.6 to 25.6°C of the dataset.

Partitioning to Organic Tissue

We will consider biological processes and effects in some detail in Chapters 8 and 12, but it is important to consider briefly how contaminants move from other compartments into biota. Relatively hydrophobic substances frequently have a strong affinity for fatty tissues (i.e., those containing high K_{ow} compounds). Therefore, such contaminants can be sequestered and can accumulate in organisms. In other words, certain chemicals are very *bioavailable* to organisms that may readily take them up from the other compartments. Bioavailability is an expression of the fraction of the total mass of a compound present in a compartment that has the potential of being absorbed by the organism. *Bioaccumulation* is the process of uptake into an organism from the abiotic compartments. *Bioconcentration* is the concentration of the pollutant within an organism above levels found in the compartment in which the organism lives. For a fish to bioaccumulate DDT, the levels found in the total fish or in certain organs (e.g., the liver) will be elevated above the levels measured in the ambient environment. In fact, DDT is known to bioconcentrate many orders of magnitude in fish. A surface water DDT concentration of 100 parts per trillion in water has been associated with 10 ppm in certain fish species (a concentration of 10,000 times!). Thus the straightforward equation for the *bioconcentration factor* (BCF) is the quotient of the concentration of the contaminant in the organism and the concentration of the contaminant in the host compartment. For a fish living in water, for example, the BCF is:

$$BCF = \frac{C_{organism}}{C_w} \qquad \text{Equation 5–12}$$

The BCF is applied to an individual organism that represents a genus or some other taxonomical group. However, considering the whole food chain and trophic transfer processes, in which a compound builds up as a result of predator/prey relationships, the term *biomagnification* is used. Some compounds that may not appreciably bioconcentrate within lower trophic state organisms may still become highly concentrated. For example, even if plankton have a small BCF (e.g., 10), if subsequently higher-order organisms sequester the contaminant at a higher rate, by the time the contaminant is taken up by top predators (e.g., alligators, sharks, panthers, and humans), they may suffer from the continuum of biomagnification, with contaminant levels in the top predator levels many orders of magnitude higher than what is found in the abiotic compartments. Thus, a distinction between a BCF and a bioaccumulation factor (BAF) is that the BCF only compares the organism's contaminant concentration to the concentration of the contaminant in water, whereas the BAF also considers the organism's food.

For a substance to bioaccumulate, bioconcentrate, and biomagnify, it must be at least somewhat persistent. If an organism's metabolic and detoxification processes are able to degrade the compound readily, it will not be present (at least in high concentrations) in the organism's tissues. However, if an organism's endogenous processes degrade a compound into a chemical species that is itself persistent, the metabolite or degradation product will bioaccumulate, and may bioconcentrate, and biomagnify. Finally, cleansing or *depuration* will occur if the organism that has accumulated a contaminant enters an abiotic environment that no longer contains the contaminant. However, some tissues have such strong affinities for certain contaminants that the persistence within the organism will remain long after the source of the contaminant is removed. For example, the piscivorous birds, such as the Common Loon (*Gavia immer*), decrease the concentrations of the metal mercury in their bodies by translocating the metal to feathers and eggs; every time the birds molt or lay eggs they undergo mercury depuration. Unfortunately, when the birds continue to ingest mercury that has bioaccumulated in their prey (fish), they often have a net increase in tissue Hg concentrations because the bioaccumulation rate exceeds the depuration rate.[30]

Bioconcentration can vary considerably in the environment. The degree to which a contaminant builds up in an ecosystem, especially in biota and sediments, is related to the compound's persistence. For example, a highly persistent compound, if nothing else, lasts longer in the environment, so there is a greater opportunity for uptake, all other factors being equal. In addition, persistent compounds often possess chemical structures that are also conducive to sequestration by fauna. Such compounds are generally quite often lipophilic, have high K_{ow} values, and usually low vapor pressures. This means that they may bind to the organic molecules in living tissues and may resist elimination and metabolic process, so that they build up over time. However, the bioaccumulation and bioconcentration can vary considerably, both among biota and within the same species of biota. For example, the pesticide mirex has been shown to exhibit a bioconcentration factor of 2600 to as high as 51,400 in pink shrimp and fathead minnows, respectively. The pesticide endrin has shown an even larger interspecies variablility in BCF values, with factors ranging from 14 to 18,000 recorded in fish after continuous exposure. Intraspecies BCF ranges may also be high. For example, oysters exposed to very low concentrations of the organometallic compound, tributyl tin, exhibit BCF values ranging from 1000 to 6000.[31]

Even the same compound in a single medium, such as a lake's water column or sediment, will show large BCF variability among species of fauna in that compartment. An example is the so-called "dirty dozen" compounds. This is a group of *persistent organic pollutants* (POPs) that have been largely banned, some for decades, but that are still found in environmental samples throughout the world. As might be expected from their partitioning coefficients, they have concentrated in sediment and biota.

The worst combination of factors is when a compound is persistent in the environment, builds up in organic tissues, and is toxic. Such compounds are referred to as *persistent bioaccumulating toxic* substances (PBTs). Recently, the United Nations Environmental Programme (UNEP) reported on the concentrations of the persistent and toxic compounds. Each region of the world was evaluated for the presence of these compounds. For example, the North American report[32] includes scientific assessments of the nature and scale of environmental threats posed by persistent toxic compounds. The results of these assessments are summarized in Table 5.6 (organic compounds) and Table 5.7 (organometallic compounds). In the United States, mining and mineral extraction activities contribute the large quantity of PBTs, with energy production the second largest source categories (see Figure 5.17). Organometallic compounds, especially lead and its compounds, comprise the lion's share of PBTs in the United States; the second largest quantity is represented by another metal, mercury, and its compounds.

The sources of PBTs are widely varied. Many are intentionally manufactured to serve some public need, such as the control of pests that destroy food and spread disease. Other PBTs are generated as unintended by-products, such as the products of incomplete combustion. In either case, there are often measures and engineering controls available that can prevent PBT releases, rather than having to deal with them after they have found their way into the various environmental compartments.

Concentration-Based and Fugacity-Based Transport Models

Let us now combine these phase and compartmental distributions into a simple fugacity-based, chemodynamic transport model. Such models are classified into three types.

Level 1 Model

This model is based on an equilibrium distribution of fixed quantities of contaminants in a closed environment (i.e., conservation of contaminant mass). The model assumes that there is no chemical or biological *degrada-*

TABLE 5.6
Summary of Persistent and Toxic Organic Compounds in North America, Identified by the United Nations as Highest Priorities for Regional Actions

Compound	*Properties*	*Persistence/Fate*	*Toxicity*[32]*
Aldrin 1,2,3,4,10,10-Hexachloro-1,4,4a,5,8,8a-hexahydro-1,4-endo,exo-5,8-dimethanonaphthalene ($C_{12}H_8Cl_6$).	Solubility in water: 27 μg L^{-1} at 25°C; Vapor pressure: 2.31 × 10^{-5} mm Hg at 20°C; log K_{ow}: 5.17–7.4.	Readily metabolized to dieldrin by both plants and animals. Biodegradation is expected to be slow, and it binds strongly to soil particles and is resistant to leaching into groundwater. Classified as moderately persistent with $T_{1/2}$ in soil ranging from 20 to 100 days.	Toxic to humans. Lethal dose for an adult estimated to be about 80 mg kg^{-1} body weight. Acute oral LD_{50} in laboratory animals is in the range of 33 mg kg^{-1} body weight for guinea pigs to 320 mg kg^{-1} body weight for hamsters. The toxicity of aldrin to aquatic organisms is quite variable, with aquatic insects being the most sensitive group of invertebrates. The 96-h LC_{50} values range from 1–200 μg L^{-1} for insects, and from 2.2–53 μg L^{-1} for fish. The maximum residue limits in food recommended by the World Health Organization (WHO) varies from 0.006 mg kg^{-1} milk fat to 0.2 mg kg^{-1} meat fat. Water quality criteria between 0.1 to 180 μg L^{-1} have been published.
Dieldrin 1,2,3,4,10,10-Hexachloro-6,7-epoxy-1,4,4a,5,6,7,8,8a-octahydroexo-1,4-endo-5,8-dimethanonaphthalene ($C_{12}H_8Cl_6O$).	Solubility in water: 140 μg L^{-1} at 20°C; vapor pressure: 1.78 × 10^{-7} mm Hg at 20°C; log K_{ow}: 3.69–6.2.	Highly persistent in soils, with a $T_{1/2}$ of 3–4 years in temperate climates, and bioconcentrates in organisms.	Acute toxicity for fish is high (LC_{50} between 1.1 and 41 mg/L) and moderate for mammals (LD_{50} in mouse and rat ranging from 40 to 70 mg kg^{-1} body weight). Aldrin and dieldrin mainly affect the central nervous system, but there is no direct evidence that they cause cancer in humans. The maximum residue limits in food recommended by WHO varies from 0.006 mg kg^{-1} milk fat and 0.2 mg kg^{-1} poultry fat. Water quality criteria between 0.1 to 18 μg L^{-1} have been published.

TABLE 5.6 *(continued)*

Compound	*Properties*	*Persistence/Fate*	*Toxicity*[32]*
Endrin 3,4,5,6,9,9-Hexachloro-1a,2,2a,3,6,6a,7,7a-octahydro-2,7:3,6-dimethanonaphth[2,3-b]oxirene ($C_{12}H_8Cl_6O$).	Solubility in water: 220–260 $\mu g\,L^{-1}$ at 25 °C; vapor pressure: 7 × 10-7 mm Hg at 25°C; log K_{ow}: 3.21–5.34	Highly persistent in soils ($T_{1/2}$ of up to 12 years have been reported in some cases). Bioconcentration factors of 14 to 18,000 have been recorded in fish, after continuous exposure.	Very toxic to fish, aquatic invertebrates, and phytoplankton; the LC_{50} values are mostly less than 1 $\mu g\,L^{-1}$. The acute toxicity is high in laboratory animals, with LD_{50} values of 3–43 $mg\,kg^{-1}$, and a dermal LD_{50} of 5–20 $mg\,kg^{-1}$ in rats. Long-term toxicity in the rat has been studied over 2 years. and a NOEL of 0.05 $mg\,kg^{-1}$ bw/day was found.
Chlordane 1,2,4,5,6,7,8,8-Octachloro-2,3,3a,4,7,7a-hexahydro-4,7-methanoindene ($C_{10}H_6Cl_8$).	Solubility in water: 180 $\mu g\,L^{-1}$ at 25°C; vapor pressure: 0.3 × 10^{-5} mm Hg at 20°C; log K_{ow}: 4.4–5.5.	Metabolized in soils, plants, and animals to heptachlor epoxide, which is more stable in biological systems and is carcinogenic. The $T_{1/2}$ of heptachlor in soil is in temperate regions 0.75 to 2 years. Its high partition coefficient provides the necessary conditions for bioconcentrating in organisms.	Acute toxicity to mammals is moderate (LD_{50} values between 40 and 119 $mg\,kg^{-1}$ have been published). The toxicity to aquatic organisms is higher, and LC_{50} values down to 0.11 $\mu g\,L^{-1}$ have been found for pink shrimp. Limited information is available on the effects in humans, and studies are inconclusive regarding heptachlor and cancer. The maximum residue levels recommended by the WHO are between 0.006 $mg\,kg^{-1}$ milk fat and 0.2 $mg\,kg^{-1}$ meat or poultry fat.
Dichlorodiphenyltri-chloroethane (DDT) 1,1,1-Trichloro-2,2-bis-(4-chlorophenyl)-ethane ($C_{14}H_9Cl_5$).	Solubility in water: 1.2–5.5 $\mu g\,L^{-1}$ at 25°C; vapor pressure: 0.02 × 10^{-5} mm Hg at	Highly persistent in soils with a $T_{1/2}$ of about 1.1 to 3.4 years. It also exhibits high bioconcentration factors (in the order of 50,000 for fish and	Lowest dietary concentration of DDT reported to cause egg shell thinning was 0.6 $mg\,kg^{-1}$ for the black duck. LC_{50} of 1.5 mg/L for largemouth bass and 56 $mg\,L^{-1}$ for guppy have been reported. The acute toxicity of DDT for mammals is moderate

	20°C; log K_{ow}: 6.19 for pp-DDT, 5.5 for pp-DDD and 5.7 for pp-DDE.	500,000 for bivalves). In the environment, the parent DDT is metabolized mainly to DDD and DDE.	with an LD_{50} in the rat of 113–118 mg kg^{-1} body weight. DDT has been shown to have an estrogen-like activity and possible carcinogenic activity in humans. The maximum residue level in food recommended by the WHO, ranges from 0.02 mg kg^{-1} milk fat to 5 mg kg^{-1} meat fat. Maximum permissible DDT residue levels in drinking water (WHO) is 1.0 μg L^{-1}.
Toxaphene Polychlorinated bornanes and camphenes ($C_{10}H_{10}Cl_8$).	Solubility in water: 550 μg L^{-1} at 20°C; vapor pressure: 0.2–0.4 mm Hg at 25°C; log K_{ow}: 3.23–5.50.	Half-life in soil from 100 days up to 12 years. It has been shown to bioconcentrate in aquatic organisms (BCF of 4247 in mosquito fish, and 76,000 in brook trout).	Highly toxic in fish, with 96-hour LC_{50} values in the range of 1.8 μg L^{-1} in rainbow trout to 22 μg L^{-1} in bluegill. Long-term exposure to 0.5 μg L^{-1} reduced egg viability to zero. The acute oral toxicity is in the range of 49 mg kg^{-1} body weight in dogs to 365 mg kg^{-1} in guinea pigs. In long-term studies NOEL in rats was 0.35 mg kg^{-1} bw/day, LD_{50} ranging from 60 to 293 mg kg^{-1} bw. For toxaphene, there exists strong evidence of the potential for endocrine disruption. Toxaphene is carcinogenic in mice and rats and is of carcinogenic risk to humans, with a cancer potency factor of 1.1 mg kg^{-1}/day for oral exposure.
Mirex 1,1a,2,2,3,3a,4,5,5a,5b,6-Dodecachloroacta-hydro-1,3,4-metheno-1H-cyclobuta[cd]pentalene ($C_{10}Cl_{12}$).	Solubility in water: 0.07 μg L^{-1} at 25°C; vapor pressure: 3×10^{-7} mm Hg at 25°C; log K_{ow}: 5.28.	Among the most stable and persistent pesticides, with a $T_{1/2}$ in soils of up to 10 years. Bioconcentration factors of 2600 and 51,400 have been observed in pink	Acute toxicity for mammals is moderate, with an LD_{50} in rat of 235 mg kg^{-1} and dermal toxicity in rabbits of 80 mg kg^{-1}. Mirex is also toxic to fish and can affect their behavior (LC_{50} [96 hr] from 0.2 to 30 mg L^{-1} for rainbow trout and bluegill, respectively). Delayed mortality of

TABLE 5.6 *(continued)*

Compound	*Properties*	*Persistence/Fate*	*Toxicity*[32]*
		shrimp and fathead minnows, respectively. Capable of undergoing long-range transport due to its relative volatility i.e. vapor pressure lowering (VPL[1]) = 4.76 Pa; K_H = 52 Pa m^3 mol^{-1}.	crustaceans occurred at 1 μg L^{-1} exposure levels. There is evidence of its potential for endocrine disruption and possibly carcinogenic risk to humans.
Hexachlorobenzene (HCB) (C_6H_6).	Solubility in water: 50 μg L^{-1} at 20°C; vapor pressure: 1.09 × 10^{-5} mm Hg at 20°C; log K_{ow}: 3.93– 6.42.	Estimated "field half-life" of 2.7–5.7 years. HCB has a relatively high bioaccumulation potential and long $T_{1/2}$ in biota.	LC_{50} for fish varies between 50 and 200 μg L^{-1}. The acute toxicity of HCB is low with LD50 values of 3.5 mg/g for rats. Mild effects on the [rat] liver have been observed at a daily dose of 0.25 mg HCB/kg bw. HCB is known to cause liver disease in humans (porphyria cutanea tarda) and has been classified as a possible carcinogen to humans by IARC.
Polychlorinated biphenyls (PCBs) ($C_{12}H_{(10-n)}Cl_n$, where n is within the range of 1–10).	Water solubility decreases with increasing chlorination: 0.01 to 0.0001 μg L^{-1} at 25°C; vapor pressure: 1.6–0.003 × 10^{-6} mm Hg at 20°C; log K_{ow}: 4.3–8.26.	Most PCB congeners, particularly those lacking adjacent unsubstituted positions on the biphenyl rings (e.g., 2,4,5-, 2,3,5- or 2,3,6-substituted on both rings) are extremely persistent in the environment. They are estimated to have $T_{1/2}$ ranging from 3 weeks to 2 years in air and,	LC_{50} for the larval stages of rainbow trout is 0.32 μg L^{-1} with a NOEL of 0.01 μg L^{-1}. The acute toxicity of PCB in mammals is generally low and LD_{50} values in rat of 1 g/kg bw. IARC has concluded that PCBs are carcinogenic to laboratory animals and probably also for humans. They have also been classified as substances for which there is evidence of endocrine disruption in an intact organism.

		with the exception of mono- and dichlorodiphenyl, more than 6 years in aerobic soils and sediments. PCBs also have extremely long $T_{1/2}$ in adult fish; for example, an 8-year study of eels found that the $T_{1/2}$ of CB153 was more than 10 years.	
Polychlorinated dibenzo-p-dioxins (PCDDs) and **Polychlorinated dibenzofurans (PCDFs)** ($C_{12}H_{(8-n)}Cl_nO_2$) and PCDFs ($C_{12}H_{(8-n)}Cl_nO$) may contain between 1 and 8 chlorine atoms. Dioxins and furans have 75 and 135 possible positional isomers, respectively.	Solubility in water: in the range 550–0.07 ng L^{-1} at 25°C; vapor pressure: 2–0.007 × 10^{-6} mm Hg at 20°C; log K_{ow}: in the range 6.60–8.20 for tetra- to octa-substituted congeners.	PCDD/Fs are characterized by their lipophilicity, semivolatility and resistance to degradation ($T_{1/2}$ of TCDD in soil of 10–12 years). Long-range transport is generally by sorption to aerosols. They are also known for their ability to bioconcentrate and biomagnify under typical environmental conditions.	The toxicological effects reported refers to the 2,3,7,8-substituted compounds (17 congeners) that are agonist for the aryl hydrocarbon receptor (AhR). All the 2,3,7,8-substituted PCDDs and PCDFs plus dioxin-like PCBs (DLPCBs) (with no chlorine substitution at the ortho positions) show the same type of biological and toxic response. Possible effects include dermal toxicity, immunotoxicity, reproductive effects and teratogenicity, endocrine disruption, and carcinogenicity. At the present time, the only consistent effect associated with dioxin exposure in humans is chloracne. The most sensitive groups are gestating and neonatal infants. Effects on the immune systems in the mouse have been found at doses of 10 ng/kg bw/day, while reproductive effects were seen in rhesus monkeys at 1–2 ng/kg bw/day. Biochemical effects have been seen in rats down to 0.1 ng/kg bw/day. In a

TABLE 5.6 *(continued)*

Compound	*Properties*	*Persistence/Fate*	*Toxicity*[32]*
			reevaluation of the exposures to dioxins, furans (and planar PCB), the WHO decided to recommend a range of 1–4 TEQ $pg\,kg^{-1}$ body weight, although more recently the acceptable intake value has been set monthly at 1–70 toxic equivalency quotient (TEQ) $pg\,kg^{-1}$ body weight
Atrazine 2-Chloro-4-(ethlamino)-6-(isopropylamino)-s-triazine ($C_{10}H_6Cl_8$).	Solubility in water: 28 mg/L at 20°C; vapor pressure: 3.0×10^{-7} mm Hg at 20°C; log K_{ow}: 2.34.	Does not adsorb strongly to soil particles and has a lengthy $T_{1/2}$ (60 to >100 days). Atrazine has a high potential for groundwater contamination despite its moderate solubility in water.	Oral LD_{50} is 3090 $mg\,kg^{-1}$ in rats, 1750 $mg\,kg^{-1}$ in mice, 750 $mg\,kg^{-1}$ in rabbits, and 1000 $mg\,kg^{-1}$ in hamsters. The dermal LD_{50} in rabbits is 7500 $mg\,kg^{-1}$ and greater than 3000 $mg\,kg^{-1}$ in rats. Atrazine is practically nontoxic to birds. The LD_{50} is greater than 2000 $mg\,kg^{-1}$ in mallard ducks. Atrazine is slightly toxic to fish and other aquatic life. Atrazine has a low level of bioaccumulation in fish. Available data regarding atrazine's carcinogenic potential are inconclusive.[1]
Hexachlorocyclohexane (HCH) 1,2,3,4,5,6-hexachlorocyclohexane (mixed isomers) ($C_6H_6Cl_6$).	γ-HCH (lindane): solubility in water: 7 $mg\,L^{-1}$ at 20°C; vapor pressure: 3.3×10^{-5} mm Hg at 20°C; log K_{ow}: 3.8.	Lindane and other HCH isomers are relatively persistent in soils and water, with half-lives generally greater than 1 and 2 years, respectively. HCHs are much less bioaccumulative than other organochlorines	Lindane is moderately toxic for invertebrates and fish, with LC_{50} values of 20–90 $\mu g\,L^{-1}$. The acute toxicity for mice and rats is moderate, with LD_{50} values in the range of 60–250 $mg\,kg^{-1}$. Lindane reported to have no mutagenic potential in a number of studies, but did have an endocrine-disrupting activity.

		of concern, because of their relatively low lipophilicity. On the contrary, their relatively high vapor pressures, particularly of the α-HCH isomer, determine their long-range transport in the atmosphere.	
Chlorinated Paraffins (CPs) Polychlorinated alkanes ($C_xH_{(2x-y+2)}Cl_y$). Manufactured by chlorination of liquid *n*-alkanes or paraffin wax and contain from 30% to 70% chlorine. The products are often divided in three groups depending on chain length: short (C_{10}–C_{13}), medium, (C_{14}–C_{17}) and long (C_{18}–C_{30}) chain lengths.	Properties largely dependent upon the chlorine content. Solubility in water: 1.7 to 236 $\mu g L^{-1}$ at 25°C; $\log K_{ow}$: in the range from 5.06 to 8.12.	May be released into the environment from improperly disposed metal-working fluids or polymers containing chlorinated paraffins. Loss of chlorinated paraffins by leaching from paints and coatings may also contribute to environmental contamination. Short-chain CPs with less than 50% chlorine content seem to be degraded under aerobic conditions. The medium- and long-chain products are degraded more slowly. CPs are bioaccumulated and both uptake and elimination are faster for	Acute toxicity of CPs in mammals is low; reported oral LD_{50} values ranging from 4–50 g/kg bw, although in repeated dose experiments, effects on the liver have been seen at doses of 10–100 mg kg^{-1} bw/day. Short-chain and mid-chain grades have been shown, in laboratory tests, to have toxic effects on fish and other forms of aquatic life after long-term exposure. The NOEL appears to be in the range of 2–5 $\mu g L^{-1}$ for the most sensitive aquatic species tested.

TABLE 5.6 *(continued)*

Compound	*Properties*	*Persistence/Fate*	*Toxicity*[32]*
		the substances with low chlorine content.	
Chlordecone or **Kepone** 1,2,3,4,5,5,6,7,9,10,10-dodecachlorooctahydro-1,3,4-metheno-2H-cyclobuta(cd)pentalen-2-one ($C_{10}Cl_{10}O$).	Solubility in water: 7.6 mg L^{-1} at 25°C; vapor pressure: less than 3 × 10^{-5} mm Hg at 25°C; log K_{ow}: 4.50	Estimated $T_{1/2}$ in soils is between 1 and 2 years, whereas in air is much higher, up to 50 years. Not expected to hydrolyze, biodegrade in the environment. Also, direct photodegradation and vaporization from water and soil is not significant. General population exposure to chlordecone mainly through the consumption of contaminated fish and seafood.	Workers exposed to high levels of chlordecone over a long period (more than one year) have displayed harmful effects on the nervous system, skin, liver, and male reproductive system (likely through dermal exposure to chlordecone, although they may have inhaled or ingested some as well). Animal studies with chlordecone have shown effects similar to those seen in people, as well as harmful kidney effects, developmental effects, and effects on the ability of females to reproduce. There are no studies available on whether chlordecone is carcinogenic in people. However, studies in mice and rats have shown that ingesting chlordecone can cause liver, adrenal gland, and kidney tumors. Very highly toxic for some species such as Atlantic menhaden, sheepshead minnow, or Donaldson trout with LC_{50} between 21.4 and 56.9 mg·L^{-1}.
Endosulfan 6,7,8,9,10,10-hexachloro-1,5,5a,6,9,9a-hexahydro-6,9-methano-2,4,3-benzodioxathiepin-3-oxide ($C_9H_6Cl_6O_3S$).	Solubility in water: 320 μg L^{-1} at 25°C; vapor pressure: 0.17 × 10^{-4} mm Hg at 25°C; log K_{ow}: 2.23–3.62.	Moderately persistent in soil, with a reported average field $T_{1/2}$ of 50 days. The two isomers have different degradation times in soil ($T_{1/2}$ of 35 and 150 days for α- and	Highly to moderately toxic to bird species (Mallards: oral LD_{50} 31–243 mg kg^{-1}), and it is very toxic to aquatic organisms (96-hour LC_{50} rainbow trout 1.5 μg L^{-1}). It has also shown high toxicity in rats (oral LD_{50}: 18–160 mg kg^{-1}, and dermal: 78–359 mg kg^{-1}). Female rats appear to be 4–5 times

		β-isomers, respectively, in neutral conditions). It has a moderate capacity to adsorb to soils and it is not likely to leach to groundwater. In plants, endosulfan is rapidly broken down to the corresponding sulfate; on most fruits and vegetables, 50% of the parent residue is lost within 3 to 7 days.	more sensitive to the lethal effects of technical-grade endosulfan than male rats. The α-isomer is considered to be more toxic than the β-isomer. There is a strong evidence of its potential for endocrine disruption.
Pentachlorophenol (PCP) (C_6Cl_5OH).	Solubility in water: 14 mg L^{-1} at 20°C; vapor pressure: 16 × 10^{-5} mm Hg at 20°C; log K_{ow}: 3.32–5.86.	Photodecomposition rate increases with pH ($T_{1/2}$ 100 hr at pH 3.3 and 3.5 hr at pH 7.3). Complete decomposition in soil suspensions takes >72 days; other authors report $T_{1/2}$ in soils of about 45 days. Although enriched through the food chain, it is rapidly eliminated after discontinuing the exposure ($T_{1/2}$ 10–24 h for fish).	Acutely toxic to aquatic organisms. Certain effects on human health. 24-h LC_{50} values for trout were reported as 0.2 mg L^{-1}, and chronic toxicity effects were observed at concentrations down to 3.2 μg L^{-1}. Mammalian acute toxicity of PCP is moderate-high. LD_{50} oral in rat ranging from 50 to 210 mg kg^{-1} bw have been reported. LC_{50} ranged from 0.093 mg/L in rainbow trout (48 h) to 0.77–0.97 mg/L for guppy (96 h) and 0.47 mg/L for fathead minnow (48 h).

TABLE 5.6 *(continued)*

Compound	*Properties*	*Persistence/Fate*	*Toxicity*[32]*
Hexabromobiphenyl (HxBB) ($C_{12}H_4Br_6$). A congener of the class polybrominated biphenyls (PBBs).	Solubility in water: 11 $\mu g L^{-1}$ at 25°C; vapor pressure: mm Hg at 20°C; log K_{ow}: 6.39.	Strongly adsorbed to soil and sediments and usually persists in the environment. Resists chemical and biological degradation. Found in sediment samples from the estuaries of large rivers and has been identified in edible fish.	Few toxicity data are available from short-term tests on aquatic organisms. The LD_{50} values of commercial mixtures show a relatively low order of acute toxicity (LD_{50} range from >1 to 21.5 $g kg^{-1}$ body weight in laboratory rodents). Oral exposure of laboratory animals to PBBs produced body weight loss, skin disorders, nervous system effects, and birth defects. Humans exposed through contaminated food developed skin disorders, such as acne and hair loss. PBBs exhibit endocrine-disrupting activity and possible carcinogenicity to humans.
Polybrominated diphenyl ethers (PBDEs)[2] ($C_{12}H_{(10-n)}Br_nO$, where n = 1–10). As in the case of PCBs the total number of congeners is 209, with a predominance in commercial mixtures of the tetra-, penta- and octa-substituted isomers.	Vapor pressure: 3.85 up to 13.3 × 10^{-3} mm Hg at 20–25°C; log K_{ow}: 4.28–9.9.	Biodegradation does not seem to be an important degradation pathway, but that photodegradation may play a significant role. Have been found in high concentrations in marine birds and mammals from remote areas. The half-lives of PBDE components in rat adipose tissue vary between 19 and 119 days, the higher values being for the more highly brominated congeners.	Lower (tetra- to hexa-) PBDE congeners likely to be carcinogens, endocrine disruptors, and/or neurodevelopmental toxicants. Studies in rats with commercial penta BDE indicate a low acute toxicity via oral and dermal routes of exposure, with LD_{50} values >2000 $mg kg^{-1}$ bw. In a 30-day study with rats, effects on the liver could be seen at a dose of 2 $mg kg^{-1}$ bw/day, with a NOEL at 1 $mg kg^{-1}$ bw/day. The toxicity to Daphnia magna has also been investigated, and LC_{50} was found to be 14 $\mu g L^{-1}$ with a NOEC of 4.9 $\mu g L^{-1}$. Although data on toxicology is limited, they have potential endocrine-disrupting properties, and there are concerns over the health effects of exposure.

Polycyclic Aromatic Hydrocarbons (PAHs) A group of compounds consisting of two or more fused aromatic rings.	Solubility in water: 0.00014–2.1 $mg\,L^{-1}$ at 25°C; $\log K_{ow}$: 4.79–8.20	Persistence of the PAHs varies with their molecular weight. The low molecular weight PAHs are most easily degraded. The reported $T_{1/2}$ of naphthalene, anthracene, and benzo(e)pyrene in sediment are 9, 43, and 83 hours, respectively, whereas for higher molecular weight PAHs, $T_{1/2}$ are up to several years in soils and sediments. The BCFs in aquatic organisms frequently range between 100 and 2000, and it increases with increasing molecular size. Due to their wide distribution, the environmental pollution by PAHs has aroused global concern.	Acute toxicity of low PAHs is moderate with an LD_{50} of naphthalene and anthracene in rat of 490 and 180,00 $mg\,kg^{-1}$ body weight, respectively, whereas the higher molecular weight PAHs exhibit higher toxicity, and LD_{50} of benzo(a)anthracene in mice is 10 $mg\,kg^{-1}$ body weight. In *Daphnia pulex*, LC_{50} for naphthalene is 1.0 $mg\,L^{-1}$, for phenanthrene 0.1 $mg\,L^{-1}$, and 0.1 $mg\,L^{-1}$, and for benzo(a)pyrene is 0.005 $mg\,L^{-1}$. The critical effect of many PAHs in mammals is their carcinogenic potential. The metabolic actions of these substances produce intermediates that bind covalently with cellular DNA. IARC has classified benz[a]anthracene, benzo[a]pyrene, and dibenzo[a,h]anthracene as probably carcinogenic to humans. Benzo[b]fluoranthene and indeno[1,2,3-c,d]pyrene were classified as possible carcinogens to humans.
Phthalates Includes a wide family of compounds. Among the most common contaminants are: Dimethylphthalate	Properties of phthalic acid esters vary greatly depending on the alcohol moieties.	Ubiquitous pollutants, in marine, estuarine, and freshwater sediments, sewage sludges, soils, and food. Degradation ($T_{1/2}$) values generally range	Acute toxicity of phthalates is generally low: the oral LD_{50} for DEHP is about 25–34 g/kg, depending on the species; for DBP reported LD_{50} values following oral administration to rats range from 8 to 20 g/kg body weight; in mice, values are approximately 5 to 16 g/kg

TABLE 5.6 *(continued)*

Compound	*Properties*	*Persistence/Fate*	*Toxicity*[32]*
(DMP), diethylphthalate (DEP), dibutylphthalate (DBP), benzylbutylphthalate (BBP), di(2-ethylhexyl) phthalate (DEHP) ($C_{24}H_{38}O_4$) and dioctylphthalate (DOP).	The $\log K_{ow}$ rates range from 1.5 to 7.1.	from 1–30 days in freshwaters.	body weight. In general, DEHP is not toxic for aquatic communities at the low levels usually present. In animals, high levels of DEHP damaged the liver and kidney and affected the ability to reproduce. There is no evidence that DEHP causes cancer in humans, but they have been reported as endocrine-disrupting chemicals. The EPA proposed a Maximum Admissible Concentration (MAC) of $6\,\mu g\,L^{-1}$ of DEHP in drinking water.
Nonyl- and Octyl-phenols NP: $C_{15}H_{24}O$; OP: $C_{14}H_{22}O$.	NP's: $\log K_{ow} = 4.5$ and OP's $\log K_{ow} = 5.92$.	NP and OP are the end degradation products of alkylphenol ethoxylates (APEs) under both aerobic and anaerobic conditions. Therefore, the major part is released to water and concentrated in sewage sludges. NP and OP are persistent in the environment, with $T_{1/2}$ of 30–60 years in marine sediments, 1–3 weeks in estuarine waters, and 10–48 hours in the atmosphere. Due to their persistence they can bioaccumulate to a	Acute toxicity values for fish, invertebrates and algae ranging from 17 to $3000\,\mu g\,L^{-1}$. In chronic toxicity tests the lowest NOEC are $6\,\mu g\,L^{-1}$ in fish and $3.7\,\mu g\,L^{-1}$ in invertebrates. The threshold for vitellogenin induction in fish is $10\,\mu g\,L^{-1}$ for NP and $3\,\mu g\,L^{-1}$ for OP (similar to the lowest NOEC). Alkylphenols are endocrine-disrupting chemicals also in mammals.

		significant extent in aquatic species. However, excretion and metabolism is rapid.	
Perfluorooctane Sulfonate ($C_8F_{17}SO_3$).	Solubility in water: 550 mg L^{-1} in pure water at 24–25°C; the potassium salt of PFOS has a low vapor pressure, 3.31×10^{-4} Pa at 20°C. Due to the surface-active properties of PFOS, the Log K_{ow} cannot be measured.	Does not hydrolyze, photolyze, or biodegrade under environmental conditions. It is persistent in the environment and has been shown to bioconcentrate in fish. It has been detected in a number of species of wildlife, including marine mammals. Animal studies show that PFOS is well absorbed orally and distributes mainly in the serum and the liver. The half-life in serum is 7.5 days in adult rats and 200 days in Cynomolgus monkeys. The half-life in humans is, on average, 8.67 years (range 2.29–21.3 years, SD = 6.12).	Moderate acute toxicity to aquatic organisms, the lowest LC_{50} for fish is a 96-hour LC_{50} of 4.7 mg L^{-1} to the fathead minnow (*Pimephales promelas*) for the lithium salt. For aquatic invertebrates, the lowest EC_{50} for freshwater species is a 48-hour EC_{50} of 27 mg L^{-1} for *Daphnia magna*, and for saltwater species, a 96-hour LC_{50} value of 3.6 mg L^{-1} for the Mysid shrimp (*Mysidopsis bahia*). Both tests were conducted on the potassium salt. The toxicity profile of PFOS is similar among rats and monkeys. Repeated exposure results in hepatotoxicity and mortality; the dose-response curve is very steep for mortality. PFOS has shown moderate acute toxicity by the oral route with a rat: LD_{50} of 251 mg kg^{-1}. Developmental effects were also reported in prenatal developmental toxicity studies in the rat and rabbit, although at slightly higher dose levels. Signs of developmental toxicity in the offspring were evident at doses of 5 mg kg^{-1}/day and above in rats administered PFOS during gestation. Significant decreases in fetal body weight and significant increases in external and visceral anomalies, delayed ossification,

TABLE 5.6 *(continued)*

Compound	*Properties*	*Persistence/Fate*	*Toxicity*[32]*
			and skeletal variations were observed. A NOAEL of 1 mg kg^{-1}/day and a LOAEL of 5 mg kg–1/day for developmental toxicity were indicated. Studies on employees conducted at PFOS manufacturing plants in the United States and Belgium showed an increase in mortality resulting from bladder cancer and an increased risk of neoplasms of the male reproductive system, the overall category of cancers and benign growths, and neoplasms of the gastrointestinal tract.

[1] The vapor pressure lowering, according to Raoult's Law is: $P = XP^o$, where P^o = the vapor pressure of the pure substance and X = the substance's mole fraction. Thus, the vapor pressure of a dissolved contaminant is proportional to the molecular percentage of that contaminant in solution. Therefore, DDT in a solution where other solutes are present will exhibit a lowered vapor pressure.

[2] See the Preface for a discussion potential endocrine and wildlife effects associated with marine ecosystems.

[3] The growing interest in PBDEs was demonstrated recently when fire retardants containing them were banned in California, based upon international health concerns. These concerns centered around the detection of PBDEs in humans, including mothers' milk. Also, in several studies, PBDE concentrations in milk has been shown to be on the rise. This elevation of PBDE concentrations in milk is happening at the same time that PCB and DDT concentrations are falling. This may indicate that new and continuing sources of PBDEs are present, while the bans on PCBs and DDT are generally residues that are taking time to degrade. See, for example, D. Meironytè, K. Norèm, and A. Bergman, 1999, "Analysis of polybrominated diphenyl ethers in Swedish human milk: A time-related trend study, 1972–1997," *Journal of Toxicology and Environmental Health A*, 58(6), 329–341; and F. Rahman, K. Langford, M. Scrimshaw, and J. Lester, 2001, "Polybrominated diphenyl ether (PBDE) flame redardants," *Science of the Total Environment*, 275 (1–3), pp. 1–17.

Source: United Nations Environmental Programme, 2002, "Chemicals: North American Regional Report," Regionally Based Assessment of Persistent Toxic Substances, Global Environment Facility.

TABLE 5.7
Summary of Persistent and Metallic Compounds in North America, Identified by the United Nations as Highest Priorities for Regional Actions

Compound	*Properties*	*Persistence/Fate*	*Toxicity*[32†]
Compounds of Tin (Sn) Organotin compounds comprise mono-, di-, tri-, and tetra butyl and triphenylene tin compounds. They conform to the following general formula, $(n\text{-}C_4H_9)_nSn\text{-}X$ and $(C_6H_5)_3Sn\text{-}X$, where X is an anion or a group linked covalently through a hetero-atom.	Organotin species have K_{ow} values ranging from 3.19 to 3.84. In seawater and under normal conditions, tributyl tin exists as three species: hydroxide, chloride, and carbonate.	Under aerobic conditions, tributyl tin takes 30 to 90 days to degrade, but in anaerobic soils may persist for more than 2 years. Due to low water solubility it binds strongly to suspended material and sediments. Tributyl tin is lipophilic and accumulates in aquatic organisms. Oysters exposed to very low concentrations exhibit BCF values ranging from 1000 to 6000.	Tributyl tin is moderately toxic, and all breakdown products are even less toxic. Its impact on the environment was discovered in the early 1980s in France with harmful effects in aquatic organisms, such as shell malformations of oysters, imposex in marine snails, and reduced resistance to infection (e.g., in flounder). Gastropods react adversely to very low levels of tributyl tin (0.06–2.3 $\mu g\,L^{-1}$). Lobster larvae show a nearly complete cessation of growth at just 1.0 $\mu g\,L^{-1}$ tributyl tin. In laboratory tests, reproduction was inhibited when female snails exposed to 0.05–0.003 $\mu g\,L^{-1}$ tributyl tin developed male characteristics. Large doses of tributyl tin have been shown to damage the reproductive and central nervous systems, bone structure, and the liver bile duct of mammals.

TABLE 5.7 *(continued)*

Compound	*Properties*	*Persistence/Fate*	*Toxicity*[32†]
Compounds of Mercury (Hg) The main compound of concern are the methylated mercury species (e.g., CH_3HgCl-methyl-mercuric chloride; or C_2H_6Hg-dimethyl mercury).	Mercury compounds vary significantly in physico-chemical properties.[1] For example, $HgCl_2$ is water soluble ($69\,g\,L^{-1}$), but HgO is only slightly soluble ($50\,\mu g\,L^{-1}$) in water. The organic forms of mercury generally are only slightly soluble in water, e.g., at 21°C, CH_3HgCl has an aqueous solubility of $0.1\,g\,L^{-1}$ and	Mercury released into the environment can either stay close to its source for long periods, or be widely dispersed on a regional or even worldwide basis. Not only are methylated mercury compounds toxic, but highly bioaccumulative as well. The increase in mercury as it rises in the aquatic food chain results in relatively high levels of mercury in fish consumed by humans. Ingested elemental mercury is only 0.01% absorbed, but methyl mercury is nearly 100% absorbed from the gastrointestinal tract. The biological $T_{1/2}$ of Hg is 60 days.	Long-term exposure to either inorganic or organic mercury can permanently damage the brain, kidneys, and developing fetus. The most sensitive target of low-level exposure to metallic and organic mercury from short-, or long-term exposures is likely the nervous system.

	C_2H_6Hg has an aqueous solubility of 1 g L^{-1}.		
Compounds of Lead (Pb) Alkyl lead compounds may be confined to tetramethyl lead (TML, $Pb(CH_3)_4$) and Tetraethyl lead (TEL, $Pb(C_2H_5)_4$).	Solubility in water: 17.9 mg L^{-1} (TML) and 0.29 mg L^{-1} (TEL) at 25°C; vapor pressure: 22.5 and 0.15 mm Hg at 20°C for TML and TEL, respectively.	Under environmental conditions, dealkylation produces less alkylated forms and finally inorganic Pb. However, there is limited evidence that under some circumstances, natural methylation of Pb salts may occur. Minimal bioaccumulations have been observed for TEL in shrimps (650), mussels (120), and plaice (130), and for TML in shrimps (20), mussels (170), and plaice (60).	Exposure to Pb and its compounds have been associated with cancer in the respiratory and digestive systems of workers in lead battery and smelter plants. However, tetra-alkyl lead compounds have not been sufficiently tested for the evidence of carcinogenicity. Acute toxicity of TEL and TML are moderate in mammals and high for aquatic biota. LD_{50} (rat, oral) for TEL is 35 mg Pb kg^{-1} and 108 mg Pb kg^{-1} for TML. LC_{50} (fish, 96 hrs) for TEL is 0.02 mg kg^{-1} and for TML is 0.11 mg kg^{-1}.

[1] See the National Research Council's report: *Toxicological Effects of Methylmercury*, 2003, National Academic Press; and International Programme of Chemical Safety, World Health Organization, 1992, Methylmercury: Environmental Health Criteria 101.

Source: United Nations Environmental Programme, 2002, "Chemicals: North American Regional Report," Regionally Based Assessment of Persistent Toxic Substances, Global Environment Facility.

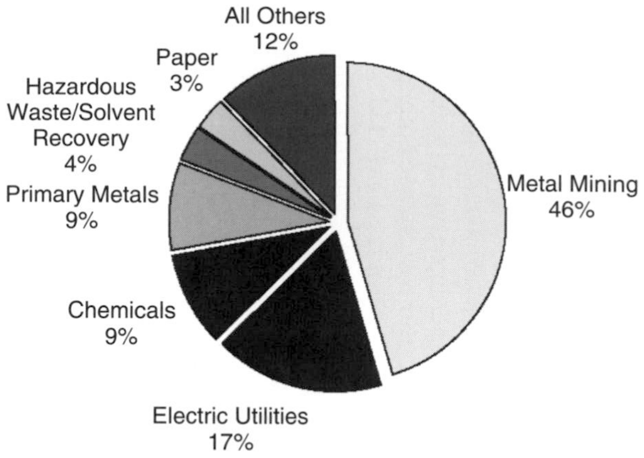

FIGURE 5.17. Total U.S. releases of contaminants in 2001, as reported in the Toxic Release Inventory (TRI). Total releases = 2.8 billion kg. *Note*: Off-site releases include metals and metal compounds transferred off-site for solidification/stabilization and for wastewater treatment, including to publicly owned treatment works. Off-site releases do not include transfers and disposal sent to other TRI facilities that reported the amount as an on-site release. (Source: U.S. Environmental Protection Agency.)

tion, no *advection*,[33] and no transport among compartments (such as sediment loading or atmospheric deposition to surface waters).

A Level 1 calculation describes how a given quantity of a contaminant will partition among the water, air, soil, sediment, suspended particles, and fauna, but does not take into account chemical reactions. Early Level 1 models considered an area of 1 km^2, with 70% of the area covered in surface water. Larger areas are now being modeled (e.g., about the size of the state of Ohio).

Level 2 Model

This model relaxes the conservation restrictions of Level 1 by introducing direct inputs (e.g., emissions) and advective sources from air and water. It assumes that a contaminant is being continuously loaded at a constant rate into the control volume, allowing the contaminant loading to reach steady state and equilibrium between contaminant input and output rates. Degradation and bulk movement of contaminants (advection) is treated as a loss term. Exchanges between and among media are not quantitified.

Since the Level 2 approach is a simulation of a contaminant being continuously discharged into numerous compartments and that achieves a steady-state equilibrium, the challenge is to deduce the losses of the contaminant due to chemical reactions and advective (nondiffusive) mechanisms.

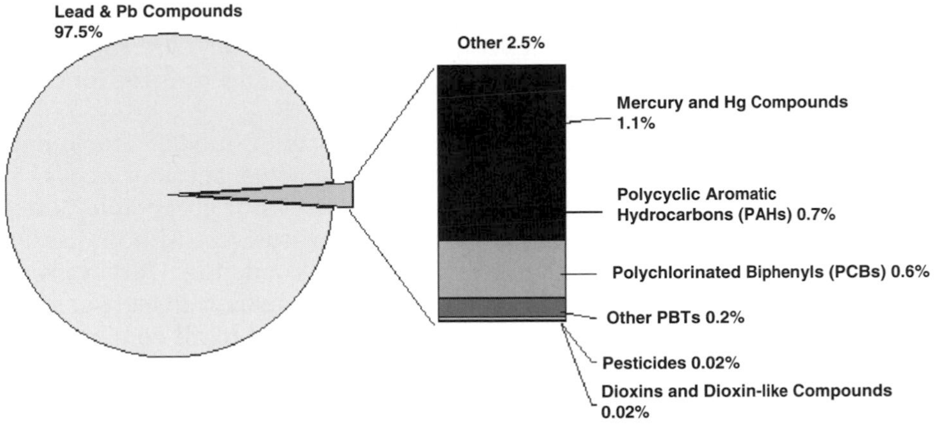

FIGURE 5.18. Total releases of persistent bioaccumulating toxic substances (PBTs) in 2001, as reported in the Toxic Release Inventory (TRI). Total releases = 206 million kg. *Note*: Off-site releases include metals and metal compounds transferred off-site for solidification/stabilization and for wastewater treatment, including to publicly-owned treatment works. Off-site releases do not include transfers and disposal sent to other TRI facilities that reported the amount as an on-site release. (Source: U.S. Environmental Protection Agency.)

Reaction rates are unique to each compound and are published according to reactivity class (e.g., fast, moderate, or slow reactions), which allows modelers to select a class of reactivity for the respective contaminant to insert into transport models. The reactions are often assumed to be first-order, so the model will employ a first-order rate constant for each compartment in the environmental system (e.g., $x\,\mathrm{mol\,hr^{-1}}$ in water, $y\,\mathrm{mol\,hr^{-1}}$ in air, $z\,\mathrm{mol\,hr^{-1}}$in soil). Much uncertainty is associated with the reactivity class and rate constants, so it is best to use rates published in the literature based upon experimental and empirical studies, wherever possible.

Advection flow rates in Level 2 models are usually reflected by residence times in the compartments. These residence times are commonly set at one hour in each medium, so the advection rate (G_i) is the volume of the compartment divided by the residence time (t):

$$G_i = Vt^{-1} \qquad \text{Equation 5–13}$$

Level 3 Model

This model is the same as Level 2, but does not assume equilibrium between compartments, so each compartment has its own fugacity. Mass balance applies to the whole system and each compartment within the system. It includes mass transfer coefficients, rates of deposition and resuspension of contaminant, rates of diffusion (discussed later), soil runoff, and area covered. All of these factors are aggregated into an intermedia transport term (D) for each compartment.

The assumption of equilibrium in Levels 1 and 2 models is a simplification, and often a gross oversimplification of what actually occurs in environmental systems. When the simplification is not acceptable, kinetics must be included in the model. Numerous diffusive and nondiffusive transport mechanisms are included in Level 3 modeling. For example, values for the various compartments' unique intermedia transport velocity parameters (in length per time dimensions) are applied to all contaminants being modeled (these are used to calculate the *D values* discussed in the next sections).

Kinetics versus Equilibrium

Since Level 3 models do not assume equilibrium conditions, a word about chemical kinetics is in order at this point. Chemical kinetics is the description of the rate of a chemical reaction.[34] This is the rate at which the reactants are transformed into products. This may take place by abiotic or by biological systems, such as microbial metabolism. Since a rate is a change in quantity that occurs with time, the change we are most concerned with is the change in the concentration of our contaminants into new chemical compounds:

$$\text{Reaction rate} = \frac{\text{change in product concentration}}{\text{corresponding change in time}} \qquad \text{Equation 5–14}$$

and,

$$\text{Reaction rate} = \frac{\text{change in reactant concentration}}{\text{corresponding change in time}} \qquad \text{Equation 5–15}$$

In environmental degradation, the change in product concentration will be decreasing proportionately with the reactant concentration, so, for contaminant X, the kinetics looks like:

$$\text{Rate} = -\frac{\Delta(X)}{\Delta t} \qquad \text{Equation 5–16}$$

The negative sign denotes that the reactant concentration (the parent contaminant), is decreasing. It stands to reason then that the degradation product Y resulting from the concentration will be increasing in proportion to the decreasing concentration of the contaminant X, and the reaction rate for Y is:

$$\text{Rate} = -\frac{\Delta(Y)}{\Delta t} \qquad \text{Equation 5–17}$$

By convention, the concentration of the chemical is shown in parentheses to indicate that the system is not at equilibrium. $\Delta(X)$ is calculated as the difference between an initial concentration and a final concentration:

$$\Delta(X) = \Delta(X)_{\text{final}} - \Delta(X)_{\text{initial}} \qquad \text{Equation 5–18}$$

So, if we were to observe the chemical transformation[35] of one isomer of the compound butane to different isomer over time, this would indicate the kinetics of the system, in this case the homogeneous gas-phase reaction of *cis*-2-butene to *trans*-2-butene (see Figure 5.19 for the isomeric structures). The transformation is shown in Figure 5.20. The rate of reaction at any time is the negative of the slope of the tangent to the concentration curve at that specific time (see Figure 5.21).

For a reaction to occur, the molecules of the reactants must meet (collide). High concentrations of a contaminant are more likely to collide than low concentrations. Thus, the reaction rate must be a function of the concentrations of the reacting substances. The mathematical expression of this function is known as the "rate law." The rate law can be determined experimentally for any contaminant. Varying the concentration of each reactant independently and then measuring the result will give a concentration curve. Each reactant has a unique rate law (this is one of a contaminant's physicochemical properties).

Let us consider the reaction of reactants A and B that yield C (A + B → C), where the reaction rate increases in accord with the increasing concentration of either A or B. This means that if we triple the amount of A, the rate of this whole reaction triples. Thus, the rate law for such a reaction is:

$$\text{Rate} = k[\text{A}][\text{B}] \qquad \text{Equation 5–19}$$

However, let us consider another reaction, X + Y → Z, in which the rate is only increased if the concentration of X is increased (changing the Y concentration has no effect on the rate law). In this reaction, the rate law must be:

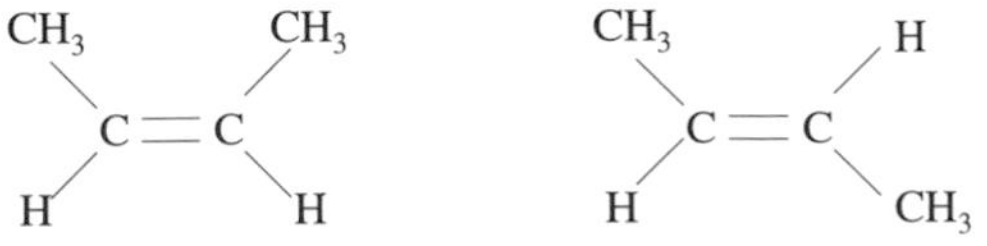

Figure 5.19. Two isomers of butene: *cis*-2-butene (left) and *trans*-2-butene (right).

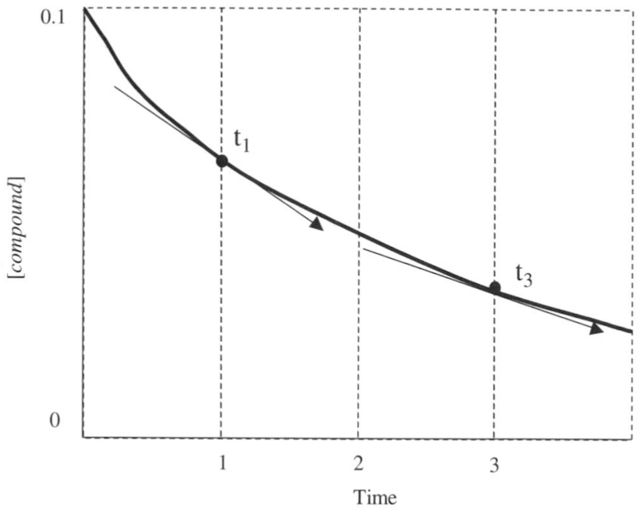

Figure 5.20. The kinetics of the transformation of a compound. The rate of reaction at any time is the negative of the slope of the tangent to the concentration curve at that time. The rate is higher at t_1 than at t_3. This rate is concentration-dependent (first-order).

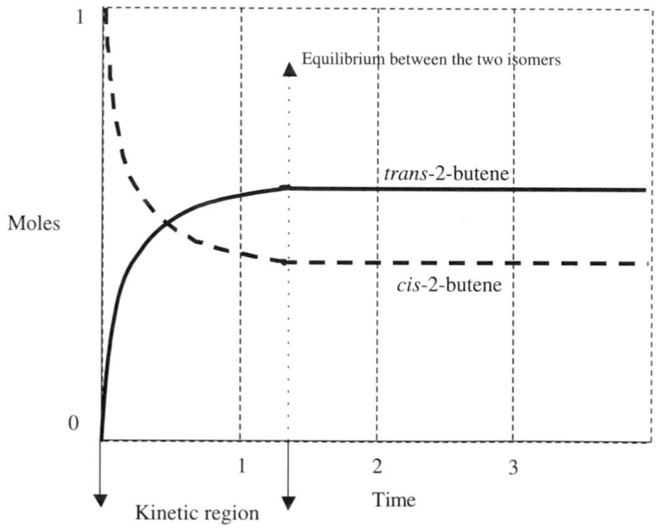

Figure 5.21. Change in respective moles of two butene isomers. Equilibrium at about 1.3 time units. The concentrations of the isomers depend on the initial concentration of the reactant (*cis*-2-butene). The actual time that equilibrium is reached depends on environmental conditions, such as temperature and other compounds present, however, at a given temperature and conditions, the ratio of the equilibrium concentrations will be the same, no matter the amount of the reactant at the start.

$$\text{Rate} = k[\text{X}] \qquad \text{Equation 5–20}$$

Thus, the concentrations in the rate law are the concentrations of reacting chemical species at any specific point in time during the reaction. The rate is how fast the reaction is occurring at that time. The constant k in Equations 5–19 and 5–20 is the *rate constant*, which is unique for every chemical reaction and is a fundamental physical constant for a reaction, as defined by environmental conditions (e.g., pH, temperature, pressure, type of solvent). The rate constant is defined as the rate of the reaction when all reactants are present in a 1 molar (M) concentration, so the rate constant k is the rate of reaction under conditions standardized by a unit concentration.

We can demonstrate the rate law by drawing a concentration curve for a contaminant that consists of an infinite number of points at each instant of time, then an instantaneous rate can be calculated along the concentration curve. At each point on the curve the rate of reaction is directly proportional to the concentration of the compound at that moment in time. This is a physical demonstration of *kinetic order.* The overall kinetic order is the sum of the exponents (powers) of all the concentrations in the rate law. For the rate k[A][B] the overall kinetic order is 2. Such a rate describes a second-order reaction, because the rate depends on the concentration of the reactant raised to the second power. Other decomposition rates are like k[X] and are first-order reactions, because the rate depends on the concentration of the reactant raised to the first power.

The kinetic order of each reactant is the power that its concentration is raised in the rate law. Thus k[A][B] is first-order for each reactant, and k[X] is first-order X and zero-order for Y. In a zero-order reaction, compounds degrade at a constant rate and are independent of reactant concentration.

Further, if we plot the change in the number of moles with respect to time, we would see the point at which kinetics ends and equilibrium begins. This simple example applies to any chemical kinetics process, but the kinetics is complicated in the "real world" by the ever-changing conditions of ecosystems, tissues, and human beings.

The progression from the original reactant to the products in an environmental chemical reaction is often described as the change of a "parent compound" into "chemical daughters" or "progeny." For example, pesticide kinetics often concerns itself with the change of the active ingredient in the pesticide to its "degradation products."

Selecting Units of Mass and Concentrations in Chemodynamics

Chemodynamic models are based on the amount of a contaminant in a specified environmental compartment, so it is necessary to understand the numerous ways that a contaminant may be expressed in terms of mass and concentration in those media.

Generally, weight per weight units can be expressed as dimensionless ratios of the mass of the contaminant and the mass of the medium. For example, in both scientific and lay publications, a contaminant may be expressed as parts per million (ppm), such as a finding of 10 ppm hexachlorobenzene (HCB) in drinking water. Such concentration units suffer from the need to assume that the contaminant and the media have the same densities. As we learned earlier, this is seldom the case.

Weight to weight units may also be expressed with the respective weight units of both contaminant and the media. This is quite common for sediment and soil, such as the finding of 100 mg of HCB per kg soil ($10\,mg\,kg^{-1}$). This alleviates the need to assume equivalent densities for the contaminant and the soil. It also allows for more precision in expressing the media's characteristics. Soils and sediment have very large ranges in their densities. For example, if soil A has twice the density of soil B, and each contains 10 mg of HCB per liter of soil ($10\,mg\,L^{-1}$), this would be adjusted so that the HCB concentration of A by weight would be twice the concentration of the HCB concentration in soil B. If the density of soil A is 2 and the density of soil B is 1, their respective weight per weight concentrations in this example would be $20\,mg\,kg^{-1}$ and $10\,mg\,kg^{-1}$.

For most contaminants in water and atmospheric media, weight per volume concentrations are commonly reported. The contaminant is described as a solute in solution in these fluids (air and water). The milligram per liter ($mg\,L^{-1}$) or microgram per liter ($\mu g\,L^{-1}$) are common units for conventional pollutants in water, such as biochemical oxygen demand (BOD) and nutrients. The highly toxic contaminants, like PCBs and dioxins, may be reported in lower concentration units, such as nanograms or picograms per liter ($ng\,L^{-1}$ and $pg\,L^{-1}$, respectively). In air, the common units are $mg\,m^{-3}$ or $\mu g\,m^{-3}$ for conventional pollutants, such as the U.S. National Ambient Air Quality Standard Pollutants, or oxides of sulfur (SO_x), oxides of nitrogen (NO_x), carbon monoxide (CO), ozone (O_3), fine particulate matter ($PM_{2.5}$), and lead (Pb). For air toxics, units may be reported as $ng\,m^{-3}$ or $pg\,m^{-3}$. As we will see in our modeling examples, it is sometimes best to use the same units, wherever possible, so that all media concentrations may be reported as grams per cubic meter ($g\,m^{-3}$), even though they have been reported in the literature by their respective conventions. A common conversion in environmental models is $1000\,L = 1\,m^{-3}$.

Molar concentrations are seldom reported in the general literature, including government reports, but are frequently encountered in the scientific literature, especially in chemistry journals. The *mole* is a fundamental measure of the quantity of a chemical element or compound. One mole of an element or compound equals the element's atomic weight or the compound's molecular weight expressed in actual mass units. In the SI system of measurements, the mole is understood to be the gram-molecular weight of a substance. So a mole of oxygen is equal to 16 grams. Incidentally, one gram-mole of any substance has a specified number of particles (e.g., atoms, molecules, or ions) that is equal to Avogadro's number (N_A), or 6.022×10^{23}. Speaking of Avogadro, his hypothesis states equal volumes of all gases at the same temperature and pressure will contain an equal number of gas molecules. At standard temperature and pressure conditions (0°C and 1 atm), one gram-mole of any gas occupies a volume of 22.4 L and contains 6.022×10^{23} molecules.

The fundamental chemical unit for environmental contaminants is moles (mol) per volume (e.g., mol per liter, cubic centimeter, or cubic meter). Environmental scientists and engineers often prefer these units because they give quantities that characterize chemical reactions. For example, a chemical reaction in the environment occurs because 1 mol of a certain *reactant* will always react with a given number of moles of another reactant, depending upon the nature of the reactions under specified conditions. Consider, for example, the simple combustion (i.e., oxidation) of methane, yielding carbon dioxide and water:

$$CH_4 + 2O_2 \rightarrow CO_2 + 2H_2O \qquad \text{Reaction 5–1}$$

In this reaction, 1 mol of methane is reacting with 2 moles molecular oxygen to yield 1 mol carbon dioxide and 2 mol water.

Actually, the methane oxidation will usually show subscripts or parenthetical notations:

$$CH_{4gas} + 2O_{2gas} \rightarrow CO_{2gas} + 2H_2O_{gas} \qquad \text{Reaction 5–2}$$

or,

$$CH_4(g) + 2O_2(g) \rightarrow CO_2(g) + 2H_2O(g) \qquad \text{Reaction 5–3}$$

This means that all of the reactants and products are gaseous. This is also known as a *homogenous reaction*, since the reactants are the same phase. An example of a heterogeneous reaction is the reaction between 1 mol solid calcium carbonate and 2 mol aqueous hydrochloric acid to yield 1 mol aqueous calcium chloride, 1 mol carbon dioxide gas, and 1 mol water (understood to be aqueous):

$$CaCO_{3s} + 2HCl_{aq} \rightarrow CaCL_{2aq}CO_{2gas} + H_2O \qquad \text{Reaction 5–4}$$

This reaction is *heterogeneous* because the reactants consist of more than one phase (solid and aqueous). An interesting phenomenon in environmental chemodynamics is the "reaction" of the same chemical to yield itself, only in a different phase. For example, natural surface waters control pH by the exertion of partial pressure of carbon dioxide gas and the production of aqueous carbon dioxide by microbes living in the waters. This tends to move the carbon dioxide dissolved in the water toward equilibrium (reversible reaction, so the arrows point in both directions) with the atmospheric carbon dioxide:

$$(CO_2)_{aq} \leftrightarrow (CO_2)_{gas} \qquad \text{Reaction 5–5}$$

The formula weight (FW) of a compound is the sum of all atomic weights of the elements comprising the compound. Molecular weight (MW) is the same as the FW and is more commonly used in environmental science literature.

Concentration Example 1

How many moles are in 0.01 g benzene?

Solution

Benzene (C_6H_6) has an atomic number of 42 and an average molecular weight of 78.11, so the number of moles will equal the mass divided by the molecular weight:

$$n = m/\mathrm{MW} = 0.001\,\mathrm{g}/(78.11\,\mathrm{g\,mol^{-1}}) = 1.28 \times 10^{-4}\,\mathrm{mol}$$

The *equivalent weight* (EW) of a contaminant or any other solute is the amount of that substance, in grams, that supplies one gram-mole (6.022 × 10^{23}) of the reacting units. The EW is particularly useful in chemodynamics, for example when salts and ions are present in a water body. It is the formula weight divided by the reaction's change in electrical charge (or valence). So in acid-base reactions, the acid supplies one gram-mole of hydrogen ions (H^+). Likewise, the base supplies one gram-mole of hydroxyl ions (OH^-). In redox (reduction-oxidation) reactions where electrons are gained and lost, the equivalent of the substance is one gram-mole of electrons. In a redox reaction, an equivalent of reactant either gives up or accepts 1 mol of electrons. The calculation of EW is thus straightforward, i.e. the molecular weight divided by change in oxidation number as a result of the chemical reaction:

$$EW = \frac{MW}{\Delta \text{ oxidation number}} \qquad \text{Equation 5–21}$$

Equivalents are commonly reported as milliequivalents (meq) or microequivalents (μeq) per liter (meq L^{-1} or μeq L^{-1}), e.g. the milliequivalents of ions, such as carbonates and hydroxides per unit volume of water that account for a water body's alkalinity.

A common mistake in water chemistry is describing high pH and alkaline conditions as synonymous. Although the two characteristics are often related, meaning that a highly alkaline water will experience an increase in pH, the two terms are different. While pH is simply an expression of the molar concentration of H^+ (negative log), alkalinity is an expression of a combination of equivalents of ions (e.g., 20 μeq L^{-1} bicarbonate, 30 μeq L^{-1} carbonate, and 10 μeq L^{-1} sulfate). Sometimes, the alkalinity is converted to equivalents for one species, such as calcium carbonate, expressed as μeq L^{-1} $CaCO_3$.

Concentration Example 2

What is the equivalent weight of sulfuric acid (H_2SO_4) in one of the reactions that occurs following acid rain deposited to surface waters?

$$H_2SO_4 + H_2O \rightarrow 2H^+ + SO_4^- + H_2O$$

Solution

The MW of sulfuric acid is about 98. Since H_2SO_4 changes from a neutral molecule to the hydrogen and sulfate ions, each with a charge, EW will be the molecular weight divided by 2:

$$EW = \frac{MW}{\Delta \text{ oxidation number}} = 98/2 = 49.$$

So the equivalent weight of sulfuric acid in this reaction is 49.

Concentration-Based Mass Balance Model Example[36]

Consider a hypothetical contaminant's transport in a single compartment (surface water). A factory has released the chemical to an estuary with an average depth of 5 m that covers an area of 2 million m^2. The flow rate of water into and out of the estuary is 24,000 m^3 per day. Sediment enters the estuary at a rate of 1 L min^{-1}. Of this, 60% settles to the sediment at the bottom of the estuary, and 40% remains suspended and is part of the estuary's outflow. The half-life of the chemical is 300 days. Its evaporation rate gives the chemical a mass transfer coefficient of 0.24 m day^{-1}. The chemical's molecular mass is 100 g mol^{-1}. Its K_{AW} is 0.01. Its particle-to-water coefficient (K_{PW}) is 6000, and its bioconcentration factor (i.e., partitioning from the water to the biota) is 9000. The particle (suspended solids) concentration in the water column is 25 ppm by volume. The volume of aquatic fauna in the estuary is 10 ppm. The factory is releasing the contaminant into the estuary at a rate of 1 kg per day. The background inflow concentration of the contaminant is 10 $\mu g\,L^{-1}$.

Calculate the steady state (constant) concentration of the contaminant in the estuary's water, particles, and fauna. Include loss rates in the calculations.

Solution

Set the total concentration of the contaminant in the water as an unknown value. This will allow us later to calculate this value by its difference from the total and other known values. Convert all units to g h^{-1} for the mass balance.

Contaminant Input

$$\text{Discharge rate } (1\text{ kg day}^{-1}) = \text{nearly } 42\,\text{g h}^{-1}.$$

Inflow rate is the flow rate of the estuary times the concentration of the contaminant in the water column = $[(24{,}000\,\text{m}^3\,\text{day}^{-1})/(24\,\text{h day}^{-1})][(10\,\mu\text{g L}^{-1})(10^{-6}\,\text{g}\,\mu\text{g}^{-1})(1000\,\text{L m}^{-3})] = 10\,\text{g h}^{-1}$.

So the total input of the contaminant is $42 + 10 = 52\,\text{g h}^{-1}$.

Partitioning between Compartments

The total volume of water in the estuary is the average depth times area ($5\,\text{m} \times 2{,}000{,}000\,\text{m}^2) = 10^7\,\text{m}^3$.

However, the total volume contains 25 ppm particles and 10 ppm fauna, or

$$\text{Particle volume} = 25 \times 10^{-6} \times 10^7\,\text{m}^3 = 250\,\text{m}^3$$

and,

$$\text{Fauna volume} = 10 \times 10^{-6} \times 10^7\,\text{m}^3 = 100\,\text{m}^3$$

Since the dissolved fraction of the contaminant concentration is $C_{dissolved}$, then the concentration of the contaminant dissolved in the water must be:

$$10^7 \cdot C_{dissolved}$$

And the particle concentration is:

$$250 \cdot K_{PW} \cdot C_{dissolved} = (250 \times 6000) C_{dissolved} = 1.5 \times 10^6 C_{dissolved}$$

And the fauna concentration is:

$$100 \cdot K_{BW} \cdot C_{dissolved} = (100 \times 9000) C_{dissolved} = 9 \times 10^5 C_{dissolved}$$

Or for water, particles, and fauna total:

$$C_{dissolved} (10 + 1.5 + 0.9) \times 10^6 = 12.4 \times 10^6 C_{dissolved}$$

Recall that the total volume must be 10^7 C_W, so we can use the ratio of the quantities in parentheses for the mass balance:

$$C_{dissolved} = 10/12.4 = 0.81 C_W$$

$$\text{Sorbed particle concentration} = 1.5/12.4 = 0.12 C_W$$

$$\text{Bioconcentration} = 0.9/12.4 = 0.07 C_W$$

Thus, 81% of the contaminant is dissolved in the estuary's surface water, 12% is sorbed to particles, and 7% is in the fauna tissue.

The concentration of the contaminant on the particles is therefore $K_{PW}C_{dissolved}$ or 0.81 $K_{PW}C_W$ = 0.81 × 6000 = 4860 C_W. And, the concentration of the contaminant in fauna tissue is 0.81 $K_{BW}C_W$ = 0.81 × 9000 = 7290 C_W.

Outflow

The outflow rate is 24,000 m^3 day^{-1} = 1000 $m^3 h^{-1}$, so the rate of transport of the dissolved contaminant is 1000 $C_{dissolved}$ $g h^{-1}$ or 810 C_W $g h^{-1}$.

Sorption is constantly occurring, so there will also be outflow of the contaminant attached to particles. Let us assume that the fauna remain in the estuary, or at least that there is no net change in contaminant mass concentrated in the biotic tissue: 40% of the sediment's 1 $L min^{-1}$ leaves the estuary; therefore the 0.4 $L min^{-1}$ = 24 $L h^{-1}$ of particles containing 4860 C_W $g m^{-3}$. Since, 24 $L h^{-1}$ = 0.024 $m^3 h^{-1}$, there will be 4860 × 0.024 = 117 C_W gh^{-1} contaminant leaving the estuary on the sediment.

Reaction

The product of the estuary water volume, concentration, and rate constant gives the reaction rate. Since the half-life is 300 days (7200 hours), the rate constant is: $= \frac{Ln(2)}{7200} = 9.6 \times 10^{-5}\,h^{-1}$.

Thus, the reaction rate is $10^7 \times C_W \times 9.6 \times 10^{-5} = 960\ C_W\ gh^{-1}$.

Sedimentation

Since the concentration of the contaminant sorbed to particles is 4860 C_W and the particle deposition (sedimentation) rate is 60% of the 1 L min^{-1} of sediment entering the estuary (i.e., 0.6 L min^{-1} = 36 L h^{-1} = 0.036 $m^3 h^{-1}$), the contaminant deposition rate is 4860 × 0.036 C_W = 175 C_W gh^{-1}.

Vaporization

The vaporization (evaporation) rate equals the product of the gas's mass transfer coefficient, the estuary's surface area, and the contaminant concentration in water. Thus, for our contaminant, the evaporation rate = (0.24 m day^{-1})(day 24 h^{-1})(2 × 10^6 m^2)(0.81 C_W) = 16,200 C_W gh^{-1}.

We will assume that no diffusion is taking place between the air and water (i.e., the air contains none of our hypothetical contaminant). If the atmosphere were a source of the contaminant, we would need to add another input term.

Combined Process Rates

If we assume steady-state conditions, we can now combine the calculated rates and set up an equality with our discharge rate (input rate):

$$\text{Discharge rate} = \text{Sum of all process rates}$$

$$\text{Discharge rate} = \text{Dissolved Outflow} + \text{Sorbed Outflow} + \text{Reaction} + \text{Sedimentation} + \text{Vaporization}$$

$$52 = 810C_W + 117C_W + 960C_W + 175C_W + 16{,}200C_W$$

$$52 = 18{,}262\,C_W$$

$$C_W = 52/18{,}262 = 0.0028\,g\,m^{-3} = 0.0028\,mg\,L^{-1} = 2.8\,\mu g\,L^{-1}$$

So, returning to our calculated rates and substituting C_W, our model shows that the estuary has the following process rates for the hypothetical contaminant:

Process	*Rate (gh^{-1})*	*Percent of Total*
Outflow dissolved in water (810 × 0.0028)	2.3	4%
Outflow sorbed to suspended particles (117 × 0.0028)	0.33	1%
Reaction (960 × 0.0028)	2.7	5%
Sedimentation (175 × 0.0028)	0.49	1%
Vaporization (16,200 × 0.0028)	45.4	89%

Our model thus tells us that the largest loss of the contaminant is to the atmosphere. Our contaminant is behaving as a volatile compound. Dissolution and chemical breakdown are also important processes in the mass balance. Sorption and sedimentation are also occurring, but they account for far less of the contaminant mass than does volatilization. This means that our contaminant is sufficiently water soluble, sorptive, reactive, and volatile that any monitoring or cleanup must account for all compartments in the environment.

To complete our model, let us consider the contaminant concentration in each environmental compartment:

$$\text{Contaminant dissolved in water} = (0.81)(0.0028\,\text{g}\,\text{m}^{-3}) = 0.0023\,\text{g}\,\text{m}^{-3} = 2.3\,\mu\text{g}\,\text{L}^{-1}$$

The concentration on particles is 4860 times the dissolved concentration:

$$\text{Contaminant sorbed to particles} = (4860)(0.0023\,\text{g}\,\text{m}^{-3}) = 11\,\text{g}\,\text{m}^{-3} = 11\,\text{mg}\,\text{L}^{-1}$$

Solid-phase media, like soil, sediment, and suspended matter, are usually expressed in weight to weight concentrations. If we assume a particle density of $1.5\,\text{g}\,\text{cm}^{-3}$, the concentration on particles is about $7.3\,\text{mg}\,\text{kg}^{-1}$.

Also, the suspended solids fraction of contaminants in surface waters are expressed with respect to water volume. Since particles make up 0.000025 of the total volume of the estuary, our contaminant's concentration is $(2.5 \times 10^{-5})(11\,\text{mg}\,\text{L}^{-1}) = 0.000275\,\text{mg}\,\text{L}^{-1}$, or about $275\,\text{ng}\,\text{L}^{-1}$ of the water column.

The concentration in the fauna is 7290 times the dissolved concentration:

Contaminant concentrated in fauna tissue = $(7290)(0.0023\,\text{g}\,\text{m}^{-3}) = 17\,\text{g}\,\text{m}^{-3} = 17\,\text{mg}\,\text{L}^{-1}$, which is about equal to $17\,\text{mg}\,\text{kg}^{-1}$ tissue. Since the fauna volume makes up 10^{-5} of the total volume of the estuary, our contaminant's concentration is $(10^{-5})(17\,\text{mg}\,\text{L}^{-1}) = 1.7 \times 10^{-4}\,\text{mg}\,\text{L}^{-1}$ or $17\,\mu\text{g}\,\text{L}^{-1}$ of the water column.

Since the total mass loading (input) is $52\,\text{g}\,\text{h}^{-1}$, we have maintained our mass balance. The concentration in each media is an indicator of the relative affinity that our contaminant has for each environmental compartment.

What if the contaminant were less soluble in water and had a higher bioconcentration rate? Reviewing our calculations, you will find that if the contaminant were less soluble, then less mass would be available to be sorbed or bioconcentrated. Keep in mind, however, that this is a mathematical phenomenon and not necessarily a physical one. Yes, the dissolved fraction is used to calculate the mass that moves to the particles and biota, but remember that the coefficients are based on empirical information. Thus, the bioconcentration factor that we were given would increase to compensate for the lower dissolved concentration. That is what makes modeling interesting and complex. When one parameter changes, the other parameters must be adjusted. Few systems in the environment are truly independent!

Fugacity, Z Values, and Henry's Law

Before we move on to model the partitioning of contaminants among the environmental media, let us revisit the relationships of Henry's Law constants to equilibrium. We have seen that the relative chemical concentrations of a substance in the various compartments and physical phases is predictable from partition coefficients. The more one knows about the affinities of a compound for each phase, the better is one's ability to predict how much and how rapidly a chemical will move. This chemodynamic behavior as expressed by the partition coefficients can be viewed as a potential, that is, at the time equilibrium is achieved among all phases and compartments, the chemical potential in each compartment has been reached.[37]

Chemical concentration and fugacity are directly related via the *fugacity capacity constant* (known as the Z value):

$$C_i = Z_i \cdot f \qquad \text{Equation 5–22}$$

Where C_i = Concentration of substance in compartment i (mass per volume)

Z_i = Fugacity capacity (time^2 per length^2)

f = Fugacity (mass per length per time^2)

And, at equilibrium, the fugacity of the system of all environmental compartments is:

$$f = \frac{M_{total}}{\sum_i (Z_i \cdot V_i)} \qquad \text{Equation 5–23}$$

Where M_{total} = Total number of moles of a substance in all of the environmental system's compartments
V_i = Volume of compartment i where the substance resides

If we assume that a chemical substance will obey the ideal gas law (which is usually acceptable for ambient environmental pressures), then fugacity capacity is the reciprocal of the gas constant (R) and absolute temperature (T). Recall that the ideal gas law states:

$$\frac{n}{V} = \frac{P}{RT} \qquad \text{Equation 5–24}$$

Where n = Number of moles of a substance
P = Substance's vapor pressure

Then,

$$P = \frac{n}{V} \cdot RT = f \qquad \text{Equation 5–25}$$

And,

$$C_i = \frac{n}{V} \qquad \text{Equation 5–26}$$

Therefore,

$$Z_{air} = \frac{1}{RT} \qquad \text{Equation 5–27}$$

This relationship allows for predicting the behavior of the substance in the gas phase. The substance's affinity for other environmental media can be predicted by relating the respective partition coefficients to the Henry's Law constants. For water, the fugacity capacity (Z_{water}) can be found as the reciprocal of K_H:

$$Z_{water} = \frac{1}{K_H} \qquad \text{Equation 5–28}$$

This is the dimensioned version of the Henry's Law constant (length^2 per time^2).

Fugacity Example 1

What is the fugacity capacity of toluene in water at 20°C?

Solution

Since Z_{water} is the reciprocal of the Henry's Law constant, which is 6.6 $\times 10^{-3}$ atm m^3 mol^{-1} for toluene, then Z_{water} must be 151.5 mol atm^{-1} m^{-3}.

The fugacity capacity for sediment is directly proportional to the contaminant's sorption potential, expressed as the solid-water partition coefficient (K_d), and the average sediment density ($\rho_{sediment}$). Sediment fugacity capacity is indirectly proportional to the chemical substance's Henry's Law constant:

$$Z_{sediment} = \frac{\rho_{sediment} \cdot K_d}{K_H} \qquad \text{Equation 5–29}$$

Fugacity Example 2

What is the fugacity capacity of toluene in sediment with an average density of 2400 kg m^{-3} at 20°C in sediment where the K_d for toluene is 1 L kg^{-1}?

Solution

Since $Z_{sediment} = \frac{\rho_{sediment} \cdot K_d}{K_H}$,

then $Z_{sediment} = \frac{(2400\,\text{kg m}^{-3}) \cdot (1\,\text{L kg}^{-1}) \cdot (1\,\text{m}^3)}{(6.6 \times 10^{-3}\,\text{atm m}^3\,\text{mol}^{-1}) \cdot (1000\,\text{L})}$

which for toluene, then $Z_{sediment}$ must be 3.6×10^{-4} mol atm^{-1} m^{-3}.

Note that if the sediment had a higher sorption capacity, for example 1.5 L kg^{-1}, the fugacity capacity constant would be higher

(50% times greater, in this case). Conversely, fugacity would decrease by a commensurate amount with increased sorption capacity. This makes physical sense if one keeps in mind that fugacity is the tendency to escape from the medium (in this case, the sediment) and move to another (surface water). If the sediment particles are holding the contaminant more tightly due to higher solid-water partitioning, the contaminant is less prone to leave the sediment. And if the solid-water partitioning is reduced, or sorption is reduced, the contaminant is more free to escape the sediment and be transported to the water.

The nature of the substrate and matrix material (e.g., texture, clay content, organic matter content, and pore fluid pH) can have a profound effect on the solid-water partition coefficient and, consequently, the $Z_{sediment}$ value.

For biota, particularly fauna and especially fish and other aquatic vertebrate, the fugacity capacity is directly proportional to the density of the fauna tissue (ρ_{fauna}), and the chemical substance's *bioconcentration factor* (BCF), and inversely proportional to the contaminant's Henry's Law constant:

$$Z_{fauna} = \frac{\rho_{fauna} \cdot BCF}{K_H} \qquad \text{Equation 5–30}$$

Fugacity Example 3

What is the fugacity capacity of toluene in aquatic fauna which have a BCF of 83 L kg^{-1} and tissue density of 1 g cm^{-3} at 20°C?

Solution

Since $Z_{fauna} = \dfrac{\rho_{fauna} \cdot BCF}{K_H}$,

then $Z_{fauna} = \dfrac{(1\,\mathrm{g\,cm^{-3}}) \cdot (83\,\mathrm{L\,kg^{-1}}) \cdot (1000\,\mathrm{cm^3}) \cdot (\mathrm{kg})}{(6.6 \times 10^{-3}\,\mathrm{atm\,m^3\,mol^{-1}}) \cdot (1\,\mathrm{L}) \cdot (1000\,\mathrm{g})}$,

then Z_{fauna} is $0.013\,\mathrm{mol\,atm^{-1}\,m^{-3}}$.

As in the case of the sediment fugacity capacity, a higher bioconcentration factor means that the fauna's fugacity capacity increases and the actual fugacity decreases. Again, this is logical, since the organism is sequestering the contaminant and keeping if from leaving if the organism has a large BCF. This is a function of both the species of organism and the characteristics of the contaminant and the environment where the organism resides. Factors like temperature, pH, and ionic strength of the water and metabolic conditions of the organism will affect BCF and Z_{fauna}. This also helps to explain why published BCF values may have large ranges.

The total partitioning of the environmental system is merely the aggregation of all of the individual compartmental partitioning. So the moles of the contaminant in each environmental compartment (M_i) are found to be a function of the fugacity, volume, and fugacity capacity for each compartment:

$$M_i = Z_i \cdot V_i \cdot f \qquad \text{Equation 5–31}$$

Comparing the respective fugacity capacities for each phase or compartment in an environmental system is useful for a number of reasons. First, if one compartment has a very high fugacity (and low fugacity capacity) for a contaminant, and the source of the contaminant no longer exists, then one would expect the concentrations in that medium to fall rather precipitously with time under certain environmental conditions. Conversely, if a compartment has a very low fugacity, measures (e.g., *in situ* remediation, or removal and abiotic chemical treatment) may be needed to see significant decreases in the chemical concentration of the contaminant in that compartment. Second, if a continuous source of the contaminant exists, and a compartment has a high fugacity capacity (and low fugacity), this compartment may serve as a conduit for delivering the contaminant to other compartments with relatively low fugacity capacities. Third, by definition, the higher relative fugacities of one set of compartments compared to another set in the same ecosystem allow for comparative analyses and estimates of sources and sinks (or "hot spots") of the contaminant, which is an important part of fate, transport, exposure, and risk assessments.

Fugacity Example 4

What is the equilibrium partitioning of 1000 kg of toluene discharged into an ecosystem of $5 \times 10^9\,m^3$ air, $9 \times 10^5\,m^3$ water, and $4.5\,m^3$ aquatic fauna, with the same K_H, *BCF*, K_d, and densities for fauna and sediment used in the three previous examples? Assume the temperature is 20°C and the vapor pressure for toluene is 3.7×10^{-2} atm.

Solution

The first step is to determine the number of moles of toluene released into the ecosystem. Toluene's molecular weight is 92.14, so converting the mass of toluene to moles gives us:

$$\frac{(1000\,\text{kg})\cdot(1000\,\text{g})\cdot(1\,\text{mol})}{(1\,\text{kg})\cdot(92.14\,\text{g})} = 10{,}853\,\text{mol}$$

The fugacity capacities for each phase are:

$$Z_{air} = \frac{1}{RT} = \frac{1}{0.0821\text{L}\cdot\text{atm}\cdot\text{mol}^{-1}\cdot\text{K}\cdot 293°\text{K}}\cdot\frac{1000\text{L}}{\text{m}^3}$$
$$= 41.6\,\text{mol}\,\text{atm}^{-1}\text{m}^{-3}$$

$$Z_{water} = \frac{1}{K_H} = \frac{1}{6.6\times10^{-3}\,\text{atm}\,\text{m}^3\text{mol}^{-1}} = 151.5\,\text{mol}\,\text{atm}^{-1}\text{m}^{-3}$$

$$Z_{fauna} = \frac{\rho_{fauna}\cdot BCF}{K_H} = \frac{(1\,\text{g}\,\text{cm}^{-3})\cdot(83\,\text{L}\,\text{kg}^{-1})\cdot(1000\,\text{cm}^3)\cdot(\text{kg})}{(6.6\times10^{-3}\,\text{atm}\,\text{m}^3\text{mol}^{-1})\cdot(1\,\text{L})\cdot(1000\text{g})}$$
$$= 0.013\,\text{mol}\,\text{atm}^{-1}\text{m}^{-3}$$

The ecosystem fugacity can now be calculated:

$$f = \frac{M_{total}}{\sum_i (Z_i \cdot V_i)} = \frac{10{,}843\,\text{mol}}{41.6\cdot5\times10^9 + 151.5\cdot9\times10^5 + 0.013\cdot4.5}$$
$$= 5.2\times10^{-8}\,\text{atm}$$

The moles of toluene in each compartment are:

$$M_{air} = 5.2\times10^{-8}\cdot5\times10^9\cdot41.6 = 10{,}816\,\text{mol}$$

$$M_{water} = 5.2\times10^{-8}\cdot9\times10^5\cdot151.5 = 7.1\,\text{mol}$$

$$M_{fauna} = 5.2\times10^{-8}\cdot4.5\cdot0.013 = 3.0\times10^{-9}\,\text{mol}$$

So, the mass of toluene at equilibrium will be predominantly in the air.

The toluene concentration of the air is 10,816 mol divided by the total air volume of $5 \times 10^9 \text{m}^3$. Since toluene's molecular weight is 92.14 grams per mol, then this means the air contains 996,586 grams of toluene, and the air concentration is $199 \mu\text{g m}^{-3}$.

The toluene concentration of the water is 7.1 mol divided by the total water volume of $9 \times 10^5 \text{m}^3$. So the water contains about 654 grams of toluene, and the water concentration is $727 \mu\text{g m}^{-3}$. However, water concentration is usually expressed on a per liter basis, or 727 ng L^{-1}.

The toluene concentration of the aquatic fauna is 3.0×10^{-9} mol divided by the total tissue volume of 4.5m^3. The fish and other vertebrates contain about 276 ng of toluene, and the tissue concentration is 0.06ng m^{-3}.

Thus, even though the largest amount of toluene is found in the air, the highest concentrations are found in the water.

Applying this information allows us to explore fugacity-based, multi-compartmental environmental models. The movement of a contaminant through the environment can be expressed with regard to how equilibrium is achieved in each compartment. The processes driving this movement can be summarized as transfer coefficients or compartmental rate constants, known as *D values*.[38] By first calculating the Z values, as we did for toluene in the previous examples, and then equating the inputs and outputs of the contaminant to each compartment, we can derive D value rate constants. The actual transport process rate (N) is the product of fugacity and the D value:

$$N = Df \qquad \text{Equation 5–32}$$

And, since the contaminant concentration is Zf, we can substitute and add a first-order rate constant k to give us a first-order rate D value (D_R):

$$N = V[c]k = (VZk)f = D_R f \qquad \text{Equation 5–33}$$

Although the concentrations are shown as molar concentrations (i.e., in brackets), they may also be represented as mass-per-volume concentrations, which will be used in our example.[39]

We will discuss the diffusive and nondiffusive transport processes later, but those processes that follow Fick's Laws, or diffusive processes, can also be expressed with their own D values (D_D), which is expressed by the mass transfer coefficient (K) applied to area A:

$$N = KA[c] = (KAZ)f = D_D f \qquad \text{Equation 5–34}$$

Nondiffusive transport (bulk flow or advection) within a compartment with a flow rate (G) has a D value (D_A) is expressed as:

$$N = G[c] = (GZ)f = D_A f \qquad \text{Equation 5–35}$$

This means that when a contaminant is moving through the environment, while it is in each phase it is affected by numerous physical transport and chemical degradation and transformation processes. The processes are addressed by models with the respective D values, so that the total rate of transport and transformation is expressed as:

$$f(D_1 + D_2 + \ldots D_n) \qquad \text{Equation 5–36}$$

Very fast processes have large D values, and these are usually the most important when considering the contaminant's behavior and change in the environment.

Fugacity-Based Mass Balance Model Example[40]

Use a fugacity approach to determine the partitioning of the hypothetical example used earlier in the concentration-based model example, assuming an average temperature of 25°C.

Solution

Let us visualize the mass transport of our hypothetical contaminant among the compartments based upon the results of our concentration-based model (see Figure 5.2). We will use units of mol m^{-3} Pa^{-1} for our Z values.

$$Z_{air} = \frac{1}{RT} = 4.1 \times 10^{-4} \text{mol m}^{-3}\text{Pa}^{-1}$$

We can derive the Z_{water} from Z_{air} and the given K_{AW} (0.01):

$$Z_{water} = \frac{Z_{air}}{K_{AW}} = \frac{4.1 \times 10^{-4}}{0.01} = 4.1 \times 10^{-2} \text{mol m}^{-3}\text{Pa}^{-1}$$

The $Z_{particles}$ value can be derived from Z_{water} and the given K_{PW} (6000):

$$Z_{particles} = Z_{water} \cdot K_{PW} = (4.1 \times 10^{-2})(6000) = 246 \,\text{mol}\,\text{m}^{-3}\text{Pa}^{-1}$$

The Z_{fauna} value can be derived from Z_{water} and the given K_{BW} (9000):

$$Z_{fauna} = Z_{water} \cdot K_{BW} = (4.1 \times 10^{-2})(9000) = 369 \,\text{mol}\,\text{m}^{-3}\text{Pa}^{-1}.$$

So the weighted total Z value (Z_{WT}) for the ecosystem is the sum of these Z values, which we can weigh in proportion to their respective volume fractions in the ecosystem:

$$\begin{aligned} Z_{WT} &= Z_{water} + (2.5 \times 10^{-4} Z_{particles}) + (10^{-5} Z_{fauna}) \\ &= (4.1 \times 10^{-2}) + (2.5 \times 10^{-4})(246) + (10^{-5})(369) = 1.06 \times 10^{-1} \,\text{mol}\,\text{m}^{-3}\text{Pa}^{-1} \end{aligned}$$

The D values (units of mol $\text{Pa}^{-1}\text{h}^{-1}$) can be found from the respective flow rates (G) given or calculated in the concentration model example, and the respective Z values:

Outflow in water:

$$D_1 = G_{water} \cdot Z_{water} = 1000 \times 4.1 \times 10^{-2} = 41 \,\text{mol}\,\text{Pa}^{-1}\text{h}^{-1}$$

Outflow sorbed to particles:

$$D_2 = G_{particle} \cdot Z_{particle} = (0.024) \cdot (246) = 5.9 \,\text{mol}\,\text{Pa}^{-1}\text{h}^{-1}$$

Reaction (using rate constant calculated from half-life of contaminant given in the concentration-based model example):

$$D_3 = V Z_{WT} k = (10^7 \times 1.06 \times 10^{-1})(9.6 \times 10^{-5}) = 101.8 \,\text{mol}\,\text{Pa}^{-1}\text{h}^{-1}$$

Sedimentation:

$$D_4 = G_{sed} Z_{particle} = (0.036) \cdot (246) = 8.9 \,\text{mol}\,\text{Pa}^{-1}\text{h}^{-1}$$

Vaporization:

The hypothetical contaminant's given mass transfer coefficient (k_M) is 0.24 m day^{-1} or 0.01 m h^{-1} (a fairly volatile substance). This mass transfer takes place across the entire surface area of the estuary (A):

$$D_5 = k_M A Z_{water} = (0.01) \cdot (2 \times 10^6)(4.1 \times 10^{-2}) = 820 \,\text{mol}\,\text{Pa}^{-1}\text{h}^{-1}$$

Overall mass balance:

Now, we can apply these D values to express the overall mass balance of the system according to the contaminant's fugacity in water

(f_{water}). Recall that the contaminant's molecular mass is $100\,\text{g}\,\text{mol}^{-1}$, and that we calculated the total input of the contaminant to be $52\,\text{g}\,\text{h}^{-1}$. Thus, the input rate is $0.052\,\text{mol}\,\text{h}^{-1}$:

$$\text{Contaminant input} = f_{water}\Sigma D_i$$

So,

$$0.052 = f_{water}D_1 + f_{water}D_2 + f_{water}D_3 + f_{water}D_4 + f_{water}D_5$$

$$0.052 = f_{water}977.6$$

This means that f_{water} equals 5.3×10^{-5}.

Further, we can now calculate the concentrations in all of the media from the derived Z values and the contaminant's f_{water}:

$$\text{Contaminant dissolved in water} = Z_{water} \cdot f_{water}$$
$$= (4.1 \times 10^{-2}) \cdot (5.3 \times 10^{-5}) = 2.2 \times 10^{-6}\,\text{mol}\,\text{m}^{-3} = 2.2 \times 10^{-4}\,\text{g}\,\text{m}^{-3}$$

$$\text{Contaminant sorbed to suspended particles} = Z_{particle} \cdot f_{water}$$
$$= (246) \cdot (5.3 \times 10^{-5}) = 1.3 \times 10^{-1}\,\text{mol}\,\text{m}^{-3} = 13\,\text{g}\,\text{m}^{-3}\ \text{particle}$$

$$\text{Contaminant in fauna tissue} = Z_{fauna} \cdot f_{water}$$
$$= (369) \cdot (5.3 \times 10^{-5}) = 2.0 \times 10^{-1}\,\text{mol}\,\text{m}^{-3} = 20\,\text{g}\,\text{m}^{-3}\ \text{tissue.}$$

The concentration derived from the fugacity model are very close to those we derived from the concentration-based model, taking into account rounding. This bears out the relationship between contaminant concentration and the Z and D values.

This model demonstrates the interrelationships within, between and among compartments. In fact, the concentration and fugacity of the contaminant are controlled by the molecular characteristics of the contaminant and the physicochemical characteristics of the environmental compartment. For example, our hypothetical example contaminant's major "forcing function" was the K_{AW}, or the mass transfer coefficient for the contaminant leaving the water surface and moving to the atmosphere. In other words, this is one of a number of *rate limiting* steps that determines where the contaminant ends up.

To demonstrate how one physicochemical characteristic can significantly change the whole system's mass balance, let us reduce the contaminant's mass transfer from a K_{AW} value of 0.24 to $0.024\,\text{m}\ \text{day}^{-1}$ $(0.001\,\text{m}\,\text{h}^{-1})$. Thus, for our new contaminant, the evaporation rate = $2.4\,\text{m}\ \text{day}^{-1}$(day

$24\,h^{-1})(2 \times 10^6\,m^2)(0.81\ C_W) = 1620\ C_W\,g\,h^{-1}$. So the combined process rates will again be the sum of all process rates:

$$\text{Discharge rate} = \text{Dissolved Outflow} + \text{Sorbed Outflow} + \text{Reaction} + \text{Sedimentation} + \text{Vaporization}$$

$$52 = 810C_W + 117C_W + 960C_W + 175C_W + 1620C_W$$

$$52 = 3628C_W$$

$$C_W = 52/3682 = 0.014\,g\,m^{-3} = 0.014\,mg\,L^{-1} = 14\,\mu L^{-1}$$

The modeled results for the estuary's process rates for the hypothetical contaminant will change to:

Process	*Rate (gh^{-1})*	*Percent of Total*
Outflow dissolved in water (810 × 0.014)	11.3	22%
Outflow sorbed to suspended particles (117 × 0.014)	1.6	3%
Reaction (960 × 0.014)	13.4	26%
Sedimentation (175 × 0.014)	2.5	5%
Vaporization (1620 × 0.014)	22.7	44%

Comparing these values to those derived from the concentration-based modeling approach shows that the change in one parameter, or decreasing the mass transfer of our pollutant to 10% of the original contaminant's vapor pressure, has led to a much more even distribution of the contaminant in the environment. While the air is still the largest repository for the contaminant at equilibrium, its share has fallen sharply (by 45%). And the fractions dissolved in water and degraded by chemical reactions account for a much larger share of the mass balance (increasing by 18% and 21%, respectively). Sorption and sedimentation's importance has also increased.

Thus, each environmental system will determine the relative importance of the physical and chemical characteristics. The partitioning coefficients will represent the forcing functions accordingly. For example, if a contaminant has a very high bioconcentration factor, even small amounts will represent high concentrations in the tissues of certain fish. Often, the molecular characteristics of a contaminant that cause it to have a high sorption potential will also render it more lipophilic, so the partitioning between the organic and aqueous phases will also be high. Conversely, the high molecular weight and chemical structures of these same molecules may render them less volatile, so that the water to air partitioning may be low. This is not always true, as some very volatile substances are also highly lipophilic (and have high octanol-water partition coefficients) and are quite readily bioconcentrated (having high BCF values). The halogenated solvents are such an example.

Also, it is important that all of these partitioning events are taking place simultaneously. So a contaminant may have an affinity for a suspended particle, but the particle may consist of organic compounds, including those of living organisms, so sorption, the organic-aqueous phase, and bioconcentration partitioning are all taking place together at the same time on the particle. The net result may be that the contaminant stays put on the particle. Researchers are interested in which of these (and other) mechanisms is most accountable for the fugacity. In the real-life environment, however, it often suffices to understand the net effect. That is why there are so many "black boxes" in environmental models.[41] We may have a good experiential and empirical understanding that under certain conditions a contaminant will move or not move, will change or not change, or will elicit or not elicit an effect. We will not usually have a complete explanation of why these things are occurring, but we can be confident that the first principles of science as expressed by the partitioning coefficients will occur unless there is some yet-to-be-explained other factor affecting them. In other words, we will have to live with an amount of uncertainty, but scientists are always looking for ways to increase certainty.

Models, therefore, although nowhere nearly perfect, are important tools for estimating the movement of contaminants in the environment. They do not obviate the need for sound measurements. In fact, measurements and models are highly complementary. Compartmental model assumptions must be verified in the field. Likewise, measurements at a limited number of points depend on models to extend their meaningfulness. Having an understanding of the basic concepts of a contaminant transport model, we are better able to explore the principle mechanisms for the movement of contaminants throughout the environment.

How Contaminants Move in the Environment

As we have seen, mechanics is the field of physics concerned with the motion and the equilibrium of matter, describing phenomena using Newton's Laws. Motion and equilibrium in the environment fall generally within the province of fluid mechanics. Things move at all scales, from molecular to global. Molecular diffusion within sediments, for example, can be an important contaminant transport mechanism. At the other end of the scale, large air masses may be able to transport gases and aerosols in bulk for thousands of kilometers from their sources.

To ensure mass balance, the flux of a contaminant is equal to the mass flux plus the dispersion flux, diffusion flux, as well as source and sink terms. Sources can be the result of a one-time or continuous release of a chemical from a reservoir, or result from desorption of the chemical along the way. Sinks can be the result of sorption and surface processes. This means that even if the source contribution is known, there will be sorption occurring in soil, sediment, and biota that will either remove the chemical from the fluid, or under other environmental conditions, the chemical will be desorbed from the soil, sediment, or biota. Thus, these interim sources and sinks must be considered in addition to the initial source and final sinks (i.e., the media of the chemical's ultimate fate).

Equipped with our understanding of mass balances and partitioning, we are now ready to investigate three important physical processes responsible for the transport of a contaminant: advection, dynamic dispersion, and diffusion.[42]

Advection

Perhaps the most straightforward contaminant transport process is advection ($J_{Advection}$), the transport of dissolved chemicals with the water or airflow. In terms of total volume and mass of pollutants moved, advection accounts for the lion's share. In fact, another name for advection is bulk transport. During advection, a contaminant is moved along with the fluid or, in the language of environmental science, the *environmental medium*. The contaminant is merely "hitching a ride" on the fluid as it moves through the environment. Environmental fluids move within numerous matrices, such as the flow of air and water between soil particles, within sediment, in unconsolidated materials underground, and in the open atmosphere. Surface water is also an environmental medium in which advection occurs.

Advection is considered a passive form of transport because the contaminant moves along with the transporting fluid. That is, the contaminant moves only because it happens to reside in the medium. Advection occurs within a single medium and among media. The rate and direction of transport is completely determined by the rate and direction of the flow of the media.

Single-Compartment Advection

The simplest bulk transport within one environmental medium or compartment is known as homogeneous advection, where only one fluid is carrying the contaminant. The three-dimensional rate of homogeneous, advective transport is simply the product of the fluid medium's flow rate and the concentration of the contaminant in the medium:

$$N = QC \qquad \text{Equation 5–37}$$

Where Q is the flow rate of the fluid medium (e.g., $m^3 sec^{-1}$) and C is the concentration of the chemical contaminant being transported in the medium (e.g., $\mu g\, m^{-3}$). Therefore, the units for three-dimensional advection are mass per time (e.g., $\mu g\, sec^{-1}$). There is much variability in these rates, so different units will be used for different media. For example, atmospheric transport and large surface waters, like rivers, move large-volume plumes relatively rapidly, while groundwater systems move very slowly.

Advection Example 1

Groundwater is flowing at $10\, m^3 sec^{-1}$ with a benzene concentration $5 \times 10^{-3}\, \mu g\, L^{-1}$. What is the three-dimensional rate of advection?

Solution

Applying Equation 5–37:

$$N = \frac{(10\, m^3\, sec^{-1})(5 \times 10^{-3}\, \mu g\, L^{-1})(1000\, L)}{m^3} = 50\, \mu g\, sec^{-1}$$

Heterogeneous advection refers to those cases where there is a secondary phase present inside the main advective medium. For example, the presence of particulate matter (i.e., suspended solids) in advecting river water, or particles carried by wind.

Heterogeneous advection involves more than one transport system within the compartment. For example, the contaminant may be dissolved in the water *and* sorbed to solids that are suspended in the water. Thus, not only the concentration of the dissolved fraction of the contaminant must be known, but also the concentration of chemical in and on the solid particles.

Advection Example 2

A river system's homogeneous advection of dissolved chromium is $500\, \mu g\, sec^{-1}$. In addition, suspended particles are moving in the river at a rate of $0.001\, m^3 sec^{-1}$. Analyses have shown that the suspended particles have an average chromium concentration of $500\,mg\,L^{-1}$. What is the heterogeneous (total) advective flow of chromium in the river?

Solution

We can treat the suspended particles as a homogenous, advective transport and add this to the dissolved fraction for the total stream load:

$$N = \frac{(0.001\,m^3\,sec^{-1})(500\,mg\,L^{-1})(1000\,L)}{m^3} = 500\,mg\,sec^{-1}$$

Total advective transport of Cr = $500\,mg\,sec^{-1}$ + $50\,\mu g\,sec^{-1}$ = $500.5\,mg\,sec^{-1}$.

This example illustrates that heterogeneous advection is a common transport mechanism for highly lipophilic compounds that are often sorbed to particles as compared to dissolved in water. Metals can form both lipophilic and hydrophilic forms, depending upon their speciation. Many organics, however, such as the PAHs and PCBs, are relatively insoluble in water, so most of their advective transport is by attaching to particles. In fact, lipophilic organics are likely to have orders-of-magnitude greater concentrations in suspended matter than are dissolved in the water.

Solutes in the groundwater also move in the general direction of groundwater flow, or, via advection, with minor control by diffusion. The zone of saturation's pore pressures are different from atmospheric pressure due to *head*. Flow is produced through the pore spaces where there is sufficient difference in head at one location versus another, so the advection follows this *hydraulic gradient* (calculation provided in Figure 5.22). Thus, dissolved contaminants in groundwater are predominantly transported by advection.

Another example of advective transport in the atmosphere is *deposition* of contaminants. The sorption of contaminants to the surface of atmospheric water droplets, in what is known as *wet deposition*, and sorption to solid particles, or *dry deposition*. The process where these contaminants are delivered by precipitation to the earth is advection.

Rather than three-dimensional transport, many advective models are represented by the one-dimensional mass *flux* equation for advection, which can be stated as:

$$J_{Advection} = \bar{v}\eta_e C \qquad \text{Equation 5–38}$$

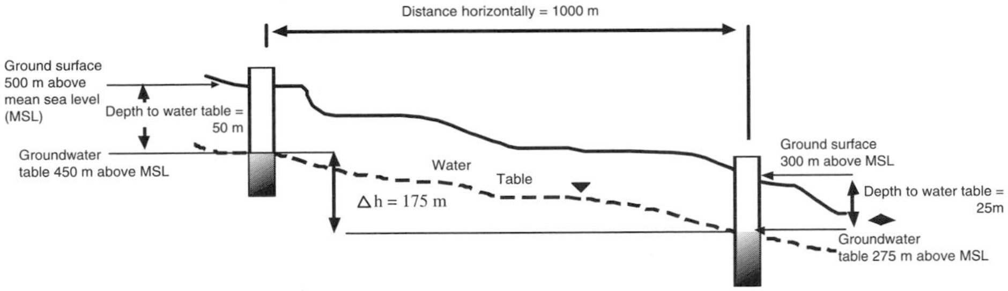

FIGURE 5.22. The hydraulic gradient (K), is the change in hydraulic head (h) over a unit distance. In this case, the horizontal distance is 1000 m. The (h) is the difference between the upper h (450 m) and the lower h (275 m), thus (h) = 175 m. So, K = 175 m/1000 m, or 0.175 (dimensionless).

Where, $\bar{v}$ = average linear velocity ($\mathrm{m\,s^{-1}}$)
η_e = effective porosity (percent, unitless)
C = chemical concentration of the solute ($\mathrm{kg\,m^{-3}}$)

Probably the most common application of the flux term is in two dimensions:

$$J_{Advection} = \bar{v}C \qquad \text{Equation 5–39}$$

Two-dimensional fluxes are an expression of the transport of a contaminant across a unit area. The rate of this transport is the flux density (see Figure 5.23), which is the contaminant mass moving across a unit area per time. In most environmental applications, fluid velocities vary considerably in time and space (e.g., think about calm versus gusty wind conditions). Thus, estimating flux density for advection in a turbulent fluid usually requires a time integration to determine average concentrations of the contaminant. For example, a piece of air monitoring equipment may collect samples every minute, but the model or calculation calls for an hourly value, so the 60 values are averaged to give one integrated concentration of the air pollutant.

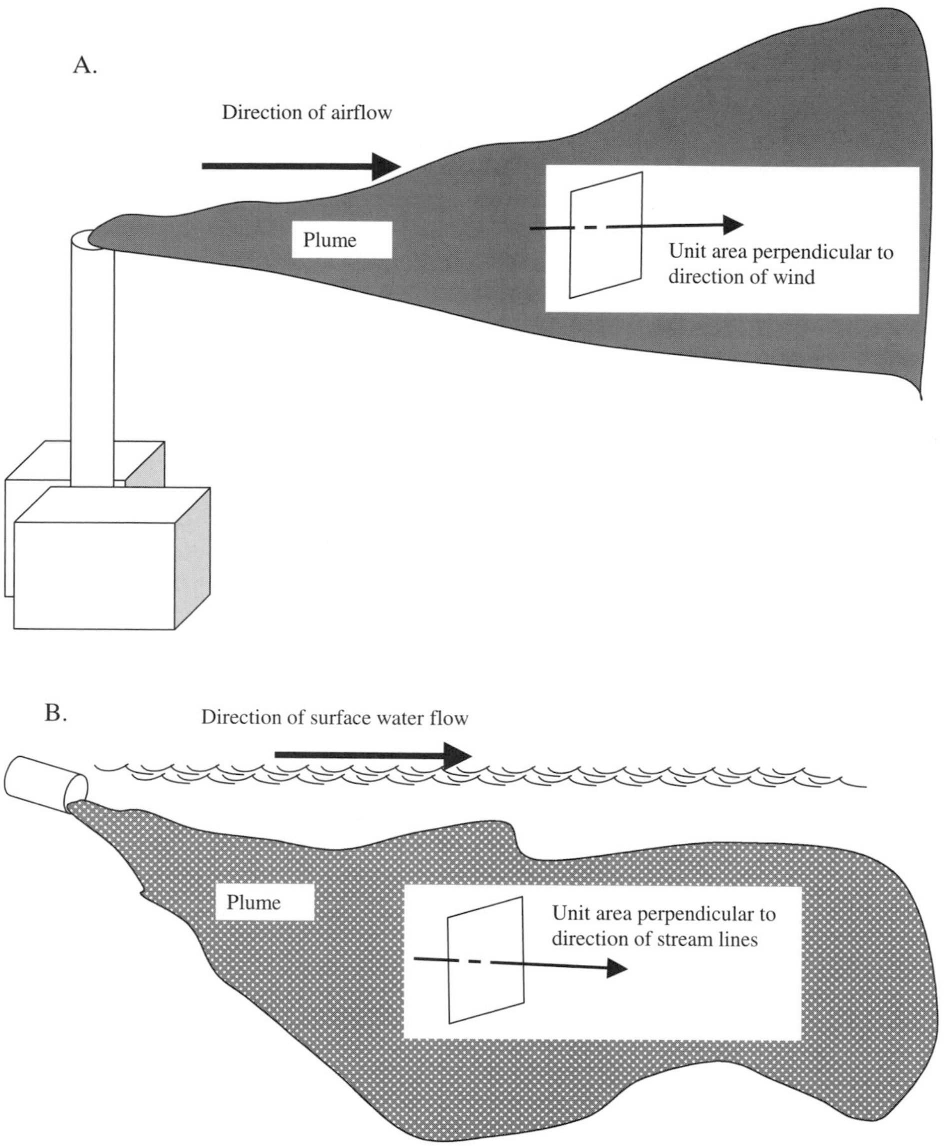

FIGURE 5.23. Determining flux density using an imaginary cross-sectional area across which contaminant flux is calculated in the atmosphere (A) and in surface waters (B).

Advection Example 3

The concentration of the pesticide dieldrin is $15\,\text{ng}\,\text{L}^{-1}$ in a stream with a velocity of $0.1\,\text{m}\,\text{sec}^{-1}$. What is the average two-dimensional flux density of the dieldrin as it moves downstream?

Solution

Applying Equation 5–39:

$$\text{dieldrin} = 15\,\text{ng}\,\text{L}^{-1} = 0.015\,\text{ng}\,\text{m}^{-3}$$

$$J_{Advection} = \overline{v}C = (0.1\,\text{m}\,\text{sec}^{-1})(0.015\,\text{ng}\,\text{m}^{-3}) = 0.0015\,\text{ng}\,\text{m}^{-2} = 1.5\ \text{picograms m}^{-2}$$

Dispersion

Numerous dispersion processes are at work in environmental chemodynamic systems. As is the case for diffusion (to be discussed in the next section), the type of dispersion can vary according to scale. Contaminant transport literature identifies two principal types: hydrodynamic dispersion and mechanical dispersion. However, these are actually not mutually exclusive terms. In fact, mechanical dispersion is a factor in dynamic dispersion. See Figures 5.24 and 5.25 for a computationally combined advective and dispersive air transport system.

Aerodynamic and Hydrodynamic Dispersion

The process of a contaminant plume's spread into multiple directions longitudinally is known as dynamic dispersion. If in air, the spreading is known as *aerodynamic dispersion*, and if in water it is *hydrodynamic dispersion*. This spreading results form physical processes that affect the velocity of different molecules in an environmental medium. For example, in aquifers the process is at work when the contaminant transverses the flow path of the moving groundwater. This results from two physical mechanisms: molecular diffusion and mechanical dispersion. Molecular diffusion, which we will discuss in detail, can occur under both freely flowing and stagnant fluid systems, while mechanical dispersion is of most importance in flowing systems. The units of dynamic dispersion d_d are area per time (e.g., $\text{cm}^2\,\text{sec}^{-1}$ for groundwater). Dynamic dispersion is expressed as:

$$d_d = av_x + D_e \qquad \text{Equation 5–40}$$

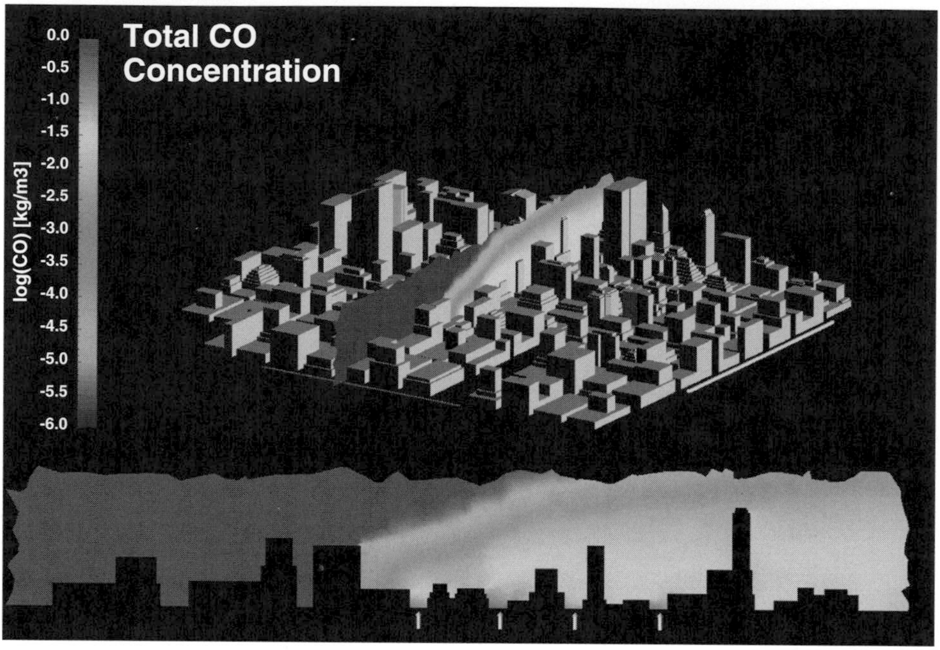

FIGURE 5.24. Profile computational fluid dynamic model depicting an air pollution (carbon monoxide) plume along 59th Street in New York City. Much of the plume is caused by advection by wind through the urban canyons. Dispersion accounts for much of the transport within the street canyons. The vertical profile at the bottom of the figure indicates the dispersion taking place above the buildings as the plume is advected horizontally. The source of the carbon monoxide is a line along the street. (Source: A. Huber, U.S. Environmental Protection Agency.)

Where, a = dispersivity of the porous medium (cm)
v_x = average linear groundwater velocity ($cm\,sec^{-1}$)
D_e = diffusion coefficient of the contaminant ($cm^2\,sec^{-1}$)

Therefore, mechanical dispersion is the result of the tortuosity of flow paths within an environmental medium. It is especially important in soil and other unconsolidated materials that render circuitous paths through which the fluid must travel. When the fluid moves through spaces in the porous media, the fluid cannot move in straight lines, so it tends to spread out longitudinally and vertically. This is what makes mechanical dispersion the dominant mechanism causing hydrodynamic dispersion at the fluid velocities that are often encountered in aquifers and soil.

Since dispersion is the mixing of the pollutant within the fluid body (e.g., aquifer, surface water, or atmosphere), a basic question is in order. Is

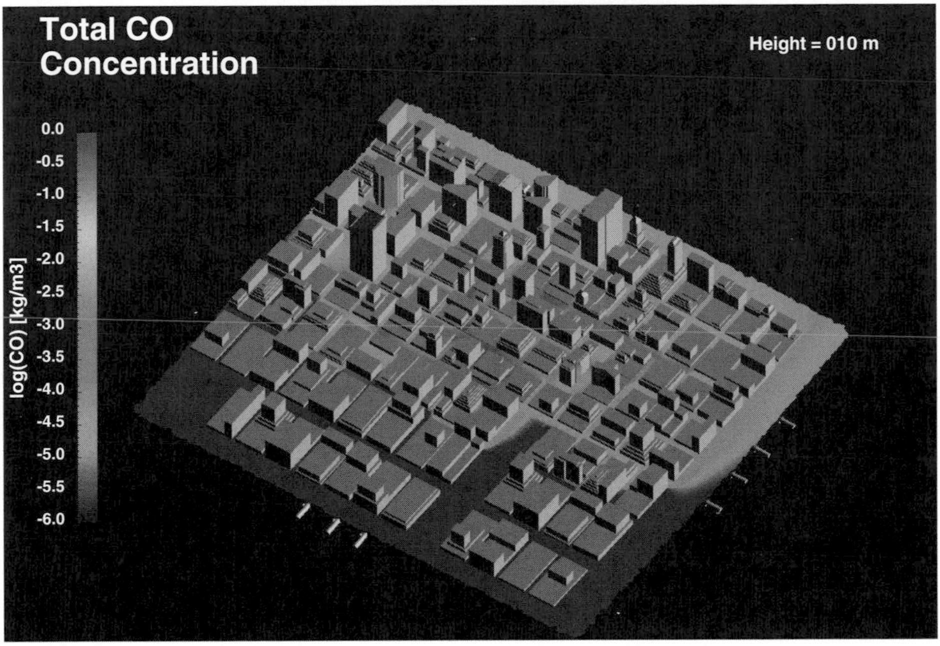

FIGURE 5.25. Plan view of computational fluid dynamic model depicting an air pollution plume along 59th Street in New York City. (Source: A. Huber, 2003, U.S. Environmental Protection Agency.)

it better to calculate the dispersion from physical principles, using a deterministic approach, than to estimate the dispersion using statistics, including probabilities and random distributions? The Eularian model derives the mass balance from conditions in a differential volume. A Lagrangian model applies the statistical theory of turbulence, assuming that turbulent dispersion is a random process described by a distribution function. The Lagrangian model follows the individual random movements of molecules released into the plume, using statistical properties of random motions that are characterized mathematically. This mathematical approach therefore estimates the movement of a volume of chemical (particle)[43] from one point in the plume to another distinct point during a unit time. In other words, the Lagrangian model estimates the path each particle takes during this time, that is, an ensemble mean field relates to the particle displacement probabilities:

$$\overline{[c]}(x, y, z, t) = M_{Total}P(Dx, t) \qquad \text{Equation 5–41}$$

Where, Dx = $x_2 - x_1$ = particle displacement (see Figure 1.2)
$P(Dx_2,t)$ = probability that the point x_2 will be immersed in the dispersing media at time t
M_{Total} = total mass of particles released at x_1
$[c]$ = mean concentration of all released particles = mass of particles the plume $dx \cdot dy \cdot dz$ around x_2

Gaussian dispersion models assume a normal distribution of the plume.

In a deterministic approach, the dispersion includes mixing at all scales. For example, in soil or other unconsolidated material, at the microscopic scale, the model accounts for frictional effects as the fluid moves through pore spaces, the path length around unconsolidated material (the tortuosity), and the size of the pores. At the larger scales, characteristics of strata and variability in the permeability of the layers must be described. A deterministic dispersion flux would then be:

$$J_{Dispersion} = \underline{D} \cdot grad\,C \qquad \text{Equation 5–42}$$

$J_{Dispersion}$ = mass flux of solute due to dispersion ($kg\,m^{-2}\,s^{-1}$)
$\underline{D}$ = dispersion tensor ($m\,s^{-1}$)
C = concentration of chemical contaminant ($kg\,m^{-3}$)

The $\underline{D}$ includes coefficients for each direction of dispersion, or longitudinally, horizontally, and vertically (D_{xx}, D_{xy}, D_{xz}, D_{yy}, D_{yz}, D_{zz}).

Diffusion

In diffusion, contaminants and other solutes move from higher to lower concentrations in a solution. For example, if a sediment contains methyl mercury (CH_3Hg) in concentrations of $100\,ng\,L^{-1}$ at a depth of 3 mm and at $10\,ng\,L^{-1}$ depth of 2 mm, diffusion would account for the upward transport of the CH_3Hg. Diffusion is described by *Fick's Laws*. The First Law says that the flux of a solute under steady-state conditions is a gradient of concentration with distance:

$$J_{Diffusion} = -D\frac{dC}{dx} \qquad \text{Equation 5–43}$$

Where D is a diffusion coefficient (units of area/time), [c] is the molar concentration of the contaminant, and x is the distance between the points of contaminant concentration measurements (units of length). Note that the concentration can also be expressed as mass per fluid volume (e.g., $mg\,L^{-1}$), in which case, flux is expressed as:

$$J_{Diffusion} = -D\frac{dC}{dx} \qquad \text{Equation 5–44}$$

The concentration gradient can also appear in the form:

$$J_{Diffusion} = -d_o i_c \qquad \text{Equation 5–45}$$

Where, d_o is again the proportionality constant, and:

$$i_c = \frac{\partial C}{\partial x} \qquad \text{Equation 5–46}$$

As in our methyl mercury example, at the beginning of this section, the negative sign denotes that the transport is from greater to lesser contaminant concentrations. Fick's Second Law comes into play when the concentrations are changing with time. The change of concentrations with respect to time is proportional to the second derivative of the concentration gradient:

$$\frac{\partial C}{\partial t} = \frac{\partial^2 C}{\partial x^2} \qquad \text{Equation 5–47}$$

All of the diffusions expressed in these equations are one-dimensional, but three-dimensional forms are available and used in models.

Two types of diffusion are important to the transport of contaminants: molecular diffusion and turbulent or eddy diffusion. Each Fickian process operates at its own scale. At the molecular level, in surface waters and atmospheric systems, diffusion dominates as a transport mechanism only in a very thin boundary layer between the fluid and the media. However, in sediments, sludge, and groundwater, this can be an important transport mechanism. Since the concentration gradient (i_c) is the change in concentration (e.g., in units of $kg\,m^{-3}$) with length (in meters), the units of i_c are $kg\,m^{-4}$. Diffusion is therefore analogous to the physical potential field theories (i.e., flow is from the direction of high potential to low potential, such as from high pressure to low pressure). This gradient is observed in all phases of matter: solid, liquid or gas. Molecular diffusion, then, is really only a major factor of transport in porous media, such as soil or sediment, and can be ignored if other processes, such as advection, lead to a flow greater than $2 \times 10^{-5}\,m\,s^{-1}$.[44] However, it can be an important process for source characterization, since it may be the principal means by which a contaminant becomes mixed in a quiescent container (such as a drum, a buried sediment, or a covered pile), or at the boundaries near clay or artificial liners in landfill systems.

Turbulent motion in fluids is characterized by the formation of eddies of various sizes. These eddies can be modeled according to Fick's First Law (concentration gradients), so that the same equations in this chapter applied to molecular diffusion may also be used to estimate the transport of contaminants by eddy diffusion. Like molecular diffusion, eddy diffusion can be modeled in one, two, or three dimensions. One-dimensional models assume that the diffusion coefficient (D) does not change with respect to direction. However, D must be adjusted to the model. This must be done when D is expected to vary with spatial location and time (which it always does, but if the change is not significant, it may be ignored). The coefficient may also be *anisotropic*; that is, it may vary in different directions or vertically in the air or water.

Diffusion Example

A restaurant is operating in a building that was formerly used as a gas station. The "foot print" of the restaurant is $200\,m^2$. The basement of the restaurant is unfinished with a dirt floor. A buried gasoline tank nearby has recently been found to be leaking fluids into the soil and groundwater. Vapors of hydrocarbons have been measured by a reputable environmental audit firm. The soil air 3 meters beneath the basement floor has a concentration of $2\,\mu g\,cm^{-3}$ total hydrocarbons (THC). If the gasoline's diffusion coefficient is $0.01\,cm^2\,sec^{-1}$ in this particular soil, and assuming that the basement air is well mixed (i.e., ventilated), what is the flux density of the vapor and the rate of vapor penetration into the basement by molecular diffusion?

Solution

Since the air in the basement is well mixed, the basement air will contain much lower vapor concentrations than in the soil air. Calculating a one-dimensional flux, the vertical concentration (upward on the *z* axis) is:

$$\frac{dC}{dz} = (2 \times 10^{-6}\,g\,cm^{-3})/300\,cm) = 6.7 \times 10^{-9}\,g\,cm^{-4}$$

The flux density is:

$$J_{Diffusion} = -D\frac{dC}{dx} = (10^{-2}\,cm^2\,sec^{-1}) \times (6.7 \times 10^{-9}\,g\,cm^{-4})$$

$$= 6.7 \times 10^{-11}\,g\,cm^{-2}\,sec^{-1}$$

Applying the flux density to the 200 m^2 ($2 \times 10^6\,cm^2$), the penetration of the vapor into the restaurant is:

$$(6.7 \times 10^{-11}\,g\,cm^{-2}\,sec^{-1}) \times (2 \times 10^6\,cm^2) \times (3600\,sec\,hr^{-1}) \times (24\,hr\,day^{-1})$$
$$= \underline{11.5\,g\,day^{-1}}$$

This is a high rate of penetration for a toxic vapor. It may even pose a fire hazard, especially if our assumption of complete ventilation is not met.

Overall Effect of the Fluxes, Sinks, and Sources

We have focused primarily on the principal *physical* processes that determine transport, but chemical degradation processes are also at play in determining the environmental fate of contaminants. Recall that one of the five laws dictating fluid dynamics mentioned at the beginning of this discussion included conservation of mass. This is true, of course, but the molecular structure of the chemical may very well change. The change depends on the chemical characteristics of the compound (e.g., solubility, vapor pressure, reactivity, and oxidation state) and those of the environment (e.g., presences of microbes, redox potential, ionic strength, and pH). The chemical degradation can be as simple as a first-order decay process (i.e., the degradation of the contaminant concentration C):

$$\frac{\partial C}{\partial t} = -\lambda c \qquad \text{Equation 5–48}$$

The degradation (λ) terms are applied to each chemical. The factors and conditions that drive the λ terms will be considered in detail in Chapter 7, "Chemical Reaction in the Environment." The new degradation products call for an iterative approach to the transport and fate of each degradation product to be described. As a new compound is formed, it must go through the same scrutiny for each transport step. This is even more critical if the degradates are toxic. Some are even more toxic than the parent compound.

A model of the expected total flux representing the fate (J_{Fate}) of the contaminant can therefore be:

$$J_{Fate} = J_{Desorption} + J_{Diffusion} + J_{Dilution} + J_{Dispersion} + J_{Advection} - J_{Sorption} - \lambda[c] \qquad \text{Equation 5–49}$$

This describes the general components and relationships of pollutant transport, and should help the scientist or engineer to select the appropriate model for the chemical and environmental needs dictated by each project.

Combining Transport and Degradation Processes Using Half-Lives and Rate Constants

In addition to elevated but highly variable BCFs, the compounds in Tables 5.5 and 5.6 are also quite *persistent* in the abiotic components of environmental compartments. Persistence is often expressed as the chemical half-life ($T_{1/2}$) of a contaminant. The greater the $T_{1/2}$, the more persistent the compound. Persistence is dependent upon the molecular structure of the compound, such as the presence of aromatic rings, certain functional groups, isomeric structures, and especially the number and types of substitutions of hydrogen atoms with halogens (specifically chlorines and bromines). Persistence potential also depends on the contaminant's relationship to its media. Compound $T_{1/2}$ values are commonly reported for each compartment, so it is possible for a compound to be highly persistent in one medium, yet relatively reactive in another.

Half-lives and rate constants represent identically ordered decay processes and are inversely related to one another. For example, first-order decay can be expressed in terms of concentration versus time, concentration versus distance, and as biodegradation rates. The first-order rates are:

$$\text{Rate constant} = \frac{0.693}{T_{1/2}}, \text{ and half-life} = \frac{0.693}{\text{Rate constant}}. \quad \text{Equation 5–50}$$

Thus, a half-life of 2 years is the same as a first-order rate constant of 0.35 year^{-1}, and a half-life of 10 years = a first-order rate constant of 0.0693 (i.e., a slower rate constant is inversely related to a longer half-life). Equation 5–50 provides a valuable method to estimate the rate at which a contaminant plume will be attenuated, and it is commonly used in groundwater studies.

Concentration-versus-time constants are known as point decay rates (k_{point}), which are derived from a single concentration value-versus-time plot and can be used to estimate the length of time that a plume will last. Bulk attenuation rates (k), derived from concentration-versus-distance plots, are used to see if the contaminant plume is expanding. Biodegradation rates (λ), which are specific to the contaminant and exclude dispersion and other transport mechanisms, can show trends in plume growth or shrinkage. The uses of these rate constants are summarized in Table 5.8.

The synergy of physical, chemical, and biological processes can be demonstrated by an equation[45] that considers transport (i.e., advection and dispersion) and decay (i.e., abiotic decay and biodegradation):

TABLE 5.8
How to Use Attenuation Rate Constants

	Point Decay Rate Constant (k_{point})	*Bulk Attenuation Rate Constant (k)*	*Biodegradation Rate Constant (λ)*
USED FOR:	*Plume Duration Estimate.* Used to estimate time required to meet a remediation goal at a particular point within the plume. If wells in the source zone are used to derive k_{point}, then this rate can be used to estimate the time required to meet remediation goals for the entire site. k_{point} should not be used for representing biodegradation of dissolved constituents in groundwater models (use λ as described in the right hand column).	*Plume Trend Evaluation.* Can be used to project how far along a flow path a plume will expand. This information can be used to select the sites for monitoring wells and plan long-term monitoring strategies. Note that k should not be used to estimate how long the plume will persist except in the unusual case where the source has been completely removed, as the source will keep replenishing dissolved contaminants in the plume.	*Plume Trend Evaluation.* Can be used to indicate if a plume is still expanding, or if the plume has reached a dynamic steady state. First calculate λ, then enter λ into a fate and transport model and run the model to match existing data. Then increase the simulation time in the model and see if the plume grows larger than the plume simulated in the previous step. Note that λ should not be used to estimate how long the plume will persist except in the unusual case where the source has been completely removed.
REPRESENTS:	Mostly the change in source strength over time with contributions from other attenuation processes such as dispersion and biodegradation. k_{point} is not a biodegradation	Attenuation of dissolved constituents due to all attenuation processes (primarily sorption, dispersion, and biodegradation).	The biodegradation rate of dissolved constituents once they have left the source. It does not account for attenuation due to dispersion or sorption.

TABLE 5.8 *(continued)*

	Point Decay Rate Constant (k_{point})	Bulk Attenuation Rate Constant (k)	Biodegradation Rate Constant (λ)
	rate as it represents how quickly the source is depleting. In the rare case where the source has been completely removed (for a discussion of source zones, see Wiedemeier et al., 1999), k_{point} will approximate k.		
HOW TO CALCULATE:	Plot natural log of concentration versus time for a single monitoring point and calculate k_{point} = slope of the best-fit line.[1] This calculation can be repeated for multiple sampling points and for average plume concentration to indicate spatial trends in k_{point} as well. 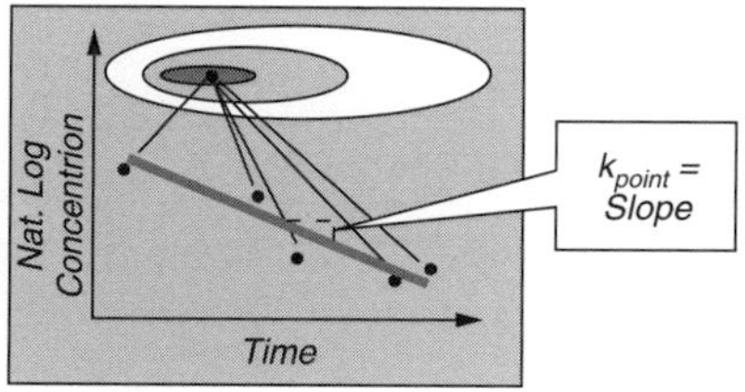	Plot natural log of concentration versus distance. If the data appear to be first-order, determine the slope of the natural log-transformed data by: 1. Transforming the data by taking natural logs and performing a linear regression on the transformed data, or 2. Plotting the data on a semi-log plot, taking the natural log of the y intercept minus the natural log of the x intercept and dividing by the distance between the two points. Multiply this slope by the contaminant velocity (seepage velocity divided by the retardation factor R) to get k.	Adjust contaminant concentration by comparison to existing tracer (e.g., chloride, tri-methyl benzenes) and then use method for bulk attenuation rate;[2] or calibrate a groundwater solute transport computer model that includes dispersion and retardation (e.g., BIOSCREEN, BIOCHLOR, BIOPLUME III, MT3D) by adjusting λ; or use the method of Buscheck and Alcantar[3] (plume must be at steady-state to apply this method). Note this method is a hybrid between k and λ as the Buscheck and Alcantar method removes the effects of longitudinal dispersion, but does

Note this calculation *does not* account for any changes in attenuation processes, particularly dual-equilibrium desorption (availability) which can reduce the apparent attenuation rate at lower concentrations.[4]

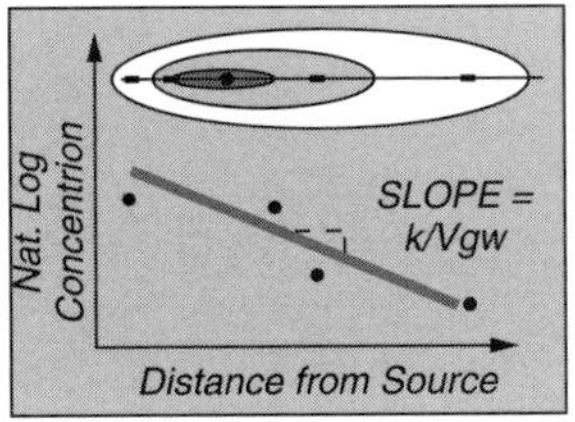

not remove the effects of transverse dispersion from their λ.

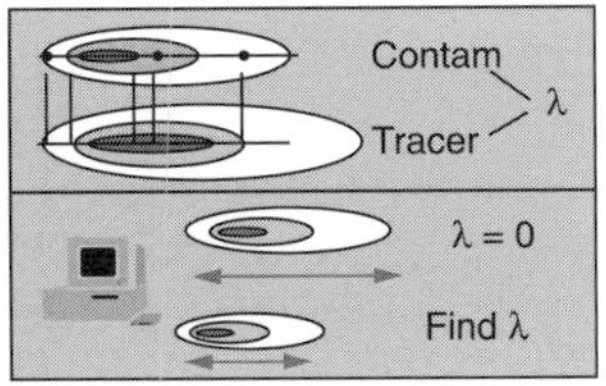

HOW TO USE: *To estimate plume lifetime:*

The time (t) to reach the remediation goal at the point where K_{point} was calculated is:

$$t = \frac{-\mathrm{Ln}\left[\frac{C_{goal}}{C_{stat}}\right]}{K_{point}}$$

To estimate if a plume is showing relatively little change:
Pick a point in the plume but downgradient of any source zones. Estimate the time needed to decay these dissolved contaminants to meet a remediation goal as these contaminants move downgradient:

$$t = \frac{-\mathrm{Ln}\left[\frac{C_{goal}}{C_{stat}}\right]}{K}$$

Calculate the distance L that the dissolved constituents will travel as they are decaying using V_s as the seepage velocity and R is the retardation factor for the

To estimate if a plume is showing relatively little change:
Enter λ in a solute transport model that is calibrated to existing plume conditions. Increase the simulation time (e.g., by 100 years, or perhaps to the year 2525), and determine if the model shows that the plume is expanding, showing relatively little change, or shrinking.

TABLE 5.8 *(continued)*

	Point Decay Rate Constant (k_{point})	*Bulk Attenuation Rate Constant (k)*	*Biodegradation Rate Constant (λ)*
		contaminant: $L = \frac{V_s}{R} \cdot t$ If the plume currently has not traveled this distance L then this rate analysis suggests the plume may expand to that point. If the plume has extended beyond point L, then this rate analysis suggests the plume may shrink in the future. Note that an alternative (and probably easier method) is to merely extrapolate the regression line to determine the distance where the regression line reaches the remediation goal.	
TYPICAL VALUES:	Reid and Reisinger[5] indicated that the mean point decay rate constant for benzene from 49 gas station sites was 0.46 per year (half-life of 1.5 years). For MTBE they reported point decay rate constants of 0.44 per year (half-life of 1.6 years). In contrast, Peargin[7] calculated rates from wells that were screened in areas with residual NAPL; the	For many BTEX plumes, *k* will be similar to biodegradation rates λ (on the order of 0.001 to 0.01 per day) as the effects of dispersion and sorption will be small compared to biodegradation.	For BTEX compounds (i.e., benzene, toluene, ethyl benzene, and xylenes), 0.1–1%/day (half-lives of 700 to 70 days).[6] Chlorinated solvent biodegradation rates may be lower than BTEX biodegradation rates at some sites.

	mean decay rate for MTBE was 0.04 per year (half life of 17 years) the rate for benzene was 0.14 per year (half life of 5 years). The following median point decay rate constants can be used: 0.33 per year (2.1 year half-life) for 159 benzene plumes at service station sites in Texas; and 0.15 per year (4.7 year half-life) for 37 TCE plumes around the United States	For more information about biodegradation rates for a variety of compounds, see Wiedemeier et al., 1999; and Suarez and Rifai, 1999.

[1] American Society for Testing and Materials, 1998. Standard Guide for Remediation of Ground Water by Natural Attenuation at Petroleum Release Sites. E 1943–98, West Conshohocken, PA. www.astm.org.
[2] Wiedemeier, T.H., J.T. Wilson, D.H. Kampbell, R.N. Miller, and J.E. Hansen, 1995. Technical Protocol for Implementing Intrinsic Remediation with Long-Term Monitoring for Natural Attenuation of Fuel Contamination Dissolved in Groundwater (Revision 0), Air Force Center for Environmental Excellence, Brooks AFB, TX, November 1995.
[3] Buscheck, T.E., and C.M. Alcantar, 1995. "Regression Techniques and Analytical Solutions to Demonstrate Intrinsic Bioremediation." In, *Proceedings* of the 1995 Battelle International Conference on In-Situ and On Site Bioreclamation, R.E. Hinchee and R.F. Olfenbuttel eds., Battelle Memorial Institute, Butterworth-Heinemann, Boston, MA.
[4] Kan, A.T., G. Fu, M. Hunter, W. Chen, C.H. Ward, and M.B. Tomson, 1998. Irreversible Sorption of Neutral Hydrocarbons to Sediments: Experimental Observations and Model Predictions," *Environmental Science and Technology*, 32:892–902.
[5] Reid, J.B., and H.J. Reisinger, 1999. Comparative MtBE versus Benzene Plume Length Behavior BP Oil Company Florida Facilities. Prepared by Integrated Sciences & Technology, Marietta, Georgia for BP Oil Company, Cleveland, Ohio.
[6] Suarez, M.P., and H.S. Rifai, 1999. Biodegradation Rates for Fuel Hydrocarbons and Chlorinated Solvents in Groundwater, *Bioremediation Journal*, 3(4):337–362, 1999.
[7] Peargin, T.R., 2002. Relative Depletion Rates of MTBE, Benzene, and Xylene from Smear Zone Non-Aqueous Phase Liquid. In *Bioremediation of MTBE, Alcohols, and Ethers*. Editors V.S. Magar, J.T. Gibbs, K.T. O'Reilly, M.R. Hyman, and A. Leeson. Proceedings of the Sixth International In Situ and On-Site Bioremediation Symposium. San Diego, California, June 4–7, 2001. Battelle Press. 67–74.
Source: U.S. Environmental Protection Agency, 2003, C. Newell, H. Rifai, J. Wilson, J. Connor, J. Aziz, and M. Suarez, "Ground Water Issue: Calculation and Use of First-Order Rate Constants for Monitored Natural Attenuation Studies," Ada, Okla.

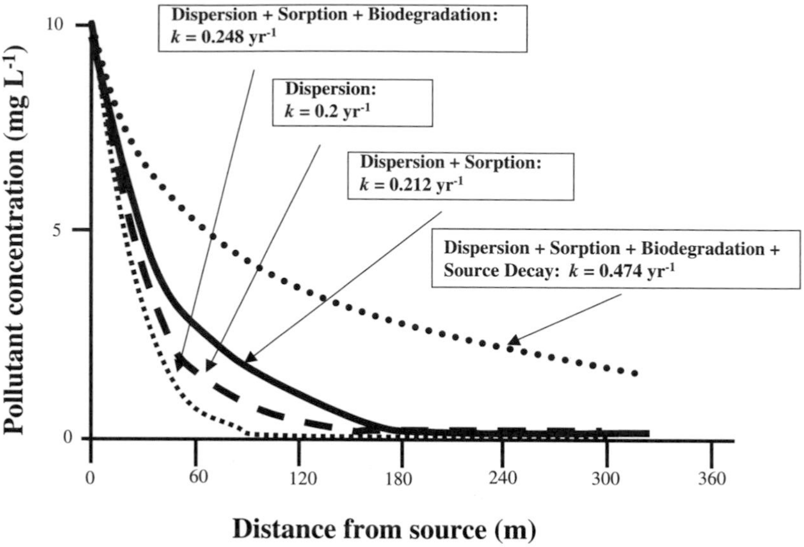

FIGURE 5.26. Effect of incremental contaminant attenuation factors on bulk rate changes to a groundwater plume. (Adapted from: U.S. Environmental Protection Agency, 2003, C. Newell, H. Rifai, J. Wilson, J. Connor, J. Aziz, and M. Suarez, "Ground Water Issue: Calculation and Use of First-Order Rate Constants for Monitored Natural Attenuation Studies," Ada, Okla.)

$$C(x,t) = \frac{C_o}{2}\exp\left[\frac{x}{2\alpha_x}\left(1-\sqrt{1+\frac{4\lambda\alpha_x}{v}}\right)\right] \operatorname{erfc}\left(\frac{x - vt\sqrt{1+\frac{4\lambda\alpha_x}{v}}}{2\sqrt{\alpha_x vt}}\right)\operatorname{erf}\left(\frac{Y}{2\sqrt{\alpha_y x}}\right) \qquad \text{Equation 5–51}$$

Where, C = contaminant concentration
C_0 = initial contaminant concentration
α_x = longitudinal dispersivity
α_y = transverse or horizontal dispersivity
λ = biodegradation rate
t = time
v = retarded velocity of groundwater $\left(v = \frac{\text{seepage velocity}}{\text{redardation factor}}\right)$, where the retardation is due to sorption.
Y = source width.

Figure 5.26 shows the results when the following values are inserted into the equation: $v_s = 100\,ft\ year^{-1}$; R = 5; Y = 40 ft; t = 10 years; and $\alpha_y = 0.1\alpha_x$. Models also need a value for source thickness (b), which is assumed to be 10 ft.

This shows that mechanical processes alone or combined underestimate the actual attenuation of contaminant concentrations as compared to when they are combined with decay factors (i.e., source decay and biodegradation). The rate doubles when the decay factors are considered (and the half-life is halved).

Notes and Commentary

1. See J. Leete, 2001, "Groundwater Modeling in Health Risk Assessment," Chapter 17 in *A Practical Guide to Understanding, Managing, and Reviewing Environmental Risk Assessment Reports,* edited by S. Benjamin and D. Belluck, Lewis Publishers, Boca Raton, Fla.
2. The tri-state area of Ohio, West Virginia, and Pennsylvania is the site of a hazardous waste incinerator operated by Waste Technologies Industries (WTI). One controversy surrounding the incinerator is an elementary school that is about 300 m from the incinerator's stack. The public meetings in the late 1980s and early 1990s showed that many citizens were concerned about where the plumes from the stack and vents would travel, so a model of the entire town and surrounding terrain was built at the U.S. EPA Fluid Modeling Facility in North Carolina. The model was placed in a wind tunnel and smoke was released to track the movement of the plumes under various conditions. The incinerator is still operating, but a number of groups, including Greenpeace USA, are protesting. For photos of the incinerator and tow of East Liverpool, Ohio, visit the Greenpeace website at: www.greenpeaceusa.org/wti/witphotostext.htm.
3. The compartmental or box models, such as the one in Figure 5.2, are being enhanced by environmental scientists and chemical engineers. Much of the information in this figure can be attributed to discussions with Yoram Cohen, a chemical engineering professor at UCLA, and Ellen Cooter, a National Oceanic and Atmospheric Administration modeler on assignment to the U.S. EPA's National Exposure Research Laboratory in Research Triangle Park, N.C.
4. Fugacity models are valuable in predicting the movement and fate of environmental contaminants within and among compartments. This discussion is based on work by one of the pioneers in this area, Don MacKay and his colleagues at the University of Toronto. See, for example, D. MacKay and S. Paterson, 1991, "Evaluating the Fate of Organic Chemicals: A Level III Fugacity Model," *Environmental Science and Technology,* Vol. 25, pp. 427–436.

5. W. Lyman, 1995, "Transport and Transformation Processes," Chapter 15 in *Fundamentals of Aquatic Toxicology: Effects, Environmental Fate, and Risk Assessment*, 2nd Edition, edited by G. Rand, Taylor and Francis, Washington, D.C.
6. Professor Daniel Richter of Duke University's Nicholas School of the Environment has spoken eloquently on this subject.
7. See J. Westfall, 1987, "Adsorption Mechanisms in Aquatic Surface Chemistry," in *Aquatic Surface Chemistry*, Wiley-Interscience, New York, N.Y.
8. L. Keith and D. Walters, 1992, *National Toxicology Program's Chemical Solubility Compendium*, Lewis Publishers, Chelsea, Mich.
9. http://ntp-db.niehs.nih.gov/htdocs/Chem_Hs_Index.html.
10. W. Lyman, 1995, "Transport and Transformation Processes," Chapter 15 in *Fundamentals of Aquatic Toxicology: Effects, Environmental Fate, and Risk Assessment*, 2nd Edition, edited by G. Rand, Taylor and Francis, Washington, D.C.
11. R. Meister, ed., 1992, *Farm Chemicals Handbook '92*, Meister Publishing Company, Willoughby, Ohio.
12. National Park Service, 1997, U.S. Department of the Interior, Environmental Contaminants Encyclopedia, O-Xylene Entry: http://www.nature.nps.gov/toxic/xylene_o.pdf.
13. See D. Mackay and F. Wania, 1995, "Transport of Contaminants to the Arctic: Partitioning, Processes, and Models," *The Science of the Total Environment*, Vol. 160, no. 161, pp. 25–28.
14. P. Lioy, C. Weisel, J. Millette, S. Eisenreich, D. Vallero, J. Offenberg, B. Buckley, B. Turpin, M. Zhong, M. Cohen, C. Prophete, I. Yang, R. Stiles, G. Chee, W. Johnson, S. Alimokhtari, C. Weschler, and L. Chen, 2002, "Characterization of the Dust/Smoke Aerosol that Settled East of the World Trade Center (WTC) in Lower Manhattan after the Collapse of the WTC 11 September 2001." *Environmental Health Perspectives*, Vol. 110, no. 7, pp. 703–14.
15. R. Lewis and S. Gordon, 1996, Sampling of Organic Chemicals in Air, in *Principles of Environmental Sampling*, edited by the American Chemical Society, Washington, D.C., pp. 401–470.
16. R. Williams, B. Ryan, S. Hern, L. Kildosher, K. Hammerstron, and C. Witherspoon, 2000, *Personal Exposures to Polycyclic Aromatic Hydrocarbons Associated in the NHEXAS Maryland Pilot*, presented at the 11th Annual International Society of Exposure Analysis, Charleston, S.C., November 7.
17. American Society for Testing and Materials, 1995, "Standard Practice for Sampling and Selection of Analytical Techniques for Pesticides and Polychlorinated Biphenyls in Air," *Annual Book of ASTM Standards*, Designation: D4861-94a.
18. U.S. Food and Drug Administration, 1999, *Pesticide Analytical Manual*, Vol. I, 3rd Edition.
19. I. Allegrini, A. DeSantis, A. Febo, C. Perrino, and M. Possanzini, 1985, "Annual Denuders to Collect Reactive Gases: Theory and Application," in *Proceedings: Annual Workshop on Methods for Acidic Deposition Measurements*, Raleigh, N.C.

20. D. Pui, C. Lewis, C. Tsai, and B. Liu, 1990, "A Compact Coiled Denuder for Atmospheric Sampling," *Environmental Science and Technology*, Vol. 24, no. 3, pp. 307–312.
21. I. Allegrini, A. DeSantis, A. Febo, C. Perrino, and M. Possanzini, 1985, "Annual Denuders to Collect Reactive Gases: Theory and Application," in *Proceedings: Annual Workshop on Methods for Acidic Deposition Measurements*, Raleigh, N.C.
22. For example, see J. Bowyer and J. Pleil, 1995, "Supercritical Fluid Extraction and Soxhlet Extraction of Organic Compounds from Carpet Samples," *Chemosphere*, Vol. 31, no. 3, pp. 2905–2918; and J. Egea-Gonzalez, M. Costro-Cano, J. Martinez-Vidal, and M. Martinez-Galera, 1997, "Analyses of Procymidone and Vinclozolin in Greenhouse Air," *International Journal of Environmental Analytical Chemistry*, Vol. 67, pp. 143–155.
23. For a complete description of the sampling and analytical methods employed, and a complete list of references, see E. Swartz, L. Stockburger, and D. Vallero, 2003, "Polyaromatic Hydrocarbons and Other Semi-Volatile Organic Compounds Collected in New York City in Response to the Events of 9/11," *Environmental Science and Technology*, Vol. 37(16), pp. 3537–3546.
24. A. Vette, M. Landis, R. Williams, D. La Posta, M. Kantz, J. Fillippelli, L. Webb, T. Ellstad, and D. Vallero, 2002, "Concentration and Composition of PM at Ground Zero and Lower Manhattan Following the Collapse of the WTC." U.S. Environmental Protection Agency, American Association for Aerosol Research Annual Meeting.
25. See, for example, T. Nielsen, 1988, "The decay of cyclopenteno (c, d) pyrene and benzo (a) pyrene in the atmosphere," *Atmospheric Environment*, Vol. 22, pp. 2249–2254; I. Kavouras, P. Koutrakis, M. Tsapakis, E. Lagoudaki, E. Stephanou, D. von Bear, and P. Oyola, 2001, "Source apportionment of urban aliphatic and polynuclear aromatic hydrocarbons (PAHs) using multivariate methods," *Environmental Science and Technology*, Vol. 35, pp. 2288–2294; and M. Tsapakis, E. Lagoudaki, E. Shephanou, I. Kavouras, P. Koutrakis, P. Oyola, and D. von Baer, 2002, "The composition and sources of PM 2.5 organic aerosol in two urban areas of Chile," *Atmospheric Environment*, Vol. 36, pp. 3851–3863.
26. G. Grimmer, J. Jacob, and K. Naujack, 1981b, "Profile of the polycyclic aromatic hydrocarbons from lubricating engine oils. Inventory by GC/MS–PAH in environmental materials, Part 1." *Fresenius. Zietschrift Analytical Chemistry*, Journal of Analytical Chemistry Vol. 309, pp. 13–19.
27. W. Rogge, L. Hildemann, M. Mazurek, G. Cass, and B. Simoneit, 1993, "Sources of fine organic aerosol: Noncatalyst and catalyst equipped automobiles and heavy-duty diesel trucks," *Environmental Science and Technology*, Vol. 27, pp, 636–651.
28. See H. Sakamoto, A. Matsuzawa, R. Itoh, and Y. Tohyama, 2000, "Quantitative analysis of styrene dimer and trimers migrated from disposable lunch boxes," *Journal of the Food Hygiene Society of Japan*, Vol. 41, pp. 200–205; and Y. Kawamura, K. Nishi, T. Maehara, and T. Yamada, 1998, "Migration of

styrene dimers and trimers from polystyrene containers into instant foods," *Journal of the Food Hygiene Society of Japan*, Vol. 39, pp. 390–398.

29. M. Mazurek, B. Simoneit, G. Cass, and H. Gray, 1987, "Quantitative high-resolution gas chromatography and high-resolution gas chromatography/mass spectrometry analyses or carbonaceous fine aerosol particles," *International Journal of Environmental Analytical Chemistry*, Vol. 29, pp. 119–139.
30. N. Schoch and D. Evers, 2002, "Monitoring Mercury in Common Loons," New York Field Report, 1998–2000, Report BRI 2001–01 submitted to U.S. Fish Wildlife Service and New York State Department of Environmental Conservation, BioDiversity Research Institute, Falmouth, Maine.
31. United Nations Environmental Programme, 2002, "Chemicals: North American Regional Report," Regionally Based Assessment of Persistent Toxic Substances, Global Environment Facility.
32. United Nations Environmental Programme, 2002.

* Chemical half-life = $T_{1/2}$
Lethal dose to 50% of tested organism = LD_{50}
Lethal concentration to 50% of tested organism = LC_{50}
Bioconcentration factor = BCF
No observable effect level = NOEL
No observable effect concentration = NOEC

† Chemical half-life = $T_{1/2}$
Lethal dose to 50% of tested organism = LD_{50}
Lethal concentration to 50% of tested organism = LC_{50}
Bioconcentration factor = BCF
No observable effect level = NOEL
No observable effect concentration = NOEC

33. Advection, i.e., the transport process that moves a contaminant solely by mass motion, is discussed in detail later in this chapter. Thus, in a Level 1 model the contaminant is not only assumed to be nonreactive chemically, but the contaminant exists in a windless and streamless systems. Also, the medium that receives the contaminant does not matter because the contaminant is assumed to reach equilibrium instantaneously when it is distributed in the environment.
34. Although "kinetics" in the physical sense and the chemical sense arguably can be shown to share many common attributes, for the purposes of this discussion, it is probably best to treat them as two separate entities. Physical kinetics, as discussed in previous sections in Chapter 2, is concerned with the dynamics of material bodies and the energy in a body owing to its motions. Chemical kinetics addresses rates of chemical reactions. The former is more concerned with mechanical dynamics, the latter with thermodynamics.
35. This example was taken from J. Spencer, G. Bodner, and L. Rickard, 2003, *Chemistry: Structure and Dynamics*, 2nd Edition, John Wiley & Sons, New York, N.Y. pp. 381–385.
36. This example is based upon guidance from D. MacKay and S. Paterson, 1993, "Mathematical Models of Transport and Fate," in *Ecological Risk Assessment*,

edited by G. Suter, Lewis Publishers, Chelsea, Mich.; and from D. MacKay, L. Burns, and G. Rand, 1995, "Fate Modeling," Chapter 18 in *Fundamentals of Aquatic Toxicology: Effects, Environmental Fate, and Risk Assessment*, 2nd Edition, edited by G. Rand, Taylor and Francis, Washington, D.C.

37. A major source of information in this section is from H. F. Hemond and E. J. Fechner-Levy, 2000, *Chemical Fate and Transport in the Environment*, Academic Press, San Diego, Calif.
38. The source of the D value discussion is D. MacKay, L. Burns, and G. Rand, 1995, "Fate Modeling," Chapter 18 in *Fundamentals of Aquatic Toxicology: Effects, Environmental Fate, and Risk Assessment*, 2nd Edition, edited by G. Rand, Taylor and Francis, Washington, D.C.
39. This is the case throughout this text. Bracketed values indicate molar concentrations, but these may always be converted to mass-per-volume concentration values.
40. This example, is based upon guidance from D. MacKay and S. Paterson, 1993.
41. Engineers and scientists refer to processes that are not well defined, but that are predictable, to be "black boxes." We may not know fully why or how the processes work, but we know that the *do* work. For example, scientists may know a lot about how much mercury is emitted from industrial sources. They may even have reliable methods for measuring mercury contamination in water, sediment, and biota. However, the mechanisms and processes that the emitted mercury undergoes before reaching the water, sediment, and biota are all too often "black boxes." This brings to mind the old cartoon by Sidney Harris depicting a feverishly working, excited scientist standing in front of a chalkboard covered with symbology, equations, and steps. Between the final equations and the conclusion, however, was the statement, ". . . a miracle happens . . ." That, my friends, is the ultimate "black box!"
42. The presentation, "Groundwater Modelling: Theory of Solute Transport," by Professor W. Schneider, Technische Unversität Hamburg-Harburg, was a source for some of the equations used in this section.
43. Science is not always consistent with its terminology. The term *particle* is used in many ways. In dispersion modeling, the term particle usually means a theoretical point that is followed in a fluid. The point represents the path that the pollutant is expected to take. Particle is also used to mean aerosol in atmospheric sciences. Particle is also commonly used to describe unconsolidated materials, such as soils and sediment. The present discussion, for example, accounts for the effects of these particles (e.g., frictional) as the fluid moves through unconsolidated material. The pollutant PM, particle matter, is commonly referred to as *particles*. Even the physicist's particle-wave dichotomy comes into play in environmental analysis, as the behavior of light is important in environmental chromatography.
44. This value is taken from W.A. Tucker and L.H. Nelkson, 1982, "Diffusion Coefficients in Air and Water." *Handbook of Chemical Property Estimation Techniques*, McGraw-Hill, New York, N.Y. Flows this low are not uncommon in some groundwater systems or at or in clay liners in landfills.

45. This equation, known as the Domenico solution, is found in C. Newell, H. Rifai, J. Wilson, J. Connor, J. Aziz, and M. Suarez, 2003, "Ground Water Issue: Calculation and Use of First-Order Rate Constants for Monitored Natural Attenuation Studies," U.S. Environmental Protection Agency, Ada, Okla. The example and associated graphics are also taken from this source.

CHAPTER 6

Fundamentals of Environmental Chemistry

The previous chapters introduced the factors and processes responsible for the physical movement and transport of substances within and among environmental compartments. While this discussion focused on physical factors and processes, numerous chemical principles had to be considered, especially the transition of contaminant reactions from kinetics to equilibrium conditions in the environment. Discussions of environmental fate and transport must always consider both physical processes associated with compartmental and phase partitioning and simultaneous reactions and chemical processes. Thus, environmental physics cannot be considered in lieu of environmental chemistry—since that is what occurs in the environment. Any complete discussion of the physical process of solubility, for example, must include a discussion of chemical phenomenon *polarity*. Further, any discussion of polarity must include a discussion of electronegativity (as we have discussed in previous chapters). Likewise, discussions of sorption and air-water partitioning must consider both chemical and physical processes. But that is the nature of environmental science; all concepts are interrelated.

This interconnectedness is evident in Table 6.1, which lists some of the most important processes involved in the fate of environmental contaminants. The previous chapters addressed about of half of these (roughly from advection through adsorption). This chapter and the next will highlight the remaining processes, with insights into the basic chemical processes at work in the environment.

We will begin with an introduction to basic environmental chemistry, then will apply these concepts to inorganic substances before moving to organic chemistry. After this introduction, attention will be paid to some of the most important chemical concepts as they relate to the transformation of substances after they are released into the environment.

In discussing environmental chemistry, it is important to bear in mind that the chemical processes include both the chemical characteristics of the

TABLE 6.1
Physical, Chemical, and Biological Processes Important to the Fate and Transport of Contaminants in the Environment

Process	*Description*	*Physical Phases Involved*	*Major Mechanisms at Work*	*Outcome of Process*	*Factors Included in Process*
Advection	Transport by turbulent flow; mass transfer	Aqueous, gas	Mechanical	Transport due to mass transfer	Concentration gradients, porosity, permeability, hydraulic conductivity, circuitousness or tortuosity of flow paths
Dispersion	Transport from source and spreading	Aqueous, gas	Mechanical	Contaminant concentration gradient-driven	Concentration gradients, porosity, permeability, hydraulic conductivity, circuitousness or tortuosity of flow paths
Diffusion (molecular)	Fick's Laws (concentration gradient)	Aqueous, gas, solid	Mechanical	Contaminant concentration gradient-driven transport	Concentration gradients
Liquid separation	Various fluids of different densities and viscosities are separated within a system	Aqueous	Mechanical	Contaminant recalcitrance due to formation of separate gas and liquid phases (e.g., gasoline in water separates among benzene, toluene, and xylene, as well as other hydrocarbons)	Polarity, solubility, K_d, K_{ow}, K_{oc}, coefficient of viscosity, density

Density stratification	Distinct layers of differing densities and viscosities	Aqueous	Physical/ Chemical	Recalcitrance or increased mobility in transport of lighter fluids (e.g., LNAPLs) that float at water table in soil and groundwater, or at atmospheric pressure in surface water	Density (specific gravity)
Migration along flow paths	Faster through large holes and conduits, e.g., path between sand particles in an aquifer; laminar at low velocities, turbulent at highest velocities	Aqueous, gas	Mechanical	Increased mobility through fractures	Porosity, flow path diameters
Sedimentation	Heavier compounds settle first (the geologic term "competence" is the inverse of sedimentation)	Solid	Chemical, physical, mechanical, varying amount of biological	Recalcitrance due to deposition of denser compounds	Mass, density, viscosity, fluid velocity, turbulence (R_N)
Filtration	Retention in mesh (i.e. size exclusion)	Solid	Chemical, physical, mechanical, varying amount of biological	Recalcitrance due to sequestration, destruction and mechanical trapping of compounds in soil micropores	Surface charge, soil, particle size, sorption, polarity

TABLE 6.1 *(continued)*

Process	*Description*	*Physical Phases Involved*	*Major Mechanisms at Work*	*Outcome of Process*	*Factors Included in Process*
Volatilization	Phase partitioning to vapor	Aqueous, gas	Physical	Increased mobility as vapor phase of contaminant migrates to soil gas phase and atmosphere	Vapor pressure ($P°$), concentration of contaminant, solubility, temperature
Dissolution	Co-solvation, attraction of water molecule shell	Aqueous	Chemical	Various outcomes due to formation of hydrated compounds (with varying solubilities, depending on the species)	Solubility, pH, temperature, ionic strength, activity
Absorption	Retention in solid matrix (three-dimensional incorporation of contaminant)	Solid	Chemical, physical, varying amount of biological	Partitioning of lipophilic compounds into soil organic matter	Polarity, surface charge, Van der Waals attraction, electrostatics, ion exchange, solubility, K_d, K_{ow}, K_{oc}, coefficient of viscosity, density
Adsorption	Retention on solid surface (two-dimensional sorption)	Solid	Chemical, physical, varying amount of biological	Recalcitrance due to ion exchanges and charge separations	Polarity, surface charge, Van der Waals attraction, electrostatics, ion exchange, solubility, K_d, K_{ow}, K_{oc}, coefficient of viscosity, density

Complexation	Reactions with matrix (e.g., soil compounds like humic acid) that form covalent bonds	Solid	Chemical, varying amount of biological	Recalcitrance and transformation due to reactions with soil organic compounds to form residues (bound complexes)	Available oxidants/reductants, soil organic matter content, pH, chemical interfaces, available O_2, electrical interfaces, temperature
Oxidation/ Reduction	Electron loss and gain	All	Chemical, physical, varying amount of biological	Destruction or transformation due to mineralization of simple carbohydrates to CO_2 and water from respiration of micro-organisms	Available oxidants/reductants, soil organic matter content, pH, chemical interfaces, available O_2, electrical interfaces, temperature
Ionization	Complete solvation leading to separation of compound into cations and anions	Aqueous	Chemical	Dissolution of salts into ions	Solubility, pH, temperature, ionic strength, activity
Hydrolysis	Reaction of water molecules with contaminants	Aqueous	Chemical	Various outcomes due to formation of hydroxides (e.g., aluminum hydroxide) with varying solubilities, depending on the species	Solubility, pH, temperature, ionic strength, activity

TABLE 6.1 *(continued)*

Process	*Description*	*Physical Phases Involved*	*Major Mechanisms at Work*	*Outcome of Process*	*Factors Included in Process*
Photolysis	Reaction catalyzed by electromagnetic (EM) energy (sunlight)	Gas (major phase)	Chemical, physical	Photo-oxidation of compounds with hydroxyl radical upon release to the atmosphere	Free radical concentration, wavelength, and intensity of EM radiation
Biodegradation	Microbially mediated, enzymatically catalyzed reactions	Aqueous, solid	Chemical, biological	Various outcomes, including destruction and formation of daughter compounds (degradation products), intracellularly and extracellularly	Microbial population (count and diversity), pH, temperature, soil moisture, acclimation potential of available microbes, nutrients, appropriate enzymes in microbes, available and correct electron acceptors (i.e. oxygen for aerobes; nitrate, ferric Fe, sulfate, and others for anaerobes)
Activation	Metabolic, detoxification process that renders a compound more toxic	Aqueous, gas, solid, tissue	Biochemical	Phase 1 or 2 metabolism, e.g., oxidation may produce toxic metabolites. For example, an	Available detoxification and enzymatic processes in cells

				aromatic compound may be converted to epoxides and diols that are carcinogenic	
Metal Catalysis	Presence of metals, e.g., Cu, Mg, Fe, Ni, speed up chemical reactions	Aqueous gas	Chemical	Chemical transformation, e.g., metal-ion catalysis via coordination with a functional group on a substrate, if the group can be hydrolyzed. For example, such a process enhances the breakdown of some pesticides, e.g., parathion. This increases the functional group's electrophilic potential and enhances nucleophilic attack	Concentration of contaminant and metal. Can speed up reactions listed in this table, e.g., oxidation/reduction, hydrolysis and biodegradation, metal-nucleophile coordination and completation (also lowers pH of aqueous systems)

compartment (e.g., air, water, soil, sediment, or biota) and those of the contaminant. One must constantly link the contaminant to where it resides in the environment. Applying chemical concepts thus requires that we recall the discussions of partitioning and balances from the previous chapters as they apply to the substrate and matrix, as well as to the contaminants. Both the media and the contaminant, obviously, are composed of matter and obey the conservation laws.

During the course of our discussions, we also mentioned a number of elements, as well as numerous examples of inorganic and organic compounds. Let us now take a few steps back to focus on the chemical processes of how and why a substance changes, beginning with the fundamental chemical concepts of abiotic, or nonliving, processes.

Basic Concepts of Environmental Chemistry

Environmental chemistry is the discipline that concerns itself with how chemicals are formed, how they are introduced into the environment, how they change after being introduced, where they end up in organisms and other receptors, and the effects they have (usually the damage they do) once they get there. To cover these concepts, environmental chemistry must address the processes in effect in every environmental compartment. This is evident by the diverse subdisciplines within environmental chemistry, including water chemistry (further divided into drinking water chemistry, wastewater chemistry, stream chemistry, etc.), atmospheric chemistry, soil chemistry, sediment chemistry, and environmental biochemistry. There are even fields such as environmental physical chemistry (including environmental photochemistry), environmental analytical chemistry (including environmental separation sciences and chromatography), and environmental chemical engineering (including fields addressing environmental thermodynamics).

The foremost subdivision, however, is between inorganic and organic chemistry. Thus, after an introduction to concepts and principles inherent to all of environmental chemistry, we will consider inorganic and organic contaminants separately.

Foundations

The *element* is a material substance that has decomposed chemically to its simplest form. These are what appear on the Periodic Table of Elements (Appendix 7). Elements may be further broken down only by nuclear reactions, where they are released as *subatomic particles*. Such particles are important sources of pollution and often are environmental contaminants. An *atom* is the smallest part of an element that can be part of a chemical reaction. The *molecule*, which may also be an atom, is the smallest subdivision of an element that is able to exist as a natural state of matter.

The *nucleus* of an atom, consisting of *protons* and *neutrons* (hydrogen has only one proton in its nucleus), account for virtually all of the atomic mass, or the atomic mass unit (amu). The term *nucleon* is inclusive of protons and neutrons (i.e., the particles comprising the atom's nucleus). An *amu* is defined as one-twelfth of the mass of carbon (C^{12}), or 1.66×10^{-27} kg. The atomic weight of an element listed in most texts and handbooks is the relative atomic weight, which is the total number of nucleons in the atom. For example, oxygen (O) has an atomic mass of 16. The atomic number (Z) is the number of protons in the nucleus. The chemical nomenclature for atomic weight A and number of element E is in the form:

$$ {}^{A}_{Z}E \qquad \text{Equation 6–1} $$

However, since an element has only one atomic number, Z is usually not shown. For example, the most stable form of carbon is seldom shown as ${}^{12}_{12}C$, and is usually indicated as ${}^{12}C$.

Elements may have different atomic weights if they have different numbers of neutrons (the number of electrons and protons of stable atoms must be the same). The elements' forms with differing atomic weights are known as *isotopes*. All atoms of a given element have the same atomic number, but atoms of a given element may contain different numbers of neutrons in the nucleus. An element may have numerous isotopes. Stable isotopes do not undergo natural radioactive decay, whereas radioactive isotopes involve spontaneous radioactive decay, as their nuclei disintegrate. This decay leads to the formation of new isotopes or new elements. The stable product of an element's radioactive decay is known as a radiogenic isotope. For example, lead (Pb; $Z = 82$) has four naturally occurring isotopes of different masses (${}^{204}Pb$, ${}^{206}Pb$, ${}^{207}Pb$, ${}^{208}Pb$). Only the isotope ${}^{204}Pb$ is stable. The isotopes ${}^{206}Pb$ and ${}^{207}Pb$ are daughter (or progeny) products from the radioactive decay of uranium (U), while ${}^{208}Pb$ is a product from thorium (Th) decay. Owing to the radioactive decay, the heavier isotopes of lead will increase in abundance compared to ${}^{204}Pb$.

The kinds of chemical reactions for all isotopes of the same element are the same. However, the rates of reactions may vary. This can be an important factor, for example, in dating material. Such processes have been used to ascertain the sources of pollution. (See Discussion Box "Engineering Technical Note: Source Apportionment, Receptor Models, and Carbon Dating.")

Radiogenic isotopes are useful in determining the relative age of materials. The length of time necessary for the original number of atoms of a radioactive element in a rock to be reduced by half (*radioactive half-life*) can range from a few seconds to billions of years. Scientists use these "radioactive clocks" to date material[1] by:

1. Extracting and purifying the radioactive parent and daughter from the relevant rock or mineral;
2. Measuring variations in the masses of the parent and daughter isotopes; and
3. Combining the abundances with the known rates of decay to calculate an age.

Radiogenic isotopes are being increasingly used as *tracers* of the movement of substances through the environment. Radiogenic isotope tracer applications using Pb, strontium (Sr), and neodymium (Nd), among others, make use of the fact that these are heavy isotopes, in contrast to lighter isotopes such as hydrogen (H), oxygen (O), and sulfur (S). Heavy isotopes are relatively unaffected by changes in temperature and pressure during transport and accumulation, variations in the rates of chemical reactions, and the coexistence of different chemical species available in the environment. Chemical reactions and processes involving Pb, for example, will not discriminate among the naturally occurring isotopes of this element on the basis of atomic mass differences (^{204}Pb, ^{206}Pb, ^{207}Pb, ^{208}Pb).

Long-term monitoring data are frequently not available for environmental systems, so indirect methods, like radiogenic isotope calculations must be used. For example, in sediments, chronological scales can be determined by the distribution of radioactive isotopes in the sediment, based upon the isotopes' half-lives.[2] The age of the sediment containing a radioactive isotope with a known half-life can be calculated by knowing the original concentration of the isotope and measuring the percentage of the remaining radioactive substance. For this process to work the chemistry of the isotope must be understood, the half-life known, and the initial amount of the isotope per unit substrate accurately estimated. The only change in concentration of the isotope must be entirely attributable to radioactive decay, with a reliable means for measuring the concentrations. The effective range covers approximately eight half-lives. The four isotopes meeting these criteria (i.e. cesium-137 (^{137}Cs), beryllium-7 (^{7}Be), ^{14}C, and ^{210}Pb) are being used to estimate how sediment has moved (e.g. by deposition and lateral transport) for the last 150 years.

The radio-dating process is analogous to an hourglass (see Figure 6.1), where the number of grains of sand in the top reservoir represents the parent isotope and the sand in the bottom reservoir represents the daughter isotopes. A measurement of the ratio of the number of sand grains in the two reservoirs will give the length of time that the sand has been flowing, which represents the process of radioactive decay. For deposited material like sediment, the counting begins when the sediment particle is deposited (t_0) and the exchange between the water and particle ceases. As the sediment particles are subsequently buried, the parent isotope decays to the daughter products.

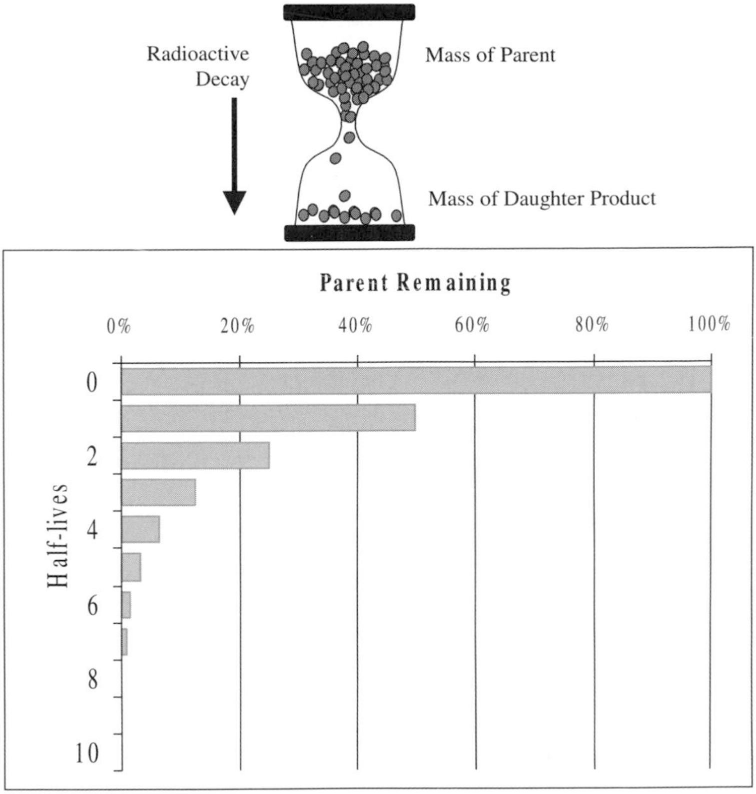

FIGURE 6.1. Radio-dating of environmental material, such as sediments, is a function of the radioactive decay of specific isotopes in the environmental compartment. The hourglass analogy holds, where the number of grains of sand in the top reservoir represents the parent isotope and the sand in the bottom reservoir represents the daughter isotopes. A measurement of the ratio of the number of sand grains in the two reservoirs will give the length of time that the sand has been flowing (radioactive decay). (Adapted from: U.S. Geological Survey, 2003, "Short-Lived Isotopic Chronometers: A Means of Measuring Decadal Sedimentary Dynamics," FS-073-98.)

Engineering Technical Note: Source Apportionment, Receptor Models, and Carbon Dating

When the results of air pollution measurements are interpreted, one of the first questions asked by scientists, engineers, and policy makers is where did it come from? Sorting out the various sources of pollution is known as *source apportionment*. A number of tools are used to try to locate the sources of pollutants. A widely used approach is the "source-receptor model," or as it is more commonly known, the *receptor model*.

Receptor models are often distinguished from the atmospheric and hydrologic dispersion models. For example, dispersion models usually start from the source and estimate where the plume and its contaminants is heading (see Figure 5.1). Conversely, receptor models are based upon measurements taken in the ambient environment and from these observations, make use of algorithms and functions to determine pollution sources. One common approach is the mathematical "back trajectory" model. Often, chemical co-occurrences are applied. It may be that a certain fuel is frequently contaminated with a conservative and, ideally, unique element. Some fuel oils, for example, contain trace amounts of the element vanadium. Since there are few other sources of vanadium in most ambient atmospheric environments, its presence is a strong indication that the burning of fuel oil is a most likely source of the plume. The model, if constructed properly, can even quantify the contribution. If measurements show that sulfur dioxide (SO_2) concentrations are found to be $10\,\mu g\,m^{-3}$ in an urban area, and vanadium is also found at sufficient levels to indicate that home heating systems are contributing a certain amount of the SO_2 to the atmosphere, the model will correlate the amount of SO_2 coming from home heating systems. If other combustion sources, such as cars and power plants, also have unique trace elements associated with their SO_2 emissions, further SO_2 source apportionment can occur, so that the total may look something like Table 6.2.

Receptor models need tracers that are sufficiently sensitive and specific to identify sources. One very promising development for such tracers is the comparison of carbon isotopes. Since combustion involves the oxidation of organic matter, which always contains carbon, it stands to reason that if there were a way to distinguish "old carbon" from "new carbon," we would have a reliable means of differentiating fossil fuels from *biogenic* hydrocarbon sources (e.g., volatile organic carbons released from coniferous trees, including pinene). As the name implies, fossil fuels are made up of carbon deposited long ago, and until now, the carbon has been sequestered. During that time the ratio of the isotopes of carbon has changed. The ratios can tell us, then, whether the carbon we are measuring had been at first sequestered a few years ago or many thousands of years ago.

Naturally-occurring radioactive carbon (^{14}C) is present at very low concentrations in all biotic (living) matter. The ^{14}C concentrations result from plants' photosynthesis of atmospheric carbon dioxide (CO_2), which contains all of the natural isotopes of carbon. However,

TABLE 6.2
Hypothetical Source Apportionment of Measured Sulfur Dioxide Concentrations

Source	*Distance from Measurement (km)*	*SO_2 Concentration Contributed to Ambient Measurement ($\mu g\,m^{-3}$)*	*Percent Contribution to Measured SO_2*
Coal-fired electric generating station	25	3.0	30
Coal-fired electric generating station	5	2.0	20
Mobile sources (cars, trucks, trains, and planes)	0–10	1.5	15
Oil refinery	30	1.5	15
Home heating (fuel oil)	0–1	1.0	10
Unknown	Not applicable	1.0	10
Total		10.0	100

no ^{14}C is found in fossil fuels since all of the carbon has had sufficient time to undergo radioactive decay. Studies have begun to take advantage of this dichotomy in ratios. For example, they have begun to address an elusive contributor to particulate matter (PM), or *biogenic* hydrocarbons. In the summer months, biogenic aerosols are formed from gas-to-particle atmospheric conversions of volatile organic compounds (VOCs) that are emitted by vegetation.[3] New methods for estimating the contribution of biogenic sources of VOCs and PM are needed, because current estimates of the importance of biogenic aerosols as contributors to total summertime PM have very large ranges (from negligible to dominant). There are large uncertainties in both the conversion mechanisms, and the amount and characteristics of biogenic VOC emissions.

The good news seems to be that direct experimental estimates can be gained by measuring the quantity of ^{14}C in a PM sample. The method depends on the nearly constant fraction of ^{14}C relative to ordinary carbon (^{12}C) in all living and recently living material, and its absence in fossil fuels. The fine fraction of PM ($PM_{2.5}$) summertime samples are available from numerous locations in the United States, from which ^{14}C measurements can be conducted. Some recent studies have shown that the carbonaceous biogenic fraction may be contributing as much as one-half of the particles formed from VOCs!

The method for measuring and calculating the isotope ratios is straightforward. The percent of modern carbon (pMC) equals the percentage of ^{14}C in a sample of unknown origin relative to that in a sample of living material, and this pMC is about equal to the percentage of carbon in a sample that originated from nonfossil (i.e., biogenic) sources. So, for sample x:[4]

$$pMC_x = \frac{(^{14}C/^{13}C)_X}{0.95 \cdot (^{14}C/^{13}C)_{SRM\ 4990B}} \times 100 \qquad \text{Equation 6–2}$$

Where the numerator is the ratio measured in the $PM_{2.5}$ sample, and the denominator is the ratio measured using the method specified by the National Institute of Standards and Testing (NIST) for modern carbon.[5] Further:

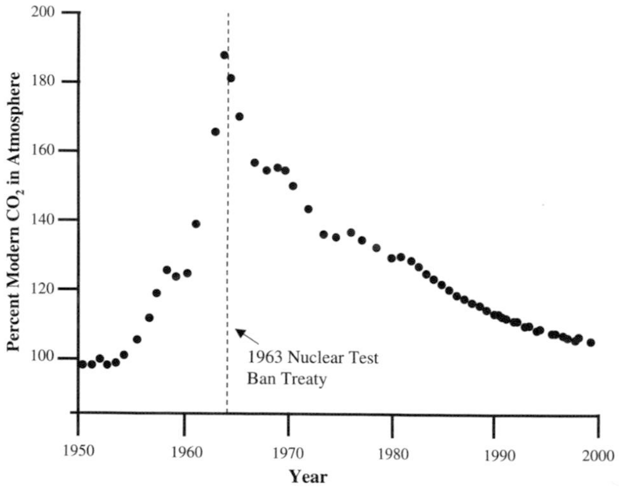

FIGURE 6.2. Biospheric ^{14}C enhancement of atmospheric modern carbon as a result of radiocarbon additions from nuclear testing and nuclear power generation. The plot indicates the time record of ^{14}C in the biosphere. The ^{14}C content of northern hemisphere biomass carbon was doubled in 1963, but since the cessation of atmospheric nuclear testing, the excess ^{14}C is now nearing natural, cosmic ray background levels. Fraction of modern carbon relative standard uncertainties are typically <0.5%. (Sources: National Institute of Standards and Technology, 2002, "A Critical Evaluation of Interlaboratory Data on Total, Elemental, and Isotopic Carbon in the Carbonaceous Particle Reference Material," NIST SRM 1649a, 107 (3); and C. Lewis, G. Klouda, and W. Ellenson, 2003, "Cars or Trees: Which Contribute More to Particulate Matter Air Pollution?" U.S. Environmental Protection Agency, Science Forum, Washington, D.C.)

$$pMC_{FossilFuel} = 0 \qquad \text{Equation 6–3}$$

Thus, for a sample x, the biogenic fraction is:

$$\%BiogenicC_x = \frac{pMC_x}{pMC_{Biognic}} \times 100 \qquad \text{Equation 6–4}$$

The 0.95 correction is needed to address the increasing atmospheric concentrations of radiocarbon due to nuclear weapons testing in the 1950s and 1960s (see Figure 6.2), and to calibrate the measurements with the standard used for radiocarbon dating (i.e., wood from 1890).[6] Although the levels have dropped since the 1963 test ban treaty, they are still elevated above the pre-1950s background level.

Expressions of Chemical Characteristics

The gravimetric fraction of an element in a compound is the fraction by mass of the element in that compound. This is found by a gravimetric (or ultimate) analysis of the compound. The empirical formula of a compound provides the relative number of atoms in the compound. The empirical formula is found by dividing the gravimetric fractions (percent elemental composition) by atomic weights of each element in the compound, and dividing all of the gravimetric fraction-to-atomic weight ratios by the smallest ratio.

Empirical Formula Development Example

An air sampling stainless steel canister was evacuated by the local fire department and brought to the environmental laboratory for analysis. The person who brought in the sample said that the sample was taken near a site where a rusty 55-gallon drum was found by some children in a creek near their school. The children and neighbors reported an unpleasant smell near the site where the drum was found.

The gravimetric analysis of the gas in the canister indicated the following elemental compositions:

Carbon: 40.0%

Hydrogen: 6.7%

Oxygen: 53.3%

Solution

First, divide the elemental percentage compositions by the respective atomic weights:

$$C: \frac{40.0}{12} = 3.3$$

$$H: \frac{6.7}{1} = 6.7$$

$$O: \frac{53.3}{16} = 3.3$$

Next, divide every ratio by the smallest ratio (3.3):

$$C: \frac{3.3}{3.3} = 1$$

$$H: \frac{6.7}{3.3} = 2$$

$$O: \frac{3.3}{3.3} = 1$$

So, the empirical formula is CH_2O or HCHO. This is formaldehyde, a toxic substance.

Preliminary Interpretation

The challenge of formaldehyde is that it comes from many sources, including emissions from factories and automobiles and even natural sources. However, since the drum seems to be a likely source, the liquid contents should be analyzed (and a search for additional drums should begin immediately—illegal dumping often is not limited to a single unit).

The first likelihood is that the liquid is formalin, a mixture that contains formaldehyde. The high vapor pressure of the formaldehyde may be causing it to leave the solution and move into the air.

Since children are in the area and there may be a relatively large amount of the substance, steps must be taken to prevent exposures and to remove the formaldehyde.

The Periodic Table

The Periodic Table (Appendix 7) follows the Periodic Law, which states that the properties of elements depend on the atomic structure and vary systematically according to atomic number. The elements in the table are arranged according to increasing atomic numbers from left to right.

An element shares many physicochemical properties with its vertical neighbors, but differs markedly from its horizontal neighbors. For example, oxygen (O) will chemically bind and react similarly to sulfur (S) and selenium (Se), but behaves very differently from nitrogen (N) and fluorine (F). Elements in the horizontal rows, known as periods, grow increasingly different with the distance moved to the left or right. So, O differs physically and chemically more from boron (B) than O does from F, and O is a very different from lithium (e.g., O is a nonmetal and Li is a light metal).

The groups (vertical columns) are designated by numerals (often Roman numerals). For example O is a group VIA element and gold (Au) is in group IB. The A and B designations are elemental families. Elements within families share many common characteristics. Within families, elements with increasing atomic weights become more metallic in their properties.

Metals (elements to the left of the periodic table) form positive ions (*cations*), are reducing agents, have low electron affinities, and have positive valences (oxidation numbers). Nonmetals (on the right side of the table) form negative ions (*anions*), are oxidizing agents, have high electron affinities, and have negative valences. Metalloids have properties of both metals and nonmetals. However, two environmentally important metalloids, arsenic (As) and antimony (Sb), are often treated as heavy metals in terms of fate, transport, and toxicity.

Some common periodic table chemical categories are:

- Metals—Every element except the nonmetals
- Heavy Metals—Metals near the center of the table
- Light Metals—Groups I and II
- Alkaline Earth Metals—Group IIA
- Alkali Metals—Group IA
- Transition Metals—All group VIII and B families
- Actinons—Elements 90–102
- Rare Earths—Lanthanons (Lanthanides), Elements 58–71
- Metalloids—Elements separating metals and nonmetals, Elements 5, 14, 32, 33, 51, 52, and 84
- Nonmetals—Elements 2, 5–10, 14–18, 33–36, 52–54, 85, and 86
- Halogens—Group VIIA
- Noble Gases—Inert elements, Group 0

It is important to keep in mind that every element in the table has environmental relevance. In fact, at some concentration, every element except those generated artificially by fission in nuclear reactors, are found in soils. Thus, it would be absurd to think of how to "eliminate" them. This is a common misconception, especially with regard to heavy metal and metalloid contamination. For example, mercury (Hg) and lead (Pb) are known to be important contaminants that cause neurotoxic and other human health effects and environmental pollution. However, the global mass balance of these metals does not change, only their locations and forms (i.e., *speciation*). Protecting health and ecological resources is thus a matter of reducing and eliminating exposures and changing the form of the compounds of these elements so that they are less mobile and less toxic. The first place to start such a strategy is to consider the oxidation states, or valence, of elements.

Electromagnetic Radiation, Electron Density, Orbitals, and Valence

Quantum mechanics tells us that the energy of a photon of light can cause an electron to change its energy state, so that the electron is disturbed from its original state. Let us briefly consider how *electromagnetic radiation* (EMR) is related to atomic structure. Much that is known about atomic structure, especially an atom's arrangement of electrons around its nucleus, is from what scientists have learned about the relationship between matter and different types of EMR. One principle is that EMR has properties of both a *particle* and a *wave*. Particles have a definite mass and occupy space; that is, they conform to the classic description of matter. Waves have no mass but hold energy with them as they travel through space. Waves have four principle characteristics: speed (v), frequency (ν), wavelength (λ), and amplitude. These are demonstrated in Figure 6.3.

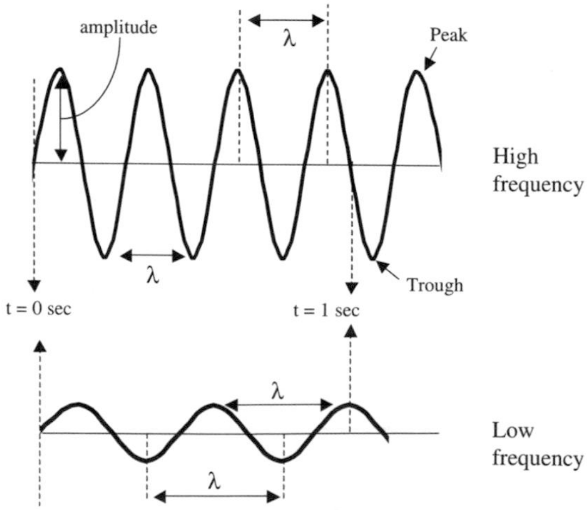

FIGURE 6.3. Electromagnetic radiation. The amplitude of the wave in the top chart is higher than that in the lower chart. The bottom wave is 2.5 cycles per seconds (2.5 hertz or 2.5 Hz). The top wave is 3.5 hertz, so the bottom wave has a 1 Hz lower frequency than the top wave.

Measuring ν in cycles per second (hertz, Hz) and λ in meters, the product gives the velocity of the wave moving through space:

$$\mathrm{v} = \nu\lambda \qquad \text{Equation 6–5}$$

For example, if a certain light's λ is 10^{-7} m and its ν is 10^{15} Hz, then the velocity of that light is $10^{8}\,\mathrm{m\,s^{-1}}$.

Field Manual Entry: Physical Contaminants—Electromagnetic Radiation

Most of the time, the term *environmental contaminant* calls to mind some chemical compound. However, contaminants may also come in the form of biological or physical agents. Biological contaminants may be pathogenic bacteria or viruses that adversely affect health, or introduced species (e.g., the zebra mussel or kudzu) that harm ecosystems. Physical agents are often the least likely to come to mind. A common physical contaminant is energy. Life depends on energy, but like most resources, when it comes in the wrong form and quantity, it may be harmful. Electromagnetic radiation (EMR) is comprised of wave functions that are propagated by simultaneous periodic variations in electrical and magnetic field intensities. Natural and many anthropogenic sources produce EMR energy in the form of waves, which are oscillating energy fields that can interact with an organism's cells. The waves are described according to their wavelength and frequency, and the energy that they produce.

Wave frequency is the number of oscillations that passes a fixed point per unit of time, measured in cycles per second (cps). 1 cps = 1 hertz (Hz). Thus, the shorter the wavelength, the higher the frequency. For example, the middle of the amplitude modulated (AM) radio broadcast band has a frequency of one million hertz (1 MHz) and a wavelength of about 300 m. Microwave ovens use a frequency of about 2.5 billion hertz (2.45 GHz) and a wavelength of 12 cm. So the microwave, with its shorter wavelength has a much higher frequency.

An EMR wave is made of packets of energy called photons. The energy in each photon is directly proportional to the frequency of the wave: the higher the frequency, the more energy there will be in each photon. Cellular material is affected in part by the intensity of the field and partly by the quantity of energy in each photon.

At low frequencies, EMR waves are known as electromagnetic fields, and at high frequencies EMR waves are referred to as electromagnetic radiations. Also, the frequency and energy determines whether an EMR will be ionizing or nonionizing radiation. Ionizing radiation consists of high frequency electromagnetic waves (e.g., X-rays and gamma rays), having sufficient photon energy to produce ionization (producing positive and negative electrically-charged atoms or parts of molecules) by breaking bonds of molecules. The general term *nonionizing radiation* refers to the portion of the electromagnetic spectrum where photon energies are not strong enough to break bonds. This segment of the spectrum includes ultraviolet (UV) radiation, visible light, infrared radiation, radio waves, and microwaves, along with static electrical and magnetic fields. Even at high intensities, nonionizing radiation cannot ionize atoms in biological systems, but such radiation has been associated with other effects, such as cellular heating, changes in chemical reactions and rates, and the induction of electrical currents within and between cells.

Not every EMR effect causes harm to an organism, such as when a mammal may respond to EMR by increasing blood flow in the skin in response to slightly greater heating from the sun. Effects may even be beneficial, such as the feeling of warmth of direct sunshine even when the air is very cold. EMR may also induce positive health effects, such as the sun's role in helping the body produce vitamin D. Unfortunately, certain direct or indirect responses to EMR may lead to adverse effects, including skin cancer.

The data supporting UV as a contaminant are stronger than that associated with more subtle fears that sources like high-energy power transmission lines and cell phones may be producing health effects. The World Health Organization (WHO) is addressing the health concerns raised about exposure to radio frequency (RF) and microwave fields, intermediate frequencies (IF), extremely low frequency (ELF) fields, and static electric and magnetic fields. Intermediate frequency (IF) and radio frequency (RF) fields produce heating and the induction of electrical currents, so it is highly plausible that these processes are also occurring to some extent in cells exposed to IF and RF fields. Fields at frequencies above about 1 MHz primarily cause heating by transporting ions and water molecules through a medium. Even very low energy levels generate a small amount of heat, but this heat is carried away by the body's normal thermoregulatory processes. However, some studies indicate that exposure to fields too weak to cause heating may still produce adverse health consequences, including cancer and neurological disorders (e.g., memory loss).

Since, electrical currents already exist in the body as a normal part of the biochemical reactions and metabolic process, the fear is that should electromagnetic fields induce sufficiently high currents, the additive effects may overload the system and engender adverse biological effects.

Extremely low frequency (ELF) electric fields exist when a charge is generated, but hardly any of the electric field penetrates into the human body. At very high field strengths they can feel like one's skin is "crawling" or one's hair is raised. Some studies, however, have associated low level ELF electric fields with elevated incidence of childhood cancer or other diseases, while other studies have not been able to establish a relationship. The WHO is recommending that more focused research be conducted to improve health risk assessments. ELF magnetic fields also exist whenever an electric current is flowing. However, unlike the ELF electric fields, magnetic fields readily penetrate an organism's tissue with virtually no attenuation. Again the epidemiology is mixed, with some studies associating ELF fields with cancer, especially in children, and others finding no such association.

The primary action in biological systems by these static electrical and magnetic fields is by inducing electrical currents and charges in cells and tissues. Reliable data are needed to come to any conclusions about chronic effects associated with long-term exposure to static magnetic fields at levels found in the working environment.[7]

The challenge of EMR is similar to that of chemical contamination. The exposure and risks associated with this hazard is highly uncertain. The key decision is whether sufficient scientific evidence exists to encourage limits on certain types of activities (either as an individual or as a public agency) where adverse effects may be occurring. Just walk in a mall, on a campus, or into a restaurant. Chances are you will see a number of people using cell phones. Drive through a neighborhood. Chances are you will see some overhead power lines. What level of certainty do you need before you change your behavior? What certainty needs to exist before you recommend that the government protect us from ourselves?

An atom's electrons occupy orbitals where the electrons contain various amounts of energy. Electrons vary in the spatial orientations and average distances from the nucleus, so that electrons occupying inner orbitals are closer to the nucleus (see Figure 6.4). The electron's velocity or position is constantly changing. Therefore, if we can measure the electron's position we will not know its velocity, and if we measure the electron's velocity, we will not be able to know its position. We are uncertain about the electron's simultaneous position and velocity. This is the basis for the Heisenberg

Uncertainty Principle, which tells us that the more we know about an electron's position, the less we can know about its velocity. Further, to keep an electron from escaping from an atom, the electron must maintain a minimum velocity, which corresponds back to the uncertainty of the electron's position in the atom. Since the uncertainty of the position is actually the whole atom, because the electron can be anywhere in the atom, chemists refer to an electron as a "cloud of electron density" within the atom rather than describing an electron as a finite particle. The electron orbitals are envisioned as regions in space where the electrons are statistically most likely to be located. We do not know where an electron is, but we know where it might be—that is, somewhere in the electron cloud.

The Schrödinger Model applies three coordinates to locate electrons, known as quantum numbers. The coordinates are principal (n), angular (ℓ), and magnetic (m_l) quantum numbers. These characterize the shape, size, and orientation of the electron cloud orbitals of the atom. The principal quantum number n gives the size of the orbital. A relative size of $n = 2$ is larger than a cloud with the size $n = 1$. Energy is needed to excite an electron to make it move from a position closer to the nucleus (e.g., $n = 1$) to a position further from the nucleus ($n = 2$, 3, or higher). So n is an indirect expression of an orbital's energy level.

The angular quantum number ℓ maps the shape of the cloud. A spherical orbital has an $\ell = 0$. Polar shaped orbitals have $\ell = 1$. Cloverleaf orbitals have $\ell = 2$. See Figure 6.4 for renderings of these shapes. The magnetic quantum number m_ℓ describes the orientation of the orbital in space. Orbitals with the same value of n form a shell. Within a shell, orbitals are divided into subshells labeled by their ℓ value. The commonly used two-character description of shells, such as $2p$ or $3d$, exemplifies the shell and subshell. For example, $2p$ indicates the shell ($n = 2$) and the subshell (p). Subshells are indicted by:

$$s: \ell = 0$$

$$p: \ell = 1$$

$$d: \ell = 2$$

$$f: \ell = 3$$

$$g: \ell = 4$$

$$h: \ell = 5$$

The number of subshells in any shell will equal the n for the shell. For example, the $n = 2$ shell contains two subshells, $2s$ and $2p$, and $n = 5$ shell contains five subshells, $5s$, $5p$, $5d$, $5f$, and $5g$ orbitals.

The electrons occupying the outermost shell are known as *valence electrons*. Valence is the number of bonds that an element can form, which is related to number of electrons in the outermost shell. The arrangement of the electrons in the outermost, or valence, determines the ultimate chemical behavior of the atom. The outer electrons become involved in transfer to and sharing with shells in other atoms, or the forming of new compounds and ions. Note that the number of valence electrons in an "A" group in the periodic table (except helium, whose shell is filled with 2) is equal to the group number. So sodium (Na), a Group 1A element, has one valence electron. Carbon (C) is a Group 4A compound, so it has four valence electrons. Chlorine (Cl), fluorine (F), and the other halogens of Group 7A have seven valence electrons.

The noble gases, except He, in Group 8A, have eight electrons in the outermost shell. The noble gases are actually the only elements that exist as individual atoms, because the noble gases have no valence electrons. At the opposite end, the fact that carbon has four electrons in its outermost shell is important. It has just as many electrons to gain or to lose (i.e., 4 is just as close to 8, for a newly filled shell, as it is to 0, for the loss of a shell), so there are many ways for it to reach chemical stability. This is one of the reasons that so many subtly, but profoundly different compounds, such as the hundreds of thousands of organic compounds, are in existence. Therefore, most atoms combine by chemical bonding to other atoms, creating molecules.

The outermost electrons tell the story of how readily an element will engage in a chemical reaction and the type of reaction that will occur. The oxidation number is the electrical charge assigned to an atom. It assumes that all bonding of that atom will be ionic (which we know is not the case). The sum of the oxidation numbers is equal to the net charge. Table 6.3 shows the oxidation numbers of certain atoms that form contaminants and nutrients in the environment. Table 6.4 gives the oxidation numbers for environmentally important radicals, i.e. groups of atoms that combine and behave as a single chemical unit. An atom will gain or lose valence electrons to form a stable ion that has the same number of electrons as the noble gas nearest the atom's atomic number. For example, Na with a single valence electron and a total of 11 electrons, will tend to lose an electron to form Na^+, the sodium cation. This ion has the same number of electrons (10 total, 8 in its outer shell) as the nearest noble gas, neon (Ne). Fluorine, with 7 valence electrons and 9 total electrons, tends to gain (accept) an electron to form a 10-electron fluorine anion, F^- that, like the sodium ion, has the same number of electrons as Ne (10).

Noble gases have an octet (i.e., group of 8) of electrons in their valence shells, meaning that the tendency of an atom to gain or to lose its valence electrons to form ions in the noble gas arrangement is called the "octet rule."[8] Chemical species of atoms are particularly stable when their outermost shells contain 8 electrons.

Recalling the previous discussion regarding polarity with respect to solubility, the relative ability of an atom to draw electrons in a bond toward itself is known as that atom's *electronegativity* (EN). If we think about the

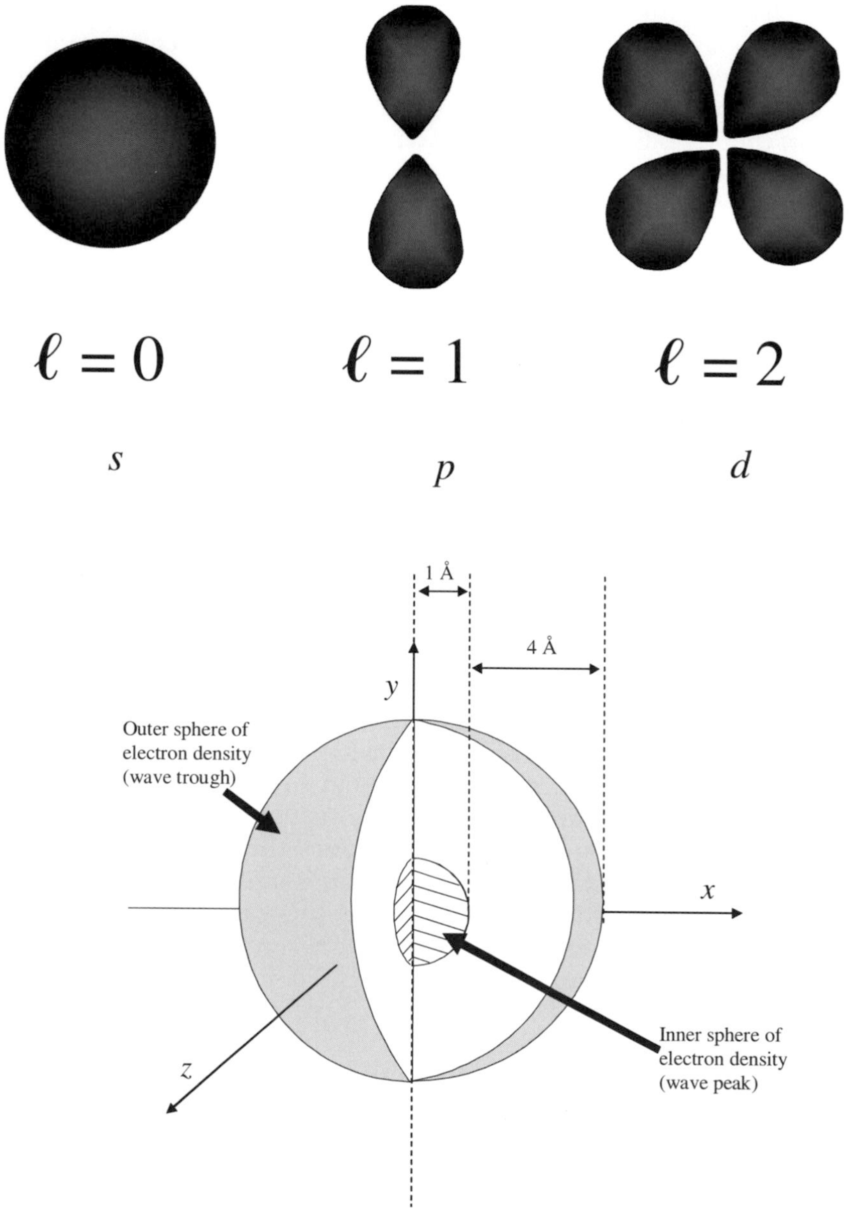

FIGURE 6.4. Two-dimensional orbital shapes, showing three angular quantum numbers (ℓ) and subshells *s*, *p*, and *d* (top). The three-dimensional 2*s* orbital (bottom) is shown as a cutaway view into the atom depicting the peak (inner sphere) and trough (outer concentric sphere) of the electron wave. The orbital is two concentric spheres of electron densities. (Adapted from: G. Loudon, 1995, *Organic Chemistry*, 3rd Edition, Benjamin/Cummings Publishing, Redwood City, Calif.)

TABLE 6.3
Oxidation Numbers for Environmentally Important Atoms

Atom	*Chemical Symbol*	*Oxidation Number(s)*
Aluminum	Al	+3
Antimony	Sb	−3, +3, +5
Arsenic	As	−3, 0, +3, +5
Barium	Ba	+2
Boron	B	+3
Calcium	Ca	+2
Carbon	C	+2, +3, +4, −4
Chlorine	Cl	−1
Chromium	Cr	+2, +3, +6
Cobalt	Co	+2, +3
Copper	Cu	+1, +2
Fluorine	F	−1
Gold	Au	+1, +3
Hydrogen	H	+1
Iron	Fe	+2, +3
Lead	Pb	+2, +4
Lithium	Li	+1
Magnesium	Mg	+2
Manganese	Mn	+2, +3, +4, +6, +7
Mercury	Hg	0, +1, +2
Nickel	Ni	+2, +3
Nitrogen	N	−3, +2, +3, +4, +5
Oxygen	O	−2
Phosphorus	P	−3, +3, +5
Plutonium	Pu	+3, +4, +5, +6
Potassium	K	+1
Radium	Ra	+2
Radon	Rn	0 (noble gas)
Selenium	Se	−2, +4, +6
Silver	Ag	+1
Sodium	Na	+1
Sulfur	S	−2, +4, +6
Tin	Sn	+2, +4
Uranium	U	+3, +4, +5, +6
Zinc	Zn	+2

forms of chemical binding as a two-dimensional entity rather than linear, the types of bonds can appear as in Figure 6.5. Thus, the binding in sodium chloride is ionic, the metal sodium's bond to a neighboring sodium atom is metallic, and molecular chlorine is covalent. However, moving into the interior of the triangle shows bonds that have characteristics of ionic, metallic, and/or covalent. For example, binding lithium chloride possesses char-

TABLE 6.4
Oxidation Numbers for Environmentally Important Radicals

Radical	*Chemical Symbol*	*Oxidation Number(s)*
Acetate	$C_2H_3O_2$	−1
Acrylate	$CHCO_2$	−1
Ammonium	NH_4	+1
Bicarbonate	HCO_3	−1
Borate	BO_3	−3
Carbonate	CO_3	−2
Chlorate	ClO_3	−1
Chlorite	ClO_2	−1
Chromate	CrO_4	−2
Cyanide	CN	−1
Dichromate	Cr_2O_7	−2
Hydroperoxide	HO_2	−1
Hydroxide	OH	−1
Hypochlorite	ClO	−1
Nitrate	NO_3	−1
Nitrate	NO_2	−1
Perchlorate	ClO_4	−1
Permanganate	MnO_4	−1
Phosphate	PO_4	−3
Sulfate	SO_4	−2
Sulfite	SO_3	−2
Thiocyanate	SCN	−1
Thiosulfate	S_2O_3	−2

acteristics of ionic and covalent, but not metallic bonds. Conversely, trisodium bismuth's binding is ionic and metallic, but not covalent. More recently, the two-dimensional approach has been made to be more quantitative by the so-called "bond-type diagrams."

Elements combine to form compounds. Two-element compounds are known as binary compounds. Three-element compounds are ternary or tertiary compounds. The representation of the relative numbers of each element is a chemical formula. Compounds are formed according to the law of definite proportions. That is, a pure compound must always be composed of the same elements that are always combined in a definite proportion by mass. Also, compounds form consistently with the law of multiple proportions, which states that when two elements combine to make more than one compound, the combining mass of each element must always be as small as integer ratios to one another. The sum of all oxidation numbers must equal zero in a stable, neutral compound. The simplest example is water. Oxygen's −2 valence is balanced by the two hydrogen's +1 valences.

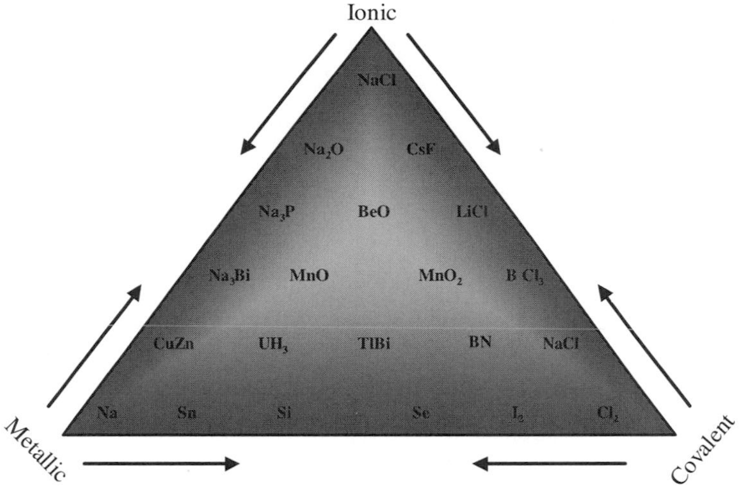

FIGURE 6.5. Diagram assuming that three different bond types, covalent, metallic, and ionic, form a two-dimensional relationship according to the likelihood of elements to form bonds. Elements closest to the corners have those types of bonds almost exclusively, while those away from the corners have a combination of bond types. (Source: W. Jolly, 1974, *The Principles of Inorganic Chemistry*, McGraw-Hill, New York, N.Y.)

Compound Formation Example

Is $PbC_2H_3O_2$ a valid compound?

Answer

Consulting Tables 6.3 and 6.4, we find that lead (Pb) has two common oxidation numbers (+2 and +4), and that acetate ($C_2H_3O_2$) has an oxidation number of −1, so the molecular formula given is *not* valid.

Lead acetate is $Pb(C_2H_3O_2)_2$. Two atoms of acetate are needed to balance the +2 valence of Pb. The molecule is often called lead (II) acetate to show that in this instance it is the "divalent" form of lead that has reacted with the acetate radical. Incidentally, lead (II) is a suspected human carcinogen, based upon experiments conducted on laboratory animals. It is also very acutely toxic and may be fatal if swallowed, and is harmful if inhaled or absorbed through the skin. Like other lead compounds, long-term exposure may harm the central nervous system, blood, and gastrointestinal tract.

Lab Notebook Entry: Importance of Valence State in Arsenic Contamination and Treatment[9]

The metalloid arsenic (As) is a constituent of rocks and soils, as well as in industrial products and processes, including wood preservatives, paints, dyes, metals, pharmaceuticals, pesticides, herbicides, soaps, and semiconductors. *Anthropogenic* sources of arsenic include mining and smelting operations; agricultural applications; and waste disposal. Arsenic is one of the most important contaminants, as shown in Figure 6.6. In the United States, As is a *contaminant of concern* (COC) at 568 hazardous waste sites or 47% of the 1209 sites on the National Priority Listing (NPL). The contamination occurs in every environmental medium (Table 6.5).

Like every element on the period table, As cannot be "destroyed" per se, but its form can be altered to make it less mobile, more treatable, and less toxic. The most important consideration for most heavy metal and metalloid contamination is the particular valence state of the element, and the concomitant compounds formed. Arsenic occurs with valence states of –3, 0, +3 (arsenate, As[III]), and +5 (arsenate, As[V]). Because the valence states –3 and 0 occur only rarely in the environment, most attention is paid to As(III) and As(V) contamination. Arsenic forms inorganic and organic compounds. The likelihood of transport, affinity for various environmental compart-

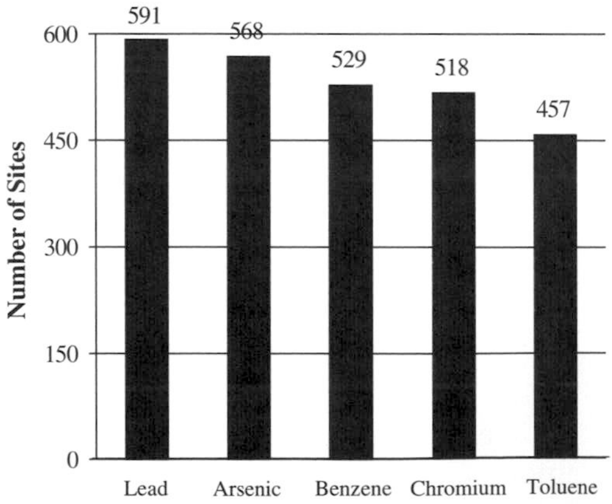

FIGURE 6.6. Five most commonly found contaminants at high priority waste sites in the United States (National Priority Listing sites). (Source: U.S. Environmental Protection Agency, 2002, "Proven Alternatives for Aboveground Treatment of Arsenic in Groundwater," Engineering Forum Issue Paper, EPA-542-S-02-002 [Revised], www.epa.gov/tio/tsp/download/arsenic_issue_paper.pdf.)

TABLE 6.5
Environmental Media Where Arsenic Is a Contaminant of Concern at High-Priority U.S. Hazardous Waste Sites Remediated under the Superfund Law

Media	*Number of Sites*
Groundwater	380
Soil	372
Sediment	154
Surface Water	86
Debris	77
Sludge	45
Solid Waste	30
Leachate	24
Liquid Waste	12
Air	8
Residuals	1
Other	21

Source: U.S. Environmental Protection Agency, 2002, "Proven Alternatives for Aboveground Treatment of Arsenic in Groundwater," Engineering Forum Issue Paper, EPA-542-S-02-002 [Revised], www.epa.gov/tio/tsp/download/arsenic_issue_paper.pdf.

ments, and toxicity of arsenic varies according to valence state and chemical form. In most environmental and public health situations As(III) is more toxic to humans and four to ten times more soluble in water than As(V). Various As compounds differ in toxicity and solubility.

Arsenic can change its valence state and chemical form in the environment. Some conditions that could affect arsenic valence and speciation include pH, redox potential, presence of complexing ions, such as those of sulfur, iron, and calcium, and microbial activity. As sorption is also affected by valence. Clay content, carbonaceous materials, and oxides of iron, aluminum, and manganese in soil and sediment affect sorption.

Much controversy has followed the attempt to revise the U.S. drinking water standard for As (known as the *maximum contaminant level* or MCL). In 2001, the U.S. Environmental Protection Agency (EPA) published a revised MCL for arsenic in drinking water that requires public water suppliers to maintain arsenic concentrations $\leq 0.010\,mg\,L^{-1}$ by the year 2006. The MCL was formerly $0.050\,mg\,L^{-1}$. Treatment goals for arsenic at groundwater cleanups can be based on MCLs, background contaminant levels, or risk.

The newly lowered treatment goals for arsenic present multiple technical challenges for the aboveground treatment of groundwater, and will likely result in higher treatment costs. Some sites that might not have needed groundwater treatment to remove arsenic under the former MCL of 0.050 $mg L^{-1}$ may now need to treat groundwater for arsenic to meet the revised MCL. At some sites, the plume of groundwater containing arsenic at concentrations greater than 0.010 $mg L^{-1}$ could be significantly larger in volume and geographic extent than the plume containing arsenic at greater than 0.050 $mg L^{-1}$. Site-specific conditions will determine if new arsenic treatment systems need to be designed, or if existing systems need to be retrofitted to treat the larger volumes. In addition, treatment of groundwater to lower arsenic concentrations can sometimes require the use of multiple technologies in sequence. For example, a site with an existing metals precipitation/coprecipitation system may need to add another technology such as ion exchange to achieve a lower treatment goal.

In some cases, a lower treatment goal might be met by changing the operating parameters of existing systems. For example, changing the type or amount of treatment chemicals used, replacing spent treatment media more frequently, or changing treatment system flow rates can reduce As concentrations in the treatment system effluent. The downside is that such changes may increase operating costs from the use of additional treatment chemicals or media, use of more expensive treatment chemicals or media, and from disposal of increased volumes of treatment residuals.

Oxidation of As(III) to As(V) can improve the performance of the treatment technologies. Chlorine, potassium permanganate, aeration, peroxide, ozone, and photo-oxidation have been used to convert As(III) to As(V). Numerous arsenic treatment systems use oxidation as a pretreatment step to improve performance. Some also add oxidation to subsequent stages. For example, greensand filtration includes oxidation and adsorption of arsenic in one unit operation. Oxidation can either be a pretreatment step or an intrinsic part of another technology, but it is usually not deployed alone as an arsenic treatment.

The U.S. EPA has compared the effectiveness of various steps in removing and treating As contamination (Table 6.6). Arsenic in drinking water, industrial wastewater, surface water, mine drainage, and leachate is often removed using the same technologies as those used to treat groundwater. "Treatment trains" consist of two or more technologies used together, either integrated into a single process or operated as a series of treatment technologies. For example, at one site a treatment train of reverse osmosis followed by ion exchange was used to remove arsenic from surface water. A common treatment train used

TABLE 6.6
Summary of the Effectiveness of Arsenic Treatment Technologies in Achieving Drinking Water Standards

Technology	*Number of Applications Identified[a] (Number with Performance Data)*			*Total Number of Applications Identified (Number with Performance Data)*	*Number of Applications Achieving <0.050 mg/L Arsenic*	*Number of Applications Achieving <0.010 mg/L Arsenic*
	Bench Scale	*Pilot Scale*	*Full Scale*			
Precipitation/Coprecipitation	NC	24 (22)	45 (30)	69 (52)	35	19
Adsorption	NC	8 (5)	15 (9)	23 (14)	12	7
Ion Exchange	NC	0	7 (4)	7 (4)	3	2
Membrane Filtration	6 (0)	25 (2)	2 (2)	33 (4)	4	2

[a]Applications were identified through a search of available technical literature. The number of applications include only those identified during the preparation of the EPA paper, and are not comprehensive.
NC = Data not collected
Source: U.S. Environmental Protection Agency, 2002, "Proven Alternatives for Aboveground Treatment of Arsenic in Groundwater," Engineering Forum Issue Paper, EPA-542-S-02-002 [Revised], www.epa.gov/tio/tsp.

for arsenic-contaminated water includes an oxidation step to change arsenic from As(III) to its less soluble As(V) state, followed by precipitation/coprecipitation and filtration to remove the precipitate. Some treatment trains are used when no single technology is capable of treating all contaminants in a particular medium. For example, at the Saunders Supply Company Superfund Site in Virginia, an aboveground system consisting of metals precipitation, filtration, and activated carbon adsorption was used to treat groundwater contaminated with As and pentachlorophenol (PCP). In this treatment train the precipitation and filtration processes were used for treating arsenic and the activated carbon adsorption process was used to treat PCP. In many instances, engineering judgment must be used to identify the right technology for treating arsenic. For example, at the Higgins Farm Superfund Site in New Jersey, a treatment train consisting of air stripping, metals precipitation, filtration, and ion exchange was used to treat groundwater contaminated with As, nonhalogenated volatile organic compounds (VOCs) and halogenated VOCs. The precipitation, filtration, and ion exchange processes were installed to remove arsenic from the wastewater, while the air stripping process was expected to treat the VOCs but have only a negligible effect on the As concentration.

Precipitation/coprecipitation is frequently used to treat As-contaminated water, and is capable of treating a large range of influent concentrations to the revised MCL for arsenic. This process can treat As with water characteristics or contaminants other than arsenic, such as hardness or heavy metals. Systems using this technology generally require skilled operators; so precipitation/coprecipitation will be most cost-effective at a large scale where labor costs can be spread over a large quantity of treated water generated. The effectiveness of adsorption and ion exchange for arsenic treatment is more readily than precipitation/co-precipitation to be made less efficient by conditions and contaminants other than arsenic. Sorption and ion exchange can treat As to the revised MCL. Small capacity systems using these technologies tend to have lower operating and maintenance costs, and require less operator expertise than is needed for precipitation systems. Thus, adsorption and ion exchange tend to be used more often when arsenic is the exclusive contaminant treated, for relatively smaller systems, and as a "polishing" technology for the effluent from larger systems.

Again, understanding elemental contamination valence state changes is an essential part of protecting the public health and environment.

Organic Chemistry

Carbon is an amazing element. It can bond to itself and to other elements in a myriad of ways. In fact, it can form single, double, and triple bonds with itself. This makes for millions of possible organic compounds. An organic compound is a compound that includes at least one carbon-to-carbon or carbon-to-hydrogen covalent bond.

Most of the hazardous compounds are organic.[10] Generally, the majority organic compounds are more lipophilic (fat soluble) and less hydrophilic (water soluble) than most inorganic compounds. However, there are large ranges of solubility for organic compounds, depending upon the presence of polar groups in their structure. For example, adding the alcohol group to n-butane to produce 1-butanol increases the solubility several orders of magnitude. Also, many inorganic compounds, e.g., some oxides, can be very hydrophobic.

Organic compounds can be classified in two basic groups: aliphatics and aromatics. Hydrocarbons are the most fundamental type of organic compound. They contain only the elements carbon and hydrogen. Aliphatic compounds are classified into a few chemical families. Each carbon normally forms four covalent bonds. Alkanes are hydrocarbons that form chains with each link comprised of the carbon. A single link is CH_4, methane. The carbon chain length increases with the addition of carbon atoms. For example, ethane's structure is:

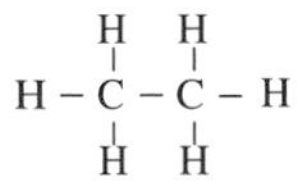

And the protypical alkane structure is:

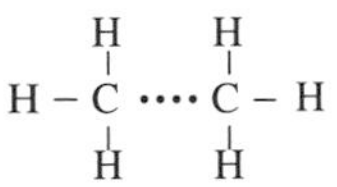

The alkanes contain a single bond between each carbon atom, and include the simplest organic compound, methane (CH_4) and its derivative "chains," such as ethane (C_2H_6) and butane (C_4H_{10}). Alkenes contain at least one double bond between carbon atoms. For example, 1,3-butadiene's structure is $CH_2{=}CH{-}CH{=}CH_2$. The numbers "1" and "3" indicate the position of the double bonds. The alkynes contain triple bonds between carbon atoms, the simplest being ethyne, $CH{\equiv}CH$, which is commonly known as acetylene (the gas used by welders).

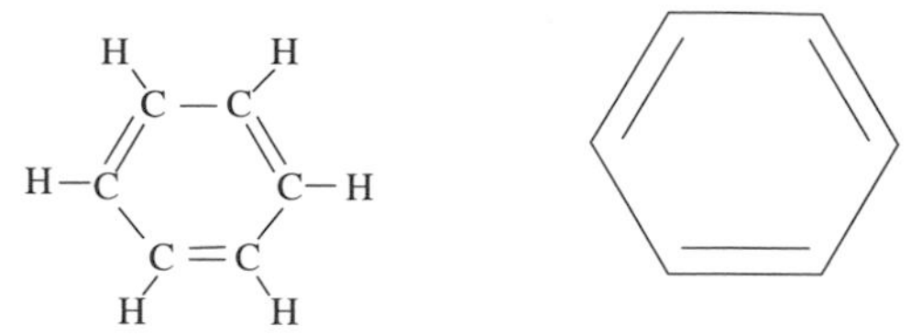

FIGURE 6.7. Benzene structure. The benzene ring on the right is the commonly used condensed form in aromatic compounds.

The aromatics are all based upon the six-carbon configuration of benzene (C_6H_6). The carbon-carbon bond in this configuration shares more than one electron, so that benzene's structure (Figure 6.7) allows for resonance among the double and single bonds; that is, the actual benzene bonds flip locations. Benzene is the average of two equally contributing resonance structures.

The term *aromatic* comes from the observation that many compounds derived from benzene were highly fragrant, such as vanilla, wintergreen oil, and sassafras. All aromatic compounds contain one or more benzene rings. The rings are planar, that is, they remain in the same geometric plane as a unit. However, in compounds with more than one ring, such as the highly toxic polychlorinated biphenyls (PCBs), each ring is planar, but the rings that may be bound together may or may not be planar. This is actually a very important property for toxic compounds. It has been shown that some planar aromatic compounds are often more toxic than their nonplanar counterparts, possibly because living cells may be more likely to allow planar compounds to bind to them and to produce nucleopeptides that lead to biochemical reactions associated with cellular dysfunctions, such as cancer or endocrine disruption.

Both the aliphatic and aromatic compounds can undergo substitutions of the hydrogen atoms. These substitutions render new properties to the compounds, including changes in solubility, vapor pressure, and toxicity. For example, halogenation (substitution of a hydrogen atom with a halogen) often makes an organic compound much more toxic. For example, trichoroethane is a highly carcinogenic liquid that has been found in drinking water supplies, whereas nonsubstituted ethane is a gas with relatively low toxicity. This is also why one of the means for treating the large number of waste sites contaminated with chlorinated hydrocarbons and aromatic compounds involves dehalogenation techniques.

The important functional groups that are part of many organic compounds are shown in Table 6.7.

Structures of organic compounds can induce very different physical and chemical characteristics, as well as change the bioaccumulation and toxicity of these compounds. For example, the differences between the

TABLE 6.7
Structures of Organic Compounds

Chemical Class	*Functional Group*
Alkanes	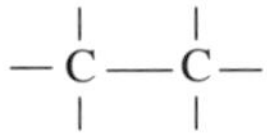
Alkenes	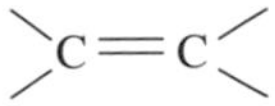
Alkynes	
Aromatics	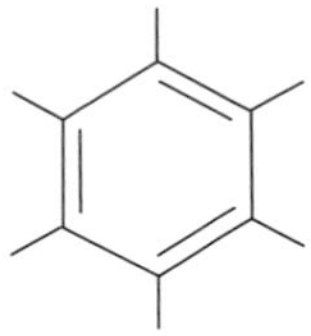
Alcohols	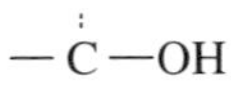
Amines	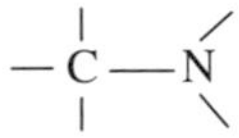
Aldehydes	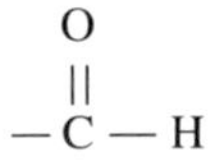
Ether	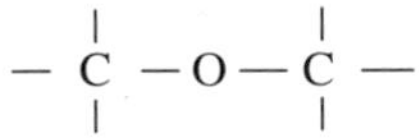
Ketones	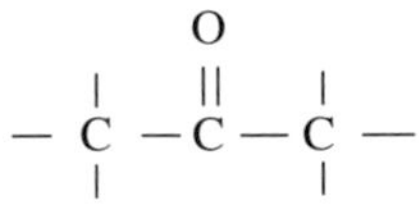
Carboxylic acids	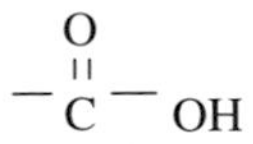
Alkyl halides[11]	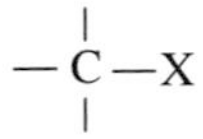
Phenols (aromatic alcohols)	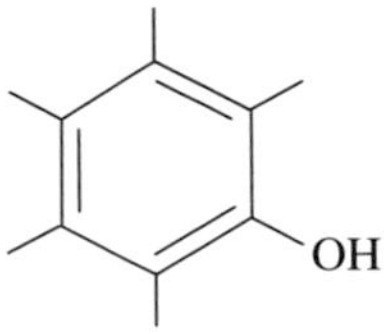

TABLE 6.7 *(continued)*

Substituted aromatics (substituted benzene derivatives):

Nitrobenzene

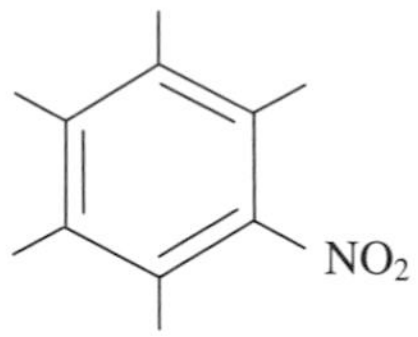

Monosubstituted alkylbenzenes

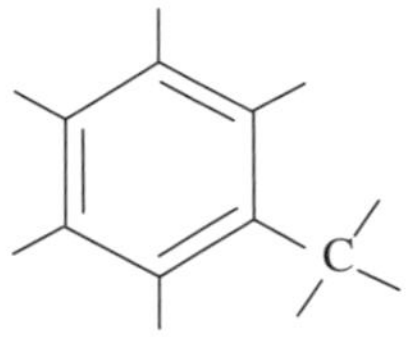

Toluene (Simplest monosubstituted alky benzene)

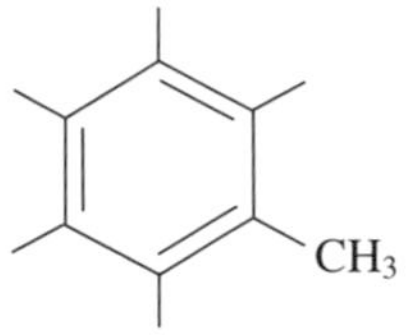

Polysubstituted alkylbenzenes

1,2-alkyl benzene (also known as ortho or *o*-...)

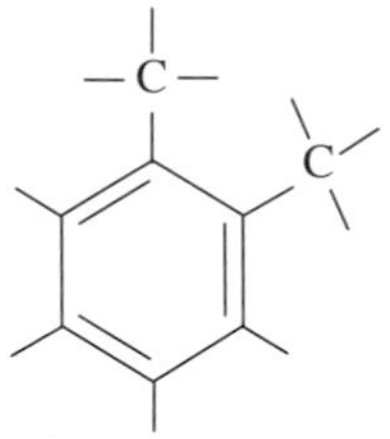

1,2-xylene or *ortho*-xylene (*o*-xylene)

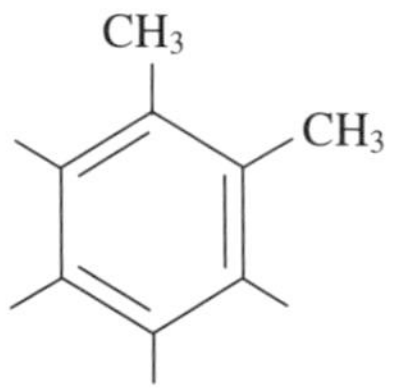

1,3-xylene or *meta*-xylene (*m*-xylene)

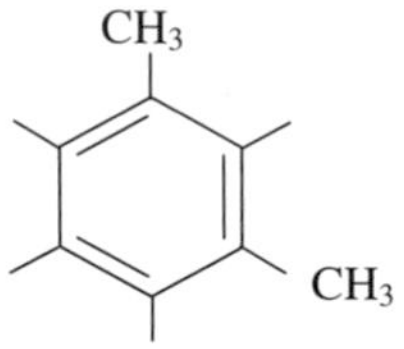

TABLE 6.7 *(continued)*

Substituted aromatics (substituted benzene derivatives):	
1,4-xylene or *para*-xylene (*p*-xylene)	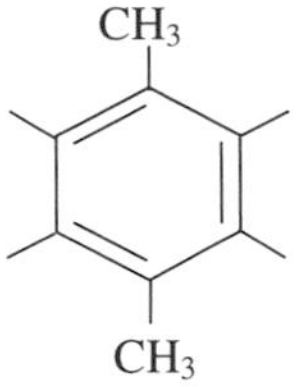
Hydroxyphenols do not follow general nomenclature rules for substituted benzenes:	
Catechol (1,2-hydroxiphenol)	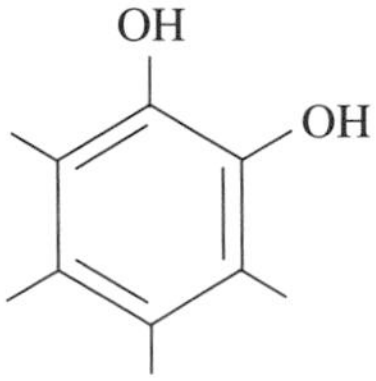
Resorcinol (1,3-hydroxiphenol)	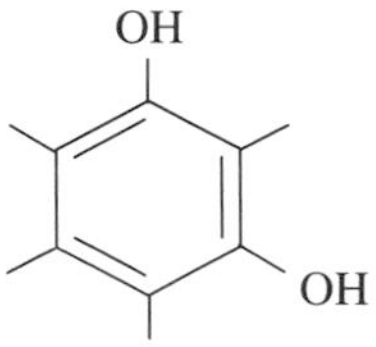
Hydroquinone (1,4-hydroxiphenol)	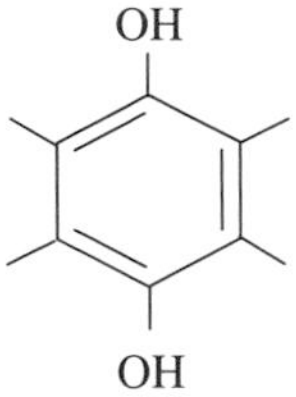

estradiol and a testosterone molecule may seem small, but they cause significant differences in the growth and reproduction of animals. The very subtle differences between an estrogen and an androgen, or the female and male hormones, respectively, can be seen in these structures (see Figure 6.8). But look at the dramatic differences in sexual and developmental changes that these "slightly different" compounds induce in organisms!

Incremental changes to a simple compound, such as ethane can make for large differences (see Table 6.8). Replacing two or three hydrogens with chlorine atoms makes for differences in toxicities between the nonhalogenated form and the chlorinated form. The same is true for the simplest aromatic, benzene. Substituting a methyl group for one of the hydrogen

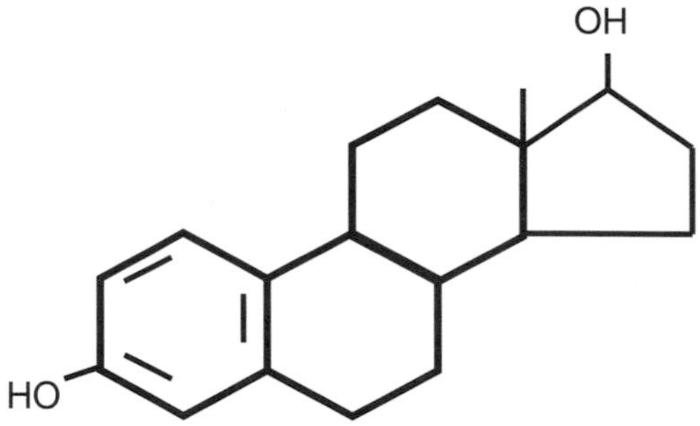

Estradiol

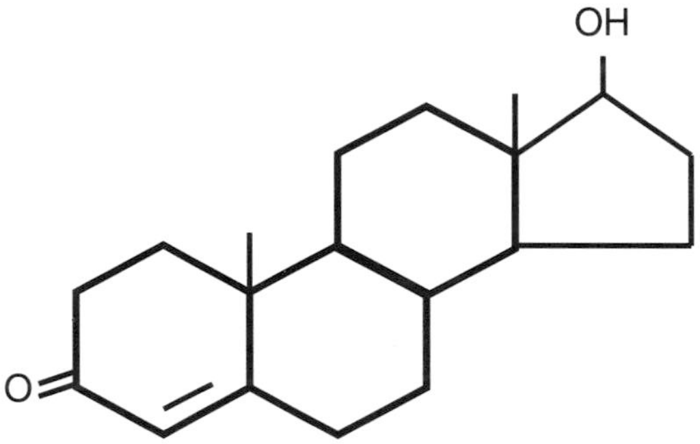

Testosterone

FIGURE 6.8. Structures of two hormones, estradiol and testosterone.

atoms forms toluene. But if instead you replace a hydrogen with a hydroxyl group, you get phenol. Replace the remaining five hydrogen atoms on the ring you get pentachlorophenol (PCP), the wood preservative that has been banned because of its toxicity, including cancer.

The lessons here are many. There are uncertainties in using surrogate compounds to represent whole groups of chemicals (since a slight change can change the molecule significantly). However, this points to the importance of "green chemistry," as well as computational chemistry and toxicology, as tools to prevent dangerous chemicals reaching the marketplace and the environment before they are manufactured. Subtle differences in molecular structure can render molecules safer, while maintaining the characteristics that make them useful in the first place, including their market value.

TABLE 6.8
Incremental Differences in Molecular Structure Leading to Changes in Physicochemical Properties and Hazards

Compound	*Physical state @ 25°C*	*$-\log S$ Solubility in H_2O @ 25°C ($mol\,L^{-1}$)*	*$-\log P^0$ Vapor Pressure @ 25°C (atm)*	*Worker Exposure Limits (parts per million)*	*Regulating Agency*
Methane, CH_4	Gas	2.8	–2.4	25	Canadian Safety Association
Tetrachloromethane (carbon tetrachloride), CCl_4	Liquid	2.2	0.8	2 short-term exposure limit (STEL) = 60 min	National Institute of Occupation Health Sciences (NIOSH)
Ethane, C_2H_6	Gas	2.7	–1.6	None (simple asphyxiant)	Occupational Safety and Health Administration (OSHA)
Trichloroethane, $C_2H\,Cl_3$	Liquid	2.0	1.0	450 STEL (15 min)	OSHA
Benzene, C_6H_6	Liquid	1.6	0.9	5 STEL	OSHA
Phenol, C_6H_6O	Liquid	0.2	3.6	10	OSHA
Toluene C_7H_8	Liquid	2.3	1.4	150 STEL	U.K. Occupational and Environmental Safety Services

Organic Structure Example 1

If the aqueous solubility as expressed as $-\log P^o$ (in $mol\,L^{-1}$) of 1-butanol, 1-hexanol, 1-octanol, and 1-nonanol are 0.1, 0.9, 2.4, and 3.1, respectively, what does this tell you about the length of carbon chains and the solubility of alcohols? (Remember that this expression of solubility is a negative log!)

Answer and Discussion

Recall that solubility in Table 6.8 is expressed as a negative logarithm, so lengthening the carbon chain *decreases polarity and, therefore, aqueous solubility*, since "like dissolves like" and water is very polar. Thus, as shown in the Figure 6.9, butanol is orders of magnitude more hydrophilic than is nonanol.

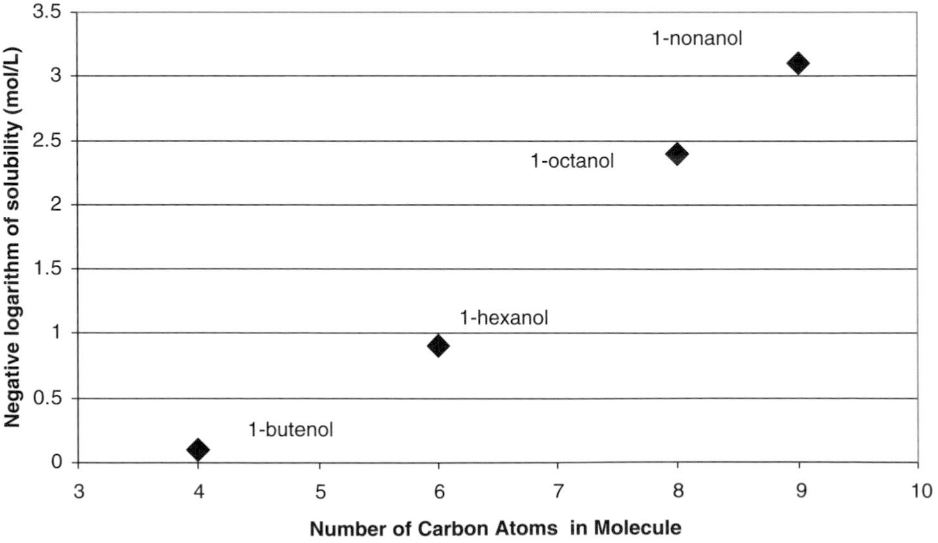

FIGURE 6.9. Aqueous solubility of four alcohols.

Organic Structure Example 2

Consider the polarity of the four alcohols in Figure 6.9 above. If n-butane, n-hexane, n-octane, and n-nonane's aqueous solubilities, as expressed as $-\log P^{o}$ are, respectively: 3.0, 3.8, 5.2, and 5.9 mol L^{-1}, what effect does the substitution of a hydroxyl functional group for the hydrogen atom have on an alkane's polarity? Can this effect also be observed in aromatics? (Hint: Compare P^{o} for benzene, toluene to phenol in Table 6.8 above; and recall that phenol is a hydroxylated benzene.)

Solution

There is a direct relationship (see Figure 6.10) between the increase in polarity and hydrophilicity when alkanes are hydroxylated into alcohols. This is why alcohols are miscible in water. This can be an important fact, especially in anaerobic treatment processes where microbes reduce organic compounds and in the process generate alcohols (and ultimately methane and water). Hydroxylation of an aromatic, as indicated by comparing the solubilities of benzene and phenol, also increases polarity and hydrophilicity.

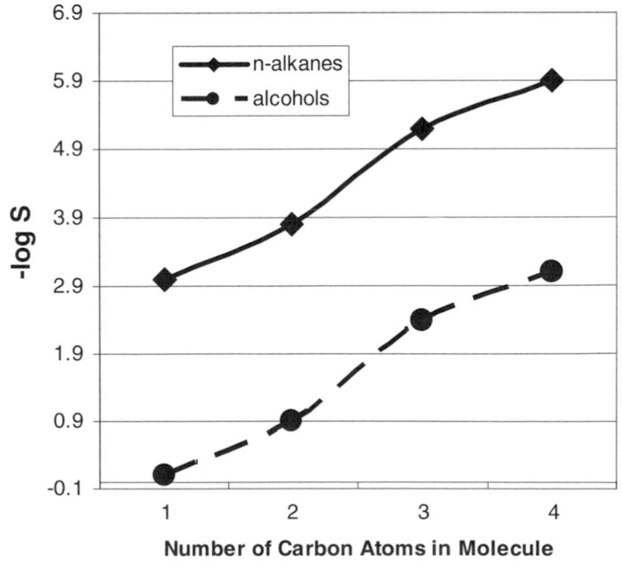

FIGURE 6.10. Aqueous solubility of selected aliphatic compounds.

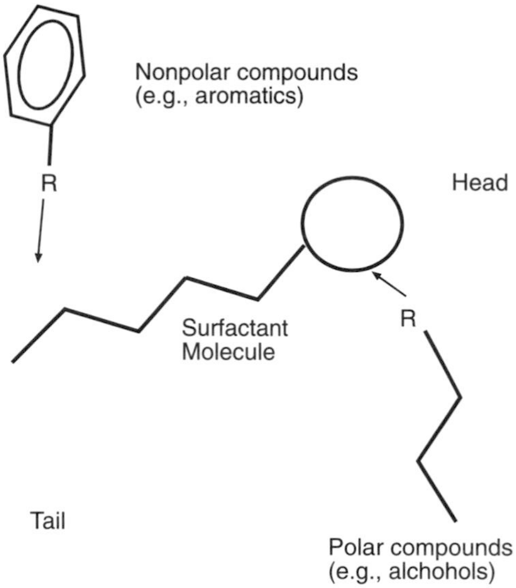

FIGURE 6.11. Structure of a surfactant prototype. The head of the molecule is relatively polar and the tail is relatively nonpolar, so in the aqueous phase the more hydrophilic compounds will react with the surfactant molecule at the head position, while the more lipophilic compounds will react at the tail position. The result is that a greater amount of the organic compounds will be in aqueous solution when surfactants are present.

It should be noted that solubility may be modified in environmental systems by using compounds known as "surfactants" (see Figure 6.11). Surfactants can be very effective in soil-washing, flushing technologies, and bioremediation of contaminated sites, by solubilizing adsorbed hydrophobic compounds in soils. The increased solubility and dispersion of hydrocarbons and aromatic compounds with very low aqueous solubility enhances desorption and bioavailability. Unfortunately, a widely used group of surfactants, such as alkylphenolethoxylates (APEs), have been banned in Europe because their breakdown products can be very toxic to aquatic organisms. The APEs have been of particular concern because of their hormonal effects, especially their estrogenicity. Therefore, one must keep in mind not only the physical and chemical advantages of such compounds, but also any ancillary risks that they may introduce. This is a classic case of competition among values (i.e., easier and more effective cleanup versus additive health risks from the cleanup chemical).

Isomers

Isomers are compounds with identical chemical formulae but different structures. They are very important in environmental organic chemistry, because even slightly different structures can evoke dramatic differences in chemical and physical properties. Isomers, then, may exhibit different chemodynamic behavior and different toxicities. For example, the three isomers of pentane (C_5H_{12}) are shown in Figure 6.12. The difference in structure accounts for significant physical differences. For example, the boiling points for *n*-pentane, isopentane, and neopentane at 1 atm are 36.1°C, 27.8°C, and 9.5°C, respectively. Thus, neopentane's lower boiling point means that this isomer has a higher vapor pressure and is more likely to enter the atmosphere than the other two isomers under the same environmental conditions.

Optical isomers or chiral forms of the same compound are those that are mirror images to each other. These differences may make one, often the left-handed form, virtually nontoxic and easily biodegradable, yet the right-handed form may be toxic and persistent.

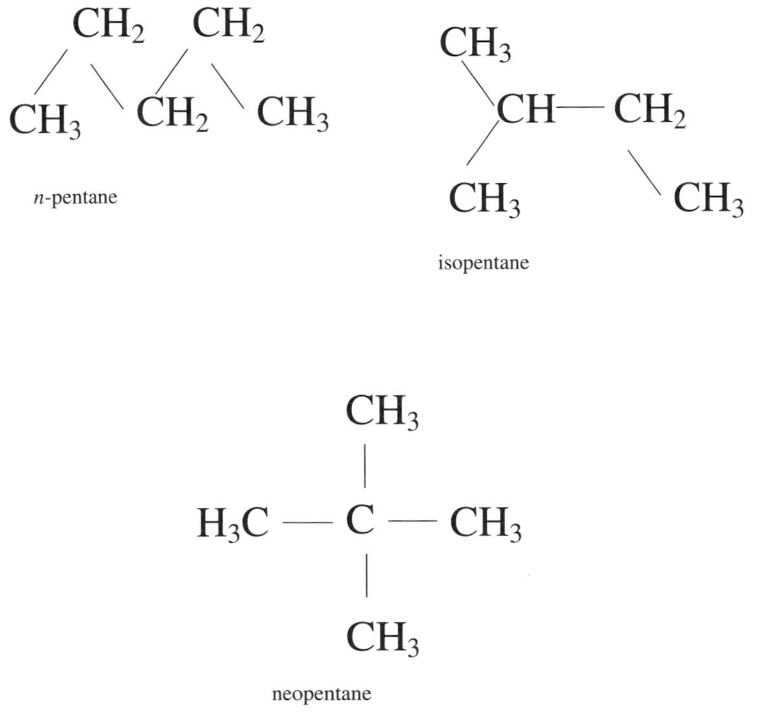

FIGURE 6.12. Isomers of pentane.

Notes and Commentary

1. U.S. Geological Survey, 2003, "Radiogenic Isotopes and the Eastern Mineral Resources Program of the U.S. Geological Survey."
2. U.S. Geological Survey, 2003, "Short-Lived Isotopic Chronometers: A Means of Measuring Decadal Sedimentary Dynamics," FS-073-98.
3. C. Lewis, G. Klouda, and W. Ellenson, 2003, "Cars or Trees: Which Contribute More to Particulate Matter Air Pollution?" U.S. Environmental Protection Agency, Science Forum, Washington, D.C.
4. This is the carbon component of a fine particulate sample ($PM_{2.5}$), such as those measured at ambient air monitoring stations. The ratios are calculated according to the National Bureau of Standards, Oxalic Acid Standard Reference Method SRM 4990B.
5. National Bureau of Standards, Oxalic Acid Standard Reference Method SRM 4990B.
6. The defined reference standard for ^{14}C is 0.95 times the ^{14}C specific activity of the original NBS Oxalic Acid Standard Reference Material (SRM 4990B), adjusted to a ^{13}C delta value of −19.09 ‰. This is "modern" carbon. It approximates wood grown in 1890 that was relatively free of CO_2 from fossil sources. Due to the anthropogenic release of radiocarbon from nuclear weapons testing and nuclear power generation, oxalic acid from plant material grown after World War II is used currently to standardize ^{14}C measurements contains more ^{14}C than 1890 wood.
7. The primary source of information for EMR is: World Health Organization, 1998, Fact Sheet N182: Electromagnetic Fields and Public Health.
8. Obviously, for the atoms near He, it is the "duet rule."
9. The source for this discussion is the U.S. Environmental Protection Agency, 2002, "Proven Alternatives for Aboveground Treatment of Arsenic in Groundwater," Engineering Forum Issue Paper, EPA-542-S-02-002 (Revised), http://www.epa.gov/tio/tsp.
10. This is true in terms of the actual number of chemical compounds. By far, most contaminants are organic. However, in terms of the total mass of reactants and products in the biosphere, inorganic compounds represent a greater mass. For example, most hazardous waste sites are contaminated with organic contaminants, but large-scale waste represented by mining, extraction, transportation, and agricultural activities have larger volumes and masses of metals and inorganic substances.
11. The letter "X" commonly denotes a halogen, e.g., fluorine, chlorine, or bromine, in organic chemistry. However, in this text, which is an amalgam of many scientific and engineering disciplines, where "x" often means an unknown variable and horizontal distance on coordinate grids, this rule is sometimes violated. Note that when consulting manuals on the physicochemical properties of organic compounds, such as those for pesticides and synthetic chemistry, the "X" usually denotes a halogen.

CHAPTER 7

Chemical Reactions in the Environment

Five categories of chemical reactions take place in the environment or in systems that ultimately lead to contamination, such as closed systems where toxic chemicals, like pesticides, are synthesized before being used and released into the ambient environment, or thermal systems where precursor compounds form new contaminants like dioxins and furans. The categories of chemical reactions are:

1. *Synthesis or combination*:

$$A + B \rightarrow AB \qquad \text{Reaction 7–1}$$

In combination reactions, two or more substances react to form a single substance. Two types of combination reactions are important in environmental systems: formation and hydration.

Formation reactions are those where elements combine to form a compound. Examples include the formation of ferric oxide and the formation of octane:

$$4Fe(s) + 3O_2(g) \rightarrow 2Fe_2O_3(s) \qquad \text{Reaction 7–2}$$

$$8C(s) + 9H_2(g) \rightarrow C_8H_{18}(l) \qquad \text{Reaction 7–3}$$

Hydration reactions involve the addition of water to synthesize a new compound, for example, when calcium oxide is hydrated to form calcium hydroxide, and when phosphate is hydrated to form phosphoric acid:

$$CaO(s) + H_2O(l) \rightarrow Ca(OH)_2(s) \qquad \text{Reaction 7–4}$$

$$P_2O_5(s) + 3H_2O(l) \rightarrow 2H_3PO_4(aq) \qquad \text{Reaction 7–5}$$

2. *Decomposition*, often referred to as *degradation* when discussing organic compounds in toxicology, environmental sciences, and engineering:

$$AB \rightarrow A + B \quad \text{Reaction 7–6}$$

In decomposition, one substance breaks down into two or more new substances, such as in the decomposition of carbonates. For example, calcium carbonate breaks down into calcium oxide and carbon dioxide:

$$CaCO_3(s) \rightarrow CaO(s) + CO_2(g) \quad \text{Reaction 7–7}$$

3. *Single replacement*, or *single displacement*:

$$A + BC \rightarrow AC + B \quad \text{Reaction 7–8}$$

This commonly occurs when one metal ion in a compound is replaced with another metal ion, such as when trivalent chromium replaces monovalent silver:

$$3AgNO_3(aq) + Cr(s) \rightarrow Cr(NO_3)_3(aq) + 3Ag(s) \quad \text{Reaction 7–9}$$

4. *Double replacement* (also *metathesis* or *double displacement*):

$$AB + CD \rightarrow AD + CB \quad \text{Reaction 7–10}$$

In metathetic reactions, cations and anions trade places. They are commonly encountered in metal precipitation reactions, such as when lead is precipitated in the reaction of a lead salt with an acid like potassium chloride:

$$Pb(ClO_3)_2(aq) + 2KCl(aq) \rightarrow PbCl_2(s) + 2KClO_3(aq) \quad \text{Reaction 7–11}$$

5. *Complete* or *efficient combustion* (thermal oxidation):

$$(CH)_x + O_2 \rightarrow CO_2 + H_2O \quad \text{Reaction 7–12}$$

Combustion is the combination of O_2 in the presence of heat (as in burning fuel), producing CO_2 and H_2O during complete combustion of organic compounds, such as the combustion of octane:

$$C_8H_{18}(l) + 17O_2(g) \rightarrow 8CO_2(g) + 9H_2O(g) \quad \text{Reaction 7–13}$$

Complete combustion may also result in the production of molecular nitrogen (N_2) when nitrogen-containing organics are burned, such as in the combustion of methylamine:

$$4CH_3NH_2(l) + 9O_2(g) \rightarrow 4CO_2(g) + 10H_2O(g) + 2N_2(g) \quad \text{Reaction 7–14}$$

Incomplete combustion can produce a variety of compounds. Some are more toxic than the original compounds being oxidized.

The alert reader will note at least two observations about these categories. First, all are kinetic, as denoted by the one-directional arrow ($\rightarrow$). Second, in the environment, many processes are incomplete, such as the common problem of incomplete combustion and the generation of new compounds in addition to carbon dioxide and water.

With respect to the first observation, indeed, many equilibrium reactions take place. However, as mentioned in previous discussions, getting to equilibrium requires a kinetic phase. So, upon reaching equilibrium, the kinetic reactions (one-way arrows) would be replaced by two-way arrows ($\leftrightarrow$). Changes in the environment or in the quantities of reactants and products can invoke a change back to kinetics.

Incomplete reactions are very important sources of environmental contaminants. As mentioned, these reactions generate *products of incomplete combustion* (PICs), such as carbon monoxide (CO), *polycyclic aromatic hydrocarbons* (PAHs), dioxins, furans, and hexachlorobenzene. (See Discussion Box "Engineer's Technical Note: Formation of Dioxins and Furans.")

Engineer's Technical Note: Formation of Dioxins and Furans

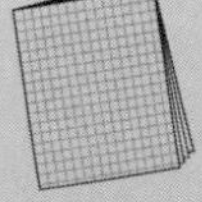

Chlorinated dioxins have 75 different forms, and there are 135 different chlorinated furans, owing to the number and arrangement of chlorine atoms on the molecules. The compounds can be separated into groups that have the same number of chlorine atoms attached to the furan or dioxin rings. Each form varies in its chemical, physical, and toxicological characteristics (see Figure 7.1).

Dioxins are highly toxic compounds that are created unintentionally during combustion processes. The most toxic form is the 2,3,7,8-tetrachlorodibenzo-p-dioxin (TCDD) isomer. Other isomers with the 2,3,7,8 chlorine substitutions are also considered to have higher toxicity than the dioxins and furans with different chlorine atom arrangements.

What is currently known about the conditions needed to form these compounds has been derived from studying full-scale municipal

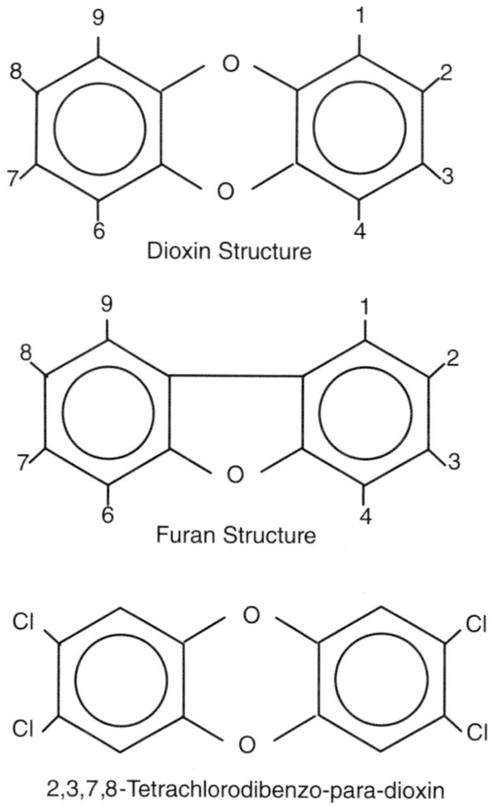

FIGURE 7.1. Molecular structures of dioxins and furans. Bottom structure is of the most toxic dioxin congener, tetrachlorodibenzo-*para*-dioxin (TCDD), formed by the substitution of chlorine for hydrogen atoms at positions 2, 3, 7, and 8 on the molecule.

solid waste incinerators, and the experimental combustion of fuels and feeds in the laboratory. Most of the chemical and physical mechanisms identified by these studies can relate to combustion systems in which organic substances are combusted in the presence of chlorine (Cl). Incinerators of chlorinated wastes are the most common environmental sources of dioxins, accounting for about 95% of the volume.

The emission of dioxins and furans from combustion processes may follow three general physicochemical pathways (see Figure 7.2). The first pathway occurs when the feed material going to the incinerator contains dioxins or furans, and a fraction of these compounds survives thermal breakdown mechanisms and passes through to be

FIGURE 7.2. Release of products of incomplete combustion (gases and particles) from stacks and vents. (Photo Credit: U.S. Department of Energy, Energy Information Administration.)

emitted from vents or stacks. This is not considered to account for a large volume of dioxin released to the environment, but it may account for the production of dioxin-like, coplanar polychlorinated biphenyls (PCBs).

The second process is the formation of dioxins and furans from the thermal breakdown and molecular rearrangement of precursor compounds, such as the chlorinated benzenes, chlorinated phenols (such as pentachlorophenol, PCP) and PCBs, which are chlorinated aromatic compounds with structural resemblances to the chlorinated dioxin and furan molecules. Dioxins appear to form after the precursor has condensed and adsorbed onto the surface of particles, such as fly ash. This is a heterogeneous process, where the active sorption sites on the particles allow for the chemical reactions, which are catalyzed by the presence of inorganic chloride compounds and ions sorbed to the particle surface. The process occurs within the temperature range 250–450°C, so most of the dioxin formation under the precursor mechanism occurs away from the high temperature zone in the incinerator, where the gases and smoke derived from combustion of the organic materials have cooled during conduction through flue ducts, heat exchanger and boiler tubes, air pollution control equipment, or the vents and the stack.

The third means of synthesizing dioxins is *de novo* within the so-called "cool zone" of the incinerator, wherein dioxins are formed from moieties different from those of the molecular structure of dioxins, furans, or precursor compounds. Generally, these can include a wide range of both halogenated compounds like polyvinylchloride (PVC), and nonhalogenated organic compounds like petroleum products, nonchlorinated plastics (polystyrene), cellulose, lignin, coke, coal, and inorganic compounds like particulate carbon, and hydrogen chloride gas. No matter which *de novo* compounds are involved, however, the process needs a chlorine donor (a molecule that "donates" a chlorine atom to the precursor molecule). This leads to the formation and chlorination of a chemical intermediate that is a precursor. The reaction steps after this precursor is formed can be identical to the precursor mechanism discussed in the previous paragraph.

De novo formation of dioxins and furans may involve even more fundamental substances than those moieties mentioned above. For example, dioxins may be generated[1] by heating of carbon particles absorbed with mixtures of magnesium-aluminum silicate complexes when the catalyst copper chloride ($CuCl_2$) is present (see Figure 7.3). The *de novo* formation of chlorinated dioxins and furans from the oxidation of carbonaceous particles seems to occur at around 300°C. Other chlorinated benzenes, chlorinated biphenyls, and

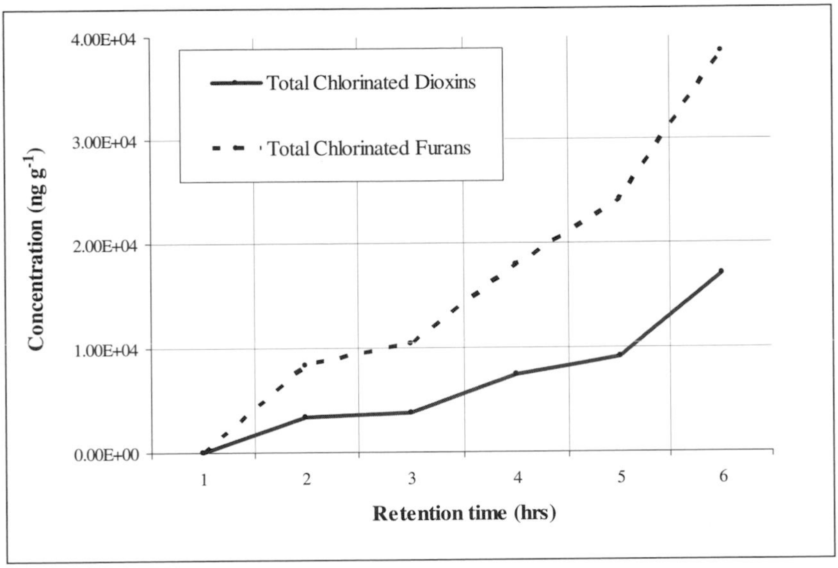

FIGURE 7.3. *De novo* formation of chlorinated dioxins and furans after heating Mg Al silicate, 4% charcoal, 7% Cl, 1% $CuCl_2 \cdot H_2O$ at 300°C. (Sources: L. Stieglitz, G. Zwick, J. Beck, W. Roth, and H. Vogg, 1989, "On the de-novo synthesis of PCDD/PCDF on the fly ash of municipal waste incinerators" *Chemosphere* 18, pp. 1219–1226; and L. Stieglitz, G. Zwick, J. Beck, J. Bautz, and W. Roth, 1989, "Carbonaceous particles in fly ash—a source for de-novo-synthesis of organochloro compounds," *Chemosphere* 19, pp. 283–290.)

chlorinated naphthalene compounds are also generated by this type of mechanism.

Other processes generate dioxin pollution. A source that has been greatly reduced in the last decade is the paper production process, which formerly used chlorine bleaching. This process has been dramatically changed, so that most paper mills no longer use the chlorine bleaching process. Dioxin is also produced in the making of PVC plastics, which may follow chemical and physical mechanisms similar to the second and third processes discussed above.

Since dioxins and dioxin-like compounds are lipophilic and persistent, they accumulate in soils, sediments, and organic matter and can persist in solid and hazardous waste disposal sites.[2] These compounds are semivolatile, so they may migrate away from these sites and be transported in the atmosphere either as aerosols (solid and liquid phase) or as gases (the portion of the compound that volatilizes).

Therefore, the engineer must take great care in removal and remediation efforts not to unwittingly cause releases from soil and sediments via volatilization or perturbations, such as landfill and dredging operations.

At the most basic level, only a few types of chemical reactions dominate in the environment. These include ionization, acid-base, precipitation, and oxidation-reduction (redox).

Environmental Ionic Reactions

Salts and Solutions

When a salt is dissolved in water, it dissociates into ionic forms. Notwithstanding their variability in doing so, under the right conditions all complexes can become dissolved in water. Ions that are dissolved in a solution can react with one another, and can form solid complexes and compounds. Actually, a compound does not exist in water. For example, when the salts calcium sulfate and sodium chloride are added to water, it is commonly held that $CaSO_4$ and NaCl are in the water. However, what is really happening is that the metals and nonmetals are associated with one another as:

$$CaSO_4(s) \Leftrightarrow Ca^{2+}(aq) + SO_4^{2-}(aq) \qquad \text{Reaction 7–15}$$

and

$$NaCl(s) \Leftrightarrow Na^{+}(aq) + Cl^{-}(aq) \qquad \text{Reaction 7–16}$$

Thus, all four of the dissociated ions are free and no longer associated with each other; that is, the Na, Ca, Cl, and SO_4 ions are "unassociated" in the water. The Na and the Cl are no longer linked to each other as they were before the compound was added to the water.

Even though atoms are neutral, in the process of losing or gaining electrons, they become electrically charged, that is, they become *ions*. An atom that loses one or more electrons is positively charged, and known as a *cation*. For example, the potassium atom loses one electron and becomes the monovalent potassium cation:

$$K - e^{-} \rightarrow K^{+} \qquad \text{Reaction 7–17}$$

When the mercury atom loses two electrons, it becomes the divalent mercury cation:

$$Hg - 2e^- \rightarrow Hg^{2+} \qquad \text{Reaction 7–18}$$

When the chromium atom loses three electrons, it becomes the trivalent chromium cation:

$$Cr - 3e^- \rightarrow Cr^{3+} \qquad \text{Reaction 7–19}$$

Conversely, an atom that gains electrons becomes a negatively charged ion, known as an *anion*. For example, when chlorine gains an electron it becomes the chlorine anion:

$$Cl + e^- \rightarrow Cl^- \qquad \text{Reaction 7–20}$$

When sulfur gains two electrons, it becomes the divalent sulfide anion:

$$S + 2e^- \rightarrow S^{2-} \qquad \text{Reaction 7–21}$$

Note that the Greek prefixes (mono-, di-, tri-, etc.) denoting the valence are the number of electrons that the ion differs from neutrality.

Reactions between ions, known as *ionic reactions*, frequently occur as ions of water-soluble salts can react in aqueous solution to form salts that are nearly insoluble in water. This causes them to separate into insoluble precipitates:

$$\text{(Ions in Solution 1)} + \text{(Ions in Solution 2)} \rightarrow \text{Precipitate} + \text{Unreacted ions} \qquad \text{Reaction 7–22}$$

Ecologist's Field Manual Entry: Anions in the Nitrogen Biogeochemical Cycle

Discussions of contaminant ions are often about the positively charged cations, especially those associated with heavy metals like lead, mercury, cadmium, strontium, selenium, and chromium, as well as the toxic metalloid arsenic. However, anions are also important contaminants. Unlike most of the toxic heavy metals, though, the anions are often also essential to the survival and growth of biota in the biochemical cycles of essential elements, especially those of nitrogen, phosphorus, and sulfur. Let us consider the essential element, N.

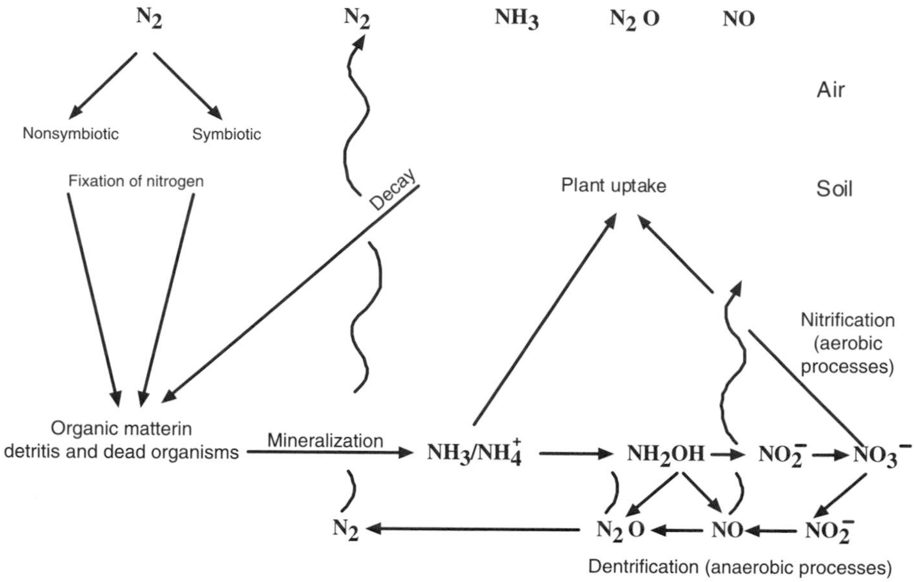

FIGURE 7.4. Biochemical nitrogen cycle.

Nitrates and Nitrites

The nitrogen cycle (Figure 7.4) includes three principal soluble forms: the cation ammonium (NH_4^+), and the anions nitrate (NO_3^-) and nitrite (NO_2^-). Nitrates and nitrites combine with various organic and inorganic compounds. Once taken into the body, NO_3^- is converted to NO_2^-. Since NO_3^- is soluble and readily available as a nitrogen source for plants (e.g., to form plant tissue such as amino acids and proteins), farmers are the biggest users of NO_3^- compounds in commercial fertilizers (although even manure can contain high levels of NO_3^-).

Ingesting high concentrations of nitrates, such as in drinking water, can cause serious short-term illness and even death. This serious illness in infants is due to the reduction of nitrate to nitrite by the body, which can interfere with the oxygen-carrying capacity of the blood, known as *methemoglobinemia*. When nitrates compete successfully against molecular oxygen, especially in small children, the blood carries methemoglobin (as opposed to healthy hemoglobin), giving rise to clinical symptoms. At 15–20% methemoglobin, children can experience shortness of breath and blueness of the skin (i.e., clinical cyanosis). At 20–40% methemoglobin, hypoxia will result. This acute condition can deteriorate a child's health rapidly over a period of days, especially if the water source continues to be used. Long-term, elevated exposures to nitrates and nitrites can cause an increase in the kidneys' production of urine (diuresis), increased starchy deposits, and hemorrhaging of the spleen.[3]

Solubility and Electrolytes

The aqueous solubility of contaminants ranges from completely soluble in water to virtually insoluble. Solubility equilibrium is the phenomenon that keeps substances in solution. Solubility and precipitation are, in a way, two sides of the same coin. There is truth in the old chemists' pun, "If you're not part of the solution, you're part of the precipitate!" Solubilities typically are quantitatively expressed in mass of solute per volume of solvent (e.g., mg L^{-1}), and sometimes progressively expressed by the adjectives "soluble," "slightly soluble," "sparingly soluble" or "insoluble."

In the solid phase, a salt is actually a collection of ions in a lattice, where the ions are surrounded by one another. However, when the salt is dissolved in water, the ions become surrounded by the water, rather than by the other ions. Each ion now has its own coordinating water envelope or "hydration sphere," or a collection of water molecules surrounding it. An ion-association reaction is an ion-ion interaction between ions in an electrolyte (i.e., ion-containing) solution.[4] When the salt lattice enters the water, then, the ions assemble into couplets of separate, oppositely charged ions: cations and anions, or the so-called "ion pairs." The pairs are held together by electrostatic attraction. Ion-association is the reverse of dissociation where the ions separate from a compound into free ions. Ions (or molecules) surrounded by water exist in the aqueous phase, so ionic compounds that are soluble in water break apart (i.e., dissociate) into their ionic components: anions and cations (see Figure 7.5).

Solutions may contain nonelectrolytes, strong electrolytes, and/or weak electrolytes. Nonelectrolytes do not ionize. They are nonionic molecular compounds that are neither acids nor bases. Sugars, alcohols, and most other organic compounds are nonelectrolytic. Some inorganic compounds are also nonelectrolytes.

Weak electrolytes only partially dissociate in water. Most weak electrolytes dissociate less than 10%: greater than 90% of these substances remain undissociated. Organic acids, such as acetic acid, are generally weak electrolytes.

An example of dissociation is strontium carbonate dissolved in water:

$$SrCO(s) \rightarrow Sr^{2+}(aq) + CO_3^{2-}(aq) \qquad \text{Reaction 7–23}$$

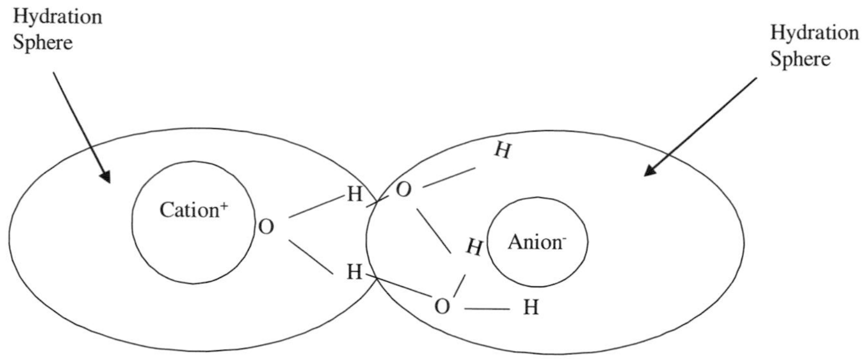

FIGURE 7.5. Ion pairs. (Source: V. Evangelou, 1998, *Environmental Soil and Water Chemistry: Principles and Applications*, John Wiley & Sons, New York, N.Y.)

Conversely, the reverse reaction forms a solid; that is, it returns from the solution to again form the lattice of ions surrounding ions. This is a precipitation reaction. In our Sr example, the carbonate species is precipitated:

$$Sr^{2+}(aq) + CO_3^{2-}(aq) \rightarrow SrCO_3(s) \qquad \text{Reaction 7–24}$$

The *ionic product* (Q) is a measure of the ions present in the solvent. The *solubility product constant* (K_{sp}) is the ionic product when the system is in equilibrium. Solubility is often expressed as the specific K_{sp}, the equilibrium constant for dissolution of the substance in water. Since K_{sp} is another type of chemical equilibrium, it is a state of balance of opposing reversible chemical reactions that proceed at constant and equal rates, resulting in no net change in the system (hence the symbol, $\leftrightarrow$). Like sorption, Henry's Law, and other equilibrium constants, solubility follows Le Chatelier's Principle, which states that in a balanced equilibrium, if one or more factors change, the system will readjust to reach equilibrium. K_{sp} values and the resulting solubility calculations for some important reactions are shown in Table 7.1. K_{sp} constants for many reactions can be found in engineering, hydrogeology, and chemistry handbooks.

Since $SrCO_3$ is a highly insoluble salt (aqueous solubility = $6 \times 10^{-3}\,mg\,L^{-1}$), its equilibrium constant for the reaction is quite small:

$$K_{sp} = [Sr^{+2}][CO_3^{2-}] = 1.6 \times 10^{-9} \qquad \text{Reaction 7–25}$$

The small K_{sp} value of the constant reflects the low concentration of dissolved ions. As the number of dissolved ions approaches zero, the compound is becoming increasingly insoluble in water. This does not mean that that

TABLE 7.1
Solubility Product Constant versus Solubility for Four Types of Salts

Salt	*Example*	*Solubility Product, K_{sp}*	*Solubility, S*
AB	$CaCO_3$	$[Ca_{2+}][CO_3^{2-}] = 4.7 \times 10^{-9}$	$(K_{sp})^{1/2} = 6.85 \times 10^{-5}$ M
AB_2	$Zn(OH)_2$	$[Zn^{2+}][OH-]^2 = 4.5 \times 10^{-17}$	$(K_{sp}/4)^{1/3} = 2.24 \times 10^{-6}$ M
AB_3	$Cr(OH)_3$	$[Zn^{3+}][OH-]^3 = 6.7 \times 10^{-31}$	$(K_{sp}/27)^{1/4} = 1.25 \times 10^{-8}$ M
A_3B_2	$Ca_3(PO_4)_2$	$[Ca^{2+}]^3[PO_4^{3-}]^2 = 1.3 \times 10^{-32}$	$(K_{sp}/108)^{1/5} = 1.64 \times 10^{-7}$ M

Source: U.S. Army Corps of Engineers, 2001, *Engineering and Design: Precipitation/Coagulation/Flocculation*, EM 1110-1-4012, Chapter 2.

an insoluble product cannot ever be dissolved, but that chemical treatment is needed to bring about dissolution. In this case, strontium carbonate requires the addition of an acid to *solubilize* the Sr^{+2} ion.

To precipitate a compound, the product of the concentration of the dissolved ions in the equilibrium expression must exceed the value of the K_{sp}. The concentration of each of these ions does not need to be the same. For example, if $[Sr^{+2}]$ is 1×10^{-5} molar, the carbonate ion concentration must exceed 0.0016 molar for precipitation to occur, because $(1 \times 10^{-5}) \times (1.6 \times 10^{-4}) = 1.6 \times 10^{-9}$ (i.e., the K_{sp} for strontium carbonate).

Dissociation and Precipitation Reaction Example

An environmental analytical chemist adds 100 mL of 0.050 M NaCl to 200 mL of 0.020 M $Pb(NO_3)_2$. Will the lead chloride that is formed precipitate?

Solution

Calculate the *ion product* (Q) and compare it to the K_{sp} for the reaction:

$$PbCl_2(s) \rightarrow Pb^{2+}(aq) + 2Cl^-(aq)$$

When the two solutions are mixed, the unassociated ions are formed as:

$$[Pb^{2+}] = 0.2L \times 2.0 \times 10^{-2} M/0.3L = 1.3 \times 10^{-2} M,$$

and,

$$[Cl^-] = 0.1L \times 5.0 \times 10^{-2} M/0.3L = 1.7 \times 10^{-2} M.$$

The value for the ion product is calculated as:

$$Q = [Pb^{2+}][Cl^-]^2 = [1.3 \times 10^{-2}][1.7 \times 10^{-2}]^2 = 3.8 \times 10^{-7}$$

The K_{sp} for this reaction is 1.6×10^{-5}.

$Q < K_{sp}$, so no precipitate will be formed. If the ion product were greater than the K_{sp} a precipitate would have formed.

Environmental conditions can affect solubility. One such instance is the *common ion effect*. Compared to a solution in pure water, an ion's solubility is decreased in an aqueous solution that contains a common ion (i.e., one of the ions that make up the compound). This allows a precipitate to form if the K_{sp} is exceeded. For example, soluble sodium carbonate (Na_2CO_3) in solution with strontium ions can cause the precipitation of strontium carbonate, since the carbonate ions from the sodium salt are contributing to their overall concentration in solution and reversing the solubility equilibrium of the "insoluble" compound, strontium carbonate:

$$Na_2CO_3(s) \rightarrow 2Na^+(aq) + CO_3^{2-}(aq) \qquad \text{Reaction 7–26}$$

$$SrCO_3(s) \rightarrow Sr^+(aq)^{2+} CO_3^{2-}(aq) \qquad \text{Reaction 7–27}$$

Also, a complexing agent or *ligand* may react with the cation of a precipitate, enhancing the solubility of the compound. In addition, several metal ions are weakly acidic and readily hydrolyze in solution. For example, hydrolyzing ferric ion (Fe^{3+}):

$$Fe^{3+} + H_2O \rightarrow Fe(OH)^{2+} + H^+ \qquad \text{Reaction 7–28}$$

When such metal ions hydrolyze, they produce a less soluble complex. The solubility of the salt is inversely related to the pH of the solution, with solubility increasing as the pH decreases. The minimum solubility is found under acidic conditions when the concentrations of the hydrolyzed species approach zero.

Ionization Example

What exactly is the pH scale? Why do pH values range from 0 to 14?

Solution and Discussion

Water ionizes. It does not exist only as molecular water (H_2O), but also includes hydrogen (H^+) and hydroxide (OH^-) ions:

$$H_2O \leftrightarrow H^+ + OH^- \qquad \text{Reaction 7–29}$$

The negative logarithm of the molar concentration of hydrogen ions, i.e. $[H^+]$ in a solution (usually water in the environmental sciences), is referred to as pH. This convention is used because the actual number of ions is extremely small. Thus pH is defined as:

$$pH = -\log_{10}[H^+] = \log_{10}([H^+]^{-1}) \qquad \text{Equation 7–1}$$

The brackets refer to the molar concentrations of chemicals, and in this case it is the ionic concentration in moles of hydrogen ions per liter. The reciprocal relationship of molar concentrations and pH means that the more hydrogen ions you have in solution, the lower your pH value will be.

Likewise, the negative logarithm of the molar concentration of hydroxide ions, $[OH^-]$ in a solution, is pOH:

$$pOH = -\log_{10}[OH^-] = \log_{10}([OH^-]^{-1}) \qquad \text{Equation 7–2}$$

The relationship between pH and pOH is constant:

$$-K = [H^+][OH^-] = 10^{-14} \qquad \text{Equation 7–3}$$

When expressed as a negative log, one can see that the pH and pOH scales are reciprocal to one another, and that they both range from 0 to 14. Thus, a pH 7 must be neutral (just as many hydrogen ions as hydroxide ions).

Upon further investigation, the log relationship means that for each factor pH unit change there is a factor of 10 change in the molar concentration of hydrogen ions. Thus, a pH 2 solution has 100,000 times more hydrogen ions than neutral water (pH 7), or $[H^+] = 10^{12}$ versus $[H^+] = 10^7$, respectively.

Environmental Acid and Base Chemistry

For our purposes, an acid is any substance that causes hydrogen ions to be produced when dissolved in water. Conversely, a base is any substance that produces hydroxide ions when dissolved in water. Acids are proton donors, and bases are proton acceptors (this is known as the Brönsted-Lowry model). Acids are electron-pair acceptors, and bases are electron-pair donors (following the Lewis model). Actually, the H^+ is a bare proton, having lost its electron, so it is highly reactive. In reality, the acid produces a hydronium ion:

$$[H^+(H_2O)] = H_3O^+ \qquad \text{Equation 7–4}$$

When acids react with bases, a double-replacement reaction takes place, resulting in neutralization. The products are water and a salt. One mole of acid neutralizes precisely one mole of base. Being electrolytes, a strong acid dissociates and ionizes 100% into H_3O^+ and anions. These anions are the acid's specific conjugate base. A strong base also dissociates and ionizes completely. The ionization results in hydroxide ions and cations, known as the base's conjugate acid. Weak acids and weak bases dissociate less than 100% into the respective ions.

There are four strong acids generally important to environmental chemistry: hydrochloric acid (HCl), nitric acid (HNO_3), sulfuric acid (H_2SO_4), and perchloric acid ($HClO_4$). Many weak acids are important in environmental chemistry, such as carbonic acid, acetic acid, and phosphoric acid.

Strong bases include sodium hydroxide (NaOH) and potassium hydroxide (KOH). Weak bases include ammonia (NH_3), which dissolves in the surface waters to become ammonium hydroxide (NH_4OH) and organic amines (i.e., compounds with the radical: —NH).

Nonmetal oxides, such as carbonate (CO_2) and sulfate (SO_2), are generally acidic, for example, forming carbonic acid and sulfuric acid, respectively, in water, while metal oxides like those of calcium (e.g., CaO) and magnesium (e.g., MgO), are generally basic. These two metal oxides, for example, form calcium hydroxide (CaOH) and magnesium hydroxide (MgOH) in water, respectively.

The principal conventional, environmental metric for acidity and basicity used in environmental contamination assessments of surface and groundwater, as well as soil, is pH. The "p" in pH represents the negative log and the "H" represents the hydrogen ion or hydronium ion molar concentration:

$$pH = -\log[H^+] \text{ and } [H^+] = 10^{-pH} \qquad \text{Equation 7–5}$$

The autoionization of water into its hydrogen ions and hydroxide ions is an equilibrium constant: the water dissociation equilibrium constant (K_w). At

25°C, the molar concentration of the product of these ions is 1.0×10^{-14}, that is:

$$K_w = [H^+][OH^-] = 1.0 \times 10^{-14} \qquad \text{Equation 7–6}$$

Thus, pH ranges from 0 to 14, with 7 being neutral. Values below 7 are acidic and values above 7 are basic.

Strong Acid/Base Example 1

A laboratory is using an aqueous solution of hydrochloric acid, which is 0.05 *M*. What is the hydrogen ion and chlorine ion molar concentration of the solution?

Solution

Since this is a strong acid, it should ionize and dissociate completely. Thus,

$[H^+] = 0.05$ and $[Cl^-] = 0.05$, or none of the associated acid remains.

Strong Acid/Base Example 2

What is the pH of the solution above?

Solution

Since $[H^+] = 0.05$, and $pH = -\log[H^+] = -\log 0.05\,M = 1.3$

Strong Acid/Base Example 3

What is the $[OH^-]$ of the solution above?

Solution

$$K_w = [H^+][OH^-] = 1.0 \times 10^{-14}$$

$$\text{So, } [OH^-] = \frac{K_w}{[H^+]} = \frac{10^{-14}}{0.05} = 2.0 \times 10^{-13}$$

Thus, even with a very high relative concentration of hydrogen ions in the acidic solution, there is still a small amount of hydroxide ion concentration.

Characterizing an acid or base as strong or weak has nothing to do with the concentration (i.e., molarity) of the solution, and everything to do with the extent to which the acid or base dissociates when it enters water. In other words, whether at a concentration of 6.0 *M* or at 0.00001 *M*, sulfuric acid will completely ionize in the water, but acetic acid will not completely ionize at any concentration. To demonstrate this, Table 7.2 shows the pH for a number of acids and bases, all with the same molar concentration. The strongest acids are at the top and the strongest bases are at the bottom. The weak acids and bases are in the middle of the table.

Most environmental acid-base reactions involve weak substances. In fact, the amount of ionization in most environmental reactions, especially those in the ambient environment (as opposed to those in chemical engineering and laboratory reactors), are quite weak, usually well below 10%

TABLE 7.2
The Experimentally Derived pH Values for 0.1 *M* Solutions of Acids and Bases at 25°C

Compound	*pH*
HCl	1.1
H_2SO_4	1.2
H_3PO_4	1.5
CH_3COOH	2.9
H_2CO_3 (in saturated solution)	3.8
HCN	5.1
NaCl	6.4
H_2O (distilled)	7.0
$NaCH_3CO_2$	8.4
$NaSO_3$	9.8
NaCN	11.0
NH_3 (aqueous)	11.1
$NaPO_4$	12.0
NaOH	13.0

Source: J. Spencer, G. Bodner, and L. Rickard, 2003, *Chemistry: Structure and Dynamics*, 2nd Edition, John Wiley & Sons, New York, N.Y.

dissociation.[5] For every 1000 molecules of a weak acid, then, only a few, say 50, molecules of the acid will dissociate into hydronium ions in the water. Thus, taking acetic acid as an example, the acid-base equilibrium reaction is:

$$CH_3COOH(aq) + H_2O(l) \Leftrightarrow H_3O^+(aq) + CH_3COO^-(aq) \qquad \text{Reaction 7–30}$$

The acetate ion (CH_3COO^-) is the reaction's conjugate base, and the hydronium ion is the active acid chemical species. Because the reaction is in equilibrium, all species exist together at a constant molar concentration ratio between the ions and the molecular acid. Therefore, we can establish another equilibrium constant for acid reactions: the acid constant (K_a):

$$K_a = \frac{[H_3O^+][CH_3COO^-]}{[CH_3COOH]} \qquad \text{Equation 7–7}$$

At 25°C, acetic acid's K_a is 1.8×10^{-5}. If the percent dissociation in an acid reaction is known, the product of this percentage and the initial acid concentration will give the molar concentration of hydrogen ions, $[H^+]$. For example, if a 0.1 *M* solution of cyanic acid (HOCN) is 2.8% ionized, the $[H^+]$ can be found. We know that HOCN is a weak acid because the percent ionization is less than 100. In fact, it is 97.2% below 100%. This means that the hydrogen ion molar concentration is: $[H^+] = 2.8\% \times 0.1\,M = 0.0028$ or 2.8×10^{-3}. Published K_a constants show that the HOCN constant at 25°C is 3.5×10^{-3}, so environmental conditions, likely temperature, are slightly affecting the pH of the solution. Remember that all equilibrium constants are temperature dependent.

Weak bases follow exactly the same protocol as weak acids, with a base equilibrium constant, K_b. Some important acid and base equilibrium constants are provided in Tables 7.3 and 7.4.

Because the atmosphere contains relatively large amounts of carbon dioxide (on average about 350 ppm), the CO_2 becomes dissolved in surface water and in soil water (because CO_2 is a common soil gas). Thus, one of the most important environmental acid-base reactions[6] is the dissociation of CO_2:

$$CO_2 + H_2O \Leftrightarrow H_2CO_3^* \qquad \text{Reaction 7–31}$$

The asterisk (*) denotes that this compound is actually the sum of two compounds, the dissolved CO_2 and the reaction product, carbonic acid H_2CO_3.

Since the carbonic acid that is formed is a diprotic acid (i.e., it has two hydrogen atoms), an additional equilibrium step reaction occurs in water. The first reaction forms bicarbonate and hydrogen ions:

$$H_2CO_3^* \Leftrightarrow HCO_3^- + H^+ \qquad \text{Reaction 7–32}$$

TABLE 7.3
Equilibrium Constants for Selected Environmentally Important Weak Monoprotic Acids and Bases at 25°C

Monoprotic Acid	*Dissociation Reaction*	K_a
Hydrofluoric acid	$HF + H_2O \Leftrightarrow H_3O^+ + F^-$	7.2×10^{-4}
Nitrous acid	$HNO_2 + H_2O \Leftrightarrow NO^+_2 + H_3O^+$	4.0×10^{-4}
Lactic acid	$CH_3CH\ (OH)\ CO2H + H_2O \Leftrightarrow CH_3CH\ (OH)\ CO2^- + H_3O^+$	1.38×10^{-4}
Benzoic acid	$C_6H_5CO_2H + H_2O \Leftrightarrow C_6H_5CO^-_2 + H_3O^+$	6.4×10^{-5}
Acetic acid	$HC_2H_3O_2 + H_2O \Leftrightarrow C_2H_3O^-_2 + H_3O^+$	1.8×10^{-5}
Propionic acid	$CH_3CH_2CO_2H + H_2O \Leftrightarrow CH_3CH_2CO^-_2 + H_3O^+$	1.3×10^{-5}
Hypochlorous acid	$HOCl + H_2O \Leftrightarrow OCl^- + H_3O^+$	3.5×10^{-8}
Hypobromous acid	$HOBr + H_2O \Leftrightarrow OBr^- + H_3O^+$	2×10^{-9}
Hydrocyanic acid	$HCN + H_2O \Leftrightarrow CN^- + H_3O^+$	6.2×10^{-10}
Phenol	$HOC_6H_5 + H_2O \Leftrightarrow OC_6H^-_5 + H_3O^+$	1.6×10^{-10}
Base	*Dissociation Reaction*	K_b
Dimethylamine	$(CH_3)_2NH + H_2O \Leftrightarrow (CH_3)_2NH^+_2 + OH^-$	5.9×10^{-5}
Methylamine	$CH_3NH_2 + H_2O \Leftrightarrow CH_3NH^+_3 + OH^-$	7.2×10^{-4}
Ammonia	$NH_3 + H_2O \Leftrightarrow NH^+_3 + OH^-$	1.8×10^{-5}
Hydrazine	$H_2NNH_2 + H_2O \Leftrightarrow H_2NNH^+_3 + OH^-$	1.2×10^{-6}
Analine	$C_6H_5NH_2 + H_2O \Leftrightarrow C_6H_5NH^+_3 + OH^-$	4.0×10^{-10}
Urea	$H_2NCONH_2 + H_2O \Leftrightarrow H_2NCONH^+_3 + OH^-$	1.5×10^{-14}

Source: A. Casparian, 2000, "Chemistry," in *How to Prepare for the Fundamentals of Engineering (FE/EIT) Exam*, M. Olia (editor), Barron's Educational Series, Inc., Hauppauge, N.Y.

TABLE 7.4
Equilibrium Constants for Selected Environmentally Important Polyprotic Acids at 25°C

Acid	*Dissociation Reactions*	K_{a1}	K_{a2}	K_{a3}
Sulfuric acid	$H_2SO_4 + H_2O \Leftrightarrow HSO^-_4 + H_3O^+$	1.0×10^3		
	$HSO^-_4 + H_2O \Leftrightarrow SO^{2-}_4 + H_3O^+$		1.2×10^{-2}	
Hydrogen sulfide	$H_2S + H_2O \Leftrightarrow HS^- + H_3O^+$	1.0×10^{-7}		
	$HS^- + H_2O \Leftrightarrow S^{2-} + H_3O^+$		1.3×10^{-13}	
Phosphoric acid	$H_3PO_4 + H_2O \Leftrightarrow HPO^-_4 + H_3O^+$	7.1×10^{-3}		
	$HPO^-_4 + H_2O \Leftrightarrow HPO^{2-}_4 + H_3O^+$		6.3×10^{-8}	
	$HPO^{2-}_4 + H_2O \Leftrightarrow PO^{3-}_4 + H_3O^+$			4.2×10^{-13}
Carbonic acid	$H_2CO_3 + H_2O \Leftrightarrow HCO^-_3 + H_3O^+$	4.5×10^{-7}		
	$HCO^-_3 + H_2O \Leftrightarrow CO^{2-}_3 + H_3O^+$		4.7×10^{-11}	

Source: A. Casparian, 2000, "Chemistry," in *How to Prepare for the Fundamentals of Engineering (FE/ET) Exam*, M. Olia (editor), Barron's Educational Series, Inc., Hauppauge, NY.)

Followed by a reaction that forms carbonate and hydrogen:

$$HCO_3^- \Leftrightarrow CO_3^{2-} + H^+ \qquad \text{Reaction 7–33}$$

Each of the two-step reactions has its own acid equilibrium constant (K_{a1} and K_{a2}, respectively). These K_{a1} and K_{a2} constants for other diprotic acids are shown in Table 7.3. For a triprotic acid, there would be three unique constants. Note that the constants decrease substantially with each step. In other words, most of the hydrogen ion production occurs in the first step.

Numerous reactions can be predicted from the relative strength of acids and bases, since their strength results from how well the protons via the hydronium ion is transferred from the acid and the electron is transferred via the hydroxide ion from the base. If an acid is weak, its conjugate base must be strong, and if an acid is strong, its conjugate base must be weak. Likewise, if a base is weak, its conjugate acid must be strong, and if the base is strong, its conjugate acid must be weak. Tables 7.3 and 7.4 show actual K_a and K_b constants, however, many handbooks and manuals report them as pK_a and pK_b. Since the "p" denotes negative logarithm, the larger the pK_a, the weaker the acid, and the larger the pK_b, the weaker the base.

An example of how the K_a is an indicator of relative strength of reactants and products is that of hydrocyanic acid:

Stronger base (↘ OH^-) **Weaker base** (↙ CN^-)

$$HCN + OH^- \Leftrightarrow CN^- + H_2O$$

Stronger acid (↗ HCN) **($K_a = 6.2 \times 10^{-10}$)**

Weaker acid (↖ H_2O) **($K_a = 1.8 \times 10^{-16}$)**

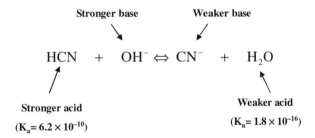

The ratio of the K_a constant values of the two acids is a direct way to quantify the equilibrium. In the hydrocyanic acid instance, the ratio is

$$\frac{6.2 \times 10^{-10}}{1.8 \times 10^{-16}} \cong 4 \times 10^{6}. \qquad \text{Reaction 7–34}$$

This large quotient indicates that the equilibrium is quite far to the right. So if HCN is dissolved in a hydroxide solution (e.g., NaOH), the resulting reaction will produce much greater amounts of the cyanide ion (CN^-) than the amount of both the hydroxide ion (OH^-) and molecular HCN. Conversely, in an aqueous solution of sodium cyanide (NaCN), the water will only react with a tiny amount of CN^-.

Among the properties of water, one of the most important environmentally is that it can behave as either an acid or a base: water is an "amphoteric" compound. That is one of the reasons that water is sometimes shown as HOH. When water acts as a base its $pK_b = -1.7$. When water acts as an acid its $pK_a = 15.7$.

Many contaminant reactions occur in water (even in the water within the cells, tissues, and organs of organisms), so the relationship between conjugate acid-base equilibrium and pH is important. The Henderson-Hasselbach equation states this relationship:

$$pK_a = pH + \log\frac{[HA]}{[A^-]} \qquad \text{Equation 7–8}$$

Thus, a review of Equation 7–8 tells us that when the pH of an aqueous solution equals the pK_a of an acidic component, the concentrations of the conjugate acids and bases must be equal (since the log of 1 = 0). If pH is 2 or more units lower than pK_a, the acid concentration will be greater than 99%. Conversely, when pH is greater than pK_a by 2 or more units, the conjugate base concentration will account for more than 99% of the solution.[7]

The Henderson-Hasselbach relationship means that mixtures of acidic and nonacidic compounds can be separated with a pH adjustment, which has strong implications for treating contaminated wastes. The application of this principal is also important to the transformation of environmental contaminants in the form of weak organic acids or bases, because these compounds in their nonionized form are much more lipophilic, meaning they will be absorbed more easily through the skin than when they exist in ionized forms. As a general rule, the smaller the pK_a for an acid and the larger the pK_b for a base, the more extensive will be the dissociation in aqueous environments at normal pH values, and the greater a compound's electrolytic nature.

Acid Rain Example

Recall from the Chapter 1 introduction to acid rain (Environmental Field Log Entry: Discussion: Normal Rain *Is* "Acid Rain") that dissolved CO_2 causes most rainfall to be slightly acidic:

$$CO_2(aq) + H_2O(l) \Leftrightarrow H^+(aq) + HCO_3^-(aq)\ K_a = 4.3 \times 10^{-7} \qquad \text{Reaction 7–35}$$

Now armed with a better understanding of K_a and K_b, as well as the Henderson-Hasselbach relationship, we can see how acid rain is exacerbated by human activities. Acidic precipitation contains strong acids: H_2SO_4 and HNO_3, mainly from the combustion of fossil fuels that contain sulfur (the air contains molecular nitrogen that is oxidized during internal combustion in engines, "mobile sources," and in any high-temperature furnaces, like those in power plants). The sulfuric acid is a stepped process. The contaminant released from stacks is predominantly sulfur dioxide when the elemental sulfur is oxidized:[8]

$$S + O_2 \rightarrow SO_2(g) \qquad \text{Reaction 7–36}$$

The sulfur dioxide in turn is oxidized to sulfur trioxide:

$$SO_2(g) + \frac{1}{2}O_2(g) \rightarrow SO_3(g) \qquad \text{Reaction 7–37}$$

The sulfur trioxide then reacts with atmospheric water (vapor, clouds, on particles) to form sulfuric acid:

$$SO_3(g) + H_2O \rightarrow H_2SO_4(aq) \qquad \text{Reaction 7–38}$$

The net result is to increase the acidity of the rain, which is a threat to aquatic life, as well as metallic and carbonate materials (including artwork, statues, and buildings). Near pollution sources, rainwater pH can be found to be less than 3 (i.e., 10,000 times more acidic than neutral!).

What are the molar concentrations of $[H^+]$ and $[OH^-]$ of rainwater at pH 3.7 at 25°C? When SO_2 dissolves in water, sulfurous acid (H_2SO_3, $K_{a1} = 1.7 \times 10^{-2}$, $K_{a2} = 6.4 \times 10^{-8}$) is formed. What is the reaction when sulfurous acid donates a proton to a water molecule? What is the Brönsted-Lowry acid and base in this reaction?

Surface waters have a natural buffering capacity, especially in regions where there is limestone, which gives rise to dissolved calcium carbonate (e.g., central Kansas is less at risk of acid rain's effects than are the Finger Lakes of New York).[9] What is the reaction of a minute amount of acid rain containing sulfuric acid reaching a lake containing carbonate (CO_3^{2-}) ions?

Solution

Since, $pH = -\log[H^+]$, then $[H^+] = 10^{-pH} = 10^{-3.7} = 2.0 \times 10^{-4}\,M$ in aqueous solution at 25°C.

The sulfurous acid proton donation reaction is:

$$H_2SO_3(aq) + H_2O(l) \Leftrightarrow HSO_3^-(aq) + H_3O^+(aq) \quad \text{Reaction 7–39}$$

H_3O^+ is the stronger acid (i.e., $K_{a1} < 1$) and HSO_3^- is the stronger base.
Regarding the buffered water system, CO_3^{2-} is the conjugate base of the weak acid HCO_3^-, so the former can react with the strong acid H_3O^+ in the sulfuric acid solution:

$$CO_3^{2-}(aq) + H_3O^+(aq) \Leftrightarrow HCO_3^-(aq) + H_2O(l) \quad \text{Reaction 7–40}$$

The $K = 1/K_{a2}$ of H_2CO_3 (making for a large K)
Similarly, HSO_4^- also reacts with CO_3^{2-}:

$$CO_3^{2-}(aq) + HSO_4^-(aq) \Leftrightarrow SO_4^{2-}(aq) + HCO_3^-(aq) \quad \text{Reaction 7–41}$$

The $K = (K_{a2}$ of $H_2SO_4)/(K_{a2}$ of $H_2CO_3)$ *(thus, large K)*

So, a HCO_3^-/CO_3^{2-} buffer is produced.

An excess of acid rain will consume all the CO_3^{2-} and HCO_3^-, converting all to H_2CO_3 (and completely eliminating the buffer).
These are important and representative reactions of the challenging global problem of acid rain.

Hydrolysis

All of the acid-base reactions involve chemical species reacting with water. It is worth noting that even some so-called neutral compounds have acidic or basic properties. For example, metal acetates ($MeC_2H_3O_{2x}$) can dissolve in water and actually react with the water to form weak acids and hydrogen ions. The resulting solutions are generally slightly basic (pH > 7), since the acetate ion is a conjugate base of the weak acid. This process is known as *hydrolysis* (i.e., *-lysis*, or breaking apart, of water molecules). Likewise, some compounds produce weak bases when dissolved in water. For example, when ammonium chloride (NH_4Cl) is dissolved in water, the solution becomes more acidic. The NH_4^+ cation serves as the conjugate base of ammonium hydroxide (NH_4OH), or molecular ammonia (NH_3) is the conjugate base of the NH_4^+ cation. Oxide gases (e.g., carbon dioxide and sulfur dioxide) can be hydrolyzed to form hydronium ions and anions:

$$CO_2(g) + H_2O(l) \Leftrightarrow H^+(aq) + HCO_3^-(aq) \qquad \text{Reaction 7–42}$$

and acids:

$$SO_2(g) + H_2O(l) \Leftrightarrow H_2SO_4 \qquad \text{Reaction 7–43}$$

These reactions also take place in organic compounds, where the organic molecule, RX, reacts with water to form a covalent bond with OH and cleaves the covalent bond of the leaving group (X) in RX, which displaces X with the hydroxide ion and an ion formed from the leaving group:

$$RX + H_2O \rightarrow ROH + H^+ + X^- \qquad \text{Reaction 7–44}$$

Amides, epoxides, carbonates, esters, organic halides, nitriles, urea compounds, and esters of organophosphate compounds are functional groups that are susceptible to hydrolysis. The process involves an electron-rich nucleus seeker (i.e., a nucleophile) attacking an electron-poor electron seeker (i.e., an electrophile) to displace the leaving group (such as a halogen). This is why hydrolysis is one of the methods of dechlorination, a detoxification process for hazardous chlorinated hydrocarbons.

Environmental engineers use hydrolysis to eliminate or to reduce the toxicity of hazardous contaminants by abiotic transformation and biotransformation, especially bacterial. For example, the highly toxic methyl isocynate, infamous as the contaminant that led to the loss of life and health in the Bhopal, India incident, can be transformed hydrolytically.

Two factors are particularly enhancing to hydrolysis. Microbial mediation, including enzymatic activity can catalyze hydrolytic reactions. This occurs both in the ambient environment, such as in the hydrolysis of inorganic and organic compounds by soil bacteria, as well as in engineered systems, such as acclimating those same bacteria to the treatment of chlorinated organic compounds in solid and liquid wastes.

The second factor is pH. Hydrolysis can be affected by specific acid and base catalysis. In acid catalysis, this H^+ ion catalyzes the reaction; and in base catalysis the OH^- ion serves as the catalyst. The effect of temperature on hydrolysis can be profound. Each 10°C incremental temperature increase results in a hydrolysis rate constant change by a factor of 2.5.[10]

Metal (Me) hydrolysis is a special case. Cations in water act like Lewis acids in that they are prone to accept electrons, while water behaves like a Lewis base because it makes its oxygen's two unshared electrons available to the cations. When strong water-metal (acid-base) interactions take place, H^+ dissociates and hydronium ions form in a prototypical reaction:

$$Me^{n+} + H_2O \Leftrightarrow MeOH^{(n+1)} + H^+ \qquad \text{Reaction 7–45}$$

Although reactions such as these have obvious applications to contaminant transformations in surface water, soil water, and groundwater, they occur in any medium where water is present. Water is present in the atmosphere, so hydrolysis occurs in clouds and fog, as well as in the water fraction of hygroscopic nuclei. Water is also present in all living things, so hydrolysis is a common process in metabolism (particularly in the first metabolic phase, as will be discussed in detail in Chapter 9, "Contaminant Hazards") and other organic processes. Thus, hydrolysis is important in numerous environmental and toxicological processes.

Photolysis

The sun's electromagnetic radiation at ultraviolet and visible wavelengths can induce chemical reactions directly and indirectly. Direct photolysis is the process where sunlight adds the activation energy needed to transform a compound. Indirect photolysis is the process by which an intermediate compound is energized, which in turn transfers energy to another compound.

Contaminants are photochemically degraded in both atmospheric and aquatic environments. Photolysis can combine or interchange with other processes, such as in the degradation pathways for chlorinated organic contaminants. For example, the degradation pathway for 1,4-dichlorobenzene in air is a reaction with photochemically-generated OH^- radicals and oxides of nitrogen. However, in soil and water, the degradation is mainly microbial biodegradation, leading to very different end products (see Figure 7.6).

Photodegradation is addressed in detail in Chapter 8's Discussion Box "Engineering Technical Note: Removing Endocrine Disruptors from Drinking Water—An Alternative Treatment Scheme Using Ultraviolet Light."

Precipitation Reactions in Environmental Engineering

Dissolved ions may react with one another to form a solid phase compound under environmental conditions of temperature and pressure. Salts are compounds that form when metals react with nonmetals, such as sodium chloride (NaCl). They may also form from cation and anion combinations, such as ammonium nitrate (NH_4NO_3). Precipitation is both a physical and chemical process, wherein soluble metals and inorganic compounds change into insoluble metallic and inorganic salts. In other words, the dissolved forms become solids. Such reactions in which soluble chemical species become insoluble products are known as *precipitation reactions*.

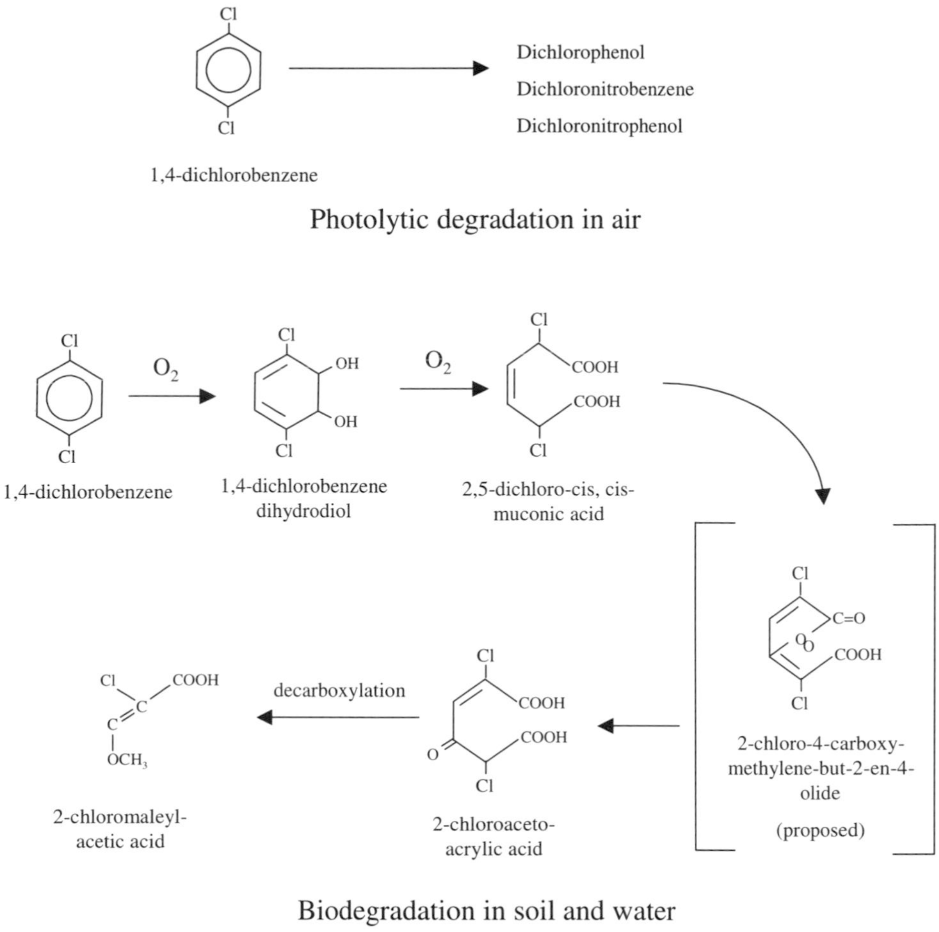

FIGURE 7.6. Different 1,4-dichlorobenzene reactions according to environmental media. (Adapted from: Agency for Toxic Substances and Disease Registry, 1998, "Toxicological Profile for 1,4-Dichlorobenzene," http://www.atsdr.cdc.gov/toxprofiles/tp10.html.)

Chemical precipitation occurs within a defined pH and temperature range unique for each metallic salt. Usually in such reactions, an alkaline reagent is added to the solution, thereby raising the solution pH. The higher pH often decreases the solubility of the metallic constituent, bringing about the precipitation (see Figure 7.7). For example, when caustic soda (NaOH) serves as the precipitating agent to lower the amount of soluble nickel, the product is the much less water soluble species nickel hydroxide precipitate (recall that "s" denotes a solid precipitate):

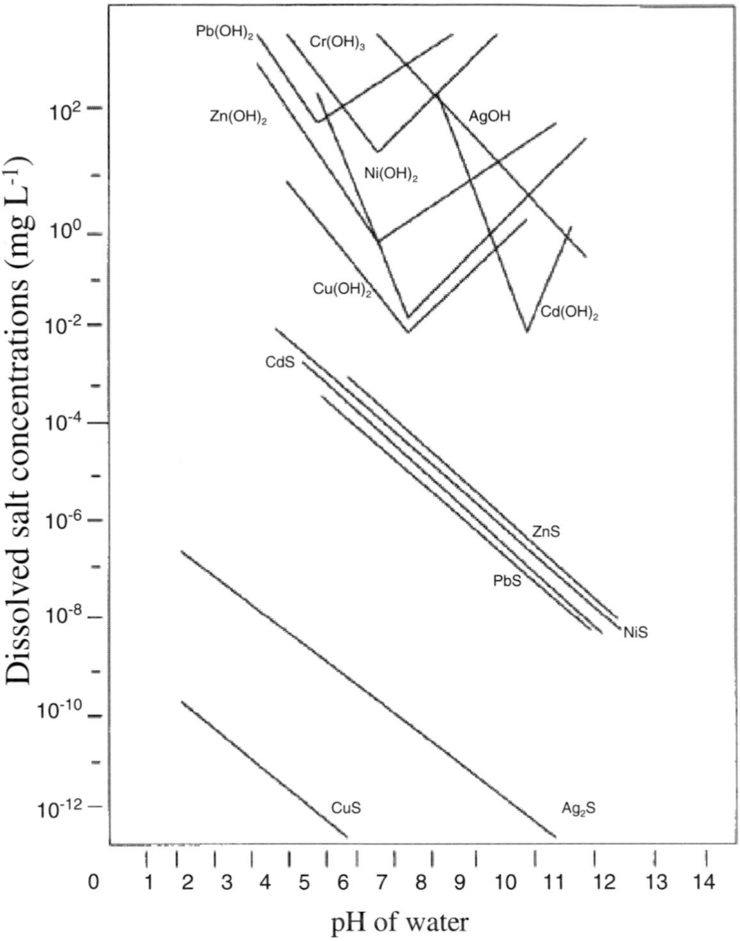

FIGURE 7.7. Effect of pH on the solubility of metal hydroxides and sulfides. (Source: U.S. Environmental Protection Agency, 1980, *Summary Report: Control and Treatment Technology for the Metal Finishing Industry; Sulfide Precipitation*, Report No. EPA 625/8-80-003, Washington, D.C.)

$$Ni^{2+} + NaOH \Leftrightarrow Na^{+} + Ni(OH)_2(s) \qquad \text{Reaction 7–46}$$

The precipitation forms small particles or colloids. If colloidal, their mass may be so small that they do not readily fall but remain suspended. In treating metal-contaminated water, for example, these colloids may have to be coagulated, flocculated, settled, clarified, or filtered from the suspension, even though they are no longer in solution. One such process is the precipitation/coagulation/filtration system ("P/C/F system") shown in Figure 7.8.

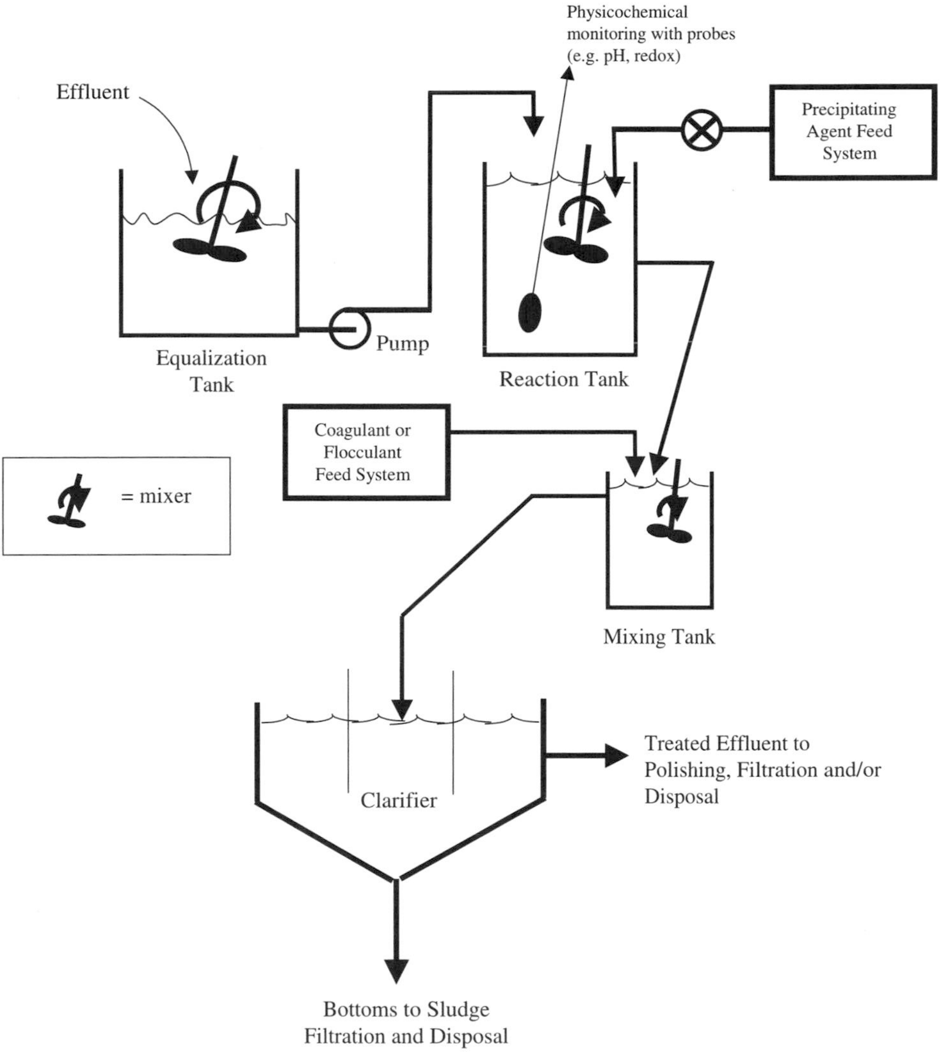

FIGURE 7.8. Precipitation/coagulation/filtration system for treating metal-contaminated water. (Adapted from: U.S. Department of the Army, 2001, *Engineering and Design: Precipitation/Coagulation/Flocculation*, Report No. EM 1110-1-4012, Washington, D.C.)

Engineer's Technical Note: Case Study—Acidic Metal Contamination[11]

Scientists have recognized for the past hundred years that acidic drainage from mines can greatly stress and even destroy ecosystems. For example, the U.S. Forest Service estimates that more than 50,000 mining sites in the western United States are releasing low pH wastes to the environment. Acidic drainage is defined as any flow into surface waters that adversely affects an aquatic ecosystem and drinking water supplies. To clean up the most egregious abandoned mine sites and to control acidic drainage from active mines, the application of physical and chemical principles must be applied. For example, the U.S. Geological Survey (USGS) is conducting research on acidic metal contamination in the Pinal Creek Basin in Arizona (see Figure 7.9) to provide information on the behavior of toxic substances in surface waters and groundwater. To ascertain information on the fate of acidic contaminants, measurements of both the soil (e.g., alluvium) and water, including stream contaminant concentrations, are needed. Of course, the interactions among stream flows and groundwater must also be understood.

Pinal Creek Basin has been the source of copper mining since 1882, starting with underground mining. Since 1948, open pit surface mining has been ongoing. The tailings[12] dominate the local landscape in some areas (Figure 7.10).

Tributaries drain the hilly and mountainous region surrounding the study area. The aquifer underlying the area consists of unconsolidated stream alluvium and fill material. The alluvium ranges from 300 to 800 meters in width, and 50 meters in depth. Most of the unconsolidated material is large-grained sand and gravel. In the northern part of Pinal Creek Basin, the aquifer becomes shallower in the direction of water flow and Inspiration Dam (see Figure 7.9). The direction of groundwater flow is toward Roosevelt Lake, an important water supply reservoir for the City of Phoenix.

The acidic mine conditions result from the oxidation of sulfide minerals exposed during mining operations and from ore processing. Unlined surface-water impoundments that received these acidic waters have become continuous contaminant sources. For example, Webster Lake, a large source, has had a pH range between 2 and 3 since the 1970s until the lake was drained in 1988. Dissolved iron and sulfate concentrations exceeded 2000 and 19,000 $mg\,L^{-1}$, respectively. The contaminated groundwater plume is extensive and has migrated approximately 25 km. Table 7.5 shows that the low pH conditions of the plume has allowed for high concentrations of sulfate, calcium,

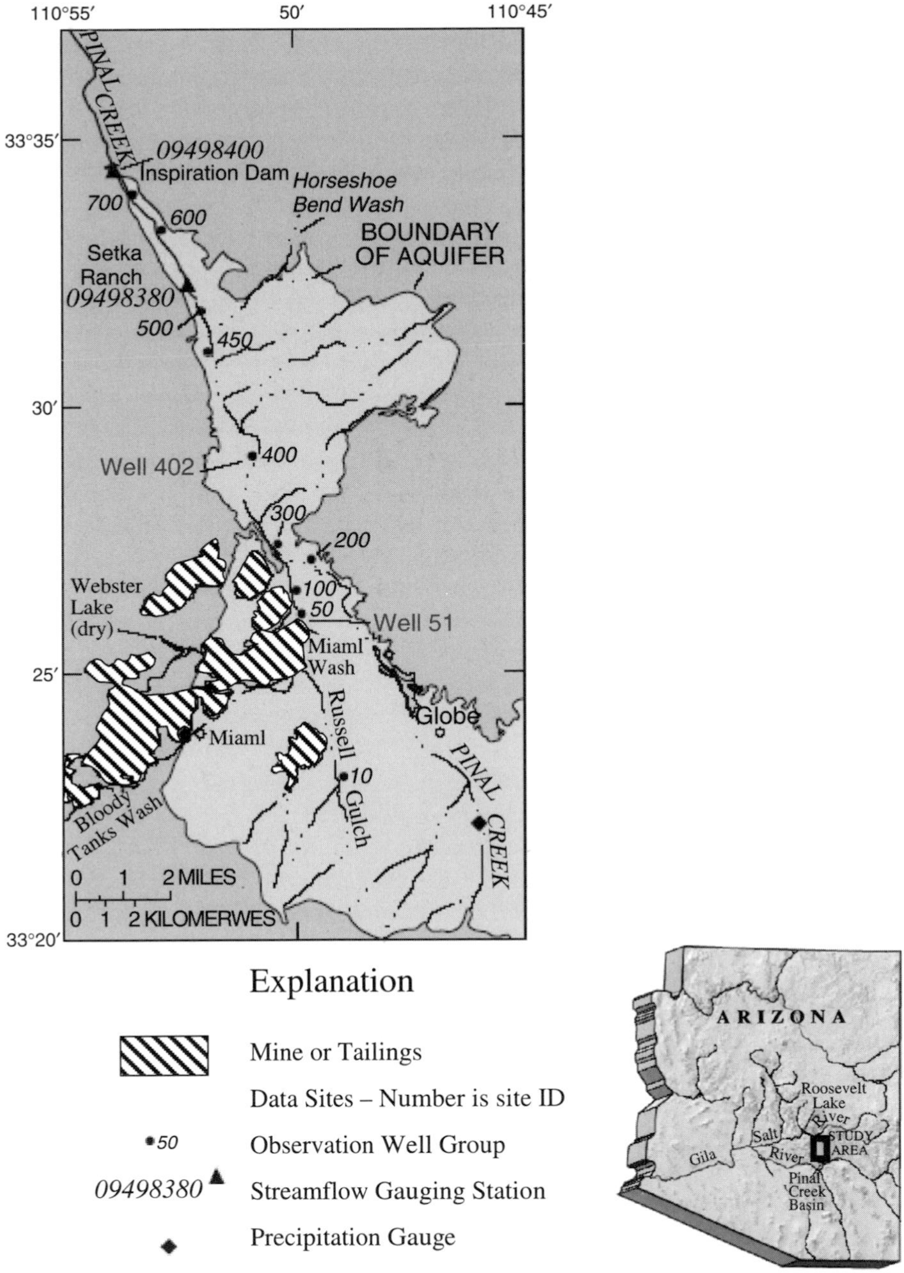

FIGURE 7.9. Groundwater and surface water measurement locations near Pinal Creek Basin, Arizona. (Source: U.S. Geological Survey.)

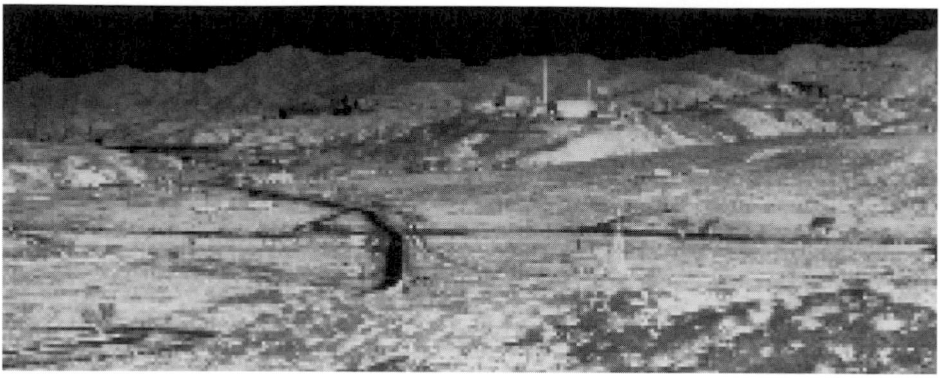

FIGURE 7.10. Mine tailings cover about 27 square kilometers in Pinal Creek Basin. (Source: U.S. Geological Survey.)

iron, manganese, copper, aluminum, and zinc. In 1996, concentrations of dissolved iron were 200 $mg\,L^{-1}$ at well 51 and 190 $mg\,L^{-1}$ at well 402.

Precipitation Reactions Used in Remediation

Acidic groundwater moving downgradient can be neutralized mainly through reaction with calcium carbonate. Such remediation has allowed the groundwater pH to increase to between 5 and 6. When this occurs, the metals (especially Fe, Cu, and Zn) precipitate or coprecipitate[13] to adsorb to mineral surfaces in the aquifer. However, some metals, such as Mn, become dissolved and more mobile. When groundwater meets surface water, the surface water becomes more basic, pH $\cong 8$, as the water equilibrates with the atmosphere. At this interface the manganese carbonates and oxides precipitate as crusts in the streambed.

The chemistry of the waters has changed as acidic sources have been removed. For example, concentrations of dissolved Fe in water from well 51 decreased from 3200 $mg\,L^{-1}$ in 1984 to 1700 $mg\,L^{-1}$ in 1990. By 1994, dissolved Fe concentrations in well 51 had dropped to 200 milligrams per liter (see Figure 7.11).

Key Lessons Learned

The Pinal Creek Basin study has shown that dissolved-metal contaminants in a groundwater plume move much more slowly than the

TABLE 7.5
Water Quality Measurements from Wells and Surface Water in the Pinal Creek Basin, Spring 1996
[Values are in milligrams per liter except for pH; <, less than; —indicates no data]

	Sample location						
	10	*51*	*302*	*402*	*503*	*702*	*Pinal Creek at Inspiration Dam*
Sample date	June 1	June 6	June 6	June 2	June 8	May 31	May 31
pH	6.8	3.9	4.0	4.2	5.3	6.9	7.8
Calcium	56	300	210	250	460	470	390
Magnesium	17	97	38	73	110	88	90
Sodium	26	77	66	58	62	57	63
Potassium	2.0	8.2	4.4	5.3	5.3	3.9	6.4
Alkalinity	199	—	—	—	27	165	90
Total dissolved inorganic carbon	—	20	20	24	24	—	—
Sulfate	36	1,900	1,200	1,400	1,800	1,200	1,500
Chloride	26	41	45	54	49	48	47
Fluoride	.3	7.6	3.4	4.9	2.9	.4	3.1
Silica	26	88	75	73	67	43	30
Iron	<.003	200	140	190	.026	<.009	<.130
Manganese	.001	13	<7.1	24	75	1.1	45.5
Aluminum	<.005	29.1	13.7	10.6	2.61	<.005	.250
Cadmium	<.001	.069	<.049	.011	.022	<.003	.007
Copper	.02	24	11	12	.67	<.03	<.03
Cobalt	<.003	—	—	.41	.87	.014	.21
Nickel	<.01	.45	.24	.42	.80	<.03	.49
Strontium	.37	1.3	.81	1.0	1.5	1.4	1.4
Zinc	.008	2.9	1.5	2.1	1.9	<.009	.35

Source: U.S. Geological Survey

plume water itself. Neutralization reactions (e.g., carbonates) slow the rate of advance of the plume's acidic front and most of the dissolved metals to about 15% of the advective or bulk rate of transport. This is important, since many models assume that contaminants move at the advective rate. (i.e. those that do not include a retardation factor).

The oxidation and precipitation of Fe is predominantly caused by the dissolution of another metal's oxides, in this instance Mn. This reaction is a source of additional acidity to the plume and a source of dissolved Mn. The dissolved Mn travels ahead of the acidic front (i.e., a kind of metallic "squall line") in the neutralized groundwater. The drop in dissolved Cu, Co, Ni, and Zn downgradient was caused by the pH-dependence of these metals' adsorption to iron hydroxides in the aquifer.

Gas exchanges between the groundwater plume and the atmosphere will affect the chemical reactions. For example, CO_2 and O_2 exchange from the plume across the water table are affected by the neutralization reactions involving carbonate minerals and from the oxidation of dissolved Fe near the water table (see Figure 7.11). After contaminated groundwater reaches Pinal Creek, gas exchange makes for a rapid increase in pH and dissolved oxygen in the surface waters. The ongoing surface water flow to and from groundwater beneath the stream (i.e., the "hyporheic zone" in Figure 7.12) enhances contact between the flowing water and the sediment, thereby stimulating the

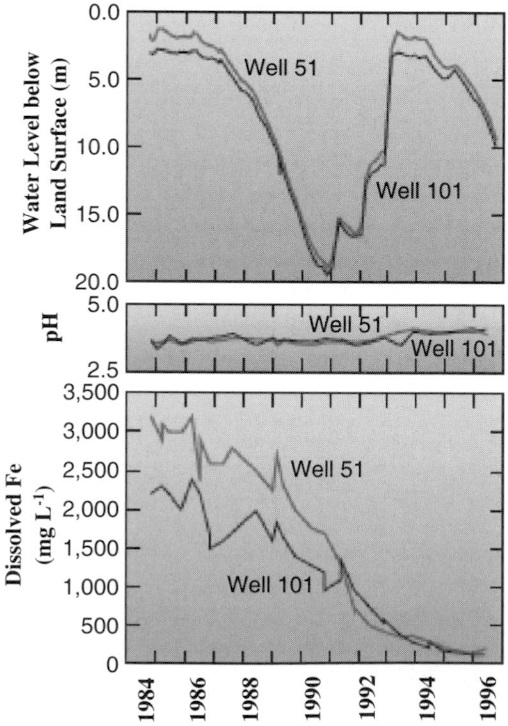

FIGURE 7.11. Levels of groundwater, pH, and iron concentrations in groundwater adjoining Miami Wash, 1984–1996. (Source: U.S. Geological Survey.)

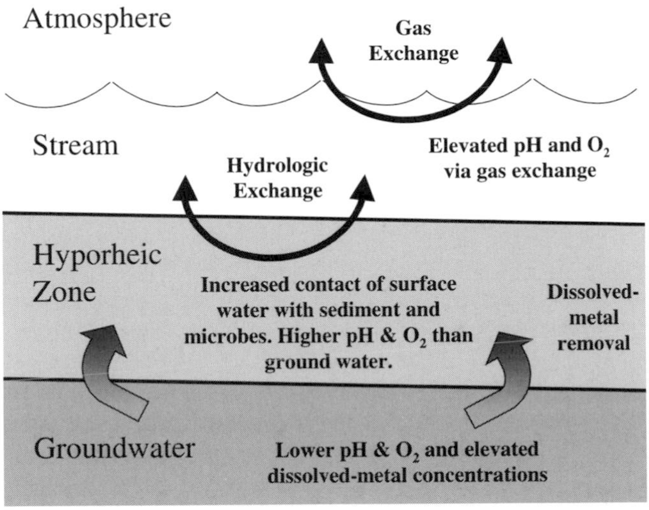

FIGURE 7.12. Exchanges and reactions between environmental compartments. In the perennial reach, some of the stream water continually moves into and out of an area in shallow groundwater known as the *hyporheic zone*. This process helps enhance removal of dissolved metals from stream flow. (Adapted from: U.S. Geological Survey.)

precipitation of dissolved Mn, which coats sediment particles with oxides of Mn. In addition, Mn-oxidizing bacteria enhance chemical reactions in the hyporheic zone. The bacteria are activated in this zone due to the increased levels of dissolved molecular oxygen by aeration in the moving stream compared to the O_2 levels in the more quiescent groundwater. About 20% of the dissolved Mn loading leaving the drainage basin is being eliminated by this approach.

Precipitation of Mn in the hyporheic zone impedes the mobility of other metals, such as dissolved Ni and Co, due to the excellent sorption by Mn oxides to these metals. This means that the loading of dissolved Ni and Co from the basin has been decreased predominantly by hyporheic exchange.

Oxidation-Reduction Reactions

An oxidation-reduction (i.e., *redox*) reaction is the simultaneous electron loss (oxidation) of one substance with electron gain (reduction) by another in the same reaction. Any one of the following actions is redox:

Oxidation	*Reduction*
Substance loses ("donates") electrons	Substance gains ("captures") electrons
Substance gains oxygen atoms	Substance loses oxygen atoms
Substance loses hydrogen atoms	Substance gains hydrogen atoms

It stands to reason from the simultaneous nature of redox reactions that each oxidation-reduction reaction is, in fact, two "half-reactions." An oxidation reaction example is the loss of two electrons by a calcium (Ca) atom to form the divalent calcium cation:

$$Ca \rightarrow Ca^{2+} + 2e^{-} \qquad \text{Reaction 7–47}$$

The companion reduction reaction example is the gain of electrons by divalent calcium to form elemental, zero-valence calcium.

$$Ca^{2+} + 2e^{-} \rightarrow Ca \qquad \text{Reaction 7–48}$$

The electronegativity relationships responsible for these changes were discussed previously in Chapter 4, "Environmental Equilibrium, Partitioning and Balances" as one of the physical principals driving polarity and solubility.

An example is the formation of two air pollutants, sulfur dioxide and nitric oxide by acidifying molecular sulfur:

$$S(s) + NO_3^{-}(aq) \rightarrow SO_2(g) + NO(g) \qquad \text{Reaction 7–49}$$

The oxidation half-reactions for this reaction are:

$$\text{Unbalanced: } S \rightarrow SO_2 \qquad \text{Reaction 7–50}$$

$$\text{Balanced: } S + 2H_2O \rightarrow SO_2 + 4H^{+} + 4e^{-} \qquad \text{Reaction 7–51}$$

The reduction half-reactions for this reactions are:

$$\text{Unbalanced: } NO_3^{-} \rightarrow NO \qquad \text{Reaction 7–52}$$

$$\text{Balanced: } NO_3^{-} + 4H^{+} + 3e^{-} \rightarrow NO + 2H_2O \qquad \text{Reaction 7–53}$$

The overall balanced oxidation-reduction reactions are:

$$4NO_3^{-} + 3S + 16H^{+} + 6H_2O \rightarrow 3SO_2 + 16H^{+} + 4NO + 8H_2O \qquad \text{Reaction 7–54}$$

$$4NO_3^{-} + 3S + 4H^{+} \rightarrow 3SO_2 + 4NO + 2H_2 \qquad \text{Reaction 7–55}$$

Of course, oxidation-reduction reactions are not all bad. In fact, they are part of essential metabolic and respiratory processes, and they are used in reducing toxicity and in treating wastes. For example, redox is employed in water treatment by adding a chemical-oxidizing or reducing agent under controlled pH. This reaction raises the valence of one reactant and lowers the valence of the other. Thus redox removes compounds that are "oxidizable," such as ammonia, cyanides, and certain metals like selenium, manganese, and iron. It also removes other "reducible" metals like mercury, chromium, lead, silver, cadmium, zinc, copper, and nickel. Thus oxidizable cyanide destruction and the reduction of chromium from Cr^{6+} to Cr^{3+} are two examples where the toxicity of inorganic contaminants can be greatly reduced by redox.[14]

Biological Redox Reactions

Redox reactions are very important in natural environmental processes. Many biologically mediated processes that degrade contaminants are redox reactions that involve the transfer of electrons from an organic contaminant to an electron acceptor. In these cases, oxygen is the electron acceptor if the process is aerobic (i.e., sufficient molecular oxygen is available). In anaerobic microbial processes, nitrate, Fe^{3+}, sulfate, and carbon dioxide can be the electron acceptors. The redox reactions for the commonly encountered contaminants benzene and substituted derivatives of benzene (i.e., xylene and ethylbenzene) are shown in Table 7.6.

The microbes will vary in dominance as the dissolved oxygen levels change. With falling dissolved oxygen concentrations, anaerobic bacteria will begin to dominate in degrading organic wastes. The transfer of electrons is the means by which the microbes receive energy from the food (i.e., the organic waste).

In the next chapter, we will consider some of the important biological concepts in contamination. However, since the reactions discussed in this chapter, especially redox, apply to the bioenergetics of microbes, and biological treatment is a large part of the arsenal used to respond to contaminated water and soil, let us consider a few examples at this time.

TABLE 7.6
Oxidation-Reduction Reactions for Aromatic compounds.

Type	*Reaction*	*Electron Acceptor*
Benzene Redox Reactions:		
Oxidation	$C_6H_6 + 12H_2O \rightarrow 6CO_2 + 30H^+ + 30e^-$	
Reduction	$7.5O_2 + 30H^+ + 30e^- \rightarrow 15H_2O$	Oxygen
Reduction	$6NO_3^- + 36H^+ + 30e^- \rightarrow 3N_2 + 18H_2O$	Nitrate
Reduction	$15Mn^{4+} + 30e^- \rightarrow 15Mn^{2+}$	Manganese
Reduction	$30Fe^{3+} + 30e^- \rightarrow 30Fe^{2+}$	Iron
Reduction	$6SO_2^- + 37.5H^+ + 30e^- \rightarrow 3.75H_2S + 15H_2O$	Sulfate
Reduction	$3.75CO_2 + 30H^+ + 30e^- \rightarrow 3.75CH_4 + 7.5H_2O$	Methanogenic bacteria
Overall	$C_6H_6 + 7.5O_2 \rightarrow 6CO_2 + 3H_2O$	Oxygen
Overall	$C_6H_6 + 6H^+ + 6NO_3^- \rightarrow 6CO_2 + 3N_2 + 6H_2O$	Nitrate
Overall	$C_6H_6 + 15Mn^{4+} + 12H_2O^- \rightarrow 6CO_2 + 30H^+ + 15Mn^{2+}$	Manganese
Overall	$C_6H_6 + 30Fe^{3+} + 12H_2O^- \rightarrow 6CO_2 + 30H^+ + 30Fe^{3+}$	Iron
Overall	$C_6H_6 + 3.75SO_4^{2-} + 7.5H^+ \rightarrow 6CO_2 + 3.75H_2S + 3H_2O$	Sulfate
Overall	$C_6H_6 + 4.5H_2O \rightarrow 2.25CO_2 + 3.75CH_4$	Methanogenic bacteria
Xylene and Ethylbenzene Redox Reactions:		
Oxidation	$C_8H_{10} + 16H_2O \rightarrow 8CO_2 + 42H^+ + 42e^-$	
Reduction	$10.5O_2 + 42H^+ + 42e^- \rightarrow 21H_2O$	Oxygen
Reduction	$8.4NO_3^- + 50.4H^+ + 42e^- \rightarrow 4.2N_2 + 25.2H_2O$	Nitrate
Reduction	$21Mn^{4+} + 42e^- \rightarrow 21Mn^{2+}$	Manganese
Reduction	$42Fe^{3+} + 42e^- \rightarrow 42Fe^{2+}$	Iron
Reduction	$5.25SO_2^- + 52.5H^+ + 42e^- \rightarrow 5.25H_2S + 21H_2O$	Sulfate
Reduction	$5.25CO_2 + 42H^+ + 42e^- \rightarrow 5.25CH_4 + 10.5H_2O$	Methanogenic bacteria
Overall	$C_8H_{10} + 10.5O_2 \rightarrow 8CO_2 + 5H_2O$	Oxygen
Overall	$C_8H_{10} + 8.4H^+ + 8.4NO_3^- \rightarrow 8CO_2 + 4.2N_2 + 9.2H_2O$	Nitrate
Overall	$C_8H_{10} + 21Mn^{4+} + 16H_2O^- \rightarrow 8CO_2 + 42H^+ + 21Mn^{2+}$	Manganese
Overall	$C_8H_{10} + 42Fe^{3+} + 16H_2O^- \rightarrow 8CO_2 + 42H^+ + 42Fe^{2+}$	Iron
Overall	$C_8H_{10} + 5.25SO_4^{2-} + 10.5H^+ \rightarrow 8CO_2 + 5.25H_2S + 5H_2O$	Sulfate
Overall	$C_8H_{10} + 5.5H_2O \rightarrow 2.75CO_2 + 5.25CH_4$	Methanogenic bacteria

(Source: U.S. Environmental Protection Agency, 2003, *Bioplume III Natural Attenuation Decision Support System*, Users Manual, Version 1.0, Washington, DC.)

Bioremediation Example 1

What is the difference between biostimulation and bioaugmentation in biological treatment of hazardous chemical contaminants? Which approach do hazardous waste remediation engineers prefer?

Answer

The process of adding nutrients to contaminated sites, such as nitrogen, phosphorus, oxygen, and other elements and compounds that serve as electron acceptors to stimulate the activity of microbial populations, is known as *biostimulation*.[15] *Bioaugmentation* is the process of adding microbes to the subsurface environment. The microbes can come from "seed" microbes that are taken from the contaminated environment and then mixed with these elements and compounds in a reactor, and then reintroduced to the contaminated soil or water, especially groundwater. Other times, special targeted and cultivated strains with known abilities to degrade certain compounds are injected into contaminated soil and groundwater.

Many bioremediation experts favor biostimulation because almost every needed microbe is available in the subsurface, where indigenous species have likely already developed enzyme production systems that will break down the contaminants, while the exogenous species that were augmented elsewhere may not survive the new, hostile environment after introduction.

Bioremediation Example 2

The Town of Eagles public works department has asked you whether bioremediation is likely to work in cleaning up two sites with the following characteristics:

Site ID	*Contaminants Detected to Date*	*Soil Type*	*Microbial Enzyme Type*[1]	*Extent of Contamination*
1	Short-chain, non-halogenated hydrocarbons and amines	Sandy	Primary	Highly concentrated within a 50 m^2 area
2	Multi-chlorinated C_{13}–C_{20}	Clayey	Cometabolism	Widely dispersed and

hydrocarbons and PCBs	unevenly distributed pockets of contamination over a 0.5 km^2 area

[1]Explained on pages 377–9.

Give reasons for expecting neither, either, or both sites to be amenable to biological treatment methods.

Answer and Discussion

This is pretty much a "no-brainer" decision. Site 1 appears to be very amenable to *in situ* bioremediation. Shorter chained, unsubstituted hydrocarbons and low molecular weight amine compounds have been shown to be very good candidates for microbial treatment. The highly permeable, sandy soils allow for the injection of nutrients and other additives to the contaminated plume. The microbes would use the contaminant as a food source, since they have developed enzymes for this purpose. The well-defined, tight boundaries of the plume make it more manageable for physical and chemical processes, such as aeration and water extraction.

Site 2 is not promising. The high molecular weight, substituted compounds resist degradation. The clay soil has low permeability, so biostimulation will be difficult and costly. Since the microbes do not use the compounds as a food source and only degrade them fortuitously via cometabolism, other compounds will have to be added so that electron accepting and donating will not be the limiting step. ΩFinally, since the waste is erratically distributed over a large area, numerous pumping and vacuum systems would have to be in place. Unfortunately, the site may not only be unsuitable for bioremediation, but probably will not be amenable to any *in situ* process. This means major excavation and transport to treatment facilities, unless on-site systems can be used. Generally, soils contaminated with polychlorinated biphenyls (PCBs) must be removed and treated, often thermally.

Unfortunately, such "no-brainer" decisions are seldom encountered in environmental situations. For example, even with short-chained, unsubstituted hydrocarbons that should be amenable to microbial remediation approaches, other factors such as the inability of bacteria to survive and thrive is not usually straightforward. This can be attributed to numerous reasons, e.g., deficiency of micronutrients, limits in gas exchange, and the presence of compounds with specific toxicity to the bacteria. Any of these factors can be the rate limiting step. That is why models employed to predict bioremediation must include sensitivity analyses to determine which factors, in a given situation, will have the greatest effects on engineering success. Remember, in environmental problem solving, "everything matters."

Bioremediation Example 3

A database has been compiled during site characterizations for certain soil microbial classes to document intrinsic bioremediation (natural attenuation) of a chlorinated hydrocarbon. The data show that the biodegradation of this compound occurs in direct proportion to the compound's concentration; that is, it follows first-order kinetics.

Calculate the biodegradation rate constant for this compound if the highest measured concentration is $90\,\mu g\,L^{-1}$ upgradient (at point A) and $450\,ng\,L^{-1}$ (corrected for dilution) downgradient (at point B 1330 m south of point A) in a groundwater plume moving at $10^{-5}\,m\,sec^{-1}$.

Answer and Discussion

The first-order biodegradation rate $= \ln\frac{C_d}{C_u}(Dv)$ Equation 7–9

Where C_d = Highest downgradient concentration of compound
C_u = Highest upgradient concentration of compound
D = Distance traveled
v = Plume velocity

So, the empirically derived first-order rate constant is:

$$\ln\frac{(450\text{ ng L}^{-1})/(9\times10^{4}\text{ ng L}^{-1})}{1330\text{ m}/10^{-5}\text{ m s}^{-1}} = \ln 2.8\times10^{-1}\text{ sec}^{-1} = -1.3\text{ sec}^{-1}$$

This is the first-order degradation rate for this compound. The microbes are accommodating!

Bioremediation Example 4

What is the shape of a first-order degradation rate compared to zero-order and second-order rates? What relevance do these differences have for remediation?

Answer

Figure 7.13 shows the prototypical decay curves. Note that the beginning and ending amounts of contamination are the same, but the rates or kinetics of the three systems are different.

The mathematics of the three orders of decay kinetics has been derived from laboratory research and small-scale studies (so-called "mesoscale" research done in bathtub-sized systems). While very useful in understanding the theory and first principles of degradation, especially biodegradation, they must be adapted to the heterogeneous conditions of the field, especially for *in situ* bioremediation and the expected growth and metabolism of microbes that catalyze contaminants.

Reactions follow the "rate law," that is, the general reaction configuration is:

$$\mathrm{aA + bB \rightarrow gG + hH}, \text{ the rate} = k[\mathrm{A}]^{\mathrm{x}}[\mathrm{B}]^{\mathrm{y}} \qquad \text{Reaction 7–56}$$

So, k represents the rate constant for molar concentrations of the reactants, and x and y represent the reaction rate order for the reactants. Summing all of the reactant orders gives the "overall order" of the reaction. A homogeneous (i.e. same physical state) abiotic example is the decomposition of the gas dinitrogen pentoxide to the gases nitrogen dioxide and oxygen:

$$2N_2O_5 \leftrightarrow 4NO_2 + O_2 \qquad \text{Reaction 7–57}$$

Laboratory studies have shown that rate law for this reaction is first-order:

$$\text{Rate} = \mathrm{k}[N_2O_5]^1 \text{ or simply } k[N_2O_5]. \qquad \text{Reaction 7–58}$$

This means that for any first-order reaction for any chemical species, the rate is:

$$\text{Rate} = \frac{-d[A]}{dt} = \frac{\Delta[A]}{\Delta t} + k[A] \qquad \text{Equation 7–10}$$

This means that in a first-order reaction, the doubling concentration of chemical species A should lead to a doubling of the reaction rate,

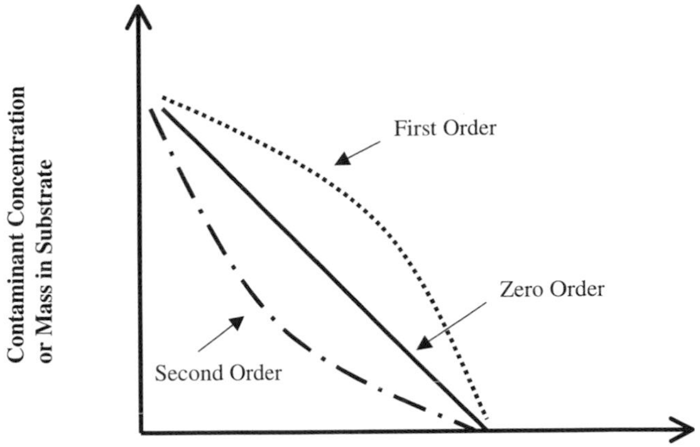

FIGURE 7.13. Prototypical decay curve for microbial degradation of organic contaminants in natural environments based upon substrate contamination and microbial biomass. The second-order rate likely results from first-order kinetics being affected by microbial population density. (Adapted from: S. Suthersan, 1997, *Remediation Engineering*, CRC Press, Inc., Boca Raton, FL).

or a tenfold increase in the concentration of A will mean ten times the reaction rate. Integrating the rate equation above yields:

$$\ln[A] = \ln[A]_0 - k\underline{\Delta t} \qquad \text{Equation 7–11}$$

Similarly, integrating the rate equation for a second-order reaction yields:

$$\frac{1}{[A]} = \frac{1}{[A]_0} + k\Delta t \qquad \text{Equation 7–12}$$

However, living systems are more complicated that this. When the substrate is not limiting, a lot of contaminant (food) is available to the microbes, and the contaminant is degraded as a function of the logarithmic growth of the microbes, following zero-order kinetics (constant log growth and log decay).

When the rate of degradation of the chemical contaminant becomes directly proportional with the concentration of the contaminant, the decay follows first-order kinetics.

Actually, one of the more realistic kinetics scenarios is second order; that is, when the first-order kinetics is related to microbial population density. Generally, though, over time the realistic[16] biodegradation rate may cycle through various orders over the life of a remediation project, as shown in Figure 7.14.

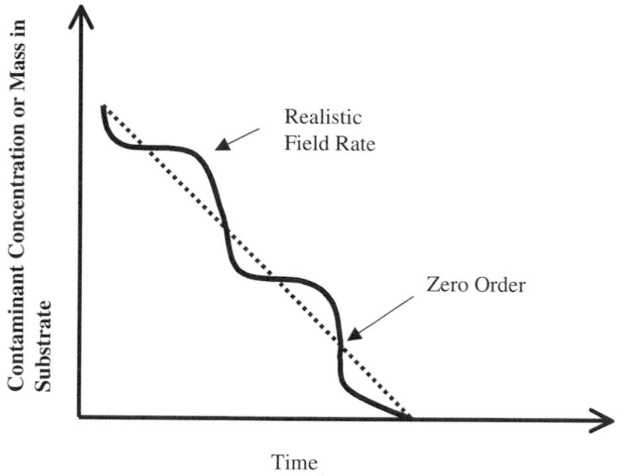

FIGURE 7.14. Rate changes within an overall zero-order decay rate. (Adapted from: S. Suthersan, 1997, *Remediation Engineering*, CRC Press, Inc., Boca Raton, FL.)

Modelers often consider microbial degradation rates to be nonlinear reactions, possibly because of the dearth of information available supporting stepwise degradation, especially in natural systems. Thus, they have developed biodegradation models to fit the results, expressed as:

$$\underline{-d[\mathrm{A}]} = k[\mathrm{A}]^{\mathrm{n}} \frac{-d[\mathrm{A}]}{dt} = \mathrm{k}[\mathrm{A}]^{\mathrm{n}} \qquad \text{Equation 7–13}$$

Where n is the rate curve fitting parameter.

Numerous reasons are needed to explain why bioremediation does not follow the theoretical decay expected if degradation depended only on total concentration of the contaminant chemical species. For one thing, it is usually not total soil concentrations that are (readily) available to the microbes, but really the water-soluble fraction that comes in contact with the microbes. For nonaqueous phase liquids (NAPLs), which comprise many of the organic contaminants, much of the contaminant is thus not in a soluble form. Many contaminants have strong affinity for soil particles due to their sorption coefficients, they resist diffusion, and they may be physically encapsulated within the soil matrices. The engineer may need to consider conditioning the soil and pore water (e.g., with surfactants) to overcome some of these factors that limit contact between microbes and chemical contaminants, but taking into account any changes in the overall mobility of the contaminants that could result from the conditioning measures.

Bioremediation Example 5

How does the life cycle of bacteria help to explain the changes in rate orders of *in situ* bioremediation?

Answer

Figure 7.15 provides an idealized biodegradation growth curve, which has well-defined stages. The extent and duration of each stage will of course vary according to the microbial species and environmental conditions.

Thus, engineered systems bring the chemical contaminant into contact with the microbes and enhance the degradation environment, such as by pumping nutrients and air (or pure oxygen) into the groundwater. Engineered systems may also add activated microbes (if indigenous microbes are not already breaking down the chemical) into the vadose or saturated zones. As these conditions change, the microbes undergo a series of stages.[17]

Lag phase: Upon initial exposure of the microbes to the chemical contaminant, a period of time is needed for the organisms to become acclimated.

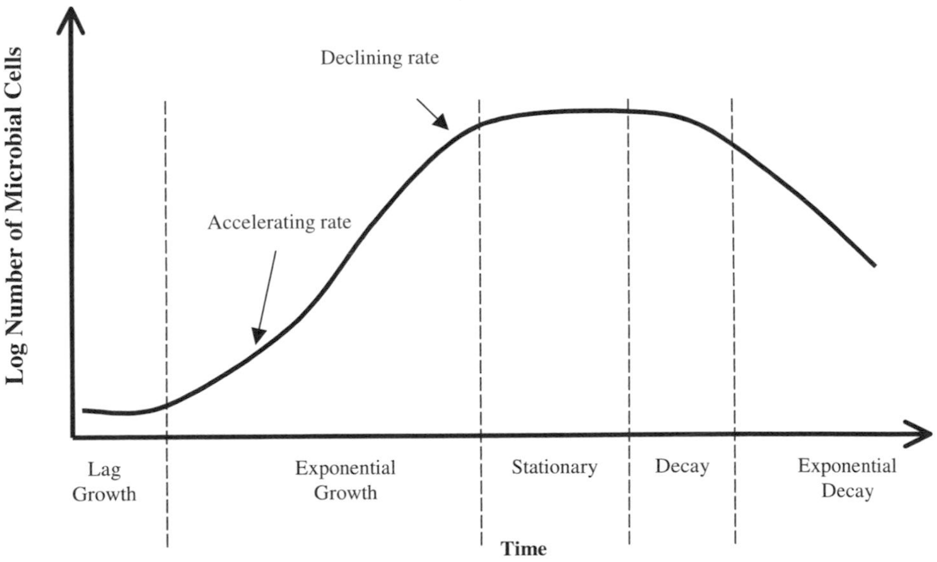

FIGURE 7.15. Prototypical growth and decay curve for bacteria.

Accelerated growth rate phase: Following acclimation, the microbes propagate at an increasing rate. There are two major processes that allow for the degradation of chemical compounds by microbes:

- The most effective is when the organisms use the contaminant as a food source for their growth, metabolism, and reproduction. This is accomplished by the microbe's ability to produce enzymes that catalyze the chemical contaminant as a carbon source, which is why bioremediation is often a good choice for organic contaminants. In addition, the chemical contaminant is also a source of electrons that the microbe needs to extract for energy; that is, the chemical is the microbe's oxygen acceptor during respiration. The microbe produces enzymes that hasten the process of breaking chemical bonds and transferring electrons from the contaminant to an electron acceptor, such as oxygen for aerobic respiration, and metals (e.g., iron and manganese) and inorganic chemicals (e.g., nitrates and sulfates) for anaerobic respiration.
- A second, less effective process is known as *secondary utilization*. Microbes can transform chemical contaminants, although the reaction provides no direct benefit to the microbial cell. Probably the most common, or at least best-understood, secondary utilization process is *cometabolism*, wherein the microbes break down chemicals coincidentally with enzymes that they normally synthesize for metabolism or detoxification. A case is the methane-oxidizing bacteria that happen to degrade chlorinated hydrocarbons, benzene, phenol, and toluene by producing enzymes needed to transfer electrons to methane (i.e., the bacterium's normal electron acceptor). The normal methane oxidation enzymes auspiciously degrade the chemical contaminants, even though the chemicals cannot serve as the primary food source for the bacteria.

Exponential growth phase: The cell mass and the number of cells are growing exponentially by binary fission.

Declining growth phase: Cell mass and numbers of cells continue to grow, but at a decreasing rate. This is usually due to depletion of the food and electron source (i.e., the contaminant, we hope). Other limiting factors could also come into play: some of the newly-generated chemicals ("degradates") could inhibit growth of the microbes due to their toxicity.

Stationary phase: Cell decay is about equal to cell propagation during this time.

Decay: Cell decay now exceeds cell propagation.

Exponential death phase: Cells are dying exponentially as cells no longer grow or propagate. We hope this is because the food source is entirely used up.

The cycle will repeat again if the microbes are again introduced to another slug of pollutants, for example, as a new batch of contaminated soil is introduced in the reactor (*ex situ*) or when the new mixes of nutrients and microbes are sent to another part of the aquifer (*in situ*).

Bioremediation Example 6

400 kg of a chemical contaminant in a water solution must be treated using a reactor system. How long will it take to reach 25 kg of the chemical, if its rate constant is $10^{-4}\,sec^{-1}$ and the reaction is first-order? How much longer will it take to get below 1 kg of the contaminant? What might this tell you about optimal loading to a reactor to achieve treatment efficiencies?

Answer and Discussion

First, determine how many half-lives it takes to get from the original mass of 400 kg to the target mass of 25 kg. Remember that a half-life is the time it takes for a concentration of a reactant to reach one-half of its initial value, so in this case it takes four half-lives to reach 25 kg.

Next, use the equation for a reaction with a first-order half-life ($t_{1/2}$):

$$t_{1/2} = \ln(2)(k)^{-1}$$

Where k = the rate constant.

So, $t_{1/2} = (0.693)(10^{-4})^{-1} = 6930\,sec$

and

$4\ t_{1/2} = 27{,}720\,sec$ or <u>7.7 hours</u> to destroy 375 kg of the initial 400 kg of the contaminant.

To reach 1 kg from the 25, will take an additional five $t_{1/2}$, so if the first-order reaction continues,[18] it will take 34,650 sec or 9.6 hours for the remaining mass of the contaminant in the solution to fall below 1 kg.

The total time needed to go from 400 kg to <1 kg of the waste is more than 17 hours. However, an engineer would likely want to achieve the higher removal rates found at higher mass (and concentrations). If this were not the only waste source for the chemical contaminant, it would be better to keep adding wastes to the reactor. In other words, if you treat at the first half-life (400 to 200 kg) continuously, it only takes 2 hours to destroy 200 kg of the contaminant. Conversely, treating at the ninth half-life (about 1.6 to 0.8 kg), it takes 2 hours to destroy 0.8 kg of the contaminant![19]

Environmental Metal Chemistry

While metals follow the general principles of chemical reactions, it is worth highlighting them, since they account for serious contamination and adverse environmental and health effects. Metals are Lewis acids since they react as electron-pair acceptors with Lewis bases (electron-pair donors), forming ion pairs, metal complexes, donor-acceptor complexes, and coordination compounds. A metal (Me) reacts with a ligand (L) to form stable compounds that are categorized as metal-ligands (Me-L), which at equilibrium are:

$$\mathrm{Me} + \mathrm{L} \Leftrightarrow \mathrm{MeL} \qquad \text{Reaction 7–59}$$

With the equilibrium constant:

$$K_{\mathrm{MeL}} = \frac{[\mathrm{MeL}]}{[\mathrm{Me}][\mathrm{L}]} \qquad \text{Equation 7–14}$$

Metal cations tend to prefer polarizable, large, and low electronegative anions, such as the sulfides, or smaller, more highly electronegative anions, such as the oxides. Environmental sinks for metals and metal compounds consist of detritus (silt and fine-sand components of sediment), oxides, humic acids, carbonates, and other organic matter, such as soil organic matter. When alkaline cations are present in solutions, metal concentrations may fall because of the competition of the alkali substances for sorption sites in soil or sediment. Redox changes, such as a decrease in molecular O_2 in water, can speciate metals and change their solubility. Under reduced conditions, metals will become sulfide complexes (e.g., sulfides of Hg, Pb, and Cd), organic complexes (e.g., Ni and Fe-chelates and ligands), chloride complexes (e.g., chlorides of Mn), and hydroxide complexes (e.g., hydroxides of oxides of Cr, which may also be the result of the higher pH induced by these conditions). Changes in pH can lead to the solution and dissolution of metals. For example, lower pH values will both increase the solution of metal carbonates and hydroxides, and make for greater amounts of desorption due to the metal cations competition with OH^- ions. Biodegradation is also an important factor in metal reactions and mobility. Metals may be transferred to and from biota, soil, and sediment into the aqueous phase as they change forms during various organisms' uptake and decomposition. Several of these processes may take place simultaneously, such as when mercury is alkylated in reduced conditions in sediment and water to form methylmercury. The methylmercury is taken up by aquatic organisms more rapidly than the inorganic, more oxidized forms.

An interesting dilemma for environmental engineers is deciding when it is most prudent to take action, such as the physical removal of metal-contaminated sediments, versus when it may be wiser to leave them in place and take measures to ensure long-term stability (e.g., capping and encapsulating contaminated soil or sediment). Some problems with dredging mercury-laden sediments have been associated with changing them from a relatively low-activity, quiescent, and reduced environment at the bottom of a lake to a highly mixed, oxidized environment in the water column, making the mercury compounds more mobile and increasing some of the chemical form's bioavailability.

The acute toxicity of metals varies among types of organisms. Most organisms are most sensitive to mercury and silver, but the acute toxicity sequence (e.g., the concentration at which one-half of the test organisms die within a specified time period, i.e. lethal dose-50 or LD_{50}, which will be discussed in Chapter 9, "Contaminant Hazards"), varies considerably:[20]

Mammals:[21] Ag, Hg, Cd > Cu, Pb, Co, Sn, Mn, Zn, Ni, Fe, Cr > Sr > Al
Fungi: Hg > Pb > Cu > Cd > Cr > Ni > Zn
Flowering Plants:[22] Ag > Hg > Cu > Cd > Cd > Cr > Ni > Co > Zn > Fe > Ca
Algae:[23] Hg > Cu > Cd > Fe > Cr > Zn > Ni > Co > Mn
Protozoa:[24] Hg, Pb > Ag > Cu, Cd > Ni, Co > Mn > Zn

Long-term toxicity has also been observed. For example, tin is not listed in the sequence above, but it has been associated with endocrine effects in many organisms, especially gastropods and other aquatic species. Also, metals include some of the most significant metal neurotoxins (e.g., Hg, Pb, and Mn), hepatotoxins, and nephrotoxins (e.g., Cd), as well as carcinogens (e.g., Cr^{6+}).

Sometimes, metals are part of organic molecules, known as *organometallic compounds*. Many are important to environmental chemistry, such as tetraethyl-lead and tributyl-tin. The lead organic compound was commonly added to gasoline to raise octane ratings in the twentieth century, but it has been banned in most areas of the world after studies showed that lead was being distributed widely and causing neurological and other health effects. The tin organic compound has recently been associated with hormonal mimicry. In many ways, organometallics behave in ways similar to those of other organic compounds.

Engineer's Technical Note: Oxidation-Reduction Reactions Applied to Environmental Probes

The principles of oxidation-reduction reactions are often used to design devices used to measure environmental factors. For example, as we will see in Chapter 8 when we discuss oxygen-depleting contaminants, probes are needed to measure the amount of molecular oxygen dissolved in water. The simplest type of probe makes use of a galvanic process, i.e. one that results when two dissimilar metals are placed in an electrolyte. The galvanic cell consists of a container filled with an electrolyte solution and which contains an anode (positive electrode) and cathode (negative electrode). As shown in Figure 7–16, the positive ions (cations) are attracted to the cathode and the negative ions (anions) are attracted to the anode.

So, to construct a probe to measure molecular oxygen, let us use the metals lead (Pb) and silver (Ag) as our electrodes.
At the Pb electrode (the cathode), the half reactions in this process are:

$$Pb + 2OH^- \rightarrow PbO + H_2O + 2e^- \qquad \text{Reaction 7–61}$$

This means that electrons are freed and travel to the Ag electrode (the anode), where the following reaction occurs:

$$2e^- + \tfrac{1}{2}O_2 + H_2O \rightarrow 2OH^- \qquad \text{Reaction 7–62}$$

We can combine the half reactions and balance the electrons:

$$Pb + 2OH^- \rightarrow PbO + H_2O + 2e^- \qquad \text{Reaction 7–63}$$

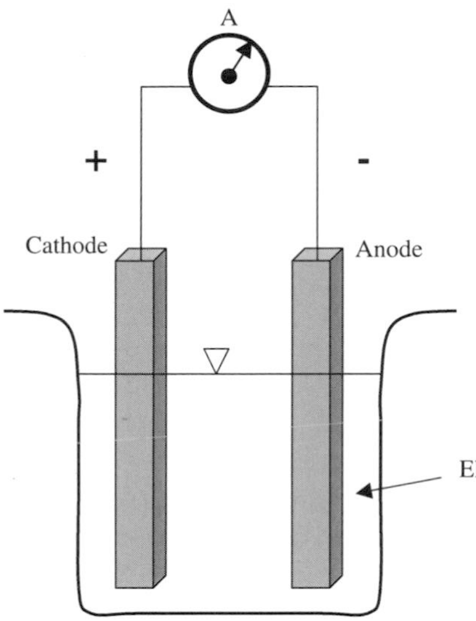

FIGURE 7.16. Galvanic cell consisting of two dissimilar metal electrodes in an electrolyte, i.e. an aqueous solution that contains ions. An ammeter between the electrodes measures the electrical flow.

$$2e^- + \tfrac{1}{2}O_2 + H_2O \rightarrow 2OH^- \qquad \text{Reaction 7–64}$$

This now gives the overall galvanic reaction:

$$Pb + \tfrac{1}{2}O_2 \rightarrow PbO \qquad \text{Reaction 7–65}$$

Thus, the oxidation-reduction reaction provides a known amount of molecular oxygen. For every two moles of lead oxide produced, one mole of oxygen must be present in the reaction. And, if no free oxygen is present in the water, no PbO is produced. Also the half reactions tell us that all we have to do to measure the amount of dissolved oxygen (DO) in a water sample is to measure the number of electrons moving between the electrodes and calibrate these electrons to the amount of DO in the sample. We do this my placing a microammeter, an instrument than can detect a very small current, in the electrolyte solution. The measured electric current can be calibrated to the DO concentration of the water sample.

The mass produced DO probes miniaturize the galvanic cell system (Figure 7–17). The electrodes are insulated from each other with plastic coatings and are covered with a permeable membrane. A few mL of electrolyte solution lie between the membrane and the electrodes and the electron flow (current) in the solution is measured with a microammeter. The amount of oxygen moving through the membrane is proportional to the concentration of DO in the water sample. According to the half reactions and the overall galvanic reaction, the higher the concentration of DO, the greater the pull of oxygen molecules and electrons through the membrane, so the amount of electricity measured between the two electrodes is directly proportional to the concentration of molecular oxygen in solution.

Similar galvanic probes give measurements of composite oxidation–reduction reactions taking place in a water sample, such as measures of conductivity. Thus, the principles of oxidation-reduction can be applied to measure water quality.

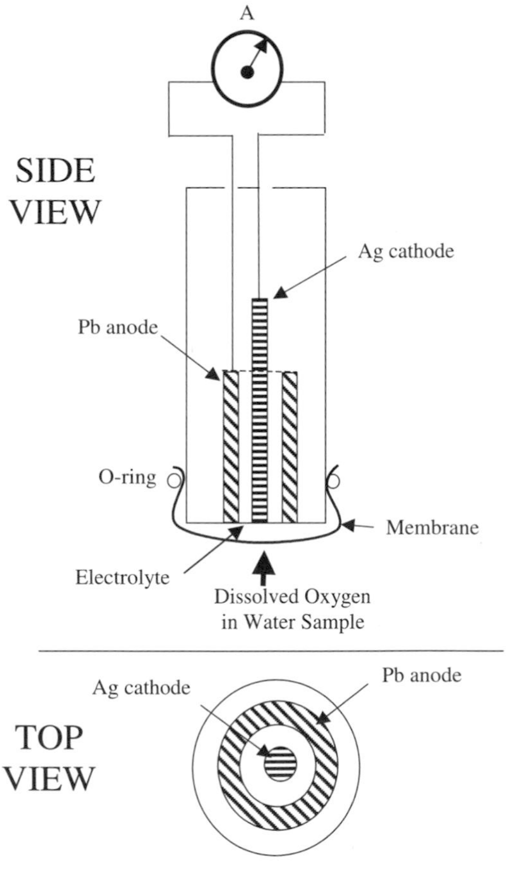

FIGURE 7.17. Prototype of a dissolved oxgen probed based on the galvanic process.

Notes and Commentary

1. L. Stieglitz, G. Zwick, J. Beck, H. Bautz, and W. Roth, 1989, "On the De-Novo Synthesis of PCDD/PCDF on the Flyash of Municipal Incinerators," *Chemosphere*, Vol. 18, pp. 1219–1226.
2. For discussion of the transport of dioxins, see C.J. Koester and R.A. Hites, 1992, "Wet and Dry Deposition of Chlorinated Dioxins and Furans," *Environmental Science and Technology*, Vol. 26, pp. 1375–1382; and R.A. Hites, 1991, "Atmospheric Transport and Deposition of Polychlorinated Dibenzo-P-Dioxins and Dibenzofurans," EPA 600/3-91/002, Research Triangle Park, N.C.
3. U.S. Environmental Protection Agency (EPA), 1999, *National Primary Drinking Water Regulations: Technical Fact Sheets*, Washington, D.C., http://www.epa.gov/OGWDW/hfacts.html.
4. The nineteenth-century Swedish chemist, Svante Arrhenius, is credited with establishing the relationship between electrical and chemical properties of molecules. In the 1884 seminal work, *Investigation on the Galvanic Conductivity of Electrolytes*, he observed that particular chemical compounds (later to be known as *electrolytes*) conduct electricity when they are dissolved in water, while other chemicals do not. He also saw that certain chemicals are involved in seemingly instantaneous reactions, while others took much longer to react. Finally, he observed that particular chemical compounds showed extremely strange colligative properties, while others were consistent with Raoult's Law, which states that the solvent's vapor pressure in an ideal solution is equal to the product of the mole fraction of the solvent and the vapor pressure of the pure solvent. The four colligative properties of solutions are the elevation of boiling point, the depression of freezing point, the decreasing of vapor pressure, and osmotic pressure. Arguably, Arrhenius's most important concept to environmental chemistry is the "activity constant," the relationship between the actual number of ions in a solution to the number of ions when all molecules have become dissociated. Perhaps, Arrhenius's contribution to electrochemistry should be characterized as "perfecting" the understanding, since others saw the relationship. In fact, a fellow Swede, Jöns Jacob Berzelius in 1817 wrote that the electric eel elicited electrical current "by an organic chemical process."
5. Note that by this definition, water itself is a weak acid in that it ionizes (autoionizes into 10^{-14} molar concentration of ions) into hydroxide and hydronium ions, which is the hydrogen ion bound to a water molecule. The importance of water's ionization in virtually all biological processes should not be underestimated. At 25°C, there are 55.35 mol water per liter. So, since half of the ions are hydronium ions, this means:

$$\frac{1.0 \times 10^{-7}\,M\ \mathrm{H_3O^+}}{55.35M\ \mathrm{H_2O}} = 1.8 \times 10^{-9} \text{ hydronium ions per water molecule!}$$

Even this small ratio provides enough H^+ given the amount of water available in the hydrological cycle and the highly reactive nature of each hydrogen ion.

6. For an excellent discussion of carbon dioxide equilibrium in water, see H.F. Hemond and E.J. Fechner-Levy, 2000, *Chemical Fate and Transport in the Environment*, Academic Press, San Diego, Calif.
7. U.S. Environmental Protection Agency, 1992, "Dermal Exposure Assessment: Principles and Applications," Interim Report, EPA 600/8-91/011B, Washington, D.C.
8. Fossil fuels, particularly coal, contain varying amounts of sulfur (S). This results from the fact that "fossil" in fossil fuels is predominantly *paleo*-plant life. Most plants contain S as a nutrient, so some remains when the plants are fossilized as coal or crude oil. Actually, the fossilizaiton process can concentrate the S content as other compounds are volatilized, oxidized, reduced, and otherwise react during the protracted time between deposition of the plant material and sedimentation processes that take place to form the fuel. So-called "high-sulfur" coals, for example, are found in specific geographic areas underlain by certain rock strata. The Energy Information Agency of the U.S. Department of Energy [1996, *U.S. Coal Reserves: A Review and Update*, DOE/EIA-0529(95), Washington, D.C.] classifies coal-bearing areas of the United States according to coal types. These types are actually representative of the coals' S-content (see Figures 7–18 and 7–19). This is the reason that areas with plentiful amounts of coal in the nation's interior region (e.g., southern Illinois) must import coal from the western regions of the contiguous United States (e.g., Wyoming). This also explains the rancor and ongoing debates among politicians, energy producers, and coal mining interests regarding who is or should be responsible for acid rain. It is a very costly enterprise to ship

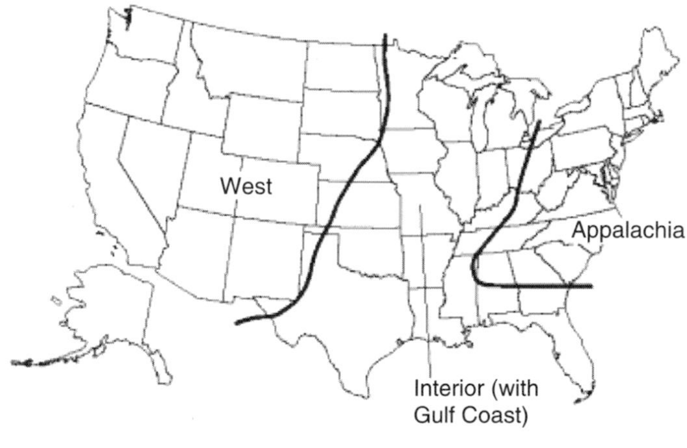

FIGURE 7.18. Demonstrated reserve base (DRB) of coal and estimated recoverable coal reserves in the United States by coal producing region as of January 1, 1997, referenced in Figure 7.19. The coal is classified by rank in quality (anthracite is highest quality in terms of caloric value and sulfur content, lignite is lowest).

coal a thousand miles in unit trains when, for example, many of the Illinois power plants sit on top of two large, albeit high-sulfur, coal fields.

9. The calcium carbonate ($CaCO_3$) acts as a buffer, which is a substance that helps to resist a pH change. For example, a well-buffered soil, like that in Kansas, has the capacity to neutralize acid deposition. Conversely, areas like the Finger Lakes region of New York, where the geology and soils do not provide buffers the carbonates, are much more sensitive to acidification.
10. R. Knox, D. Sabatini, and L. Canter, 1993, *Subsurface Transport and Fate Processes*, Lewis Publishers, Boca Raton, Fla.
11. The principal source for this case study is U.S. Geological Survey, 1997, "Research on Acidic Metal Contaminants in Pinal Creek Basin near Globe, Arizona," USGS Fact Sheet FS-005-97, Reston, Va.
12. Tailings are piles of mining and mineral processing waste, usually crushed rock and minerals.
13. Coprecipitation is the physicochemical proess whereby elements are incorporated into other compounds as the elements precipitate from solution. For example, a heavy metal may be incorporated into metal oxide minerals at the

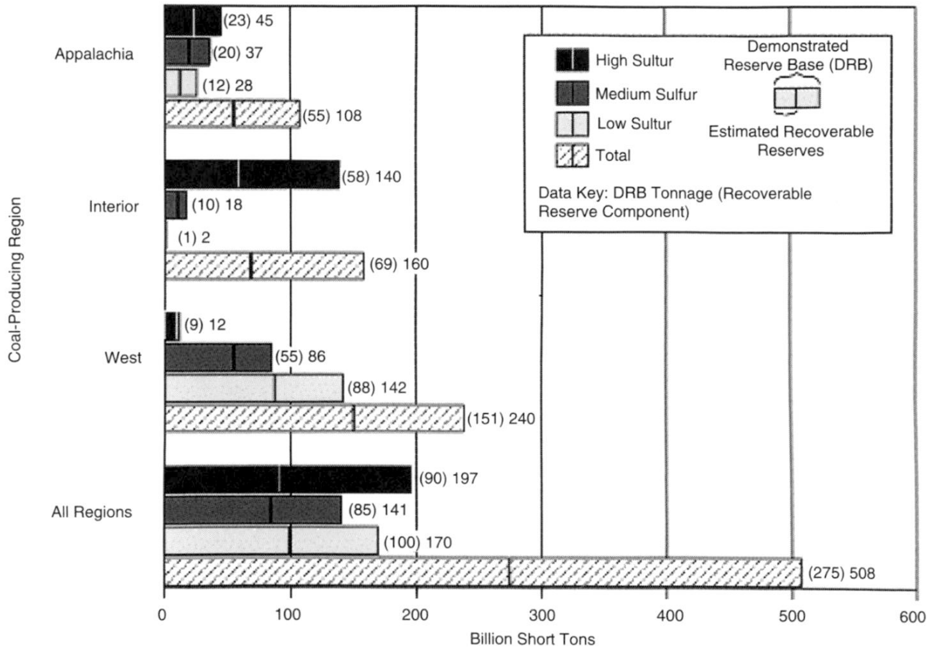

FIGURE 7.19. Coal-bearing regions of the United States [Source: U.S. Department of Energy, Energy Information Agency, 1999, *U.S. Coal Reserves: A Review and Update*, DOE/EIA-0529(95), Washington, D.C.; See: *http://www.eia.doe.gov/cneaf/coal/reserves/chapter1.html* for more information.]

same time that the metal or metalloid is precipitation from contaminated water.

14. Redox reactions are controlled in closed reactors with rapid mix agitators. Oxidation-reduction probes are used to monitoring reaction rates and product formation. The reactions are exothermic and can be very violent when the heat of reaction is released, so care must be taken to use only dilute concentrations, along with careful monitoring of batch processes.
15. See S.S. Suthersan, 1997, *Remediation Engineering: Design Concepts*, CRC Press, Boca Raton, Fla., pp. 143–144.
16. Ibid.
17. See D. Grasso, 1993, *Hazardous Waste Site Remediation: Source Control*, CRC Press, Boca Raton, Fla., pp. 13–16.
18. Remember that reaction rates can be affected by concentrations. For example, when much reactant is available the reaction is not rate limited by concentration, but as the mass drops in the solution, the microbes may transition to non-exponential growth. You were not told whether this is a biological or abiotic treatment approach.
19. We owe a debt to the economists, including Malthus, for the concept of the Law of Diminishing Returns, which we see at work here.
20. See E. Nieboer and D. Richardson, 1980, "The Replacement of the Nondescript Term 'Heavy Metals' by a Biologically and Chemically Significant Classification of Metal Cations," *Environmental Pollution Serial B*, Vol. 1, p. 3.
21. The study animals were rat, mouse, and rabbit.
22. The study plant was barley.
23. The study algae was *Chlorella vulgaris*.
24. The study organism was *Paramecium*.

CHAPTER 8

Biological Principles of Environmental Contamination

The Cell

The fundamental building block in physics is the particle. For chemistry, it is the atom. For chemical reactions, the basic unit is the element. The fundamental unit in living systems is the cell.[1] Whether it is a self-contained organism like that of bacteria or algae, or a part of a complex organism like a human being, the cell is where the biochemical processes—for good or bad—take place. When operating effectively, cells are the factories that turn nutrients and energy into biomass through the processes of photosynthesis, metabolism, and ion exchanges in microbes and plants. In animals, the cell is the location of metabolism and respiration. These mechanisms unfortunately are often disrupted by environmental contaminants. They are also the processes that convert complex chemical contaminants into simpler, less toxic forms through the process of *biotransformation*.

Most cells, whether in unicellular organisms or in complex, higher animals, consist of common structures (see Figure 8.1), but cells can have large ranges in size and function (Figure 8.2). Cells were first identified in the seventeenth century when the English physicist Robert Hooke observed cork under a microscope and detected cellular structures in the wood. Since modern concepts of the organism were still nascent ideas, Hooke hypothesized that the cells were containers of the "noble juices" and "fibrous threads" of a living organism. Hooke also limited his hypotheses to plants and did not include animals. It was not until 1838 when botanist Matthias Schleiden and zoologist Theodor Schwann characterized the similarities between cells in plants and animals. In 1847, Schwann's research was synthesized into a paper where he stated that blood, skin, bone, muscle, and other tissue are composed of cells. The German pathologist, Rudolf Virchow is credited, in 1858, with espousing modern cell theory; in *Cellurpathologic* (1839), Virchow stated that an "animal appears as a sum of vital units, each of which bears in itself the complete characteristics of life." These units are what are now known as cells.

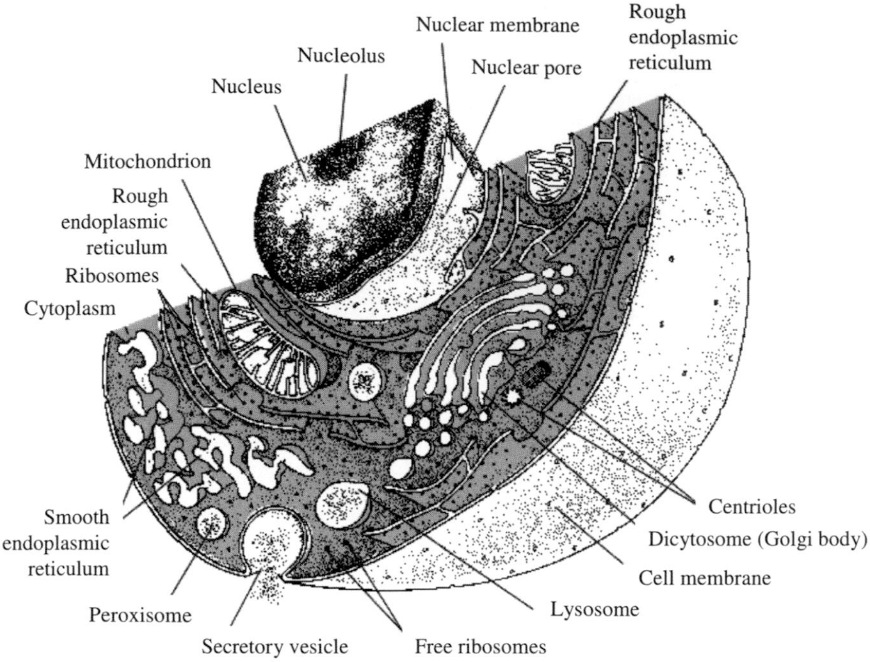

FIGURE 8.1. Structures of the cell. (Source: National Institutes of Medicine, National Institute of General Medical Science; based on a figure in H. Curtis, 1983, *Biology*, 4th Edition, Worth Publishers, New York, N.Y.)

Biochemistry can be traced to Antoine Lavoisier, the eighteenth-century French scientist who explained the role of oxygen in the metabolism of food to provide energy in both plants and animals, established the composition of water and other compounds, and introduced methods of measuring aspects of chemical reactions, thereby laying the foundation for modern chemistry. By the nineteenth century, important compounds were being identified, including organic pigments, such as hemoglobin in blood and chlorophyll in plants (see discussion later in Chapter 11, "Contaminant Sampling and Analysis" on chlorophyll as an environmental indicator). Two fundamental cell types exist: prokaryotic and eukaryotic. The more primitive cell type, prokaryotic cells, has no membrane around its nuclear region, and include bacteria, mycoplasma, and simple blue-green algae, or cyanobacteria. By contrast, eukaryotic cells have double membranes separating the nucleus from the cytoplasm, and numerous internal membranes to set apart their organelles. All animal and plant cells are eukaryotic. Prokaryotic and eukaryotic cells play important roles in the fate of environmental contaminants.

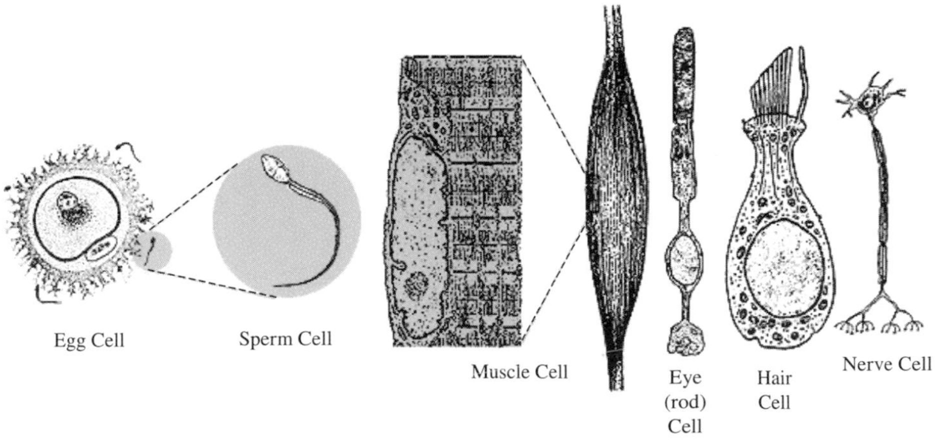

FIGURE 8.2. Differentiation and specialization in eukarotic cells. Shown here, for example, are human cells. Following its fertilization by a sperm, the human egg cell divides many times over, generating numerous types of specialized cells structured in ways to perform needed functions in the organism. An extreme example is a single nerve cell that must be a meter in length to connect the toe to the spine. (Source: National Institutes of Health, National Institute of General Medical Science.)

Prokaryotic organisms commonly produce only exact duplicates of themselves, but higher eukaryotic organisms' cells can be differentiated into diverse cell types. Prokarytic cells, then, have the advantage of simple needs for nutrients. This allows for their being able to break down contaminants via biotransformation. This is also a factor in why engineers are able to acclimate bacteria and other prokaryotes in processes to treat hazardous wastes and wastewater. In addition, prokaryotes can resist adverse environmental conditions, grow rapidly, and divide geometrically. Thus, they are ideal for environmental treatment scenarios.

Production of proteins is the principal chemical compound output of all cells. Eukarotes start producing proteins in the nucleus, the large, dense structure within the cell. In the twentieth century, the rod-like bundles of deoxyribonucleic acid (DNA) in the nucleus, known as *chromosomes* (see Figure 8.3), were linked to heredity. The nucleus provides the cellular information through chemical messaging systems, including polypeptides (see Figure 8.4). Genes, comprised of DNA, direct the formation of cells, or what kind they are and what types will be made and differentiated in the organism. The nucleus is thus the location of all messages regarding reproduction and cell division. Molecular DNA consists of bases linked to form a double helix structure. Two bases are joined together by chemical bonds and attached to chains of chemically bonded sugar and phosphate mole-

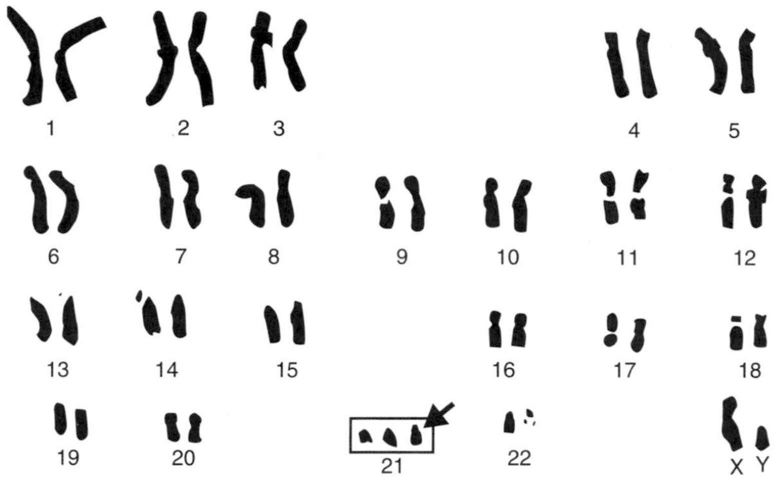

FIGURE 8.3. Differences in size and banding pattern allow the 24 chromosomes to be distinguished from each other, an analysis called a karyotype. For example, in this micrograph, in persons with Downs syndrome, cells contain a third copy of chromosome 21, which is diagnosed by karyotype analysis. (Source: Oak Ridge National Laboratory, 2004, Chromosome FAQs, http://www.ornl.gov/sci/techresources/Human_Genome/posters/chromosome/faqs.shtml.)

cules. A nucleotide is a unit of DNA that is made up of one sugar molecule, one phosphate molecule, and one base. Only four bases exist: adenine (A), thymine (T), guanine (G), and cytosine (C). The base A is always joined to T. The base G is always linked to C. Thus the sequence of bases on one side of the helix (e.g., AGCGT) complements and establishes the sequence (TCGCA) on the other side of the helix (see Figure 8.5). This sequencing allows for billions of possible messages. Unfortunately, it is also the errors in such sequencing that lead to many of the adverse outcomes resulting from exposures to environmental contaminants, such as cancer and birth defects.

The "Bio" Terms

Armed with an understanding of cellular structure and function, let us consider the numerous means by which contaminants are influenced by biological processes in the environment. Many terms in the environmental sciences include the prefix *bio*. Often, this is meant to distinguish a process initiated, mediated, and sited in living systems, especially at molecular and cellular levels. For example, the chemistry may be identical or very different in *abiotic* systems (e.g., sand and air), than in *biotic* systems (e.g., a forest, a tree, a leaf, a leaf cell, or a receptor molecule on the leaf cell). Such

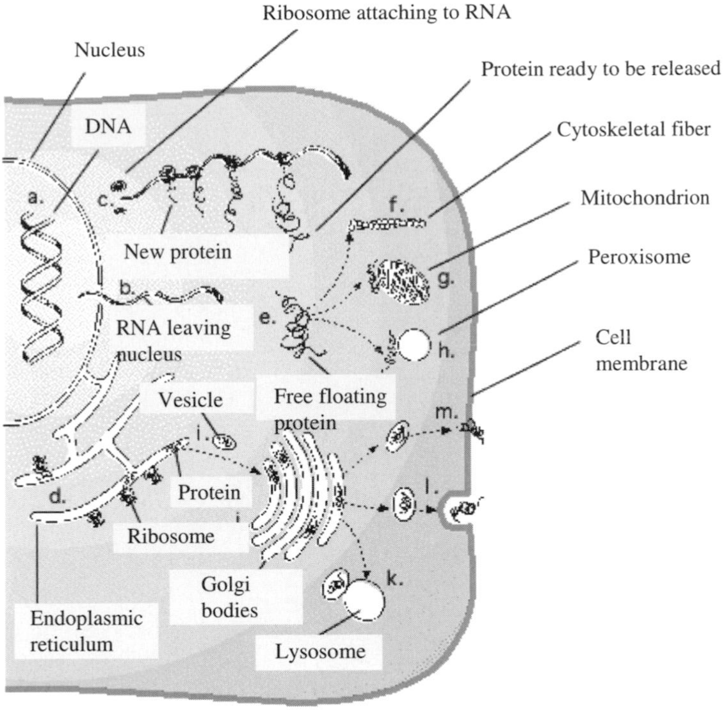

FIGURE 8.4. Protein production begins in the nucleus at the DNA (a). A coded message for a protein leaves the nucleus in the form of RNA (b) and goes to either free ribosomes (c) or to ribosomes bound to the endoplasmic reticulum (d). When released from a free ribosome, a protein (e) can become incorporated into cytoskeletal fibers (f) or into such organelles as a mitochondrion (g) or a peroxisome (h). Proteins made in the endoplasmic reticulum leave in a vesicle (i) and migrate to the Golgi apparatus (j). Proteins are sorted in the Golgi and are then carried in vesicles to lysosomes (k), or are secreted (l) or incorporated into the cell's surface membrane (m). (Source: National Institutes of Health, National Institute of General Medical Science; based on a figure in J. Darnell, H. Lodish, and D. Baltimore, 1986, *Molecular Cell Biology*, Scientific American Books, New York, N.Y.)

"bio" terms include bio-effective dose (bio-exposure), bio-uptake, bioactivation, bioaccumulation, biosequestration, bioconcentration, biotransformation, biodegradation, biomagnification, and biodepuration (elimination).

Since most of these terms are expressions or parts of the function of risk, let us first consider some of the basic elements of toxicology and risk.

Bio-Uptake and Bioaccumulation

Once the contaminant enters an organism it is taken up through several processes, especially accumulation, metabolism, and excretion. These

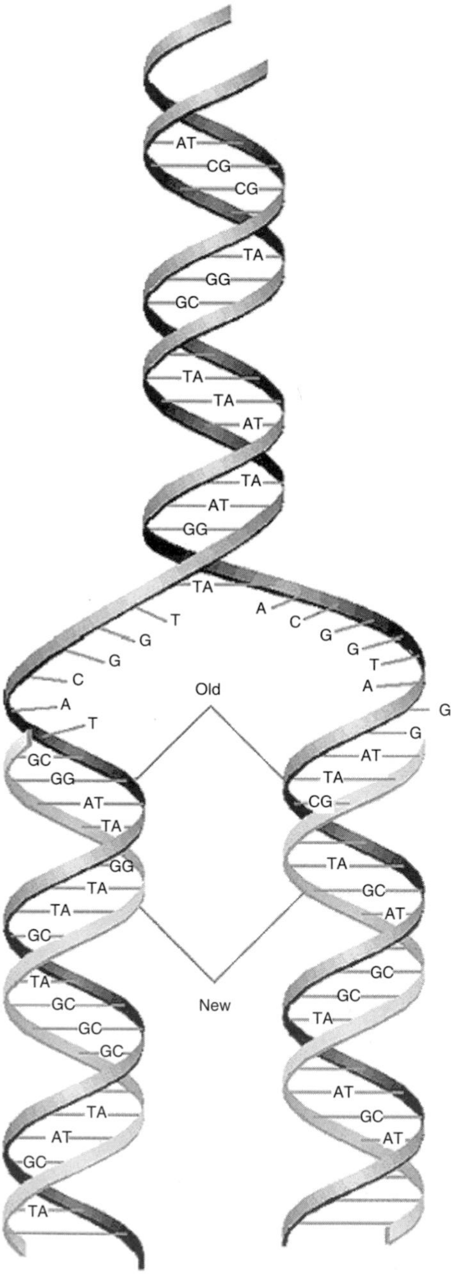

FIGURE 8.5. Structure and replication of deoxyribonucleic acid. (Source: National Institutes of Health, National Institute of General Medical Science from J.D. Watson, J. Tooze, and D.T. Kurtz, 1983, *Recombinant DNA: A Short Course*, W.H. Freeman and Company, New York, N.Y.)

processes will end by moving or changing the contaminant chemically. All organisms share the pharmacokinetic processes of absorption, distribution, and excretion. Bioaccumulation is a function of these three processes. However, the type of chemicals able to be processed, the time that each mechanism takes, and the ultimate change to the compound after uptake vary significantly among species, or even strains of the same species. Thus bioaccumulation is a "species-dependent" factor.

The mass of the contaminant that ultimately is accumulated by an organism is known as the organism's *body burden.* Bioaccumulation is another equilibrium condition. As shown in Figure 8.6, the organism goes through a stage, at times even before birth, where it begins uptake of a contaminant. The rate of uptake is greater than the rate of elimination during the toxicokinetics phase. Eventually, the accumulation reaches equilibrium with its surrounding environment, so that the body burden remains constant. Through treatment or with the elimination of the source and release of the contaminant from its fatty tissues or other storage sites (e.g., the liver), the process of biodepuration may result in a reduced body burden.

Once an environmental contaminant has been taken up by an organism, for an adverse effect—or any effect, for that matter—to occur, the contaminant must interact with its cells. The interaction sites may be on the cell's surface, such as when an endocrine disruptor mimics a hormone by

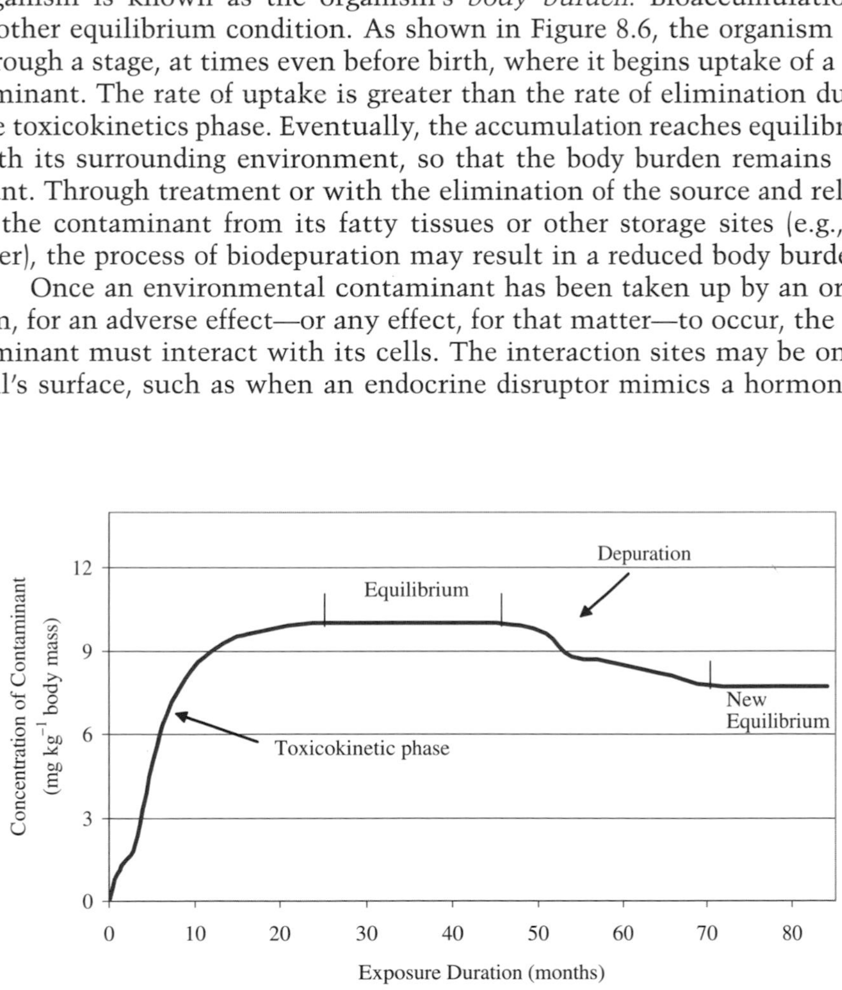

FIGURE 8.6. Bioaccumulation in an organism. During the toxicokinetic stage, uptake of the contaminant is greater than elimination. At equilibrium, the uptake and elimination processes are equal. During detoxification or depuration, elimination is greater than uptake, so the body burden of the organism is reduced until a new equilibrium is established.

linking with a hormone receptor site on the cells surface. The interaction may also occur within a cell, such as when a carcinogenic contaminant enters a cell's nucleus and interferes with normal DNA sequencing. The contaminant may also interact within an organism's extracellular spaces. Thus, for plants, a contaminant may interact with root cells, stomata, vascular tissue, and cuticle tissues. Animal interactive sites include skin and stomach tissue. The lung for land animals and the gills for fish are also sites that interact with contaminants that have been taken up.

If the dose-response and biological gradient relationships hold, and they usually do, the intensity of an adverse effect from exposure to an environmental contaminant must depend on the concentration of the contaminant. If a contaminant persists in an organism, it is more likely to elicit toxicity, particularly if the contaminant stays at the ultimate site of action. For example, if a neurotoxic contaminant is stored in fat reserves, but does not find its way to the central nervous system or any other nerve site, the organism will not exhibit neural dysfunction. However, once the neurotoxin is released and distributed to a nerve site, the neurotoxicity will be manifested. And if the contaminant finds its way to a nerve cell, the longer it remains at this site of action, the more likely the cell will be damaged. Since the endogenous target molecule in the cell is the site of action, a contaminant's chemical reactions with that molecule represents the initiation of toxicity in the organism. For example, a dioxin molecule may react with a receptor molecule on the cell's surface. This reaction may signal the feminine or masculine responses (e.g., hair growth, or testis or ova development) much as a hormone would do (see Figure 8.7). In other words, the dioxin and the natural hormone both bind to the cell's receptor. They are both *ligands*, or molecules that travel through the bloodstream as chemical messengers that will bind to a target cell's receptor. Or, the new polypeptide that is formed from this receptor-contaminant interaction may react with DNA in the nucleus. The former reaction is an example of an endocrine response, while the latter may lead to mutagenicity or cancer.

The same contaminant can elicit different responses. In our example above, dioxins have been shown to be both endocrine disruptors by binding with or interfering with cellular receptors, and they are carcinogenic and mutagenic because of the reactions that they or their metabolites have with the DNA molecule. Contaminants may also react with a wide range of molecules besides receptors and DNA, including lipids and microfilamental proteins. Contaminants may also enter into catalytic reactions, where enzymes are involved. Enzymes are important in the metabolism of cells, whether in a unicellular bacterium or a multicellular human being.

Absorption is the process whereby a contaminant moves from the site of exposure (e.g., the skin, lung tissue, or stomach) to the circulatory system. The principal mechanism for transferring a contaminant that has entered

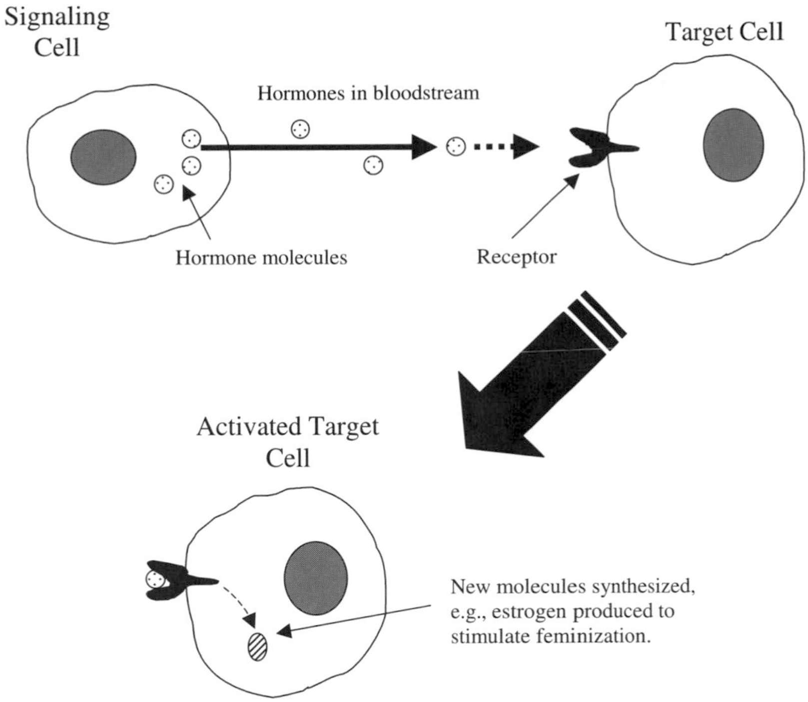

FIGURE 8.7. Schematic of the process for endocrine signals between cells. The signaling cell releases hormones into the bloodstream that reach the receptor of the target cell. When the receptor binds to the hormone, new molecules are synthesized in the activated target cell.

an organism is diffusion, or movement of the contaminant from high to low concentrations. Most contaminants travel across epithelial barriers to find their way to blood capillaries via diffusion. Thus, if a contaminant mass is high enough (i.e., sufficient rate of exposure) and the chemical can be readily dissolved into the bloodstream, then absorption will occur. Absorption also depends on the area of exposure, the type of epithelial layers, the microcirculation intensity in the subepithelial regions, and the properties of the contaminant.[2]

It is possible for some contaminants to be eliminated even before being absorbed. This process is known as "presystemic elimination" and can take place while the contaminant is being transferred from the exposure site (e.g., the outer layer of the skin or the gastrointestinal, or GI, tract. As a contaminant moves through the GI mucosal cells, lungs, or liver, much of the contaminant may be eliminated. For example, the heavy metal manganese (Mn) can be eliminated during uptake by the liver, even before it is absorbed into the bloodstream. Presystemic elimination, however, does not necessarily mean that an organism experiences no adverse effect. In fact, Mn exposure can damage the liver without ever being absorbed into the bloodstream. This is also one of the complications of biomarkers (which will be discussed later), since the body is protected against Mn toxicity by low rates of absorption or by the liver's presystemic Mn elimination.[3]

Distribution[4] is the step where contaminants move from the point of entry or absorption to other locations in an organism. The principal mechanism for distribution is circulation of fluids. The absorbed contaminant first moves through cell linings of the absorbing organ, for example, the skin or GI tract. After this, the contaminant enters that organ's interstitial fluid, or the fluid that surrounds cells. About 15% of the human body mass is interstitial fluids. The contaminant may continue to be distributed into intracellular fluids, which account for about 40% of body mass. Contaminant movement to more remote locations occurs in blood plasma (about 8% of body mass). Interstitial and intracellular fluids are stationary, or remain in place, so while the contaminant resides in these fluids they are not mechanically transported. Only after entering the bloodstream does distribution become rapid. A contaminant can leave the interstitial fluids by entering cells of local tissue, by flowing into blood capillaries and the blood circulatory system, and by moving into the lymphatic system.

A contaminant's distribution is largely influenced by its affinity for binding to proteins, such as albumin, in the blood plasma. When a contaminant binds to these proteins it is no longer available for potential cell interactions. In the bloodstream, only the bound fraction of the contaminant is in equilibrium with the free contaminant. Only the free (unbound) fraction may pass through the capillary membranes. The portion of the contaminant that is bound to proteins, therefore, determines the contaminant's biological half-life and toxicity. Passive diffusion of the toxicant to and from fluids is the result of the contaminant's concentration gradient. The diffusive processes follow the same Fickian principles as those discussed in previous chapters. The apparent volume of distribution (V_D) is the total volume of fluids (units = liters) in the body to which the contaminant has been distributed:

$$V_D = \frac{m}{C_{plasma}} \qquad \text{Equation 8–1}$$

Where, m is the mass or pharmacological dose (mg) of the contaminant, and C_{plasma} is the concentration of the contaminant in the plasma ($mg\,L^{-1}$).

Contaminants distributed exclusively in the blood will have higher values of V_D, while those distributed to several fluid types (blood and the interstitial and intracellular fluids) will be more diluted and would have lower V_D values. These values can be influenced by a contaminant's rates of sequestration, biotransformation, and elimination. The value is a good indication of just how widely a contaminant is distributed within an organism. It is also a key factor in calculating the contaminant *body burden* (mg):

$$\text{Body burden} = C_{\text{plasma}} \cdot V_D \qquad \text{Equation 8–2}$$

Contaminant Distribution Example

Bob is exposed to 30 mg Contaminant A and has a blood plasma concentration of $3\,\text{mg}\,\text{L}^{-1}$. Cindy is exposed to 9 mg of Contaminant B but has a plasma concentration of $3\,\text{mg}\,\text{L}^{-1}$. What are the volumes of distribution and body burdens for each person?

Solution

The volume of distribution is the quotient of the dose and the concentration in the plasma, thus Bob's $V_D = 30/3 = 10\,\text{L}$ of contaminant A and Cindy's $= 9/3 = 3\,\text{L}$ of Contaminant B.

Since body burden is the product of the plasma concentration and the volume of distribution, Bob's body burden $= 3 \times 10 = 30\,\text{mg}$ of A, and Cindy's body burden $= 3 \times 3 = 9\,\text{mg}$ of B.

Thus, Contaminant B is distributed less than A (only 30%). Also, this has caused Bob to have a greater body burden of A than Cindy does of B. It is important to keep in mind, however, that numerous factors can affect distribution and body burden. For example, the sex and age of a person can influence how rapidly a contaminant is distributed. In fact, if men on average distribute these contaminants 3.3 times more rapidly than women, then A and B could be the same contaminant (all other factors, such as age, being equal).

The route of exposure is an important factor that can affect the concentration of the parent contaminant or its metabolites within the blood or lymph regions. This can be important since the degree of biotransformation, storage, elimination, and, ultimately, toxicity can be influenced by the time and path taken by the contaminant within the body. For example, if the contaminant goes directly to the liver before it travels to other parts of the body, most of the contaminant mass can be biotransformed rapidly. This means that "downstream" blood concentrations will be muted or entirely eliminated, which obviates any toxic effects. This occurs when contaminants become absorbed through the gastrointestinal (GI) tract. The absorbed contaminant mass that enters the vascular system of the GI tract is carried by the blood directly to the liver via the portal system. Blood from the liver subsequently travels to the heart and then on to the lung, before being distributed to other organs. Thus, contaminants that enter from the GI tract are immediately available to be biotransformed or excreted by the liver and eliminated by the lungs. This is known as the "first pass effect." For example, if the first-pass biotransformation of a contaminant is 75% via the oral exposure route, the contaminant-blood concentration is only about 25% of that of a comparable dose administered intravenously.

The routes of exposure follow the same principles discussed previously. For example, respiratory exposures to contaminant gases are a function of gas diffusion. Recall that Fick's Law, as given in Equation 5–43, expresses gas flux as:

$$J_{Diffusion} = -D\frac{dC}{dx} \qquad \text{Equation 8–3}$$

This may be reordered, and values added[5] for the contaminant and the lung:

$$J_{Diffusion} = -D \times \frac{S}{MW^{1/2}} \times \frac{A}{d} \times (p_a \times p_b) \qquad \text{Equation 8–4}$$

Where, $J_{Diffusion}$ = diffusion rate (mass per length^2 per time)
D = diffusion coefficient for the contaminant (area per time)
S = solubility of the contaminant gas in the blood (mass per volume)
MW = molecular weight of the contaminant (dimensionless)
A = surface area of membrane in contact with the contaminant (length^2)
d = membrane thickness (length)
p_a = partial pressure of contaminant gas in inhaled air (pressure units)
p_b = partial pressure of contaminant gas in blood (pressure units)

The Fickian relationship shows that so long as p_a is larger than p_b, the diffusion rate is positive and the contaminant is taken up (i.e., is more likely to reach the target organ). As the partial pressure in the blood increases and becomes greater than that in the air, the gradient reverses and the contaminant moves out of the lung. Also, note that for a highly soluble compound, the rate of diffusion is rapid. Obviously, the slowest processes (smallest variable in the numerators, largest variable in the denominators) will be rate limiting. Aerosols (particles) will effectively diffuse if the contaminant is lipophilic. Particle size is a major limiting factor and is inversely proportional to dose. Currently, particles with diameters ≤2.5 m are considered to be most effective in passing by the nasopharyngeal region and penetrating to the tracheobronchial region and being deposited in alveoli. Larger particles are filtered physically and are considered to be less problematic.

Fundamental chemical principles apply to the oral route. For example, the pH varies among the fluids found in different organs, lowest in the stomach (pH near 1.0) and highest in some urines (pH about 7.8). Blood is also slightly basic, with a pH of 7.4, while the small intestines are slightly acidic (pH about 6.5). This means that the acid-base relationships described in our discussions of chemical reactions are very important to the oral exposure route. For example, lipophilic organic acids and bases will be absorbed by passive diffusion only when they are not in an ionized form, so the Henderson-Hasselbach equation (Equation 7–8) is a determinant in the amount of organic acids absorbed.

Contaminants absorbed through the inhalation or dermal routes will enter the blood and go directly to the heart and systemic circulation. Therefore, the contaminant is distributed to other organs of the body before it finds its way to the liver and is not subject to this first-pass effect. Also, a contaminant entering the lymph of the intestinal tract will not first travel to the liver. Rather, the contaminant will slowly enter the circulatory system. The proportion of a contaminant that moves via lymph is much smaller than that amount carried in the blood. The contaminant blood concentration also depends on the rate of biotransformation and excretion. Some contaminants are rapidly biotransformed and excreted, while others are slowly biotransformed and excreted.[6]

Disposition is a mechanism that integrates the processes of distribution, biotransformation, and elimination. Disposition (kinetic) models describe how a contaminant moves within the body with time. The disposition models are named for the number of compartments of the body where a contaminant may be transported. Important compartments include, blood, fat (adipose) tissue, bone, liver, kidneys, and brain.

Kinetic models may be a one-compartment open model, a two-compartment open model or a multiple-compartment model. The one-compartment open model (Figure 8.8) describes the disposition of a substance that is introduced and distributed instantaneously and evenly in the body, and eliminated at a rate and amount that is proportional to the amount left in the body. This is known as a "first-order" rate, and represented as the logarithm of concentration in blood as a linear function of time.

The half-life of the chemical that follows a one-compartment model is simply the time required for half the chemical to no longer be found in the plasma. Only a few contaminants adhere to simple, first-order conditions of the one-compartment model.

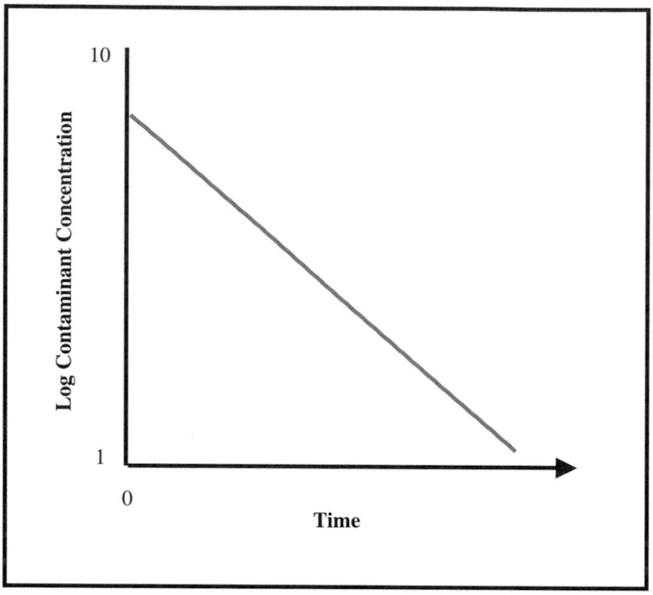

FIGURE 8.8. One-compartment toxicokinetic model. (Adapted from: National Library of Medicine, 2003, Toxicokinetics Tutor Program.)

For most chemicals, it is necessary to describe the kinetics in terms of at least a two-compartment model (Figure 8.9). This model assumes that the contaminant enters and distributes in the first compartment, usually the blood. From there, the contaminant is distributed to another compartment, from which it can be eliminated, or it may return to the first compartment.

Concentration in the first compartment declines continuously over time. Concentration in the second compartment rises, peaks, and subsequently declines as the contaminant is eliminated from the body.

A half-life for a chemical whose kinetic behavior fits a two-compartment model is often referred to as the "biological half-life." This is the most commonly used measure of the kinetic behavior of a trace contaminant.

Frequently, the kinetics of a chemical within the body cannot be adequately described by either of these models since there may be several peripheral body compartments that the chemical may go to, including long-term storage. In addition, biotransformation and elimination of a chemical may not be simple processes but subject to different rates as the blood levels change.

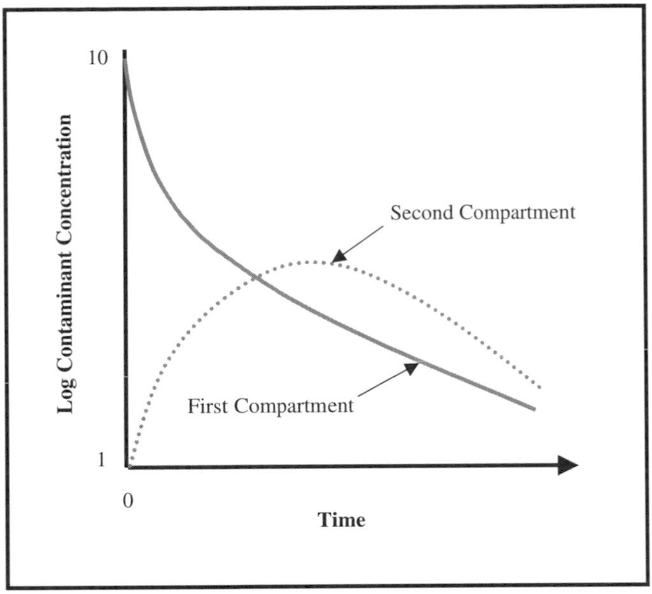

FIGURE 8.9. Two-compartment toxicokinetic model. (Adapted from: National Library of Medicine, 2003, Toxicokinetics Tutor Program.)

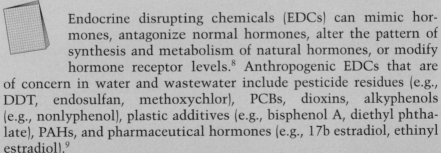

Engineering Technical Note: Removing Endocrine Disruptors from Drinking Water—An Alternative Treatment Scheme Using Ultraviolet Light[7] Contributed by Erik Rosenfeldt and Karl Linden, Duke University

Endocrine disrupting chemicals (EDCs) can mimic hormones, antagonize normal hormones, alter the pattern of synthesis and metabolism of natural hormones, or modify hormone receptor levels.[8] Anthropogenic EDCs that are of concern in water and wastewater include pesticide residues (e.g., DDT, endosulfan, methoxychlor), PCBs, dioxins, alkyphenols (e.g., nonlyphenol), plastic additives (e.g., bisphenol A, diethyl phthalate), PAHs, and pharmaceutical hormones (e.g., 17b estradiol, ethinyl estradiol).[9]

Recent research has shown that many EDCs are present in the environment at levels capable of negatively affecting wildlife. One of the first EDCs heavily researched was DDT.[10] Throughout the 1980s, exposure to this pesticide was associated with abnormal sexual differentiation in seagulls, thinning and cracking of bald eagle eggs,[11] and a sharp decrease in the numbers of male alligators in Lake Apopka,

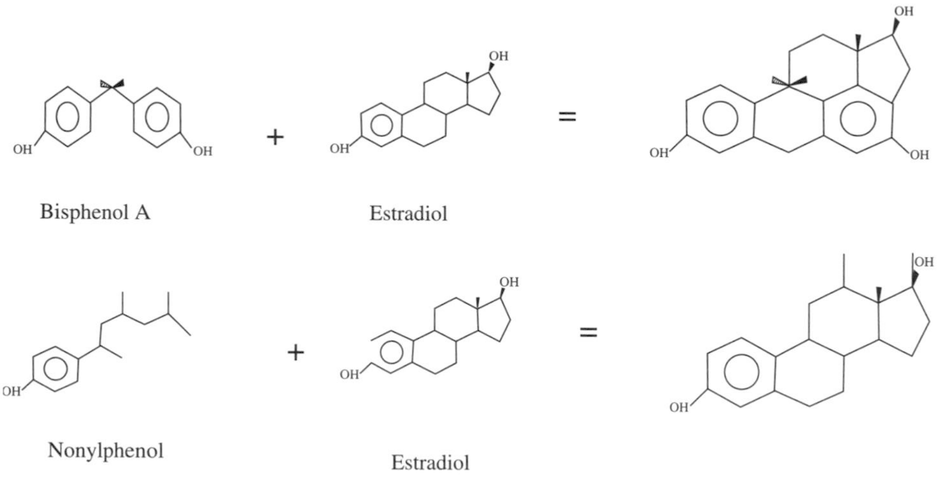

FIGURE 8.10. Comparison of the structure of bisphenol A and nonylphenol with estradiol, showing their overlap in the combined structures.

Florida, with feminization and loss of fertility found in the remaining males.[12] Since then, other pesticides and chemicals have been associated with endocrine-related abnormalities in wildlife, including the inducement of feminine traits, such as secretion of the egg-laying hormone, vitellogenin, in males of numerous aquatic species downstream for treatment.[13] Birds and terrestrial animals are also affected by EDCs.[14] Recently, these problems have found their way to humans, exposed to halogenated compounds and pesticides.[15] A recent nationwide survey of pharmaceuticals in U.S. surface water found EDCs at $ng\,L^{-1}$ levels in 139 stream sites throughout the United States. Several of these EDCs were found in concentrations high enough to be reported in units of $\mu g\,L^{-1}$ levels, including nonylphenol ($40\,\mu g\,L^{-1}$), bisphenol A ($12\,\mu g\,L^{-1}$), and ethinyl estradiol ($0.831\,\mu g\,L^{-1}$).[16] Many of these compounds are extremely persistent in the environment, so their removal before entering environmental media is paramount to reducing exposures.

The search for the specific chemical structure moiety responsible for inducing the estrogenic response is the subject of quite a lot of research in the field of endocrine disruption. Many sources have postulated that phenolic rings are a major structural reason for the estrogenicity of EDCs.[17] Figure 8.10 shows how two known EDCs compare structurally with estrogen, the hormone they are thought to mimic.

Determining the Estrogenicity of EDCs

Several bioassays are currently being developed and tested for their ability to predict the estrogenicity of various compounds. These assays work in various ways, but all have the common goal of identifying compounds that will cause responses similar to estrogen in various organisms. Some of these bioassays include the Yeast Estrogen Screen (YES), Human cell reporter gene construct (ER-CALUX), MCF-7 cell proliferation (E-Screen), Vitellogenin induction in fish, and developmental studies of fish with specific endpoints.

For example, the YES is an assay based on yeast cells modified to harbor the human estrogen receptor. When activated, this receptor binds to the estrogen response element of some plasmid DNA that is engineered to produce S-galactosidase. When estrogens are present, S-galactosidase is excreted by the cells into the culture medium where it reacts and liberates a red dye. The resulting color change is measured with a spectrophotometer, and the responses have been calibrated based on the response of actual estrogen. This method has been widely used to determine the "estrogenicity" (in terms of ability to bind with the estrogen receptor and produce a response) of many compounds and mixtures of compounds of known and unknown composition. Table 8.1 displays the relative binding affinity for several suspected EDCs as compared with estrogen (17-β-estradiol).

Environmental Fate of Endocrine-Disrupting Compounds

By examining the physical and chemical properties of EDCs, it is possible to examine where in the environment a threat from the chemicals will occur. Table 8.2 displays physical data and major uses of three EDCs of particular concern to human health: bisphenol A (BPA), 17-β-estradiol (E2), and 17-β-ethinyl estradiol (EE2). These three compounds are xeno-estrogens, or natural or synthetic compounds that act to mimic the effect of estrogens.

The low vapor pressures of BPA and EE2 mean that they are not generally found in the atmosphere unless they are sorbed to particles. Since EE2 and E2 have similar structures, the vapor pressure for E2 is also expected to be low. Due to their hydrophobic nature, these contaminants will more readily associate with organic solvents and particles in a liquid water phase. However, portions of the compounds do exist in the aqueous phase, and this proportion can be greater at higher pH values, especially for BPA. Also, due to the hormonal nature of these compounds, their effects can be felt at extremely low concentrations (on the order of $ng\,L^{-1}$). Thus, treatment technologies to

TABLE 8.1
Relative Binding Affinity Compared to Estrogen (YES Assay)

Test Compound	*Relative Estrogenic Potency*
17-β-estradiol (E2)	1.0
17-β-ethinylestradiol (EE2)	0.7
Diethylstilbestrol (DES)	1.1
Nonylphenol (NP)	7.2×10^{-7}
Bisphenol A (BPA)	6.2×10^{-5}

Source: Data from E. Silva, N. Rajapakse, and A. Kortenkamp, 2002, "Something from 'Nothing': Eight Estrogenic Chemicals Combined at Concentrations below no Observable Effect in Centrations (NOECs) Produce Significant Mixture Effects," *Environmental Science and Technology*, Vol. 36, no. 8, pp. 1751–1756; and L. Folmar, 2002, "A Comparison of the Estrogenic Potencies of Estradiol, Ethynylestradiol, Diethylstilbestrol, Nonylphenol and Methoxychlor In Vivo and In Vitro," *Aquatic Toxicology*, Vol. 60, pp. 101–110.

TABLE 8.2
Physical Data for and Major Products Containing BPA, EE2, and E2

Compound	*Melting Point (°C)*	*Vapor Pressure (mm Hg)*	*Solubility ($mg\,L^{-1}$)*	*Log K_{ow}*	*Uses*
BPA	153	4×10^{-8}	129 (25°C)	3.32	Plasticizer (adhesives, paints, CDs, baby bottles)
EE2	183	6×10^{-9}	11.3 (27°C)	3.67	Synthetic estrogen (birth control pills)
E2	178.5	NA	3.6 (27°C)	4.01	Natural estrogen

NA = not abailable.
Source: Physical data from Chemfinder.com. (http://chemfinder.cambridgesoft.com/reference/chemfinder.asp); source for EE2 vapor pressure is K.M. Lai, K.L. Johnson, M.D. Scrimshaw, and J.N. Leiber, 2000, "Binding of waterborne steroid estrogens to solid phases in river and estuarine systems," *Environmental Science and Technology*, 34, 3890–3894.

remove these contaminants in drinking water to levels below their active concentrations must be found and utilized in order to protect human health. Endocrine Disruptor Degradation Example 1 examines the possible impacts of an industrial spill of an EDC, even when a viable treatment scheme exists to protect against such an accident.

Endocrine Disruptor Degradation Example 1

A chemical plant that produces polycarbonate for baby bottles recently spilled 1 ton of BPA into a wastewater stream with a flow of one million gallons per day (1 MGD) that discharges its effluent into the Ohio River. The plant has the capability to feed $10\,mg\,L^{-1}$ Powdered Activated Carbon (PAC) into the wastewater stream and enough holding capacity to achieve 4 hours of contact time. Adding PAC to water can remove endocrine disruptors, and the removed rate is dose-dependent (see Figure 8.11). To remove this PAC and other solids, the plant also has the capability to filter solid particles with diameters down to 1 μm from their wastewater in an emergency. If it is assumed that the spill is evenly dispersed throughout one day, and equilibrium conditions are achieved in the water stream, what is the final concentration of the water being discharged from the plant into the Ohio River? Also, according to the YES assay, how "estrogenic" is the wastewater stream due to the BPA?

Solution
The answer follows three steps.

Step 1: Find the concentration of BPA in the waste stream before any treatment.

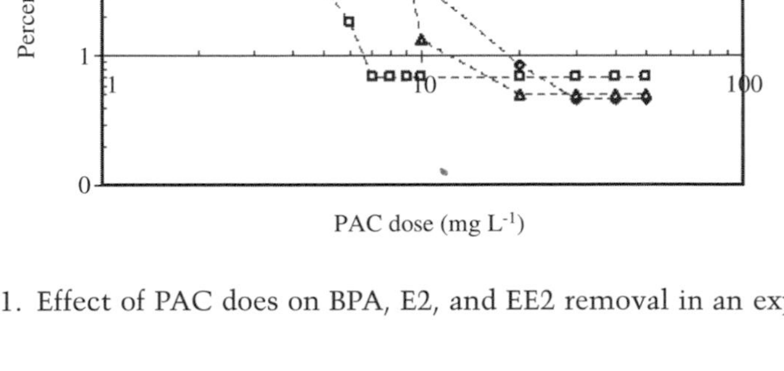

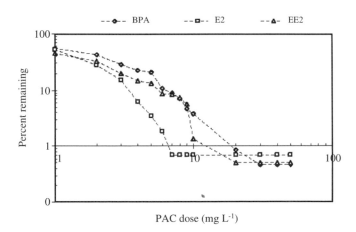

FIGURE 8.11. Effect of PAC does on BPA, E2, and EE2 removal in an experimental water.

If 1 ton of solid BPA is spilled into 1 million gallons (assume even dispersion through the waste stream for 1 day), upon unit conversion, a concentration of 237 $mg L^{-1}$ would be achieved if all BPA is dissolved in water. However, Table 8.2 shows that the solubility of BPA in water is only 129 $mg L^{-1}$, implying that this is the maximum concentration of BPA in water at 25°C. The rest of the BPA remains as solid particles in the water (assumed greater than 1 μm in diameter).

Step 2: Determine the concentration after PAC addition and filtration.
For the conditions given (10 $mg L^{-1}$ PAC with a contact time of 4 hours removing BPA), approximately 4% of the original concentration of BPA remains in solution. Also, the filtration step will remove all PAC, plus any undissolved BPA, implying the final concentration in the wastewater stream will be approximately 5.16 $mg L^{-1}$.

Step 3: How "estrogenic" is this stream?
According to the YES data given in Table 8.1, BPA displays a relative potency of 6.2×10^{-5} as compared to estrogen. This means the concentration of 5.16 $mg L^{-1}$ displays the "estrogenic" response of 0.32 $\mu g L^{-1}$ of 17-β-estradiol. This is equivalent to an estrogen concentration capable of inducing estrogenic responses in all of the bioassays.

Note that the wastewater will be substantially diluted when it enters the Ohio River. However, it is naïve to assume that a wastewater stream dumping into a larger water body will disperse widely, a conclusion supported by a U.S. Geological Survey study examining wastewater discharge from Las Vegas into Lake Meade, which used the vitellogenin bioassay to show elevated levels of EDCs greatly affecting male carp, fish that prefer sheltering near large underwater objects, including wastewater effluent pipes.[18]

Treatment of EDCs in Drinking Water: UV Applications

Because of the proven ability of EDCs to interfere with the normal endocrine function of so many aquatic species at low concentrations, and their presence in waters used as drinking water sources, inclusion of a treatment technology capable of removing or destroying EDCs in a drinking water treatment train may be imperative to the goal of protecting human health. Current treatment technologies that have been tested for their efficacy regarding removal or degradation of EDCs include conventional biological treatment, chlorination, activated carbon (GAC, PAC), membranes, and several oxidative techniques, with mixed success. Several recent reports have caused great concern, for they have indicated that chlorination, a treatment process utilized by nearly every water utility in the United States, may react with certain EDCs to produce products that exhibit greater estrogenic activity than their parent compounds. These studies were performed regarding the chlorination of bisphenol A and nonylphenol,[19] two persistent EDCs.

A novel approach to removing synthetic estrogens involves using emerging ultraviolet light (UV)–based water treatment technology, currently used to disinfect microbial contaminants in drinking water. Ultraviolet radiation water treatment has proved very effective in removing threats presented by pathogenic organisms, and is being installed in many treatment facilities throughout the world. The use of UV radiation for destruction of contaminants, an area of increasing interest, may also present a viable alternative for effective treatment of EDCs in water supplies throughout the world.

UV Basics

Two types of mercury-based UV lamps are traditionally used for water treatment. Low pressure (LP) lamps emit radiation at a wavelength of 253.7 nm (monochromatic UV radiation), near the peak germicidal efficiency for inactivating most microorganisms. Medium pressure (MP) lamps emit radiation spread out over a broader wavelength spectrum (polychromatic UV radiation). The "pressure" refers to the mercury vapor pressure inside the lamp. The differences in the pressures of the gas inside the lamps cause the difference in the output spectrum of the two lamps. Typically LP lamps have a pressure of less then 0.013 atm, and MP lamps a pressure of about 1.3 atm. Figure 8.12 displays the relative emission spectra of low and medium pressure lamps, as well as the relative absorbance of three EDCs.

Using UV radiation to treat chemicals in drinking water is completed via two mechanisms: direct and indirect photolysis. Direct photolysis involves the direct absorption of UV radiation by the chemical to be treated, and the excitation and destruction of the compound. Indirect photolysis involves the absorption of UV radiation by a chemical other than that being treated, formation of an excited intermediate, and a degradation reaction between this intermediate species and the compound.

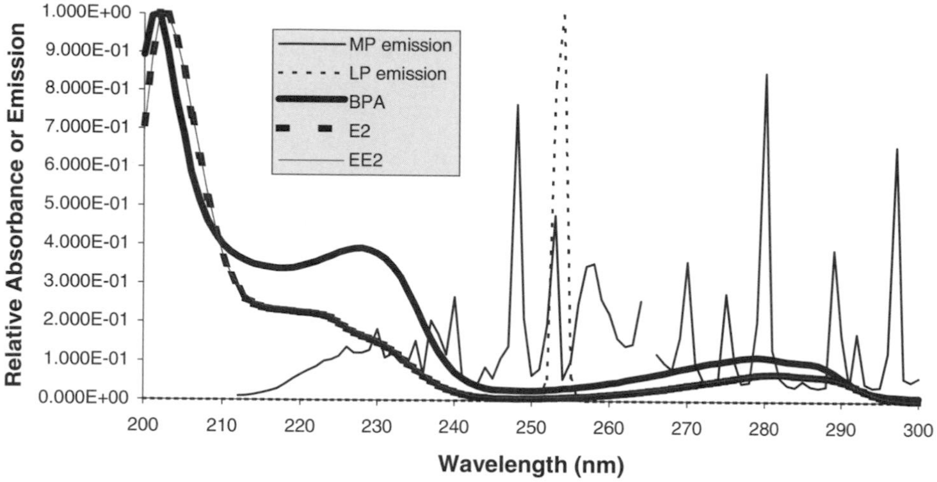

FIGURE 8.12. Relative emission spectra for MP and LP UV lamps, with relative absorbance spectra of BPA, E2, and EE2 overlain.

Direct Photolysis

Direct UV photolysis is governed by two main parameters, the molar absorption coefficient, and the quantum yield. Both parameters are chemical specific and describe the interaction of the chemical with UV radiation.

The molar absorption coefficient describes the amount of radiation at a specific wavelength that a compound within solution will absorb. With inverse molar concentration units per area ($M^{-1} cm^{-1}$), the UV absorbance due to a solution of the compound at a specific concentration is described by using the molar absorption coefficient. Figure 8.12 displays the relative absorption at each wavelength from 200 to 300 nm of BPA, E2, and EE2. All three compounds exhibit a multimodal absorption spectrum over this range, and each exhibits an absorption minimum at approximately 250 nm. This is significant, because LP lamps emit UV radiation only at approximately 254 nm, which corresponds closely to the minimum absorption of each contaminant, while MP lamps emit radiation throughout the UV range.

The first law of photochemistry states that only radiation that is absorbed can produce a photochemical effect. Thus, direct UV treatment of contaminants is effective only if the UV radiation emitted by a UV lamp is absorbed by the contaminant. The emission spectrum for the MP lamps overlaps much of the major absorbance features of the contaminants under study. Therefore, it is expected that an MP lamp will destroy the contaminants more rapidly simply because more radiation is absorbed by the compounds.

Another important factor in understanding direct UV treatment is the quantum yield (Φ). The quantum yield is a measure of the photon efficiency of a photochemical reaction. It is defined as the number of moles of reactant removed per Einstein (mole of photons) absorbed by the chemical.[20] There are no simple rules to predict reaction quantum yields from chemical structure, so Φ values need to be determined experimentally for each compound. Additionally, the wavelength dependence of Φ must be considered when using polychromatic radiation sources. Quantum yield values can be approximated as wavelength independent, at least over the wavelength range of a given absorption band, corresponding to one mode of excitation.[21] If, as in the case of our EDCs, multiple light absorption bands are displayed, quantum yields may have to be determined for various wavelengths to predict accurately the transformation rate of a given compound.

Indirect Photolysis: UV/H_2O_2 Advanced Oxidation Technology

When hydrogen peroxide (H_2O_2) is added to the solution before irradiation with UV, the direct photolysis process for the target compound is augmented by an indirect degradation process through the production of the hydroxyl radical ($\cdot OH$). Addition of UV energy in the presence of H_2O_2 is known as an advanced oxidation process (AOP). AOPs can be generated via a number of scenarios including vacuum UV in water, ozone, ozone/peroxide, UV/ozone/peroxide, UV/TiO_2, UV/NO_3, fenton processes, and photo-fenton processes. AOPs are characterized by the formation of a highly reactive, oxidative-intermediate species, such as the hydroxyl radical.

When UV radiation hits a H_2O_2 molecule, the molecule splits apart into two OH radicals.

$$H_2O_2 + h\nu \rightarrow 2\cdot OH \qquad \text{Reaction 8–1}$$

Although the stoichiometry of this reaction implies two radicals per parent H_2O_2 molecule, due to recombining and inefficiencies in the process, only one OH radical is formed per photon of light absorbed. Therefore, the quantum yield of the process is unity. This means that in the bulk solution, for every mole of photon of light absorbed by

H_2O_2, one mole of hydroxyl radical is formed. Once the hydroxyl radical is formed, it will rapidly undergo an oxidation reaction with almost any species present, including the contaminant of interest. OH radicals will also react quickly with carbonate species (HCO_3^-, CO_3^{-2}), natural organic matter (NOM), other organic compounds present, chloride ion, and even H_2O_2. Given this nonselective nature of OH radicals, water quality must be accounted for when determining the effectiveness of the process towards degrading a specific contaminant of concern. Table 8.3 displays the second-order rate constants of OH radical with several organic contaminants of concern, as well as with carbonate species and NOM.

Modeling the UV/H_2O_2 Process

Because of the unselective nature of the OH radical, the concentration of the species can often be considered constant and relatively low (10^{-14} – 10^{-12} M) when compared to the levels of other species in the water. Using these assumptions, the steady-state model for destruction involving the OH radical has been developed. This model assumes the OH radical concentration at a constant level throughout the process, thus reducing the second-order rate equation:

$$\frac{d[M]}{dt} = k[\cdot OH][M] \qquad \text{Equation 8–5}$$

to a pseudo–first-order rate equation

$$\frac{d[M]}{dt} = k'[M] \qquad \text{Equation 8–6}$$

TABLE 8.3
Second-Order Rate Constants of OH Radical with Several Organic Contaminants and Inorganic Species[22]

Compound	*Second-Order Rate Constants*	*Source*
Atrazine ($M^{-1}s^{-1}$)	3×10^9	Acero (2000)
MTBE ($M^{-1}s^{-1}$)	1.6×10^9	Huber (2003)
Ethinyl estradiol ($M^{-1}s^{-1}$)	9.8×10^9	Huber (2003)
HCO_3^- ($M^{-1}s^{-1}$)	8.5×10^6	Buxton (1988)
CO_3^{2-} ($M^{-1}s^{-1}$)	3.9×10^8	Buxton (1988)
DOM (L (mg C)$^{-1}$ s^{-1})	2.5×10^4	Larson and Zepp (1988)
H_2O_2 ($M^{-1}s^{-1}$)	2.7×10^7	Buxton (1988)[20]

where [M] is the molar concentration of the compound that is degraded, and k′ is the product of the second-order rate constant and the steady-state OH radical concentration.

The steady-state OH radical concentration is influenced by many parameters in the UV/H_2O_2 process, including intensity of the UV radiation, concentration of H_2O_2, and water quality. Equation 8–7 is used to calculate $[OH]_{ss}$ for a low pressure lamp with a known hydrogen peroxide concentration.

$$[OH]_{ss} = \frac{I_{ave}\Phi\varepsilon[H_2O_2]}{\sum k_s[S]} \qquad \text{Equation 8–7}$$

where I_{ave} is the average UV irradiance (Einsteins per second, i.e., $Es\ s^{-1}$).

Φ is the quantum yield of OH radical formation from H_2O_2 ($1\ mol\ Es^{-1}$)

ε is the molar absorption coefficient of H_2O_2 ($17.9\ M^{-11}\ cm^{-1}$ at 254 nm)

$[H_2O_2]$ is the initial concentration of hydrogen peroxide (M)

$\Sigma k_s[S]$ is the sum of the second-order rate constants times the concentration of all scavenger species present

Endocrine Disruptor Degradation Example 2 is an opportunity to use this model to examine the degradation of an EDC in a natural water using the UV/H_2O_2 process.

Endocrine Disruptor Degradation Example 2

Given the second-order rate constant for the reaction between EE2 and OH radical in Table 8.3, find the time required to degrade EE2 by 2 logs (99%), using LP UV/H_2O_2 process (average irradiance = $0.015\ mEs\ s^{-1}$, $[H_2O_2]_i = 15\ mg\ L^{-1}$) in a water described by the water quality parameters given in Table 8.4.

Solution

The solution consists of three steps.

Step 1: Find the OH radical steady-state concentration.
First, the molar concentration of all scavenger species must be known. The scavengers in this case are H_2O_2, DOM, HCO_3^-, and CO_3^{2-} (the initial concentration of EE2 can be neglected because it is significantly

less than the concentration of other organics in the water). HCO_3^- and CO_3^{2-} are calculated using the pH and alkalinity. A simplified version of the alkalinity equation is:

$$Alk = [OH^-] - [H^+] + [HCO_3^-] + [CO_3^{-2}] \qquad \text{Equation 8–8}$$

And the carbonate species are related through acid/base chemistry (discussed in detail in Chapter 7):

$$K_a = \frac{[CO_3^{2-}][H^+]}{HCO_3^-} \qquad \text{Equation 8–9}$$

Where K_a is the second acid dissociation constant for the carbonate system ($K_a = 10^{-10.3}$).

By manipulating these equations and solving for the molar concentrations of the other species, the concentrations of the scavenging species are as follows:

$$[H_2O_2] = 4.4e-4\ M,\ DOM = 4.92\ mg\,L^{-1}, [HCO_3^-] = 5.0e-4\ M, [CO_3^{-2}] = 5.6e-7\ mg\,L^{-1}$$

The second-order OH radical rate constants for all of these species can be found in Table 8.3, so Equation 8–6 can be solved to find the steady-state OH radical concentration of 8.5×10^{-13} M.

Step 2: Integrate the pseudo–first-order rate equation.
To find the time necessary for a reaction to occur, an integrated rate expression must be found. In this case, separating the variable and integrating both sides of Equation 8–6 yields the following integrated rate equation.

$$\frac{C}{C_o} = e^{-k,t} \qquad \text{Equation 8–10}$$

Step 3: Solve the integrated rate equation to find the time needed for 2 log removal.
Two logs of removal implies 99% removal, so if 0.01 is input for the left-hand side of Equation 8–10 and k_{EE2} from Table 8.3 (9.8×10^9) is multiplied by $[\cdot OH]_{ss}$ to find k′, a time of 553 seconds, or 9 minutes and 13 seconds, is needed to achieve the desired removal.

Figure 8.13 shows the destruction kinetics of EE2 as a function of time for this system.

TABLE 8.4
Water Quality Parameters for a Natural Water

H_2O_2 (ppm) MW = 34 g/mol	$[EE2]_i$ ($\mu g L^{-1}$)	DOM (excluding EE2) mg/L	pH	Alkalinity ($mg L^{-1}$ as $CaCO_3$)
15	50	4.92	7.35	24.8

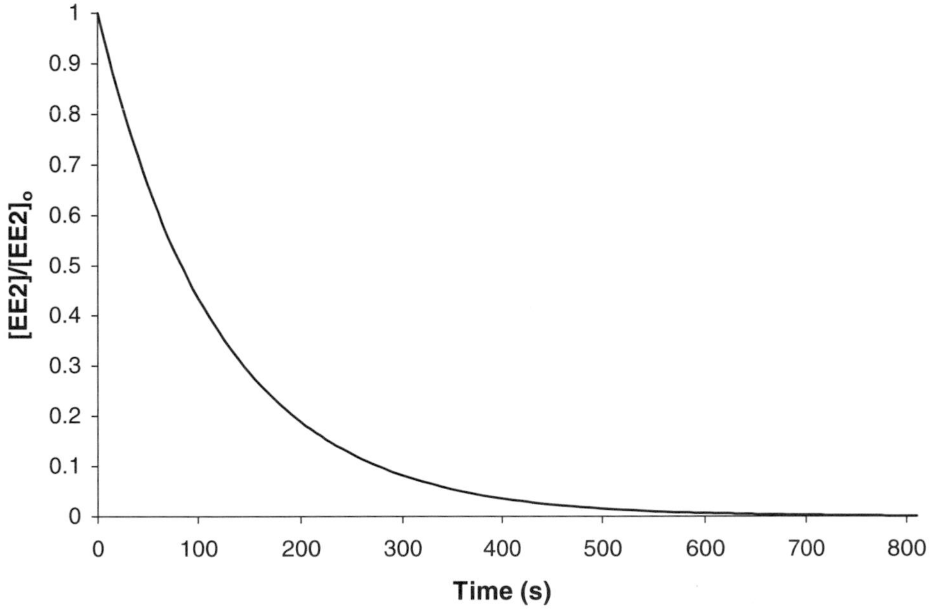

FIGURE 8.13. Destruction of EE2 over time by the UV hydrogen peroxide process, as modeled with the steady state OH radical model.

As a final note, complete mineralization of organic contaminants can be achieved by utilizing the UV/H_2O_2 advanced oxidation process. However, complete transformation to mineral acids, H_2O and CO_2 takes long exposure times and high concentrations of H_2O_2. As is the case in most chemical treatment situations, incomplete destruction of the contaminants will occur with the UV/H_2O_2 processes. As such, a variety of destroyed but not mineralized by-products will remain in the treated water. These products are likely to be more polar and smaller than the original pollutant. Both the identities and the toxicity of these compounds must be determined to evaluate the true effectiveness of any degradation process. The ultimate question that needs to be answered when determining the effectiveness of the treatment process for destruction of EDCs is, "Does this treatment process solve the problem, exacerbate it, or cause new problems?" By utilizing various bioassays, including the YES assay, the E-Screen assay, or a developmental fish assay, future research will attempt to examine the relative toxicity of the by-products of the destruction processes, in an effort to determine the effectiveness of the UV treatment process not only in regards to destruction of EDCs, but ultimately in regards to protecting the water supply from these contaminants and the possibility of the estrogenic behavior of their degradation products.

Oxygen-Depleting Contaminants

Oxygen, particularly in its molecular form O_2, is essential to most life forms, all higher-order organisms, and all environmental systems. In fact, one of the key indicators of an environmental system's condition is the availability of ample O_2 concentrations. When O_2 concentrations are depleted, the system becomes *anoxic* and less capable of sustaining diverse populations. We may be tempted to think of contaminants as only those agents that elicit direct toxic responses, however, certain contaminants are particularly harmful because they use up resources in the environment at the expense of other organisms. Thus it is important to consider the processes that lead to falling O_2 levels.

An interesting aspect of O_2 depletion is the unevenness with which it affects ecosystems. For example, lowering dissolved O_2 in surface waters will first adversely affect more sensitive aquatic species, such as trout, salmon, and other game fish. The lower O_2 concentrations may actually benefit "rough" fish like carp and buffalo, until the waters are almost completely devoid of dissolved O_2.

Also, the effect is different for various life stages: at lower oxygen levels, the fish may live but not be able to reproduce effectively. Figure 8.14 shows that the combination of vulnerabilities during these different life stages will determine the ability of an aquatic population to thrive, or at least survive (in this instance, in saltwater). For example, below 4 $mg\,L^{-1}$ dissolved O_2 an adult trout may not suffer acute effects, but trout larvae or young-of-the-year fish may not survive. Dissolved oxygen criteria apply to both continuous and cyclically depressed oxygen levels. If O_2 concentrations are continuously above the chronic criterion for growth (about 4.8 $mg\,L^{-1}$ in many systems), the aquatic life at that location should not be harmed. When dissolved oxygen conditions at a site fall below the juvenile/adult survival criterion (2.3 $mg\,L^{-1}$), there is not a sufficient amount of oxygen to protect aquatic organisms. Thus, when conditions lead to persistently

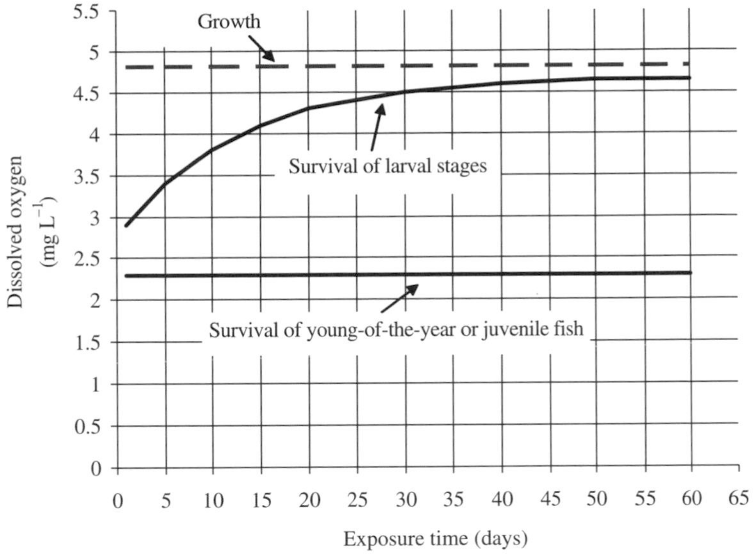

FIGURE 8.14. Summary of dissolved molecular oxygen (O_2) criteria for persistent exposure for a fish population. Shown are the lower bound limits on protective O_2 concentrations. The chronic growth limit may be violated for a specific number of days provided the chronic larval recruitment limit is not violated. (Source: U.S. Environmental Protection Agency.)

depressed oxygen levels, and DO conditions are between the growth and survival levels, the duration and intensity of these depressed oxygen levels require ongoing monitoring to ensure that the aquatic ecosystem remains healthy.[23]

It is important to know the sources of oxygen depletion, which can be direct depleters, such as contaminants that react with oxygen chemically. They may also be indirect depleters, or those that allow for large growth of bacteria or algae that in turn uses up the oxygen. The total amount of oxygen used chemically and biochemically is known as chemical oxygen demand (COD). The amount used by microbes is known as biochemical oxygen demand (BOD).

Naturally occurring organic matter, including organic wastes from sewage treatment plants, improperly operating septic systems, and runoff from agricultural and residential areas, are actually the energy sources (i.e., "food") for water-borne bacteria. Bacteria decompose these organic materials using dissolved oxygen, thereby decreasing the DO present for aquatic life. BOD is the amount of oxygen that bacteria will consume in the process of decomposing organic matter under aerobic conditions. The BOD is measured by incubating a sealed sample of water for five days and measuring the loss of oxygen by comparing the O_2 concentration of the sample at time = 0 (just before the sample is sealed) to the concentration at time = 5 days (i.e., BOD_5). Samples are commonly diluted before incubation to prevent the bacteria from depleting all of the oxygen in the sample before the test is complete.[24]

Chemical oxygen demand (COD) does not differentiate between biologically available and inert organic matter, and it is a measure of the total quantity of oxygen required to oxidize all organic material into carbon dioxide and water. COD values always exceed BOD values for the same sample. Sometimes COD measurements are conducted simply because they require only a few hours compared to the 5 days for BOD.

If effluent with high BOD concentrations reaches surface waters, it may diminish DO to levels lethal to some fish and many aquatic insects. As the water body re-aerates as a result of mixing with the atmosphere and by algal photosynthesis, O_2 is added to the water, and the oxygen levels will slowly increase downstream. The drop and rise in DO concentrations downstream from a source of BOD is known as the DO sag curve, because the concentration of dissolved oxygen "sags" as the microbes deplete it. Thus, the falling O_2 concentrations fall with both time and distance from the point where the high BOD substances enter the water (see Figure 8.15).

Biomarkers of Contaminants

When a contaminant interacts with an organism, substances like enzymes are generated as a response. Thus, measuring such substances in fluids and tissues can provide an indication or "marker" of contaminant exposure and biological effects resulting from the exposure. The term *biomarker* includes any such measurement that indicates an interaction between an environmental hazard and a biological system.[25] In fact, biomarkers may indicate any type of hazard—chemical, physical, and biological. An exposure biomarker is often an actual measurement of the contaminant itself or any chemical substance resulting from the metabolism and detoxification processes that take place in an organism. For example, measuring total lead (Pb) in the blood may be an acceptable exposure biomarker for people's exposures to Pb. However, other contaminants are better reflected by measuring chemical by-products.

Exposure biomarkers are also useful as an indication of the contamination of fish and wildlife in ecosystems. For example, measuring the activity of certain enzymes, such as ethoxyresorufin-*O*-deethylase (EROD), in fish *in vivo* biomarker, indicates that the organism has been exposed to planar halogenated hydrocarbons, PAHs, or other similar contaminants. The mechanism for EROD activity in the fish is the receptor-mediated induction of cytochrome P450-dependent mono-oxygenases when exposed to these contaminants.[26]

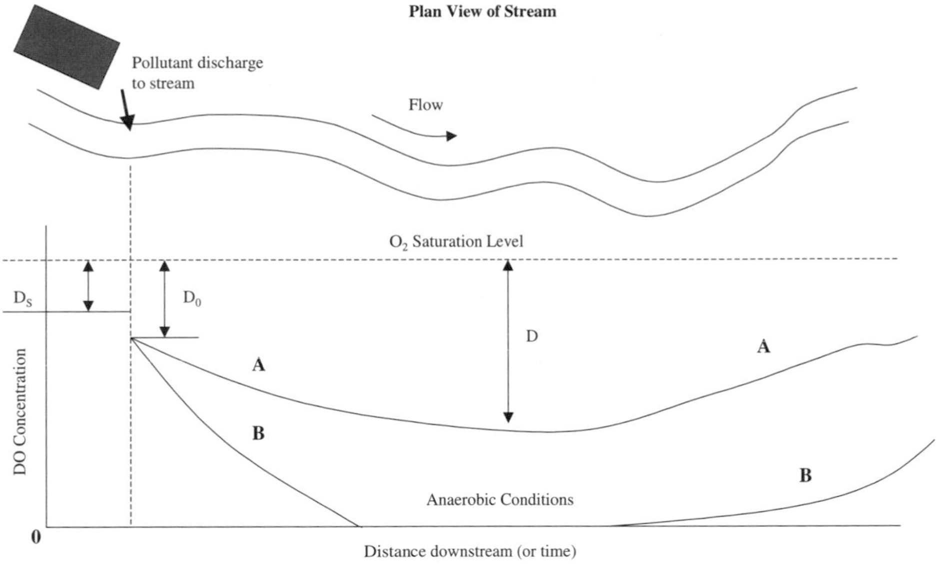

FIGURE 8.15. Dissolved oxygen sag curve downstream from an oxygen-depleting contaminant source. The concentration of dissolved oxygen in Curve A remains above 0, so although the available oxygen is reduced, the system remains aerobic. Curve B shows a sag where dissolved oxygen falls to 0 and anaerobic conditions result. D_S is the background oxygen deficit before the pollutants enter the stream. D_0 is the oxygen deficit after the pollutant is mixed. D is the deficit for contaminant A, which may be measured at any point downstream. This indicates both the distance and time of microbial exposure to the source. For example, if the stream's average velocity is $5\,\text{km}\,\text{h}^{-1}$, D measured 10 km downstream also represents 2 hours of microbial activity to degrade the pollutant.

Laboratory Notebook Entry: Measuring and Assessing a Biological Agent—*Stachybotrys*

Fungi comprise the kingdom of organisms that includes about 250,000 species, only about 200 of which have been identified as pathogenic.[27] Molds are fungi that live on numerous surfaces, including indoor walls and fixtures, as well as outdoors in soil, on plants, and on detritus (see Figure 8.16). Over 1000 species of molds have been found in indoor environments. Mold growth is usually increased with increasing temperature and humidity under environmental conditions, but this does not mean that molds cannot grow in colder conditions. Species may

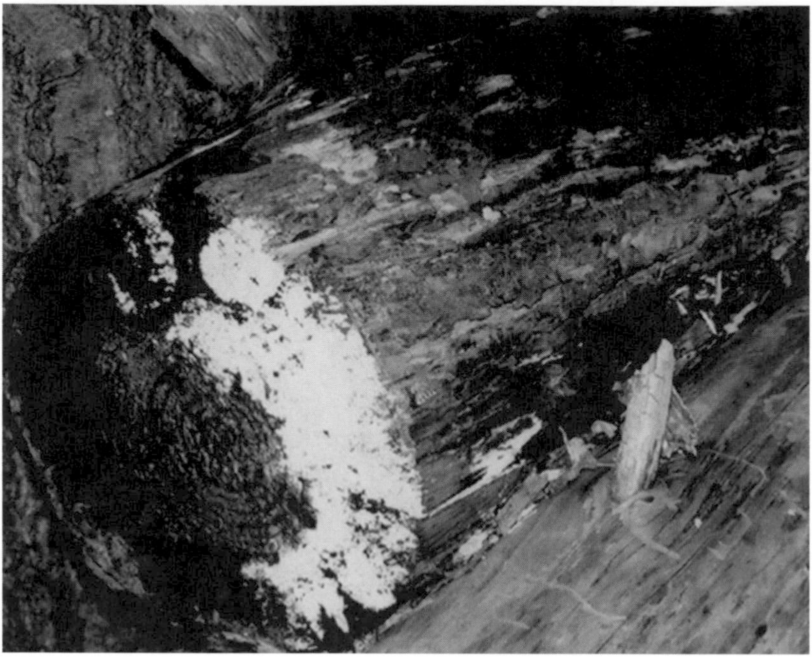

FIGURE 8.16. Mold growing outdoors on firewood. Both white and black molds are present in this photo. (Source: U.S. Environmental Protection Agency, 2002, *A Brief Guide to Mold, Moisture and Your Home*, EPA 402-K-02-003, Washington, D.C.)

be of wide range of colors and often elicit particles and gases that render odors, often referred to as "musty." Like other fungi, molds reproduce by producing spores that are emitted into the atmosphere. Living spores (Figure 8.17) are disseminated to colonize growth wherever conditions allow. Most ambient air contains large amounts of so-called *bioaerosols*, or particles that are part of living or once-living organisms. In this instance, the bioaerosols are live mold spores, meaning that inhalation is a major route of exposure.

Indoor sources of molds include leaking pipes and fixtures, damp spaces such as those in basements and crawl spaces, heating, and air conditioning and ventilation (HVAC) systems, especially those that allow for condensation from temperature differentials between surfaces and ambient air, kitchens, and showers.

Some molds produce toxic substances called mycotoxins. There is much uncertainty related to possible health effects associated with inhaling mycotoxins over a long time periods. Extensive mold growth

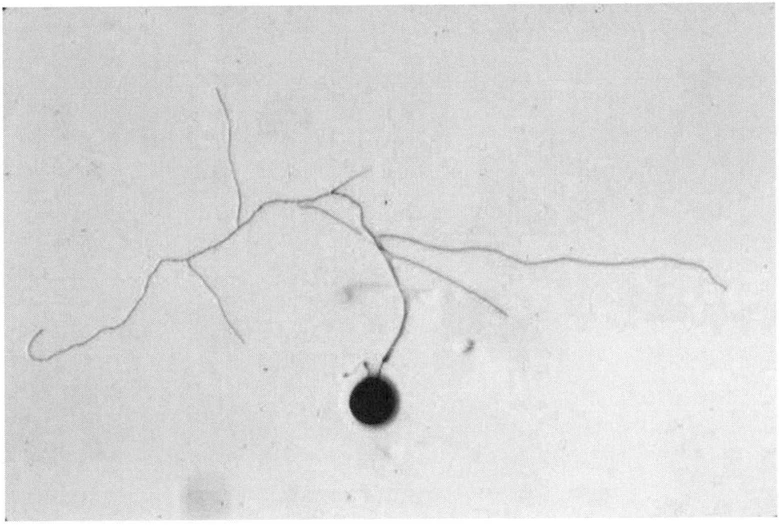

FIGURE 8.17. Optical micrograph of a *Glomus intraradices* spore that has begun to germinate. This is an example of a beneficial fungal type known as arbuscular mycorrhizal (AM) fungi, which colonize plant root cells to establish symbiosis, i.e., a mutual association where the fungi hasten plant growth, and improve drought and disease resistance by increasing the plant's root mass thereby improving nutrient transport to the plant cells. (Source: U.S. Department of Agriculture, 2004, Microbial Biophysics and Residue Chemistry Unit, "AM Fungi," http://www.arserrc.gov/mbb/AMFungi.htm.)

may cause nuisance odors and health problems for some people. It can damage building materials, finishes, and furnishings, and in some cases, cause structural damage to wood.

Sensitive persons may experience allergic reactions, similar to common pollen or animal allergies, flu-like symptoms, and skin rash. Molds may also aggravate asthma. Rarely, fungal infections from building-associated molds may occur in people with serious immune diseases. Most symptoms are temporary and eliminated by correcting the mold problem, although much variability exists on how people are affected by mold exposure. Particularly sensitive subpopulations include:

- Infants and children
- Elderly people
- Pregnant women

- Individuals with respiratory conditions or allergies and asthma
- Persons with weakened immune systems (for example, chemotherapy patients, organ or bone marrow transplant recipients, and people with HIV infections or autoimmune diseases)

Persons cleaning mold should wear gloves, eye protection, and a dust mask or respirator to protect against breathing airborne spores (Figure 8.18). A professional experienced in mold evaluation and remediation, such as an industrial hygienist, may need to be consulted to address extensive mold growth in structures. It is important to correct large mold problems as soon as possible by first eliminating the source of the moisture and removing contaminated materials, cleaning the surfaces, and finally drying the area completely.

If visible mold is present, then it should be remediated, regardless of what species are present and whether samples are taken. In specific instances, such as cases where health concerns are an issue, litigation is involved, or the source of contamination is unclear, sam-

FIGURE 8.18. Mold remediation worker using the recommended personal protection equipment, including an N-95 respirator, goggles, and gloves. (Source: U.S. Environmental Protection Agency, 2002, *A Brief Guide to Mold, Moisture and Your Home*, EPA 402-K-02-003, Washington, D.C.)

pling may be considered as part of a building evaluation. Sampling is needed in situations where visible mold is present and there is a need to have the mold identified. A listing of accredited labs can be found at http://www.aiha.org/LaboratoryServices/html/lists.htm.

Environmental investigations must be interpreted in the context of medical and epidemiological information for infectious diseases from environmental sources. For example, finding *Legionella* colonization of a water supply serving an immuno-compromised population has potential health significance notwithstanding cases of infection. Health hazards from exposure to environmental molds and their metabolites relate to four broad categories of chemical and biological characteristics: (1) irritants, (2) allergens, (3) toxins, and, rarely, (4) pathogens. Risks from exposure to a particular mold species vary depending on a number of factors. Uncertainty is increased with the lack of information on specific human responses to well-defined mold contaminant exposures. In combination, these knowledge gaps make it impossible to set simple exposure standards for molds and related contaminants.

A useful method for interpreting microbiological results is to compare the kinds and levels of organisms detected in different environments. Usual comparisons are indoors to outdoors or complaint areas to areas where no complaints have been made. Specifically, in buildings without mold problems, the qualitative diversity of airborne fungi indoors and outdoors are expected to be similar. On the other hand, dominance of one or a few species of fungi indoors and their absence outdoors may indicate a moisture problem and degraded air quality. Also, the consistent presence of certain fungi species, including *Stachybotrys chartarum, Aspergillus versicolor*, or various *Penicillium* species in counts above background concentrations, may indicate the conditions conducive to their growth (i.e., moisture and ventilation problems). Generally, indoor mold types should be similar and levels should be no greater than outdoor and background areas. Analysis of bulk material or dust samples can also be compared to results of similar samples collected from reasonable comparison areas.

Total bacterial levels indoors versus outdoors may not be as useful as with fungi, because bacteria reservoirs exist in both.However, specific strains of bacteria that are present may help in apportioning potential building-related sources. More information is available at:

- *Field Guide for the Determination of Biological Contamination* (stock #227-RC-96), American Industrial Hygiene Association (AIHA), http://www.aiha.org
- *Report of Microbial Growth Task Force* (stock #458-EQ-01), AIHA, http://www.aiha.org

- Listing of AIHA Laboratory Quality Assurance Program Environmental Microbiology Laboratory Accreditation Program (LQAP EMLAP) accredited laboratories, AIHA, http://www.aiha.org
- Bioaerosols: Assessment and Control, American Conference of Governmental Industrial Hygienists (ACGIH), http://www.acgih.org
- *Standard and Reference Guide for Professional Water Damage Restoration, Institute of Inspection, Cleaning, and Restoration Certification,* IICRC S500, http://www.iicrc.org
- *Mold Remediation in Schools and Commercial Buildings,* EPA 402-K-01-001, Environmental Protection Agency, http://www.epa.gov/iaq/molds/index.html
- Draft Guideline for Environmental Infection Control in Healthcare Facilities (especially sections I.C.3, I.C.4, I.F, II.C.1, and Appendix B), Centers for Disease Control (CDC), http://www.cdc.gov/ncidod/hip/enviro/env_guide_draft.pdf
- EPA and FEMA (Federal Emergency Management Agency) Flood Clean-Up Guidelines: http://www.epa.gov/iaq/pubs/flood.html and http://www.fema.gov/hazards/floods/
- Centers for Disease Control and Prevention (CDC): http://www.cdc.gov/nceh/airpollution/mold/default.htm
- California Indoor Air Quality Program: http://www.cal-iaq.org/iaqsheet.htm
- New York City Department of Health "Guidelines on Assessment and Remediation of Fungi in Indoor Environments": http://www.nyc.gov/html/doh/html/epi/moldrpt1.html
- American College of Occupational and Environmental Medicine guideline, "Adverse Human Health Effects Associated with Molds in the Indoor Environment": http://www.acoem.org/guidelines/pdf/mold-10-27-02.pdf

Accelerated Biodegradation: Bioremediation

Extracting a microbe from the environment and exposing it to a target contaminant under controlled conditions is one means of breaking the contaminant down into less toxic components. This is the goal of *bioremediation.* Microbes, or even higher organisms like plants (phytoremediation) and animals, can reduce the potential toxicity of chemical contaminants by transforming, degrading, and immobilizing these compounds in the environment. Environmental scientists and engineers know a great deal about the pathways for organic degradation and the degradation mechanisms. Such treatment processes include cometabolism, anaerobic biotransformations of highly chlorinated solvents, and alternate electron acceptors, which are used frequently in controlled bioremediation efforts. Biological treatment methods will be addressed in more detail in Chapter 11, "Contaminant Sampling and Analysis."

Biocriteria: A New Way to Determine Environmental Quality

The traditional way of dealing with pollution is to measure the chemical concentrations of contaminants and, if they are outside of the healthy range (e.g., elevated contamination), to take action. However, other ways are available to assess environmental quality. The presence, condition, and diversity of plants, animals, and other living things can be used to assess the health of a specific ecosystem, such as a stream, lake, estuary, wetland, or forest. Such organisms are referred to as *biological indicators.*

An indicator is in a sense an "integrated" tool that incorporates highly complex information in an understandable manner. A well-known bioindicator is the famous canary in the coal mine. Miners were aware that if they hit a vein that contained "coal gas" (actually high concentrations of methane) they had little time to evacuate before inhalation of the gas would lead to death. However, they realized that due to its small mass, a smaller animal would succumb to the toxic effects before a human would be affected. The miners did not really care so much *how* it worked (i.e., the dose-response relationships and routes that will be discussed in the next chapter), they only cared *that* it worked. Actually, the canary is an example of a *bioassay*, which is a test of toxicity or other adverse effect on one or a few organisms to determine the overall expected effect on a system.

An ecological indicator can be a single measure, an index that embodies a number of measures, or a model that characterizes an entire ecosystem or components of that ecosystem. An indicator integrates the physical, chemical, and biological aspects of ecological condition. It is used to determine status and to monitor or predict trends in environmental conditions and possible sources of contamination and stress on systems.

Biocriteria are metrics of a system's biological integrity. A system must be able to support communities of organisms in a balanced manner.[27] One means of determining biological integrity is to compare the current condition of an ecosystem to that of pristine or undisturbed conditions (see Figure 8.20). The *threshold* is the condition below which a system suffers from dysfunction or impairment, such as a minimum concentration of oxygen or a maximum concentration of a toxic contaminant. A reference condition is frequently associated with biological integrity. However, few systems have not been in some way affected by humans, so the "pristine" system is rare indeed, and an environmental scientist will more often refer to a reference system as one that is "minimally impaired," or one with high biological integrity. Ecosystems and environmental compartments can be degraded by chemical contamination as well as by physical changes that alter habitats, such as the withdrawal of irrigation water from aquifers and surface waters, overfishing, and overgrazing, and by introducing opportunistic exotic species. Biota are selectively sensitive to all forms of pollution (as in the difference between game and rough fish discussed earlier).

Estimating biological integrity requires the application of direct or indirect evaluations of a system's attributes. Indirect evaluations can have the advantage of being cheaper than the direct approaches, but they will not often be as robust. An attribute of natural systems to be protected, such as a fish population, is an example of an assessment endpoint; whereas an attribute that is quantified with actual measurements, such as age classes of the fish population, is known as a measurement endpoint. Reliable and representative assessment and measurement endpoints are needed to reflect a system's biological integrity.

Arguably the most widely used metric for biological integrity is the Index of Biotic Integrity (IBI), which consists of 12 attributes in three major groups: species richness and composition, trophic structure, and abundance and condition of fish and other aquatic organisms.

Species richness is a measure of the number of different species of organisms in an ecosystem. *Composition* is the classification of the types of species in an ecosystem (e.g., a system contains four species of top predators, 30 species of first order consumers, 300 species of producers, etc.). The *trophic structure* of an ecosystem is the means by which energy flows within it; that is, from producers (plant life) progressing to higher-order organisms (consumers) up to the top predators. Decomposers break down the organic compounds in the remains of other organisms into simpler compounds, ultimately to inorganic substances (this process is known as "mineralization"). *Abundance* is the total number of species in the ecosystem (e.g., if Ecosystem A and B both have 10,000 total number of animals, but A has 20 different species, while B has 200 different species, then B has greater species abundance than A). The condition of the ecosystem is an expression of both its trophic structure and how well the ecosystem is functioning. So, a system's ecological integrity is a measure of the ecosystem's condition. The metrics for integrity must include physical, chemical, and biological attributes that are compared to an ideal (e.g., "unimpaired") condition.[28] The elements of the biosphere are essential to the protection of biological integrity (see Table 8.5).

The ecosystem processes follow the hierarchy of a system's organization, including its various structures and functions, so the metabolism of individual organisms are at one extreme. Population processes, such as reproduction, recruitment, dispersal, and speciation are next, while at the highest level of organization, or communities or ecosystems, processes include nutrient cycling, interspecies interactions, and energy flows. Only a representative amount of biota needs to be sampled. Such selections must aggregate an optimal number of attributes with sufficient precision and

TABLE 8.5
Components of Biological Integrity

Biospheric Elements	*Ecosystem Processes*
Genetics	Mutation, recombination
Individual	Metabolism, growth, reproduction
Population/species	Age-specific birth and death rates Evolution/speciation
Asssemblage (community and ecosystem)	Interspecies interactions Energy flow
Landscape	Water cycle Nutrient cycles Population sources and sinks Migration and dispersal

Source: U.S. Environmental Protection Agency.

FIGURE 8.19. A benthic invertebrate. (Source: U.S. Environmental Protection Agency.)

sampling efficiency to provide robust indicators of ecosystem health. For example, *benthic* aquatic invertebrates (Figure 8.19) living at the bottom of surface water systems can be very powerful bioindicators since they live in the water for all or most of their lives and remain only in areas suited to their survival (i.e., higher quality conditions). Benthic invertebrates are also relatively easy to collect and identify in the laboratory. They have limited mobility and differ in their ability to tolerate different kinds of pollution, so they are good "sentries" of biological integrity. Since benthic invertebrates can live for more than one year and are limited in their mobility,

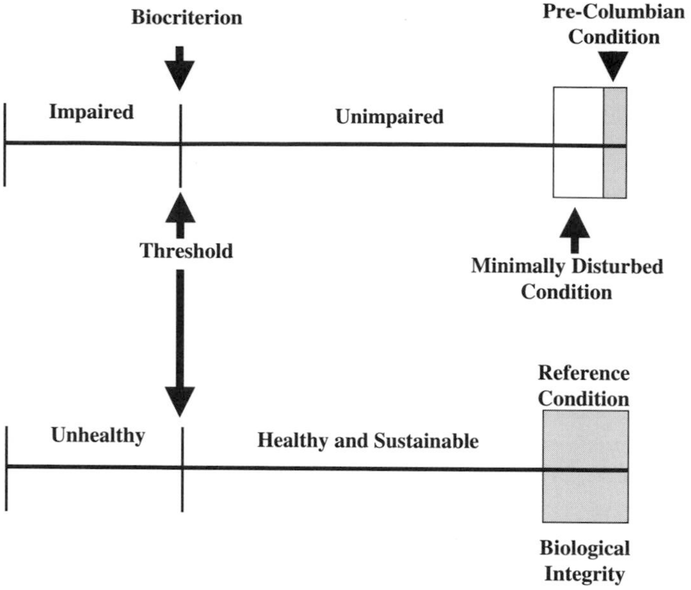

FIGURE 8.20. Need to have biocriteria that match actual ecosystem integrity. (U.S. Environmental Protection Agency, 2003, *Biological Indicators of Watershed Health*, http://www.epa.gov/bioindicators/html/about.html.)

they can be ideal "integrators" of surface water conditions. These and other "sentry" organisms, analogous to the "canary in the coal mine," integrate or "index" environmental quality. When the correct diversity, productivity, and abundance of representative organisms are present, the bio-indicators are telling us that the system is healthy.

Biological effects at the cellular level range from acute cellular toxicity to changes in the cellular ribonucleic and deoxyribonucleic acid structures, leading to cellular (and tissue) mutations, including cancer. The cells are also homes to chemical signaling processes, such as those in the stimulus-response systems in microbes and plants, as well as the endocrine, immune, and neural systems in animals.

Notes and Commentary

1. The principal source for the cell discussion is the National Institutes of Health, National Institute of General Medical Science.
2. The source of this and the following discussions on contaminant toxicity mechanisms is Z. Gregus and C. Klaasen, 1996, "Mechanisms of Toxicity," in *Casarett and Doull's Toxicology: The Basic Sciences of Poisons*, 5th Edition,

edited by C. Klaasen, McGraw-Hill, New York, N.Y. The whole edition is an excellent source of information on most aspects of toxicology.

3. For example, see J. Greger, 1998, "Dietary Standards for Manganese: Overlap between Nutritional and Toxicological Studies," *Journal of Nutrition*, Vol. 128, no. 2, pp. 368S–371S.
4. The general source for the distribution and toxicokinetic modeling discussion is the National Library of Medicine's Toxicokinetics Tutor program.
5. The following discussions on pharmacological subjects are based on discussions in S. Zakrewski, 1991, *Principles of Environmental Toxicology*, American Chemical Society, Washington, D.C. This is an excellent introduction to toxicology as it applies to public health and environmental assessments.
6. There is another important exposure pathway, i.e. nasal. Recent discussions among scientists and engineers within Duke University's Pratt School of Engineering and Medical School seem to indicate an almost direct link between nasal, olfactory, and brain absorption of some volatile contaminants (e.g. nitric oxide, NO).
7. This engineering technical note was prepared by Erik Rosenfeldt, M.S. and Dr. Karl Linden, of Duke University's Department of Civil and Environmental Engineering. The Duke University ultraviolet (UV) research laboratory studies pathogen disinfection and UV photochemical treatment of drinking water, through direct UV and advanced oxidation processes. For more information, contact Mr. Rosenfeldt at ejr3@duke.edu or Dr. Linden at kglinden@duke.edu, or visit the website for the International Ultraviolet Association at http://www.iuva.org. The figures in the technical note, except Figure 8.7 were developed by Rosenfeldt and Linden, Figure 8.7 was drawn by the writer.
8. C. Sonnenschein and A.M. Soto, 1998, "An Updated Review of Environmental Estrogen and Androgen Mimics and Antagonists," *Journal of Steroid Biochemistry and Molecular Biology*, 65, (1–6), pp. 143–150.
9. U.S. Environmental Protection Agency, 2001, "Removal of Endocrine Disruptor Chemicals Using Drinking Water Treatment Processes," EPA/625/R-00/015, Washington, D.C.
10. D. Fry and C. Toone, 1981, "DDT–Induced Feminization of Gull Embryos," *Science* 213, 922–924.
11. S.N. Weimeyer, T.G. Lamont, C.M. Burck, C.R. Sindelar, F.J. Gramlich, J.D. Fraser, and M.A. Byrd, 1984, "Organochlorine, Pesticide, Polychlorobiphenyl, and Mercury Residues in Bald Eagle Eggs—1969–79—and Their Relationships to Shell Thinning and Reproduction," *Archives of Environmental Contamination and Toxicology*, Vol. 13, no. 5, p. 529–549.
12. L.J. Guillette, T.S. Gross, G.R. Masson, J.M. Matter, H.F. Percival, and A.R. Woodward, 1994, "Developmental Abnormalities of the Gonad and Abnormal Sex-Hormone Concentrations in Juvenile Alligators from Contaminated and Control Lakes in Florida," *Environmental Health Perspectives*, Vol. 102, no. 8, p. 680–688.

13. See, for example, C.E. Purdom, P.A. Hardiman, V.J. Bye, N.C. Eno, C.R. Tyler, and J.P. Sumpter, 1994, "Estrogenic Effects from Sewage Treatment Works," *Chemistry and Ecology*, 8, pp. 275–285; and S. Jobling, D. Sheahan, J. A. Osborne, P. Matthiessen, and J.P. Sumpter, 1996, "Inhibition of Testicular Growth in Rainbow Trout (Oncorhynchus Mikiss) Exposed to Estrogenic Alkylphenolic Chemicals," *Environmental Toxicology and Chemistry*, 15, (2), pp. 194–202.
14. G.A. Fox, 2001, "Effects of Endocrine Disrupting Chemicals on Wildlife in Canada: Past, Present, and Future," *Water Quality Research Journal of Canada*, 36, (2), pp. 233–251.
15. See E.K. Sheiner, E. Sheiner, R.D. Hammel, G. Potashuit, and R. Carel, 2003, "Effect of Occupational Exposures on Male Fertility: Literature Review," *Industrial Health*, 41, (2), pp. 55–62; P.S. Guzelian, 1982, "Comparative Toxicology of Chlordecone (Kepone) in Humans and Experimental Animals," *Annual Reviews of Pharmacology and Toxicology*, Vol. 22, 89–113; and T.B. Hayes, A. Collins, M. Lee, M. Mendoza, N. Noriega, A.A. Strart, and A. Vonk, 2002, "Hermaphroditic, Demasculinized Frogs after Exposure to the Herbicide Atrazine at Low Ecologically Relevant Doses," *Proceedings of the National Academy of Sciences of the United States of America*, 99, (8), pp. 5476–5480.
16. D.W. Koplin, E.T. Furlong, M.T. Meyer, E.M. Thurman, S.D. Zangg, L.B. Barber, and H.T. Buxton, 2002, "Pharmaceuticals, Hormones, and Other Organic Wastewater Contaminants in U.S. Streams, 1999–2000: A National Reconnaissance," *Environmental Science and Technology*, 36, (11), pp. 1202–1211.
17. See T.W. Schultz, J.R. Seward, and G.D. Sinks, 2000, "Estrogenicity of Benzophenones Evaluated with a Recombinant Yeast Assay: Comparison of Experimental and Rules-Based Predicted Activity," *Environmental Toxicology and Chemistry*, Vol. 19, no. 2, pp. 301–304; T.W. Schultz, G.D. Sinks, and M.T.D. Cronin, 2002, "Structure-Activity Relationships for Gene Activation Oestrogenecity: Evaluation of a Diverse Set of Aromatic Chemicals," *Environmental Toxicology*, Vol. 17, no. 1, pp. 14–23; Y. Tabira, N. Makoto, A. Daisuke, Y. Yakate, Y. Jahara, J. Shinmyozu, M. Noguchi, M. Takatsuki, and Y. Shimohigashi, 1999, "Structural Requirements of Para-alkylphenols to Bind to Estrogen Receptor," *European Journal of Biochemistry*, Vol. 262, no. 1, pp. 240–245; and C.L. Waller, T.I. Oprea, K. Chae, H.K. Park, K.S. Korach, S.C. Laws, T.E. Wiese, W.R. Kelce, and L.E. Gray, 1996, "Ligand-Based Identification of Environmental Estrogens," *Chemical Research in Toxicology*, Vol. 9, no. 8, pp. 1240–1248.
18. H.E. Bevans, S.L. Goodbred, J.F. Miesher, S.A. Watkins, T.S. Gross, N.D. Denslow, and T. Choeb, 1996, U.S. Geological Surry, Synthetic Organic compounds and carp endocrinology and histology, Las Vegas, Wash and Las Vegas and Callville bays of Lake Mead Nevada, 1992 and 1995. Water-Resources Investigations Report 96–4266.
19. J.Y. Hu, G.H. Xie, and T. Aizawa, 2002, "Products of Aqueous Chlorination of 4-nonylphenol and Their Estrogenic Activity," *Environmental Toxicology and Chemistry*, Vol. 21, no. 10, pp. 2034–2039.

20. R. G. Zepp, 1978, "Quantum Yields for Reaction of Pollutants in Dilute Aqueous Solution,"*Environmental Science and Technology*, 12, p. 327.
21. J.R. Bolton, 2001, *Glossary of Terms in Photocatalysis and Radiation Catalysis*, paper presented at International Union of Pure and Applied Chemistry, 41st General Assembly, Brisbane, Australia, July 2001)
22. J.R. Bolton, 2001, *Glossary of Terms in Photo Catalysis and Radiation Catalysis*, Poster presented at the International Union of Pure and Applied Chemistry, 41st General Assembly, Brisbane, Australia, July 2001.
23. J.L. Acero, K. Stemmler, and U. Von Gunten, 2000, "Degradation Kinetics of Atrazine and Its Degradation Products with Ozone and OH Radicals: A Pre-Dictive Tool for Drinking Water Treatment," *Environmental Science and Technology*, 34(4), 591–597; Huber et al., 2003, "Oxidation of Pharmaceuticals During Ozonation and Advanced Oxidation Processes," *Environmental Science and Technology*, 37(5), p. 1016; G.V. Buxton et al., 1988, "Critical Review of Data Constants for Reactions of Hydrated Electrons, Hydrogen Atoms, and Hydroxyl Radicals in Aqueous Solutions," *J. Phys. Chem. Ref. Data*, 17, 513–886; and, R. Larson and R. Zepp, (1988). "Reactivity of the Carbonate Radical with Aniline Derivatives," *Environmental Toxicology and Chemistry*, 7, pp. 265–274.
24. U.S. Environmental Protection Agency, 2000, "Fact Sheet: Dissolved Oxygen (Saltwater): Cape Cod to Cape Hatteras," EPA-822-F-99-009, Washington, D.C.
25. State of Georgia, 2003, *Watershed Protection Plan Development Guidebook*.
26. National Research Council, 1989, *Biologic Markers in Reproductive Toxicology*, National Academy Press, Washington, D.C.
27. See T. Bucheli and K. Fent, 1995, "Induction of Cytochrome P450 as a Biomarker for Environmental Contamination in Aquatic Ecosystems," *Critical Reviews in Environmental Science and Technology*, Vol. 25, pp. 201–268; and J. Stegeman and M. Hahn, 1994, "Biochemistry and Molecular Biology of Monooxygenases: Current Perspectives on Forms, Functions, and Regulation of Cytochrome P450 in Aquatic Species," in *Aquatic Toxicology: Molecular, Biochemical, and Cellular Perspectives*, edited by D. Malins and G. Ostrander, CRC Press, Boca Raton, Fla.
28. S. Reid, 2002, "State of the Science on Molds and Human Health, Centers for Disease Control and Prevention," Statement for the Record before the Subcommittees on Oversight and Investigations and Housing and Community Opportunity, U.S. House of Representatives, Washington, D.C.
29. J. Karr and D. Dudley, 1981, "Ecological Perspectives on Water Quality Goals," *Environmental Management*, Vol. 5, pp. 55–68.
30. Much has been written about how to compare the actual condition of an ecosystem or habitat to some standard condition. Few "natural" or even "unimpaired" sites in fact exist. Human activities have affected even the most remote corners of the earth. For example, PCBs, DDT, and other persistent pollutants have been found in polar regions as a result of long-range transport mechanisms, usually via the atmosphere.

    One means of comparing a site's ecological condition is to establish so-called "reference sites." Such sites would have to be measured in the same manner as that of the sites to which they will be compared. The U.S. federal government, for example, has established Long Term Ecological Monitoring (LTEM) sites precisely for this purpose. For example, the National Park Service routinely collects data from each of its LTEM sites, including mapping soils, geology, meteorology, and vegetation; measuring air and water quality; conducting inventories of vertebrates, vascular plants, including distributions and status; and, providing prototypes of various types of ecosystems (i.e., one prototype LTEM site for each of 10 biomes).

Part III

Contaminant Risk

CHAPTER 9

Contaminant Hazards

Environmental Toxicology

Toxicology is the study of poisons. Environmental science concerns itself with surrounding conditions and interrelationships that affect organisms. Thus, environmental toxicology is a specialty within specialties. It is the science of harmful agents that surround organisms. Often, it is most concerned with the human environment, so it addresses the contaminants that affect or may affect humans. However, much attention is also devoted to toxicology in ecosystems, known as *ecological toxicology,* or simply *ecotoxicology.*

The word *poison* is used infrequently in environmental matters. The reason may be that the term has been strongly associated with immediate and acute effects after exposure, usually ingestion but also inhalation and dermal contact, to relatively low doses of a substance. In lay terms, most people would consider arsenic to be a poison, but there would be less unanimity of opinion on whether nicotine is a poison. In fact, to most toxicologists, both can be "poisons." As the famous scientist Paracelsus said in the sixteenth century, "Dose alone makes a poison. . . . All substances are poisons, there is none which is not a poison. The right dose differentiates a poison and a remedy."[1]

Paracelsus' quote illuminates a number of toxicological concepts. Let us consider two. First, the poisonous nature, or the toxicology, of a substance must be related to the circumstances of exposure. What is the age of the person exposed? What is that person's existing health status? What is the chemical and physical form of the contaminant? Is the agent part of a mixture, or is it a pure substance? How was the person exposed: from food, drink, air, or through the skin? These and other characterizations of a contaminant must be known to determine the extent and degree of harm. The second concept highlighted by Paracelsus is that dose is related to response. This is what scientists refer to as a biological gradient, or a *dose-response* relationship: the more poison, the more harm.

The classification of harm is an expression of a contaminant's *hazard,* which is a component of risk. A hazard is expressed as the potential for unacceptable outcome. A hazard can be expressed in numerous ways (see Table 9.1). For chemical or biological agents, the most important hazard is the potential for disease or death (measured by epidemiologists as *morbidity* and *mortality*, respectively). Thus, the hazards to human health are referred to collectively in the medical and environmental sciences as *toxicity*. Toxicology is chiefly concerned with these health outcomes and their potential causes.

TABLE 9.1
Four Types of Hazards Important to Hazardous Wastes, as Defined by the Resource Conservation and Recovery Act (RCRA)

Hazard Type	*Criteria*	*Physical/Chemical Classes in Definition*
Corrosivity	A substance with an ability to destroy tissue by chemical reactions.	Acids, bases, and salts of strong acids and strong bases. The waste dissolves metals, other materials, or burns the skin. Examples include rust removers, waste acid, alkaline cleaning fluids, and waste battery fluids. Corrosive wastes have a pH of <2.0 or >12.5. The U.S. EPA waste code for corrosive wastes is "D002."
Ignitability	A substance that readily oxidizes by burning.	Any substance that spontaneously combusts at 54.3°C in air or at any temperature in water, or any strong oxidizer. Examples are paint and coating wastes, some degreasers, and other solvents. The U.S. EPA waste code for ignitable wastes is "D001."
Reactivity	A substance that can react, detonate, or decompose explosively at environmental temperatures and pressures.	A reaction usually requires a strong initiator (e.g., an explosive like TNT, i.e., trinitrotoluene), confined heat (e.g., saltpeter in gunpowder), i.e. explosive reactions with water (e.g., Na). A reactive waste is unstable and can rapidly or violently react with water or other substances. Examples include wastes from cyanide-based plating operations, bleaches, waste oxidizers, and waste explosives. The U.S. EPA waste code for reactive wastes is "D003."

TABLE 9.1 *(continued)*

Hazard Type	*Criteria*	*Physical/Chemical Classes in Definition*
Toxicity	A substance that causes harm to organisms. Acutely toxic substances elicit harm soon after exposure (e.g., highly toxic pesticides causing neurological damage within hours after exposure). Chronically toxic substances elicit harm after a long period of time of exposure (e.g., carcinogens, immunosuppressants, endocrine disruptors, and chronic neurotoxins).	Toxic chemicals include pesticides, heavy metals, and mobile or volatile compounds that migrate readily, as determined by the Toxicity Characteristic Leaching Procedure (TCLP), or a "TC waste." TC wastes are designated with waste codes "D004" through "D043."

Toxicity Example 1

Review the criteria in Table 9.2. Which criteria are more likely to be measures of chronic toxicity? Which are likely to be indicators of acute toxicity? How might variations in species' response come into play?

Solution and Discussion

Generally, if the criterion contains the word *lethal*, it is an indicator of an acute effect. Phytotoxicity is also an indicator of acute toxicity, since plants generally are stressed within a growing season. This is actually a "welfare" cost as opposed to an ecological or human health stress, since much of the interest in phytotoxicity is related to crops and standing timber.[2] However, there have been associated studies of crop damage and ecological risk assessment. The United States' Environmental Monitoring and Assessment Program, for example, includes "agroecosystems" as one of the seven ecological resources that it monitors. A recent example of crop damage as it relates to ecological and human health is research on tropospheric ozone. Agricultural researchers were interested in crop stress, but ecologists and human exposure researchers were interested in how these stresses relate to ecosystem condition and human health effects.

Bioconcentration is often a long-term process, so it may be an indicator of chronic exposure.

TABLE 9.2
Biologically-Based Classification Criteria for Hazardous Waste

Criterion	*Description*
Bioconcentration	The process by which living organisms concentrate a chemical to levels exceeding the surrounding environmental media (e.g., water, air, soil, or sediment).
Lethal Dose (LD)	A dose of a chemical calculated to expect a certain percentage of mortality in a population of an organism (e.g., minnow) exposed through a route other than respiration (dose units are mg [contaminant] kg^{-1} body weight). The most common metric from a bioassay is the lethal dose 50 (LD_{50}), wherein 50% of a population exposed to a chemical is killed.
Lethal Concentration (LC)	A calculated concentration of a chemical in the air that, when respired for four hours (i.e., exposure duration = 4 h) by a population of an organism (e.g., rat) will kill a certain percentage of that population. The most common metric from a bioassay is the lethal concentration 50 (LC_{50}), wherein 50% of a population exposed to a chemical is killed. (Air concentration units are mg [chemical] L^{-1} air.)
Phytotoxicity	The chemical's ability to elicit biochemical reactions that harm flora (plant life).

Source: P. Aarne Vesilind, J. Jeffrey Peirce, and Ruth F. Weiner, 1993, *Environmental Engineering*, 3rd Edition, Butterworth-Heinemann, Boston, Mass.

Toxicity Testing

There is a need to set criteria to determine if a waste exhibits any of the hazardous characteristics.[3] This includes the potential to contaminate groudwater. The "extraction procedure" (EP) was the original test developed by the U.S. EPA to establish whether a waste was hazardous by virtue of its toxicity and its likelihood to leach. Because the Resource Conservation and Recovery Act (RCRA) defines a hazardous waste as a waste that presents a threat to human health and the environment when the waste is "improperly managed," the government identified the set of assumptions that would allow for the means for a waste to be disposed if the waste is not subject to controls as mandated by Subtitle C of RCRA. This so-called "mismanagement scenario" was designed to simulate a "plausible worst case" of mismanagement. Under a worst-case scenario, a potentially hazardous waste is assumed to be disposed along with municipal solid waste in a landfill with actively decomposing substances overlying an aquifer. When the government developed the mismanagement scenario, it recognized that not all wastes would be managed in this manner, but that a dependable set of assumptions would be needed to ensure that the hazardous waste definition is implemented. Thus, the U.S. federal government took a conservative approach.

The conservative assumption of mismanagement drove the EP. This led to selecting drinking water that has leached from a landfill as the most likely pathway for human exposure. The EP defined the toxicity of a waste by measuring the potential for finding potentially toxic substances in the waste that have leached and migrated to contaminate groundwater and surface water (and ultimately sources of potable water).

The specific EP called for the analysis of a liquid waste or liquid waste extract to see if it contained unacceptably high concentrations of any of 14 toxic constituents identified in the National Interim Primary Drinking Water Standards,[4] because at the time that the EP was being developed these were the only official health-based federal standards available.

Following the worst-case scenario, the solid waste (following particle size reduction, if necessary) was extracted using organic acids (acids likely to be found in a landfill containing decomposing municipal wastes). To simulate the likely dilution and degradation of the toxic constituents as they would migrate from the landfill to a water source, the drinking water standards were multiplied by a "dilution and attenuation factor" (DAF) equal to 100, which the government considered to represent a substantial hazard.

The amendments to RCRA, known as the Hazardous and Solid Waste Amendments of 1984 (HSWA) redirected the government to broaden the toxicity characteristic (TC) and to reevaluate the EP, especially to see if the EP adequately addressed the mobility of toxic chemicals under highly variable environmental conditions. The Congress was specifically concerned that the leaching medium being used was not sufficiently "aggressive" to identify a wide range of hazardous wastes, but mainly focused on metals (particularly in their elemental form) and did not give enough attention to wastes that contain hazardous organic compounds. So in 1986, a new procedure was developed.

The Toxicity Characteristic Leaching Procedure (TCLP) was designed to provide replicable results for organic compounds and to yield the same type of results for inorganic substances as those from the original EP test. The government added 25 organic compounds to the test (see Table 9.3). These additions were based upon the availability of chronic toxicity reference levels. The U.S. EPA applied a subsurface fate and transport model to confirm whether the dilution and attenuation foctor (DAF) of 100 used by the EP test was still adequate. That is, any waste that does not dilute and attenuate 100-fold as the pollutants migrate to the groundwater fails the test: $DAF = C_L/C_{RW}$; where C_L is the concentration of the contaminant in

TABLE 9.3
Toxicity Characteristic Chemical Constituent Regulatory Levels for 39 Hazardous Chemicals

Contaminant	*Regulatory Level ($mg\ L^{-1}$)*	*EPA Identification Number*
Arsenic	5.0	D004
Barium	100.0	D005
Cadmium	1.0	D006
Chromium	5.0	D007
Lead	5.0	D008
Mercury	0.2	D009
Selenium	1.0	D010
Silver	5.0	D011
Endrin	0.02	D012
Lindane	0.4	D013
Methoxychlor	10.0	D014
Toxaphene	0.5	D015
2,4-D	10.0	D016
2,4,5 TP (Silvex)	1.0	D017
Benzene	0.5	D018
Carbon tetrachloride	0.5	D019
Chlordane	0.03	D020
Chlorobenzene	100.0	D021
Chloroform	6.0	D022
o-Cresol	200.0	D023
m-Cresol	200.0	D024
p-Cresol	200.0	D025
Cresol	200.0	D026
1,4-Dichlorobenzene	7.5	D027
1,2-Dichloroethane	0.5	D028
1,1-Dichloroethylene	0.7	D029
2,4-Dinitrotoluene	0.13	D030
Heptachlor (and its hydroxide)	0.008	D031
Hexachloroethane	3.0	D032
Hexachlorobutadiene	0.5	D033
Hexachloroethane	3.0	D034
Methyl ethyl ketone	200.0	D035
Nitrobenzene	2.0	D036
Pentachlorophenol	100.0	D037
Pyridine	5.0	D038
Tetrachloroethylene	0.7	D039
Trichloroethylene	0.5	D040
2,4,5-Trichlorophenol	400.0	D041
2,4,6-Trichlorophenol	2.0	D042
Vinyl chloride	0.2	D043

the leachate ($mg\,L^{-1}$) and C_{RW} is the concentration of the contaminant in the receiving groundwater ($mg\,L^{-1}$). So, if the leachate contains $10\,mg\,L^{-1}$ of chemical X and the receiving groundwater contains $0.11\,mg\,L^{-1}$ of chemical X, the waste fails the test because the DAF is slightly below 100. The TCLP begins with the same mismanagement assumptions as those that established the EP. The test procedure is the same as that of the EP, except that the TCLP allows the use of two extraction media. Which specific medium used in the test is dictated by the alkalinity of the solid waste. The liquid extracted from the waste is analyzed for the 39 toxic constituents listed in Table 9.3, and the concentration of each contaminant is compared to the TCLP standards specific to each contaminant.

Toxicity Example 2

Explain why the EP procedure was replaced by the TCLP. What are the major differences in the two tests?

Solution and Discussion

The extraction procedure (EP) dealt only with inorganic wastes. Obviously, organic wastes like pesticides, chlorinated benzenes, and solvents were a growing concern for people living around waste sites and facilities that generate, transport, and store hazardous substances. Thus, as a first step, a new list of representative organic compounds (most of which are chlorinated) was added to the eight heavy metals in what was called the Toxicity Characteristic Leaching Procedure (TCLP) list. Since organics generally behave differently in the environment than do inorganic substances, the leaching procedure (under a "mismanagement scenario") had to be applied differently (i.e., did the dilution and attenuation factor change for any or all of these newly listed substances?).

Hazardous Waste Characteristics

The concept of risk is expressed as the likelihood (statistical probability) that harm will occur when a receptor (e.g., human, a part of an ecosystem or even a commodity, e.g., a national monument)[5] is exposed to that hazard. Thus, an example of a toxic hazard is a *carcinogen* (a cancer-causing chemical), and an example of a toxic risk is the likelihood that a certain population will have an *incidence* of a particular type of cancer after being exposed to that carcinogen (e.g., the population risk that one person out of a million will develop lung cancer when exposed to a certain dose of carcinogen X for a certain period of time).

Other hazards besides *toxicity* are also important to hazardous waste engineering. The outcome may relate to environmental quality, such as an ecosystem stress, loss of important habitats, and decreases in the size of the population of sensitive species. Outcomes related to public and personal safety are also important, such as a substance's potential to ignite, its corrosiveness, its flammability, or its explosiveness. Finally, a substance may be a "public welfare hazard" that damages property values or physical materials, expressed for example as its corrosiveness or acidity. The so-called hazard may be inherent to the substance, but more than likely, the hazard depends on the situation and conditions where the exposure may occur. The substance is most hazardous when a number of conditions exist simultaneously; witness the hazard to firefighters using water in the presence of oxidizers.

The challenge to the environmental professional is how to remove or modify the characteristics of a substance that render it hazardous, or to relocate the substance to a situation where it has value. An example of the former would be the dehalogenation of chlorinated benzenes to transform them into compounds that can be used as solvents in manufacturing or laboratories. An example of the latter is the so-called "adopt a chemical" programs in laboratories where solvents and reagents left over in one laboratory are made available to other laboratories.[6]

Risk assessment sounds like a very technical term. It can be, but risk assessment is really something that people do constantly. Human beings decide throughout each day whether the risk from particular behaviors is acceptable or whether the potential benefits of a behavior do not sufficiently outweigh the hazards associated with that behavior. Classic examples may include one's decision whether to drink coffee that contains the alkaloid, caffeine. The benefits include the morning "jump-start," but the potential hazards include induced cardiovascular changes in the short term, and possible longer-term hazards from chronic caffeine intake.

A particular type of hazard important to environmental protection is what is known as "hazardous waste." Hazardous waste is an important classification of contaminants because it has been specifically defined and targeted for action by government agencies. Section 1004(5) of the Resource Conservation and Recovery Act (RCRA)[7] defines a hazardous waste to be a solid waste that may "pose a substantial present or potential threat to human health and the environment when improperly treated, stored, transported, or otherwise managed." RCRA made the U.S. EPA responsible for defining which specific solid wastes would be considered hazardous waste either by identifying the characteristics of a hazardous waste or by "listing" particular hazardous wastes. Thus, a solid waste is "hazardous" if:[8]

1. The waste is officially "listed" as a hazardous waste on one of the four U.S. EPA groupings. (Note: The engineer should check frequently whether any of the wastes of concern have been listed, since the lists are updated periodically by the federal government as new data and research are published.):

a. *F List*—Chemicals that are generated via non\specific sources by chemical manufacturing plants to produce a large segment of chemicals. A solvent must comprise at least 10% of the waste prior to use.
b. *K List*—Wastes from 17 specific industries that use specific chemical processes (e.g., veterinarian drug or wood preservative manufacturing). The processes included on the K List are very specifically defined by regulation, so the engineer involved in work related to chemical manufacturing processes is well advised to investigate all past, present, and possible processes to determine whether they fall onto this list.
c. *P List*—Acutely hazardous, technical grade (i.e., approximately 100% composition and sole active ingredient) chemicals discarded by commercial operations.
d. *U List*—Toxic, but not acutely hazardous, technical grade chemicals discarded by commercial operations that are also classified as corrosive, ignitable, reactive or toxic (see Table 9.1).

2. Based upon testing, the waste is found to be corrosive, ignitable, reactive or toxic (see Table 9.1).
3. The generator of the waste reports and declares that the waste is "hazardous" based upon its proprietary information or other knowledge about the waste. (Note: It is always good ethics and good business practice to exercise full disclosure in matters related to potential hazards, including those for chemicals that are not listed per se by the enforcement agencies. It is also sound legal practice, since it would be quite embarrassing and potentially damaging to an engineer's career if information was available to the company documenting a hazard, but this was not disclosed until legal proceedings.)

Mixtures of any listed hazardous waste with other wastes will require that the engineer manage all of the mixture as a listed hazardous waste. Spills of listed wastes that impact soils and other unconsolidated material are also regulated as the listed hazardous waste. If a listed hazardous waste is spilled, the environmental professional must immediately notify the appropriate state agency or the U.S. EPA to determine how best to manage the impacted material that contains the listed waste. The so-called "characteristic waste" may not appear on one of the EPA lists; it is considered hazardous if it exhibits one or more of the following of the characteristics described in Table 9.1.

Other classifications have been applied to hazardous wastes. For example, biologically-based criteria have been used to characterize the hazard and ability of chemicals to reach and affect organisms (including those in Table 9.2).

Hazards are frequently concerned with human health hazards and risks, however, hazards may also pose threats to other "receptors," especially those associated with ecosystems. Identification of ecological hazards is part of any ecological risk assessment.

Hazard Assessment Example 1

If a waste has a pH of 1.9, what type of hazard exists?

Answer and Discussion

The federal government has explicitly defined four types of environmental hazards: corrosivity, reactivity, ignitability, and toxicity (see Table 9.1 for the actual standards). Other federal, state, and local agencies define various other types of hazards, but not all are directly related to environmental hazards, such as those to ensure safety (e.g., fire hazards) and to promote hygiene (e.g., biological hazards).

Corrosivity is the hazard of destroying tissues, so corrosives lie at the extreme end of the pH scale. Thus, any chemical with a pH < 2.0 is considered by the U.S. Environmental Protection Agency to be corrosive. It is also the hazard for chemicals with pH > 12.5. These are extremely acidic and extremely basic, respectively. This is a classification defined by the federal government. However, from a scientific perspective, a substance may be deemed "corrosive" by many kinds of chemical reactions, not simply acid and base reactions. However, in the strict confines of hazardous waste management, pH is the metric used to determine corrosivity.

An interesting follow-up question is what is the target for the hazard? Are we mainly concerned about corrosiveness because the wastes destroy materials (known as a "welfare" hazard) contributing to engineering failures (i.e., corrosion fatigue), or are they an indication of the hazard to humans handling or possibly exposed to the substance (e.g., first responders)?

Hazard Assessment Example 2

When does the fire code in your state require that you construct a separate, detached building on your site to store "water reactive" substances and pyrophoric gases?

Solution and Discussion

Whether maintaining facilities during a site cleanup or for general operations at a manufacturing facility, environmental professionals should be aware of the hazards of all substances used. Yes, although it may seem counterintuitive, many hazardous substances are often used to clean up hazardous wastes![9]

In addition, you must understand the state laws pertaining to generation, storage, and transportation of hazardous substances wherever you practice. In this instance, storage is codified in fire codes. For example, in North Carolina, the storage requirements for hazardous substances are provided in the *North Carolina Fire Prevention Code*,[10] which categorizes substances into various hazardous classifications. The larger the class number, the more imminently dangerous the substances. Table 2703.8.2 describes the external containment requirements and Table 2703.1.1 gives the "maximum allowable quantities" of each substance classification.

According to the North Carolina fire code, Class 3 water reactive substances are those that react explosively with water, even without added heat or confinement. Class 2 water reactives will potentially react explosively when they come in contact with water. Class 1 water reactives do not react violently.

Class 3 water reactive substances stored in excess of 1 ton (908 kg) and Class 2 water reactive substances stored in excess of 25 tons (22,700 kg) must be stored in specially designed detached buildings. Class 1 water reactives do not need special external containment.

Pyrophoric gases are those that autoignite at air temperatures less than or equal to −11°C. The NC fire code requires a specially designed detached building to store in excess of 2000 cubic feet ($56.64\,m^8$) of these compounds.

Segregating these substances prevents the opportunity for reactions, often which can be violent and explosive. Thus, if a separate building is dedicated to them, only personnel with a need to use them will come into contact with them. This means that the likelihood of incidental contact is greatly reduced, especially incompatible activities, such as the very common use of water in a laboratory or in an industrial setting, or the possibility of breaking a piece of tubing with building renovations, such as the installation of electrical wiring or plumbing.

Bio-Effective Dose

Dose is the amount, often mass, of a contaminant administered to an organism ("applied dose"), the amount of the contaminant that enters the organism ("internal dose"), the amount of the contaminant that is absorbed by an organism over a certain time interval ("absorbed dose"), or the amount of the contaminants or its metabolites that reach a particular "target" organ

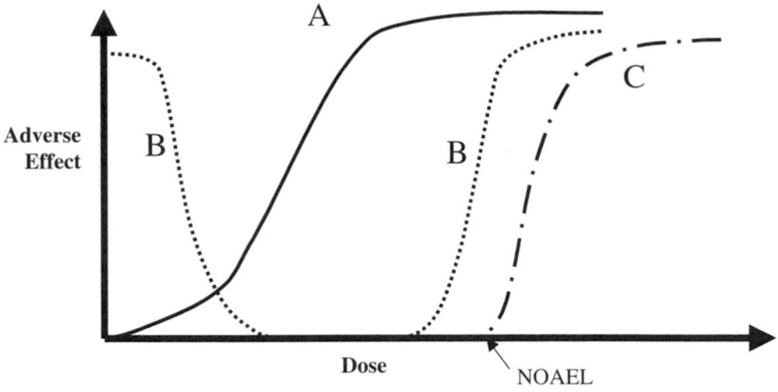

FIGURE 9.1. Three prototypical dose-response curves. Curve A represents the no-threshold curve, which expects a response (e.g., cancer) even if exposed to a single molecule (this is the most conservative curve). Curve B represents the essential nutrient dose-response relationship, and includes essential metals, such as trivalent chromium or selenium, where an organism is harmed at the low dose due to a "deficiency" (left side) and at the high dose due to "toxicity" (right side). Curve C represents toxicity above a certain threshold (noncancer). This threshold curve expects a dose at the low end where no disease is present. Just below this threshold is the "no observed adverse effect level" or NOEAL. Another threshold value is the No Observable Effects Concentration (NOEC), which is highest concentration of a toxic substance to which an organism is exposed for its whole life or part of its life (i.e., short-term exposure) where no effect on survival is detected ($NOEC_{survival}$) or no effect on growth and reproduction is detected ($NOEC_{growth}$). (Source: D. Vallero, 2003, *Engineering the Risks of Hazardous Wastes*, Butterworth-Heinemann, Elsevier Science, Burlington, Mass.)

("biologically effective dose" or "bio-effective dose"), such as the amount of a neurotoxin (a chemical that harms the nervous system) that reaches the nerve or other nervous system cells. Theoretically, the higher the concentration of a hazardous substance that an organism contacts, the greater the expected adverse outcome. The classic demonstration of this gradient is the so-called "dose-response" curve (Figure 9.1). If one increases the amount of the substance, a greater incidence of the adverse outcome would be expected.

The three curves in Figure 9.1 represent those generally found for toxic chemicals.[11] Curve A is the classic cancer dose-response. Regulatory agencies generally subscribe to the precautionary principle that any amount of exposure to a cancer-causing agent may result in an expression of cancer at the cellular level. Thus, the curve intercepts the *x*-axis at 0. Metals can be toxic at high levels, but several are essential to the development and metabolism of organisms. Thus, Curve B represents an essential chemical (i.e., a "nutrient") that will cause dysfunction at low levels (below the minimum intake needed for growth and metabolism) and toxicity at high levels. The segment of Curve B that runs along the x-axis is the "optimal range" of an essential substance. Curve C is the classic noncancer dose-response curve. The steepness of the three curves represents the potency or severity of the toxicity. For example, Curve C is steeper than Curve A, so the adverse outcome (disease) caused by the chemical in Curve C is more potent than that of the chemical in Curve A. This simply means that the response rate is higher. However, if the diseases in question are cancer (Curve A) and a relatively less important disease for Curve C, such as short-lived headaches, the steepness simply represents a higher incidence of the disease, not greater importance.

Another aspect of the dose-response curve is that with increasing potency, the range of response decreases. In other words, as shown in Figure 9.2, a severe response represented by a steep curve will be manifested in greater mortality or morbidity over a smaller range of dose. For example, an acutely toxic contaminant's dose that kills 50% of test animals (i.e., the LD_{50}) is closer to the dose that kills only 5% (LD_5) and the dose that kills 95% (LD_{95}) of the animals. The dose difference of a less acutely toxic contaminant will cover a broader range, with the differences between the LD_{50} and LD_5 and LD_{95} being more extended than that of the more acutely toxic substance.

The shape and slope of the curve are formed according to available data. There are a number of uncertainties associated with these data. Often, the dose-response relationship is based upon comparative biology from animal studies. These are usually high-dose, short duration (at least compared to a human lifetime) studies. From these animal data, models are constructed and applied to estimate the dose-response that may be expected in humans. Thus, the curve may be separated into two regions (see Figure 9.3). When environmental exposures do not fall within the range of observation, extrapolations must be made to establish a dose relationship. Generally, extrapolations are made from high to low doses, from animal to human responses, and from one route of exposure to another. The first step in establishing a dose-response is to assess the data from empirical observations. To complete the dose-response curve, extrapolations are made either by modeling or by employing a default procedure based upon information from the compound's chemical and biochemical characteristics.

Toxicokinetics and Toxicodynamics

Following uptake and entry into an organism, a contaminant will move and change. As is often the case, we use similar or identical terms in one area of science that have different meanings in others. So far we have encountered two definitions of the term "kinetics," one for physics and one for chemistry. Let us now add a third connotation that applies to toxicology.

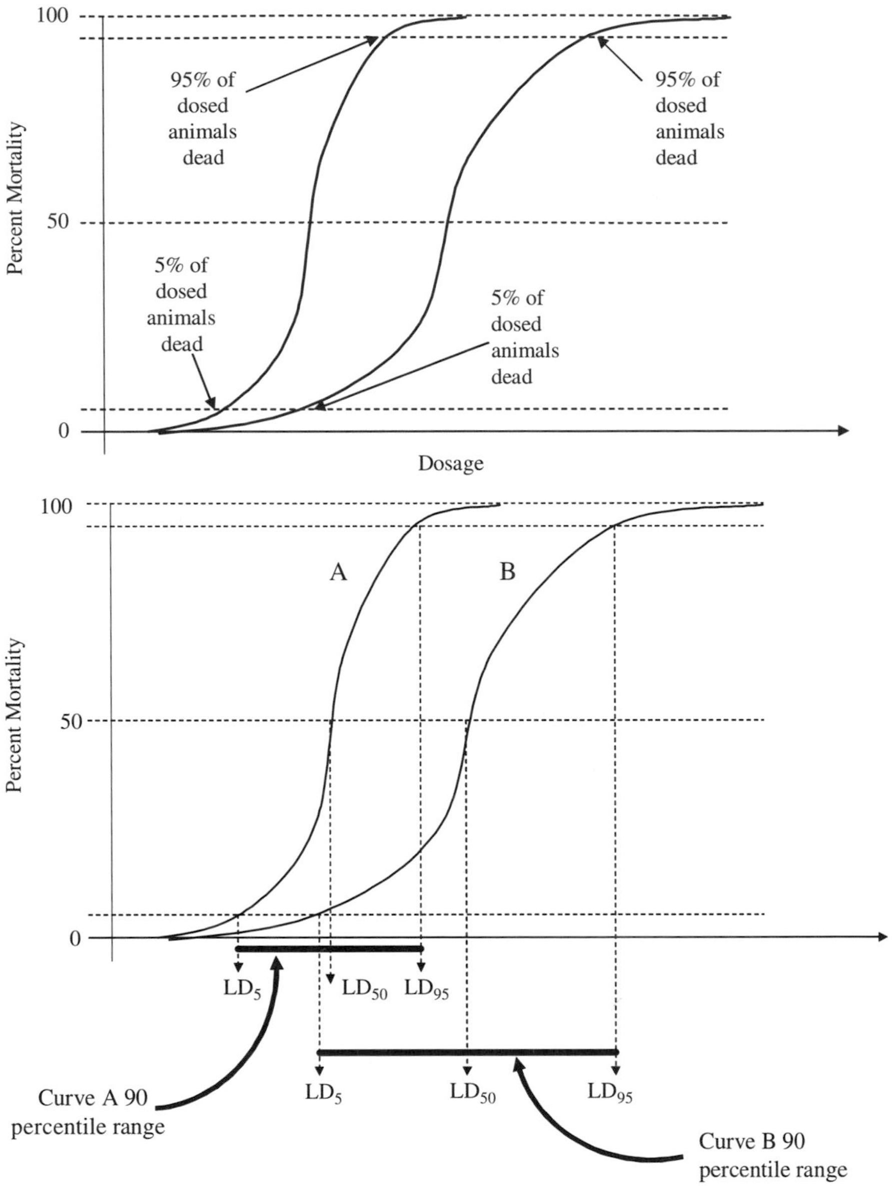

FIGURE 9.2. The greater the potency or severity of response (i.e., steepness) of the dose-response curve, the smaller the range of toxic response (90 percentile range shown in bottom graph). Thus, note that both curves have thresholds, and that curve B is less acutely toxic based upon all three reported lethal doses (LD_5, LD_{50}, and LD_{95}). In fact, the LD_5 for curve A is nearly the same as the LD_{50} for curve B, meaning that about the same dose of contaminant A kills nearly half the test animals, but that of contaminant B has only killed 5%. Thus, contaminant A is much more acutely toxic.

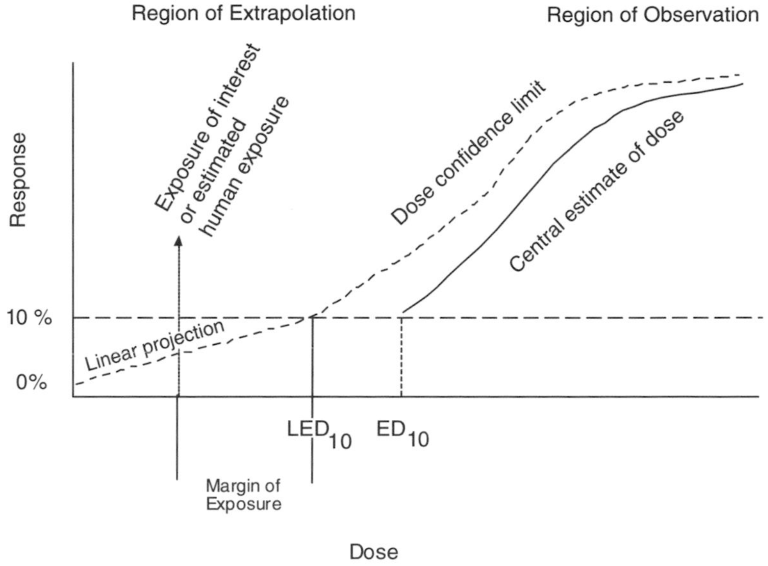

FIGURE 9.3. Dose-response curves showing the two major regions of data availability. LED_{10} = the lower 95% confidence limit on a dose associated with 10% extra risk; ED_{10} = the estimate of the dose that would lead to 10% increase in response. (Source: Based on discussions with the U.S. Environmental Protection Agency.)

In the *kinetic phase*, a contaminant may be absorbed, metabolized, stored temporarily, distributed, and excreted. In this sense, the organism is, in the terms of chemical engineering, a biological reactor or "bioreactor." The amount of contaminant that has been absorbed in the exact form of the original contaminant is known as the active parent compound. If the parent contaminant is metabolized, it will either become less toxic and excreted as a "detoxified metabolite," or it will be *activated* to become a more toxic "active metabolite." Models and studies that interpret or elucidate these processes are referred to as *toxicokinetic*. These are almost identical to the pharmacokinetic models that have been developed to ascertain the changes of drugs after they are administered to humans and animals.

Another term of physics, *dynamics*, is also important to toxicology. In the *dynamic phase* the contaminant or its metabolite undergoes interactions with a body's cells, causing a toxic response. The three major steps of *toxicodynamics* are the primary reaction, the biochemical effects, and the organism's response to these effects. Those compounds that are rapidly absorbed and slowly eliminated, such as 2,3,7,8-tetrachlorodibenzo-*p*-dioxin and the polychlorinated biphenyls (PCBs), are actually in some ways more amenable to toxicodynamic modeling because bioaccumulation, or body burden, is a reliable indicator of time-integrated exposure and absorbed dose. In other words, since the compound is so easily absorbed, its kinetic phase is straightforward (the absorbed amount is very nearly the same as the amount taken in), and its degradation is so slow that calculating how much is expected to be in an organism over time is simple. For example, let us consider two contaminants that are rapidly and completely absorbed. Contaminant A is completely eliminated within 10 hours in a human. However, only 10% of contaminant B is eliminated after iterations of metabolic reactions that take weeks, with the remaining 90% sequestered in fatty tissue. If the exposure is a single event, the person exposed to contaminant A will have a body burden of contaminant A of zero after 10 hours. By contrast, after 10 hours the person exposed to contaminant B would have a body burden of contaminant B that is nearly the same as the amount absorbed.

As discussed in Chapters 5 and 6, many environmental contaminants are hydrophobic and lipophilic. Such compounds are often particularly toxic as they build up in the organism, so the organism needs a mechanism to eliminate them. These are often enzyme-catalyzed metabolic mechanisms. There are two fundamental phases of metabolism: *phase 1 reactions* and *phase 2 reactions*. Recall from our earlier discussions in Chapter 4 that water is quite polar, and since "like dissolves like," any process that makes a molecule more polar makes it more soluble in aqueous media. Thus, to get a contaminant into bodily fluids that are predominantly water (e.g., in the cell or in systems, such as urinary and gastrointestinal tracts), the molecule must be transformed into a more polar metabolite. Thus, in phase 1 reactions, the lipophilic compounds become more water soluble by the attachment of polar groups, such as:

hydroxide	-OH
epoxide	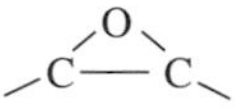
hydroxylamine	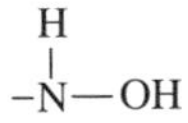
sulfhydryl	–SH

The majority of phase I reactions are catalyzed by the compound cytochrome P-450, in the endoplasmic reticulum of the cells. The liver is the site of most animals' cytochrome P-450. These particular phase 1 reactions are known as *microsomal mixed-function oxidase* reactions. The "microsomal" adjective means that that enzyme comes from an organelle; in this instance, the endoplasmic reticulum. Reactions where new polar groups are added include oxidation, reduction, hydrolysis, and hydration. However, not all phase 1 reactions result in additions, such as those reactions that remove halogens, like chlorine and fluorine. Removing halogens can be quite difficult since they lend so much stability to the compound. Consider, for example, the metabolism[12] of chlorobenzene (Figure 9.4). In all the processes, including those catalyzed by enzymes, such as oxidation, hydroxylation, and epoxidation, the chorine atoms remain intact.

After these catalyzed reactions and the additions of the polar groups, the contaminant is more easily eliminated by phase 2 metabolism. In this phase, the contaminant or its metabolites from a phase 1 reaction are bound with endogenous molecules (conjugation), further increasing the amount of water-soluble derivatives that can be eliminated in bile and urine.

The metabolism does not necessarily end with phase 2. For example, in a phase 3 reaction, the biotransformed[13] compounds generated in phase 2 may be metabolized, sometimes into toxic forms.

As mentioned, if the parent contaminant is metabolized, one of two things can happen. It may become less toxic, so the new compound is known as the detoxified metabolite, which is not only less toxic but also easier to excrete than the parent contaminant. Unfortunately, the second possible effect is that the compound is biologically *activated* into a more toxic form. A dramatic example of activation is the epoxidation of benzo(a)pyrene. In phase 1 reactions, oxidation is one of the processes that occurs in a cell to render a contaminant molecule more hydrophilic. It is believed that the ultimate carcinogen resulting from exposure benzo(a)pyrene is a form that is produced from the metabolism. In other words, the polycyclic aromatic hydrocarbon may not be the active parent contaminant in this case. The carcinogenic active metabolite is formed by three enzyme-catalyzed reactions (see Figure 9.5). First, benzo(a)pyrene is epoxidized to benzo(a)pyrene 7,8 epoxide via cytochrome P450 catalysis. Next, this compound is hydrolyzed via epoxide hydrolase to the diol structure, benzo(a)pyrene 7,8 dihydrodiol. Finally, another epoxide is formed, this time at the 9,10 position, to form the actively carcinogenic metabolite, (+(anti))benzo(a)pyrene 7,8 dihydrodiol 9,10 epoxide.

Dose-response models may be biologically-based with parameters calculated from curve-fitting of data. If data are sufficient to support a biologically-based model specific to a chemical, and significant resources are available, this is usually the model of choice. Biologically-based models require large amounts of data.

Case-specific models employ model parameters and information gathered from studies specific to a particular chemical. Often, however, neither the biologically based nor case-specific model is selected, because the necessary data or the significant costs cannot be justified.

Curve-fitting is another approach used to estimate dose-response relationships for chemicals. Such models are used when response data in the observed range are available. A so-called "point of departure for

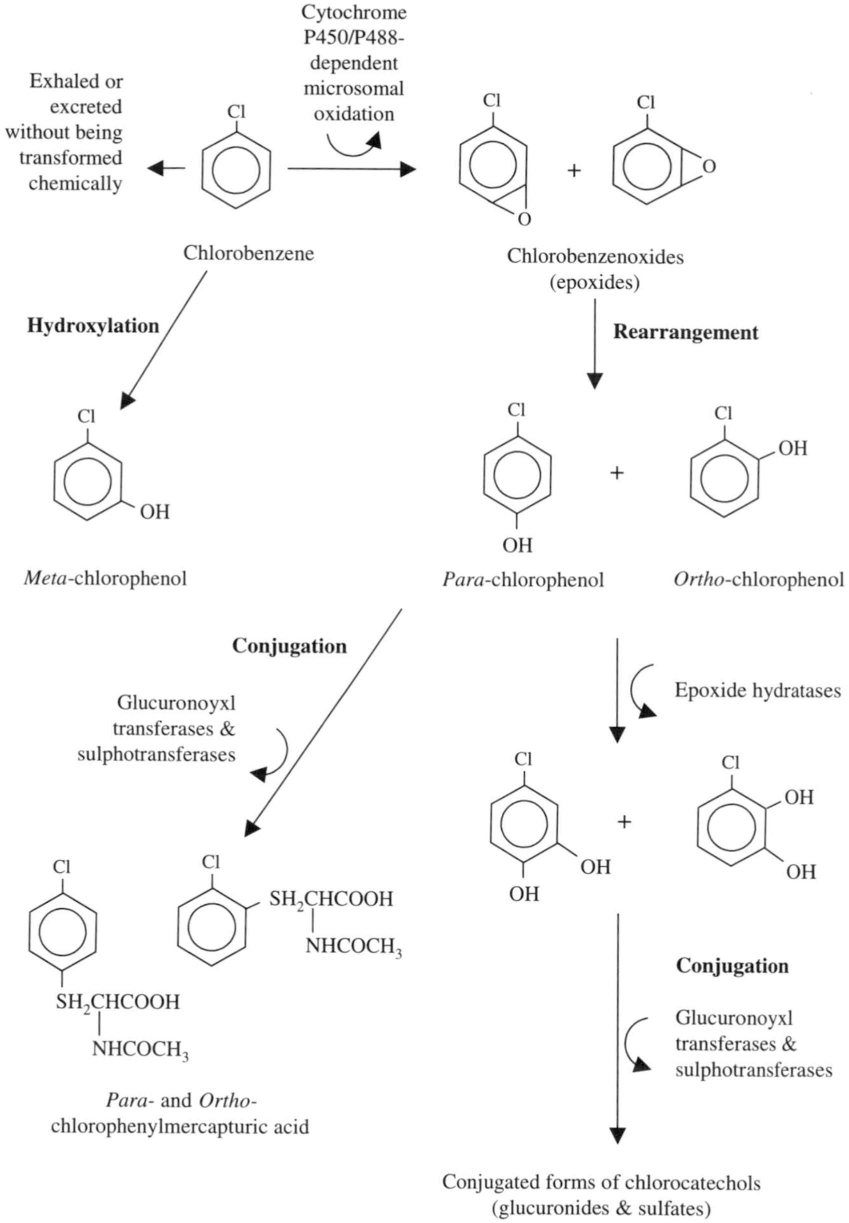

FIGURE 9.4. Proposed metabolic pathways for dichlorobenzene. (Adapted from: California Environmental Protection Agency, 2003, *Public Health Goals for Chemicals in Drinking Water: Chlorobenzene*, Sacramento, Calif.)

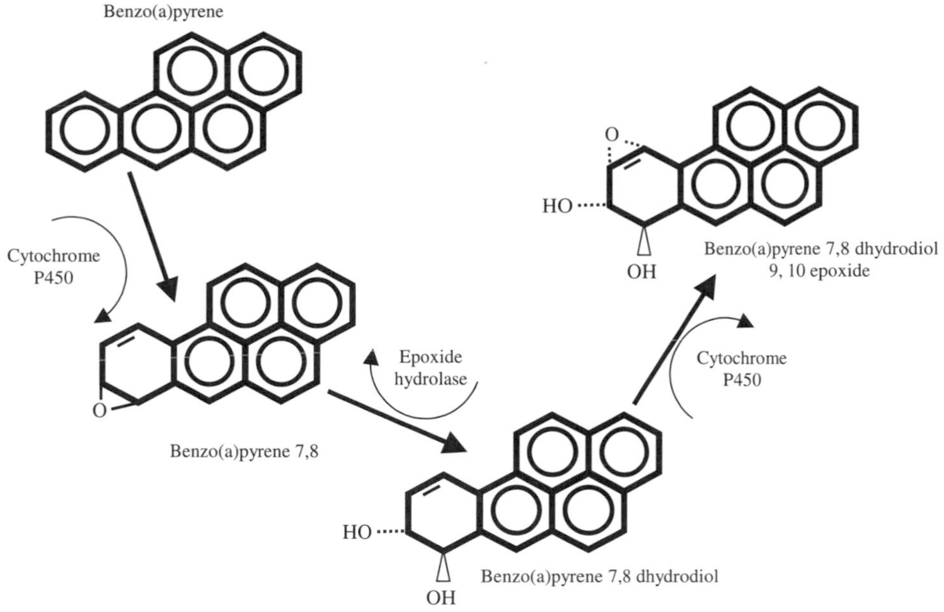

FIGURE 9.5. Biological activation of benzo(a)pyrene to form the carcinogenic active metabolite, benzo(a)pyrene 7,8 dihydrodiol 9, 10 epoxide.

extrapolation" is estimated from the curve. The point of departure is a point that is either a data point or an estimated point that can be considered to be in the range of observation, without the need for much extrapolation. The LED_{10} in Figure 9.3 is the lower 95% confidence limit on a dose associated with 10% extra risk. This is an example of such a point and, in fact, is often the standard point of departure. The central estimate in Figure 9.3 of the ED_{10} (the estimate of a 10% increased response), also may be used to describe a relative hazard and potency ranking.

Dose-Response Example 1

When can a dose-response curve that is steeper (i.e., induces a greater response) than that of another substance be of less concern than the substance with the less steep dose-response?

Solution and Discussion

The steepness of the dose-response curve is an indication of the potency or severity of the effect. However, the potency is only as

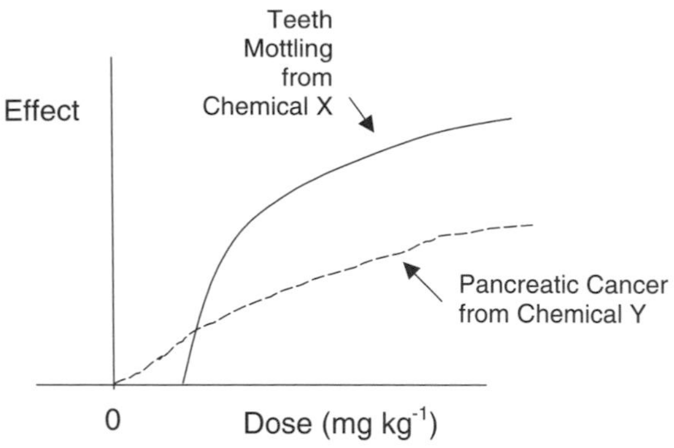

FIGURE 9.6. Hypothetical dose response curves for two different health endpoints from exposure to two different contaminants.

important as the effect. Thus, a chemical that has a very steep curve for mottling of teeth (e.g., a small amount of a fluoride compound in water leads readily to the effect) is of less concern from a public health perspective than a flatter curve for an organic solvent associated with pancreatic cancer (see the graph in Figure 9.6).

Note the immediate steepness of the chemical X curve compared to the more gentle slope for chemical Y. Also note the NOAEL for X, doses below which generate no effect. Cancer curves do not have NOAELs.

Dose-Response Example 2

What is the major difference between the dose-response curve's region of extrapolation versus the region of observation in Figure 9.7? How are models used in these two regions?

Solution and Discussion

Extrapolations may be mathematical or scientific. Mathematically, the extrapolation can be made from the region (or range) of observation to the region of extrapolation. For example, there may be points observed from epidemiological studies. The graph shows the result of four different studies of the ability of Chemical Z to produce tumors in rats.

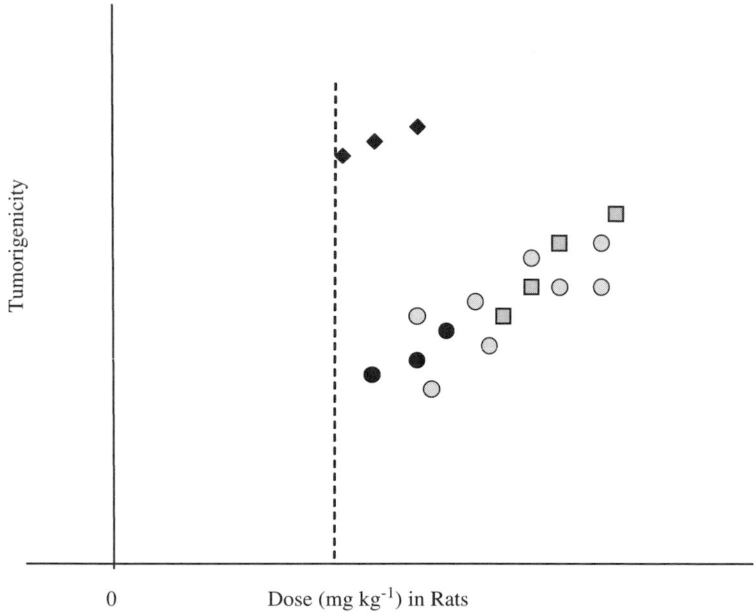

FIGURE 9.7. Studies from four hypothetical studies of tumors in rats. Each shape indicates a data point from a particular study.

Note that each of the studies has a gradient (dose-response), but that one study (the diamonds) shows a much greater tumor effect from Chemical Z. This could result from errors in study design, measurement, or other experimental flaws, or it could be that the study differs from the others (e.g., different means of dosing the animals, such as oral versus dermal; different organs tested; or the presence of a "promoter"[14] in the diet or elsewhere). Either way, the scientific extrapolation will have to determine why these studies differ. These points are interpolated to generate a curve in the region of observation, but no actual results from experiments are available below these dosages. A number of statistical methods can be used to extend the curve to the origin.

The graph also indicates a clear demarcation between the region of observation (right of the dashed line) and the region of extrapolation.

By the way, the type of data provides a means for discussing the importance of outliers in science. It is important to be able to distinguish when such data may and should be ignored, when outliers are important, and what they may tell us that other data cannot. It is also a good place to discuss scientific integrity and the appropriate use of scientific findings, especially the temptations and pitfalls of "trimming, cooking, and forging" data.[15] The bottom line is that if any data are not used, this should be clearly noted, along with the rationale (e.g., a statistical modeling technique) for their omission or selective use.

The scientific extrapolations take many forms. These include the so-called PBPK (physiologically based *pharmacokinetic*) models that extrapolate human effects from animal and other studies. There will also need to be extrapolations within human studies, including extrapolating data from adults to children (when only adults are studied), from men to women, and from occupational studies to general environmental exposures. In addition, extrapolations from one tissue to another (e.g., modeling liver response from kidney studies) and from a known route (e.g., ingestion) that has been well studied to an understudied route of exposure (e.g., inhalation of volatile contaminants when showering). Modelers must either develop algorithms for each of these extrapolations or establish defaults that are usually very conservative.

Risk is calculated by multiplying the slope of the dose-response curve by the actual contact with the substance, that is, exposure. If either term is zero, the risk is zero. The risk of even the most toxic substance is zero if there is no exposure. If there is an extremely toxic substance on the planet Jupiter, one's risk on Earth is zero. The risk will only increase if the substance finds its way to Earth or if we find our way to Jupiter. Similarly, a "nontoxic" substance, if there is such a substance, will never elicit a risk because the toxicity is zero. However, the reality of risk is always within these extremes. The environmental professional is challenged to reduce risks at both ends of the spectrum. For example, chemical engineers may apply *green chemistry* approaches to decrease the mobility or toxicity of a substance and by considering potential uses and eliminating, or at least limiting, the exposures to the substance before full-scale manufacturing.

Environmental Epidemiology

Epidemiology is the study of the distribution and determinants of states or events related to health in specific populations. It is an important resource for environmental and health risk assessment. Two key measures used to describe and analyze diseases in populations are *incidence*, the number of newly reported cases in a population during a year, and *prevalence*, the total number of cases in a population.

Epidemiology Example 1

A specific hormonal dysfunction has an incidence of 150 per million and a prevalence of 150 per million. Give at least two possible explanations for such a finding.

Answers and Discussion

The cure rate could be equal to the number of new cases each year, or the mortality rate could be 100% in every year studied, so that the numbers of new cases are the only ones that show up in the data each year. Let's hope it's the former, not the latter!

Epidemiology Example 2

Over a ten-year span, the incidence of a respiratory illness increased from 10 to 200 per million. Give two plausible explanations. What if incidence increases over 10 years, but prevalence is decreasing? Give at least one explanation.

Answer and Discussion

The twentyfold increase could be the result of an actual increase in the number of new cases, possibly from an increase in the concentration of a stressor in the environment leading to increased exposures. Another explanation could be improving detection capabilities. For example, the 200 is closer to the actual number of new cases, but physicians have become better at recognizing the symptoms associated with the disease, thus improving the nosological data. A third possibility is misdiagnosis and erroneous reporting of health statistics. For example, physicians may increase their diagnoses of a new syndrome that was previously diagnosed as something else. The syndrome incidence may not have increased, but it has become popular to so designate. These interpretations point to the need to understand the data underlying health reports.

Epidemiology Example 3

Describe four health study designs used in epidemiology. (Hint: Start with Ira B. Tager, "Current View of Epidemiologic Study Designs for Occupational and Environmental Lung Diseases," *Environmental Health Perspectives*, Vol. 108, Supp. 4, August 2000.)

Discussion

The Tager article discusses cohort and case-control studies. Investigators observe diseases and exposures over time. Two types of cohort studies are life table studies and longitudinal studies. Life table cohort studies follow traditional life table methodologies, observing the ratio of general exposures and person-times to the incidences of diseases. Longitudinal studies follow populations and strata within these populations over time to link various types of exposures and diseases to specific changes experienced by the population and subpopulations over a specified time. Longitudinal studies include time-series and panel studies, as well as "ecologic" (between group differences) studies. Time-series studies collect observations sequentially to observe changes in exposures and health outcomes (e.g., changes in asthma incidence with changes in particulate matter concentrations over time). Panel studies involve measuring subjects daily for symptoms and physiological functions and comparing these health metrics to possible exposures or ambient levels of contaminants. Longitudinal studies can be either prospective, where a group is identified and then followed for years after, or retrospective, where the group is identified and the investigators try to determine which risk factors and exposures appear to be associated with the group's present health status:

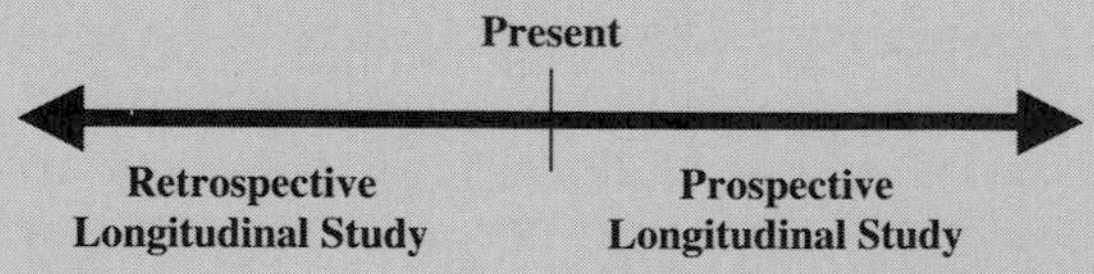

Case-control studies are usually clinical, where investigators identify two groups: people who already have the health outcome (cases), and people who do not have the outcome (controls). The two groups are studied to determine the extent to which an exposure was more prevalent in the past history in comparison to the other group. A "nested" case-control study is one that is part of a cohort study. The advantage of a nested case-control study over a regular case-control study is that the exposure measurements are obtained before the health outcome has occurred, so bias is reduced. See the Tager article on how case-control studies may be staged.

Another type of study is the so-called "cluster." A particular group of people in a tightly defined area may develop a disease at a rate much higher than that of the general population. A cluster study is actually a type of retrospective longitudinal study in that it identifies the group to be studied because the members share a particular health outcome and researchers must investigate the myriad of exposures and risk factors that could explain the outcome.

Epidemiology Example 4

What are some of the weaknesses in epidemiological data? What are some obstacles to interpreting human effects from animal data?

Answers and Discussion

No data set is perfect. One of the weaknesses of epidemiological data is the inability to control for confounders, the conditions or variables that are both a risk factor for disease and associated with an exposure of interest. An association between exposure and a confounder (a true risk factor for disease) will falsely indicate that the exposure is associated with disease. For example, if a person is exposed to Chemical X at home and develops lung cancer, one must be sure that the Chemical X is linked to the cancer, rather than a confounding condition, such as the fact that the person smoked two packs of cigarettes per day. Confounding factors may not even be known at the time the epidemiological study was designed so it was not controlled.

Similarly, not all populations respond in the same way to exposures. For example, exposure to ultraviolet (UV) light can produce more severe effects in persons with less skin pigmentation (e.g., melanin) than in persons with greater skin pigmentation. Much variability exists among subpopulations' susceptibility to particular diseases.

Another weakness has to do with the accuracy and representativeness of the data. For example, if physicians are inconsistent in disease taxonomy or in the ways that they report diseases, this will be reflected in the data. One physician may report pneumonia and another bronchitis, while a third may report acute asthma symptoms, all for an identical health episode. Spatial representation is difficult. For example, the address reported for a patient with a chronic disease may be near the health care facility where the patient has recently moved. However, the exposure or risk factors were encountered long ago and far away from the current address that is reported.

Contaminant Groupings

If a compound can exist in a number of forms, the toxicity of all these forms may be grouped together using the toxic equivalency factor (TEF) method. In other words, the engineer is more concerned about the contaminant in all of its forms, rather than each species. For example, the chlorinated dioxins have 75 different forms and there are 135 different chlorinated furans, simply by the number and arrangement of chlorine atoms on the molecules. The compounds can be separated into groups that have the same number of chlorine atoms attached to the furan or dioxin ring. Each form varies in its chemical, physical, and toxicological characteristics. The most

TABLE 9.4
Toxic Equivalency Factors for the Most Toxic Dioxin and Furan Congeners

Dioxin Congeners	*Toxic Equivalency Factor*
2,3,7,8-Tetrachlorodibenzo-*para*-dioxin	1.0
1,2,3,7,8-Pentachlorodibenzo-*para*-dioxin	0.5
1,2,3,4,7,8-Hexachlorodibenzo-*para*-dioxin	0.1
1,2,3,6,7,8-Hexachlorodibenzo-*para*-dioxin	0.1
1,2,3,7,8,9-Hexachlorodibenzo-*para*-dioxin	0.1
1,2,3,4,6,7,8-Heptachlorodibenzo-*para*-dioxin	0.01
Octachlorodibenzo-*para*-dioxin	0.001
Furan Congeners	
2,3,7,8-Tetrachlorodibenzofuran	0.1
1,2,3,7,8-Pentachlorodibenzofuran	0.5
2,3,4,7,8-Pentachlorodibenzofuran	0.05
1,2,3,4,7,8-Hexachlorodibenzofuran	0.1
1,2,3,6,7,8-Hexachlorodibenzofuran	0.1
1,2,3,7,8,9-Hexachlorodibenzofuran	0.1
2,3,4,6,7,8-Hexachlorodibenzofuran	0.1
1,2,3,4,6,7,8-Heptachlorodibenzofuran	0.01
1,2,3,4,7,8,9-Heptachlorodibenzofuran	0.01
Octachlorodibenzofuran	0.001

toxic form is the 2,3,7,8-tetrachlorodibenzo-p-dioxin (TCDD) isomer. Other isomers with the 2,3,7,8 configuration (but with additional chlorine atoms) are also considered to have higher toxicity than the dioxins and furans with different chlorine atom arrangements. It is best that the individual toxicity of each chemical in a mixture like hazardous waste is known, but this is usually quite costly and time-consuming or even impossible to determine. Thus, the TEF (Table 9.4) provides an aggregate means of estimating the risks associated with exposure to mixtures of the chlorinated dioxins and furans, as well as other highly toxic groups such as the PCBs and polycyclic aromatic hydrocarbons (PAHs). These standards can serve as benchmarks for hazardous waste cleanup target levels.

TEF Example

Using Table 9.4, how much octachlorodibenzofuran is needed to equal the toxicity of $10\,\mathrm{ng\,L^{-1}}$ tetrachlordibenzo-*para*-dioxin? Give reasons for the differences in toxicity of the compounds listed in the table.

Solution and Discussion

Since the TEQ is based upon the toxicity of the most toxic species (tetrachlordibenzo-*para*-dioxin, TCDD), which is set at 1, then all species can be compared according to their TEFs.

Octachlorodibenzofuran's TEF = 0.001, so it is one one-thousandth as toxic as TCDD. Thus it would take 1000 times the amount of TCDD to produce the same toxic effect, or: $1000 \times 10\,\mathrm{ng\,L^{-1}} = 10^4\,\mathrm{ng\,L^{-1}}$ or $10\,\mu\mathrm{g\,L^{-1}}$ octachlorodibenzofuran. That is a very high concentration of dioxin.

The structure, especially the planarity, of a compound affects its ability to bind with cellular receptors, which leads to the toxicity. Dioxin-like compounds are able to bind to the aryl hydrocarbon (Ah) hydroxylase receptor, reach the nucleus, and directly increase the transcription of genes responsive to the compounds. The compounds' binding strength to the Ah receptor correlates with their ability to induce DNA transcription. This is the foundation for TEFs. Thus, the ideal configuration is the planarity brought about by the dioxin's two oxygen bonds between the two benzene rings. The halogenation configuration in the "para" position adds to the binding affinity. However, additional chlorination above four or five Cl atoms on either the dioxin or furan configuration appears to inhibit Ah receptor binding (possibly due to steric hindrance because of the larger molecules' difficulty being absorbed through cellular membranes, i.e., they are size excluded).

Arguably, dioxins represent one of the most important groups of contaminants for which TEFs are calculated. The U.S. EPA classifies the toxicity of each individual isomer in the mixture by assigning each form a toxicity equivalency factor as it compares to the most toxic form of dioxin. The chlorinated dioxins and furans share certain similar chemical structures and biological characteristics. Scientists consider dioxins to cause toxic effects in similar ways. Dioxins usually exist as mixtures of congeners in the environment. Using TEFs, the toxicity of a mixture can be expressed in terms of its Toxicity Equivalents (TEQs), which represents the amount of TCDD that would be required to equal the combined toxic effect of all the dioxin-like compounds found in that mixture. As shown in the TEF Example, the concentration of each dioxin is multiplied by its respective TEF. The measurement and analysis of dioxins is complicated and difficult. There are a number of ways of expressing limitations in analysis and detection. If a result is reported as "nondetected," the U.S. EPA conservatively sets it to one-half of the detection level.[18] The products of the concentrations and their respective TEFs are then summed to arrive at a single TCDD TEQ value for the complex mixtures of dioxins in the sample. The TEF method is determined by first identifying all of the isomeric forms in the mixture, and multiplying the concentrations of each isomer by its corresponding TEF factor.[19] For dioxins, the products are summed to obtain the total 2,3,7,8-TCDD equivalents in the mixture. From these summed equivalents the human exposures and risks are calculated as total dioxin/furan exposures and total dioxin/furan risks (the same procedure is used for calculating total PCB or PAH exposures and risks).

For the past 30 years, a number of different sets of TEFs have been used for evaluating mixtures of dioxin compounds. No uniform set of TEFs presently exists, but two are currently in use: the International set and the World Health Organization (WHO) set. Both the International approach and the WHO approach include a total of 17 dioxin and furan compounds shown in Table 9.4 (see Chapter 7, "Chemical Reactions in the Environment," for descriptions of these compounds and how they form). The WHO approach for developing TEF values differs from the International approach for three compounds, two of which would not alone significantly change any TCDD TEQ value. However, for one compound, a pentachlorinated dioxin, the TEF for the WHO 1998 approach is twice as high (1 versus 0.5) as that of the International approach. In other words, in the International approach, the pentachlorinated dioxin is considered to have half the potency of TCDD, but WHO considers the two compounds to have equal potencies (TEF = 1). This is not simply a hypothetical problem, because if a waste has high amounts of a pentachlorinated dioxin, it is going to be considered much more hazardous using the WHO method. This very issue arose in reporting dioxin results during the World Trade Center environmental response, where it was decided that the International approach would be applied to dioxin findings.

Carcinogens

Cancer Classifications

Based on the scientific weight-of-evidence available for the hazardous chemical, the U.S. EPA classifies the substance's cancer-causing potential. Carcinogens fall into the following classifications (in descending order of strength of weight-of-evidence):

"A" Carcinogen—The chemical is a human carcinogen.
"B" Carcinogen—The chemical is a probable human carcinogen, with two subclasses:
- B1—Chemicals that have limited human data from epidemiological studies supporting their carcinogenicity;
- B2—Chemicals for which there is sufficient evidence from animal studies, but for which there is inadequate or no evidence from human epidemiological studies.

"C" Carcinogen—The chemical is a possible human carcinogen.
"D" Chemical—The chemical is not classifiable as to human carcinogenicity.
"E" Chemical—There is evidence that the chemical does not induce cancer in humans.

Cancer Example 1

Is an A Carcinogen more potent than a B Carcinogen? Which type do you believe will be more likely to meet with skepticism in a public meeting?

Answer and Discussion

The term "potency" applied to carcinogenesis (or any adverse effect, for that matter) is the gradient of the dose-response curve. In other words, potency is reflected by the steepness of a curve. Dose-response curves for suspected carcinogens have slopes with orders of magnitude differences in steepness. As we will see in the next sections on slope factors, the slope for dioxin is eight orders of magnitude steeper than the slope for chloromethane! The cancer classifications represent the state of knowledge of whether exposure of a contaminant induces cancer in humans. "A" carcionogeneity is strongly supported by human data (e.g., epidemiology and clinical studies). "B" carcinogenesis is supported by more limited human (B1) and/or animal (B2) data. However, the limited data of a B carcinogen may lead to a steeper dose-response curve than that of a strongly supported A carcinogen. That is, the data are stronger and less refutable for the A than for the B carcinogen, but the B may well be a more potent carcinogen (it is simply "understudied").

It is perilous to attempt to predict a public meeting. Some folks may be more concerned about a chemical that is associated with human data (A), but others may be impressed by the term "potency."

Cancer Example 2

What do mutagenicity and carcinogenicity have in common?

Discussion

The general commonality between a mutagen (a substance that can induce an alteration in the structure of deoxyribonucleic acid) and a carcinogen is that they both modify genetic material. That is one of the reasons that mutagenicity studies are included in the "weight of evidence" determinations in cancer classifications.

Cancer Example 3

Give an example of an E chemical.

Discussion

According to the EPA Cancer Classification System, an E Group Chemical is an agent with no associated increased incidence of neoplasms in at least two well-designed and well-conducted animal studies of adequate power and dose in different species. This is a proactive statement, that is, it is not that evidence has not been found to associate doses of the chemical with cancer, but that evidence that has been found shows that dosing with the chemical does not induce tumors or other neoplastic growth. Contrary to some opinions, relatively few chemicals have been linked to cancer. However, it does not mean that all of the E chemicals are safe, since they may be associated with other noncancer effects (e.g., mercury with neurotoxicity).

Some chemicals[16] in Group E are aldicarb, avermectin B1, bardac 22, bentazon, boron, boric acid, borax, bromoxynil, bromuconazole, bronopol, butylate, cadusafos, chlorpropham, chlorpyrifos, coumaphos, cyromazine, desmedipham, difenzoquat methyl sulfate, diflubenzuron, dinocap, diquat dibromide, disulfoton, dithiopyr, esfenvalerate, ethion, fenamiphos, fenbutatin oxide, fenitrothion, fenpropathrin, fenthion, fenvalerate, flumetsulam, flumiclorac pentyl, fluridone, flutolanil, fonofos, formetanate hydrochloride, glycophosphate, glyphosate trimesium, imazapyr, imidacloprid, maleic hydrazide, mepiquat chloride, methamidophos, methomyl, myclobutanil, naled, nicosulfuron, oxamyl, paraquat dichloride, phorate, phostebupirim, picloram diethanolamine salt, triisoproanolamine salt, potassium salt, triethylamine salt, profenofos, prohexadione, rimsulfuron, rotenone, sulfentrazone, sulfosate, tebufenozide, terbacil, terbufos, triasulfuron, and triflumizole.

The Slope Factor

Unlike the reference dose (discussed later in this chapter), which provides a "safe" level of exposure, cancer risk assessments generally assume there is no threshold. Thus, the NOAEL and LOAEL (lowest observable adverse effect level) are meaningless for cancer risk. Instead, cancer slope factors are used to calculate the estimated probability of increased cancer incidence over a person's lifetime (the so-called "excess lifetime cancer risk" or ELCR). Like the reference doses, slope factors follow exposure pathways.

Table 9.7 provides the toxicity values[17] for a number of pesticides that have been applied in the state of New York. Note that only the malathion and permethrin are designated as carcinogens, and that permethrin is two orders of magnitude more carcinogenic than malathion (i.e., the slope factor is $1.84 \times 10^{-2}\,\text{kg·d mg}^{-1}$ for permethrin versus $1.52 \times 10^{-4}\,\text{kg·d mg}^{-1}$ for malathion).

Slope Factors for Cancer

Slope factors (SFs) are expressed in inverse exposure units since the slope of the dose-response curve is an indication of risk per exposure. Thus, the units are the inverse of mass per mass per time, usually $(\text{mg kg}^{-1}\ \text{day}^{-1})^{-1} = \text{kg·d mg}^{-1}$. This means that the product of the cancer slope factor and exposure, i.e., risk, is unitless. This should make sense upon examination, because risk is a probability of adverse outcomes and, therefore, is simply a fraction or percentage. The SF is the toxicity value used to calculate cancer risks. SF values are contaminant-specific and route-specific. Thus, one must not only know the contaminant, but how a person is exposed (e.g., via inhalation, through the skin, or via ingestion). Inhalation and oral cancer slope factors are shown in Table 9.5. Note that the more potent the carcinogen, the larger the slope factor (i.e., the steeper the slope of the dose-response curve). For example, arsenic and benzo(a)pyrene are quite carcinogenic, with slope factors of 1.51 and 3.10, respectively. Their cancer potency is three orders of magnitude greater than aniline, bromoform, and chloromethane, for example.

The route of exposure can greatly influence the cancer slope. Note, for example, that the carcinogeniety of 1,2-dibromo-3-chloropropane is three orders of magnitude steeper via the oral route than from breathing vapors. Conversely, the cancer slope factor for chloroform is more than an order of magnitude greater from inhalation than from oral ingestion. Such information is important in deciding how to protect populations from exposure to contaminants. For example, if an industrial facility is releasing vinyl chloride, both inhalation and oral ingestion must be considered as possible routes of exposures for people living nearby. Both the inhalation and oral slope factors are high, i.e., 3.00×10^{-1} and $1.90\,\text{kg·d mg}^{-1}$, respectively. In addition, if the vinyl chloride finds its way to the water supply, not only the amount in food and drinking water must be considered, but also indirect inhalation routes, such as showering, since vinyl chloride is volatile and can be released and inhaled. The physical and chemical characteristics, such as vapor pressure and Henry's Law constants, of vinyl chloride coupled with its marked toxicity via multiple routes of exposure, make it a particularly onerous contaminant.

Table 9.5 also indicates that the structure of a compound greatly affects its biological activity. For example, comparing halogen substitutions indicates that the greater number of chlorine atoms on a molecule, the steeper the slope of the dose-response curve. Unsubstituted ethane is not carcinogenic (no slope factor). A single chlorine substitution in chloromethane renders the molecule carcinogenic, with a slope factor of 2.90×10^{-3}. Adding another chlorine atom to form 1,2-dichloroethane increases the slope to 9.10×10^{-2}. Completely halogenated ethane, i.e. hexachloroethane, has seen its cancer slope factor increase to 1.40×10^{-2}. Also, where the chlorine or bromine substitutions occur on the molecule will affect the cancer potential. For example, the isomers of tetracloroethane have different slope factors; 1,1,1,2-tetrachloroethane's slope factor is 1.40×10^{-2}, but 1,1,2,2-tetrachloroethane's slope factor is 2.03×10^{-1}. This seemingly small difference in molecular structure leads to an order of magnitude greater cancer potency.

Dermal exposures are generally extrapolated from the other two major routes. For example, the dermal slope factor for Aroclor 1254, the polychlorinated biphenyl (PCB) mixture (21% $C_{12}H_6Cl_4$, 48% $C_{12}H_5Cl_5$, 23% $C_{12}H_4Cl_6$,

TABLE 9.5
Cancer Slope Factors for Selected Environmental Contaminants[20]

Contaminant	*Inhalation Slope Factor* *(kg·dmg⁻¹)*	*Oral Slope Factor* *(kg·dmg⁻¹)*
Acephate	1.74×10^{-2}	8.70×10^{-3}
Acrylamide	4.55	4.50
Acrylonitrile	2.38×10^{-1}	5.40×10^{-1}
Aldrin	1.71×10^{1}	1.70×10^{1}
Aniline	5.70×10^{-3}	5.70×10^{-3}
Arsenic	1.51×10^{1}	1.50
Atrazine	4.44×10^{-1}	2.22×10^{-1}
Azobenzene	1.09×10^{-1}	1.10×10^{-1}
Benz(a)anthracene	3.10×10^{-1}	7.30×10^{-1}
Benzene	2.90×10^{-2}	2.90×10^{-2}
Benzo(a)pyrene	3.10	7.30
Benzo(b)fluoranthene	3.10×10^{-1}	7.30×10^{-1}
Benzo(k)fluoranthene	3.10×10^{-2}	7.30×10^{-2}
Benzotrichloride	1.63×10^{1}	1.30×10^{1}
Benzyl chloride	2.13×10^{-1}	1.70×10^{-1}
Beryllium	8.40	Not given
Bis(2-chloroethyl)ether	1.16	1.16
Bis(2-chloroisopropyl)ether	3.50×10^{-2}	1.10×10^{-2}
Bis(2-ethyl-hexyl)phthalate	1.40×10^{-2}	7.00×10^{-2}
Bromodichloromethane	6.20×10^{-2}	6.20×10^{-2}
Bromoform	3.85×10^{-3}	7.90×10^{-3}
Cadmium	Not given	6.30
Captan	7.00×10^{-3}	3.50×10^{-3}
Chlordane	3.50×10^{-1}	3.50×10^{-1}
Chlorodibromomethane	8.40×10^{-2}	8.40×10^{-2}
Chloroethane (Ethylchloride)	2.90×10^{-3}	2.90×10^{-3}
Chloroform	8.05×10^{-2}	6.10×10^{-3}
Chloromethane	3.50×10^{-3}	1.30×10^{-2}
Chromium(VI)	3.50×10^{-3}	Not given
Chrysene	3.10×10^{-3}	7.30×10^{-3}
DDD	2.40×10^{-1}	2.40×10^{-1}
DDE	3.40×10^{-1}	3.40×10^{-1}
DDT	3.40×10^{-1}	3.40×10^{-1}
Dibenz(a,h)anthracene	3.10	7.30
Dibromo-3-chloropropane,1,2-	2.42×10^{-3}	1.40
Dichlorobenzene,1,4-	2.20×10^{-2}	2.40×10^{-2}
Dichlorobenzidine,3,3-	4.50×10^{-1}	4.50×10^{-1}
Dichloroethane,1,2-	9.10×10^{-2}	9.10×10^{-2}
Dichloroethene (mixture),1,1-	1.75×10^{-1}	6.00×10^{-1}
Dichloromethane	7.50×10^{-3}	1.64×10^{-3}
Dichloropropane,1,2-	6.80×10^{-2}	6.80×10^{-2}
Dichloropropene,1,3-	1.30×10^{-1}	1.75×10^{-1}
Dieldrin	1.61×10^{1}	1.61×10^{1}
Dinitrotoluene, 2,4-	6.80×10^{-1}	6.80×10^{-1}

TABLE 9.5 *(continued)*

Contaminant	*Inhalation Slope Factor ($kg \cdot d\,mg^{-1}$)*	*Oral Slope Factor ($kg \cdot d\,mg^{-1}$)*
Dioxane, 1,4-	2.20×10^{-2}	1.11×10^{-2}
Diphenylhydrazine, 1,2-	7.70×10^{-1}	8.00×10^{-1}
Epichlorohydrin	4.20×10^{-3}	9.90×10^{-3}
Ethyl acrylate	6.00×10^{-2}	4.80×10^{-2}
Ethylene oxide	3.50×10^{-1}	1.02
Formaldehyde	4.55×10^{-2}	Not given
Heptachlor	4.55	4.50
Heptachlor epoxide	9.10	9.10
Hexachloro-1,3-butadiene	7.70×10^{-2}	7.80×10^{-2}
Hexachlorobenzene	1.61	1.60
Hexachlorocyclohexane, alpha	6.30	6.30
Hexachlorocyclohexane, beta	1.80	1.80
Hexachlorocyclohexane, gamma (lindane)	1.30	1.30
Hexachloroethane	1.40×10^{-2}	1.40×10^{-2}
Hexahydro-1,3,5-trinitro-1,3,5-traizine (RDX)	2.22×10^{-1}	1.11×10^{-1}
Indeno(1,2,3-cd)pyrene	3.10×10^{-1}	7.30×10^{-1}
Isophorone	9.50×10^{-4}	9.50×10^{-4}
Nitrosodi-*n*-propylamine, *n*-	7.00	7.00
Nitrosodiphenylamine, *n*-	4.90×10^{-3}	4.90×10^{-3}
Pentachloronitrobenzene	5.20×10^{-1}	2.60×10^{-1}
Pentachlorophenol	1.20×10^{-1}	1.20×10^{-1}
Phenylphenol, 2-	3.88×10^{-3}	1.94×10^{-3}
Polychlorinated biphenyls (Arochlor mixture)	3.50×10^{-1}	2.00
Tetrachlorodibenzo-*p*-dioxin, 2,3,7,8	1.16×10^{5}	1.50×10^{5}
Tetrachloroethane,1,1,1,2-	2.59×10^{-2}	2.60×10^{-2}
Tetrachloroethane,1,1,2,2-	2.03×10^{-1}	2.03×10^{-1}
Tetrachloroethene (PCE)	2.00×10^{-3}	
Tetrachloroethylene	2.03×10^{-3}	5.20×10^{-2}
Tetrachloromethane	5.25×10^{-2}	1.30×10^{-1}
Toxaphene	1.12	1.10
Trichloroethane,1,1,2-	5.60×10^{-2}	5.70×10^{-2}
Trichloroethene (TCE)	6.00×10^{-3}	1.10×10^{-2}
Trichlorophenol, 2,4,6-	1.10×10^{-2}	1.10×10^{-2}
Trichloropropane, 1,2,3-	8.75	7.00
Trifluralin	3.85×10^{-3}	7.70×10^{-3}
Trimethylphosphate	7.40×10^{-2}	3.70×10^{-2}
Trinitrotoluene, 2,4.6- (TNT)	6.00×10^{-2}	3.00×10^{-2}
Vinyl chloride	3.00×10^{-1}	1.90

Sources: U.S. Environmental Protection Agency, 2002, Integrated Risk Information System; U.S. EPA, 1994, Health Effects Summary Tables, 1994.

TABLE 9.6
Gastrointestinal Absorption Rates and Dermal Cancer Slope Factors for Selected Environmental Contaminants[22]

Contaminant	*GI Absorption*	*Dermal Slope Factor ($kg \cdot d\,mg^{-1}$)*
Acephate	0.5	1.74×10^{-2}
Acrylamide	0.5	9.00
Acrylonitrile	0.8	6.75×10^{-1}
Aldrin	1	1.72×10^{1}
Aniline	0.5	1.14×10^{-3}
Arsenic	0.95	1.58×10^{1}
Atrazine	0.5	4.44×10^{-1}
Azobenzene	0.5	2.20×10^{-1}
Benz(a)anthracene	0.5	1.46
Benzene	0.9	3.22×10^{-2}
Benzo(a)pyrene	0.5	1.46×10^{1}
Benzo(b)fluoranthene	0.5	1.46
Benzo(k)fluoranthene	0.5	1.46×10^{-1}
Benzotrichloride	0.8	1.63×10^{1}
Benzyl chloride	0.8	2.13×10^{-1}
Beryllium	0.006	Not given
Bis(2-chloroethyl)ether	0.98	1.13
Bis(2-chloroisopropyl)ether (DEHP)	0.8	8.75×10^{-2}
Bis(2-ethyl-hexyl)phthalate	0.5	2.80×10^{-2}
Bromodichloromethane	0.98	6.37×10^{-2}
Bromoform	0.75	1.05×10^{-2}
Cadmium	0.044	Not given
Captan	0.5	7.00×10^{-3}
Chlordane	0.8	4.38×10^{-1}
Chloroethane (Ethylchloride)	0.8	1.28
Chloroform	1	6.10×10^{-3}
Chloromethane	0.8	1.63×10^{-2}
Chromium(VI)	0.013	Not given
Chrysene	0.5	1.46×10^{-2}
DDD, 4,4-	0.8	3.00×10^{-1}
DDE, 4,4-	0.8	4.25×10^{-1}
DDT, 4,4-	0.8	4.25×10^{-1}
Dibenz(a,h)anthracene	0.5	1.46×10^{1}
Dibromo-3-chloropropane,1,2-	0.5	1.12×10^{-1}
Dichlorobenzene,1,4-	1	2.40×10^{-2}
Dichlorobenzidine,3,3-	0.5	9.00×10^{-1}
Dichloroethane,1,2- (EDC)	1	9.10×10^{-2}
Dichloroethene,1,1-	1	6.00×10^{-1}
Dichloropropane,1,2-	1	6.80×10^{-2}
Dichloropropene,1,3-	0.98	1.84×10^{-1}
Dieldrin	1	1.60×10^{1}
Dinitrotoluene, 2,4-	1	6.80×10^{-1}
Dioxane, 1,4-	0.5	2.20×10^{-2}

TABLE 9.6 *(continued)*

Contaminant	*GI Absorption*	*Dermal Slope Factor ($kg \cdot d\,mg^{-1}$)*
Diphenylhydrazine, 1,2-	0.5	1.60
Epichlorohydrin	0.8	1.24×10^{-2}
Ethyl acrylate	0.8	6.00×10^{-2}
Ethylene oxide	0.8	1.28
Formaldehyde	0.5	Not given
Heptachlor	0.8	5.63
Heptachlor epoxide	0.4	2.28×10^{1}
Hexachloro-1,3-butadiene	1	7.80×10^{-2}
Hexachlorobenzene	0.8	2.00
Hexachlorocyclohexane, alpha	0.974	6.47
Hexachlorocyclohexane, beta	0.907	1.99
Hexachlorocyclohexane, gamma (lindane)	0.994	1.31
Hexachloroethane	0.8	1.75×10^{-2}
Hexahydro-1,3,5-trinitro-1,3,5-traizine (RDX)	0.5	2.22×10^{-1}
Indeno(1,2,3-cd)pyrene	0.5	1.46
Isophorone	0.5	1.90×10^{-3}
Nitrosodi-*n*-propylamine, *n*-	0.475	1.47×10^{1}
Nitrosodiphenylamine, *n*-	0.5	9.80×10^{-3}
Pentachloronitrobenzene	0.5	5.20×10^{-1}
Pentachlorophenol	0.5	2.40×10^{-1}
Phenylphenol, 2-	0.5	3.88×10^{-3}
Polychlorinated biphenyls (Arochlor mixture)	0.85	2.35
Tetrachlorodibenzo-*p*-dioxin, 2,3,7,8	0.9	1.68×10^{5}
Tetrachloroethane,1,1,1,2-	0.8	3.25×10^{-2}
Tetrachloroethane,1,1,2,2-	0.7	2.86×10^{-1}
Tetrachloroethene (PCE)	1	5.20×10^{-2}
Tetrachloromethane	0.85	1.53×10^{-1}
Toxaphene	0.63	1.75
Trichloroethane,1,1,2-	0.81	7.04×10^{-2}
Trichloroethene (TCE)	0.945	1.16×10^{-2}
Trichlorophenol, 2,4,6-	0.8	2.20×10^{-2}
Trichloropropane, 1,2,3-	0.8	8.75
Trifluralin	0.2	3.85×10^{-3}
Trimethylphosphate	0.5	7.40×10^{-2}
Trinitrotoluene, 2,4,6-(TNT)	0.5	6.00×10^{-2}
Vinyl chloride	0.875	2.17

Sources: U.S. Environmental Protection Agency, 2002, *Integrated Risk Information System*; U.S. EPA, 1994, *Health Effects Summary Tables*, 1994.

TABLE 9.7
Toxicity Values for Six Pesticides Used in the Northeastern United States

Active ingredient	*Noncancer Hazards*									*Cancer Risk (the additional probability of contracting cancer over a lifetime)*	
	Acute (short-term exposure duration)			*Subchronic (intermediate exposure duration)*			*Chronic (long-term exposure duration)*				
	RfD skin ($mg\,kg^{-1}d^{-1}$)	*RfD Ingestion* ($mg\,kg^{-1}d^{-1}$)	*RfC Inhalation* ($mg\,L^{-1}$)	*RfD Skin* ($mg\,kg^{-1}d^{-1}$)	*RfD Ingestion* ($mg\,kg^{-1}d^{-1}$)	*RfC Inhalation* ($mg\,L^{-1}$)	*RfD Skin* ($mg\,kg^{-1}d^{-1}$)	*RfD Ingestion* ($mg\,kg^{-1}d^{-1}$)	*RfC Inhalation* ($mg\,L^{-1}$)	*Cancer Slope Factor (CSF)* ($kg\,d\,mg^{-1}$)	*Unit Risk Factor (UR), Dust Inhalation* ($(\mu g\,m^{-3})^{-1}$)
Malathion	0.5	0.50	0.0001	0.5	0.024	0.0001	0.024	0.024	0.0001	0.00152	0.000000434
Naled	0.01	0.01	0.0000022	0.01	0.01	0.0000023	0.002	0.002	0.0000022	NC	NC
Permethrin	1.5	0.26	0.0025	1.5	0.155	0.0025	0.05	0.05	0.00025	0.0184	0.00000626
Resmethrin	10	0.1	0.0001	10	0.1	0.0001	0.03	0.03	0.00001	NC	NC
Sumithrin	10	0.7	0.0029	10	0.7	0.0029	0.071	0.071	0.00029	NE	NE
Piperonyl Butoxide	10.0	2.0	0.00074	10	0.0175	0.00074	0.0175	0.0175	0.00007	NE	NE

Notes: RfC = Reference concentration.
RfD = Reference dose.
CSF = Cancer slope factor.
UR = Unit risk factor.
NC = No evidence of carcinogenicity.
NE = Limited evidence of carcinogenicity; no CFS established.
$mg\,kg^{-1}d^{-1}$ = mg pesticide active ingredient per kilogram human body weight per day.
$mg\,L^{-1}$ = mg pesticide active ingredient per liter of air per day.
$kg\,d\,mg^{-1} = (mg\,kg^{-1}d^{-1})^{-1}$ = risk per milligram of active ingredient per kilogram human body weight per day.
$\mu g\,m^{-3}$ = risk per microgram of active ingredient per cubic meter of air.
Source: U.S. Environmental Protection Agency.

and 6% $C_{12}H_3Cl_7$), for dermal exposure to soil or food, is 222 kg·d mg^{-1}. Keep in mind that this is the dose-response slope associated with handling or other skin contact with the contaminant, not the actual ingestion. The Aroclor 1254 dermal slope factor for exposure to water is 444 kg·d mg^{-1}. Both of these dermal slopes have been extrapolated from a gastrointestinal absorption factor of 0.9000.[21] All of the dermal slope factors shown in Table 9.6 have been extrapolated from other routes. The GI tract absorption rate is also given, since these are often used to extrapolate slope factors for dermal and other routes of exposure. Note that the larger the GI absorption decimal, the more completely the contaminant is absorbed. For complete absorption, the value equals 1.

The absorption factor is not only important for extrapolating slope factors, but it is a variable in calculating certain exposures. As we shall see later, the air (both particle and gas) and water exposure equations include an absorption factor. The dermal exposure equation does not include an absorption factor, but since dermal cancer slope factors are extrapolated from the inhalation or ingestion slopes, by extension, the absorption factor is part of the dermal risk calculations. Thus, all other factors being equal, a contaminant with a larger absorption factor will have a larger risk. This is evident when considering the pathway taken by a chemical after it enters an organism. As shown in Figure 9.8, the potential dose in a dermal exposure is what is available before coming into contact with the skin, but after this contact (i.e., the applied dose), it crosses the skin barrier and is absorbed. The absorption leads to the biological effectiveness of the contaminant when the chemical reaches the target organ, where it may elicit the effect (e.g., cancer). The absorption factor is the first determinant of the

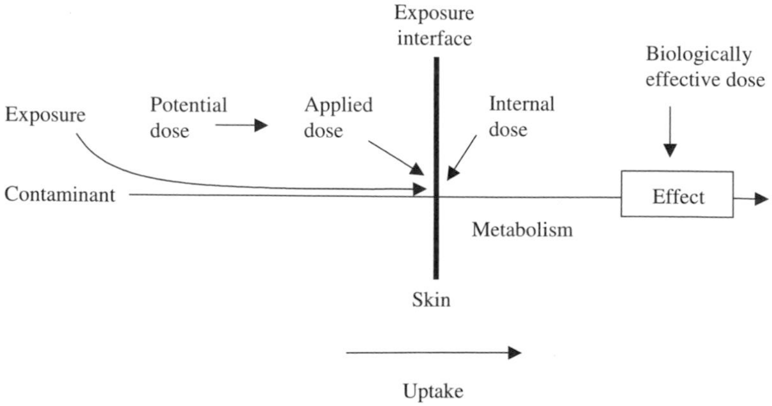

FIGURE 9.8. Pathway of a contaminant from ambient exposure through health effect. (Source: U.S. Environmental Protection Agency; and D. Vallero, 2003, *Engineering the Risks of Hazardous Wastes*, Butterworth-Heinemann, Boston, Mass.)

amount of the contaminant that reaches the target organ. For example, although the dermal slope factors for 1,4-dioxane and 1,4-dichlorobenzene are nearly the same (2.20×10^{-2} and 2.40×10^{-2}, respectively), all of the dichlorobenzene is expected to be absorbed (i.e., absorption = 1), while only half of the dioxane will be absorbed (i.e., absorption = 0.5). This means that if all other factors are equal, the risk from dichlorobenzene is twice that of dioxane.

It is worth noting the higher dermal slopes compared to inhalation slopes for some compounds. Note, for example, that the very lipophilic PCBs have a dermal slope that is an order of magnitude steeper than their inhalation slope. The absorption rate and, hence, the dermal slope is also affected by the contaminant's chemical structure. For example, trichloroethene (TCE) with its double bond between the carbon atoms has an absorption rate of 0.945, while 1,1,2-trichloroethane with only single bonds has a much lower absorption rate of 0.81, even though both have three chlorine substitutions.

When gathering information about a contaminant's toxicity, especially its cancer potential, it is important to understand the specifics of the toxic response. For example, the volatile organic compound methylene chloride, which more properly is known as dichloromethane, is not only carcinogenic, it is also a potent neurotoxin.[23] Since noncancer endpoints, such as neurotoxicity, unlike cancer endpoints have a threshold below which no effect is observed, dichloromethane has a safe level and a reference dose for that noncancer effect. Thus, since a carcinogen generally will not cause cancer everywhere in an organism, but has an affinity for certain tissues and organs, it is important to know which target organ was the basis for the slope factor. For example, the dichloromethane target organ is the liver, and the types of cancer are adenomas, carcinomas, and cancer nodules. However, dichloromethane's slope factor was based upon adenomas and carcinomas of the lung and liver. Finally, in addition to oral and inhalation routes, dichloromethane has a slope factor for the skin. Its dermal slope of 7.89×10^{-3} is very nearly the same as its inhalation slope. Interestingly, however, the dermal slope was calculated from the gastrointestinal absorption factor of 0.95.

Molecular stereochemical differences play a role in the way a contaminant interacts with a cell, and the efficiency of a contaminant's ability to "turn on" cellular responses via receptors on and in the cell. For example, *entantiomers* and *chiral* forms of the same compound can lead to dramatic differences in toxicity. Chirals are mirror-image isomers, where the arrangement dictates, for example, whether a compound is an effective pharmaceutical or a highly toxic compound. For this reason, pesticide and pharmaceutical companies must ensure that only the nontoxic, effective mirror image form is found in certain products. These steric differences, at least in part, account for the different slopes of the alpha, beta, and gamma isomers of hexachlorocyclohexane (HCH). There are eight geometric isomers of HCH. The isomers differ in the axial and equatorial positions of the chlorine atoms. One of the isomers, α-HCH, exists in two enantiomeric forms. HCH is commercially produced by photochemical chlorination of benzene. The product, technical-grade HCH, consists principally of five isomers: α–HCH (60–70%), β–HCH (5–12%), γ–HCH (10–15%), δ–HCH (6–10%), and ε–HCH (3–4%).[24] This mixture is marketed as an inexpensive insecticide, but since γ-HCH is the only isomer that exhibits strong insecticidal properties, it has been common to refine it from the technical HCH and market it under the name "lindane." However, all commercially produced lindane contains trace amounts of other HCH isomers.

All the HCH isomers are acutely toxic to mammals, and chronic exposure has been linked to a range of health effects in humans. Of the different isomers, α-HCH exhibits the most carcinogenic activity and has been classified along with technical-grade HCH as a Group B2 probable human carcinogen by the U.S. EPA.[25] As the most metabolically stable isomer, β-HCH is the predominant isomer accumulating in human tissues. All isomers of HCH have high water solubilities compared to most other chlorinated organic compounds. They also have moderately high vapor pressures when compared to other organochlorine pesticides. Thus, HCH is usually present in the environment as a gas or dissolved in water, with only a small percentage adsorbed onto particles. The compounds have fairly long lifetimes in air and can be transported long distances.

The toxicity may also be indirectly related to the variance in physical and chemical properties of HCH among the isomers. For example, the vapor pressure of α-HCH is somewhat less than that of γ-HCH. α-HCH has also been shown to be slightly more lipophilic than γ-HCH ($\log K_{ow}$ 3.8 versus 3.6). The Henry's Law constant for α-HCH is about twice as high as that of γ-HCH, so α-HCH is more likely to partition to the air. Another important difference between the isomers is the persistence of the β-isomer. β-HCH is resistant to environmental degradation. It is also more lipophilic than the other isomers. These properties may result from its significantly smaller molecular volume. Since β-HCH's bonds between H, C, and Cl at all six positions are equatorial (i.e., within the plane of the ring), the molecule is denser and small enough to be stored in the interstices of lipids in animal tissues. Thus, even though the isomers possess identical elemental composition, the difference in their molecular arrangement leads to very different physical, chemical, and biological behavior.

Chronic Noncancer Health Endpoints

Reference Dose and Reference Concentrations: "Safe" Levels of Exposure

The *reference dose* (RfD) represents the highest allowable daily exposure associated with a noncancerous disease. It is calculated from the threshold value below which no adverse effects are observed (the so-called *no observable adverse effect level* or NOAEL), along with uncertainty and modifying factors based upon the quality of data and the reliability and representativeness of the studies that produced the dose-response curve:

$$RfD = \frac{(NOAEL)}{(UF_{1...n}) \times (MF_{1...n})} \qquad \text{Equation 9–1}$$

where, RfD = Reference dose ($mg\,kg^{-1}\,d^{-1}$)
$UF_{1...n}$ = Uncertainty factors related to the exposed population and chemical characteristics (dimensionless, usually factors of 10)
$MF_{1...n}$ = Modifying factors which reflect the results of qualitative assessments of the studies used to determine the threshold values (dimensionless, usually factors of 10).

The uncertainty factors address the robustness and quality of data used to derive the RfD, especially to be protective of sensitive populations (e.g., children and the elderly). It also addresses extrapolation of animal data from comparative biological studies to humans, accounting for differences in dose-response among different species. An uncertainty factor can also be applied when the studies upon which the RfD is based are conducted with various study designs; for example, if an acute or subchronic exposure is administered to determine the NOAEL, but the RfD is addressing a chronic disease, or if a fundamental study used a *lowest observable adverse effect level* (LOAEL) as the threshold value, requiring that the NOAEL be extrapolated from the LOAEL. The modifying factors address the uncertainties associated with the quality of data used to derive the threshold values, mainly from qualitative, scientific assessments of the data. For airborne contaminants, a *reference concentration* (RfC) is used in the same way as the RfD. That is, the RfC is an estimate of the daily inhalation exposure that is likely to be without appreciable risk of adverse effects during a lifetime.

The oral chronic RfD is used with administered oral doses under long-term exposures (i.e., exposure duration >7 years) effect, while the oral subchronic RfD is applied for shorter exposures of 2 weeks to 7 years. The inhalation chronic RfD applies to long-term exposures and is derived from an inhalation chronic RfC. Likewise, the subchronic inhalation RfD is derived from the inhalation subchronic RfC. The dermal chronic RfD and subchronic RfD relate to absorbed doses under chronic exposures and subchronic exposures, respectively.

Reference Dose Example 1

Referring to Table 9.7, and assuming that the dermal health data for pesticide propenyl butoxide are excellent[26] (i.e., no modifying factor is applied) and gathered from acute studies (i.e., no uncertainly factor for acute extrapolations), some subchronic studies (UF = 10), and more limited chronic data (UF = 100), what would be the respective NOAELs for acute, subchronic, and chronic noncancer dose-response relationships via the dermal pathway?

Solution and Discussion

Studies are often conducted on one or a few species and via one pathway (or for animal studies, one means of administration, such as dermal, inhalation, gavage, etc.). If studies are not done for a particular route of exposure or pathway, extrapolations will be needed to ascertain what these studies mean, and additional error and uncertainty beyond the animal to human, or sample to population, extrapolations. In this case, the scientists feel very good about the dose-response relationship for acute effects and data quality, but are increasingly unsure about subchronic and chronic effects.

For skin, exposure to propenyl butoxide (Table 9.7) shows that acute RfD is $10\,\mathrm{mg\,kg^{-1}\,day^{-1}}$. According to Equation 9–1, since there are no modifying or uncertainty factors, the NOAEL is the same as the RfD, $10\,\mathrm{mg\,kg^{-1}\,day^{-1}}$.

For subchronic skin, the RfD is again $10\,\mathrm{mg\,kg^{-1}\,day^{-1}}$, but the NOAEL would be ten times higher (UF = 10), so it is $100\,\mathrm{mg\,kg^{-1}\,d^{-1}}$.

For chronic, the RfD is $0.0175\,\mathrm{mg\,kg^{-1}\,day^{-1}}$. Our UF of 100 means the chronic skin NOAEL is $1.75\,\mathrm{mg\,kg^{-1}\,day^{-1}}$.

A quick look at the arithmetic shows that the conservative nature of reference doses moves the concentrations of noncancer contaminants that are deemed to be "safe" well below the calculated threshold derived from the dose-response studies. In other words, the level where scientists feel comfortable that no harm will occur is lowered with decreasing confidence, so the reference dose is an expression of both outcome and certainty, while the NOAEL is an expression of outcome only as dictated by the studies conducted on a specific compound. Another way to think about reference doses is that they are concentrations at the lower end of the confidence interval around the NOAEL.

Reference Dose Example 2

Making the unlikely assumption that the same modifying and uncertainty conditions hold for inhalation as were reported for dermal, what would be the respective NOAELs for acute, subchronic, and chronic noncancer dose-response relationships of exposures to propenyl butoxide via the dermal pathway?

Discussion

First of all, the assumption is unlikely because the studies are conducted on different species, pathways, and concentrations of even the same substance, in this instance propenyl butoxide. Some laboratories and universities used only one type of administration on various compounds, so their procedures will differ from other laboratories using different approaches. Thus, Lab A may conduct many inhalation studies of rats exposed to various contaminants in a chamber. Lab B may conduct only skin administrations. Therefore, the likelihood that both the skin and lung pathway studies would have identical data quality and certainties is somewhat farfetched. But, since we are making this assumption, let us go ahead and calculate the NOAELs for propenyl butoxide from Table 9.7's RfCs and Equation 9–1.

For inhaled air, the table shows that acute RfC is $7.4 \times 10^{-4}\,mg\,L^{-1}$. Since there are no modifying or uncertainty factors, the NOAEL is the same as the RfC, or $7.4 \times 10^{-4}\,mg\,L^{-1}$.

For subchronic inhalation, the RfC is again $7.4 \times 10^{-4}\,mg\,L^{-1}$, but the NOAEL would be ten times higher (UF = 10), so it is $7.4 \times 10^{-3}\,mg\,L^{-1}$.

For chronic, the RfC is $7 \times 10^{-5}\,mg\,L^{-1}$. Our UF of 100 means the chronic skin NOAEL is $7 \times 10^{-3}\,mg\,L^{-1}$.

One of the principal reasons for concern about exposure to contaminants is disease. Arguably, most attention in this area is paid to diseases that result from long-term, chronic exposures to contaminants. The major area of focus is cancer, as is evident from the previous discussion. However, other health endpoints are increasingly gaining attention, especially endocrine disruption, neurotoxicity, and immunological disorders.

Few contaminants are mutually exclusive in their ability to elicit disease; for example, they may be carcinogenic and may cause hormonal dysfunction or damage to the central nervous system. This raises the specter of an absurd notion that a carcinogen may have as safe a level of exposure as a neurotoxin! It is important to keep in mind that toxicologists tend to be compartmental when looking at diseases. When decisions are made of whether or at what dose a contaminant causes a disease in a population, these are usually made irrespective of other diseases. For example, when scientists review the animal and human data and literature to decide the level where no neurological dysfunction appears to result, this is done exclusively for the neurotoxic effects.

Thus, consider the hypothetical neurotoxin A that has an established reference dose (RfD) at $7\,\mathrm{mg\,kg^{-1}\,day^{-1}}$. This means that even if one is exposed at this dose over time, no neurotoxicity is expected. However, if neurotoxin A is also a carcinogen, it has no safe level (i.e., no threshold or no observable adverse effect level, NOAEL). Without a threshold, an RfD cannot be calculated, since the RfD equation's numerator is the NOAEL! Thus, even though a person exposed to contaminant A at $7\,\mathrm{mg\,kg^{-1}\,day^{-1}}$ is not expected to contract neural problems, that person is still at risk of cancer.

That said, let us now consider the so-called "noncancer endpoints."

Environmental Endocrine Disruptors[27]

In our discussion in Chapter 8 of ultraviolet light as a means to treat endocrine disruptors we learned that they are chemical compounds that can interfere with the proper functions of endocrine systems in humans and other organisms. Much scientific research related to hormonally active chemicals has pertained to medicine and pharmaceuticals. Currently, scientists and engineers are taking steps to reduce the risks posed by these chemicals, which are increasingly discovered in the environment.

The Endocrine System

The endocrine system is a vital part of an organism's normal growth, development, and reproduction, so even small disturbances in endocrine function at the wrong time may lead to long-lasting, irreversible effects. An organism is particularly vulnerable to endocrine disruption during highly sensitive times of development, such as prenatal and pubescent periods, when small changes in endocrine status may have delayed consequences that may not appear until much later in adult life or even in future generations. The most infamous case of multigenerational endocrine disruption is arguably that of diethylstilbestrol (DES), a synthetic hormone that was prescribed to pregnant women from 1940 to 1971 as a treatment for the mothers to prevent miscarriages. Unfortunately, DES has been subsequently classified as a "known carcinogen." The major concern was not with the treatment of the mothers, but with the *in utero* exposure that led to a high incidence of cervical cancers in the daughters of the treated mothers. Although an endocrine disruptor may or may not also be carcinogenic, DES shows that the effects may be linked.

While DES is generally recognized as a pharmaceutical problem it is, at a minimum, an environmental indicator of the potential problems of newly introduced chemicals. Thus, an emerging concern, particularly about human and animal pharmaceuticals, is their "pass-through" into the environment. For example, drugs used in combined animal feeding operations (CAFOs), have been found in waters downstream, even after treatment. This is problematic in at least two ways. First, the drugs and their metabolites (after passing through the animals) may themselves be hormonally active or may suppress immune systems. Second, antibiotics are being introduced to animals in large quantities, giving the targeted pathogens an opportunity to develop resistance and rendering them less effective.

Even more troubling is the phenomenon of *cross-resistance.* For example, the U.S. Food and Drug Administration recently proposed withdrawing the approval of enrofloxacin in the treatment of poultry in CAFOs. Enrofloxacin is one of the antibacterials known as fluoroquinolones, which have been used to treat humans since 1986.[28] Fluoroquinolone drugs keep chickens and turkeys from dying from *Escherichia coli* (*E. coli*) infection, usually contracted from the animals' own droppings. The pharmaceutical may be an effective prophylactic treatment for *E. coli*, but another genus, *Campylobacter*, appears to be able to build resistance (see Figure 9.9). Humans consuming the poultry products contaminated with fluoroquinolone-resistant *Campylobacter* risk infection by a strain of *Campylobacter* that is increasingly difficult to treat. Worse yet, the whole class of reliable fluoroquinolone drugs is at risk of being ineffective, since the cross-resistance can carry over to drugs with similar structures. This problem has been observed in numerous classes of drugs, including synthetic penicillin. And the use of drugs is not limited to treating diseases. In fact, large quantities of antibiotics have been used as growth promoters in CAFOs, so the probability of human exposure is further increased.

An organism's glands produce chemical compounds, known as hormones, which are secreted into the bloodstream to be delivered to the cells throughout the body. Cells contain specific sites, known as receptors, made up of molecules that will bind to specific hormones. When this binding occurs, the cell has accepted the signal sent from the gland. These chemical signals transmitted by the endocrine system are needed for an organism's reproduction, development, and homeostasis (the regulation of temperature, fluid exchange, and other internal processes). The binding process "turns on" the cell. For example, when ovaries secrete the female sex hormone estrogen, this is a message to specific cells to regulate ovulation and to develop secondary sexual characteristics in girls during puberty. Likewise, the release of a male sex hormone androgen, such as testosterone, sends the message to specific cells to trigger the creation of sperm and the

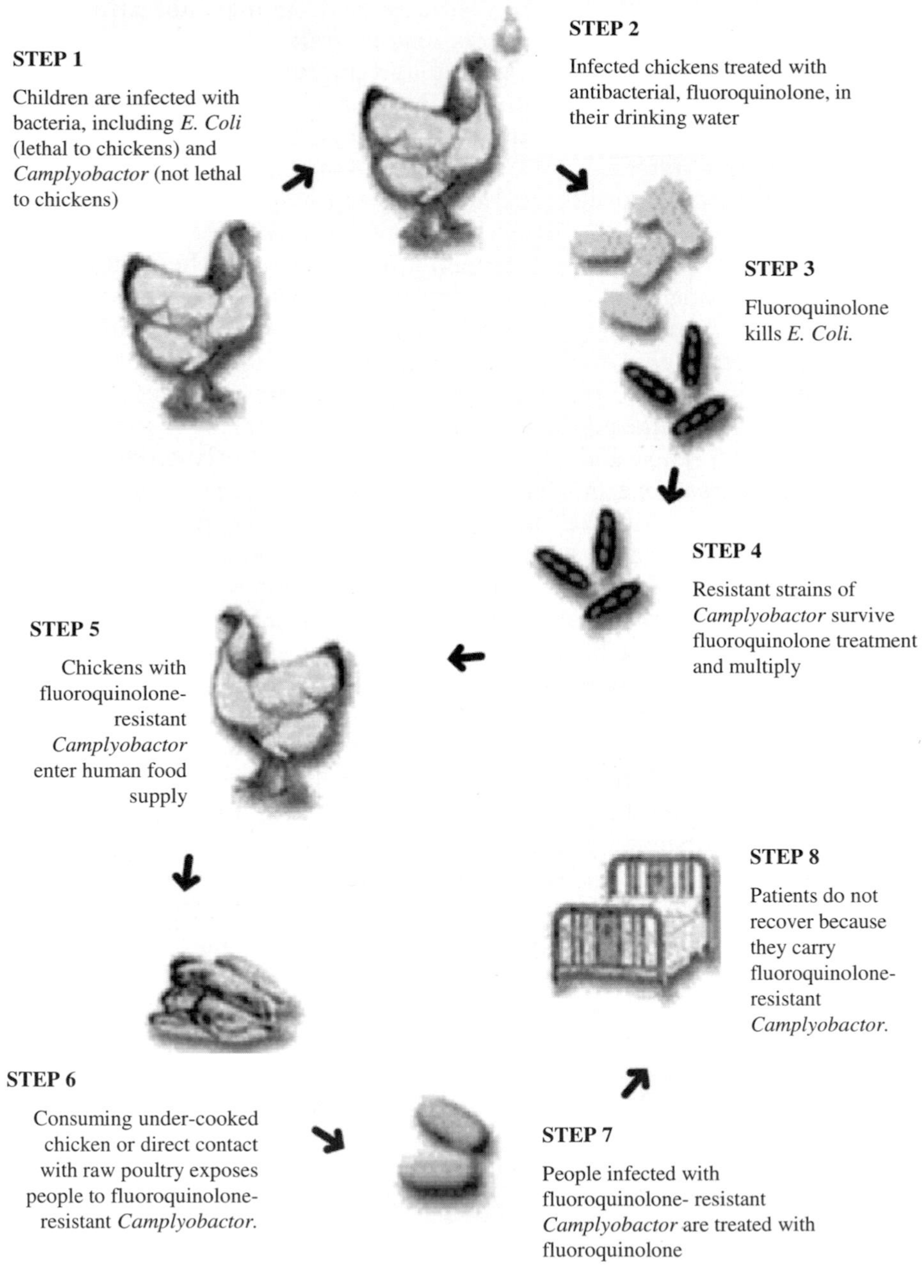

STEP 1
Children are infected with bacteria, including E. Coli (lethal to chickens) and Camplyobactor (not lethal to chickens)
STEP 2
Infected chickens treated with antibacterial, fluoroquinolone, in their drinking water
STEP 3
Fluoroquinolone kills E. Coli.
STEP 4
Resistant strains of Camplyobactor survive fluoroquinolone treatment and multiply
STEP 5
Chickens with fluoroquinolone-resistant Camplyobactor enter human food supply
STEP 6
Consuming under-cooked chicken or direct contact with raw poultry exposes people to fluoroquinolone-resistant Camplyobactor.
STEP 7
People infected with fluoroquinolone- resistant Camplyobactor are treated with fluoroquinolone
STEP 8
Patients do not recover because they carry fluoroquinolone-resistant Camplyobactor.

FIGURE 9.9. Steps in the cross-resistance of *Camplyobactor* to fluoroquinolone drugs. (Source: U.S. Food and Drug Administration, L. Bren, 2001, "Antibiotic Resistance from Down on the Farm," *FDA Veterinarian*, Vol. 16, no. 1, pp. 2–4. Graphic by R. Gordon.)

onset of secondary sexual characteristic in pubescent boys. Although sex hormones have received the most attention, endocrine disruption occurs in all glands, including the thyroid, adrenal, pituitary, and pineal glands.

Types of Environmental Endocrine Disruptors

Chemicals that act like natural hormones by binding to a cell's receptor are known as *agonists*. Conversely, chemicals that inhibit the receptor are called *antagonists*. Environmental endocrine disruptors can be either type (see Figure 8.7).

An organism's endocrine, neural, and immune systems are chemical messaging systems that are intimately interconnected. A change in one can potentially lead to changes in the others. A third type of endocrine disruption is termed *indirect*, where a substance endogenously affects the hormonal signals by interfering with the neural or immune systems. A classic example is the strong neurotoxin, mercury (Hg), which may not bind to an estrogen or androgen site, but its effect on the neurological system in turn interferes with the organism's hormonal health.

Environmental endocrine disruption was observed through much of the twentieth century.[29] Abnormal mating patterns in bald eagles on the east coast of North America were observed in the 1940s. Rachel Carson in her classic book *Silent Spring* (1962) observed that predatory birds, like eagles, accumulated chlorinated hydrocarbon pesticides. The eggshells of these birds were abnormally thin, and DDT contaminated birds were less successful in hatching than those with lower concentrations of the pesticide. During the last two decades of the twentieth century, scientists found associations between exposure to chemical compounds and changes affecting endocrine systems in humans and animals. These compounds were first called environmental estrogens, but since other hormonal effects were increasingly identified, the term environmental hormone gained usage. "Hormone mimicker" was sometimes used to describe compounds that could elicit a response like that of a natural hormone. For example, certain pesticides bind very easily to estrogen receptors, resulting in increased feminization of the organism. A hormonally active agent and an endocrine disruptor are more general classifications of any chemical that causes hormonal dysfunction.

Like other environmental contaminants, endocrine disruptors vary in physical and chemical forms. These different forms possess numerous and distinct ways that they can be broken down in the environment. Resistance to this breakdown is known as persistence. Chemicals also vary in their toxicity and their ability to build up in the food chain, termed bioaccumulation. The persistent, bioaccumulating toxics, the so-called PBT endocrine disruptors, are those that have long half-lives (the amount of time for half of the mass of the compound to break down in the environment), easily build up in the food chain, and lead to toxic effects. The PBT endocrine-disrupting compounds can be organic (known as persistent organic pollutants or POPs), inorganic (certain metals and their salts), or organometallic (including chelates and other metallic complexes, such as the butylated or phenalated forms of tin [Sn], which have been associated with endocrine effects in aquatic fauna). Table 9.8 identifies some of the important endocrine disrupting chemical groups.

Studying Endocrine Disruptors

Engineers and scientists can reduce the risks from substances that disrupt the endocrine system. Actions can be taken at various stages, from the synthesis of compounds (i.e., "green chemistry"), to the release of chemicals into the environment, to removal and treatment after release. Our understanding of endocrine disruption has been enhanced by scientific research in three major areas: laboratory studies (including animal testing and cellular receptor binding studies); epidemiology (the study of the incidence and distribution of diseases in human populations and ecosystems); and natural experiments (unplanned events, uncontrolled per se by scientists, but that allow for "before and after" comparisons). A profound endocrine disruption natural experiment was that of a large spill of dicofol, a pesticide chemically similar to DDT, into Lake Apopka, Florida. Studies following the spill indicated that the male alligators, the lake's top predators, showed marked reductions in gonad size.

Although many pesticides and some industrial chemicals have undergone toxicity testing (i.e., potential for adverse health effects), these tests are in many ways inadequate to ascertain the degree to which a substance will interact with the endocrine system. Scientific knowledge related to endocrine disruptors is evolving, but there is general scientific agreement that better endocrine screening and testing of existing and new chemicals is needed. To this end, the U.S. Government has established the Endocrine Disruptor Screening Program (EDSP) to evaluate the potential of hundreds of chemicals to cause hormonal effects (recall our discussion in Chapter 1).

Endocrine disruptors enter the environment in numerous ways. Wastes from households and medical facilities may contain hormones that reach landfills and wastewater treatment plants, where they pass through untreated or incompletely treated and enter waterways. Engineers designing treatment facilities must consider the possibility that wastes will contain hormonally active chemicals and find ways to treat them. For example, fish downstream from treatment plants have shown symptoms of endocrine disruption, so engineers must design treatment systems that eliminate the endocrine disruptors before the effluent is released into surface waters.

Manufacturers of hormonally active substances must find ways to eliminate them. One means of doing this is to change the chemical structure of compounds so that they do not bind or block receptor sites on cells.

TABLE 9.8
Some compounds found in the environment that have been associated with endocrine disruption, based on *in vitro, in vivo,* cell proliferation, or receptor-binding studies. (For full list, study references, study types and cellular mechanisms of action, see Chapter 2 of National Research Council, *Hormonally Active Agents in the Environment,* National Academy Press, Washington, DC, 2000. Source for asterisked (*) compounds is Colburn, et al, http://www.ourstolenfuture.org/Basics/chemlist.htm.)

Compound[1]	*Endocrine Effect*[2]	*Potential Source*
2,2′,3,4′,5,5′-Hexachloro-4-biphenylol and other chlorinated biphenylols	Antiestrogenic	Degradation of PCBs released into the environment
4′,7-Dihydroxy daidzein and other isoflavones, flavones, and flavonals	Estrogenic	Natural flora
Aldrin*	Estrogenic	Insecticide
Alkylphenols	Estrogenic	Industrial uses, surfactants
Bisphenol A and phenolics	Estrogenic	Plastics manufacturing
DDE (1,1-dichoro-2,2-bis(p-chlorophenyl)ethylene)	Antiandrogenic	DDT metabolite
DDT and metabolites	Estrogenic	Insecticide
Dicofol	Estrogenic or antiandrogenic in top predator wildlife	Insecticide
Dieldrin	Estrogenic	Insecticide
Diethylstilbestrol (DES)	Estrogenic	Pharmaceutical
Endosulfan	Estrogenic	Insecticide
Hydroxy-PCB congeners	Antiestrogenic (competitive binding at estrogen receptor)	Dielectric fluids
Kepone (Chlorodecone)	Estrogenic	Insecticide
Lindane (γ-hexachlorocyclohexane) and other HCH isomers	Estrogenic and thyroid agonistic	Miticide, insecticide
Lutolin, quercetin, and naringen	Antiestrogenic (e.g., uterine hyperplasia)	Natural dietary compounds
Malathion*	Thryroid antagonist	Insecticide
Methoxychlor	Estrogenic	Insecticide

TABLE 9.8 *(continued)*

Compound[1]	*Endocrine Effect*[2]	*Potential Source*
Octachlorostyrene*	Thryroid agonist	Electrolyte production
Pentachloronitrobenzene*	Thyroid antagonist	Fungicide, herbicide
Pentachlorophenol	Antiestrogenic (competitive binding at estrogen receptor)	Preservative
Phthalates and their ester compounds	Estrogenic	Plasticizers, emulsifiers
Polychlorinated biphenyls (PCBs)	Estrogenic	Dielectric fluid
Polybrominated Diphenyl Ethers (PDBEs)*	Estrogenic	Fire retardants, including *in utero* exposures
Polycyclic aromatic hydrocarbons (PAHs)	Antiandrogenic (Aryl hydrocarbon-receptor agonist)	Combustion byproducts
Tetrachlorodibenzo-*para*-dioxin and other halogenated dioxins and furans*	Antiandrogenic (Aryl hydrocarbon-receptor agonist)	Combustion and manufacturing (e.g., halogenation) byproduct
Toxaphene	Estrogenic	Animal pesticide dip
Tributyl tin and tin organometallic compounds*	Sexual development of gastropods and other aquatic species	Paints and coatings
Vinclozolin and metabolites	Antiandrogenic	Fungicide
Zineb*	Thyroid antagonist	Fungicide, insecticide
Ziram*	Thyroid antagonist	Fungicide, insecticide

[1] Not every isomer or congener included in a listed chemical group (e.g., PAHs, PCBs, phenolics, phthlates, and flavinoids) has been shown to have endocrine effects. However, since more than one compound has been associated with hormonal activity, the whole chemical group is listed here.

[2] Note that the antagonists' mechanisms result in an opposite net effect. In other words an antiandrogen feminizes and an antiestrogen masculinizes an organism.

The addition or deletion of a single atom or the arrangement of the same set of atoms (i.e., an isomer) can significantly reduce the likelihood of toxic effects elicited by a compound.

Neurotoxins

A glance at the minimum risk levels (MRLs) in Appendix 8 shows the central and peripheral nervous systems to be targets for a number of compounds.

Environmental neurotoxins can inflict nerve damage on any population, but children are arguably the most vulnerable and important exposed subpopulations. Numerous studies have linked neurological diseases and impairments in children to environmental contaminants, especially lead and carbon monoxide, and more recently to dioxins, PCBs, and numerous contaminants. Lead is the most notorious neurotoxin in many parts of the world. The former Secretary of Health and Human Services in the United States has stated, "Lead poisoning is the most common and societally devastating environmental disease of young children."[30]

Lead, like numerous other neurotoxins, accumulates and persists in tissues. Lead enters the food chain following deposition on soil, in surface waters, and on plants. Upon being entrained in the atmosphere, it may be transported thousands of miles if the lead particles are sufficiently small or if the lead speciates into volatile compounds.[31]

Elevated lead concentrations have been observed in people living in urban areas; near roads (due to past use of the now-banned lead in gasoline); near mines, smelters, and shipping facilities; as well as near other industrial sites, such as battery manufacturing operations.[32] Routes of exposure and toxicology vary. However, ingestion and inhalation routes account for a larger amount of Pb exposure than do dermal exposures. Socioeconomic conditions are also a factor in exposure; for example, the percentage of children with elevated blood-lead concentration is, on the whole, higher for children in low-income families and in African-American children compared to the general U.S. population.

Most inhaled Pb is absorbed, while 20% to 94% of lead in adults is stored in bones and teeth on average, while 73% of Pb stored in children's bodies is in their bones and teeth.[33] All Pb absorption depends on the intake of nutrients. Low intake of calcium, zinc, and iron can enhance lead absorption in the small intestines.[34] Over time, the stored Pb can be released into the bloodstream, particularly during calcium stress. This means that during pregnancy and lactation, as well as during menopause and stages of osteoporosis, women and their babies are at a particularly high risk of Pb toxicity.

Developmental neurotoxicity is concerned with harm during prenatal and perinatal stages of life. For example, *in utero* exposure to PCBs in mammals has been associated with neural dysfunction, and observations of elevated PCB exposures in pregnant women have been followed by developmental impairments and neurobehavioral delays in the children.[35]

Mercury: The Neurotoxin

Possibly the most notorious neurotoxin in the environment is the metal mercury (Hg) and its compounds. Releases of Hg to water are continuous because, like other heavy metals, it is found in many soils. However, anthropogenic sources, like mining, metal processing, and burning fossil fuels, concentrate Hg to much higher levels than are found in the natural background. After its release, Hg is converted to reduced, organic forms, such as monomethylmercury and dimethylmercury. These conversions are mediated by microbes, but they may also occur as a result of metabolism by higher organisms. In fact, methylated forms of Hg are commonly biomagnified. By the time the Hg reaches the top predator, such as a shark or an alligator, it may have increased to 10 million times the concentration of the water.[36]

Although some areas are highly contaminated, Hg is fairly ubiquitous. For example, in 1987 the mean wet weight concentration of Hg in 43 states has been observed to be as high as 520 ppb (in Walleye the freshwater game fish, *Stizostedion vitreum*),[37] and even higher in the northeastern U.S. states at 770 ppb (in Walleye).[38] Every Walleye tested in the 1998 northeast study was found to contain some detectable level of Hg.

Much of what is known about exposures and adverse effects associated with Hg comes from accidental releases. In fact, some of the most unsettling episodes have provided researchers at least in part the basis for what is known about high-end exposures and long-term effects of Hg. One such episode occurred in Minamata Bay near Kyushu, Japan. A chemical plant using Hg as a catalyst released wastes into the bay. Shellfish and other aquatic wildlife, a food staple for Minamata City residents, concentrated the Hg in the bay. This resulted in neurotoxicity; including disturbances of the senses and visual field constriction (what were to become the typical symptoms of what was to become known as "Minamata Disease"). The majority of patients also experienced coordination disturbance, speech disorders (i.e., dysarthria), hearing disturbance, problems from dimethjyl mercury in walking, and tremors.[39]

The neurotoxicity can be acute or chronic. Acute exposures to Hg (especially methylated forms) can harm the central nervous systems. It can also cause severe impairment to the kidneys, GI tract, and the cardiovascular system, as well as lethal poisoning (which recently occurred from dimethyl mercury in a laboratory setting).

Chronic exposures are more likely those associated with environmental pollution. The Minamata case is an example of chronic effects. In fact, chronic mercury neurotoxicity is steeped in history. The expression "mad as a hatter" was used by author Lewis Carroll[40] in his 1865 book,

Alice in Wonderland. Carroll's link was based on a disease endemic to nineteenth-century hat makers, who used a solution of Hg (or "quicksilver") to turn fur into felt. The hatters breathed the volatile Hg, often in the poorly ventilated areas of their workshops; so that their body burdens of Hg crept up the longer the solutions were used. Once a threshold amount of Hg was reached, they experienced neural dysfunction, such as trembling, coordination problems, slurred speech, dental problems, loss of memory, anxiety and depression. In fact, the term "mad hatter syndrome" still has currency in describing Hg toxicity.[41]

Based on data obtained from unfortunate episodes in Japan and elsewhere,[42] dose-response relationships between blood mercury levels ($<10\,\mu g\,dL^{-1}$ to $500\,\mu g\,dL^{-1}$)[43] and frequency and severity of symptoms have been established. These Hg exposures have been associated with mild symptoms that occur at the lower blood mercury levels and with deaths that occurred at levels $>300\,\mu g\,dL^{-1}$.

As in cancer and endocrine disruptor endpoints, neurotoxicity relies on extrapolations from various data sources with large ranges in quality, inconclusive results from studies, and the need to extrapolate from animal studies and high-dose episodes in human populations. This can lead to large uncertainties.

Immunotoxins

Exposure to environmental contaminants can damage the immune system. One cellular mechanism of action is involved in immune system dysfunction when the contaminant binds with or blocks the aryl hydrocarbon (Ah)-receptor. For example, PCBs, dioxins, and furans have been found to act at the Ah-receptor site of the cell.

Immunosuppression has been observed in high-dose incidents in the human population. The damage to the immune system can remain for decades after the exposure. For example, a large release of dioxins occurred in 1976 from a plant accident in Seveso, Italy. Twenty years after the event, exposed individuals have continued to show depressed levels of immunoglobin G (IgG).[44] Immunoglobulins are part of a body's system to attack antigens (i.e., foreign materials) that have entered, including pathogenic microbes. IgG and other immunoglobulins react with the antigens chemically. About three-fourths of human immunoglobulins are IgG, unless the immune system has been compromised. When these levels drop significantly, as they did in Seveso ($p < 0.0002$), people are less able to resist disease.

The immune system disorders, neurological problems, and endocrine disruption are related to one another. For example, following the Seveso dioxin release episode, all three endpoints were observed. The chronic nature of the diseases could be the result of lasting consequences of an initial exposure or continuous effects resulting from the persistent contaminant body burdens. That is, the contaminants are stored in fatty tissue reserves, bones, and other locations and released periodically over long periods of time.

Thus, chronic toxicity can take many forms, presenting a challenge to environmental professionals.

Ecological Toxicity

While much of this text is concerned with human health endpoints, toxicity can be extended to other species. The growing fields of eco-toxicology and eco-risk have several things in common with human toxicology and risk assessment, such as concern about ambient concentrations of contaminants and uptake in water, air, and soil. In some ways, however, ecological dose-response and exposure research differs from that in human systems. First, ecologists deal with many different species, some more sensitive than others to the effects of contaminants. Second, the means of calculating exposure are different, especially if one is concerned about the exposure of an entire ecosystem.

Ecosystems are quite complex. Ecologists tend to characterize them by evaluating their composition, structure, and functions. Ecosystem composition is a listing or taxonomy of every living and nonliving part of the ecosystem. Ecological structure, as the term implies, is how all of the parts of the system are linked to form physical patterns of life forms, for example, the patterns can range from single forest stands to biological associations and plant communities. A single wetland or prairie is an example of a much simpler structure compared to a multilayered forest, which consists of plant and microbial life in the detritus, herbs, saplings, newer trees, and canopy trees.

Ecosystem functions are what the ecosystems do. This includes cycles of nitrogen, carbon, and phosphorous, which allow for the biotic processes such as production, consumption, and decomposition.

Indicators of ecosystem health include:

- *Diversity*: One ecologist[45] defines *biodiversity* as the "composition, structure, and functions [that] determine, and in fact constitute, the biodiversity of an area. Composition has to do with the identity and variety of elements in a collection, and includes species lists and measures of species diversity and genetic diversity. Structure is the physical organization or pattern of a system, from habitat complexity as measured within communities to the pattern of patches and other elements at a landscape scale. Function involves ecological and evolutionary processes, including gene flow, disturbances, and nutrient cycling."

- *Productivity*: This is an expression of how economical a system is with its energy. It is a measure of how much biomass is produced from abiotic (e.g., nutrients and minerals) and biotic resources (from microbial populations to canopy plant species to top predator fauna). One common measure is "net primary productivity," which is the difference between two energy rates:

$$P_1 = k_p - k_e \qquad \text{Equation 9–2}$$

Where P_1 = Net primary productivity
k_p = Rate of chemical energy storage by primary producers
k_e = Rate at which the producers use energy (via respiration)

- *Sustainability*: How likely is it that the diversity and productivity will hold up? Even though an ecosystem appears to be diverse and highly productive, is there something looming that threatens the continuation of these conditions? For example, is an essential nutrient being leached out of the soil, or are atmospheric conditions changing that may threaten a key species of animal, plant or microbe? Sustainability is difficult to quantify precisely.

Perhaps the best way to understand some of the metrics used to characterize ecosystem health or stress is to consider some examples.

Eco-Toxicity Example 1

In 1990, you conducted an environmental assessment of microbes in a small stream at your plant. You found seven species of these critters. Your actual number count of each microbial species in the stream community was 16, 49, 69, 124, 212, 344, and 660 number of individual organisms per liter (mL^{-1}).

Find the diversity of this stream community using the Shannon-Weiner index:

$$D = -\sum_{i=1}^{m} P_i \log_2 P_i \qquad \text{Equation 9–3}$$

or

$$D = -1.44\sum_{i=1}^{m} (n_i/N)\ln(n_i/N) \qquad \text{Equation 9–4}$$

Where, D = index of community diversity
$P_i = n_i/N$
n_i = number (i.e., density) of the ith genera or species
N = total number (i.e., density) of all organisms in the sample
i = 1, 2, . . . , m
m = number of genera or species

Solution and Discussion

Construct a table to derive the values needed to find D, using Equation 9–4:

$$D = -1.44\sum_{i=1}^{m}(n_i/N)\ln(n_i/N).$$

i	n_i	n_i/N	$-1.44\ln(n_i/N)$	$-1.44(n_i/N)\ln(n_i/N)$
1	16	0.010855	6.513331	0.070701
2	49	0.033243	4.901637	0.162945
3	69	0.046811	4.408745	0.20638
4	124	0.084125	3.564653	0.299876
5	212	0.143826	2.792374	0.401617
6	344	0.233379	2.095335	0.489006
7	660	0.447761	1.157033	0.518075
Σ	1474	1		2.148599

Thus, the answer is 2.1. This doesn't tell you a whole lot. It is really most useful when comparing systems. Thus, if your stream is 2.1 and surrounding streams are all around 4, you may have a problem. Generally, D values range from about 1.5 to 4.5.

Eco-Toxicity Example 2

What would happen if all of the numbers of species doubled?

Solution and Discussion

Nothing, the index would still be 2.1. So diversity differs from the total count of organisms (abundance).

Eco-Toxicity Example 3

You conducted a follow-up study in 1995 and found that the density of these same species had changed to 2000, 25, 17, 18, 21, 40, and 11 microbes L^{-1}. How had the numbers and diversity changed in five years?

Solution and Discussion

Again, calculate D by constructing a table:

i	n_i	n_i/N	$-1.44 ln(n_i/N)$	$-1.44(n_i/N) ln(n_i/N)$
1	2000	0.93809	0.09204	0.08634
2	25	0.01173	6.40215	0.07507
3	17	0.00797	6.95751	0.05548
4	18	0.00844	6.8752	0.05805
5	21	0.00985	6.65322	0.06553
6	40	0.01876	5.72535	0.10742
7	11	0.00516	7.58437	0.03913
Σ	2132	1		0.48701

This shows that in five years, the actual number of microbes is increasing, but the diversity is far less ($D = 0.5$ versus 2.1). This may indicate that conditions favorable to one species, such as the presence of a toxic chemical, are detrimental to the other six species.

A key question to ask is whether the two studies are comparable. For example, were they conducted in the same season? (Many microbes grow better in warmer conditions, while others may compete more effectively in cooler waters.) If the studies are comparable, further investigation is needed, but this certainly is an indication that things are amiss, since Shannon values usually range from about 1.5 to 4.5!

Eco-Toxicity Example 4

What is the difference between advisory levels and allowable levels of exposure?

Discussion

The major difference is between the bad and the good! An advisory level, such as the county issuing a fishing advisory because sufficiently high concentrations of a PCB have been found in the gar in Lake X, is put in place to keep you from eating any (or two much) gar. Likewise, a swimming advisory is designed to keep you from entering waters that may contain chemicals, enteric pathogens, or other harmful agents so that you are not exposed via ingestion or dermally. In other words, you are "advised" to avoid the exposure.

Conversely, the allowable levels are telling you that even with a margin of safety and precaution, if you go about your day and are exposed to the ADI (average daily intake), scientific data and evidence tell us that you will be okay. The concept is similar to the NOAELs and RfDs discussed earlier in this chapter. Obviously, as knowledge grows and studies become more sophisticated, both advisory and allowance thresholds are adjusted, usually downwardly. For example, we may be plodding along for years with an ADI for chemical X at 10 $\mu g\,kg^{-1}$, but suddenly new research indicates that effects may occur at an order of magnitude lower concentration. If borne out, the ADI would have to be adjusted downwardly to 1 $\mu g\,kg^{-1}$. This also happens when standards, such as maximum contaminant levels (MCLs) must be changed in light of new information about a contaminant's hazard.

Similarly, we only worried about one type of fecal coliform for decades, but if a new strain of *E. coli* is found to be more virulent and toxic to humans than others, the swimming advisory would have to take this into account and the advisory would kick in at much lower counts of fecal coliform bacteria. Thus, little Johnnie is not allowed in the lake that he has swam in for years, and his parents cannot understand why. This is a classic example of the challenge of the town engineer and public health officials to explain why people cannot use their own resources. It is also a challenge to do the right things to prevent contamination so that advisories can be avoided in the first place.

Hazards come in many sizes, types and forms. The environmental professional must have an understanding of all of them to address and to prevent environmental risks.

Notes and Commentary

1. Actually Paracelsus was referring to the medieval concept of "hormesis," i.e., the beneficial application of toxic chemicals in small doses. (See W.C. Kreiger, "Paraselsus: Dose Response" in the *Handbook of Pesticide* Toxicology, 2nd Edition, 2001. Academic Press, R. Kreiger, J. Doull, and D. Ecobichon (Editors).

2. Standing timber is a "crop," so to speak, as indicated by the fact that the U.S. Forest Service is housed in the Department of Agriculture.
3. The U.S. EPA also developed standard approaches and set criteria to determine whether waste exhibited any of the hazardous characteristics. The testing procedures are generally defined and described in the *Test Methods for Evaluating Solid Waste* (SW-846). See U.S. Environmental Protection Agency, 1995, *Test Methods for Evaluating Solid Waste*, Vols. I and II (SW-846), 3rd Edition.
4. U.S. Environmental Protection Agency, 1976, *National Interim Primary Drinking Water Regulations*, EPA-570/9-76-003.
5. These receptors, respectively, represent the three types of environmental values, i.e., human health, ecological integrity, and public welfare.
6. The federal laboratories in Research Triangle Park, North Carolina, have instituted an Adopt a Chemical program that not only includes chemicals but has subsequently been extended to share glassware and other laboratory apparatus. Although much of the apparatus does not qualify as hazardous wastes per se, they may contain hazardous substances that would have to be disposed of properly if they were not adopted. One example would be the chromatographs and detectors, such as the electron capture detectors (ECDs) that contain radioactive nickel.
7. RCRA is a good place to find the definitions of hazardous waste, but certainly it is not the only one. In fact, RCRA hazardous waste has excluded some wastes that most reasonable people would consider to be hazardous. Notably, it does not include nuclear source, special nuclear, or by-products defined by the Atomic Energy Act of 1954, as amended. These wastes are known as *nuclear wastes*. Obviously, sources such as military installations and nuclear power generation facilities have both types, i.e., hazardous and nuclear wastes. That is, they have chemical (and possibly biological) wastes, as well as radioactive wastes produced as by-products of nuclear reactions.

   The definitions of waste are important, as evidenced by the recent request of the U.S. Department of Energy (DOE) to change the federal definition of nuclear wastes. The requested change would allow the Nuclear Regulatory Commission (NRC) to decide whether nuclear reprocessing wastes from weapon manufacturing (e.g., plutonium enrichment) can be transported to a nuclear waste disposal site in Yucca Mountain, Nevada. The revised definition, according to DOE, would help to clarify the distinction between "high level" and other nuclear wastes. Critics have opposed the change, arguing that it is merely a way to get around recent court mandates regarding how and where certain wastes may be disposed. For more details on both sides of the debate, see the August 22, 2003 article by S. Straglinski in the *Las Vegas Sun*.

   Another important nuance in the definition of a waste is inherent to RCRA itself. Section 3001(b)(3)(A)(ii) excludes "solid wastes from the extraction, beneficiation, and processing of ores and minerals" from the definition of hazardous wastes. Several years of lawsuits and draft regulations have resulted in long lists of specific types of wastes that would be excluded in the U.S. EPA's Mining Waste Exclusion Final Rule (Effective date: March 1, 1990). The rationale for such exclusions is the sheer volume of wastes generated by mining activities. Just visit a mine in Wyoming, West Virginia, Kentucky, or Illinois! The opposing view is that "waste is waste" and no matter where it comes from, it is still just as hazardous. As is often the case in environmental decision making, no one seems completely happy with any of the definitions, and especially the exclusions, for hazardous waste.

8. Mandated by 40CFR261 (U.S. Code of Federal Regulations).
9. For example, in soil-washing cleanup practices, strong acids may be used to help mobilize contaminants from soil particles.
10. North Carolina Building Code Council and North Carolina Department of Insurance, 2002, *Fire Prevention Code*, Tables 2703.1.1 and 2703.8.2.
11. J. Duffus and H. Worth provide an excellent introduction to the concepts of dose, hazards, and risk in their 2001 training program, "The Science of Chemical Safety: Essential Toxicology—4, Hazard and Risk," *IUPAC Educators' Resource Material*, International Union of Pure and Applied Chemistry.
12. B. Hellman, 1993, *Basis for an Occupational Health Standard: Chlorobenzene. National Institute for Occupational Safety and Health*, U.S. Department of Health and Human Services (National Institute of Occupational Safety and Health) DHHS (NIOSH) Publication No. 93-102.
13. The term *biotransformation* is limited in some usage to the chemical breakdown of complex molecules by microbes, especially the mineralization of organic molecules (i.e., making them less organic and more inorganic). However, any biological process, such as metabolism in higher-order organisms, can be included as a biotransformation process. In addition to metabolic processes, biotransformation may include binding to receptors in and on a cell.
14. The "two-hit" theory on carcinogenesis was posited by A.G. Knudson in 1971 (see A.G. Knudson, 1985, "Hereditary cancer, oncogenes, and antioncogenes," *Cancer Research*, 45(4), 1437–1443). The theory argues that cancer develops after DNA is damaged. The initial damage, known as "initiation," may, but does not always, lead to cancer. Subsequent cellular damage, caused by agents known as "promoters," changes the nature and make-up of the cell. In fact, the normal homeostasis (self-regulation) of cells is gone, so clonal cancer cells keep dividing unchecked.
15. See Chapter 3 of *Sigma Xi*, The Scientific Research Society, 1986, *Honor in Science, Sigma Xi*, Research Triangle Park, N.C.
16. U.S. Environmental Protection Agency, Office of Pesticide Programs, 1999, "List of Chemicals Evaluated for Carcinogenic Potential," Washington, D.C., August.
17. Source is D. Kincaid, 2001, "Toxicity Code," Lehman College, University of New York, Bronx, N.Y.
18. The limit of detection (LOD) is both an analytical and a sampling threshold. If an instrument can only detect down to 1 ppb, this is an analytical limitation. However, in reality, if the sample has been held for some time, or the sample must be extracted from the soil or trapping device in the field, this is a limit,

even if the laboratory can detect down to 1 ppb. In our discussion, such a laboratory would report the nondetects as 500 ppt (i.e., one-half of 1 ppb). This is certainly not the only means of dealing with nondetects. Other statistical methods for dealing with nondetects are used, but a nondetect should never be reported as 0, since one can only say with confidence that it was not seen. It may or may not be present, but we can only report what we know, and that is dictated by the LOD.

19. For example, the World Health Organization has recently assigned the value of 1 to the pentachlorodibenzo-para-dioxin, which means that the organization considers the toxicity equal to that of TCDD.
20. These values are updated periodically; if a carcinogen is not listed in the table, visit http://risk.lsd.ornl.gov/tox/rap_toxp.shtml.
21. This information was obtained from the Risk Assessment Information System of the Oak Ridge National Laboratory, 2003.
22. These values are updated periodically; if a carcinogen is not listed in the table, visit http://risk.lsd.ornl.gov/tox/rap_toxp.shtml.
23. Risk Assessment Information System, 2003, Oak Ridge National Laboratory.
24. F. Kutz, P. Wood, and D. Bottimore, 1991, "Organochlorine pesticides and polychlorinated biphenyls in human adipose tissue," *Review of Environmental Contamination Toxicology*, Vol. 120, pp. 1–82.
25. Agency for Toxic Substances and Disease Registry (ATSDR), 1997, *Toxicological Profile for Alpha-, Beta-, Gamma- and Delta-Hexachlorocyclohexane*; 205-93-0606; Research Triangle Institute, Research Triangle Park, N.C., pp. 1–239.
26. This is a very arbitrary assumption and probably incorrect, and is only used here for illustrative purposes. For updated and actual information about the quality of underlying data and models, visit the ATSDR website.
27. Sources for the endocrine disruptor section are: R. Carson, 1962, *Silent Spring*, Houghton Mifflin, Boston, Mass.; T. Colborn, D. Dumanoski, and J.P. Myers, *Our Stolen Future*, 1997, Penguin, Books, New York, N.Y.; U.S. Environmental Protection Agency, 1998, "Endocrine Disruptor Screening and Testing Advisory Committee (EDSTAC) Final Report"; S. Goodbred, R. Gilliom, T. Gross, N. Denslow, W. Bryant, and T. Schoeb, 1997, "Reconnaissance of 17β-Estradiol, 11-Ketotestosterone, Vitellogenin, and Gonad Histopathy in Common Carp of United States Streams: Potential for Contaminant-Induced Endocrine Disruption," U.S. Geological Survey Open-File Report 96–627, Sacramento, Calif.; National Academy of Sciences, 1999, *Hormonally Active Agents in the Environment*, National Academy Press, Washington, D.C.; United Nations Environment Programme (Chemicals), Regionally Based Assessment of Persistent Toxic Substance, 2002, North American Regional Report; and U.S. Environmental Protection Agency, 1997, Special Report on Environmental Endocrine Disruption: An Effects Assessment and Analysis.
28. U.S. Food and Drug Administration, L. Bren, 2001, "Antibiotic Resistance from Down on the Farm," *FDA Veterinarian*, Vol. 16, no. 1, pp. 2–4; and C. Richardson, 2000, Ontario Ministry of Agriculture and Food: http://www.gov.on.ca/OMAFRA/english/livestock/sheep/facts/info_resist.htm.

29. Endocrine disruptor Internet links: environmental endocrine disruptors are an international problem; numerous resources are updated frequently on the status of these contaminants, including:
    - http://www.epa.gov/scipoly/oscpendo/ (U.S. Environmental Protection Agency Screening Program)
    - http://iccvam.niehs.nih.gov/methods/endocrine.htm (National Institute for Environmental Health Sciences' Endocrine Disruptor Test Methods)
    - http://ehp.niehs.nih.gov/topic/endodisrupt.html (Endocrine disruptor articles online, published in the journal, *Environmental Health Perspectives*)
    - http://www.epa.gov/endocrine/ (U.S. Environmental Protection Agency's Endocrine Disruptor Research Program)
    - http://edkb.fda.gov/ (U.S. Food and Drug Administration's Endocrine Disruptor Knowledge Base, which includes numerous links to other endocrine disruptor sites)
    - http://www.cerc.cr.usgs.gov/endocrine/ (U.S. Geological Survey's Endocrine Disruptor Links)
    - http://docs.pesticideinfo.org/documentation4/ref_toxicity5.html (Pesticide Action Network's Pesticide Database, which includes links to endocrine disruptor chemical lists)
30. L. Sullivan, 1991, Speech by Secretary of U.S., Department of Health and Human Services on lead poisoning, presented at 1st Annual Conference on Childhood Lead Poisoning, Washington D.C., October 7.
31. Agency for Toxic Substances and Disease Registry, 1993, "Toxicological Profile for Lead: Final Report."
32. Environment Canada, 1995, Envirofacts. Toxic Chemicals in Atlantic Canada—Lead. EN 40-226/1-1995 E: http://www.ns.ec.gc.ca/epb/envfacts.
33. ATSDR, 1993, "Toxicological Profile for Lead: Final Report."
34. P. Mushak and A.F. Crochetti, 1996, "Lead and Nutrition," *Nutrition Today*, Vol. 31, pp. 12–17.
35. U.S. Environmental Protection Agency, 1993, Workshop Report on Developmental Neurotoxic Effects Associated with Exposure to PCBs, EPA/630/R-02/004, Research Triangle Park, N.C.
36. U.S. Environmental Protection Agency, 2001, "Mercury Update: Impact on Fish Advisories," Fact Sheet No. EPA-823-F-01-011, Washington, D.C.
37. U.S. Environmental Protection Agency, 1993, *National Study of Chemical Residues in Fish*, Vol. 1, EPA-823R-92-008a, Washington, D.C.; and D. Bahnick, C. Sauer, B. Butterworth, and D. Kuehl, 1994, "A National Study of Mercury Contamination in Fish IV: Analytical Methods and Results," *Chemosphere*, Vol. 29, no. 3, pp. 537–547.
38. Northeast States for Coordinated Air Use Management, 1998, *Northeast States and Eastern Canadian Provinces Mercury Study: A Framework for Action*, Boston, Mass.
39. H. Tokuomi, T. Okajima, J. Kanai, M. Tsunoda, Y. Ichiyasu, H. Misumi, K. Shimomura, and M. Takaba, "Minamata Disease," *World Neurology*, 1961, 2, pp. 536–545.

40. Pseudonym for the English mathematician and writer, Charles Lutwidge Dodgson.
41. Complementary Medical Association, 2003, http://www.the-cma.org.uk/index.htm.
42. Oak Ridge National Laboratory, 2003, Risk Assessment Information System, Toxicity Summary for Methyl Mercury: http://risk.lsd.ornl.gov/tox/profiles/methyl_mercury_f_V1.shtml#t3.
43. dL = deciliter, the standard volume unit for blood.
44. A. Baccarelli, P Mocarelli, D. Patterson, Jr., M. Bonzini, A. Pesatori, N. Caporaso, and M. Landi, 2002, "Immunologic Effects of Dioxin: New Results from Seveso and Comparison with Other Studies," *Environmental Health Perspectives*, Vol. 110, pp. 1169–1173.
45. R. Noss, 1990, "Indicators for Monitoring Biodiversity: A Hierarchical Approach," *Conservation Biology*, Vol. 4, no. 4, pp. 355–364. Quote is from p. 355.

CHAPTER 10

Contaminant Exposure and Risk Calculations

With our understanding of environmental hazards, especially toxicity, we must now consider the other half of the risk equation: exposure.

Exposure Assessment

Scientists conduct exposure assessments to evaluate the kind and magnitude of exposure to contaminants. Such assessments are usually site-specific for clearly identified contaminants of concern. For example, they may be conducted for an abandoned hazardous waste site or a planned industrial facility. For the former site, the list of contaminants of concern would be based upon the sampling and analysis of the various environmental compartments, while the latter would be based upon the types of chemicals to be used or generated in the construction and operation of the industrial facility. Thus the assessment considers sources of contaminants, pathways through which contaminants are moving or will be moving, and routes of exposure where the contaminants find their way to receptors (usually people, but also receptors in ecosystems, such as fish and wildlife). Table 10.1 includes some of the most important considerations in deciding on the quality of information needed to conduct an exposure assessment.

The necessary information to quantify is determined by both the characteristics of the contaminant and the route of exposure, for example, dermal exposure to DDT in soil requires permeability coefficients, soil absorption factors, surface area of body exposed, and soil adherence to the skin. Also, the target group being protected must be identified. For example, for hazardous wastes assessments, exposure is calculated for an "average individual" (i.e., a central tendency), wherein the 95% upper confidence limit (UCL) on the arithmetic mean is selected for an exposure point concentration, and central estimates (i.e., arithmetic average, 50th percentile, median) for all other exposure variables.[1]

TABLE 10.1
Questions to Be Asked When Determining the Adequacy of Information Needed to Conduct Exposure Assessments

Compartment	*Question*
Soil	If humans have access to contaminated soils, can ranges of contamination be provided on the basis of land use (i.e., restricted access, road/driveway/parking lot access, garden use, agriculture and feedlot use, residential use, playground and park use, etc.)? Have the soil depths been specified? Do soil data represent surface soil data (≤3 inches in depth) or subsurface soil data (>3 inches in depth)? If soil depth is known, but does not meet surface or subsurface soil definitions, designate the data as *soil* and specify the depth (e.g., 0–6 inches). If the soil depth is unknown, the health assessor should designate the data as unspecified soil. Has soil been defined in the data? If not, the health assessor should assume soil includes any unconsolidated natural material or fill above bedrock that is not considered to be soil and excludes manmade materials such as slabs, pavements or driveways of asphalt, concrete, brick, rock, ash, or gravel. A soil matrix may consist of pieces of each of these materials. Do soil data include uphill and downhill samples and upwind and downwind samples both on and off the site?
Sediment	Have the sediment samples been identified as grab samples or cores? (see Chapter 11) Was the depth of the samples specified? Was the sampling program designed to collect sediment samples at regular intervals along a waterway or from depositional areas or both? Do the sediment data include results for upstream and downstream samples both on- and off-site? Has sediment been defined by the samplers? (To prevent confusion between sediment and soil, assume *sediment* is defined as any solid material, other than waste material or waste sludge that lies below a water surface, that has been naturally deposited in a waterway, water body, channel, ditch, wetland, or swale, or that lies on a bank, beach, or floodway land where solids are deposited.) Have any sediment removal activities (e.g., dredging, excavation, etc.) occurred that may have altered the degree of sediment contamination (leading to a false negative). This becomes important when the following occur: 1. Sediment contamination in fishable waters is used to justify sampling and analyses of edible biota; 2. Sediment data are used to justify additional downstream sampling, particularly at points of exposure and in areas not subject to past removal activities; and 3. The significance of past exposure is assessed.

TABLE 10.1 *(continued)*

Compartment	*Question*
Surface Water	Do surface-water data include results for samples both upstream and downstream of the site? Was information obtained on the number of surface-water samples taken at each station, as well as the frequency, duration, and dates of sampling?
Groundwater	Were groundwater samples collected in the aquifer of concern? Did sampling occur both upgradient and downgradient of the site and the site's groundwater contamination plume?
All	Did the sampling design include selected hot spot locations and points of possible exposure?

Source: Agency for Toxic Substances and Disease Registry, 2003, *ATSDR Public Health Assessment Guidance Manual.*

Calculating Exposures

Generally, exposure is calculated for each route and pathway. All exposure contains a term for the concentration of the contaminant in the environment. It also includes assumptions about the population of concern, such as typical lifetimes and body weight.

The generic equation for exposure is an expression of intake or uptake of a contaminant. Intake is the step where the contaminant enters the organism, but has not yet been distributed (e.g., it has traveled through the mouth and nose and entered the lungs, but has not yet traversed the absorption barrier and is yet to enter the bloodstream). Intake is directly proportional to the chemical concentration, contact with the substance, frequency and duration of the contact, and the rate of absorption of the contaminant. Intake is also indirectly proportional to the overall body mass of the exposed organism and the averaging time of exposure. Thus, generic contaminant intake (I) can be expressed as:

$$I = \frac{C \cdot CR \cdot EF \cdot ED \cdot AF}{BW \cdot AT} \qquad \text{Equation 10–1}$$

Where C = Chemical concentration of contaminant (in units of mass per volume)

CR = Contact rate (in units of mass per time)

EF = Exposure frequency (number of events)

ED = Exposure duration (length of time of the exposure, in units of time)

AF = Absorption factor (fraction of contaminant available for toxic effect, dimensionless, ≤1)
BW = Body weight (in units of mass)
AT = Averaging time (in units of time; for chronic exposures in humans, usually = 70 years)

So if two rodents, each weighing 100 g, are each exposed to a total of 200 ng of hexachlorobenzene, but one is exposed over 30 days while the other is exposed for 300 days, the rodent exposed for 300 days would have only 10% of the intake of hexachlorobenzene, because its averaging time was 10 times longer. Likewise, if two rodents, one weighing 50 g and the other weighing 200 g, were exposed to 100 ng of hexachlorobenzene, both over a 30-day period, the 50 g rodent's intake would be twice that of the 100 g rodent, because of the smaller rodent's lesser body weight. The point is that even though the mass (dose) is the same, the other factors, in these instances averaging time and body weight, cause the exposures to differ.

All of the exposure equations for the various routes (e.g., drinking water, diet, dermal contact, and breathing) are based upon the intake equation, although physically and chemically, so the disease or health outcome of concern should not matter in calculating exposures. However, the way the exposure calculations are used to calculate risk will differ as a matter of practice and policy. For chronic risk, especially cancer risks, the exposure is an expression of chronic intake, such as the lifetime average daily dose, and for noncancer risks, the exposure is part of a formulation known as the hazard quotient, wherein the contaminant intake is divided by the safe concentration (RfD or RfC). We will now consider both applications.

The Lifetime Average Daily Dose (LADD)

Human exposures to chemicals associated with cancer and other chronic, long-term diseases are usually represented by estimates of lifetime average daily dose (LADD), which is a function of the concentration of the chemical, contact rate, contact fraction, and exposure duration per a person's body weight and life expectancy. For example, exposure from ingesting contaminated water can be calculated[2] as:

$$LADD = \frac{(C)\cdot(CR)\cdot(ED)\cdot(AF)}{(BW)\cdot(TL)} \qquad \text{Equation 10–2}$$

Where, LADD = lifetime average daily dose ($mg\,kg^{-1}\,d^{-1}$); C = concentration of the contaminant in the drinking water ($mg\,L^{-1}$); CR = rate of water consumption ($L\,d^{-1}$); ED = duration of exposure (d); AF = portion (fraction) of the ingested contaminant that is physiologically absorbed[3] (dimensionless); BW = body weight (kg); and TL = typical lifetime (d).

The LADD equations for the major routes of exposure are provided in Table 10.2.

TABLE 10.2
Equations for Calculating Lifetime Average Daily Dose (LADD) for Various Routes of Exposure

Route of Exposure	*Equation* LADD (in $mg\,kg^{-1}\,d^{-1}$) =	*Definitions*
Drinking Water	$\frac{(C)\cdot(CR)\cdot(ED)\cdot(AF)}{(BW)\cdot(TL)}$	C = concentration of the contaminant in the drinking water ($mg\,L^{-1}$) CR = rate of water consumption ($L\,d^{-1}$) ED = duration of exposure (d) AF = portion (fraction) of the ingested contaminant that is physiologically absorbed (dimensionless) BW = body weight (kg) TL = typical lifetime (d)
Inhaling Aerosols (Particulate Matter)	$\frac{(C)\cdot(PC)\cdot(IR)\cdot(RF)\cdot(EL)\cdot(AF)\cdot(ED)\cdot(10^{-6})}{(BW)\cdot(TL)}$	C = concentration of the contaminant on the aerosol/particle ($mg\,kg^{-1}$) PC = particle concentration in air ($mg\,m^{-3}$) IR = inhalation rate ($m^{-3}\,h^{-1}$) RF = respirable fraction of total particulates (dimensionless) EL = exposure length ($h\,d^{-1}$) ED = duration of exposure (d) AF = absorption factor (dimensionless) 10^{-6} is a conversion factor (kg to mg) Other variables are the same as above.
Inhaling Gas Phase Contaminants	$\frac{(C)\cdot(IR)\cdot(EL)\cdot(AF)\cdot(ED)}{(BW)\cdot(TL)}$	C = concentration of the contaminant in the gas phase ($mg\,m^{-3}$) Other variables the same as above.

TABLE 10.2 *(continued)*

Route of Exposure	*Equation* LADD (in $mg\,kg^{-1}\,d^{-1}$) =	*Definitions*
Contact with Soil-borne Contaminants	$\frac{(C)\cdot(SA)\cdot(BF)\cdot(FC)\cdot(SDF)\cdot(ED)\cdot(10^{-6})}{(BW)\cdot(TL)}$	C = concentration of the contaminant in the soil ($mg\,kg^{-1}$) SA = skin surface area exposed (cm^{-2}) BF = bioavailability (percent of contaminant absorbed per day) FC = fraction of total soil from contaminated source (dimensionless) SDF = soil deposition, the mass of soil deposited per unit area of skin surface ($mg\,cm^{-1}\,d^{-1}$) Other variables are the same as above.

Source: M. Derelanko, 1999, "Risk Assessment," *CRC Handbook of Toxicology*, edited by M.J. Derelanko and M.A. Hollinger, CRC Press, Boca Raton, Fla.

LADD Example

In the process of synthesizing pesticides over an 18-year period, a coating and plastics company has contaminated the soil on its property with high levels of vinyl chloride. Even though the plant closed two years ago, fugitive emissions of vinyl chloride vapors continue to reach the neighborhood surrounding the plant at an average concentration of $1\,mg\,m^{-3}$. As a precautionary approach, assume that people are breathing at a high ventilation rate of $4.8\,m^3\,h^{-1}$. The legal settlement allows neighboring residents to evacuate and sell their homes to the company. However, they may also stay. The neighbors have asked to compare their exposures if they leave versus if they stay, since they have already been exposed for 20 years. In other words, they are asking what would be the overall difference in LADD if they leave or stay.

Solution and Discussion

Vinyl chloride is highly volatile, so its phase distribution will be mainly in the gas phase rather than the aerosol phase. Although some of the vinyl chloride may be sorbed to particles, we will use only vapor phase LADD equation, since the particle phase is likely to be relatively small. Also, since we are using the high ventilation rate, and people do not breathe at this rate for long periods of time, the safety factor should well compensate for any residual particle-bound vinyl chloride. Also, we will assume that outdoor concentrations are the exposure concentrations. This is unlikely, however, since most people spend fewer hours per day outdoors compared to indoors. To determine how much vinyl chloride penetrates living quarters, indoor air studies would have to be conducted. For a scientist to compare exposures, indoor air measurements should be taken.

Find the appropriate equation in Table 10.2 and insert default values from Table 10.3 (factors commonly used in exposure equations), and assume that a person lives the remainder of an entire typical lifetime exposed at these levels. By convention, the longest time this would be is 70 years; if the person is now 20 years of age and has already been exposed for that time, and lives the remaining 50 years exposed at $10\,\text{mg}\,\text{m}^{-3}$:

$$\begin{aligned}\text{LADD} &= \frac{(C)\cdot(IR)\cdot(EL)\cdot(AF)\cdot(ED)}{(BW)\cdot(TL)}\\ &= \frac{(1)\cdot(4.8)\cdot(24)\cdot(0.875)\cdot(25550)}{(70)\cdot(25550)}\\ &= 1.4\ \text{mg kg}^{-1}\ \text{day}^{-1}\end{aligned}$$

If the 20-year-old leaves today, assuming no future vinyl chloride exposure, the exposure duration would be for the 20 years that the person lived in the neighborhood. Thus, only the ED term would change, or from 25,550 days or 70 years to 7300 days or 20 years. Thus, the LADD falls to 2/7 of its value:

$$\text{LADD} = 0.4\ \text{mg kg}^{-1}\ \text{day}^{-1}$$

These exposure values can now be used for comparative risk characterizations.

Although other equations are commonly used to calculate chronic exposure to contaminants, they are similar to the LADD equations included in Table 10.2. For example, the ATSDR calculates the "exposure dose" as:

$$E_D = \frac{C \cdot IR \cdot E_F}{BW} \qquad \text{Equation 10–3}$$

Where, E_D = exposure dose
C = contaminant concentration
IR = intake rate of contaminated medium
E_F = exposure factor
BW = body weight

All of the variables are the same as those for intake or LADD, except for E_F. Some exposures are intermittent or irregular, so the E_F is calculated to integrate the dose over the exposure interval. E_F is calculated as the product of the exposure frequency and the exposure duration, divided by the time period over which the dose is to be averaged. For example, if a child comes into contact with contaminated soil three times a week over a five-year period, the exposure factor would be:

$$E_F = (3\ \text{days/week} \cdot 52\ \text{weeks/year} \cdot 5\ \text{years})/(5\ \text{years} \cdot 365\ \text{days/year}) \qquad \text{Equation 10–4}$$

$$\text{The child's } E_F \approx 0.4$$

If the exposure is not intermittent, $E_F = 1$.

Keep in mind that the exposure factors, such as those in Table 10.3, are based upon average conditions and can vary among populations and

TABLE 10.3
Commonly Used Human Exposure Factors

Exposure Factor	*Adult Male*	*Adult Female*	*Child (3–12 Years of Age)*[8]
Body weight (kg)	70	60	15–40
Total fluids ingested ($L\,d^{-1}$)	2	1.4	1.0
Surface area of skin, without clothing (m^2)	1.8	1.6	0.9
Surface area of skin, wearing clothes (m^2)	0.1–0.3	0.1–0.3	0.05–0.15
Respiration/ventilation rate ($L\,min^{-1}$)—Resting	7.5	6.0	5.0
Respiration/ventilation rate ($L\,min^{-1}$)—Light activity	20	19	13
Volume of air breathed ($m^3\,d^{-1}$)	23	21	15
Typical lifetime (years)	70	70	NA
National upper-bound time (90th percentile) at one residence (years)	30	30	NA
National median time (50th percentile) at one residence (years)	9	9	NA

Sources: U.S. Environmental Protection Agency, 2003, *Exposure Factor Handbook*; and Agency for Toxic Substances and Disease Registry, 2003, *ATSDR Public Health Assessment Guidance Manual*.[7]

under various environmental conditions, such as differences in occupational versus general environmental exposures. They may also vary by age or other demographic and social factors. For example, it is not uncommon for pesticide applicators to remove articles of clothing due to hot weather. If these activities are not accounted for in the equations, or using the default value of 0.1–0.3 m^2 of skin surface area in contact with the pesticide, the exposure of this group under these conditions would be underestimated.[4]

Exposure calculations are often more complicated than the equations may indicate. For example, dietary exposures to food can be a significant route for many persistent organic compounds and heavy metals. So information about people's activities, such as the amounts and types of food they eat will affect the exposure. For example, some parts of North America have subpopulations that, on average, eat much more fish than others. So, these people may be exposed to contaminants like mercury compounds and PCBs via dietary food ingestion at much higher rates than others. Likewise, some regions have higher red meat consumption, which is a route of dioxin exposure. For example, in Canada, food ingestion is estimated to contribute 96% of the amount of human exposure to dioxins and furans.[5]

Thus, food ingestion dose (ID_f) of a contaminant can be estimated:

$$ID_f = \frac{\sum_{i=1}^{n} CL_i \times CR_i \times EF}{BW} \qquad \text{Equation 10–5}$$

Where, ID_f is in units of $mg\,kg^{-1}\,d^{-1}$

CL_i = Concentration of contaminant in food group i ($mg\,g^{-1}$);
CR_i = Consumption rate of food group i ($g\,d^{-1}$);
EF = Exposure factor (unitless);
BW = Body weight (kg);
n = Total number of food groups.

So, if we are concerned about possible emissions from an incinerator, we would want to calculate the food ingestion exposure dose for emitted contaminants. For example, we would likely want to calculate how much cadmium (Cd) people are exposed to by eating homegrown vegetables. Thus, we would apply the CR to the percentage of food that is homegrown (PH). The factors for food intake can be found at ATSDR's website: http://www.atsdr.cdc.gov/HAC/HAGM/app-e.pdf.

Calculating exposure from homegrown foods must consider the percentage of contaminated food that is homegrown, so our intake equation becomes:

$$ID_f = \frac{\sum_{i=1}^{n} CL_i \times CR_i \times EF \times PH_i}{BW} \qquad \text{Equation 10–6}$$

Where, PH_i = Percentage of food group that is homegrown (Table E.7 from the ATSDR website).

The units for CR values were converted to $g\,d^{-1}$ from published $g\,kg^{-1}\,d^{-1}$ by multiplying by 60 kg.[6]

If a garden survey provides Cd concentrations (i.e., CL) in vegetables as shown in the table, the calculation of the food ingestion exposure dose for Cd through garden crop contamination can be calculated using consumption rates and percentage of foods that are homegrown as published in the U.S. EPA's *Exposure Factors Handbook* (http://www.epa.gov/ordntrnt/ORD/WebPubs/exposure/).

The sum of each food type's exposure dose represents the garden vegetable contribution to the overall Cd exposure, i.e. $0.036\,mg\,kg^{-1}\,d^{-1}$. The same procedure would be used for every contaminant of concern. Since dietary habits can vary widely, even within the same town, it is recommended that a dietary survey be completed before estimating contaminant exposure by way of food ingestion. However, using the published values can be a good first step in assessing the food ingestion component of exposure.

TABLE 10–4
Hypothetical Exposure Doses of Cadmium to Persons Downwind from an Incinerator, Calculated Using Default Consumption Rates and Percentage of Foods That Are Homegrown as Published in the U.S. EPA's *Exposure Factors Handbook* (http://www.epa.gov/ordntrnt/ORD/WebPubs/exposure/) and from Methods Published by ATSDR, 2004, *Public Health Assessment Guidance Manual* (Update), Appendix E: Calculating Exposure Doses (http://www.atsdr.cdc.gov/HAC/PHAManual/appe.html)

Food Group	*Cd Concentration in Food Group ($mg\,g^{-1}$)*	*Consumption Rate of Food Group ($g\,d^{-1}$)*	*Home-grown Percent of Food Group of Total Food Group Consumed (%)*	*Exposure Factor (unitless)*	*Body Weight (kg)*	*Exposure Dose ($mg\,kg^{-1}\,d^{-1}$)*
Potatoes	0.02	65.6	3.8	1	70	0.0007
Dark green vegetables	0.01	10.8	4.4	1	70	0.00007
Deep yellow vegetables	0.51	8.8	6.5	1	70	0.004
Tomatoes	0.24	52.6	18.4	1	70	0.03
Other vegetables	0.01	79.0	6.9	1	70	0.0008
Total						0.036

Exposure Dose Example

Consider the ingestion of soil by an adult male working at a hazardous waste site. The soil contaminant concentration is $100\,mg\,kg^{-1}$ and a daily ingestion rate of $50\,mg\ day^{-1}$. The person is on-site five days per week, 50 weeks per year, for 30 years.

Solution

Use the specific equation for soil ingestion[9] from the ATSDR; which is the same as Equation 10–3, but with a conversion factor for soil.

$$ID_S = \frac{C \cdot IR \cdot E_F \cdot 10^{-6}}{BW} \qquad \text{Equation 10–7}$$

Where, IDs = Soil ingestion exposure dose ($mg\,kg^{-1}\ day^{-1}$)
C = Contaminant concentration ($mg\,kg^{-1}$)
IR = Soil ingestion rate ($mg\ day^{-1}$)
E_F = Exposure factor (unitless)
BW = Body weight (kg)

The conversion factor of $10^{-6}\,kg\,mg^{-1}$ converts the soil contaminant concentration (C) from $mg\,kg^{-1}$ soil to $mg\,mg^{-1}$ soil.

First, calculate the exposure factor:

$$E_F = \text{exposure frequency} \times \text{exposure duration} \div \text{exposure time}$$

$$E_F = \frac{(5\text{ days week}^{-1}) \cdot (50\text{ weeks year}^{-1}) \cdot 30\text{ years}}{(365\text{ days year}^{-1}) \cdot (70\text{ years})} = 0.29$$

$$ID_S = \frac{100\text{ mg kg}^{-1} \cdot 50\text{ mg day}^{-1} \cdot 0.29 \cdot 10^{-6}}{70\text{ kg}} = 2 \times 10^{-5}\text{ mg kg}^{-1}\text{ day}^{-1}$$

Incidentally, any worker at a hazardous waste site should be wearing personal protective equipment that would reduce the daily ingestion rate far below $50\,mg\,d^{-1}$.

E_F is dimensionless, so whether calculated as a lifetime dose or an exposure dose, the key clue is in the units, so both are expressions of exposure that can be input into risk characterization equations.[10]

Calculating Risk

In its simplest form, risk is an expression of the probability of harm from a defined activity. By extension, environmental risk is the probability that exposure to a contaminant will cause a specific harm to a part of the environment. Thus, environmental risk differs with regard to the target of the harm. Are we concerned about the health of human populations, ecosystems, or even nonliving resources, such as the pyramids in Egypt or the integrity of buildings?[11]

Risk is the product of exposure and hazard. For contaminants, it is the product of exposure to the contaminant and the slope of the dose-response curve. For carcinogens, risk is exposure times the slope factor:

$$\text{Cancer Risk} = \text{LADD} \times \text{SF} \qquad \text{Equation 10–8}$$

Thus, inserting the LADD and the inhalation slope factor of 3.00×10^{-1} from Table 9.5 into our vinyl chloride example, where the two LADD values are under consideration, the cancer risk to the neighborhood exposed for 20 years gives us $1.4\,\text{mg}\,\text{kg}^{-1}\,\text{day}^{-1} \times 0.3\,(\text{mg}\,\text{kg}^{-1}\,\text{day}^{-1})^{-1} = 0.43$. This is an incredibly high risk! The threshold for concern is often 1 in a million, while this is a probability of 43%.

Even at the shorter duration period (20 years of exposure instead of 70 years), the risk is calculated as $0.4 \times 0.3 = 0.12$. The combination of a very steep slope factor and very high lifetime exposures leads to a very high risk. Always bear in mind that risk assessment is both quantitative and qualitative. At times, simply using the slope factor gives ridiculous results. For example, if the exposure in the instance cited here were $1\,\text{mg}\,\text{kg}^{-1}\,\text{d}^{-1}$, the slope would be greater than 1! Mathematically, a probability value must lie between 0 and 1. One of the reasons for this is that the dose-response curves are derived from animal studies and other inexact sources (see Figure 9.3). Another reason is that the part of the curve used for the slope may be the "linearized" region of the curve. But, as shown in Figures 9.1, 9.2 and 9.3, dose-response curves are often sigmoidal (i.e., "S"-shaped) rather than purely linear, with a plateau of responses (in this case, tumors) at very high doses. This phenomenon is known as a "saturation effect," because doubling the dose cannot possibly double the response. In other words at some point in the curve, 51% of exposed individuals are expected to have tumors. At this point, doubling the dose cannot double the effect, because there cannot be 102% chance of tumerogenesis. This means that somewhere the linear relationship (i.e., direct relationship between dose and response) tapers off so that increasing dose still results in an increased effect, but at a decreasing rate. Modelers are interested in the specific region of the dose-response curve in Figure 9.3 from which the slope is calculated. In the curvilinear range, for example, this can be found from the tangent of the slope of the curve. So, the very steep slope factors in published tables, including those in Chapter 9, will not be valid for very high exposures, i.e., exposures greater than the linearized range of the dose-response curve.

In our exposure dose example, we can use the same formulation as we do with LADD, that is:

$$\text{Cancer Risk} = ID_S \times SF \qquad \text{Equation 10–9}$$

Recall that $ID_S = 2 \times 10^{-5}\,\text{mg}\,\text{kg}^{-1}\,\text{day}^{-1}$. If the contaminant were vinyl chloride, we would use its oral slope factor since this is an ingestion intake. Thus, the risk from ingesting soil contaminated with vinyl chloride would be $2 \times 10^{-5} \times 3.80$ or 0.000038. Although smaller than our inhalation example, this additional lifetime cancer risk remains a concern since it is a greater risk than 1 in a million.

Applying Cancer Risk Calculations to Cleanup Levels

Historically, environmental protection has been based upon two types of controls: technology-based and quality-based. Technology-based controls are determined by what is "achievable" from the current state of the science and engineering. In a way, these are feasibility-based. For example, over its history the Clean Air Act has called for "best achievable control technologies (BACT)," and more recently for "maximally achievable control technologies (MACT)." Both have reflected the reality that even though from an air quality standpoint it would be best to have extremely low levels of pollutants, technologies are not available or are not sufficiently reliable to reach these levels. In fact, mandating unproven or unreliable technologies may even exacerbate the pollution, such as in the early days of wet scrubbers on coal-fired power plants. Theoretically, the removal of sulfur dioxide could be accomplished by venting the power plant flue through a slurry of carbonate. However, the reliability of this technology was at the time unproven, so untreated emissions were released while the slurry systems were repaired. The tradeoff of the benefit of improved treatment over older methods was outweighed in these instances by the frequency of no treatment (see Figure 10.1).

Technology-based standards are a part of most environmental programs. Wastewater treatment, groundwater remediation, soil cleaning, sediment reclamation, drinking water supply, air emission controls, and hazardous waste site cleanup all are in part determined by availability and feasibility of control technologies.

Quality-based controls are those that are required to ensure that an environmental resource is in good enough condition to support a particular use. For example, a stream may need to be improved so that people can swim in it and so that it can be a source of water supply. Certain streams

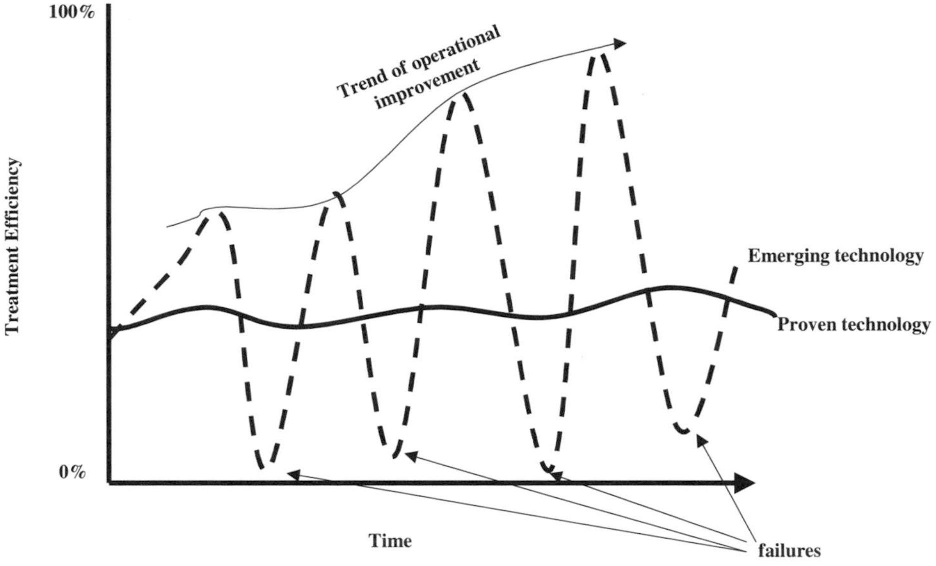

FIGURE 10.1 Hypothetical failure rate of new versus proven technologies.

may need to be protected more than others, such as the so-called "wild and scenic rivers." The parameters will vary, but usually include minimum levels of dissolved oxygen and maximum levels of contaminants. The same goes for air quality, where ambient air quality must be achieved, so that concentrations of contaminants listed as National Ambient Air Quality Standards, as well as certain toxic pollutants, are below levels established to protect health and welfare.

More recently, environmental protection has become increasingly "risk-based." Risk-based approaches to environmental protection, especially contaminant target concentrations, are designed to require engineering controls and preventive measures to ensure that risks are not exceeded. This embodies elements of both technology-based and quality-based standards. The technology assessment helps determine how realistic it will be to meet certain contaminant concentrations, while the quality of the environment sets the goals and means to achieve cleanup. Commonly, the threshold for cancer risk to a population is 1 in a million excess cancers. However, one may find that the contaminant is so difficult to remove due to its affinity for soil, that the best way to achieve the risk target is to simply fence the area in and allow no access. This is often unsatisfying, however, and the public and courts may mandate that even if costs are high and technology unreliable, a strong effort to clean up the site must be made.

Risk-based targets can be calculated by solving for the target contaminant concentration in the exposure and risk equations. Enumerating the intake equation within the risk equation gives:

$$\text{Risk} = \text{I} \times \text{SF} \qquad \text{Equation 10–10}$$

$$\text{Risk} = \frac{\text{C} \cdot \text{CR} \cdot \text{EF} \cdot \text{ED} \cdot \text{AF} \cdot \text{SF}}{\text{BW} \cdot \text{AT}} \qquad \text{Equation 10–11}$$

Solving for C:

$$\text{C} = \frac{\text{Risk} \cdot \text{BW} \cdot \text{AT}}{\text{I} \cdot \text{CR} \cdot \text{EF} \cdot \text{ED} \cdot \text{AF} \cdot \text{SF}} \qquad \text{Equation 10–12}$$

This is the target concentration for each contaminant needed to protect the population from the specified risk, such as $10^{-6.}$ In other words, this is the concentration needed to protect a population having an average body weight and over a specified averaging time from an exposure of certain duration and frequency that leads to a risk of 1 in a million.

Risk-Based Contaminant Cleanup Example 1

Again, consider the ingestion of soil by adults working at a hazardous waste site. Recall that the soil contaminant concentration is $100\,\text{mg}\,\text{kg}^{-1}$ and a daily ingestion rate of $50\,\text{mg}\,\text{day}^{-1}$. The person is on-site 5 days per week, 50 weeks per year, for 30 years. If this is the only pathway for the contaminant, how much does the soil have to be remediated to protect the workers so that their risk of cancer is below 10^{-6}, if the contaminant has a slope factor of $1 \times 10^{-2}\ (\text{mg}\,\text{kg}^{-1}\,\text{d}^{-1})^{-1}$?

Solution

Recall that the IDs $= 2 \times 10^{-5}\,\text{mg}\,\text{kg}^{-1}\ \text{day}^{-1}$. If the slope factor is $10^{-2}\ (\text{mg}\,\text{kg}^{-1}\ \text{day}^{-1})^{-1}$, the risk is the product of these two values, so it is 2×10^{-7}, already below the 10^{-6} risk, so no cleanup is needed to achieve this level.

Risk-Based Contaminant Cleanup Example 2

A well is the principal water supply for the Town of Chemtown. A study has found that the well contains $80\,\text{mg}\,\text{L}^{-1}$ tetrachloromethane (CCl_4). Assuming that the average adult in the town drinks $2\,\text{L}\,\text{d}^{-1}$ of water from the well and lives in the town for an entire lifetime, what is the lifetime cancer risk to the population if no treatment is added?

What concentration is needed to ensure that the population cancer risk is below 10^{-6}?

Solution

The lifetime cancer risk added to the Chemtown's population can be estimated using the LADD and slope factor for CCl_4. In addition to the assumptions given, we will use default values. Also, since people live in the town for their entire lifetimes, their exposure duration is equal to their typical lifetime. Thus, ED and TL terms cancel, leaving the abbreviated LADD =

$$\frac{(C)\cdot(CR)\cdot(AF)}{(BW)} \qquad \text{Equation 10–13}$$

Since the problem does not specify male or female adults, we will use the average body weight, assuming that there are about the same number of males as females. The absorption factor of CCl_4 given in Table 9.6 is 0.85, so the adult lifetime exposure is:

$$\text{LADD} = \frac{(80)\cdot(2)\cdot(0.85)}{(65)} = 4.2 \text{ mg kg}^{-1}\text{ day}^{-1}$$

Using the midpoint value between the default values $\left(\frac{15+40}{2} = 27.5 \text{ kg}\right)$ for body weight and default CR values $(1\,\text{L}\,\text{d}^{-1})$ the children lifetime exposure is:

$$\text{LADD} = \frac{(80)\cdot(1)\cdot(0.85)}{(27.5)} = 2.5 \text{ mg kg}^{-1}\text{ day}^{-1} \text{ for the first 13 years,}$$

and the adult exposure of 4.2 mg kg^{-1} day^{-1} thereafter.

Table 9.6 shows the oral SF for CCl_4 to be 1.30×10^{-1} kg day^{-1}, so the added adult lifetime risk from drinking the water is:

$$4.2 \times (1.30 \times 10^{-1}) = 5.5 \times 10^{-1}$$

And, the added risk to children is:

$$2.5 \times (1.30 \times 10^{-1}) = 3.3 \times 10^{-1}.$$

However, for children, environmental and public health agencies recommend an additional factor of safety beyond what would be used to calculate risks for adults. This is known as the "10X" rule, because a common additional risk of 10 times the amount is expected for children. Thus, with the added risk in this case, our reported risk would be 3.3. While this is statistically impossible (i.e., one cannot have a probability greater than 1 because it would mean that the outcome is more than 100% likely, which of course is impossible!). What this tells us is that the combination of a very high slope of the dose-response curve and a very high LADD leads to much needed protections, and removal of either the contaminants from the water or the provision of a new water supply. The city engineer or health department should mandate bottled water immediately.

The cleanup of the water supply to achieve risks below 1 in a million can also be calculated from the same information and reordering of the risk equation to solve for C:

$$\text{Risk} = \text{LADD} \times \text{SF} \qquad \text{Equation 10–14}$$

$$\text{Risk} = \frac{(\text{C}) \cdot (\text{CR}) \cdot (\text{AF}) \cdot (\text{SF})}{(\text{BW})} \qquad \text{Equation 10–15}$$

$$\text{C} = \frac{(\text{BW}) \cdot \text{Target Risk}}{(\text{CR}) \cdot (\text{AF}) \cdot (\text{SF})} \qquad \text{Equation 10–16}$$

Based on adult LADD, the well water must be treated so that the tetrachloromethane concentrations are below:

$$\text{C} = \frac{(65) \cdot 10^{-6}}{(2) \cdot (0.85) \cdot (0.13)} = 2.9 \times 10^{-4}\ \text{mg L}^{-1} = 290\ \text{ng L}^{-1}$$

Based on children's LADD, and the additional "10X," the well water must be treated so that the tetrachloromethane concentrations are below:

$$\text{C} = \frac{(27.5) \cdot 10^{-7}}{(1) \cdot (0.85) \cdot (0.13)} = 2.5 \times 10^{-5}\ \text{mg L}^{-1} = 25\ \text{ng L}^{-1}$$

The town has a major and important task ahead of it. They will have to remove the contaminant so that the finished water is less than six orders of magnitude below that of the untreated well water!

The cumulative risk from all routes of exposure is found by adding each risk together. Thus, after calculating the maximum contaminant concentrations, a cleanup target for each route is based on the total risk. For example, if people are exposed to a contaminant by inhalation and by dermal contact with soil, and these are expected to have the same contribution to the exposure (i.e., people's dose is the same from soil as from air), and we want to keep the total risks below 10^{-6}, one way to do so is to reduce the total risk in each route by half. Thus, the risk in the equation for inhalation is 5×10^{-7}, and the risk for soil ingestion is also 5×10^{-7}.

Oftentimes, however, it is much easier and more practical to control one route *versus* another. So, if cleaning the soil is more costly than stripping the volatile contaminants, the city may be able to reduce the exposures and risks more cost-effectively by devoting more attention to the air pathway. Also, if the soil is more contaminated (i.e., has higher concentrations), than the water, removal efficiencies may be better for the soil than the water because it is often easier to remove the first bulk of contaminants than it is to remove the final increment (see Figure 10.2). However, always keep in mind that state and federal regulations will require a certain level of cleanup in all media.

Non-Cancer Hazard and Risk Calculations

As we saw in the previous chapter, environmental health endpoints are categorized as either cancer or noncancer effects. While the calculations for either type of endpoint include functions of exposure and the dose-response, the manner in which the dose-response curve is used to ascertain risk is quite different for the two effects.

Unlike the cancer risk equation, which is principally concerned with the cancer potency as reflected in the slope of the dose-response curve, noncancer risk is concerned with the threshold below which noncancer effects are not manifested. Recall that the cancer curve has no such threshold, because any amount of carcinogen can, theoretically, result in a cancer response (i.e., the curve intercepts the x-axis and y-axis at 0). The area below this dose threshold on the x-axis (no observable adverse effects level, or NOAEL) is deemed "safe." This safe level would be one molecule below the NOAEL if the data and models used to draw the curve and establish the NOAEL were perfect. Science with its variability and uncertainty, obviously, never allows this to be the case, so when the safe dose (i.e., the reference dose, or RfD) is established, certain precautionary factors, known as modifying factors and uncertainty factors, are applied to move the safety point closer to zero. The RfD is the quotient of the NOAEL divided by the product of the uncertainty and modifying factors, so when the dose-response

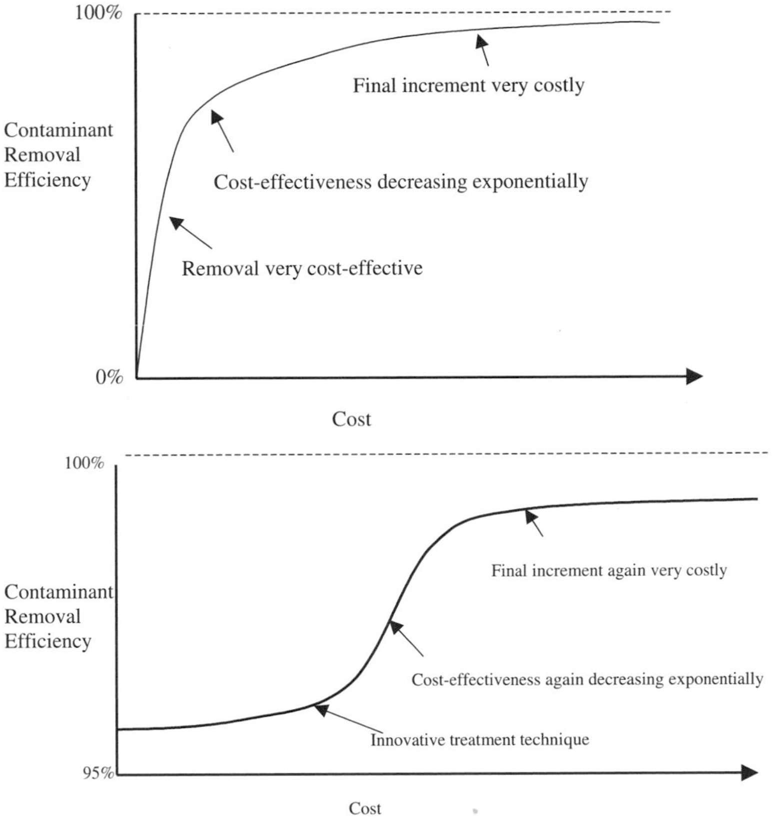

FIGURE 10.2 Prototypical contaminant removal cost-effectiveness curve. In the top diagram, during the first phase, a relatively large amount of the contaminant is removed at comparatively low costs. As the concentration in the environmental media decreases, the removal costs increase substantially. At an inflexion point, the costs begin to increase exponentially for each unit of contaminant removed, until the curve nearly reaches a steady state where the increment needed to reach complete removal is very costly. The top curve does not recognize innovations that, when implemented, as shown in the bottom diagram, can make a new curve that will again allow for a steep removal of the contaminant until it cost-effectiveness decreases. This concept is known to economists as the law of diminishing returns.

relationship has been established with large uncertainties and unreliable models, the only safe level of exposure would practicably be zero.

Sources such as those of the U.S. EPA's Integrated Risk Information System (IRIS) and the U.S. Department of Energy, Oak Ridge National Laboratory's Risk Assessment Information System (RAIS) provide updated information about noncancer health effects for both organic and inorganic contaminants. For example, noncancer information about the organic solvent tetrachloromethane (CCl_4) is available.

The CCl_4 oral chronic RfD:

- Is 7.00×10^{-4} mg kg^{-1} day^{-1}.
- Has a modifying factor of 1 (i.e., excellent information for extrapolations).
- Has an uncertainty factor of 1000 (i.e., questions about the quality of data).
- Is applied to the liver (i.e., the target organ), so the noncancer endpoint for oral route is hepatotoxicity. In this case the critical effect is liver lesions.
- Has an overall confidence rated "medium."

The CCl_4 dermal chronic RfD:

Is 4.55×10^{-4} kg^{-1} day^{-1}.
Is based on a gastrointestinal absorption factor of 0.6500.

Similarly, information regarding metals, including the toxicity information for each of their valence states, is demonstrated by the RAIS entry for chromium (Cr). For insoluble Cr^{3+} salts:

- The oral chronic RfD is 1.50 mg kg^{-1} day^{-1}.
- The oral chronic RfD has a modifying factor of 10.
- The oral chronic RfD has an uncertainty factor of 100.
- The overall confidence in the oral chronic RfD is low.
- The dermal chronic RfD is 7.00×10^{-3} mg kg^{-1} day^{-1}.
- The dermal chronic RfD is based on a gastrointestinal absorption factor of 0.0050.

For chromic acid (Cr^{6+}) mists:

- The oral chronic RfD is 3.00×10^{-3} mg kg^{-1} day^{-1}.
- The oral chronic RfD has a modifying factor of 3.
- The oral chronic RfD has an uncertainty factor of 300.
- The overall confidence in the oral chronic RfD is low.
- The inhalation chronic reference concentration (RfC) is 8.00×10^{-3} mg kg^{-1} day^{-1}.
- The inhalation chronic RfC has a modifying factor of 1.
- The inhalation chronic RfC has an uncertainty factor of 90.
- The inhalation chronic RfC study target tissue is nasal.
- The inhalation chronic RfC study critical effect is septum atrophy.
- The overall confidence the inhalation chronic RfC is low.
- The dermal chronic RfD is 6.00×10^{-3} mg kg^{-1} day^{-1}.
- The dermal chronic RfD is based on a gastrointestinal absorption factor of 0.0200.

The noncancer effect entries also include references to the studies that served as the bases for these values.

The Hazard Quotient

The hazard quotient (HQ) is the ratio of the potential exposure to a specific contaminant to the concentration at which no adverse effects are expected. The HQ is the ratio of a single contaminant exposure, over a specified time period, to a reference dose for that contaminant, derived from a similar exposure period:

$$HQ = \frac{Exposure}{RfD} \qquad \text{Equation 10–17}$$

If the calculated HQ < 1, no adverse health effects are expected to occur at these contaminant concentrations. If the calculated HQ > 1, there is a likelihood that adverse outcome can occur at these concentrations.

For example, the chromic acid (Cr^{6+}) mists dermal chronic RfD is $6.00 \times 10^{-3}\,mg\,kg^{-1}\,day^{-1}$. If the actual dermal exposure of people living near a plant is calculated (e.g., by intake or LADD) to be $4.00 \times 10^{-3}\,mg\,kg^{-1}\,day^{-1}$, the HQ is 2/3 or 0.67. Since this is less than 1, one would not expect people chronically exposed at this level to show adverse effects from skin contact. However, at this same chronic exposure, or $4.00 \times 10^{-3}\,mg\,kg^{-1}\,day^{-1}$, to hexavalent chromic acid mists via oral route, the RfD is $3.00 \times 10^{-3}\,mg\,kg^{-1}\,day^{-1}$, meaning the HQ = 4/3 or 1.3. The value is greater than 1, so we cannot rule out adverse noncancer effects, such as gastric ulcers and mucosa erosions.

The calculated HQ value cannot be translated into a probability that adverse health effects will occur (i.e., it is not actually a metric of risk). The HQ is a benchmark that can be used to estimate the likelihood of risk.[12] It is not even likely to be proportional to the risk. Thus, an HQ > 1 does not necessarily mean that adverse effects will occur.

Noncancer hazard estimates often have substantial uncertainties from a variety of sources. Scientific estimates of contaminant concentrations, exposures, and risks always incorporate assumptions to the application of available information and resources. Uncertainty analysis is the process used by scientists to characterize just how good or bad the data are in making these estimates.

The Hazard Index

The HQ values are for individual contaminants. The hazard index (HI) is the sum of more than one HQ value to express the level of cumulative noncancer hazard associated with inhalation of multiple pollutants (e.g., certain classes of compounds, such as solvents, pesticides, dioxins, fuels, etc.):

$$HI = \sum_{1}^{n} HQ \qquad \text{Equation 10–18}$$

An HI can be developed for all pollutants measured, such as the 32 compounds measured in New Jersey as part of the National Air Toxics Assessment (Figure 10.3). An HI can also be site-specific: for example, if an environmental audit shows that only CCl_4 and Cr^{6+} were detected by sampling of soil. Recall that the previously calculated Cr^{6+} dermal HQ was 0.67. The dermal chronic RfD of CCl_4 is $4.55 \times 10^{-4}\,mg\,kg^{-1}\,day^{-1}$. If the exposure is $1.00 \times 10^{-4}\,mg\,kg^{-1}\,day^{-1}$, the HQ for chronic dermal exposure to tetrachloromethane is $1.00/4.55 = 0.22$.

Thus, the HI for this site is $0.67 + 0.22 = 0.89$. Since the HI is under 1, the noncancer effect is not expected at these levels of exposure to the two compounds. However, if the chronic dermal exposure to CCl_4 had been $2.00 \times 10^{-4}\,mg\,kg^{-1}\,day^{-1}$, the HQ for CCl_4 would have been 0.44, and the HI would have been calculated as $0.67 + 0.44 = 1.11$. This is a benchmark that

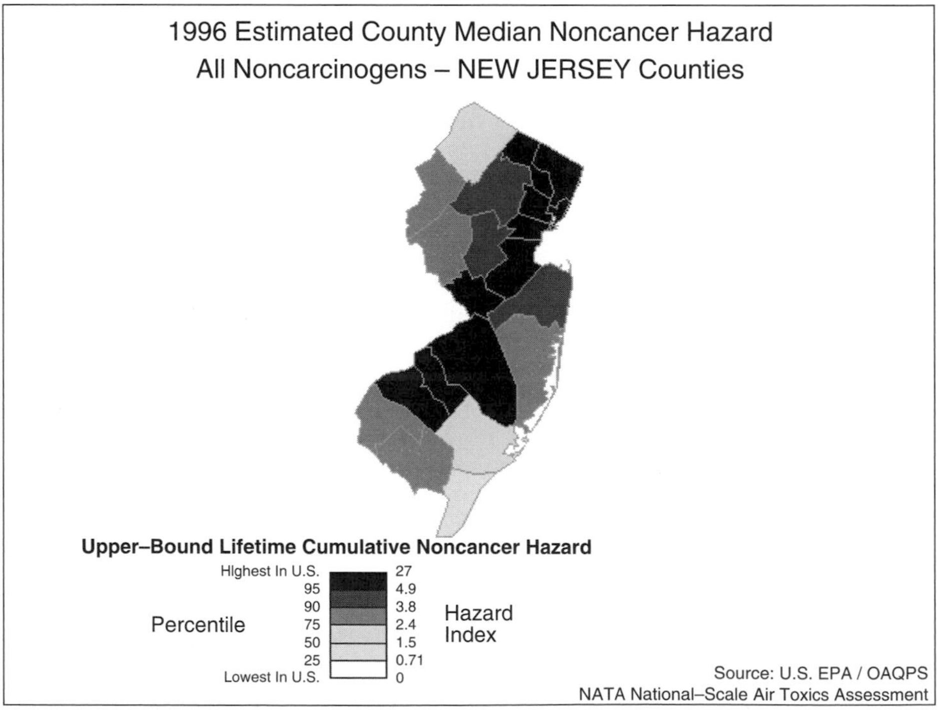

FIGURE 10.3 Noncancer hazard index for 32 air toxics included in the Clean Air Act, based upon inhalation exposure data in New Jersey from political subdivisions. Estimates do not include indoor emissions and are based on exposure estimates for the median individual within each census tract, which EPA considers to be a "typical" exposure, meaning that individuals may have substantially higher or lower exposures based on their activities. (Source: U.S. Environmental Protection Agency, National Air Toxics Assessment.)

indicates that the cumulative exposures to the two contaminants may lead to noncancer effects.

Comprehensive Risk Communication

The amount of data and information regarding contaminant concentrations, exposure, and effects can be overwhelming when presented to the public and clientele. Thus, these data must be reduced into meaningful formats. A recent example of how the information discussed in this and the previous chapters can be presented is that of the Ohio Environmental Protection Agency's Urban Air Toxic Monitoring Program,[13] which addresses potential risks in large urban areas with many industrial air pollution sources. Air quality samples were collected between 1989 and 1997 near a large industrial area in Cuyahoga County. The contaminant concentrations are typical of many urban areas, and the concentrations are expected to be lower in the future. Pollution prevention activities by industry, vehicle emission tests by motorists, and mandates in the Federal Clean Air Act will all help reduce toxics in the air. Samples were analyzed for volatile organic compounds (VOCs), heavy metals, and polycyclic aromatic hydrocarbons (PAHs). The Ohio agency conducted a risk assessment based upon both the cancer and non-cancer health risks, assuming that an individual is exposed constantly to the same concentration of the pollutant for a lifetime (i.e., ED = TL). The results of the cancer health risk assessment are provided in Table 10.5. Heavy metals contributed the majority of the cancer risk (about 66%).

Each category in the table shows the cumulative risks from exposure to all compounds detected under a specific contaminant class. The U.S. EPA has defined acceptable exposure risks for individual compounds to range from 10^{-6} to 10^{-4}. Also, it is quite possible that one or a few contaminants are contributing the lion's share of risk to each contaminant class. For example, a particularly carcinogenic PAH, like benzo(a)pyrene or dibenz(a,h)anthracene (each with an inhalation cancer slope factor of 3.10), could account for most of the risk, even if its concentrations are about the same as other PAHs. In fact, this appears to be the case when looking at the individual chemical species listed in Table 10.6 that were used to derive

TABLE 10.5
Cumulative Cancer Risk Based on Air Sampling in Cuyahoga County, Ohio, 1989–1997

Source of Cancer Risk	*Total Estimated Risk*
VOCs	0.515×10^{-4}
Heavy metals	1.21×10^{-4}
PAHs	0.123×10^{-4}
Total carcinogenic risk	1.85×10^{-4}

Source: Ohio Environmental Protection Agency, 1999.

TABLE 10.6
Individual Chemical Species Used to Calculate Cancer Risks Shown in Table 10.5

Compound	*Carcinogenic Unit Risk ($m^3\,\mu g^{-1}$)*	*Source**	*Average Concentration ($\mu g\,m^{-3}$)*	*Carcinogenic Risk*
VOCs				
Methyl chloride	1.8 E-06	HEAST	0.68	1.22 E-06
Dichloromethane	4.7 E-07	IRIS	2.06	9.70 E-07
Trichloromethane	2.3 E-05	IRIS	0.27	6.29 E-06
Benzene	8.3 E-06	IRIS	3.91	3.25 E-05
Carbon tetrachloride	1.5 E-05	IRIS	0.55	8.30 E-06
Trichloroethene	1.7 E-06	HEAST	0.55	9.42 E-07
Tetrachloroethene	9.5 E-07	HEAST	1.07	1.02 E-06
Styrene	5.7 E-07	HEAST	0.49	2.81 E-07
SUM				**5.15 E-05**
HEAVY METALS				
Arsenic	4.30 E-03	IRIS	0.00271	1.17 E-05
Cadmium	1.80 E-03	IRIS	0.00765	1.38 E-05
Chromium (total)**	1.20 E-02	IRIS	0.00800	9.60 E-05
SUM				**1.21 E-04**
		Toxic Equivalence		
PAHs				
Benzo(a)pyrene***	2.10 E-03	1	0.006	1.26 E-06
Benzo(a)anthracene	2.10 E-04	0.1	0.0048	1.01 E-06
Benzo(b)fluoranthene	2.10 E-04	0.1	0.0023	4.83 E-07
Benzo(k)flouranthene	2.10 E-04	0.1	0.0007	1.47 E-07
Chrysene	2.10 E-05	0.01	0.0047	9.87 E-08
Dibenz(a,h)anthracene	2.10 E-03	1	0.0047	8.61 E-06
Indeno [1,2,3-cd]pyrene]	2.10 E-04	0.1	0.0031	6.51 E-07
SUM				**1.23 E-05**
TOTAL CARCINOGENIC RISK				**1.85 E-04**

* HEAST = U.S. EPA's Health Effects Assessment Summary Tables; IRIS = U.S. EPA's Integrated Risk Information System.
Source for PAHs is the ATSDR's toxicological profiles for PAHs.
** Estimation based on slope factor of chromium VI.
*** Estimation based on slope factor of oral route.
Source: Ohio Environmental Protection Agency, 1999, Cleveland Air Toxics Study Report.

the risks. Likewise, the VOC cancer risk was largely determined by the concentrations of benzene, while the heavy metals, although largely influenced by Cr^{6+}, were more evenly affected by arsenic and cadmium.

The cancer risk calculations are based on the unit risk estimate (URE), which is the upper-bound excess lifetime cancer risk that may result from continuous exposure to an agent at a defined concentration. For inhalation this concentration is 1 $\mu g\,m^{-3}$ in air. For example, if the URE = 1.5×10^{-6} per $\mu g\,m^{-3}$, then 1.5 excess tumors are expected to develop per million population being exposed daily for a lifetime to 1 μg of the contaminant per cubic meter of air.

The cancer risk reported for each individual contaminant is below the level designated by federal health agencies as acceptable, and falls within the range of risks expected for large cities, with their numerous sources of toxic air contaminants (i.e., the so-called "urban soup").

The noncancer hazard index calculations are provided in Table 10.7. Noncarcinogenic health effects include developmental, reproductive, or cardiovascular health problems. Any total hazard index number below 100% (1.00) is generally regarded as a safe level of exposure.

As was the case for cancer risk, a few compounds can drive the noncancer hazard index. For example (as shown in Table 10.8), 3-chloropropene and tetrachloromethane (shown as carbon tetrachloride) account for an HI of 0.53, while all the other measured VOCs account for only 0.10. In addition, these two compounds account for almost 82% of the total noncarcinogenic risk estimates. This shows that the importance of ensuring that all potential contaminants are measured.

A wealth of information is available for every step in risk assessment, from toxicological studies supporting the hazard identification and dose-response relationships of chemical contaminants, to data needed to conduct exposure assessments, to models needed to characterize risks to populations and subpopulations.

TABLE 10.7
Cumulative Hazard Index Based on Air Sampling in Cuyahoga County, Ohio, 1989–1997

Source of Noncarcinogenic Risk	*Hazard Index (2 Significant Figures)*
VOCs	0.63
Heavy metals	0.008
PAHs	0.012
Total noncarcinogenic risk	0.65

Source: Ohio Environmental Protection Agency, 1999.

Table 10.8
Individual Chemical Species Used to Calculate Noncancer Hazard Indexes Shown in Table 10.7

Compound	*Reference Conc. (μg/m³)*	*Source*	*Average Conc. (μg/m³)*	*Hazard Percent Index (HI)*
VOCs				
Dichlorodifluoromethane	7.00 E+02	IRIS**	2.44	0.35
Trichlorofluoromethane	1.05 E+03	IRIS**	3.94	0.37
Dichloromethane	2.10 E+02	IRIS**	2.06	0.98
3-chloropropene	1.00 E+00	IRIS (RfC)	0.31	30.60
1,1,2-trichloro-1,2,2-trifluoroethane	1.05 E+05	IRIS**	0.54	5.16 E-04
Trichloromethane	3.50 E+01	IRIS**	0.27	0.78
1,1,1-trichloroethane	1.00 E+03	HEAST	2.12	0.21
Carbon tetrachloride	2.45 E+00	IRIS**	0.55	22.57
Toluene	4.00 E+01	IRIS (RfC)	7.00	1.75
Tetrachloroethene	3.50 E+01	IRIS**	1.07	3.06
Ethylbenzene	1.00 E+03	IRIS (RfC)	1.17	0.12
m+p-xylene	7.00 E+02	HEAST	4.85	0.69
Styrene	1.00 E+03	IRIS (RfC)	0.49	0.05
o-xylene	7.00 E+02	HEAST	1.27	0.18
1,2,4-trichlorobenzene	3.50 E+01	IRIS**	0.55	1.58
p-dichlorobenzene	8.00 E+02	IRIS (RfC)	0.62	0.08
SUM				**63.38**
PAHs				
Naphthalene	1.40 E+01	HEAST	0.152	1.08
Acenaphthene	2.10 E+02	IRIS**	0.027	0.01
Fluorene	1.40 E+02	IRIS**	0.028	0.02
Anthracene	1.05 E+03	IRIS**	0.004	0.00
Fluoranthene	1.40 E+02	IRIS**	0.025	0.02
Pyrene	1.05 E+02	IRIS**	0.017	0.02
SUM				**1.15**
HEAVY METALS				
Arsenic	1.05 E+00	IRIS**	0.0027	0.26
Cadmium	1.75 E+00	IRIS**	0.0077	0.44
Chroumium (total)***	1.75 E+01	IRIS**	0.008	0.05
Nickel and compounds	7.00 E+01	IRIS**	0.0289	0.04
Zinc and compounds	1.05 E+03	IRIS**	0.1904	0.02
SUM				**0.80**
TOTAL HEALTH EFFECTS PERCENTAGE				**65.33**

** Estimation based on route to route extrapolation from RfD.
Source: Ohio Environmental Protection Agency, 1999, Cleveland Air Toxics Study Report.

Reference Dose Example 1

Mathematically, why can an RfC or RfD not be calculated for a carcinogen? What characteristic of the noncancer dose-response curve is most important for calculating noncancer risk?

Answer and Discussion

Recall that the equations for the RfC and RfD (see Equation 9–1) contain the NOAEL in the numerator. Carcinogens, as a precautionary measure, have no safe level of dose or exposure (i.e., no level below which no tumorigenesis will occur), so they have no NOAEL. Thus, no reference dose or reference concentration can be calculated.

The most important characteristic of the noncancer dose-response curve is the NOAEL, since that is the point on the dose axis from which the displacement is made to the safe dose (i.e., RfD), as shown in Figure 10.4.

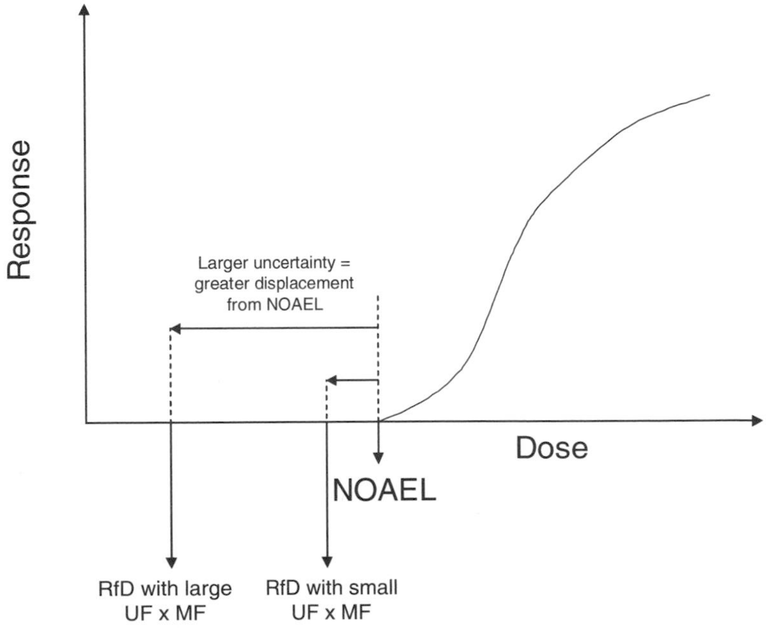

FIGURE 10.4 Effect of uncertainty on reference dose.

Reference Dose Example 2

Numerous diseases fall under the government's moniker "noncancer" risk. An important noncancer endpoint is neurotoxicity. What is the reference dose (RfD) for a neurotoxic substance if its NOAEL is $0.9\,\text{mg}\,\text{kg}^{-1}\ \text{day}^{-1}$, its data used in the neurotoxicity model is only derived from animal studies showing subchronic effects (UF = 100), and its data have certain weaknesses in quality assurance and approach (MF = 100)?

Solution and Discussion

The RfD equation is:

$$\text{RfD} = (\text{NOEAL})/(\text{UF} \times \text{MF}) = (9 \times 10^{-1})/10^{4}$$
$$= 9 \times 10^{-5}\ \text{mg kg}^{-1}\ \text{day}^{-1} = \underline{9 \times 10^{-2}\ \mu\text{g kg}^{-1}\ \text{day}^{-1}}$$

This means that a population exposed to this neurotoxin at this dosage would not be expected to show neurological disorders.

Reference Dose Example 3

Are there times when even at exposure levels below the RfD we should be concerned about a noncancer hazard?

Answer and Discussion

Yes. You should still be concerned if the compound acts synergistically with another compound. If the effects of compound A and compound B are synergistic and one or both is neurotoxic, then the RfD for exposure to both would have to be reduced. For example, if the RfD for compound A is $9 \times 10^{-2}\,\mu\text{g}\,\text{kg}^{-1}\ \text{day}^{-1}$ and the RfD for compound B is $1 \times 10^{-2}\,\mu\text{g}\,\text{kg}^{-1}\ \text{day}^{-1}$, but research shows that the neurotoxic effect is twice as potent when an organism is exposed to both, then we know that the synergistic RfD would have to be less than the lower RfD (i.e., in this case, $<9 \times 10^{-2}\,\mu\text{g}\,\text{kg}^{-1}\ \text{day}^{-1}$). Pharmacokinetic modeling would have to be applied to know just how much lower the RfD should be.

Another possibility is that the compound may elicit other noncancer health effects, such as immune and endocrine system dysfunctions. In this case, separate RfDs would have to be calculated for each disease, because the manifestation of the health effects would occur at different thresholds (i.e., NOAELs) and uncertainties in the data (new UF and MF values).

Reference Dose Example 4

What is the population risk of your town if the maximum daily exposure from the drinking water source a compound is $19\,\mu g\,kg^{-1}$ and a safe dose for that compound is considered to be $9\,\mu g\,kg^{-1}\,day^{-1}$? What does this mean for the good people of your town?

Solution and Discussion

Population risk is the quotient of maximum daily dose (MDD) and allowable daily intake (ADI). The ADI is set by regulatory agencies (especially the Food and Drug Administration, ATSDR, EPA, and state health departments) as a recommended maximum daily exposure for individuals, so we can use it here as a "safe" dose. Thus:

$$\text{Population Risk} = \text{MDD}/\text{ADI} = (19\ \mu g\ kg^{-1})/(9\ \mu g\ kg^{-1}) = 2.1$$

This means that the population being studied is twice as likely to exhibit the health effects as that of a population exposed at what the authorities consider to be a safe dose (or concentration) from their water supply. If you are the town engineer, this also means you would be expected to propose measures to reduce the concentrations of the compound in your water supply system below $9\,\mu g\,kg^{-1}$.

Risk Example

Why can risk ever be zero?

Solution and Discussion

Risk is the probability of an adverse outcome. In this instance, it is the probability that an individual, or population, or ecosystem will experience harm after being exposed to a chemical. Keep in mind that risk is specific in terms of the type of harm (i.e., a specific disease in a specific organ of a specific population within a specified time period). Thus, risk becomes very difficult to calculate for mixtures of chemicals (are the effects additive, synergistic, or protective?).

Since risk is a probability, it is dimensionless; it is a fraction. Population risk is the number of individuals with harmful outcomes (numerator) over the number of individuals in a population. Risk is calculated by multiplying the potency of the hazard (i.e., the slope of the dose-response curve) by the exposure. If the hazard is cancer, the slope of the dose-response curve is published for numerous substances. Thus, *mathematically at least*, risk can be zero if either the hazard or exposure is zero. If a substance is not carcinogenic (over time, we find that it is an "E Chemical"), no matter what the exposure, the risk of cancer will be zero, because any number multiplied by zero is zero. Similarly, if exposure is zero, risk is zero.

However, scientifically, nothing is "risk free," simply because there are so many hazards out there. For example, water is not carcinogenic, but you can drown it in, become asphyxiated by it, be burned when it is sprayed on an oxidizing chemical, and slip on it in its solid phase (ice).[14]

Notes and Commentary

1. U.S. Environmental Protection Agency, 2003, Risk Assessment Guidance for Superfund (RAGS), Vol. I: *Human Health Evaluation Manual* (Part E, Supplemental Guidance for Dermal Risk Assessment) Interim.
2. See M. Derelanko, 1999, *CRC Handbook of Toxicology*, "Risk Assessment," edited by M.J. Derelanko and M.A. Hollinger, CRC Press, Boca Raton, Fla., for equations related to all routes and pathways of exposure.
3. Default value for absorption = 1; that is, unless otherwise specified, one can assume that all of the contaminant is absorbed.
4. For controlled pesticides, state and federal agencies require that applicators be "certified." Even for less strictly controlled pesticides, label instructions advise skin covering. However, I have seen people applying pesticides in sandals, shorts, and no shirts. This occurs "down on the farm," but it may actually be more prominent at beachfront properties (e.g., maintenance people applying herbicides and insecticides to beach properties) where the "culture" dictates very little skin covering, no matter what the chore.
5. Health Canada, 1995, *Investigating Human Exposure to Contaminants in the Environment: A Handbook for Exposure Calculations*, Great Lakes Effects Program, Ottawa, Canada, p. 28.
6. Method given in: ATSDR, 2004, Public Health Assessment Guidance Manual (Update), Appendix E: Calculating Exposure Doses, http://www.atsdr.cdc.gov/HAC/PHAManual/appe.html.
7. These factors are updated periodically by the U.S. EPA in the *Exposure Factor Handbook* at www.epa.gov/ncea/exposfac.htm.
8. The definition of "child" is highly variable in risk assessment. The *Exposure Factors Handbook* uses these values for children between the ages of 3 and 12 years.
9. See ATSDR's *Public Health Assessment Guidance Manual* (http://www.atsdr.cdc.gov/HAC/HAGM).

10. This is purely mathematical. Since risk is a probability, risk is unitless. Thus, the units of the dose-response curve slope or slope factor are the inverse of those for exposure, because risk is the product of exposure and slope. This means that exposure's units must be $mg\,kg^{-1}\,day^{-1}$, and the units of the dose-response slope must be $kg{\cdot}d\,mg^{-1}$.
11. The concern for the built environment or cultural resources is known as "welfare" considerations. Welfare risk assessments are less discussed in environmental circles than health and ecological assessments, but are very important. For example, benefit/cost analyses, which are frequently part of environmental assessments and environmental impact statements, include welfare considerations. Also, language in state and federal legislation includes welfare considerations, such as the Clean Air Act's intent to protect national monuments and parks from impairment of visibility, by placing greater protections, or more stringent requirements on ambient sulfur dioxide and particulate concentrations. The logic of such welfare protections is that there is little point in having such beautiful resources if one's ability to see them is impaired. There are also indirect or secondary benefits to welfare considerations. For example, welfare protection often requires even lower levels of contaminant concentrations than do health and ecosystem protection. As a result, an additional margin of safety is provided. Your lungs will benefit from the lower levels of pollution even if the more protective regulations are the result of welfare. Thus, if the Safe Drinking Water Act invokes secondary standards (maximum contaminant levels, or MCLs) to make your water taste better, the additional treatment may lead to even better health-related standards, such as removal of heavy metals even beyond the health related MCL. Finally, aesthetics itself can be beneficial to human health and the condition of ecosystems. Many argue, and I agree, that people are intimately tied to their surroundings. The feedbacks that an aesthetically pleasing neighborhood and building provide add to the health of individuals and populations. Unfortunately, the converse is equally true. Displeasing surroundings add to stresses. The spiritual value of creation must not be underestimated, although it is difficult to quantify.

    The foregoing discussion leads to the need to improve the way the environment is "valued." Valuation is problematic in environmental assessment. The previously mentioned benefit/cost relationships are usually done following a utilitarian model espoused by John Stuart Mill. This calls for the "greatest good for the greatest number." The way such good is usually determined is by a monetized system of valuation. Monetization is biased toward resources that are readily capitalized, that is, the so-called "nonuse" values are not easily input. A resource that adds to the gross domestic product is added to the "benefit" side of the B:C ratio. To illustrate, strictly applying the monetization model to a hypothetical standing forest, made up of 50% timber (harvestable trees) and 50% old growth trees (protected forest), only the timber is included in the benefit because it has utility. Although efforts are made to try to monetize the protected growth, its actual value (e.g., to biodiversity, disease resistance, sustainability, and aesthetics) is often not represented, or at least underrepresented. There are a few brave souls working to develop better valuation models for environmental resources, but they must overcome much inertia. Some of these new approaches to environmental valuation identified by the U.S. Department of Energy, 1997, C. Ulibarri and K. Wellman, *Natural Resource Valuation: A Primer on Concepts and Techniques,* http://tis.eh.doe.gov/oepa/guidance/cercla/valuation.pdf are:

Economic Valuation Techniques	*Types of Benefits*
Market price approach	Recreational/existence value
Appraisal methods	"Fair market values" of land
Resource replacement cost	Groundwater resource values
Travel cost method	Recreational/existence values
Random utility models	Recreational/existence values
Hedonic price method	Groundwater/land value Human health/value of life
Factor income approach	Fresh water supply
Contingent valuation method	Use/nonuse values
Benefit-transfer method	Air quality/visibility
Unit-day value method	Recreational value
Ecological valuation approach	Gross primary energy value and intrinsic value

12. National Research Council, 1994, *Science and Judgment in Risk Assessment,* National Academy Press, Washington, D.C.
13. Ohio Environmental Protection Agency, 1999, Press Release, August 17, 1999.
14. In 1998, my son, who at that time was an engineering student at North Carolina State University, gave me a copy of the May 1998 edition of *Broadside,* a student publication. The issue contained an article on the hazards of dihydrogen monoxide and conducted a survey that found the majority of students wanted to ban this chemical! A very enlightening and funny website that delves even further into the hazards of dihydrogen monoxide is http://www.dhmo.org/.

Part IV

Interventions to Address Environmental Contamination

CHAPTER 11

Contaminant Sampling and Analysis

The first step in controlling contaminants is to know as precisely and accurately as possible where they are and at what concentrations they exist. Therefore, we will begin with an overview of environmental contaminant sampling and analysis.

Environmental Monitoring

The terms *monitoring* and *sampling* are frequently used interchangeably by environmental scientists, but monitoring is the more inclusive term. Environmental monitoring is dependent upon the quality of sample collection, preparation, and analysis. Sampling is a statistical term, and usually also a geostatistical term.

The environmental sample is a small portion of air, water, soil, biota, or other environmental media (e.g., paint chips, food, etc., for indoor monitoring) that represents a larger entity. For example, a sample of air may consist of a canister or bag that holds a defined quantity of air that will be subsequently analyzed. The sample is representative of a portion of an air mass. Ideally, a sufficient number of samples are collected and their results aggregated to ascertain with defined certainty the quality of an air mass. More samples will be needed for the New York City air shed than for that of a small town. However, intensive sampling is often needed for highly toxic contaminants and for sites that may be particularly critical, such as a national park, a hazardous waste site, or an "at risk" neighborhood (such as one near a chemical manufacturing facility). Like other statistical measures,[1] the sample is used to infer the condition of the larger population or larger area (in the case of geostatistics). A simple example of the representativeness of a sample is that if one wishes to characterize the amount of trichloroethane (TCA) in a lake and gathers a single 500 mL sample in the middle of the lake that contains 1 million liters of water, the sample represents only 5×10^{-7} of the lake's water volume. It also is limited in location vertically and horizontally, so there is much uncertainty. However, if 10 samples are taken at 10 spatially distributed sites, the inferences are improved. Further, if the samples were taken in each season, then there would be some improvement to the understanding of intra-annual variability. And, if the sampling is continued for several years, the inter-annual variability is even better characterized. Such sampling can be conducted on any of the environmental media.

Before the samples are collected and arrive at the laboratory, the general monitoring plan, including quality assurance provisions, must be in place. The plan describes the procedures to be employed to examine a particular site. These procedures must be strictly followed to investigate the extent of contamination of an environmental resource. The plan describes in detail the kinds of samples to be taken (e.g., real-time probes, sample bags, bottles, and soil cores), the number of samples needed (based upon statistical tests to be performed), methods for collection, sample handling, and transportation. The quality and quantity of samples are determined by data quality objectives (DQOs), which are defined by the objectives of the overall contaminant assessment plan. DQOs are qualitative and quantitative statements that translate nontechnical project goals into scientific and engineering outputs needed to answer technical questions. Quantitative DQOs specify a required level of scientific and data certainty, while qualitative DQOs express decisions goals without specifying those goals in a quantitative manner. Even when expressed in technical terms, DQOs must specify the decision that the data will ultimately support, but not the manner in which the data will be collected. DQOs guide the determination of the data quality that is needed in both the sampling and analytical efforts. The U.S. Environmental Protection Agency has listed three examples of the range of detail of quantitative and qualitative DQOs:[2]

1. *Example of a less detailed, quantitative DQO*: Determine with greater than 95% confidence that contaminated surface soil will not pose a human exposure hazard.
2. *Example of a more detailed, quantitative DQO*: Determine to a 90% degree of statistical certainty whether or not the concentration of mercury in each bin of soil is less than 96 ppm.
3. *Example of a detailed, qualitative DQO*: Determine the proper disposition of each bin of soil in real-time using a dynamic work plan and a field method able to complete analyses and report results of lead (Pb) in the soil samples within two hours of sample collection.

In other words, if all we needed to know were the seasonal change in pH near a fish hatchery, only a few samples using simple pH probes would be defined as the DQO. However, if the environmental assessment or audit calls for the characterization of year-round water quality for trout in the stream, the sampling plan's DQO may dictate that numerous samples at various spatial locations be continuously sampled for inorganic and organic contaminants, turbidity, nutrients, and ionic strength. The sampling plan must include all environmental media, such as soil, air, water, and biota, that are needed to characterize the exposure. It should explicitly point out which methods will be used. For example, if highly toxic chemicals like dioxins are being monitored, the U.S. EPA specifies particular sampling and analysis methods.[3]

The geographic area where data are to be collected is defined by distinctive physical features such as volume (e.g., of water body or air mass) or area (e.g., metropolitan city limits), the soil within the property boundaries down to a depth of 6 cm, a specific water body, length along a shoreline, or the natural habitat range of a particular animal species. Care should be taken to define boundaries. For example, Figure 11.1 indicates a study area by a grid, wherein each cell is sampled.[4]

The target population may be divided into relatively homogeneous subpopulations within each area or subunit. This can reduce the number of samples needed to meet the tolerable limits on decision errors, and to allow more efficient use of resources.

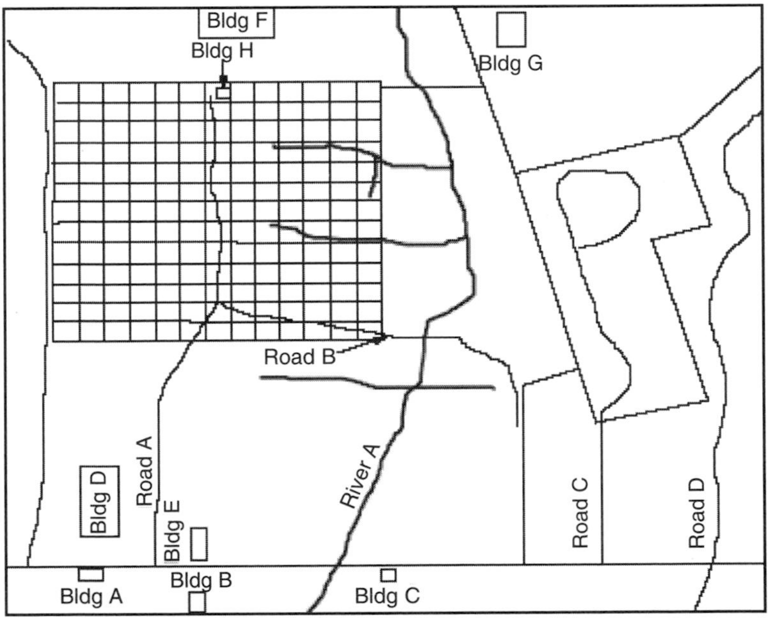

FIGURE 11.1. Environmental assessment area delineated by map boundaries. Each grid cell is sampled. (Source: U.S. Environmental Protection Agency, 2002, *Guidance for the Data Quality Objectives Process*, EPA QA/G-4, EPA/600/R-96/055, Washington, D.C.)

Time is also important in defining the monitoring needed. Conditions vary over the course of a study due to changes in weather conditions, seasons, operation of equipment, and human activities. These include seasonal changes in groundwater levels, seasonal differences in farming practices, daily or hourly changes in airborne contaminant levels, and intermittent pollutant discharges from industrial sources. Such variations must be considered during data collection and in the interpretation of results.

The EPA's *Guidance for the Data Quality Objectives Process* identifies the following examples of temporally sensitive attributes in environmental studies:

- measurement of lead in dust on windowsills may show higher concentrations during the summer when windows are raised and paint/dust accumulates on the windowsill;
- terrestrial background radiation levels may change due to shielding effects related to soil dampness;
- measurement of pesticides on surfaces may show greater variations in the summer because of higher temperatures and volatilization;
- instruments may not give accurate measurements when temperatures are colder; or
- measurements of airborne particulate matter may not be accurate if the sampling is conducted in the wetter winter months rather than the drier summer months.

Another obvious example is the difference in ambient ozone levels in the troposphere at night versus during the daytime hours, and winter versus summer. This does not mean that samples are only taken at certain dates and times, but that the differences due to temporal and seasonal factors must be documented.

Thus, the population and optimum time frame for collecting data are crucial in the monitoring plan. Feasibility should also be considered, including access to the properties, equipment acquisition and operation, and environmental conditions, times, and conditions when sampling is prohibited (e.g., freezing temperatures, high humidity, noise, and legal access to private property).

Sampling Approaches

The environmental sampling plan defines the kinds of samples to be collected. A *grab sample* is simply a measurement at a site at a single point in time. *Composite sampling* physically combines and mixes multiple grab samples (from different locations or times) to allow for physical, instead of mathematical, averaging. The acceptable composite provides a single value of contaminant concentration measurement that can be used in statistical calculations. Multiple composite samples can provide improved sampling precision and reduce the total number of analyses required compared to noncomposite sampling[5] (e.g., "grab" or integrated soil sample of x mass or y volume), the number of samples needed (e.g., for statistical significance), the minimum acceptable quality as defined by the quality assurance (QA) plan and sampling standard operating procedures (SOPs), and sample handling after collection.

A weakness of composite sampling is the false negative effect. For example, samples are collected from an evenly distributed grid of homes to represent a neighborhood exposure to a contaminant, as shown in Figure 11.2, where action is needed above the threshold of 5 $mg\,kg^{-1}$ soil. The assessment found the contaminant concentrations at the five sites to be 3, 1, 2, 12, and 2 $mg\,kg^{-1}$, and the mean contamination concentration is only 4 $mg\,kg^{-1}$, so it would be reported below the threshold level of 5 $mg\,kg^{-1}$. However, the fourth home is well above the safety level. This could also have a false positive effect. For example if the mean concentration were 6 $mg\,kg^{-1}$ in the example, the whole neighborhood may not need cleanup if the source is isolated to a confined area in the yard of home 5. Another example of where geographic composites may not be representative is in cleaning up and monitoring the success of cleanup actions. For example, if a grid is laid out over a contaminated groundwater plume (Figure 11.3), it may not take into account horizontal and vertical impervious layers, unknown sources (e.g.,

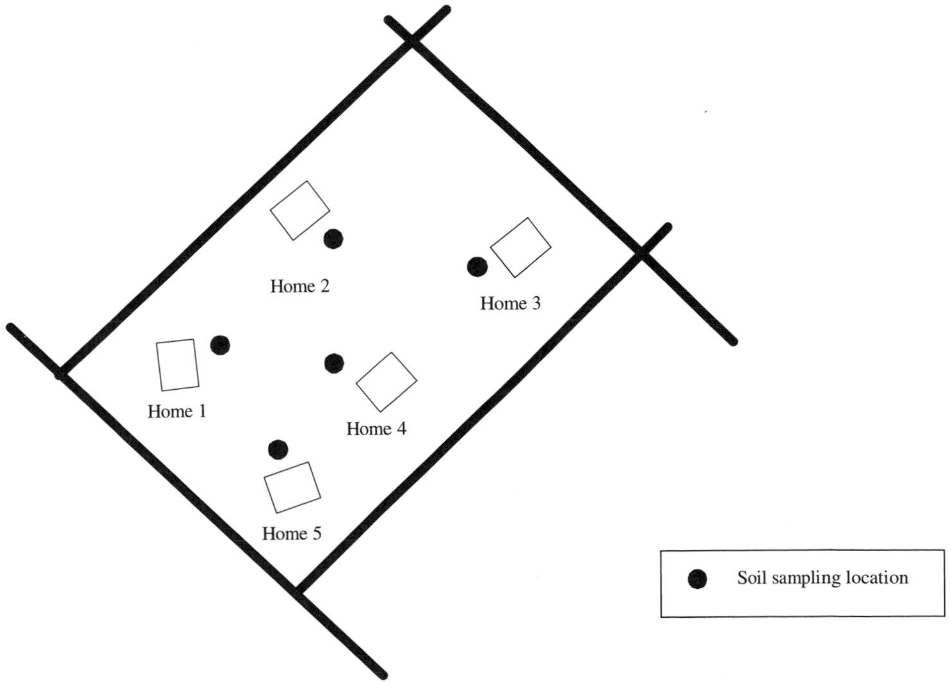

FIGURE 11.2. Composite sampling grid for a neighborhood.

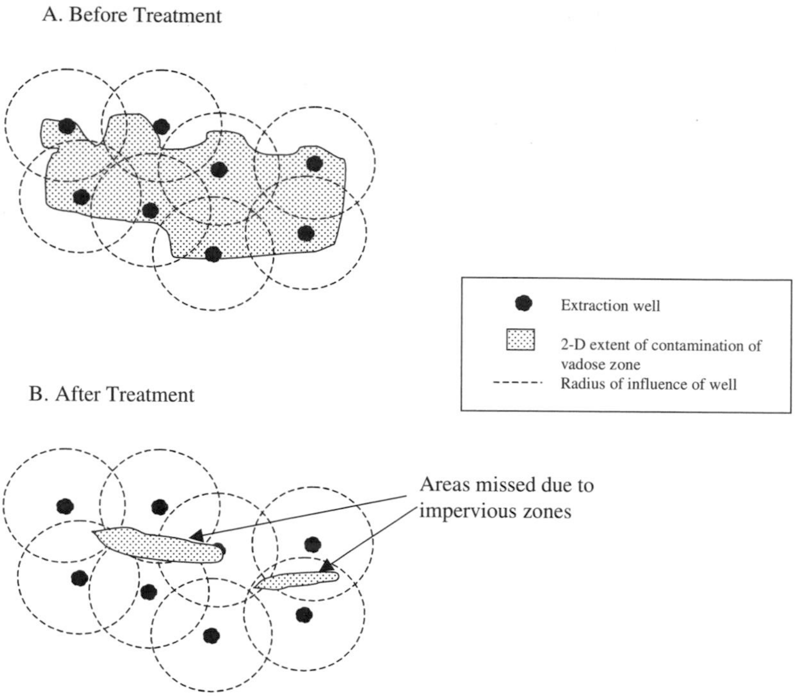

FIGURE 11.3. Extraction well locations on a geometric grid, showing hypothetical cleanup after 6 months.

buried tanks), and flow differences among strata, so that some of the plume is eliminated but pockets are left (as shown in Figure 11.3B).

Therefore, it is often good practice to assume that a contaminated site will have a heterogeneous distribution of contamination. Let us now consider some sampling methods and considerations on their use.[6]

Random Sampling

While it has the value of statistical representativeness, with a sufficient number of samples for the defined confidence levels (e.g., x samples needed for 95% confidence), random sampling may lead to large areas of the site being missed by sampling due to chance distribution of results. It also neglects prior knowledge of the site. For example, if maps show an old tank that may have stored contaminants, a purely random sample will not give any preference to samples near the tank.

Stratified Random Sampling

This method proceeds by dividing the site into areas and randomly sampling within each area, avoiding the omission problems of random sampling alone.

Stratified Sampling
Contaminants or other parameters are targeted. The site is subdivided and sampling patterns and densities varied in different areas. Stratified sampling can be used for complex and large sites, such as mining.

Grid or Systematic Sampling
The whole site is covered (see Figure 11.1). Sampling locations are readily identifiable, which is valuable for follow-up sampling, if necessary. The grid does not have to be rectilinear. In fact, rectangles are not the best polygon to use if the value is to be representative of a cell or centroid. Circles provide equidistant representation, but overlap. Hexagons are sometimes used as a close approximation to the circle, while avoiding overlap (ala the bee's honeycomb). The U.S. Environmental Monitoring and Assessment Program (EMAP) has used a hexagonal grid pattern, for example to represent large areas of the United States.

Judgmental Sampling
Samples are collected based upon knowledge of the site. This overcomes the problem of ignoring sources or sensitive areas, but it is vulnerable to bias of both inclusion and exclusion. Obviously, this would not be used for spatial representation, but for pollutant transport, plume characterization, or monitoring near a sensitive site (e.g., a day care center). Judgment is also used in "methods studies," for example, to see how well a particular approach to study or clean up contaminants works under worst case conditions. This is particularly useful if an approach has worked to the satisfaction of scientists in the laboratory or in a limited field test, but it remains to be seen how well the approach works in the "real world."

At every stage of monitoring from sample collection through analysis and archiving, only qualified and authorized persons should be in possession of the samples. This is usually assured by requiring chain-of-custody manifests, that is, everyone who takes possession of the samples signs off, and again when they are passed on to the next responsible person. Sample handling includes specifications on the temperature range needed to preserve the sample, the maximum amount of time the sample can be held before analysis, special storage provisions (e.g., some samples need to be stored in certain solvents, kept from light, etc.), and chain-of-custody provisions (again, only certain, authorized persons should be in possession of samples after collection).

The reason each person in possession of the samples is required to sign and date the chain-of-custody form before transferring the samples is because samples have evidentiary and forensic content. Any compromising of the sample integrity must be avoided, not only for scientific credibility, but for legal requirements.

Monitoring Example 1

You have received a report that shows that the level of detection (LOD) for naphthalene is $5\,\mu g\,L^{-1}$. The LOQ for the same chemical is $10\,\mu g\,L^{-1}$. The report shows that the soil concentrations of naphthalene every 10 meters from a shed to a roadway 50 m away:

$6\,\mu g\,L^{-1}$ closest to the shed
$7\,\mu g\,L^{-1}$
$8\,\mu g\,L^{-1}$
$8\,\mu g\,L^{-1}$
$9\,\mu g\,L^{-1}$
$9\,\mu g\,L^{-1}$ at the roadway

Can a gradient be calculated from these data? Why or why not? If yes, what would the average concentration gradient be? If not, how might these data be useful? (Hint: Consider what an LOD tells you about the data and the presence of naphthalene.)

What if the report showed the following concentrations?

$10\,\mu g\,L^{-1}$ closest to the shed
$11\,\mu g\,L^{-1}$
$11\,\mu g\,L^{-1}$
$12\,\mu g\,L^{-1}$
$13\,\mu g\,L^{-1}$
$13\,\mu g\,L^{-1}$ at the roadway

Can a gradient be calculated from these data? Why or why not? If yes, what is the average concentration gradient?

If these data are truly representative and accurate, is the shed the likely source of naphthalene contamination? Explain.

Solution and Discussion

Since all of the values in the first data set are above the LOD, we can say that naphthalene is present in the soil. However, since the values are below the LOQ (see Table 11.1), we cannot confidently tell whether they vary amongst themselves, so we cannot calculate a gradient.

Since all values in the second data set are at or above the LOQ, we can calculate a gradient. For example, in a span of 50 meters the naphthalene concentrations increased by $3\,\mu g\,L^{-1}$ from the shed to the roadway. Thus, the gradient is $3\,\mu g\,L^{-1}\ 50\,m^{-1} = 0.6\,\mu g\,L^{-1}\,m^{-1}$.

Since the concentrations of naphthalene are decreasing from the road to the shed, the roadway is more likely to be the location of the source rather than the shed.

Monitoring Example 2

If soil naphthalene concentrations similar to those found in Monitoring Example 1 are also found along the roadway for a distance of 50 meters where there are no buildings and the land is undeveloped, what is a likely source of the naphthalene? Why were only naphthalene results reported?

Discussion

This finding opens up more scenarios. Was there a spill, or is there a continuous source of the naphthalene at a distant site that is depositing it on the soil? An investigation is in order.

An environmental assessment of one compound may be the result of an investigation requested by a company, town, or regulatory agency due to a complaint, for example, anonymously via a "hot line" call or a formally registered complaint or inquiry from a citizen or group. Or, the analysis could also be part of a research study by a laboratory or university. Another possibility is that a whole suite of compounds was studied. In this instance, a number of polycyclic aromatic compounds were targeted, but only naphthalene was detected.

It behooves the scientist and engineer to find out which, if any, of these is the reason for only one identified contaminant.

Laboratory Analysis

Analytical chemistry is an essential part of exposure characterization. When the sample arrives at the laboratory, the next step may be "extraction." Extraction is needed for two reasons. First, the environmental sample may be in sediment or soil, where the chemicals of concern are sorbed to particles and must be freed for analysis to take place. Second, the actual collection may have been by trapping the chemicals onto sorbants. Thus, to analyze the sample, the chemicals must first be freed from the sorbant matrix. Again, dioxins provide an example. Under environmental conditions, dioxins are fat soluble and have low vapor pressures, so they may be found on particles, in the gas phase, or in the water column suspended to colloids (and very small amounts dissolved in the water itself). Therefore, to collect the gas phase dioxins, the standard method calls for trapping it on polyurethane foam (PUF). Thus, to analyze dioxins in the air, the PUF and particle matter must first be extracted, and to analyze dioxins in soil and sediment, those particles must also be extracted.

Extraction makes use of physics and chemistry. For example, many compounds can be simply extracted with solvents, usually at elevated temperatures. A common solvent extraction is the Soxhlet extractor, named after the German food chemist, Franz Soxhlet (1848–1913). The Soxhlet extractor (the U.S. EPA Method 3540) removes sorbed chemicals by passing a boiling solvent through the media. Cooling water condenses the heated solvent and the extract is collected over an extended period, usually several hours. Other automated techniques apply some of the same principals as solvent extraction but allow for more precise and consistent extraction, especially when large volumes of samples are involved. For example, supercritical fluid extraction (SFE) brings a solvent, usually carbon dioxide, to the pressure and temperature near its critical point of the solvent, where the solvent's properties are rapidly altered with very slight variations of pressure.[7] Solid phase extraction (SPE), which uses a solid and a liquid phase to isolate a chemical from a solution, is often used to clean up a sample before analysis. Combinations of various extraction methods can enhance the extraction efficiencies, depending upon the chemical and the media in which it is found. Ultrasonic and microwave extractions may be used alone or in combination with solvent extraction. For example, the U.S. EPA Method 3546 provides a procedure for extracting hydrophobic (i.e., not soluble in water) or slightly water-soluble organic compounds from particles such as soils, sediments, sludges, and solid wastes. In this method, microwave energy elevates the temperature and pressure conditions (i.e., 100–115°C and 3.4–11.9 atm) in a closed extraction vessel containing the sample and solvent(s). This combination can improve recoveries of chemical analytes and can reduce the time needed compared with the Soxhlet procedure alone.

Not every sample needs to be extracted. For example, air monitoring using canisters and bags allows the air to flow directly into the analyzer. Water samples may also be directly injected. Surface methods, such as fluorescence, sputtering, and atomic absorption, require only that the sample be mounted on specific media (e.g., filters). Also, continuous monitors, like the chemiluminescent system mentioned in the last section of this chapter, provide ongoing measurements.

Chromatography consists of separation and detection. Separation makes use of the chemicals' different affinities for certain surfaces under various temperature and pressure conditions. The first step, injection, introduces the extract to a "column." The term *column* is derived from the time when columns were packed with sorbents of varying characteristics, sometimes meters in length, and the extract was poured down the packed column to separate the various analytes. Today, columns are of two major types, gas and liquid. Gas chromatography (GC) makes use of hollow tubes ("columns") coated inside with compounds that hold organic chemicals. The columns are in an oven, so that after the extract is injected into the column, the temperature is increased, as well as the pressure, and the various organic compounds in the extract are released from the column surface differentially, whereupon they are collected by a carrier gas (e.g., helium) and transported to the detector. Generally, the more volatile compounds are released first (they have the shortest retention times), followed by the semivolatile organic compounds. Thus, boiling point is often a very useful indicator as to when a compound will come off a column. This is not always the case, since other characteristics such as polarity can greatly influence a compound's resistance to be freed from the column surface.[8] For this reason, numerous GC columns are available to the chromatographer (different coatings, interior diameters, and lengths). Rather than coated columns, liquid chromatography (LC) makes use of columns packed with different sorbing materials with differing affinities for compounds. Also, instead of a carrier gas, LC uses a solvent or blend of solvents to carry the compounds to the detector. In the high performance LC (HPLC), pressures are also varied.

Detection is the final step for quantifying the chemicals in a sample. The type of detector needed depends upon the kinds of pollutants of interest. Detection gives the "peaks" that are used to identify compounds (see Figure 11.4). For example, if hydrocarbons are of concern, GC with flame ionization detection (FID) may be used. GC-FID gives a count of the number of carbons; for example, long chains can be distinguished from short chains. The short chains come off the column first and have peaks that appear before the long-chain peaks. However, if pesticides or other halogenated compounds are of concern, electron capture detection (ECD) is a better choice.

A number of detection approaches are also available for LC. Probably the most common is light absorption. Chemical compounds absorb energy at various levels, depending upon their size, shape, bonds, and other structural characteristics. Chemicals also vary in whether they will absorb light or how much light they can absorb depending upon wavelength. Some absorb very well in the ultraviolet (UV) range, while others do not. Diode arrays help to identify compounds by giving a number of absorption ranges in the same scan. Some molecules can be excited and will fluoresce. The Beer-Lambert Law tells us that energy absorption is proportional to chemical concentration:

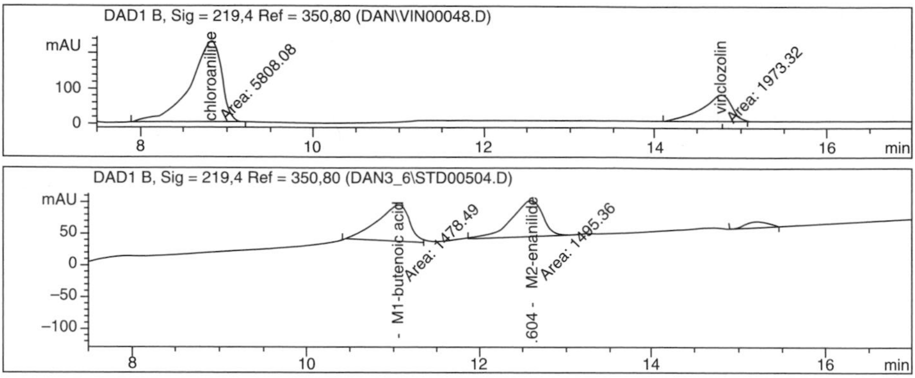

FIGURE 11.4. High-performance liquid chromatograph/ultraviolet detection peaks for standard acetonitrile solutions: $9\,mg\,L^{-1}$ 3,5-dichloroaniline and $8\,mg\,L^{-1}$ the fungicide vinclozolin (top); and its intermediate degradation products, $7\,mg\,L^{-1}$ M1 and $9\,mg\,L^{-1}$ M2 (bottom). (Source: D. Vallero, 2003, *Engineering the Risks of Hazardous Wastes,* Butterworth-Heinemann, Boston, Mass.)

$$A = eb[C] \qquad \text{Equation 11–1}$$

where, A is the absorbency of the molecule, e is the molar absorptivity (proportionality constant for the molecule), b is the light's path length, and $[C]$ is the chemical concentration of the molecule. Thus, the concentration of the chemical can be ascertained by measuring the light absorbed.

One of the most popular detection methods is mass spectrometry (MS), which can be used with either GC or LC separation. The MS detection is highly sensitive for organic compounds and works by using a stream of electrons to consistently break apart compounds into fragments. The positive ions resulting from the fragmentation are separated according to their masses. This is referred to as the "mass to charge ratio" or m/z. No matter which detection device is used, software is used to decipher the peaks and to perform the quantitation of the amount of each contaminant in the sample.

For inorganic substances and metals, the additional extraction step may not be necessary. The actual measured media (e.g., collected airborne particles) may be measured by surface techniques like atomic absorption (AA), X-ray fluorescence (XRF), inductively coupled plasma (ICP), or sputtering. As for organic compounds, the detection approaches can vary. For example ICP may be used with absorption or MS. If all one needs to know is elemental information, for example to determine total lead or nickel in a sample, AA or XRF may be sufficient. However, if speciation (i.e., knowing the various compounds of a metal) is required, then significant sample preparation is needed, including a process known as *derivatization*. Derivatizing a sample is performed by adding a chemical agent that transforms the compound in question into one that can be recognized by the detector.

This is done for both organic and inorganic compounds, such as when the compound in question is too polar to be recognized by MS.

The physical and chemical characteristics of the compounds being analyzed must be considered before visiting the field and throughout all the steps in the laboratory.

Although it is beyond the scope of this book to go into detail, it is worth mentioning that the quality of results generated about contamination depends upon the sensitivity and selectivity of the analytical equipment. Table 11.1 defines some of the most important analytical chemistry threshold values.

Environmental Chromatography Example

Your analytical laboratory has generated the following chromatogram and table from an HPLC/UV at 254 nm using a 5 μm, C_{18}, 4.6 × 250 mm column from a water sample you submitted:

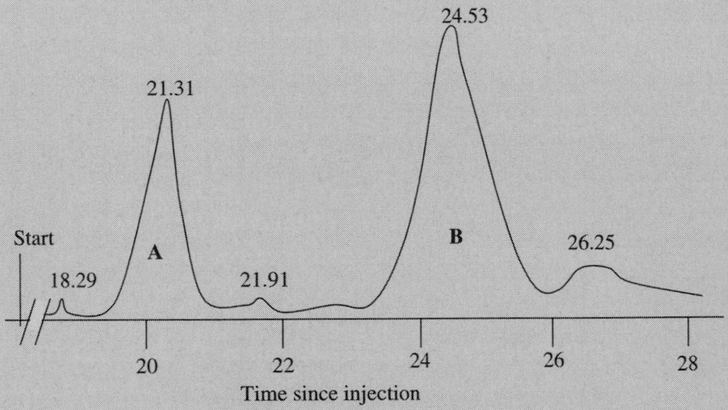

Retention Time	*Area*	*Type*	*Area/Height*	*Area %*
18.29	NA	NA	NA	0.1
21.31	NA	NA	NA	31.4
21.91	NA	NA	NA	0.2
24.53	NA	NA	NA	67.2
26.25	NA	NA	NA	1.1

Even with the missing entries in the table, you can still ascertain certain information. What are the retention times of compound A and

B? Which compound is present in a larger amount? Which compound has the higher boiling point? What would happen to the retention times of compounds A and B if the column temperature were raised? You suspect that compound B is benzo(a)pyrene. How would you find out whether this is the case?

Answers and Discussion

The retention time of compound A is 21.31 minutes, shown above of the peak and in the table's retention time column. The retention time of compound B is 24.53 minutes.

You cannot tell from this table or chromatogram which compound is present in a larger amount, since the only way to do so is to have calibration curves from known concentrations of compound A and compound B (at least three, but preferably five). For example, you would run the HPLC successively with injections of pure solutions of 0.01, 0.1, 1, 10, and 100 $\mu g L^{-1}$ concentrations of compound A, and again with pure solutions of the same concentrations of compound B. These concentrations would give peak areas associated with each known concentration. Then you could calculate (actually the HPLC software will calculate) the calibration curve. So, for example, if an peak with an area of 200 is associated with 1 $\mu g L^{-1}$ of compound A and a peak with an area of 2000 is associated with 10 $\mu g L^{-1}$ of compound A (i.e., a linear calibration curve) at 21.31 minutes after the aliquot is injected into the HPLC, then when you run your unknown sample and a peak at 21.31 minutes with an area of 1000 would mean you have about 5 $\mu g L^{-1}$ concentration of compound A in your sample. The same procedure would be followed to draw a calibration curve for compound B at a retention time of 24.53 minutes.

The reason you cannot simply look at the percent area is that each compound is physically and chemically different, and recall from the Beer-Lambert Law (Equation 11.1) that the amount of energy absorbed (in this case, the UV light) is what gives us the peak. If a molecule of compound A absorbs UV at this wavelength (i.e., 254 nm) at only 25% as that of compound B, compound A's concentration would be higher than that of compound B (because even though compound B has twice the percent area, compound A's absorbance is four times that of compound B).

Compound A has the lower boiling point since it comes off the column first. Of course, this is only true if other factors, especially polarity, are about the same. For example, if compound B has about the same polarity as the column being used, but compound A has a very different polarity, compound A will have a greater tendency to leave the column. Generally, however, retention time is a good indicator of boiling point; i.e., shorter retention times mean lower boiling points.

If the column temperature were raised, both compounds A and B would come off the column in a shorter times. Thus, the retention times of both compounds A and B would be shorter than before the temperature was raised.

To determine whether the peak at 24.53 minutes is benzo(a)pyrene, you must first obtain a true sample of pure benzo(a)pyrene to place in a standard solution. This is the same process as you used to develop the calibration curve above. That is, you would inject this standard of known benzo(a)pyrene into the same HPLC and the same volume of injection. If the standard gives a peak at a retention time at about 25 minutes, there is a good chance it is benzo(a)pyrene. As it turns out, benzo(a)pyrene absorbs UV at 254 nm and does come off an HPLC column at about 25 minutes.

The column type also affects retention time and peak area. The one used by your laboratory is commonly used for polycyclic aromatic hydrocarbons, including benzo(a)pyrene. However, numerous columns can be used for semivolatile organic compounds, so both the retention time and peak area will vary somewhat. Another concern is co-elution, i.e., two distinct compounds that have nearly the same retention times. One means of reducing the likelihood of co-elution is to target the wavelength of the UV detector. For example, the recommended wavelength for benzo(a)pyrene is 254 nm, but 295 nm is preferred by environmental chromatographers because the interference peak in the benzo(a)pyrene window is decreased at 295 nm. Another way to improve detection is to use a diode array detection system with the UV detector. This gives a number of different chromatograms simultaneously at various wavelengths. Finally, there are times when certain detectors do not work at all. For example, if a molecule does not absorb UV light (i.e., it lacks a group of atoms in a molecule responsible for absorbing the UV radiation, known as chromophores), there is no way to use any UV detector. In this case another detector, e.g., mass spectrometry, must be used.

Sources of Uncertainty

Contaminant assessments have numerous sources of uncertainty. There are two basic types of uncertainty: Type A and Type B. Type A uncertainties result from the inherent unpredictability of complex processes that occur in nature. These uncertainties cannot be eliminated by increasing data col-

TABLE 11.1
Expressions of Chemical Analytical Limits

Type of Limit	*Description*
Limit of Detection (LOD)	Lowest concentration or mass that can be differentiated from a blank with statistical confidence. This is a function of sample handling and preparation, sample extraction efficiencies, chemical separation efficiencies, and capacity and specifications of all analytical equipment being used (see IDL below).
Instrument detection limit (IDL)	The minimum signal greater than noise detectable by an instrument. The IDL is an expression of the piece of equipment, not the chemical of concern. It is expressed as a signal to noise (S:N) ratio. This is mainly important to the analytical chemists, but the engineer should be aware of the different IDLs for various instruments measuring the same compounds, so as to provide professional judgment in contracting or selecting laboratories and deciding on procuring for appropriate instrumentation for all phases of contaminant remediation.
Limit of quantitation (LOQ)	The concentration or mass above which the amount can be quantified with statistical confidence. This is an important limit because it goes beyond the "presence-absence" of the LOD and allows for calculating chemical concentration or mass gradients in the environmental media (air, water, soil, sediment, and biota).
Practical quantitation limit (PQL)	The combination of LOQ and the precision and accuracy limits of a specific laboratory, as expressed in the laboratory's quality assurance/quality control (QA/QC) plans and standard operating procedures (SOPs) for routine runs. The PQL is the concentration or mass that the engineer and scientist can consistently expect to have reported reliably.

Source: D. Vallero, 2003, *Engineering the Risks of Hazardous Wastes*, Butterworth-Heinemann, Boston, Mass.

lection or enhancing analysis. The scientist and engineer must simply recognize that Type A uncertainty exists, but must not confuse it with Type B uncertainties, which can be reduced by collecting and analyzing additional scientific data.

The first step in an uncertainty analysis is to identify and describe as many uncertainties that may be encountered. Sources of Type B uncertainty take many forms.[9] There can be substantial uncertainty concerning the numerical values of the attributes being studied (e.g., contaminant concentrations, wind speed, discharge rates, groundwater flow, and other variables). Modeling generates its own uncertainties, including errors in selecting the variables to be included in the model, such as surrogate contaminants that represent whole classes of compounds (e.g., does benzene represent the behavior or toxicity of other aromatic compounds?). The application of the findings, even if the results themselves have tolerable uncertainty, lead to the propagation of uncertainties when ambiguity arises regarding their meaning. For example, a decision rule is a statement about which alternative will be selected, such as for cleanup, based on the characteristics of the decision situation. A "decision-rule uncertainty" occurs when there are disagreements or poor specification of objectives (i.e., is our study really addressing the client's needs?).

Variability and uncertainty must not be confused. Variability consists of measurable factors that differ across populations such as soil type, vegetative cover, or body mass of individuals in a population. Uncertainty consists of unknown or not fully known factors that are difficult to measure, such as the inability to access an ideal site that would be representative because it is on private property.

Modeling uncertainties, for example, may consist of extrapolations from a single value to represent a whole population, that is, a point estimate (e.g., 70 kg as the weight of an adult male). Such estimates can be typical values for a population or an estimate of an upper end of the population's value, such as 70 years as the duration of exposure used as a "worse case" scenario. Another approach is known as the Monte Carlo technique (Figure 11.5). The Monte Carlo-type exposure assessments use probability distribution functions, which are statistical distributions of the possible values of each population characteristic according to the probability of the occurrence of each value. These are derived using iterations of values for each population characteristic. While the Monte Carlo technique may help to deal with the point estimate limitation, it can suffer from confusing variability with uncertainty.

Other data interpretation uncertainties can result from the oversimplification of complex entities. For example, assessments consist of an aggregation of measurement data, modeling, and combinations of sampling and modeling results. However, these complicated models are providing only a snapshot of highly dynamic human and environmental systems. The use of more complex models does not necessarily increase precision, and extreme values can be improperly characterized. For example, a 50th percentile value can always be estimated with more certainty than a 99th percentile value.

The bottom line is that uncertainty is always present in sampling, analysis, and data interpretation, so the monitoring and data reduction plan should be systematic and rigorous. Studies must generate data that are sufficiently precise and accurate (see discussion box: Quality in Environmental Science: Precision and Accuracy). The uncertainty analysis must be addressed for each step of the contaminant assessment process, including

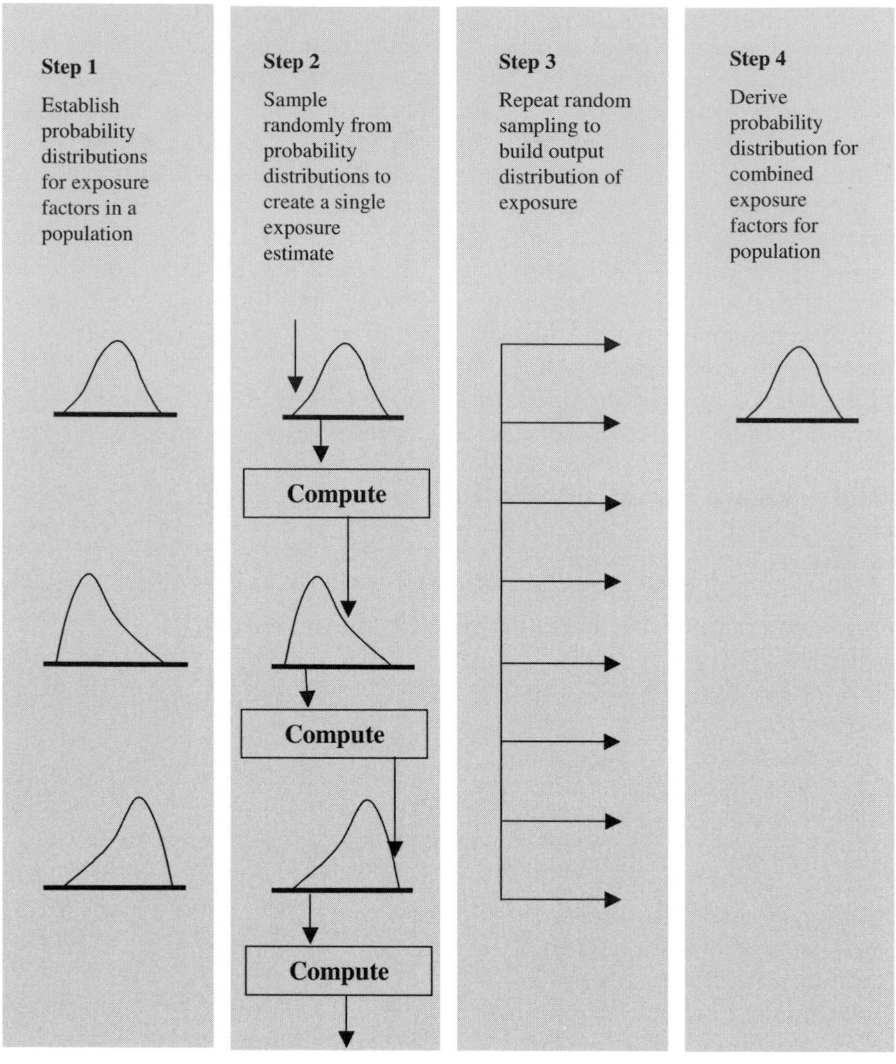

FIGURE 11.5. Principles of the Monte Carlo method for aggregating data. (Source: Adapted from Australian Department of Health and Ageing, 2002, Environmental Risk Assessment: Guidelines for Assessing Human Health Risks from Environmental Hazards.)

any propagation and enlargement of cumulative error (e.g., an incorrect pH value that goes into an index where pH is weighted heavily, and then used in another algorithm for sustainability). The characterization of the uncertainty of the assessment includes selecting and rejecting data and information ultimately used to make environmental decisions, and includes both qualitative and quantitative methods (see Table 11.2).

TABLE 11.2
Example of an Uncertainty Table for Exposure Assessment

	Effect on Exposure[a]		
Assumption	*Potential Magnitude for Over-Estimation of Exposure*	*Potential Magnitude for Under-Estimation of Exposure*	*Potential Magnitude for Over- or Under-Estimation of Exposure*
Environmental sampling and analysis			
Sufficient samples may not have been taken to characterize the media being evaluated, especially with respect to currently available soil data.			Moderate
Systematic or random errors in the chemical analyses may yield erroneous data.			Low-High
Exposure parameter estimation			
The standard assumptions regarding body weight, period exposed, life expectancy, population characteristics, and lifestyle may not be representative of any actual exposure situation.			Moderate
The amount of media intake is assumed to be constant and representative of the exposed population.	Moderate		
Assumption of daily lifetime exposure for residents.	Moderate to high		

[a] As a general guideline, assumptions marked as "low," may affect estimates of exposure by less than one order of magnitude; assumptions marked "moderate" may affect estimates of exposure by between one and two orders of magnitude; and assumptions marked "high" may affect estimates of exposure by more than two orders of magnitude.
Source: Australian Department of Health and Ageing, 2002, *Environmental Health Risk Assessment: Guidelines for Assessing Human Health Risks from Environmental Hazards*.

Quality in Environmental Science: Precision and Accuracy

Environmental decisions must be based upon reliable information. For information to be reliable, it must be of a quality matched to the needs of the decision maker. For example, a farmer needs to know whether a soil sample contains 10% or more of organic matter and if the soil pH value less than 7. A school district deciding whether to site a new elementary school wants to know the concentrations of several toxic compounds in the soil at the site to five significant figures. These two instances are both demanding that sampling and analysis be conducted, but the precision and accuracy of the results are very different. The farmer needs much less precision, but is hoping that within these defined ranges of organic matter and pH that the results will be accurate. The school district needs highly precise information that must also be sufficiently accurate.

But what do these terms really mean? Are they not often used interchangeably? Actually, precision and accuracy mean two very different things, especially to scientists and engineers. They are statistical terms.

Precision describes how refined and repeatedly an operation can be performed, such as the exactness in the instruments and methods used to obtain a result. It is an indication of the uniformity or reproducibility of a result. This can be likened to shooting arrows,[10] with each arrow representing a data point. Targets A and B in Figure 11.6 are equally precise. Assuming that the center of the target, i.e., the bulls-eye, is the "true value," data set B is more accurate than A. If we are consistently missing the bull's eye in the same direction at the same distance, this is an example of bias (systematic error). The good

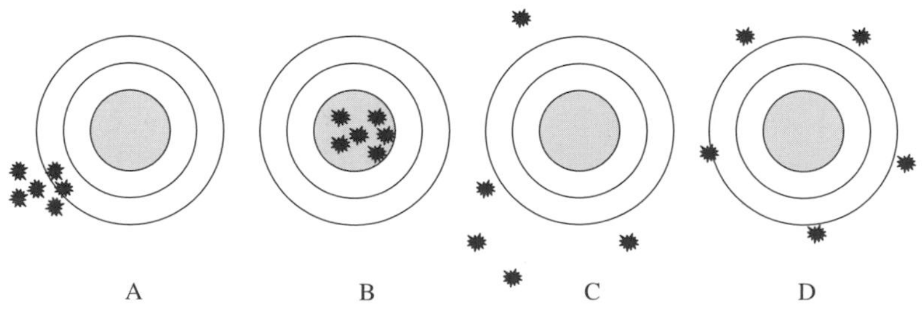

Figure 11.6. Precision and accuracy. The bulls-eye represents the true value. Targets A and B demonstrate data sets that are precise; targets B and D data sets that are accurate, and targets C and D data sets that are imprecise. Target B is the ideal data set, which is precise and accurate.

news is that if we are aware that we are missing the bulls-eye (e.g. by running known standards in our analytical equipment), we can calibrate and adjust the equipment. To stay with our archery analogy, the archer would move her sight up and to the right.

Thus, *accuracy* is an expression of how well a study conforms to some defined standard, (the true value). So, accuracy expresses the quality of what we find, while precision expresses the quality of the operation by which we obtained our finding. So, the two other scenarios of data quality are shown in Targets C and D. Thus, the four possibilities are that our data is precise but inaccurate (Target A), precise and accurate (Target B), imprecise and inaccurate (Target C), and imprecise and accurate (Target D).

At first blush, Target D, may seem unlikely, but it is really not all that uncommon. The difference between Targets B and D are simply that D has more "spread" in the data. For example the variance and standard deviation of D is much larger than that of B. However, their measures of central tendency, i.e., the means, are nearly the same. So, both data sets are giving us the right answer, but almost all of the data points in B are near the true value. None of the data points in D are near the true value, but the mean is.

Certainty in Reporting Environmental Data: Significant Figures[11]

Environmental science depends on expressing results with appropriate precision. Uncertainties in environmental measurement are introduced through human error, equipment malfunctions, and experimental bias. Data are useful only to the extent that one can be confident in their validity. This means that each figure or digit in the numerical expression of a measurement should be significant. A significant figure is a number that is considered to be correct within a specified or implied limit of error. Thus, if the height of a woman, expressed in significant figures, is written as 5.18 feet, it is assumed that only the *last figure* (i.e., the 8) may be in error. Uncertainty in the first or second figure would obviate the significance from the last figure (you have to know the number of feet, before worrying about inches). If the last figure is uncertain by a specified amount, the woman's height can be given as 5.18 ± 0.01 feet.

The significant figures of a number can be found by reading the number from the left to the right and counting all digits starting with the first digit that is not zero. The decimal point is ignored since it is determined by the particular units, not by the precision of the operation. So, the measurements 33.35 cm and 333.5 mm are equivalent, and both have four significant figures.

In addition or subtraction, any figure in the answer is significant only if each number in the problem contributes a significant figure at that decimal level (that is, the level of greatest magnitude will determine how many significant figures should be carried in the answer):

$$\begin{array}{r} 507.7812 \\ 0.00034 \\ \underline{20.31} \\ 528.09 \end{array}$$

When "rounding off" (discarding nonsignificant figures), the last significant figure is unchanged if the next figure is less than 5, and is increased by 1 if the next figure is 5 or more:

4.7349 rounded to 4.735 (four significant figures)
4.7349 rounded to 4.73 (three significant figures)
2.8150 rounded to 2.82 (three significant figures)

In multiplication and division, the number of significant figures in the answer is the same as that in the quantity with the fewest significant figures:

$$(2.0 \times 5199)/0.0834 = 5.4 \times 10^7$$

In a calculation with multiple steps, first find the number of significant figures in the answer as discussed above, and then round off each number with excess significant figures to one or more significant figures than necessary. Then round off the answer to the correct significant figures. This procedure preserves significance without too much work. For example, $x = 4.3 \times (311.8/273.1) \times [760/(784 - 2)]$.

There are two significant figures in the number 4.3; therefore, the answer will have two significant figures. Round off according to rules 1 and 2, to one extra significant figure. Note that the presence of only one significant figure in the number 2 does not mean that there is only one significant figure in the answer because 784 – 2 = 782, which has three significant figures. So,

$$x = 4.3 \times (312/273) \times (760/782)$$

Solve and round off to two significant figures,

$$x = 4.8$$

Significant Figure Example

Your laboratory reports that the total suspended solids (TSS) for Townville's influent (the untreated water) is 201 mg L^{-1}, and the effluent leaving the treatment plant is 85.99 mg L^{-1}. What is the TSS removal efficiency?

Solution

Removal efficiency can be expressed as the percent TSS removal

$$\%\ \text{TSS removal} = \frac{\text{TSS removal}}{\text{TSS influent}} \times 100\% \qquad \text{Equation 11–2}$$

$$= \frac{\text{TSS}_{\text{inf}} - \text{TSS}_{\text{eff}}}{\text{TSS influent}} \times 100\% \qquad \text{Equation 11–3}$$

$$= \frac{201 - 85.99}{201} \times 100\%$$
$$= 59.05\%$$

The effluent data are reported as four significant figures. However, since the influent data, i.e., 201, has only three significant figures, we should report the TSS removal to three significant figures. So, the correct expression of TSS removal efficiency for Townville's wastewater treatment facility is 59.1%.

Expressions of Data Precision

One means of showing the level of precision of laboratory and field measurements is to compare duplicate measurements. This is expressed by relative percent difference (RPD):

$$RPD = \frac{(C_1 - C_2)}{(C_1 + C_2)/2} \times 100\% \qquad \text{Equation 11–4}$$

Relative Percent Difference Example 1

An environmental field team has just completed a stack test for the toxic gas hydrogen sulfide (H_2S) for a manufacturing plant. This resulted in duplicate samples that, upon analysis, the H_2S stack emission concentrations were found to be 111 and 102 $\mu g\, m^{-3}$. The QA plan for the project calls for an RPD ≤10.00%. Do these duplicates meet the QA performance requirements for the H_2S stack test?

Solution

Calculate the RPD:

$$RPD = \frac{(111 - 102)}{(111 + 102)/2} \times 100\%$$
$$= 8.45\%$$

Since the RPD is less than 10.00% the results are sufficiently precise to meet the QA requirements.

Relative Percent Difference Example 2

In the same study as the previous example, a new QA requirement of an RPD ≤3.00% was added to improve precision. To address this need, the team purchased new equipment that yielded H_2S concentrations of 111 and 107 $\mu g\, m^{-3}$. Does the new equipment allow the team to meet the new precision requirements?

Solution

$$\text{RPD} = \frac{(111-107)}{(111+107)/2} \times 100\%$$
$$= 3.67\%$$

Even though the precision improved by almost 5%, the RPD is still greater than the 3.00% QA requirement, so the results are too imprecise to be used.

When more than two replicates are available, the relative standard deviation (RSD) is used:

$$\text{RSD} = \frac{s}{\overline{Y}} \times 100\% \qquad \text{Equation 11–5}$$

Where, s = the standard deviation of the replicates and $\overline{Y}$ = mean of the replicate measurements.[12]

The standard deviation, s, is calculated as:

$$s = \sqrt{\left[\sum_{n=1}^{n}\left[\frac{(Y_1-\overline{Y})^2}{n-1}\right]\right]} \qquad \text{Equation 11–6}$$

Relative Standard Deviation Example 1

What is the RSD for biochemical oxygen demand (BOD) 23, 18, 17, 20, 23 mg L^{-1} from a river water quality study?

Solution

First, calculate the sample mean: $\overline{Y} = 101 \div 5 = 20$

Next, find s:

$$s = \sqrt{\left[\sum_{n=1}^{n}\left[\frac{(Y_1-\overline{Y})^2}{n-1}\right]\right]} = \sqrt{\frac{3^2+2^2+3^2+0^2+3^2}{4}} \qquad \text{Equation 11–7}$$
$$= 7.8$$

Finally, calculate the RSD:

$$\text{RSD} = \frac{s}{\overline{Y}} \times 100\% = 7.8/20 \times 100\% = 39\%$$

Relative Standard Deviation Example 2

A known amount of the chemical being tested is often added to samples as a QA check. This is known as a "spiked" sample. The analysis of the unspiked samples is first determined, followed by an analysis after the samples are spiked.

You are measuring the amount of benz(a)anthracene in the air near a power plant. The extracts from the filters that collected the air samples are shown in Table 11.3. The table also shows that your lab technician has added $0.05\,\mathrm{mg\,L^{-1}}$ benz(a)anthracene to each sample. Calculate the sample precision for the 10 samples collected and spiked.

Solution

$$s = \sqrt{\frac{(\sum deviation)^2}{n-1}} \qquad \text{Equation 11–8}$$

$$= \sqrt{\frac{(0.37)^2}{10-1}} = 0.123 \text{ or about } 0.12$$

TABLE 11.3
Hypothetical Concentrations of Benz(a)anthracene in Sample Extracts

1 *Concentration of Spike Sample*	*2* *Concentration of Sample Only*	*3* *Recovery (1–2)*	*4* *Known Concentration Added*	*5* *Deviation from Expected (3–4)*
1.13	1.15	–0.02	0.05	–0.07
1.10	1.15	–0.05	0.05	–0.10
2.11	2.03	0.08	0.05	0.03
3.09	3.03	0.06	0.05	0.01
9.67	9.07	0.60	0.05	0.55
5.57	5.69	–0.12	0.05	–0.17
9.90	9.77	0.13	0.05	0.08
9.54	9.52	0.02	0.05	–0.03
8.90	8.85	0.05	0.05	0.00
1.11	0.99	0.12	0.05	0.07

Σ Deviation = 0.37

Expressions of Data Accuracy

Accuracy combines bias and precision to reflect how close the measured data are to the true value. Spiking is a valuable analytical tool to assure accuracy. A known amount of the substance of interest (i.e., the analyte) is added to the sample. The percent recovery (%R) for the spiked sample is found as follows:

$$\%R = \frac{S-U}{C} \times 100\% \qquad \text{Equation 11–9}$$

Where, S = the measured concentration in the spiked sample
U = the measured concentration of the unspiked sample
C = actual concentration of known addition to the sample

Accuracy Example

What is the percent recovery of a sample to measure benz(a)anthracene in soil that has a measured concentration in the spiked aliquot of 5.25 ng kg^{-1} and a measured concentration of 3.85 ng kg^{-1} in the unspiked aliquot? The soil has been spiked with a known amount equal to 2.00 ng kg^{-1} of benz(a)anthracene.

$$\%R = \frac{S-U}{C} \times 100\% = [(5.25-3.85)/2.00] \times 100\% = 70\%$$

A standard reference material (SRM) is often used to determine accuracy. An SRM contains a known analyte concentration that is certified by an outside source. An aliquot of the standard reference material is processed as a sample and processed through the complete analytical procedure used for all environmental samples. When SRM is available, the percent recovery can be calculated as

$$\%R = \frac{C_M}{C_{SRM}} \times 100\% \qquad \text{Equation 11–10}$$

Where C_M is the concentration of the SRM that is measured and C_{SRM} is the actual concentration of the SRM. So, if a certified laboratory provides us with an aliquot that they state has a concentration of 3.95 ng kg^{-1} benz(a)anthracene, but our analysis shows 3.85 ng kg^{-1}, then the %R = (3.85/3.95) × 100% = 97.47%.

As the numerator in the %R equation approaches the value of the denominator, the sample accuracy improves. Values deviate from 100% recovery because of bias and analytical imprecision, including operator and equipment deficiencies.

The overall accuracy of all of the samples in a study can be evaluated by summing their individual percent recoveries and dividing by the number of samples. This gives a mean percent recovery for the study. The variance and standard deviation of the overall recoveries can also be calculated.

Another indicator of data quality is completeness. One measure of completeness is to determine the fraction of measurements that are considered to be valid against the total number of measurements taken. In this way, percent completeness also represents operational efficiency. Representativeness is also a consideration of data, particularly if decisions are going to be made. For example, do the data represent an area beyond where the measurements were taken. This involves judgment as to whether information may be extrapolated from the measurements. Models are sometimes used to do this. Conversely, representativeness may also include interpolation, e.g., deciding how to model the areas between measurements. All such indicators of data quality hinge on sufficient precision and accuracy.

In our previous discussions of reference doses and concentrations (RfDs and RfCs, respectively) in Chapter 9, "Contaminant Hazards," recall that uncertainty factors (UFs) were applied to address both the inherent and study uncertainties upon which to establish safe levels of exposure to contaminants. These include tenfold factors, used to derive the RfD and RfC from experimental data. The UFs consider the uncertainties resulting from the variation in sensitivity among the members of the populations, including interhuman and intraspecies variability; the extrapolation of animal data to humans (i.e., interspecies variability); the extrapolation from data gathered in a study with less-than-lifetime exposure to lifetime exposure (i.e., extrapolating from acute or subchronic to chronic exposure); the extrapolation from different thresholds, such as the LOAEL rather than from a NOAEL; and the extrapolation from an incomplete database is incomplete. Note that most of these sources of uncertainty have a component associated with measurement and analysis.

The numerical value uncertainties are directly related to the quality and representativeness of the sampling design and the analytical expressions described in Table 11.2. When these values are input into models, they are known as "parameter uncertainties." Environmental studies, particularly field studies, have large amounts of variability (e.g., seasonal, year-to-year, spatial). There is also always imprecision and inaccuracy associated with the measurement and analytical equipment and systemic weaknesses in data gathering (i.e., bias).[13] It is important that these uncertainties not be ignored. Although they are realities of science, we should strive to reduce error as much as we can.

Ecologist's Field Manual: Importance of Chlorophyll as an Environmental Indicator[14]

Chlorophyll is the pigment that gives plants their green color and is an essential component of photosynthesis, whereby plants derive their energy for metabolic, growth, and reproductive processes. Scientists measure the amount of chlorophyll in water as an indirect yet reliable indicator of the amount of photosynthesizing taking place in a water body. For example, in a sample collected in a lake or pond, the photosynthetic activity of algae or phytoplankton is indicated by a chlorophyll metric. Such a measurement reflects the amount of chlorophyll pigments, both active (alive) or inactive (dead). Thus, chlorophyll can allow the distinction between different life cycles of algal growth (Figure 11.7). "Chlorophyll *A*" is a measure of the active fraction of the pigments; that is, the portion that was still actively respiring and photosynthesizing at the time of sampling.

The amount of algae found in a surface water body will have a large effect on the physical, chemical, and biological mechanisms in the water because the algae produce oxygen when light is present, and consume oxygen in the dark, so O_2 levels are affected within diurnal cycles. Algae also expend oxygen when they die and decay. In addition, the decomposition of algae results in the release of nutrients to surface waters, which may allow more algae to grow. Thus, the algal and plankton photosynthesis and respiration will affect the water body's pH and suspended solids content. In fact, in ponds and lakes, the presence of algae in the water column is the principal factor affecting turbidity measurements (e.g., Secchi disk readings). Algal proliferation can also lead to negative aesthetics, such as the "algal blooms" that show up as a greenish scum floating atop ponds and lakes in the summer, as well as the odors associated with the growth. Increasing amounts of sunlight, temperature, and available nutrients with spring warming and summer heat increase algal growth and, therefore, the chlorophyll *A* concentrations. Until limited by the availability of one or more nutrients (especially nitrogen or phosphorus), algae will continue to grow. Strong winds provide mixing of waters, leading to an

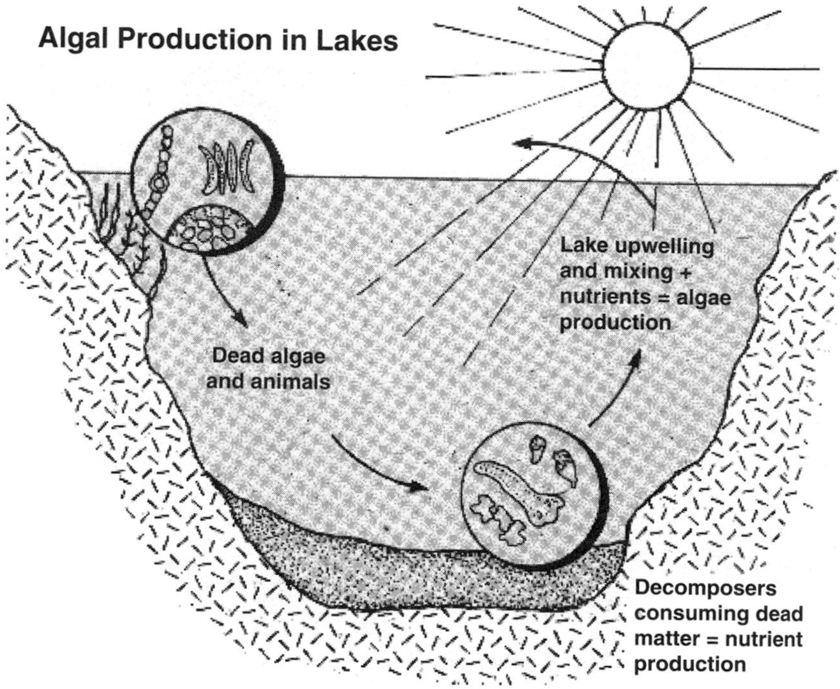

FIGURE 11.7. Algal growth cycle. (Source: State of Washington, Department of Ecology, 2003, *A Citizen's Guide to Understanding and Monitoring Lakes and Streams*.)

immediate decrease in algae concentrations as the organisms are distributed throughout the water column. But, winds may also help to release nutrients into the surface water system by agitating nutrients sequestered in bottom sediments, so that a nitrogen- or phosphorus-limited lake or pond may experience a spike in algal growth following the windy conditions.

The decreasingly available light and reduced temperatures with the onset of fall results in decreasing algal growth. However, in deep lakes where there is stratification (i.e., different temperatures at different lake levels), a fall algal bloom may occur because the lake mixes with the change of density of the layers due to the temperature differentials at various levels. This allows for more nutrients to be made available to the algae in the water body.

Algal populations, and therefore chlorophyll *A* concentrations, vary greatly with lake depth. Algae must stay within the top portion of the lake where there is sunlight to be able to photosynthesize and stay alive. As they sink below the sunlit portion of the lake, they die. Therefore, few live algae (as measured by chlorophyll *A*) are found at greater depths. Some algae, notably blue-greens (Figure 11.8), have internal "flotation devices" that allow them to regulate their depth and so remain within the top portion of the lake to photosynthesize and reproduce.

Certain algal species, especially the "blue-green" prokaryotes, produce toxins. Usually, the concentrations of toxin are too small to elicit health problems, but should the algal populations become dense, the toxins may exceed safe thresholds. For example, animals have been known to die from consuming water contaminated by algae. The blooms of these algal species usually have a characteristic bluish green sheen.

Since limiting nutrients will limit the number of algae that can grow in surface waters, the best way to address algal problems is to limit the amount of nitrogen and phosphorus entering the water. At

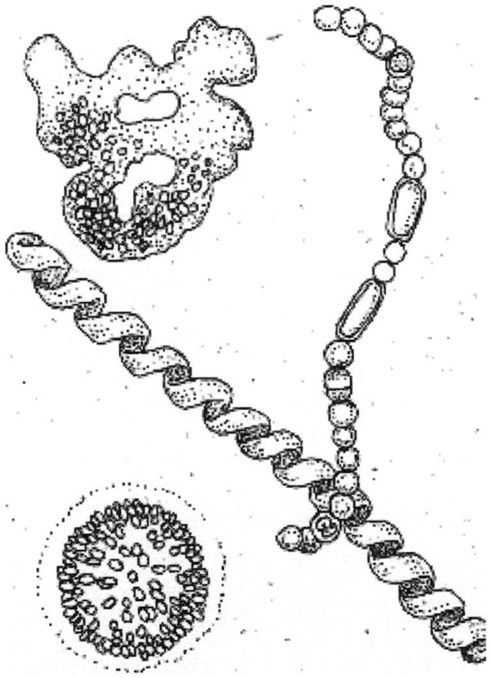

FIGURE 11.8. Blue-green algae. (Source: State of Washington, Department of Ecology, 2003, *A Citizen's Guide to Understanding and Monitoring Lakes and Streams*.)

TABLE 11.4
Chlorophyll *A* Concentrations ($\mu g\,L^{-1}$) Measured in the Top Layer (Epilimnion) of Three Washington State Lakes in June and September 1989

	Summit Lake	*Blackmans Lake*	*Black Lake*
June	1.5	3.3	7.6
September	1.5	3.9	56.2

one time, *point sources* were a principal source of such contamination, but with greater controls on wastewater treatment plants and other large sources, much of the loading of nutrients to lakes and ponds comes from *nonpoint sources*, such as runoff from farms and septic tanks. Lake management plans, for example, often include a large amount of measures to reduce the amount of nutrients reaching surface waters, including top soil erosion control programs, contour farming, minimum till agriculture, and reduced amounts of fertilizers applied to fields, as well as the banning or strong controls of septic tanks and other nutrient leaching sources in a lake watershed.

Chlorophyll *A* is reported in mass per volume units (usually $\mu g\,L^{-1}$). Many states have no water quality standard for chlorophyll *A*. Concentrations of chlorophyll *A* can vary considerably from one lake to another, even though they may be in the same region. For example, the concentrations for three lakes in the western region of Washington State are shown in Table 11.4. Black Lake would appear to have greater algal growth than do Summits and Blackmans Lakes. Also, Black Lake would appear to be temperature stratified and to experience fall mixing, allowing for an increase in algal populations in September.

Chlorophyll *A*: An Environmental Indicator

Chlorophyll *A* concentrations can be a tool to characterize a lake's trophic status. Though trophic status is not related to any water quality standard, it is a mechanism that can be used to rate a surface water body's productive state. Phytoplankton biomass in aquatic ecosystems can be simply measured as an indicator of water quality and ecosystem condition. Chlorophyll *A* has been established as an indicator of both the potential amount of photosynthesis and of the quantification of phytoplankton biomass[15] and has become a principal measure of the amount of phytoplankton present in a water body. Chlorophyll *A* is also an indirect measure of light penetration.[16] Rel-

TABLE 11.5
Model Values of Seasonal Mean and Salinity Regime-Specific Chlorophyll *A* Concentrations ($\mu g\,L^{-1}$) Characterizing Trophic Conditions to Support Acceptable Dissolved Oxygen Levels. Salinity content increases from left to right. Oligohaline waters have ≤ 0.005% salinity content, mesohaline waters have 0.005 to 0.018% salinity content, and polyhaline waters have 0.018 to 0.030% salinity content.

Season	*Tidal-Fresh*	*Oligohaline*	*Mesohaline*	*Polyhaline*
Spring	4	5	6	5
Summer	12	7	5	4

Source: Coastnet, Oregon State University Extension Sea Grant Program, 1996, *Sampling Procedures: A Manual for Estuary Monitoring.*

TABLE 11.6
Visual Inspection Criteria for Trophic Conditions of a Water Body

Algal Index Value	*Category*	*Description*
0	Clear	Conditions vary from no algae to small populations visible to the naked eye.
1	Present	Some algae visible to the naked eye but present at low to medium levels.
2	Visible	Algae sufficiently concentrated that filaments or balls of algae are visible to the naked eye. May be scattered streaks of algae on water surface.
3	Scattered Surface Blooms	Surface mats of algae scattered. May be more abundant in localized areas if winds are calm. Some odor problems.
4	Extensive Surface Blooms	Large portions of the water surface covered by mats of algae. Windy conditions may temporarily eliminate mats, but they will quickly redevelop as winds become calm. Odor problems in localized areas.

Source: Coastnet, Oregon State University Extension Sea Grant Program, 1996, *Sampling Procedures: A Manual for Estuary Monitoring.*

atively rapid methods are available for measuring the concentration of chlorophyll *A* in water samples and *in vivo*.[17] Methods are also available to measure chlorophyll *A* with remote sensing and passive multispectral signals associated with phytoplankton.[18] Chlorophyll *A* is a robust indicator of nitrogen and phosphorus enrichment.[19] Reduced water clarity and low dissolved oxygen conditions improve when excess phytoplankton or blooms, measured as chlorophyll *A*, are

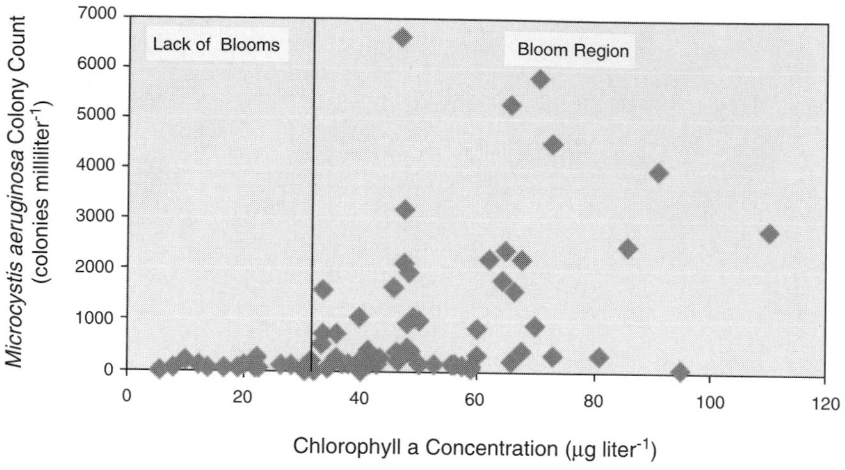

FIGURE 11.9. Colony counts of the algal species *Microcystis aeruginosa* compared to the gradient of chlorophyll *A*, measured in Chesapeake Bay. The vertical line depicts the threshold between bloom and nonbloom conditions (approximately 500 colonies mL^{-1} and $30\,\mu g\,L^{-1}$ chlorophyll *A*). (Source: Maryland Department of National Resources, 2003, unpublished data.)

lowered. Thus, chlorophyll *A* can be a robust indicator of the trophic state of a water body. An example is that modeled for lakes in Oregon (Tables 11.5 and 11.6). Along with the chlorophyll *A* readings, visual inspections of surface waters can indicate trophic conditions. Those identified for Chesapeake Bay are shown in Figure 11.9.

Chemiluminescence and Fluorescent *In-Situ* Hybridization (FISH): Monitoring the Magnitude of the Risks Associated with Environmental Contamination

Many sophisticated instruments and procedures are available to monitor the presence and concentration of contaminants in water, air, and soil. These instruments and their attendant procedures offer many opportunities to better understand the magnitude of the associated risks. For example, these instruments and techniques are used to determine the magnitude of a pollution problem, answering such questions as: How much toluene is in the soil surrounding a hazardous waste landfill? The same instruments and techniques are used to track the successes and/or failures of attempts to control the risks associated with environmental contaminants, answering such questions as: How much has this pump-and-treat technology, which was used for the past 6 months at a cost of $1,200,000, reduced the levels of toluene in the soil surrounding this hazardous waste landfill?

Let us consider two examples of measurement and monitoring instruments and procedures that are at the cutting edge of efforts to determine levels of hazardous contaminants in water, air, and soil. Chemiluminescence and fluorescent *in-situ* hybridization (FISH) are summarized as state-of-the-art examples of the science, engineering, and technology of measurement instrumentation and techniques available to address pollution problems.

Example Measurement and Monitoring Problem: Contaminated Soil

Soils at sites nationwide are contaminated with dense nonaqueous phase liquids (DNAPLs) that sink in the groundwater, nonaqueous phase liquids (NAPLs) that float within the groundwater, as well as heavy metals. These sites have become a nationwide public health and economic concern. Active soil remediation systems such as *pump and treat* and passive remediation systems such as *natural attenuation* require elaborate, expensive, decades-long monitoring for process control, performance measurement, and regulatory compliance. Under current monitoring practices, many liquid and/or soil samples must be collected from contaminated sites, packaged, transported, and analyzed in a certified laboratory, all at great expense and with great time delay to the owner of the site and to regulatory agencies at the state and federal levels. In addition, waste that is generated by sample collection and by sample analyses must be properly disposed, again at great cost. Monitoring of contaminated soils will continue into the future to ensure that public health and the environment are properly protected.

Chemiluminescence for Sensing of the Levels of Nitric Oxide (NO) Emissions from Soil

Consider the extremely complex and only partially understood biogeochemical nitrogen cycle (see Figure 7.4). The processes in soil that traditionally are suggested to contribute to the levels of NO emissions are, in order of general importance, autotrophic nitrification, respiratory denitrification, chemo-denitrification, and heterotrophic nitrification. Except for chemo-denitrification, all of these mechanisms are microbially mediated transformations performed by such bacteria as *Nitrosolobus* and *Nitrobacter* genera in autotrophic nitrification, and *Pseudomonas* and *Alcaligenes* genera in respiratory denitrification. Autotrophic nitrification and then respiratory denitrification are suggested to be the principal sources of NO in the cycle, while heterotrophic nitrification and chemo-denitrification can be important NO sources under extraordinary soil pH and other conditions.[20]

As levels of contamination change in a soil, chemiluminescence monitoring of NO emissions from contaminated soil can indicate the absence or presence, including the level, of contamination of a pollutant in the soil. This monitoring of NO emissions from soil may be used as a surrogate indicator of the level of contamination in soils during remediation and post-remediation activities at contaminated soil sites, and thus could assist in determining when expensive soil pollution remediation activities may cease.

Historically, NO concentrations in ambient air have been determined using chemiluminescence analyzers that are inexpensive, durable, accurate, and precise. For example, these analyzers are used widely by the state and federal environmental agencies to measure NO concentrations as precursors to ozone formation in cities and towns nationwide, contributing to decades of successful ambient air quality monitoring programs.

Chemiluminescence analyzers convert NO to electronically excited NO_2 (indicated as $NO_2{}^{*}$) when 0_3 is supplied internally by the analyzer, as summarized:

$$O_3 + NO \rightarrow NO_2{}^{*} + O_2 \qquad \text{Reaction 11–1}$$

These excited $NO_2{}^{*}$ molecules emit light when they move to lower energy states as:

$$NO_2{}^{*} \rightarrow NO_2 + hv(590 < \lambda < 3000\,\text{nm}) \qquad \text{Reaction 11–2}$$

The intensity of the emitted light is proportional to the NO concentration and is detected and converted to a digital signal by a photomultiplier tube that is recorded.

Dynamic test chambers and systems are available to measure the NO flux from soil. The mass balance for NO in the chamber is summarized by:

$$\frac{dC}{dt} = \left(\frac{Q[C]_0}{V} + \frac{JA_1}{V}\right) - \left(\frac{LA_2}{V} + \frac{Q[C]_f}{V}\right) + R \qquad \text{Equation 11–11}$$

Where, A = surface area of the soil
V = volume of the chamber
Q = air flow rate through the chamber
J = emission from the soil (flux)
C = NO concentration in the chamber

$[C]_o$ = NO concentration at the inlet of the chamber
$[C]_f$ = NO concentration at the outlet of the chamber
L = loss of NO on the chamber wall assumed first order in $[C]$
R = chemical production/destruction rate for NO in the chamber.

The NO emissions from soil to the headspace are calculated as J.

Using FISH to Analyze Soil Microbial Communities Exposed to Different Soil Contaminants and Different Levels of Contamination

The so-called "FISH" method identifies microorganisms by using fluorescently labeled oligonucleotide probes homologous to target strains or groups of microorganisms and viewing them by epifluorescent microscope in samples of soil studied in the laboratory and in the field. This technique was first applied to activated sludge cultures in 1994 but is continually undergoing modifications building on the understandings of procedures and oligonucleotide probes designed and applied to identify nitrifying bacteria in wastewater treatment systems. Methods of FISH application to soil samples are evolving and will continue to evolve as does every method of contaminant monitoring and measurement.[21]

Historically, the FISH techniques applied to the study of microbial communities in soil have not been as well developed as have the FISH techniques applied to the study of microbial communities in water or slurried sediment samples. The classification of active soil bacteria using FISH is a challenging research topic that appears to be developing almost exclusively outside the United States. The FISH techniques for identifying bacteria extracted from soils generally are particularly difficult to perform due to: (1) the high background fluorescence signals from soil particles; (2) the exclusion of bacteria associated with soil particles; (3) the nonspecific attachment of the fluorescent probes to soil debris; (4) probing microorganisms that are entrapped in soil solids; and (4) determining the optimal stringency of hybridization.

Other obstacles to the application of the general FISH method to soils include difficulties in sequence retrieval, finding rRNA sequences of less common organisms, nonspecific staining, low signal intensity, and target organism accessibility. In addition cells that are in the stationary phase often do not contain a sufficient cellular rRNA content to produce a detectable fluorescent image with FISH. These challenges can be overcome with the development of a variety of directed modifications to the general FISH methodologies, including altering experimental procedures for extraction and filtration of soil microbes, different selection and sequencing of oligonucleotide probes, and improving detection instrumentation, particularly the software to analyze the images obtained on a microscope

TABLE 11.7
Examples of Oligonucleotide Probes

Probe	*Target Bacteria*	*Applicability in NO Studies*
EUB338	Eubacteria	All bacteria
ALF1B	α proteobacteria	*Pseudomonas* and *Nitrobacter*
BET42A	β proteobacteria	*Nitrosomas*
GAM42A	γ proteobacteria	*Pseudomonas* (e.g., *P. putida*)

with epifluorescent capability. Example FISH probes are presented in Table 11.7.

Integration of Monitoring Techniques: Chemiluminescence and Fluorescent *In-Situ* Hybridization (FISH)

For different soils from different contaminated sites having different levels of different contaminants, microbial activity and consequently NO production will be affected during remediation and postremediation activities in the field. Consider, for example, a site where the soil is contaminated with NAPLs. At sampling locations at this site, observed NO emissions measurements that are lower than representative background levels of NO emissions from the soil could indicate depressed levels of microbial activity due to high levels of contamination that are toxic to the microorganisms in the soil. On the other hand, observed NO emission levels that are in the range of representative background levels of NO emissions from the soil could indicate normal levels of microbial activity due to acceptably low or nonexistent levels of contamination, or levels found in a successfully remediated soil or soil that was never contaminated. Depending on the type of contaminant, the level of contamination, and the physical, chemical, and microbiological characteristics of the soil itself, chemiluminescence NO emissions monitoring at different locations at the site could indicate the presence, absence, or level of contamination in the soil.

NO emissions from soil are seen as a direct indicator of microbiological activity in the soil, which in turn can suggest the presence, absence, or concentration of different contaminants in the soil. For example, laboratory measurements of NO emissions from uncontaminated soil and soil that has been contaminated with toluene have shown that the toluene-contaminated soil can produce ten times more NO than the uncontaminated soil. The additional production of NO is suggested to be the result of increased microbial activity in the contaminated soil.

These are examples of innovative ways of combining monitoring methods to determine the extent of contamination.

Notes and Commentary

1. The distinction between a "statistic" and a "parameter" is that a statistic is based upon a sample of a population, but a parameter describes a population. So if a population of an endangered species equals 1000 individuals, and every individual is tested (say, length of talons of a bird-of-prey), then that is a parameter. If, however, a population of the species is one million and a sample of 1 in 10 is taken, that is a statistic, even though both studies looked at 1000 birds. The term "parameter" also means any variable used in a model.
2. D. Crumbling, U.S. Environmental Protection Agency, 2001, *Clarifying DQO Terminology Usage to Support Modernization of Site Cleanup Practice*, EPA 542-R-01-014.
3. For sampling and analyzing dioxins and furans in soil and water, a good place to start is U.S. EPA, 1994, "Method 1613," Tetra-through octa-chlorinated dioxins and furans by isotope dilution HRGC/HRMS (Rev. B), Office of Water, Engineering and Analysis Division, Washington, D.C., as well as U.S. EPA, September 1994, "RCRA SW846 Method 8290," Polychlorinated dibenzodioxins (PCDDs) and polychlorinated dibenzofurans (PCDFs) by high resolution gas chromatograph/high resolution mass spectrometry (HRGC/HRMS), Office of Solid Waste. For air, the best method is the PS-1 high-volume sampler system described in U.S. EPA, 1999, "Method TO-9A" in *Compendium of Methods for the Determination of Toxic Organic Compounds in Ambient Air*, 2nd Edition, EPA/625/R-96/010b.
4. U.S. Environmental Protection Agency, 2002, *Guidance for the Data Quality Objectives Process*, EPA QA/G-4, EPA/600/R-96/055, Washington, D.C.
5. U.S. Environmental Protection Agency, 2003, "Test Methods: Frequently Asked Questions." http://www.epa.gov/cgi-bin/epaprintonly.cgi.
6. Australian Department of Health and Ageing, 2002, *Environmental Health Risk Assessment: Guidelines for Assessing Human Health Risks from Environmental Hazards*.
7. See M. Ekhtera, G. Mansoori, M. Mensinger, A. Rehmat, and B. Deville, 1997, "Supercritical Fluid Extraction for Remediation of Contaminated Soil," in *Supercritical Fluids: Extraction and Pollution Prevention*, edited by M. Abraham and A. Sunol, *ACSSS*, Vol. 670, pp. 280–298, American Chemical Society, Washington, D.C.
8. This is another of the many general rules of science and engineering that has important exceptions. A good indicator of a compound's retention time (i.e., the time it takes before it separates from the column) is its boiling point. The lower the boiling point, generally, the shorter the retention time will be. Specific estimates of retention time, however, must also consider polarity (i.e.,

retention time increases as the polarity of compounds approaches the polarity of the substance that makes up the coating of the column material). Remember "like dissolves like."

9. See A. Finkel, 1990, *Confronting Uncertainty in Risk Management: A Guide for Decision-Makers*, Center for Risk Management, Resources for the Future, Washington, D.C.
10 My apologies to the anonymous originator of this analogy, who deserves much credit for this teaching device. The target is a widely used way to describe precision and accuracy.
11 See: ATSDR, 2001, Public Health Assessment Guidance Manual, Appendix K, Units and Conversions, http://www.atsdr.cdc.gov/HAC/HAGM/appk.html.
12 For many environmental situations, the random error, i.e., the difference between the value that is measured and the actual value, is assumed to be distributed normally. This is the prototypical "bell curve" that is also known as a Gaussian distribution. The mean of the entire population has the symbol μ and the variance (i.e., the "spread" of values as indicated by how far the tails extend from the mean on the bell curve) is represented by σ^2. The population standard deviation is the square root of the variance, so is simply σ. The mean of the sample of the population is the "Y bar" ($\bar{Y}$) shown here, and the sample standard deviation is represented by *s*.
13. Bias is systematic error. It can result from numerous sources, including improperly calibrated equipment or faulty assumptions in a model. One way to visualize bias is to consider some everyday equipment, such as an appliance. For example, I set my alarm clock 10 minutes fast, so that when it shows 6:00 A.M. the actual time is 5:50 A.M. Statisticians would refer to the 5:50 A.M. reading as the "true value." If someone happens to use my biased clock without making the necessary adjustment that I always do (i.e., subtracting 10 minutes from the reading), they will be early until they correct the bias (e.g., look at their watch and reset my clock).
An interesting consideration is that one can reduce random error (i.e., incorrect conclusions due to chance) by increasing sampling size. However, bias is not dependent on sample size, so the correction would do nothing to help with systematic error. You would just have more values that are wrong.
14. Source: State of Washington, Department of Ecology, 2003, "A Citizen's Guide to Understanding and Monitoring Lakes and Streams," http://www.ecy.wa.gov/programs/wq/plants/management/joysmanual/chlorophyll.html.
15. See D. Flemer, 1969, "Continuous Measurement of *In Vivo* Chlorophyll of z Dinoflagellate Bloom in Chesapeake Bay," *Chesapeake* Science, Vol. 10, pp. 99–103; and D. Flemer, 1969, "Chlorophyll Analysis as a Method of Evaluating the Standing Crop of Phytoplankton and Primary Production," *Chesapeake Science*, Vol. 10, pp. 301–306.
16. For example, see C. Lorenzen, 1972, "Extinction of Light in the Ocean by Phytoplankton," *Journal of Conservation*, Vol. 34, pp. 262–267.
17. See D. Flemer, 1969, "Continuous Measurement of *In Vivo* Chlorophyll of a Dinoflagellate Bloom in Chesapeake Bay," *Chesapeake Science*, Vol. 10, pp.

99–103; and U.S. Environmental Protection Agency (EPA), 1997, *Methods for the Determination of Chemical Substances in Marine and Estuarine Environmental Matrices*, 2nd Edition, Method 446.0, EPA/600/R-97/072, U.S. EPA, Office of Research and Development, Washington, D.C.

18. L. Harding, Jr., E. Itsweire, and W. Esais, 1992, "Determination of Phytoplankton Chlorophyll Concentrations in the Chesapeake Bay with Aircraft Remote Sensing," *Remote Sensing of Environment*, Vol. 40, pp. 79–100.
19. L. Harding, Jr., and E. Perry, 1997, "Long-term Increase of Phytoplankton Biomass in Chesapeake Bay, 1950–1994," *Marine Ecology Progress Series*, Vol. 157, pp. 39–52.
20. Research in the area of nitric oxide emissions from soil includes: F. Chase, C. Corke, and J. Robinson, 1968, "Nitrifying Bacteria in the Soil," in *Ecology of Soil Bacteria*, edited by T.R.G. Gray and D. Parkinson, University of Liverpool Press, Liverpool; H. Christensen, M. Hansen, and J. Sorensen, 1999, "Counting and Size Classification of Active Soil Bacteria by Fluorescence In Situ Hybridization with an rRNA Oligonucleotide Probe," *Applied and Environmental Microbiology*, Vol. 65, no. 4, pp. 1753–1776; I. Galbally, 1989, "Factors Controlling NO Emissions from Soils," in *Exchange of Trace Gases between Terrestrial Ecosystems and the Atmosphere: The Dahlem Conference,* edited by M.O. Andreae and D.S. Schimel, John Wiley & Sons, New York, N.Y.; S. Jousset, R. Tabachow, and J. Peirce, 2001, "Nitrification and Denitrification Contributions to Soil Nitric Oxide Emissions," *Journal of Environmental Engineering*, Vol. 127, no. 4, pp. 322–338; J. Peirce and V. Aneja, 2000, "Laboratory Study of Nitric Oxide Emissions from Sludge Amended Soil," *Journal of Environmental Engineering*, Vol. 126, no. 3, pp. 225–232; and D. Rammon and J. Peirce, 1999, "Biogenic Nitric Oxide from Wastewater Land Application," *Atmospheric Environment*, Vol. 33, pp. 2115–2121.
21. Developing research in the area of FISH applications to the microbial populations in water and soil includes: G.A. Kowalchuk, J.R. Stephen, W. De Boer, J.I. Prosser, T.M. Embley, and J.W. Woldendorp, 1997, "Analysis of *B*-Proteobacteria Ammonia-oxidising Bacteria in Coastal Sand Dunes Using Denaturing Gradient Gel Electrophoresis and Sequencing of PCR Amplified 16S rDNA Fragments," *Applied and Environmental Microbiology*, Vol. 63, pp. 1489–1497; W. Manz, R. Amann, M. Wagner, and K.-H. Schleifer, 1992, "Phylogenetic Oligonucleotide Probes for the Major Subclasses of Proteobacteria: Problems and Solutions," *Systematic and Applied Microbiology*, Vol. 15, pp. 593–600; B. Nogales, E.R.B. Moore, E. Llobet-Brossa, R. Rossello-Mora, R. Amann, and K.N. Timmis, 2001, "Combined Use of 16S Ribosomal DNA and 16S RNA to Study the Bacterial Community of Polychlorinated Biphenyl-polluted Soil," *Applied and Environmental Microbiology*, Vol. 67, no. 4, pp. 1874–1884; M. Wagner, G. Rath, H.-P. Koops, J. Flood, and R. Amann, 1996, "In Situ Analysis of Nitrifying Bacteria in Sewage Treatment Plants," *Water Science and Technology*, Vol. 34, no. 1–2, pp. 237–244.

CHAPTER 12

Intervention: Managing the Risks of Environmental Contamination

With an understanding of the sources, transport, transformation, and fate of environmental contaminants, as well as the risks that they pose, we must consider how to reduce the concentrations and exposures to these chemicals after they are released. This is risk management. The first step is to plan and gather reliable environmental data, as described in the previous chapter. Next, the contaminants are characterized by applying the principles and concepts in this book, in order to prepare the wastes for disposal and decontamination. We follow discussions of these steps with brief discussions of interventions that can be made to treat contaminated media and to reduce the risks posed by contamination. We will end with a consideration of the ways to communicate what we have found and what we intend to do about the contamination.

A Template for Cleaning Up Contaminants

Environmental regulations lay out explicit steps for cleaning up contamination. Although these steps are usually codified[1] according to which environmental compartment has been contaminated, we can apply them more generally to environmental assessments and cleanups in all environmental media, including the air, water, soil, and living systems. The contaminant cleanup process is shown in Figure 12.1. The first step of a contaminant cleanup is a preliminary assessment (PA). Generally, who knows what one will find until one looks? Even if data are available, what has been reported may just be the tip of the iceberg. During the PA of a site, readily available information about a site and its surrounding area are collected to differen-

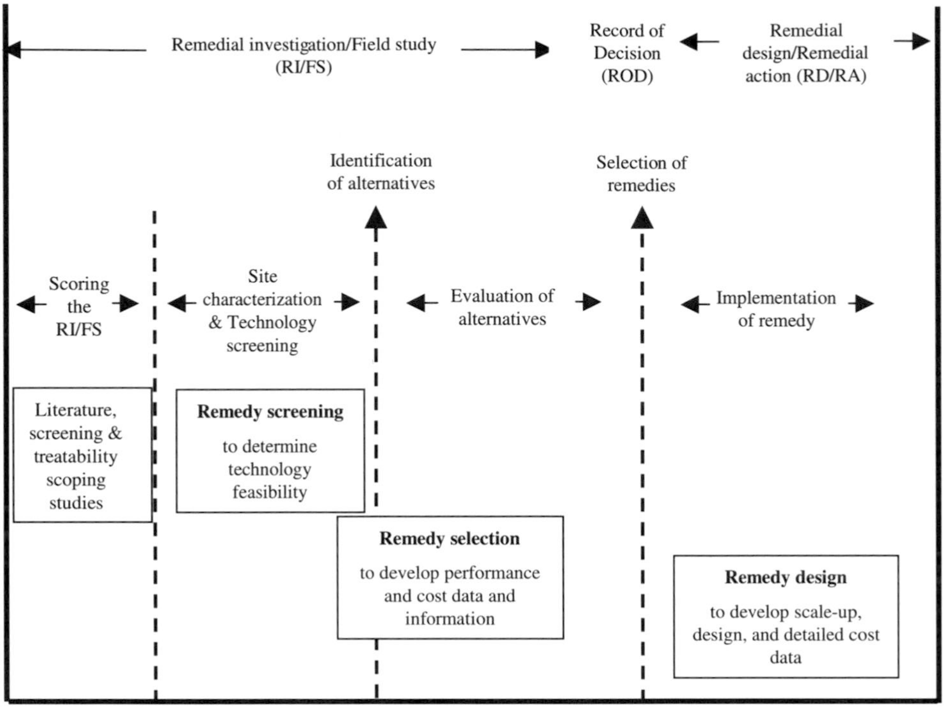

FIGURE 12.1. Steps in a contaminated site cleanup, as mandated by Superfund. (Source: U.S. Environmental Protection Agency, 1992, *Guide for Conducting Treatability Studies under CERCLA: Thermal Desorption*, EPA/540/R-92/074 B.)

tiate sites posing little of no threat to human health and ecosystems from sites that potentially pose threats and that may need further investigation.[2] Any possible *emergency response* actions may also be identified. A site inspection is performed if the PA, based on limited data, calls for one (that is why this step is often referred to as the PA/SI).[3]

In the United States, certain hazardous waste sites are considered to be of sufficient concern to be listed on the National Priority List. The listing is actually a combination of the hazard (usually toxicity) of the contaminants found at the site and the likelihood that people or ecosystems will be exposed to these pollutants. Severely polluted sites and sites that contain very toxic compounds in measurable quantities are ranked higher than those with less toxic substances in lesser quantities.

A public disclosure of the condition following contaminant treatment must be made. This final record of decision (ROD) indicates that the specific engineering remedy has been selected for the site. Like any other aspects of hazardous waste cleanup, this decision is subject to later contests (legal, scientific, or otherwise). Since public officials are not exempt from personal tort liabilities in their decisions, the ROD is usually made as a collective, agency decision based upon past and ongoing contaminant measurements, and includes provisions for monitoring for years to come to ensure that the engineered systems continue to perform according to plan. The ROD must also ensure that a plan for operating and maintaining all systems is in place, including a plan for dealing with failures and other unexpected contingencies, such as improvements in measurement techniques that later identify previously undetected pollutants.

Characterizing Contaminants in the Environment

The environmental monitoring and analysis gives important information about the various media that have been contaminated or that are vulnerable to future contamination. Before intervening, however, information is also needed to characterize the contaminants themselves. Characterization must factor in all of the physical, chemical, and biological characteristics of the contaminant with respect to the matrices and substrates (if soil and sediment) or fluids (air, water, or other solvents) where the contaminants are found.

Selection factors for treatment technologies must consider the target contaminants, the characteristics of the environmental compartment where the contaminant is found, and how to implement the treatment processes. Further, the selected approach must meet criteria for *treatability*. The comprehensive remedy must consider the effects that each action taken has had and will have on past and proceeding steps.

Eliminating or reducing concentrations of contaminants begins with an assessment of the physical and chemical characteristics of each contaminant, and matching these characteristics with the appropriate treatment technology. All of the kinetics and equilibria, such as solubility, fugacity, sorption, and bioaccumulation factors, will determine the effectiveness of destruction, transformation, removal, and immobilization of these contaminants. For example, Table 12.1 ranks the effectiveness of selected treatment technologies on organic and inorganic contaminants typically found in contaminated slurries, soils, sludges, and sediments. As shown, there can be synergies (e.g., innovative incineration approaches are available that not only effectively destroy organic contaminants but in the process also destroy the inorganic cyanide species). Unfortunately, there are also antagonisms among certain approaches, such as the very effective incineration processes for organic contaminants that transform heavy metal species into more toxic and more mobile forms. The increased pressures and temperatures are good for breaking apart organic molecules, and removing functional groups that lend them toxicity, but these same factors oxidize or in other ways transform the metals into worse forms. Thus, when mixtures of organic and inorganic contaminants are targeted, more than one technology may be required to accomplish project objectives, and care must

TABLE 12.1
Effect of the Characteristics of the Contaminant on Decontamination Efficiencies

Treatment Technology	*Organic Contaminants*					*Inorganic Contaminants*		
	PCBs	*PAHs*	*Pesticides*	*Petroleum Hydrocarbons*	*Phenolic Compounds*	*Cyanide*	*Mercury*	*Other Heavy Metals*
Conventional Incineration	D	D	D	D	D	D	xR	pR
Innovative Incineration[a]	D	D	D	D	D	D	xR	I
Pyrolysis[a]	D	D	D	D	D	D	xR	I
Vitrification[a]	D	D	D	D	D	D	xR	I
Supercritical Water Oxidation	D	D	D	D	D	D	U	U
Wet Air Oxidation	pD	D	U	D	D	D	U	U
Thermal Desorption	R	R	R	R	U	U	xR	N
Immobilization	pI	pI	pI	pI	pI	pI	U	I
Solvent Extraction	R	R	R	R	R	pR	N	N
Soil Washing[b]	pR	pR	pR	pR	pR	pR	pR	pR
Dechlorination	D	N	pD	N	N	N	N	N
Oxidation[c]	N/D	N/D	N/D	N/D	N/D	N/D	U	xN
Bioremediation[d]	N/pD	N/D	N/D	D	D	N/D	N	N

Note: PCBs—polychlorinated biphenyls
PAHs—polynuclear aromatic hydrocarbons
Primary designation
D = effectively destroys contaminant
R = effectively removes contaminant
I = effectively immobilizes contaminant
N = no significant effect
N/D = effectiveness varies from no effect to highly efficient depending on the type of contaminant within each class
U = effect not known
Prefixes: p = partial; x = may cause release of nontarget contaminant

[a] This process is assumed to produce a vitrified slag.
[b] The effectiveness of soil washing is highly dependent on the particle size of the sediment matrix, contaminant characteristics, and the type of extractive agents used.
[c] The effectiveness of oxidation depends strongly on the types of oxidant(s) involved and the target contaminants.
[d] The effectiveness of bioremediation is controlled by a large number of variables as discussed in the text.

Source: U.S. Environmental Protection Agency, 2003, *Remediation Guidance Document*, Chapter 7, EPA-905-B94-003.

be taken not to trade one problem (e.g., PCBs) for another (e.g., more mobile species of cadmium).

The characteristics of the soil, sediment, or water will affect the performance of any contaminant treatment or control. For example, sediment, sludge, slurries, and soil characteristics will influence the efficacy of treatment technologies include particle size, solids content, and high contaminant concentration (see Table 12.2).

Particle size may be the most important limiting characteristic for application of treatment technologies to sediments. Most treatment technologies work well on sandy soils and sediments. The presence of fine-grained material adversely affects treatment system emission controls,

TABLE 12.2
Effect of Particle Size, Solids Content, and Extent of Contamination on Decontamination Efficiencies

Treatment Technology	*Predominant Particle Size*			*Solids Content*		*High Contaminant Concentration*	
	Sand	*Silt*	*Clay*	*High (slurry)*	*Low (in situ)*	*Organic Compounds*	*Metals*
Conventional Incineration	N	X	X	F	X	F	X
Innovative Incineration	N	X	X	F	X	F	F
Pyrolysis	N	N	N	F	X	F	F
Vitrification	F	X	X	F	X	F	F
Supercritical Water Oxidation	X	F	F	X	F	F	X
Wet Air Oxidation	X	F	F	X	F	F	X
Thermal Desorption	F	X	X	F	X	F	N
Immobilization	F	X	X	F	X	X	N
Solvent Extraction	F	F	X	F	X	X	N
Soil Washing	F	F	X	N	F	N	N
Dechlorination	U	U	U	F	X	X	N
Oxidation	F	X	X	N	F	X	X
Bioslurry Process	N	F	N	N	F	X	X
Composting	F	N	X	F	X	F	X
Contained Treatment Facility	F	N	X	F	X	X	X

Note: F—sediment characteristic favorable to the effectiveness of the process
N—sediment characteristic has no significant effect on process performance
U—effect of sediment characteristic on process is unknown
X—sediment characteristic may impede process performance or increase cost
Source: U.S. Environmental Protection Agency, 2003, *Remediation Guidance Document*, Chapter 7, EPA-905-B94-003.

because it increases particulate generation during thermal drying, is more difficult to dewater, and has greater attraction to the contaminants (particularly clays). Clayey sediments that are cohesive also present materials handling problems in most processing systems.

Solids content generally ranges from high, usually the *in situ* solids content (30–60% solids by weight), to low, such as hydraulically dredged sediments (10–30% solids by weight). Treatment of slurries is better at lower solids contents, but this can be achieved even for high solids contents by water addition at the time of processing. It is more difficult to change a lower to a higher solids content, but evaporative and dewatering approaches, such as those used for municipal sludges, may be employed. Also, thermal and dehalogenation processes are decreasingly efficient as solids content is reduced. More water means increased chemical costs and increased need for wastewater treatment.

Elevated levels of organic compounds or heavy metals in high concentrations must also be considered. Higher total organic carbon (TOC) content favors incineration and oxidation processes. The TOC can be the contaminant of concern or any organic, since they are combustibles with caloric value. Conversely, higher metal concentrations may make a technology less favorable by increasing the contaminant mobility of certain metal species following application of the technology.

A number of other factors may affect selection of a treatment technology other than its effectiveness for treatment (some are listed in Table 12.3). For example, vitrification and supercritical water oxidation have only been used for relatively small projects and would require more of a proven track record before implementing them for full-scale sediment projects. Regulatory compliance and community perception are always a part of decisions regarding an incineration system. Land use considerations, including the amount of acreage needs are commonly confronted in solidification and solid-phase bioremediation projects (as they are in sludge farming and land application). Disposing of ash and other residues following treatment must be part of any process. Treating water effluent and air emissions must be factored into the decontamination decision-making process.

Estimating Contaminant Migration[4]

Ways to estimate potential contaminant releases (i.e., "losses," as defined by environmental regulators) from various combinations of treatment technologies are quite difficult due to the variability of chemical and physical characteristics of contaminated media (especially soils and sediments), the strong affinity of most contaminants for fine-grained sediment particles, and the limited track record or "scale-up" studies for many treatment technologies. Off-the-shelf models can be used for simple process operations, such as extraction or thermal vaporization applied to single contaminants

TABLE 12.3
Selected Factors on Selecting Decontamination and Treatment Approaches

Treatment Technology	*Implementability at Full Scale*	*Regulatory Compliance*	*Community Acceptance*	*Land Requirements*	*Residuals Disposal*	*Wastewater Treatment*	*Air Emissions Control*
Conventional Incineration		✓	✓				✓
Innovative Incineration		✓	✓				✓
Pyrolysis		✓					✓
Vitrification	✓	✓					✓
Supercritical Water Oxidation	✓						
Wet Air Oxidation							
Thermal Desorption				*	✓	✓	✓
Immobilization				✓			✓
Solvent Extraction					✓	✓	
Soil Washing					✓	✓	
Dechlorination							✓
Oxidation	✓						
Bioslurry Process	✓						✓
Composting				✓			✓
Contained Treatment Facility				✓		✓	✓

Note: ✓—the factor is critical in the evaluation of the technology
Adapted from: U.S. Environmental Protection Agency, 2003, *Remediation Guidance Document*, Chapter 7, EPA-905-B94-003.
*Author's note: Soil type can be a factor since fine textured soils are conductive to wet air oxidation.

in relatively pure systems. However, such models have not been validated for the sediment treatment technologies discussed in this chapter, because of the limited database on treatment technologies for contaminated sediments or soils.

Treatability Tests[5]

Standard engineering practice for evaluating the effectiveness of treatment technologies for any type of contaminated media (solids, liquids, or gases) is to perform a treatability study for a sample that is representative of the contaminated material. The performance data from treatability studies can aid in reliably estimating contaminant concentrations for the residues following treatment, as well as possible waste streams generated by a technology. Treatability studies may be performed at the bench-scale (in the lab) or at the pilot-scale level (e.g., a real-world study, but limited in the number of contaminants, in the spatial extent, or to a specific, highly controlled form of a contaminant, such as one pure congener of PCBs rather than the common mixtures). Most treatment technologies include post-treatment or controls for waste streams produced by the processing. The contaminant losses can be defined as the residual contaminant concentrations in the liquid or gaseous streams released to the environment. For technologies that extract or separate the contaminants from the bulk of the sediment, a concentrated waste stream may be produced that requires treatment off-site at a hazardous waste treatment facility, where permit requirements may require destruction and removal efficiencies greater than 99.9999% (i.e., the so-called rule of "six nines"). The other source of contaminant loss for treatment technologies is the residual contamination in the sediment after treatment. The disposal sites of treated wastes are subject to leaching, volatilization, and losses by other pathways. The significance of these pathways depends on the type and level of contamination that is not removed or treated by the treatment process. Various waste streams for each type of technology that should be considered in treatability evaluations are listed in Table 12.4.

Contaminant Treatment and Control Approaches[6]

Five steps in sequence define an event that results in environmental contamination of the water, air, or soil. These steps individually and collectively offer opportunities to intervene and to control the risks associated with pollution and thus protect public health and the environment. The steps that are necessary and sufficient for such pollution to occur from a contaminant source, a contaminated environmental medium, or accidental release are:

TABLE 12.4
Selected Waste Streams Commonly Requiring Treatability Studies

Contaminant Loss Stream	*Treatment Technology Type*						
	Biological	*Chemical*	*Extraction*	*Thermal Desorption*	*Thermal Destruction*	*Immobilization*	*Particle Separation*
Residual solids	✗	✗	✗	✗	✗	✗	✗
Wastewater	✗	✗	✗	✗			✗
Oil/organic compounds			✗	✗			✗
Leachate						✗[a]	
Stack gas				✗	✗		
Adsorption media			✗	✗			
Scrubber water					✗		
Particulates (filter/cyclone)				✗	✗		

[a]Long-term contaminant losses must be estimated using leaching tests and contaminant transport modeling similar to that used for sediment placed in a confined disposal facility. Leaching could be important for residual solids for other processes as well.
Source: U.S. Environmental Protection Agency, 2003, *Remediation Guidance Document*, Chapter 7, EPA-905-B94-003.

SOURCE →
RELEASE →
TRANSPORT →
EXPOSURE →
RESPONSE

As a first step, the contaminant source must be identifiable. A hazardous substance must be released from the source, be transported through the water, air, or soil environment, and reach a human, animal, or plant receptor in a measurable dose. The receptor must also have a quantifiable detrimental response in the form of death or illness. All five steps must exist for a pollution problem to exist. No pollution problem? No need to intervene!

If a problem does exist, intervention can occur at any one of these steps to control the risks to public health and to the environment. Of course, any intervention scheme and subsequent control must be justified by the environmental professional as well as the public or private client in terms of scientific evidence, sound engineering design, technological practicality, economic realities, ethical considerations, and the laws of local, state, and national governments.

Intervention at the Source of Contamination

A contaminant must be identifiable, either in the form of an industrial facility that generates waste by-products, a hazardous waste processing facility, a surface or subsurface land storage/disposal facility, or an accidental spill into a water, air, or soil receiving location. The intervention must minimize or eliminate the risks to public health and the environment by utilizing technologies at this source that are economically acceptable and based on applicable scientific principles and sound engineering designs.

In the case of an industrial facility producing hazardous waste as a necessary by-product of a profitable item, as considered here, for example, the environmental professional can take advantage of the growing body of knowledge that has become known as *life cycle analysis*.[7] In the case of a hazardous waste storage facility or a spill, the engineer must take the source as a given and search for possibilities for intervention at a later step in the sequence of steps.

Under the life cycle analysis method of intervention, the environmental manager considers the environmental impacts that could incur during the entire life cycle of (1) all of the resources that go into the product; (2) all the materials that are in the product during its use; and (3) all the materials that are available to exit from the product once it or its storage containers are no longer economically useful to society. Few simple examples exist that describe how *life cycle analysis* is conducted, but consider for now any one of a number of household cleaning products. Consider that a particular cleaning product, a solvent of some sort, must be fabricated from one of several basic natural resources. Assume for now that this cleaning product currently is petroleum-based. The engineer could intervene at this initial step in the life cycle of this product, as the natural resource is being selected, and consequently the engineer could preclude the formation of a source of hazardous waste by suggesting instead the production of a water-based solvent.

Similarly, intervention at the production phase of this product's life cycle and the suggestion of fabrication techniques can preclude the formation of a source of certain contaminants from the outset. In this case, the recycling of spent petroleum materials could provide for more household cleaning product with less or zero hazardous waste generation, thus controlling the risks to public health and the environment. Another example is that of "co-generation," which may allow for two manufacturing facilities to colocate so that the "waste" of one is a "resource" for the other. An example is the location of a chemical plant near a power generation facility, so that the excess steam generated by the power plant can be piped to the nearby chemical plant, obviating the need to burn its own fuel to generate the steam needed for chemical synthesis. Another example is the use of an alcohol waste from one plant as a source for chemical processes at another.

In life cycle analysis the product under consideration must be considered long before any switches are flipped and valves turned. For example, a particular household cleaning product may result in unintended human exposure to buckets of solvent mixtures that fumigate the air in a home's kitchen or pollute the town's sewers as the bucket's liquid is flushed down a drain. In this way, life cycle analysis is a type of systems engineering where a critical path is drawn, and each decision point considered.

Under the plan, the disposal of this solvent's containers must be considered from a long-term risk perspective. The challenge is that every potential and actual environmental impact of a product's fabrication, use, and ultimate disposal must be considered. This is seldom, if ever, a "straight line projection."

Intervention at the Point of Release

Once a contaminant source has been identified, the next step is to intervene at the point at which the waste is released into the water, air, or soil environment. This point of release could be at the end of a pipe running from the source of pollution to a receiving water body like a stream, from the top of a stack running from the source of pollution to a receiving air shed, or from the bottom-most layer of a clay liner in a hazardous waste landfill connected to surrounding soil material. Similarly, this point of release could be a series of points as a contaminant is released along a shoreline from a plot of land into a river, or through a plane of soil underlying a storage facility (i.e., a so-called "nonpoint source"). This intervention applies the physical, chemical, and microbiological processes discussed throughout this book. These processes can be used to prevent the release of contaminants into the environment.

Intervention as the Contaminant Is Transported in the Environment

Wise site selection of facilities that generate, process, and store contaminants is the first step in preventing or reducing the likelihood that they will move. For example, in surface waters, the distance from a source to a receiving body of water is a crucial factor in controlling the quantity and characteristics of waste as it is transported.

In the air, meteorology helps to determine the opportunities to control the atmospheric transport of contaminants. For example, manufacturing, transportation, and hazardous waste generating, processing, and storage facilities must be sited to avoid areas where specific local weather patterns are frequent and persistent. These avoidance areas include ground-based inversions, elevated inversions, valley winds, shore breezes, and city heat islands. In each of these venues, the pollutants become locked into air masses with little or no chance of moving out of the respective areas. Thus the concentrations of the pollutants can quickly and greatly pose risks to public health and the environment. In the soil environment the engineer has the opportunity to site facilities in areas of great depth-to-groundwater, as well as in soils (e.g., clays) with very slow rates of transport. In this way, engineers and scientists must work closely with city and regional planners early in the site selection phases.[8]

Intervention to Control the Exposure

The receptor of contamination can be a human, other fauna in the general scheme of living organisms, flora, or materials or constructed facilities. In the case of humans, the contaminant can be ingested, inhaled, or dermally contacted. Such exposure can be direct with human contact to, for example, particles of lead that are present in inhaled indoor air. Such exposure also can be indirect, as in the case of human ingestion of the cadmium and other heavy metals found in the livers of beef cattle that were raised on grasses receiving nutrition from cadmium-laced municipal wastewater treatment biosolids.

Heavy metals or chlorinated hydrocarbons similarly can be delivered to domestic animals and animals in the wild. Construction materials also are sensitive to exposure to hazardous wastes, from the "greening" of statues, through the de-zinc process associated with low pH rain events, to the crumbling of stone bridges found in nature. Isolating potential receptors from exposure to contaminants gives the engineer or planner an opportunity to control the risks to those receptors. Incidentally, controlling exposures is also possible at the "point of use." For example, a growing number of people add filters to water taps, use indoor air cleaning devices and systems, and drink bottled water.

The opportunities to control exposures to contaminants are directly associated with the ability to control the amount of hazardous pollutants delivered to the receptor through source control and siting of facilities. One solution to environmental contamination could be to increase their dilution in the water, air, or soil environments. We will discuss specific examples of this type of intervention later in this chapter.

Intervention at the Point of Response

Scientists working on finding the prevention and cure for cancer may some day enter this step and save the day. Enhancement of immune response and preventive medicine are promising, such as the role of free radicals in cancer prevention. This, unfortunately, is beyond the scope of this text.

Opportunities for intervention are grounded in basic scientific principles, engineering designs and processes, and applications of proven and developing technologies to control the risks associated with contaminants. Let us consider in more detail examples of such opportunities: (1) thermal processing; (2) microbiological treatment; and (3) landfills as long-term repositories for contaminated materials.

Thermal Processing: The Science, Engineering and Technology of Contaminant Destruction

Contaminants, if completely organic in structure are, in theory, completely destructible using principles based in thermodynamics with the engineering inputs and outputs summarized as:

$$\text{Hydrocarbons} + O_2(+\text{energy?}) \rightarrow CO_2 + H_2O(+\text{energy?}) \quad \text{Reaction 12–1}$$

Contaminants are mixed and react with oxygen, sometimes in the presence of an external energy source, and in fractions of seconds or several seconds the by-products of gaseous carbon dioxide and water are produced to exit to top of the reaction vessel while a solid ash is produced to exit the bottom of the reaction vessel.[9] Energy may also be produced during the reaction and the heat may be recovered. A derivative problem in this simple reaction could be global warming associated with the carbon dioxide.

On the other hand, if the contaminant of concern to the engineer contains other chemical constituents, in particular halogens (especially chlorine and bromine) or heavy metals, the original simple input and output relationship is modified to a very complex situation:

Hydrocarbons + O_2 (+energy?) + Cl or heavy metal(s) + H_2O + inorganic salts + nitrogen compounds + sulfur compounds + phosphorus compounds → CO_2 + H_2O (+energy?) + chlorinated hydrocarbons or heavy metal(s) inorganic salts + nitrogen compounds + sulfur compounds + phosphorus compounds

Reaction 12–2

With these contaminants the potential exists for destroying the initial contaminant, but also actually exacerbating the problem by generating more hazardous off-gases containing chlorinated hydrocarbons and/or ashes containing heavy metals (e.g., the improper incineration of certain chlorinated hydrocarbons can lead to the formation of the highly toxic chlorinated dioxins, furans, and hexachlorobenzene).

All of the thermal systems discussed below have common attributes. All require the balancing of the three "*T*s": *T*ime of incineration, *T*emperature of incineration, and *T*urbulence in the combustion chamber. The space required for the incinerator itself ranges from several square yards, to the back of a flatbed truck, to several acres to sustain a regional incinerator system.

The advantages of thermal systems include: (1) the potential for energy recovery; (2) volume reduction of the contaminant; (3) detoxification as selected molecules are reformulated; (4) the basic scientific principles, engineering designs, and technologies are well understood from a wide range of other applications including electric generation and municipal solid waste incineration; (5) application to most organic contaminants that compose a large percentage of the total contaminants generated worldwide; (6) the possibility to scale the technologies to handle a single gallon per pound (liter per kilogram) of waste; or millions of gallons per pound (liter per kilogram) of waste; and (7) land areas that are small relative to other hazardous waste management facilities such as landfarms and landfills.

Each system design must be customized to address the specific contaminants under consideration, including the quantity of waste to be processed over the project period as well as the physical, chemical, and microbiological characteristics of the waste. Laboratory testing and pilot studies matching a given waste to a given incinerator must be conducted prior to the design, citing, and construction of each incinerator. Generally, the same reaction applies to most thermal processes, i.e. gasification, pyrolysis, hydrolysis, and combustion:[10]

$$C_{20}H_{32}O_{10} + x_1O_2 + x_2H_2O \rightarrow y1C + y_2CO_2 + y_3CO + y_4H_2 + y_5CH_4 + y_6H_2O + y_7C_nH_m \qquad \text{Reaction 12–3}$$

The coefficients *x* and *y* balance the compounds on either side of the equation. In many thermal reactions, C_nH_m includes the alkanes, C_2H_2, C_2H_4, C_2H_6, C_3H_8, C_4H_{10}, C_5H_{12}, and benzene, C_6H_6. The actual reactions from test burns for commonly incinerated compounds are provided in Table 12.5.

Of all of the thermal processes, incineration is the most common process for destroying organic contaminants in industrial wastes. Incineration simply is heating wastes in the presence of oxygen to oxidize organic compounds (both toxic and nontoxic). The principal incineration steps are shown in Figure 12.2.

The mention of the word *incineration* evokes controversy in communities. Incineration alone does not remove heavy metal contamination. In fact, incineration generally increases the leachability of metals through the process of oxidation (but processes like slagging or vitrification actually reduce the mobility of many metals, by producing nonleachable, basalt-like residue). The increased leachability of metals would be problematic if the ash and other residues are to be buried in landfills or stored in piles. The leachability of metals is generally measured using the toxicity characteristic leaching procedure (TCLP) test. Incinerator ash that fails the TCLP must be disposed of in a waste facility approved for hazardous wastes. Enhanced leachability would be advantageous only if the residues are engineered to undergo an additional step to treat metals. This points to the need for a systematic approach for any contaminant treatment process.

There are a number of points during the contaminant incineration process flow where new compounds may need to be addressed. As mentioned, ash and other residues may contain elevated levels of metals, usually at much higher concentrations than the original feed as a result of thermal processes that combust hydrocarbons. These processes leave behind only the

TABLE 12.5
Balanced Combustion Reactions for Selected Organic Compounds

Chlorobenzene:	$C_6H_5Cl + 7O_2 \rightarrow 6CO_2 + HCl + 2H_2O$
TCE (tetrachloroethene):	$C_2Cl_4 + O_2 + 2H_2O \rightarrow 2CO_2 + HCl$
HCE (hexachloroethane):	$C_2Cl_6 + \frac{1}{2}O_2 + 3H_2O \rightarrow 2CO_2 + 6HCl$
"Post-chlorinated polyvinyl chloride (CPVC):	$C_4H_5Cl_3 + 4\frac{1}{2}O_2 \rightarrow 4CO_2 = 3HCl + H_2O$
Natural Gas Fuel (Methane):	$CH_4 + 2O_2 \rightarrow CO_2 + 2H_2O$
PTFE Teflon:	$C_2F_4 + O_2 \rightarrow CO_2 + 4HF$
Butyl Rubber:	$C_9H_{16} + 13O_2 \rightarrow 9CO_2 + 8H_2O$
Polyethylene:	$C_2H_4 + 3O_2 \rightarrow 2CO_2 + 2H_2O$

Wood is considered to have the composition of $C_{6.9}H_{10.6}O_{3.5}$. Therefore, the combustion reactions are simple carbon and hydrogen combustion:

$$C + O_2 \rightarrow CO_2$$
$$H + 0.25O_2 \rightarrow 0.5H_2O$$

Source: U.S. Environmental Protection Agency.

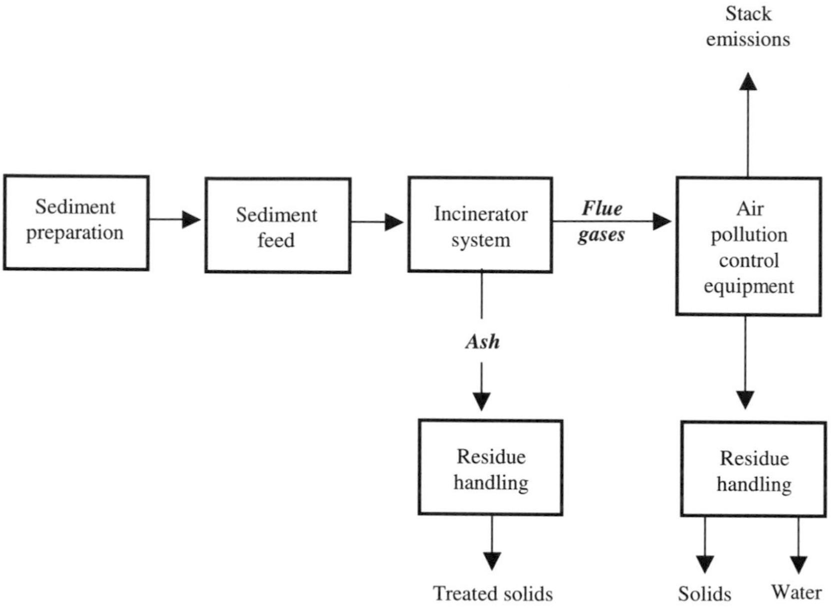

FIGURE 12.2. Steps in the incineration of contaminants. (Source: U.S. Environmental Protection Agency, 2003, *Remediation Guidance Document*, Chapter 7, EPA-905-B94-003.)

noncombustible materials, predominantly heavy metals and inorganic compounds. Certainly, a fraction of the metals and inorganic compounds are volatile and leave through the stack, but compared to most organic compounds, more of the inorganic and metallic substances remain in the ash and solid residues. Thus, the flue gases are likely to include both organic and inorganic compounds that have been released as a result of temperature-induced volatilization or newly transformed products of incomplete combustion with heat-enhanced vapor pressures than the original contaminants.

The disadvantages of hazardous waste incinerators include: (1) the equipment is capital-intensive, particularly the refractory material lining the inside walls of each combustion chamber that must be replaced as cracks form whenever a combustion system is cooled and heated (i.e., expansion/contraction cycles); (2) the operation of the equipment requires very skilled operators and is more costly when fuel must be added to the system; (3) ultimate disposal of the ash is necessary and particularly troublesome and costly if heavy metals or chlorinated compounds are found during the expensive monitoring activities; and (4) air emissions may be hazardous and thus must be controlled and continuously monitored for chemical constituents.

Given the underlying principles and disadvantages of incineration, seven general guidelines emerge:

1. Only liquid, purely organic contaminants are true candidates for combustion;
2. Chlorine-containing organic materials deserve special consideration if in fact they are to be incinerated at all; special materials used in the construction of the incinerator, long (many seconds) of combustion time, high temperatures (> 1600°C), with continuous mixing if the contaminant is in the solid or sludge form;
3. Feedstock containing heavy metals generally should not be incinerated;
4. Sulfur-containing organic material will emit sulfur oxides, (SO_x) which must be controlled (e.g., with SO_x scrubbers);[11]
5. The formation of nitrogen oxides (NO_x) can be minimized if the combustion chamber is maintained above 1100°C;[12]
6. Destruction depends on the interaction of a combustion chamber's temperature, dwell time, and turbulence; and,
7. Off-gases and ash must be monitored for chemical constituents, each residual must be treated appropriately so the entire combustion system operates within the requirements of the local, state, and federal environmental regulations, and hazardous components of the off-gases, off-gas treatment processes, and the ash must reach ultimate disposal in a permitted facility.

As an example of the specificity in matching contaminants to treatment technologies, let us consider the five general categories of incinerators that are available to destroy contaminants: (1) rotary kiln; (2) multiple hearth; (3) liquid injection; (4) fluidized bed; and (5) multiple chamber.

Rotary Kiln

The combustion chamber in a rotary kiln incinerator, as illustrated in Figure 12.3, is a heated rotating cylinder that is mounted at an angle with baffles possibly added to the inner face to provide the turbulence necessary for the contaminant destruction process to take place. Engineering design decisions, based on the results of laboratory testing of a specific contaminant, include: (1) angle of the drum; (2) diameter and length of the drum; (3) presence and location of the baffles; (4) rotational speed of the drum; and (5) use of added fuel to increase the temperature of the combustion chamber as the specific contaminant requires. The liquid, sludge, or solid hazardous waste is input into the upper end of the rotating cylinder, rotates with the cylinder-baffle system, and falls with gravity to the lower end of the cylinder. The heated upward moving off-gases are collected, monitored for chemical constituents, and subsequently treated as appropriate prior to release. Meanwhile, the ash falls with gravity to be collected, monitored for chemical constituents, and also treated as needed before ultimate disposal.

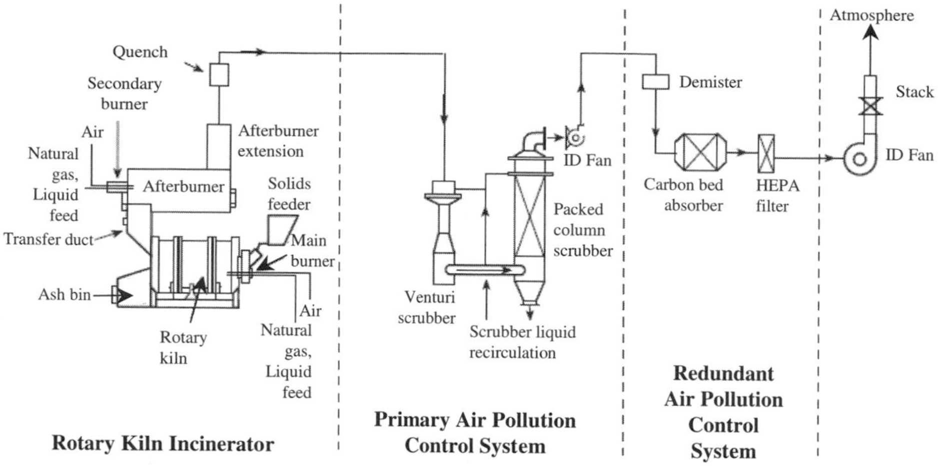

FIGURE 12.3. Rotary kiln system. (Source: J. Lee, D. Fournier, Jr., C. King, S. Venkatesh, and C. Goldman, 1997, "Project Summary: Evaluation of Rotary Kiln Incinerator Operation at Low-to-Moderate Temperature Conditions," U.S. Environmental Protection Agency.)

Recent designs of the rotary kiln system[13] consist of a primary combustion chamber, a transition volume, and a fired afterburner chamber. After exiting the afterburner, the flue gas is passed through a quench section followed by a primary APCS (air pollution control system). The primary APCS can be a venturi scrubber followed by a packed-column scrubber. Downstream of the primary APCS, a backup secondary APCS with a demister, an activated-carbon adsorber, and a high-efficiency particulate air (HEPA) filter can collect the contaminants that have not been destroyed by the incineration.

The rotary kiln is amenable to the incineration of most organic contaminants, is well suited for solids and sludges, and in special cases can be injected with liquids and gases through auxiliary nozzles in the side of the combustion chamber. Operating temperatures generally vary from 800°C to 1650°C. Engineers use laboratory experiments to design residence times of seconds for gases, and minutes or possibly hours for the incineration of solid material. In this manner, combustion conditions are tailored to the feedstock.

Multiple Hearth

In the multiple hearth illustrated in Figure 12.4, contaminants generally in solid or sludge form are fed slowly through the top of the vertically stacked hearth; in special configurations hazardous gases and liquids can be injected

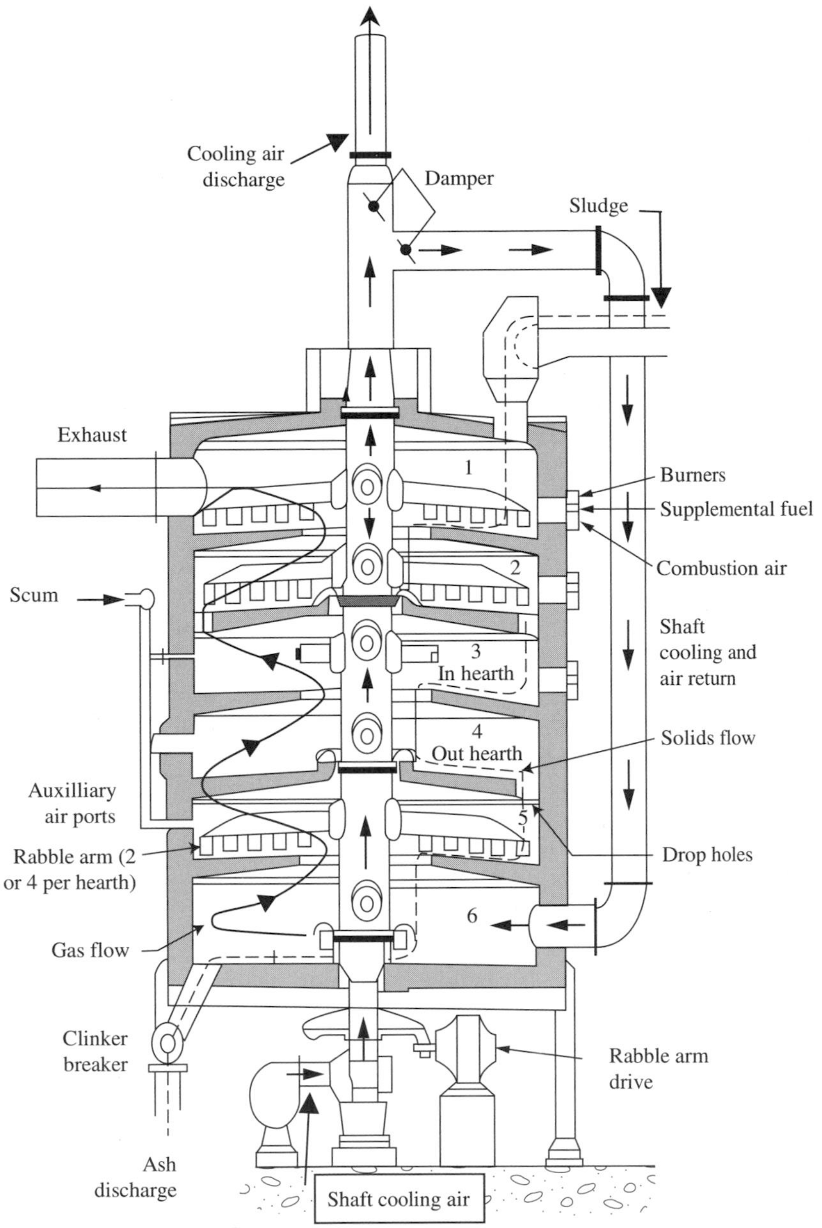

FIGURE 12.4. Multiple-hearth incineration system. (Source: U.S. Environmental Protection Agency, 1998, *Locating and Estimating Air Emissions from Sources of Benzene*, EPA-454/R-98-011, Research Triangle Park, N.C.)

through side nozzles. Multiple hearth incinerators, historically developed to burn municipal wastewater treatment biosolids (i.e., sludge), rely on gravity and scrapers working the upper edges of each hearth to transport the waste through holes from the upper hotter hearths to the lower cooler hearths. Heated upward-moving off-gases are collected, monitored for chemical constituents, and treated prior to release; the falling ash is collected, monitored for chemical constituents, and subsequently treated prior to ultimate disposal.

Most organic wastes can be incinerated using a multiple-hearth configuration. Operating temperatures vary from 300°C to 980°C, cooler than most rotary kilns. Multiple-hearth systems are designed with residence times of seconds if gases are fed into the chambers, to several hours if solid materials are placed on the top hearth and allowed to drop eventually to the bottom hearth and exiting as ash.

Liquid Injection

Vertical or horizontal nozzles spray liquid hazardous wastes into liquid injection incinerators specially designed for specific wastes or retrofitted to one of the other incinerators discussed here. The wastes are atomized through the nozzles that match the waste being handled with the combustion chamber, as determined in laboratory testing. The application obviously is limited to liquids that do not clog these nozzles, though some success has been experienced with hazardous waste slurries.[14] Operating temperatures generally vary from 650°C to 1650°C (a wider range than the rotary kiln and multiple-hearth systems. Liquid-injection systems (Figure 12.5) are designed with residence times of fractions of seconds as the

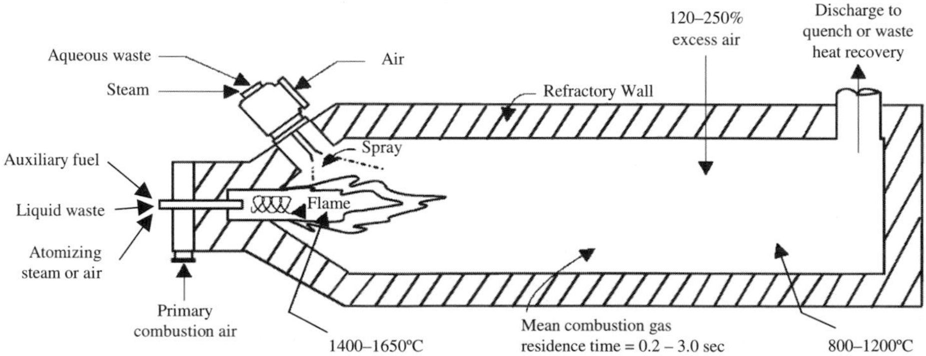

FIGURE 12.5. Prototype of liquid injection system. (Source: U.S. Environmental Protection Agency, 1998, *Locating and Estimating Air Emissions from Sources of Benzene*, EPA-454/R-98-011, Research Triangle Park, N.C.)

upwardly moving off-gases are collected, monitored for chemical constituents, and treated as appropriate prior to release to the lower troposphere. These very short residence times are allowed because all wastes are in the liquid phase. Times will increase with increasing suspended solids contents of the liquids (e.g., slurries will require longer retention times than nearly pure solutions).

Fluidized Bed

Contaminated feedstock is injected under pressure into a heated bed of agitated inert granular particles, usually sand, as the heat is transferred from the particles to the waste, and the combustion process proceeds as summarized in Figure 12.6. External heat is applied to the particle bed prior to the injection of the waste and is applied continually throughout the combustion operation as the situation dictates. Heated air is forced into the bottom of the particle bed, and during this continuous fluidizing process the particles become suspended among themselves. The openings created within the bed permit the introduction and transport of the waste into and through the bed. The process enables the contaminant to come into contact with the granular particles that maintain their heat better than, for example, the gases inside a rotary kiln. The heat maintained in the particles increases the time the contaminant is in contact with a heated element, and thus the

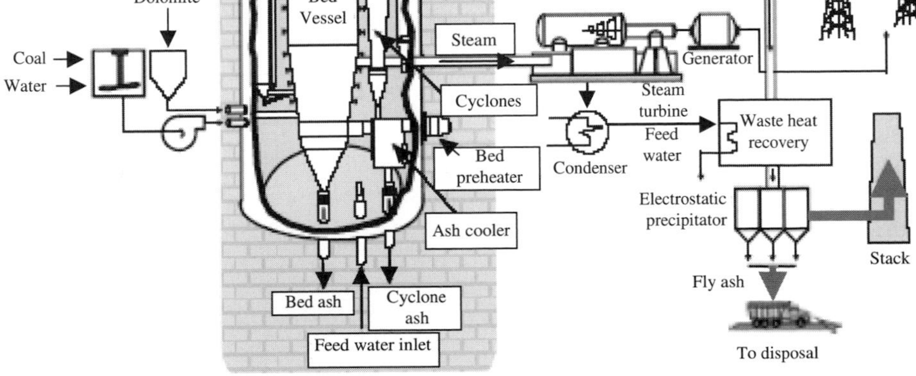

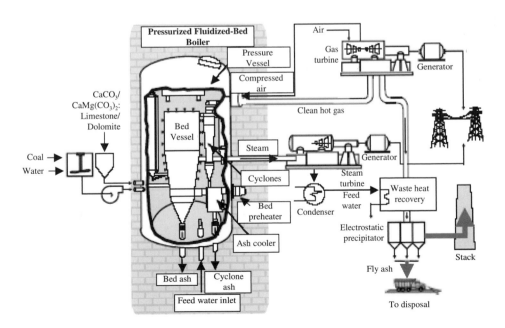

FIGURE 12.6. Pressurized fluidized bed system. (Source: U.S. Department of Energy, 1999, Tidd Pressurized Fluid Bed Combustion Demonstration Project.)

combustion process can become more complete than in the rotary kiln, generating fewer harmful by-products. Off-gases are collected, monitored for chemical constituents, and treated as needed, prior to release, and the falling ash is collected, monitored for chemical constituents, and subsequently treated prior to ultimate disposal.

Most organic wastes can be incinerated in a fluidized bed, but the system is best suited for liquids. Operating temperatures generally vary in a small range from 750°C to 900°C. Liquid injection systems are designed with residence times of fractions of seconds as the upwardly moving off-gases are collected, monitored for chemical constituents, and treated as appropriate prior to release to the lower troposphere.

Multiple Chamber

Contaminants are turned to a gaseous form on a grate in the ignition chamber of a multiple chamber system. The gases created in this ignition chamber travel through baffles to a secondary chamber where the actual combustion process takes place. Often, the secondary chamber is located above the ignition chamber to promote natural advection of the hot gases through the system. Heat may be added to the system in either the ignition chamber or the secondary chamber as required for specific burns.

The application of multiple-chamber incinerators generally is limited to solid wastes, with the waste entering the ignition chamber through a opened charging door in batch, not continuous, loading. Combustion temperatures typically hover near 540°C for most applications. These systems are designed with residence times of minutes to hours for solid hazardous wastes, as off-gases are collected, monitored for chemical constituents, and treated as appropriate prior to release to the lower troposphere. At the end of each burn period the system must be cooled so the ash can be removed prior to monitoring for chemical constituents and subsequent treatment prior to ultimate disposal.

Calculating Destruction Removal Efficiency

Federal hazardous waste incineration standards require that hazardous organic compounds meet certain destruction efficiencies. These standards require 99.99% destruction of all hazardous wastes and 99.9999% destruction of extremely hazardous wastes like dioxins.[15] The destruction removal efficiency (DRE) is calculated as:

$$\mathrm{DRE} = \frac{W_{in} - W_{out}}{W_{in}} \times 100 \qquad \text{Equation 12–1}$$

Where W_{in} = Rate of mass of waste flowing into the incinerator
W_{out} = Rate of mass of waste flowing out of the incinerator.

Destruction Efficiency Example

Calculate the DRE if during a stack test, the mass of pentachlorodioxin is loaded into the incinerator at the rate of $10\,\mathrm{mg\,min^{-1}}$ and the mass flow rate of the compound measured downstream in the stack is $2\,\mathrm{pg\,min^{-1}}$. Is the incinerator up to code for the thermal destruction of this dioxin? Calculate the DRE during the same stack test for the mass of tetrachloromethane (CCl_4) that is loaded into the incinerator at the rate of 100 liters $\mathrm{min^{-1}}$ with the mass flow rate of the compound measured downstream at $1\,\mathrm{ml\,min^{-1}}$. Is the incinerator up to code for CCl_4?

Solution

Since the difference between the input and outflow of the contaminant is 8, then this is only an 80% removal, well short of the 99.99%, and even worse in the case of a dioxin, which is an extremely hazardous waste requiring 99.9999% removal. Therefore, the incinerator is not up to code.

The CCl_4 removal is much better than that of dioxin, since 100 L are in and 0.001 are leaving, so the DRE = 99.999%. This is an acceptable or even better removal efficiency than 99.99% by an order of magnitude, so long as CCl_4 is not considered an extremely hazardous compound. If it were, then it would have to meet the rule of six nines (it only has five).

Other Thermal Processes

Incineration is frequently used to decontaminate soils with elevated concentrations of organic hazardous constituents. High-temperature incineration, however, may not be needed to treat soils provides contaminated with most volatile organic compounds (VOCs). Also, in soils with heavy metals, high temperature incineration will likely increase the volatilization of some of these metals into the combustion flue gas (see Tables 12.6 and 12.7). High concentrations of volatile trace metal compounds in the flue gas poses increased challenges to air pollution control. Thus, other thermal processes, especially thermal desorption and pyrolysis, can provide an effective alternative to incineration.

When successful in decontaminating soils to the necessary treatment levels, thermally desorbing contaminants from soils provides a number of benefits compared to incineration, including lower fuel consumption, no formation of slag, less volatilization of metal compounds, and less complicated air pollution control demands. Thus, beyond monetary costs and ease of operation, a less energy-demanding or heat-intensive system can be more advantageous in terms of actual pollutant removal efficiency.

TABLE 12.6
Conservative Estimates of Heavy Metals and Metalloids Partitioning to Flue Gas as a Function of Solids Temperature and Chlorine Content*

Metal or Metalloid	*871°C*		*1093°C*	
	Cl = 0%	*Cl = 1%*	*Cl = 0%*	*Cl = 1%*
Antimony	100%	100%	100%	100%
Arsenic	100%	100%	100%	100%
Barium	50%	30%	100%	100%
Beryllium	5%	5%	5%	5%
Cadmium	100%	100%	100%	100%
Chromium	5%	5%	5%	5%
Lead	100%	100%	100%	100%
Mercury	100%	100%	100%	100%
Silver	8%	100%	100%	100%
Thallium	100%	100%	100%	100%

* The remaining percentage of metal is contained in the bottom ash. Partitioning for liquids is estimated at 100% for all metals. The combustion gas temperature is expected to be 100°F to 1000°F higher than the solids temperature.
Source: U.S. Environmental Protection Agency, 1989, *Guidance on Setting Permit Conditions and Reporting Trial Burn Results: Vol. II*, Hazardous Waste Incineration Guidance Series, EPA/625/6-89/019, Washington, D.C.

TABLE 12.7
Metal and Metalloid Volatilization Temperatures

Metal or Metalloid	*Without Chlorine*		*With 10% Chlorine*	
	Volatility Temperature (°C)	*Principal Species*	*Volatility Temperature (°C)*	*Principal Species*
Chromium	1613	CrO_2/CrO_3	1611	CrO_2/CrO_3
Nickel	1210	$Ni\ (OH)_2$	693	$NiCl_2$
Beryllium	1054	$Be\ (OH)_2$	1054	$Be\ (OH)_2$
Silver	904	Ag	627	AgCl
Barium	84	$Ba\ (OH)_2$	904	$BaCl_2$
Thallium	721	Tl_2O_3	138	TIOH
Antimony	660	Sb_2O_3	660	Sb_2O_3
Lead	627	Pb	–15	$PbCl_4$
Selenium	318	SeO_2	318	SeO_2
Cadmium	214	Cd	214	Cd
Arsenic	32	As_2O_3	32	As_2O_3
Mercury	14	Hg	14	Hg

Source: B. Willis, M. Howie, and R. Williams, 2002, *Public Health Reviews of Hazardous Waste Thermal Treatment Technologies: A Guidance Manual for Public Health Assessors*, Agency for Toxic Substances and Disease Registry.

Pyrolysis is the process of chemical decomposition induced in organic materials by heat in the absence of oxygen. It is practicably impossible to achieve a completely oxygen-free atmosphere, so pyrolytic systems run with less than stoichiometric quantities of oxygen. Because some oxygen will be present in any pyrolytic system, there will always be a small amount of oxidation. Also, desorption will occur when volatile or semivolatile compounds are present in the feed.

During pyrolysis,[16] organic compounds are converted to gaseous components, along with some liquids, as coke, or the solid residue of fixed carbon and ash. CO, H_2, and CH_4 and other hydrocarbons are produced. If these gases cool and condense, liquids will form and leave oily tar residues (including solid and liquid phase PAHs) and water with high concentrations of total organic carbon (TOC). Pyrolysis generally takes place well above atmospheric pressure at temperatures exceeding 430°C. The secondary gases need their own treatment, such as by a secondary combustion chamber, by flaring, and by partial condensation. Particulates must be removed by additional air pollution controls, such as fabric filters or wet scrubbers.

Conventional thermal treatment methods, such as rotary kiln, rotary hearth furnace, or fluidized bed furnace, are used for waste pyrolysis. Kilns or furnaces used for pyrolysis may be of the same design as those used for combustion (i.e., incineration), as discussed earlier, but must operate at lower temperatures and with less air than in combustion.

The target contaminant groups for pyrolysis include semivolatile organic compounds (SVOCs), including pesticides, PCBs, dioxins, and PAHs. Pyrolysis allows for separating organic contaminants from various wastes, including those from refineries, coal tar, wood-preservatives, creosote-contaminated and hydrocarbon-contaminated soils, mixed radioactive and hazardous wastes, synthetic rubber processing, and paint and coating processes. Pyrolysis systems may be used to treat a variety of organic contaminants that chemically decompose when heated (i.e., "cracking"). Pyrolysis is not effective in either destroying or physically separating inorganic compounds that coexist with the organics in the contaminated medium. Volatile metals may be removed and transformed, but since their mass balance will not change, the operator must account for all chemical species of metals and metalloids (e.g. arsenic).

Emerging Thermal Technologies

Other promising thermal processes include high-pressure oxidation and vitrification.[17] *High-pressure oxidation* combines two related technologies, wet air oxidation and supercritical water oxidation, which combine high temperature and pressure to destroy organics. Wet air oxidation can operate at pressures of about 10% of those used during supercritical water oxidation, an emerging technology that has shown some promise in the treatment of PCBs and other stable compounds that resist chemical reaction.

Wet air oxidation has generally been limited to conditioning of municipal wastewater sludges, but can degrade hydrocarbons (including PAHs), certain pesticides, phenolic compounds, cyanides, and a number of organic compounds. Oxidation may benefit from catalysts.

Vitrification uses electricity to heat and destroy organic compounds and immobilize less chemically reactive contaminants. A vitrification unit has a reaction chamber divided into two sections: the upper section to introduce the feed material containing gases and pyrolysis products, and the lower section consisting of a two-layer molten zone for the metal and siliceous components of the waste. Electrodes are inserted into the waste solids, and graphite is applied to the surface to enhance its electrical conductivity. A large current is applied, resulting in rapid heating of the solids and causing the siliceous components of the material to melt as temperatures reach about 1600°C. The end product is a solid, glass-like material that is very resistant to leaching.

Microbiological Processing: The Science, Engineering, and Technology of Contaminant Biotreatment

Contaminants, if completely organic in structure are, in theory, completely destructible using principles based in microbiology with the engineering inputs and outputs summarized as:

$$\text{Hydrocarbons} + O_2 + \text{microorganisms (+energy)} \rightarrow \\ CO_2 + H_2O + \text{microorganisms (+energy?)} \qquad \text{Reaction 12–4}$$

In aerobic biotreatment processes, contaminant wastes are mixed with oxygen and aerobic microorganisms, sometimes in the presence of an external energy source[18] in the form of added nutrition for the microorganisms. In seconds, hours, or possibly days the by-products of gaseous carbon dioxide and water are produced, which exit the top of the reaction vessel while a solid mass of microorganisms is produced to exit the bottom of the reaction vessel.[19] The only presently obvious indirect effect of this simple reaction is the generation of carbon dioxide and its potential association with global climate change.

On the other hand, if the waste of concern to the engineer contains other chemical constituents, in particular chlorine or heavy metals, and if in fact the microorganisms are able to withstand and flourish in such an environment and not shrivel and die, the simple input and output relationship is modified to:

$$\text{Hydrocarbons} + O_2 + \text{microorganisms (+energy?)} + \text{Cl or heavy metal(s)} \\ + H_2O + \text{inorganic salts} + \text{nitrogen compounds} + \text{sulfur compounds} \\ + \text{phosphorus compounds} \rightarrow CO_2 + H_2O \text{ (+energy?)} \\ + \text{chlorinated hydrocarbons or heavy metal(s) inorganic salts} \\ + \text{nitrogen compounds} + \text{sulfur compounds} + \text{phosphorus compounds}$$

If the microorganisms do survive in this complicated environment, the potential exists for the transformation to a potentially more toxic molecule that contains chlorinated hydrocarbons, higher heavy metal concentrations, as well as more mobile or more toxic chemical species of heavy metals.

All of the bioreactor systems discussed in the following have some similar attributes. All rely on populations of microorganisms to metabolize organic contaminants into, ideally, the harmless by-products of $CO_2 + H_2O$ (+ energy?). In all of the systems the microorganisms must be either initially cultured in the laboratory to be able to metabolize the specific organic waste of concern, or target populations of microorganisms in the system must be given sufficient time, i.e. days, weeks, possible even years, to evolve to the point where the cumbersome food, that is the contaminant, is digestible by the microorganisms. What makes the food, i.e. the contaminant, "cumbersome" is its chemical structure, particularly, the availability of functional groups and other sites on the molecule where biochemical reactions may occur. When these sites are protected, e.g. by halogens, the microbes will have a more difficult time biodegrading them.

During all treatment processes, the input waste must be monitored and possibly controlled to maintain environmental conditions that do not upset or destroy the microorganisms in the system. These monitoring and control requirements for each of the systems include but are not limited to:

1. Temperature, possibly in the form of a heated building;
2. pH, possibly in the form of lime addition to decrease acidity;
3. Oxygen availability, possibly in the form of atmospheric diffusers that pump ambient atmosphere into the mixture of microorganisms and contaminant;
4. Additional food sources or nutrients, possibly in the form of a secondary carbon source for the microorganisms; and,
5. Changes in the characteristics of the input contaminant mixtures, including hydrocarbon availability and chemicals that may be toxic to the microorganisms, possibly including holding tanks to homogenize the waste prior to exposure to the microorganisms.

The populations of microorganisms must be matched to the particular contaminant of concern. The engineer must plan for and undertake extensive and continual monitoring and fine-tuning of each microbiological processing system during its complete operation.

The advantages of the biotreatment systems include: (1) the potential for energy recovery; (2) volume reduction of the hazardous waste; (3) detoxification as selected molecules are reformulated; (4) the basic scientific principles, engineering designs, and technologies are well understood from a wide range of other applications, including municipal wastewater treatment at facilities across the United States; (5) application to most organic contaminants, which as a group compose a large percentage of the total hazardous waste generated nationwide; (6) the possibility to scale the technologies to handle a single gallon per pound (liter/kilogram) of waste per day, or millions of gallons per pounds (liters/kilograms) of waste per day; and (7) land areas that could be small relative to other hazardous waste management facilities such as landfills.

The disadvantages of the biotreatment systems include: (1) the operation of the equipment requires very skilled operators and is more costly as input contaminant characteristics change over time and correctional controls become necessary; (2) ultimate disposal of the waste microorganisms is necessary and particularly troublesome and costly if heavy metals or chlorinated compounds are found during the expensive monitoring activities; and (3) the sometimes lengthy periods of time needed to reach microbial population sizes to biodegrade certain pollutants.

Given these underlying principles of biotreatment systems, four general guidelines are suggested whenever such systems are considered as a potential solution to any contaminant problem:

1. Only liquid organic contaminants are true candidates;
2. Chlorine-containing organic materials deserve special consideration if in fact they are to be biotreated at all, and special testing is required to match microbial communities to the chlorinated wastes, realizing that useful microbes may not be identifiable, and even if they are the reactions may take years to complete;
3. Hazardous waste containing heavy metals generally should not be bioprocessed; and,
4. Residual masses of microorganisms must be monitored for chemical constituents, and each residual must be addressed as appropriate so the entire bioprocessing system operates within the requirements of the local, state, and federal environmental regulations.

Each application of biotechnology must address the specific characteristics of the contaminant under consideration, including the quantity of waste to be processed over the planning period as well as the physical, chemical, and microbiological characteristics of the waste over the entire period of the project. Laboratory tests matching a given waste to a given bioprocessor must be conducted prior to the design and citing of the system.

Three different types of bioprocessors that generally are available to the engineer are introduced in the following sections, with accompanying text and summary diagrams: (1) trickling filter; (2) activated sludge; and (3) aeration lagoons. As a group these three types of treatment systems represent a broad range of opportunities available to engineers searching for methods to control the risks associated with contaminants.

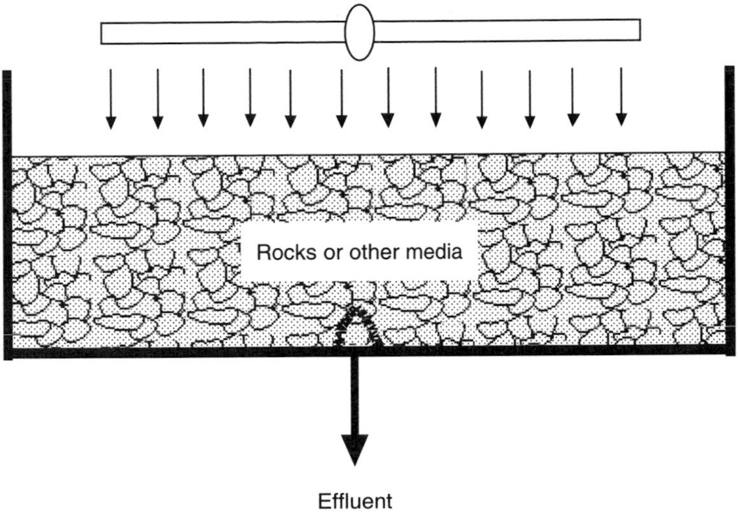

FIGURE 12.7. Trickling filter treatment system. (Source: D. Vallero, 2003, *Engineering the Risks of Hazardous Wastes*, Butterworth-Heinemann, Boston, Mass.)

Trickling Filter

The classical design of a trickling filter system, illustrated in Figure 12.7, includes a bed of fist-sized rocks or other matrices, enclosed in a rectangular or cylindrical structure, through which is passed the waste of concern. Biofilms are selected from laboratory studies and encouraged to grow on the rocks; as the liquid waste moves downward with gravity through the bed, the microorganisms comprising the biofilm are able to come into contact with the organic contaminant and food source and ideally metabolize the waste into relatively harmless CO_2 + H_2O + microorganisms (+ energy?). Oxygen is supplied by blowers from the bottom of the reactor and passes upward through the bed. The treated waste that moves downward through the bed subsequently enters a quiescent tank where the microorganisms that are sloughed off of the rocks are settled, collected, and ultimately disposed. Trickling filters are actually considered to be mixed treatment systems because aerobic bacteria grow in the upper, higher oxygen layers of the media, while anaerobes grow in the lower, more reduced regions lower in the system.

Activated Sludge

The key to the activated sludge system summarized schematically in Figure 12.8 and shown in Figure 12.9 is that the microorganisms that are available to metabolize the contaminant/food source are recycled within the system. This reuse enables this bioprocessor actually to evolve over time as the

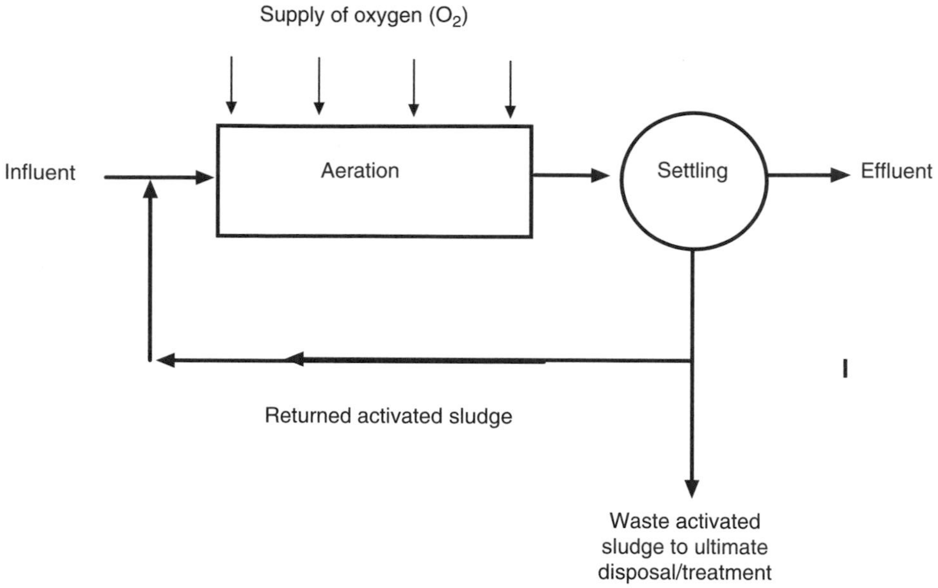

FIGURE 12.8. Activated sludge treatment system. (Source: D. Vallero, 2003, *Engineering the Risks of Hazardous Wastes*, Butterworth-Heinemann, Boston, Mass.)

microorganisms adapt to the changing characteristics of the influent contaminant; with this evolution comes the potential for the microorganisms to be more efficient at metabolizing the waste stream of concern. A ready supply of tailored and hungry microorganisms is always available to the engineer operating the facility!

A tank full of liquid waste is injected with a mass of microorganisms. Oxygen is supplied through the aeration basin as the microorganisms come in contact, sorb, and metabolize the waste ideally into $CO_2 + H_2O +$ microorganisms (+ energy?). This is an aerobic process since molecular oxygen is present (see Figure 12.9). The heavy, satisfied microorganisms then flow into a quiescent tank where the microorganisms are settled with gravity, collected, and ultimately disposed. Depending on the current operating conditions of the facility, some or many of the settled and now hungry and active microorganisms are returned to the aeration basin where they are given another opportunity to chow down. Liquid effluent from the activated sludge system may require additional microbiological or chemical processing prior to release into a receiving stream or city sewer system.

The activated sludge process in theory and in practice is a sequence of three distinct physical, chemical, and biological steps:

1. *Sorption.* The microorganisms come in contact with the food source, the organic material in the contaminant, and the food either is adsorbed to the cell walls or adsorbed through the cell walls of the

microorganisms. In either case the food is now directly available to the individual microorganisms. In a correctly operated facility, this sorption phase generally takes about 30 minutes.

2. *Growth.* The microorganisms metabolize the food and biochemically break down, or destroy, the hazardous organic molecules. This growth phase, during which individual organisms grow and multiply, may take up to hours or possibly days for complete metabolism of the hazardous constituents in the waste. Thus the design of the activated sludge system must include a basin with a detention time adequate for the correct amount of growth to take place.
3. *Settling.* Solid (the microorganisms) and supernatant liquid (the liquid remaining from the process) separation is achieved in a settling basin where the heavy and satisfied microorganisms sink to the bottom with gravity.

A critical design consideration of the activated sludge system is the loading to the aeration basin. *Loading* is defined as the food (F) to microorganism (M) ratio (F:M) at the start of the aeration basin. The planning is similar to the planning that precedes a Thanksgiving Day feast, with the trick being to make sure enough food is on hand for all of those in attendance. In the activated sludge system, the food shows up in the form of the organic constituents of the contaminant. The invited guests show up in the form of microorganisms that are returned from the settling basin to the aeration tank. With little or no control over the amount of food that may arrive during any given time period, the operating engineer must adjust the F:M ratio by adjusting the number of returned microorganisms. This balancing act between the amount of food and the numbers of microorganisms is summarized in two extreme examples suggesting ranges of F:M ratios, aeration times, and treatment efficiencies:

F to M Ratio + Aeration Time → Degree of Treatment
1. lower longer higher
(little food, lots of hungry mouths to feed, lots of time at the dinner table)
2. higher shorter lower
(smaller tanks, shortened time at the dinner table)

Sample loadings that are observed in practice range from 0.05 to greater than 2.0. The process of *extended aeration,* lasting up to greater than 30 hours, might have a loading of between 0.05 and 0.20 with an efficiency of contaminant removal in excess of 95%. The process of *conventional aeration,* closer to 6 hours for aeration, might have a loading between 0.20 and 0.50 with a treatment efficiency of possibly 90%. The process of *rapid aeration,* in the range of 1 to 3 hours for aeration, might have a loading between 1.0 and 2.0 with a removal efficiency closer to 85%. For each given problem, the engineer must design an individually activated sludge facility based on laboratory testing of a specific contaminant mixture; the engineer must operate that facility and select different loadings through time based on ongoing laboratory tests of the facility's input, process variables, and outputs.

Variations of the classic activated sludge system just summarized exist to help process very specific and difficult to treat contaminants. These variations in the design and operation of such facilities include:

1. *Tapered Aeration.* As seen in Figure 12.10, the oxygen that is supplied to the aeration basin is in greater amounts at the input end of the basin and in lesser amounts at the output end of the basin, with the goal of supplying more oxygen where it may be needed the most to address a specific contaminant loading.

FIGURE 12.9. Aeration waste treatment system. An aerobic treatment approach for breaking down toxic substances, which can also be persistent, from household or manufacturing sources. The waste is combined with recycled biomass and aerated to maintain a target dissolved oxygen (DO) content. Organisms use the organic components, expressed as biochemical oxygen demand (BOD) of waste as food, decreasing the organic levels in the wastewater. Oxygen concentrations must be controlled to maintain optimal treatment efficiencies. The system provides high concentrations of oxygen near the influent to accommodate the large oxygen demand from microbes as waste is introduced to the aeration tank. (Adapted from D. A. Vallero, *Engineering the Risks of Hazardous Wastes*, Butterworth-Heinemann, Boston, 2003; photo courtesy of D.J. Vallero, used with permission.)

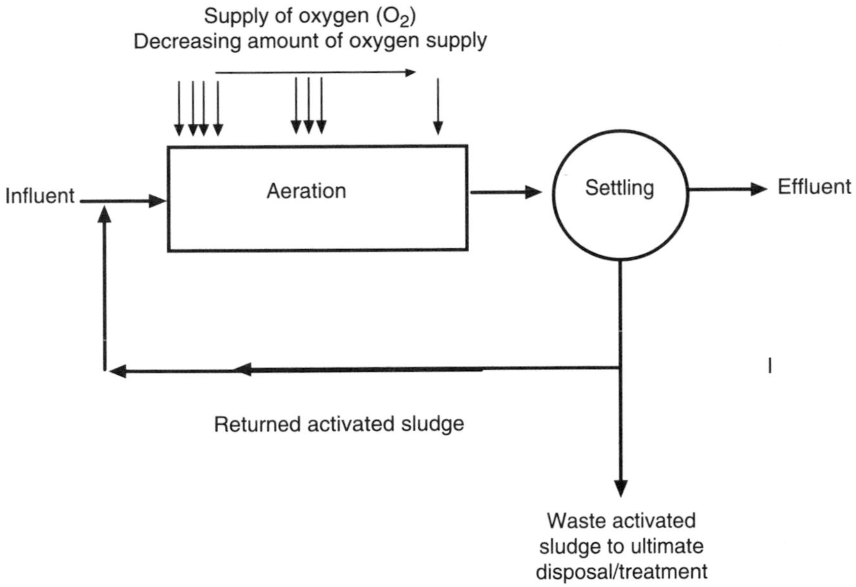

FIGURE 12.10. Tapered aeration activated sludge treatment system (greater amount of oxygen added closer to influent due to the large oxygen demand from microbes as waste is introduced to the aeration tank). (Source: D. Vallero, 2003, *Engineering the Risks of Hazardous Wastes*, Butterworth-Heinemann, Boston, Mass.)

2. *Step Aeration.* As seen in Figure 12.11, the influent oxygen *and* the contaminant is supplied to the aeration basin in equal amounts throughout the basin with the goal of matching the oxygen demand to the location where it may be needed the most for a specific contaminant problem.
3. *Contact Stabilization or Biosorption.* As seen in Figure 12.12, the sorption and growth phases of the microbiological processing system are separated into different tanks with the goal of achieving growth at higher solids concentrations, saving tank space, and thus saving money.

Anaerobic Treatment

We have stressed aerobic systems, however, some contaminants are more easily broken down anaerobically (see Figure 12.13). For example, PCBs may be treated either aerobically or anaerobically, but aerobic treatment is usually less effective for congeners that have increasing number of chlorine (Cl) atoms. PCB congeners with more than four Cl atoms are not usually treated aerobically. While aerobic biotreatment of PCBs works by cleaving the aromatic rings, anaerobic degradation removes the chlorine atoms from the PCB molecule (i.e., dehalogenation or dechlorination). The Cl atoms are

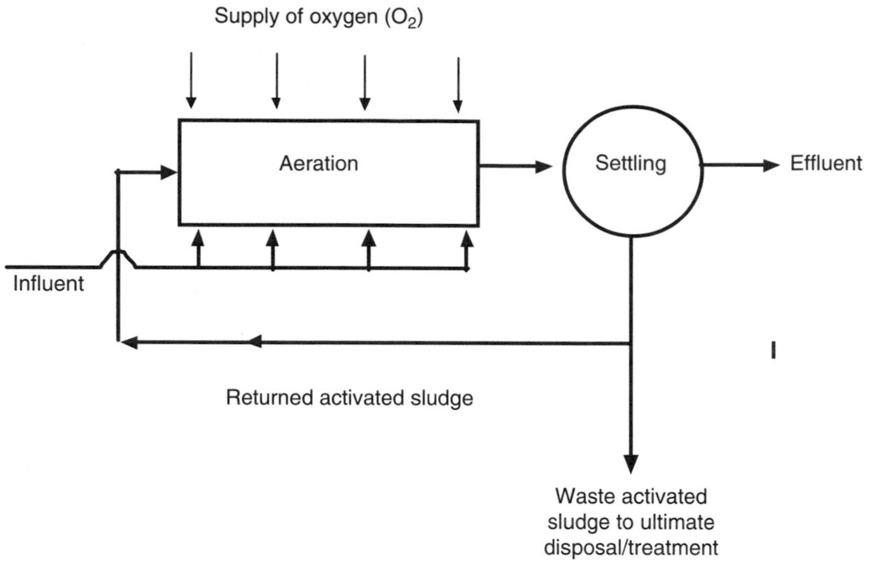

FIGURE 12.11. Step activated sludge treatment system. (Source: D. Vallero, 2003, *Engineering the Risks of Hazardous Wastes*, Butterworth-Heinemann, Boston, Mass.)

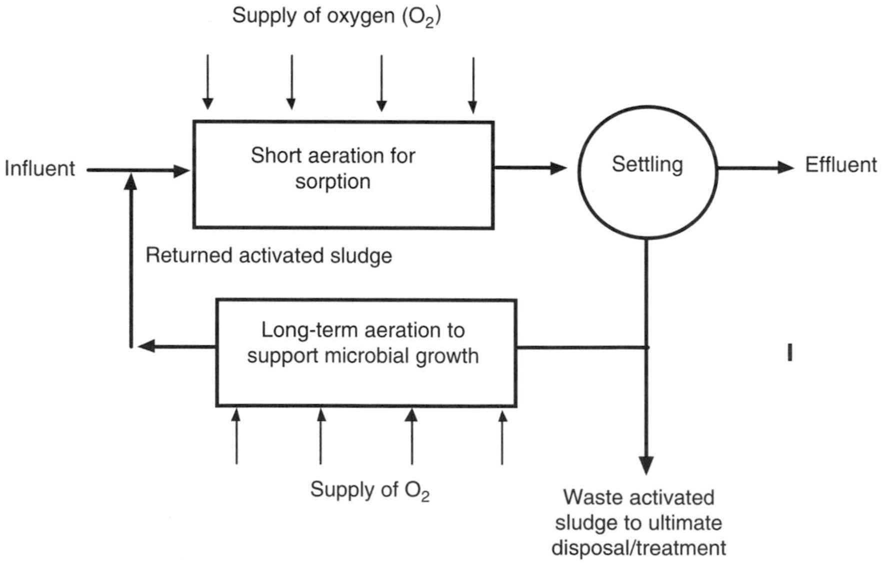

FIGURE 12.12. Contact stabilization activated sludge treatment system. (Source: D. Vallero, 2003, *Engineering the Risks of Hazardous Wastes*, Butterworth-Heinemann, Boston, Mass.)

FIGURE 12.13. Anaerobic or anoxic system. In this process, anaerobic bacteria grow by using reduction-oxidation sources other than molecular oxygen (O_2). Anaerobic systems can be used to treat industrial wastes. Other anaerobic systems (e.g., lagoons) encourage the growth of facultative bacteria, those that can grow in the presence or absence of O_2. Facultative systems can remove toxic wastes by creating a balance between bacteria and algae by modulating aerobic and anaerobic conditions to enhance chemical uptake. The anaerobic and facultative processes are enhanced in a constructed wastewater treatment system, as shown here. These processes can also take place in other systems, like landfills. Note in the photo the lighter substances that have migrated to the surface. These may be fats that have been separated physically during the treatment process, or bubbles from gases, such as methane, that are produced when the microbes degrade the wastes. Note also that although this is an anoxic chamber, a thin film layer at the surface will be aerobic because it is in contact with the atmosphere. *(Photo courtesy of D.J. Vallero, used with permission.)*

removed because the reduced, anoxic conditions allow hydrogen atoms to substitute for the Cl atoms. This renders the newly formed molecules less toxic (e.g., lase carcinogenic) and less likely to bioaccumulate (i.e., lower BCF values). Some systems are multistep. The anaerobic stage is followed by aerobic treatment. The benzene ring cleavage is more practical for the degradation product of anaerobic treatment because there are now fewer Cl atoms surrounding the PCB molecule than were present on the parent PCB. Thus, the degraded molecule is less sterically hindered and is less persistent. This process works for many halogenated organic compounds, such as those with bromine (Br) substitutions, i.e., brominated compounds like the polybrominated biphenyls (PBBs), as well as for chlorinated alkanes and aromatics.

Aeration Ponds

Ponds like the one illustrated in Figure 12.14 treat liquid and dissolved contaminants for over the long-term, from months to years. Persistent organic molecules, those not readily degraded in trickling filter or activated sludge systems, are potentially broken down by certain microbes into CO_2 + H_2O + microorganisms (+ energy?) if given enough time. The ponds are open to the weather, and ideally oxygen is supplied directly to the microorganisms from the atmosphere. Design decisions based on laboratory experiments and pilot studies include:

1. Design: Pond size: 0.5 to 20 acres
2. Design: Pond depth: 1 foot to 30 feet
3. Design: Detention time: days to months to possibly years
4. Operation: in series with other treatment systems, other ponds, or stand-alone
5. Operation: the flow to the pond is either continuous or intermittent
6. Operation: the supply of additional oxygen to the system through blowers and diffusers may be required.

Again, the critical engineering concerns in the design and operation of ponds and other biotreatment facilities are the identification and maintenance of microbial populations that metabolize the specific contaminant of concern.

Hazardous Waste Storage Landfills: Examples of the Science, Engineering, and Technology of Long-Term Storage of Contaminated Media

The four stages in the life of long-term storage facilities, (1) siting, (2) design, (3) operation, and (4) post-closure management, offer the engineer myriad

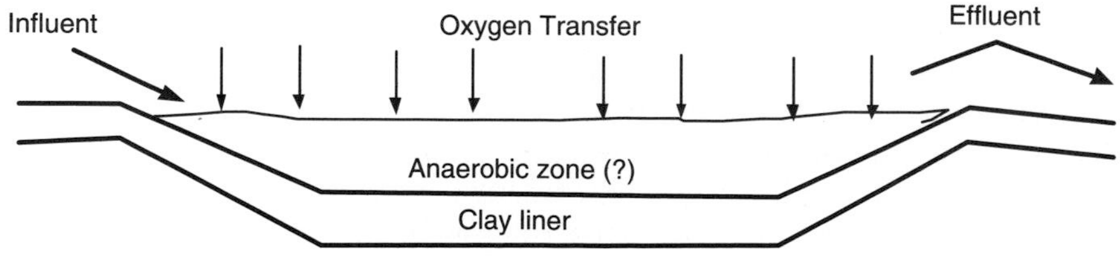

FIGURE 12.14. Aeration pond. (Source: D. Vallero, 2003, *Engineering the Risks of Hazardous Wastes*, Butterworth-Heinemann, Boston, Mass.)

opportunities to intervene and control the risks associated with contaminants. At each stage of a landfill's life, any intervention scheme must be justified by the engineer in terms of the science, engineering, and technological aspects of the project. However, as recent U.S. history indicates, economic realities and public perceptions as well as the laws of local, state, and national governments drive the decision-making process throughout all stages of landfill considerations. Few if any individuals are willing to accept a landfill in their "backyards," and thus the over-used expression "NIMBY," short for "not in my back yard."[20]

Siting

Engineering determinations are extremely important at the beginning of any process to site any facility, whether that facility is a local shopping center or a hazardous waste landfill. The balance is to identify sites of land that are at least acceptable from a scientific and engineering standpoint. In no particular order, the engineer can overlay a regional landmass with at least the seven site selection criteria discussed in this section. Of course these criteria are region-specific and must be identified on a case-by-case basis. However, the goal is always that, after the selection criteria for a specific region have been established and overlaid upon the region, at least one area within the region remain standing as a contender for the landfill site.

Historically, the release and transport of hazardous chemicals from a landfill has included waste moving along the surface of the earth or into groundwater supplies beneath the earth. The location of a hazardous waste landfill must consider these release and transport possibilities; thus the landfill must be sited well above historically high groundwater tables and well away from surface streams and lakes. Once the horse is out of the barn, so to speak, the damage is done as surface and subsurface drinking water supplies are jeopardized.

Climatology must also be considered when options are screened to identify sites for a hazardous waste landfill. Intensive rain events can damage the integrity of any waste barrier system found in any landfill anywhere; thus the landfill must be located outside the paths of reoccurring storms. Hurricane paths in North Carolina and tornado alleys in the Midwest offer vivid examples, but microclimates exist throughout the United States that result in deluges that could and do assist in the migration of contaminants from landfill impoundments to a receiving surface or subsurface body of water.

The geology of the region is similarly important as the engineer searches for a site to construct a hazardous waste landfill. The potential construction site must be stable in geological time; thus areas of active and dormant faults must be avoided. The vertical soil profile must be composed of soil materials that are generally impervious to liquid migration; thus sandy soils and cracked bedrock must be avoided. With permeabilities ranging up to 10^{-6} cm sec^{-1} and with thickness exceeding hundreds of meters, natural clay deposits could provide the most promising materials on which to site a hazardous waste landfill.

The ecology of the region poses particularly troublesome difficulties during the site selection process. Areas of low fauna and flora densities are preferred while natural wilderness areas sensitive ecotones, wildlife refuges, and migration routes should be avoided. Areas supporting endangered species must also be avoided.

Transportation routes to and from a potential site raise the questions of local human receptors exposed to contaminants if a roadside spill occurs while the waste is in transit. The need also exists for an all-weather highway that helps support adequate emergency responses and evacuations should such accidental spills or catastrophic events occur at the landfill. Thus existing or possible additional transportation routes must be part of the site selection process.

Alternatives for land resource utilization also must be considered as locations are screened to identify potential sites for a landfill within a region. The long-term storage facility should only occupy land that has low alternative land use value. No sense putting the landfill where a golf course and housing development could go. Recreational areas must be avoided to help limit the accessibility of the site to the general public.

Environmental health often is the primary concern in siting these types of facilities. The landfill must be located away from drinking water wells, surface drinking water supplies, and populated areas. The goal is to avoid placing drinking water supplies and receptors in close proximity to the landfill.

The real challenge to the engineer involved in the site selection process is that, with numerous and often conflicting site selection criteria, few if any acceptable sites may be identified within a given region at the end of the day. For example wildlife refuges have few if any human inhabitants and thus siting a landfill in the refuge could maximize the distance from the landfill to potential human receptors of the contaminant. But one site selection criterion is the avoidance of wildlife areas! An engineer often is faced with a "darned if we do, darned if we don't" situation.

A tool that can be very helpful in site selection is the geographic information system (GIS). The GIS allows the planner or engineer to input data from tables and text to create layers of visual outputs. For example, if a landfill siting requires that it be located within 10 km of the town border, that 40 hectares (ha) of land are be available, that it be no closer than 0.7 km to a water body and within 0.4 km of an existing road, the GIS can identify land parcels that satisfy these siting criteria. This GIS map outputs are also valuable in meetings and technical communications (See Figure 12.15).

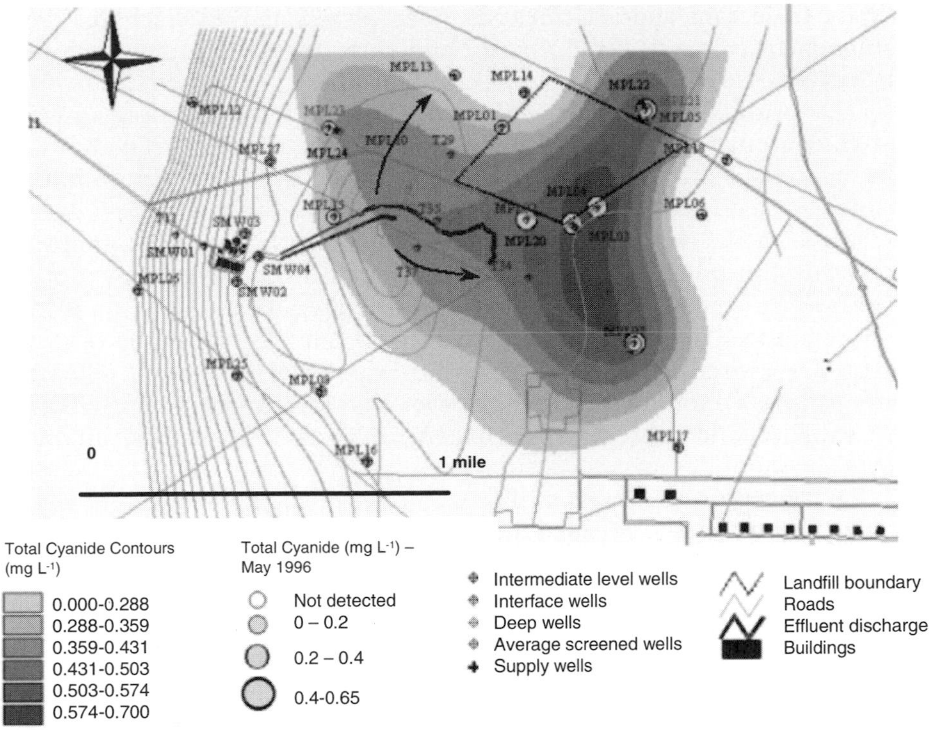

FIGURE 12.15. A map generated with a geographic information system (GIS). A GIS map can be a powerful tool for interpreting site data and for presenting results. Different layers of environmental and physiographic data and information can be turned off and on, allowing for many different scenarios based on the analytical parameters, site physical features, hydrogeological data, and contaminant concentration data. This example map displays cyanide levels based upon comprehensive site data. By generating sequential visualizations of monitoring, for the site. GIS maps can also help to monitor plume movement, plume size, and changes in contaminant migration directions.
Source: Naval Facilities Engineering Service Center, 2002, Guide to Optimal Groundwater Monitoring, Port Hueneme, California.

Design

The engineer can control the risks associated with the long-term storage of hazardous waste by incorporating sound engineering design considerations into any and all of five levels of safeguards to be found in modern proposed and existing landfill designs. Starting from the top and going down through a landfill, these five potential levels of safeguard include: (1) a cover to prevent water from entering the landfill; (2) solidification of the hazardous waste; (3) a primary barrier to liquid release with leachate collection and treatment as appropriate; (4) a secondary barrier to liquid release with leachate collection and treatment as appropriate; and (5) discharge wells downgradient from the site to pump and treat any contaminated "horse" that has escaped the barn.

The landfill must be covered to prevent the movement of rainwater into and through the impoundment. The cap illustrated in Figure 12.16 must be constructed with layers of materials. The first layer should be topsoil that is graded to promote the controlled runoff of all storm events. The soil is seeded with grasses having short root systems to promote the evapotranspiration of rain that falls on the landfill. The second layer of the cap should be composed of an impermeable material that also is graded to promote controlled runoff and prevent erosion of the cap, while also preventing movement of the rainwater into the depths of the landfill. The third layer of the cap should be a sand lens that is graded to promote the collection, exhaust, and subsequent treatment of gases that may be produced within the landfill.

The waste within the landfill must be solidified to help preclude movement of any waste within the landfill. The first level of the solidification process is the mixing of all liquid and sludge wastes with sorbent material

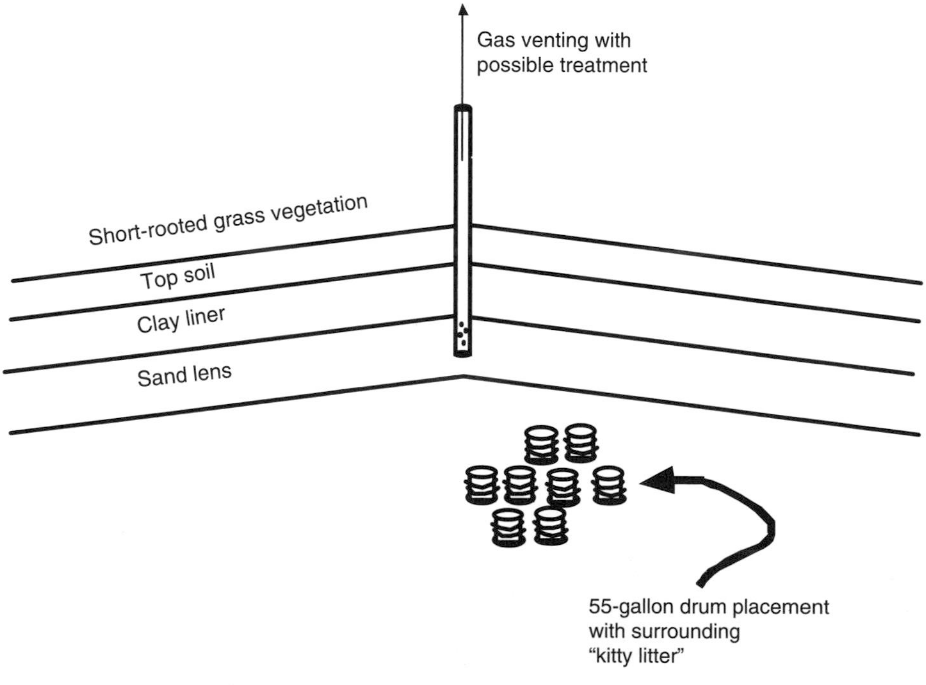

FIGURE 12.16. Engineered cover for a hazardous waste landfill. (Source: D. Vallero, 2003, *Engineering the Risks of Hazardous Wastes*, Butterworth-Heinemann, Boston, Mass.)

prior to burial. In practice, the sorbent material quite often is an oven-dried clay taken from a nearby natural clay formation. The material is identical to kitty litter products used in households and to oil-dry products used in auto repair shops. The second level of solidification generally is the filling of painted 55-gallon drums with the kitty litter/waste mixture. The third level of solidification is the surrounding of all of the 55-gallon drums with more kitty litter at the time of burial.

Leachate collection systems are illustrated in Figure 12.17. Many regulatory agencies require two or three pairs of these systems as design redundancies to protect the integrity of a landfill. A primary leachate collection and treatment system must be designed like the bottom of the landfill bathtub. This leachate collection system must be graded to promote the flow of liquid within the landfill from all points in the landfill to a central collection point where the liquid can be pumped to the surface for subsequent monitoring and treatment. Crushed stone and perforated pipes are used to channel the liquid along the top layer of this compacted clay liner to the pumping location(s).

Immediately below the primary leachate collection is a secondary leachate collection available in case the primary system fails. This leachate collection system also must be graded to promote the flow of liquid within the landfill from all points in the landfill to a central collection point where the liquid can be pumped to the surface for subsequent treatment. The secondary system typically is constructed of a flexible membrane liner (FML) material, at least 2 mm thick, and an unbending plastic garbage bag.

The final barrier to liquid waste migration from the hazardous waste landfill must be a field of monitoring and extraction wells. The monitoring

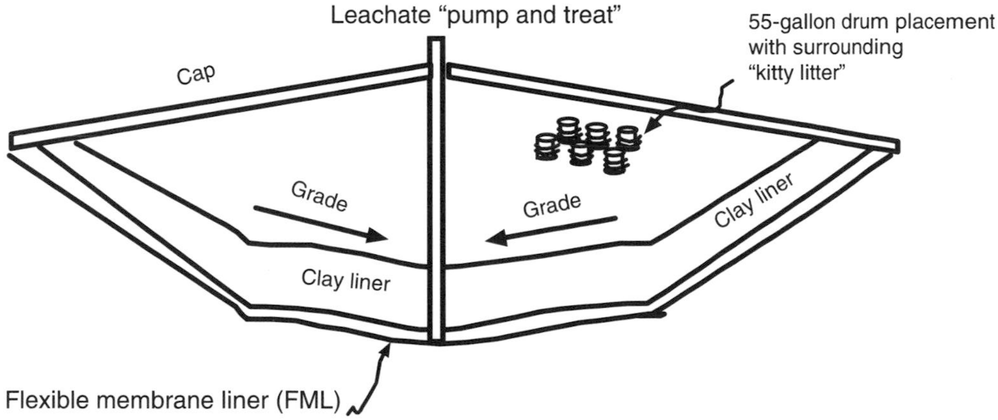

FIGURE 12.17. Leachate collection system for a hazardous waste landfill. (Source: D. Vallero, 2003, *Engineering the Risks of Hazardous Wastes*, Butterworth-Heinemann, Boston, Mass.)

wells are located upgradient and downgradient from the site, as seen in Figure 12.18. The upgradient monitoring wells provide a method to identify background concentrations of the constituents in the groundwater against which to compare the information collected from the downgradient monitoring wells. If a chemical substance has been detected downgradient from the landfill that has not been detected in the upgradient monitoring wells, or if a chemical substance is detected at higher levels downgradient from the landfill, then the landfill has sprung a leak. The downgradient pump-and-treat wells then can be used to extract groundwater at rates that prohibit any additional transport of contaminant through the soil underlying the landfill. The entire process often becomes similar to finding a needle in a haystack; thus the location of the wells in the field becomes of paramount importance to the success of this monitoring and pump-and-treat system at the landfill.

Operation

As the landfill enters its operational phase, the phase when waste is actually buried in the facility, the engineer has additional opportunities to help control the risks associated with the contaminants. Any leachate that is collected from the liner system(s) must continually be monitored and treated as appropriate. The groundwater monitoring wells must be operated continually, with liquid samples collected and analyzed for chemical con-

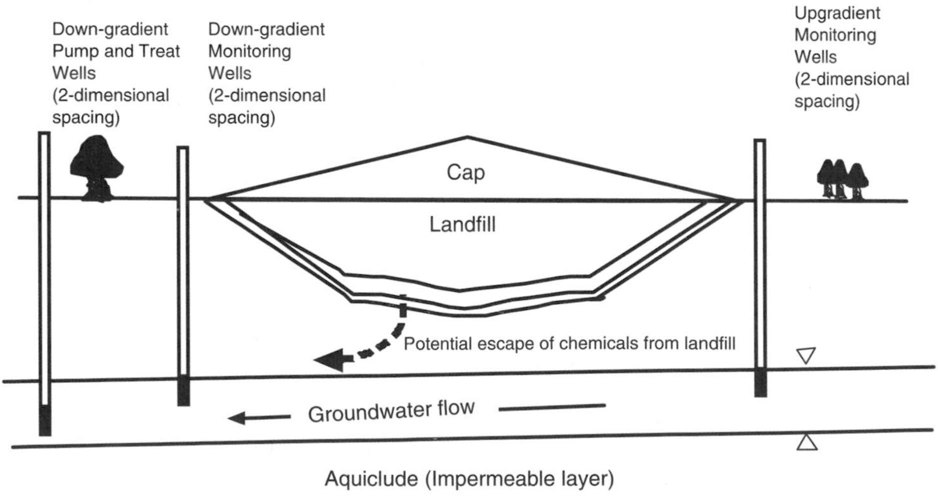

FIGURE 12.18. Monitoring and pump-and-treat wells surrounding a landfill. (Source: D. Vallero, 2003, *Engineering the Risks of Hazardous Wastes*, Butterworth-Heinemann, Boston, Mass.)

stituents, and with subsequent operation of the downgradient pump-and-treat wells as appropriate. The location of the solidified waste canisters must be three-dimensionally mapped to promote excavation at a future time if and when advancing science, engineering, and recovery technologies provide economical reprocessing and recycling of the buried waste materials.

Post-Closure Management

Once the landfill is full of solidified hazardous waste the engineer can further control the risks associated with the waste by conducting very important post-closure management procedures. The filled landfill must be covered with the cap discussed earlier in this chapter. The liquids from the leachate and monitoring wells must be analyzed continually, and the pump-and-treat wells must be maintained continually and used when necessary. Most importantly, access to the site must be limited to those people responsible for the post-closure management of the facility; all other people and animals must be denied entry.

Ex Situ and *In Situ* Treatment

Contaminated soil and sediment must often first be removed and then treated off-site, also known as *ex situ* treatment. Contaminated soil may be excavated and transported to kilns or other high-temperature operations, where the contaminated soil is mixed with combustible material. High sand content soils may even be part of an asphalt mix. Contaminated soil may also be distributed onto an impermeable surface, allowing the more volatile compounds to evaporate. Microbial biodegradation can be accelerated and enhanced by adding nutrients and moisture to the soil. A faster process, thermal desorption, entails heating the soil to evaporate the contaminants and capture the compounds, and then burning them in a vapor-treatment device.

Generally, groundwater is treated by drilling recovery wells to pump contaminated groundwater to the surface. Commonly used groundwater treatment approaches include air stripping, filtering with granulated activated carbon (GAC), and air sparging. Air-stripping transfers volatile compounds from water to air. Groundwater is allowed to drip downward in a tower filled with a permeable material through which a stream of air flows upward. Another method bubbles pressurized air through contaminated water in a tank. Filtering groundwater with GAC entails pumping the water through the GAC to trap the contaminants. In air sparging, air is pumped into the groundwater to aerate the water and to increase the volatilization of contaminants due to the increased partial pressures exerted by the oxygen. Most often, a soil venting system is combined with an air sparging system for vapor extraction.

Contaminants can also be treated where they are found without first removing them, known as *in situ* remediation. Bioremediation makes use of living microorganisms to break down toxic chemicals or to render the chemicals less hazardous. This is often done by using bacteria, and in some instances algae and fungi, that are already living in the soil, sediment, or water. These microbes are exposed to incrementally increasing amounts of the chemical, so that the organisms adapt to using the chemical as an energy (food) source. This process is known as acclimation. The acclimated microbes can then be taken from the laboratory and applied to the waste either in a treatment facility or *in situ* in the field. The most passive form of bioremediation is *natural attenuation*, where no engineering intervention is used, and the contaminants are allowed to be degraded by resident microbes over time. The only role for the engineer is to monitor the soil and groundwater to measure the rate at which the chemicals are degrading. Natural attenuation can work well for compounds that are found in the laboratory to break down under the conditions found at the site. For example, if a compound is degraded under reduced, low pH conditions in the laboratory, it may also degrade readily in soils with these same conditions (e.g., in deeper soil layers where bacteria have adapted to these conditions naturally).

Plant life may also be used to reduce the amount of contamination. In *phytoremediation*, contaminated areas are seeded, and as the plants grow their roots extract the chemicals from the soil. The harvested plants are either treated on-site, for example by composting, or transferred to a treatment facility. In reality, both microbial and macrophytic processes occur simultaneously. Poplar trees can help to treat areas contaminated with agricultural chemicals. Plants, such as grasses and field crops, have even been used to treat the very persistent polychlorinated biphenyls (PCBs), wood preservatives, and petroleum. Plants have also been used to extract heavy metals and radioactive substances from contaminated soil. Bioremediation has been used successfully to treat numerous other organic and inorganic compounds.

Various methods are available for treating substances after they have been released into the environment. Eliminating the wastes before they are released, however, is the best means of reducing risks to humans and other organisms.

Answering a few seemingly straightforward questions can illustrate the complexity of a comprehensive remediation effort. The questions are also a type of "final exam" that assesses what we have learned in Chapters 1 through 12.

Contaminant Treatment Question 1

Review the table belows, which shows the phase distribution of organic contaminants following a hypothetical tanker truck spill of 40,000 liters of gasoline into a medium sand aquifer with a water table 5 meters beneath the ground surface.

Phase	*Volume of Aquifer Material Contaminated (m^3)*	*Percent of Total Volume of Aquifer Contamination*	*Volume of Gasoline (L)*	*Percent of Total Volume of Gasoline*
Free mobile, nonaqueous	2500	0.6	24,700	61.8
Sorbed to soil	85,000	20.3	11,200	28.0
Dissolved in water	330,000	79.0	400	1.0
Gaseous (vapor phase)	Not measured	Unknown	Unknown	Unknown

Give reasons why the relatively small percentage of gasoline in solution represents the largest fraction of the extent of contamination.

Answer and Discussion

First, characterize the contaminant. Gasoline is not a single compound but a mixture of many aliphatic and aromatic compounds. A few of the alkanes and even some of the aromatics (e.g. benzene) have relatively high water solubility, while a large fraction of the larger chains and aromatics have comparatively low solubilities and end up as NAPLs in this aquifer. In addition, some gasolines have metallic and inorganic additives. Since these NAPLs, by definition, are relatively insoluble in water, they are not carried easily through the sandy soil.

Conversely, a small percentage of the gasoline mixture is soluble in water. This transport is rapid in a sandy aquifer with high values of hydraulic conductivity.

An important question to ask is how much time has elapsed between these measurements and the spill. If it is only a few days, we may see migration of the sorbed and nonaqueous phases. The amount of sorbed organics represents most of what remains in the aquifer. This is a function of the sorption coefficient of each hydrocarbon molecule, the availability of particles and micelles to serve as surfaces, and physical phenomena, such as the double-layer effect and the ionic strength of the soil.

Contaminant Treatment Question 2

Explain why the NAPLs represent such a small portion of the plume, yet they comprise over 60% of the gasoline spilled.

Answer and Discussion

NAPLs by definition are nonaqueous, so they are typically hydrophobic, and they are unlikely to move readily with the groundwater without other physical conditions forcing this to happen, such as surface sorption and the existence of surface-active substances like surfactants in the pore water.

There is also the possibility of measurement error, which understates the amount of NAPLs. The dense NAPLs (known as DNAPLs) are heavier than water, so they tend to sink in the aquifer, while the light NAPLs (i.e., LNAPLs) are less dense than water, and tend to float along the top of the aquifer at the water table. The monitoring devices may not have taken into account the lack of mixing of the NAPLs. Thus, if a measurement were taken at the center of the aquifer, it is possible that most of the NAPLs were missed.

Contaminant Treatment Question 3

Why was vapor phase not measured?

Answer and Discussion

It could be that the response team only had soil and water monitoring devices available when they took the measurements. Volatile compounds in the air are measured using canisters or bags that are evacuated, allowing air to enter. Measurements are also made using traps, such as activated carbon that is extracted and analyzed in the lab.

It is also possible that by the time the team took the measurements, most of the mass of volatile organic compounds (those with vapor pressure $>10^{-2}$ kP) had already evaded into the atmosphere. Depending on the actual mixture of gasoline, it may be that there was not a large fraction that was volatile. The assessment team should ask the manufacturer to provide an assay of the actual storage tanks of gasoline at the refinery or distributor.

Contaminant Treatment Question 4

The total volume of gasoline measured represents only 90% of the total spill. We need a complete mass balance. Where is the rest?

Answer and Discussion

Since the water table is 5 m below the surface, it could be found in pockets in the soil above the vadose zone. Soil is very heterogeneous in texture and porosity, so even though it is classified as medium sand, there may be clay lenses and other areas of differential conductivity and permeability. The gasoline may have run off on the surface and became volatilized before the measurements were taken. Fractions of the more volatile materials may have evaded between the times of cleanup and measurement. Since this was a spill, it is likely that the first responders sprayed water and other fire-retarding substances onto the tank and may have used containment devices to prevent runoff. These devices may not have been included in the phase distribution totals. It is also highly likely that these values are estimates or values based on models that take into account contaminant and hydrogeologic conditions. The model may not have allowed for the quantitation of certain compounds or certain conditions.

Contaminant Treatment Question 5

What are the most common treatment approaches used at hazardous wastes sites?

Answer and Discussion

As shown in Figure 12.19, four major types of treatment accounted for 89% of the processes used through 1991 at abandoned hazardous waste sites in the U.S. About 36% have employed thermal processes (thermal desorption, and on-site and off-site incineration). Solidification accounted for 26%, soil vapor extraction was used at 17% of the sites, while bioremediation was used at less than 10% of the sites.[21] Although not specifically a treatment process per se, landfills and containment structures account for a third of all remedial actions.

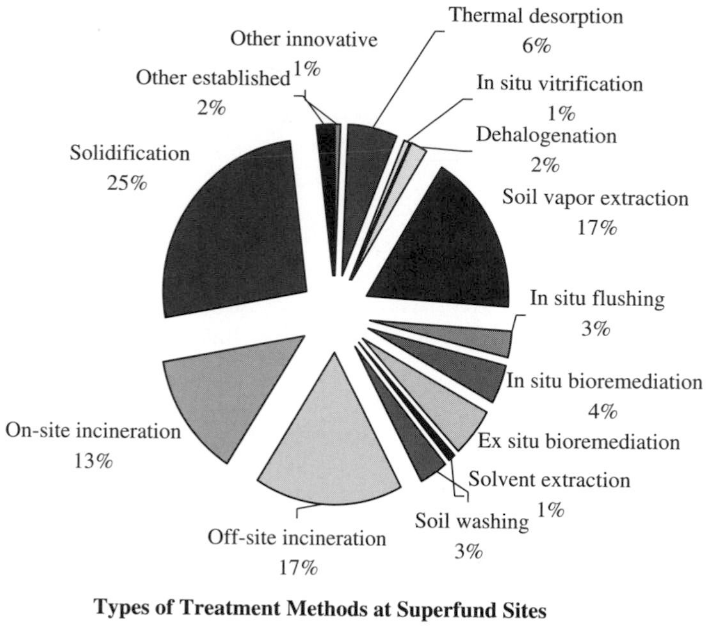

FIGURE 12.19. Treatment types at abandoned hazardous wastes sites in the United States (Source: U.S. Environmental Protection Agency, 1992, *Innovative Treatment Technologies: Semiannual Status Report*, Washington, D.C.)

Contaminant Treatment Question 6

What are some of the likely remedial steps that need to be taken when you find hazardous waste leaking from a landfill?

Answer and Discussion

Containment of the waste implies that you must reduce the flow from the landfill to levels required to eliminate contaminant transport. This usually includes installing impermeable barriers with extremely low hydraulic conductivity around the site. These barriers can be synthetic geomembranes, trenches backfilled with extremely low permeability clay (e.g., bentonite) slurries, or combinations of the two.

To reduce infiltration of water into the landfill, a cap (usually a layer of clay materials) is installed. One of the most challenging steps will be to deal with the migration of waste-laden water from beneath the landfill. Sometimes this requires complete excavation of all source materials in the landfill that are stockpiled safely while lining is installed (again synthetic and/or clays) at the bottom of the hole. After the bottom liner is in place, the contaminated materials may be returned to the landfill.

Monitoring wells must continue to be operated upstream and downstream, triangulating the site, to ensure that migration has been halted and that the liners continue to operate as designed. Even well-designed linings can fail for numerous, often "low-tech" reasons, such as the invasion of burrowing animals. With apologies to the Captain and Tennille,[22] there should be no "muskrat love" in and around your newly engineered landfill.

Contaminant Treatment Question 7

What are some likely differences between a former municipal landfill and an active industrial hazardous waste storage landfill?

Answer and Discussion

There are numerous differences. The municipal landfill will contain, we hope, mostly nonhazardous materials, but the wastes will be much more diverse than most industrial landfills. Although the overall toxicity of the industrial landfill's contents is probably much higher than the old city dump, from an engineering standpoint, the more toxic wastes may be easier to deal with because the engineering controls can be more specific. For example, if concentrations of chlorobenzene in the municipal landfill and industrial landfill are in the parts per billion (ppb) and parts per thousand, respectively, but the municipal landfill also has chlorinated solvents, pesticides, and aromatic compounds in the ppb range, the industrial landfill engineering solutions can be targeted completely at what is best to decontaminate the chlorobenzene. However, the city engineer will have to worry about a very diverse mix of contaminants.

Also, the "matrix" is likely to be much more heterogeneous for the municipal landfill (stuff is coming from homes, businesses, small industries, illegal dumping, etc.), so the contaminants may reside in everything from grass clippings and yard wastes to small containers to household items. Each of these matrices will have its own partitioning coefficients and kinetics (e.g., grass clippings may be similar to soil organic matter, but contaminants mixed in with latex paint may be similar to an industrial slurry).

Another major difference between these two landfills is that one is inactive and the other is active. This difference can be both good and bad. The inactive site has the advantage of less change in the source term for the contaminant. In other words, it may be a source of very nasty substances, but at least it is no longer changing from day to day, except for the degradation of the compounds within the source.

On the other hand, the active site has the advantage that laws, such as the Resource Conservation and Recovery Act, require that detailed manifest reports be generated for all incoming wastes. The engineer should consult these reports to ascertain how the contaminant mix has changed over time. For example, if the reports show that PCBs were accepted 15 years ago, these persistent compounds are in all likelihood still in the landfill and are potentially being transported from the site where they can contaminate the groundwater. Even if your monitoring wells do not indicate that PCBs are in the leachate, this "sword of Damocles" stands precariously, waiting to present itself in the form of groundwater contamination.

Contaminant Treatment Question 8

You have been asked to oversee the removal and transportation of contaminated soil for *ex situ* treatment. This requires careful planning. What important factors must be considered prior to excavation of the soil?

A detailed excavation plan includes:

- Digging boundaries (vertical and horizontal)
- Locations of stockpiled materials
- Access for heavy equipment and trucks
- Access for light-duty vehicles and persons
- Maintenance and supply of equipment (including facilities for fueling, lubrication, repair, and upgrades)
- Actual methods for excavating material (from the large coverage of dipper shovels, draglines, clamshells, tractor loaders, and backhoes,[23] to hand shovels and even minute probes for the fine details of research and monitoring)

- Contingency plans for emergencies, spills, unexpected releases, unexpected findings (e.g., archeological, historical, and institutional), and dangerous conditions, such as buried power lines and pipelines, especially those that may not have been properly mapped
- Management and control of excavations, including division of labor and "command and control" of all personnel (site engineer, workers, foreman, official representatives of government, and others with a need to be on-site)
- Access restricted to only those who need to be on-site and the means for such restriction, including security
- Decision points on working conditions, including postponements and changes due to weather conditions and the potential for unsafe conditions
- Adherence to fire and other safety code provisions as defined by state and local authorities
- Keeping of manifests and chain of custody information for all material excavated and moved off-site, including such reports at the ultimate treatment site.

Contaminant Treatment Question 9

Now that you have an excavation plan in place you are asked to estimate the amount of material that needs to be moved. Fortunately, you have a reliable environmental assessment of the site showing that heptachlorodioxin and heptachlorofuran are the only toxic species detected in the soil. The soil is contaminated to a depth of 15 cm. It is a very large removal area, about 300 m by 600 m. Estimate the amount of soil that needs to be treated off-site.

Answer and Discussion

$$\text{Area of removal} = 300\,\text{m} \times 600\,\text{m} = 1.8 \times 10^5\,\text{m}^2$$

$$\text{Depth of removal} = 15\,\text{cm} = 1.5 \times 10^{-1}\,\text{m}$$

Thus, the amount removed is:

$$(1.8 \times 10^5\,\text{m}^2) \times (1.5 \times 10^{-1}\,\text{m}) = 2.7 \times 10^4\,\text{m}^3.$$

Contaminant Treatment Question 10

Your equipment and grade conditions, as well as the type of equipment selected in your excavation plan, take out soil to a depth of 9 inches. How much soil will you end up removing compared to the actual extent of contaminated soil? Why do you need to remove this much soil?

Answer and Discussion

$$9\,\text{in} = 22.5\,\text{cm} = 0.225\,\text{m depth.}$$

Thus, the amount removed is:

$$(1.8 \times 10^5\,\text{m}^2) \times (2.25 \times 10^{-1}\,\text{m}) = 4 \times 10^4\,\text{m}^3$$

This is quite a bit more than the amount of contaminated soil. However, it may give you a factor of safety, since contaminant migration through soil is highly variable, depending on soil characteristics and environmental conditions. After all, these are dioxins and furans, so the client and public will need to be assured that you are "getting it all." But it will cost more than if you had removed only to the extent of contamination.

By the way, it is a good idea to take periodic samples of the contaminants at various soil depths as excavation continues. If the gradient that we expected at the outset is different, especially if it is less steep (i.e., the contamination is continuing to lower depths than we expected), we will need to revise our excavation plan to remove all contaminated soil. This means that we may want to make sure that our original assumptions about the type and size of the dioxin source are still valid. For example, if we find a buried tank or a "hot spot" of high dioxin levels at depth, this must be factored into our removal (and treatment) planning, especially if we now have reason to believe that additional contaminants of concern are present. It is likely that the *ex situ* treatment of these dioxins, such as thermal treatment, will be good for most organics, since dioxins are among the most difficult to break down. However, these processes will not remove metals, so a type of pre-thermal treatment (or post-thermal treatment, depending on the stoichiometry and soil type) will need to be added. The other thing to keep in mind about metals and organic chemistry is that metal catalysis is possible, so kinetics and degradation pathways will change under certain conditions, which means that temperature, pressure, sorption, and other conditions may have to be adapted in the ultimate treatment processes.

Notes and Commentary

1. The major template used in this discussion is that of three major pieces of U.S. legislation: the Superfund Law, or the Comprehensive Environmental Response, Compensation, and Liability Act of 1980, the Superfund Amendment and Reauthorization Act of 1986, and the Resource Conservation and Recovery Act of 1976, as well as their amendments.
2. U.S. EPA website (May 2003), http://www.epa.gov/superfund/whatissf/sfproces/pasi.htm.
3. See the EPA publication, *Guidance for Performing Preliminary Assessments under CERCLA*, September 1991, PB92-963303, EPA 9345.0-01A, and the electronic scoring program "PA-Score" found at http://www.epa.gov/superfund/resources/pascore for additional information on how to conduct a preliminary assessment (PA).
4. U.S. Environmental Protection Agency, 2003, *Remediation Guidance Document*, Chapter 7, EPA-905-B94-003.
5. *Ibid.*
6. The principal sources for this section are discussions with two of my colleagues and mentors, Ross E. McKinney and J. Jeffrey Peirce. Dr. McKinney, who before his recent retirement spent most of his illustrious career at the University of Kansas, is among the most highly recognized authorities in biological treatment. Dr. Peirce, of Duke University's Pratt School of Engineering, contributed to my recent book, *Engineering the Risks of Hazardous Wastes* (Chapter 4 is a direct source for this section). He is a leader in environmental engineering. Each contributed much of what I have been able to convey in this book, going back to my studies at Kansas and Duke, and my ongoing pursuit of their wisdom as of this writing.
7. An interesting recent publication that will introduce the reader to life-cycle analysis is J.K. Smith and J.J. Peirce, 1996, "Life Cycle Assessment Standards: Industrial Sectors and Environmental Performance," *International Journal of Life Cycle Assessment*, Vol. 1, no. 2, pp. 115–118.
8. This goes beyond zoning. Obviously, the engineer should be certain that the planned facility adheres to the zoning ordinances, land use plans, and maps of the state and local agencies. However, it behooves all of the professionals to collaborate, preferably before any land is purchased and contractors are retained. Councils of Government (COGs) and other "A-95" organizations can be rich resources when considering options on siting. They can help avoid the need for problems long before implementation, to say nothing of contentious zoning appeal and planning commission meetings and perception problems at public hearings!
9. Numerous textbooks address the topic of incineration in general and hazardous waste incineration in particular. For example, see C.N. Haas and R.J. Ramos, 1995, *Hazardous and Industrial Waste Treatment*, Prentice-Hall, Englewood Cliffs, N.J.; C.A. Wentz, 1989, *Hazardous Waste Management*, McGraw-Hill, New York, N.Y.; and J.J. Peirce, R.F. Weiner, and P.A. Vesilind,

1998, *Environmental Pollution and Control*, Butterworth-Heinemann, Boston, Mass.

10. Biffward Programme on Sustainable Resource Use, 2003, *Thermal Methods of Municipal Waste Treatment*, http://www.biffa.co.uk/pdfs/massbalance/Thermowaste.pdf.
11. Scrubbers are air pollution control devices that separate soluble gases, like SO_x from a gas mixture (e.g., flue gases). Most scrubbers bring the gas mixture from the bottom and while the gases rise they are sprayed with a liquid (e.g., ammonia) from the top. The liquid dissolves the SO_x and is collected at the bottom of the scrubber. This "scrubber sludge" must be handled properly.
12. At high temperatures, nitrogen dioxide (NO_2) is rapidly reduced to nitric oxide (NO). If methane or another reductant is in excess, the NO is further reduced to N_2 (nitrogen gas or molecular nitrogen). A challenge for the engineer is to deal with any products of incomplete combustion (PICs), such as CO and the polycyclic aromatic hydrocarbons (PAHs) that can form. Thus, a second step is usually needed to destroy the PICs. However, this second step may allow NO_x compounds to form because it involves oxidation (molecular nitrogen + oxygen $\rightarrow NO_x$). So, the engineer must balance the process so that neither NO_x nor PICs are released at concentrations above emission standards. That is also why sufficient monitoring (#7 on this list) is so crucial. One needs to know if the reactions are changing due to operation and maintenance of the combustion facility.
13. J. Lee, D. Fournier, Jr., C. King, S. Venkatesh, and C. Goldman, 1997, "Project Summary: Evaluation of Rotary Kiln Incinerator Operation at Low-to-Moderate Temperature Conditions," U.S. Environmental Protection Agency, EPA/600/SR-96/105, Cincinnati, OH.
14. A slurry is a liquid that contains suspended solids in amounts sufficient to change the liquid's viscosity so that it behaves more like a non-Newtonian fluid (see Figures 3.3 and 3.4).
15. "Extremely hazardous waste" (EHW) is a legal term. For example, in California an EHW <50 mg kg^{-1} LD_{50} oral toxicity, <43 mg kg^{-1} dermal toxicity, or <100 ppm LC_{50} inhalation toxicity. It also includes substances that are carcinogenic, that react violently with water, and that are very likely to persist and bioaccumulate.
16. Federal Remediation Technologies Roundtable, 2002, *Remediation Technologies Screening Matrix and Reference Guide*, 4th Edition.
17. A principal source for all of the thermal discussions is U.S. Environmental Protection Agency, 2003, *Remediation Guidance Document*, Chapter 7, EPA-905-B94-003.
18. Addition of food other than the waste itself is why Reaction 12–4 and subsequent reactions show the parenthetical "(+ energy?)." In many systems, if the microbes adapt sufficiently no external energy is needed because the microbes use the waste as their entire food source.
19. For decades books have been published that focus on the current understandings of the science, engineering, and technology of biological waste treatment.

See, for example, G. Tchobanoglous and F. Burton, 1991, *Wastewater Engineering, Metcalf and Eddy*, McGraw-Hill, New York, N.Y. and its most recent edition: D.H. Stessel, G. Tchobanoglous, and F. Burton, 2002, *Wastewater Engineering: Treatment and Reuse*, 4th Edition. McGraw-Hill, New York, N.Y.; A. Gaudy and E. Gaudy, 1988, *Elements of Bioenvironmental Engineering*. Engineering Press, San Jose, Calif.; and J. Peirce, R. Weiner, and P. Vesilind, 1998, *Environmental Pollution and Control*, Butterworth-Heinemann, Boston, Mass. For a particular focus on the biotreatment of hazardous wastes see, for example, C. Haas and R. Ramos, 1995, *Hazardous and Industrial Waste Treatment*. Prentice-Hall, Englewood Cliffs, N.J., and C. Wentz, 1989, *Hazardous Waste Management*, McGraw-Hill, New York, N.Y.

20. A more complete discussion of hazardous waste storage facilities appears in a wide range of textbooks, including C. Haas and R. Ramos, 1995, *Hazardous and Industrial Waste Treatment*. Prentice-Hall, Englewood Cliffs, N.J., and C. Wentz, 1989, *Hazardous Waste Management*, McGraw-Hill, New York, N.Y.
21. U.S. Environmental Protection Agency, 1992, *Innovative Treatment Technologies: Semiannual Status Report*, Washington, D.C.
22. Look them up. They were huge in the 1970s!
23. Backhoes are by far the most frequently used excavating equipment in removing contaminated soils and unconsolidated materials at hazardous waste sites, because they can limit the depth of excavation to boom size, can exceed excavation depths of 13 m, can remove materials efficiently and quickly from below grade, and can also be used to trench and skim surfaces, so they are also useful for grading and for drainage work beyond excavation.

CHAPTER 13

Environmental Decisions and Professionalism

As we have seen throughout this text, data are important but are meaningless at best, and dangerous at worst, if they are not interpreted properly. The professional's credo is different from that of the general public, or even that of the entrepreneur. We cannot live by *caveat emptor*, or "Let the buyer beware!" The professional credo must be *credat emptor*, "Let the buyer trust!" This mandate goes beyond avoiding actions that are clearly wrong. We must actively seek ways to do what is right. And we must ask, who is the buyer, and who is our client? Certainly, the company, agency, or holders of contracts are our clients, but in the environmental realm, our clients are vast. They are all of the people who may be affected by our advice, those now living and those of future generations. The "environment" in all of its connotations is our client.

Our advice and actions must be seen through the prism of sustainability. As such, our actions must not simply solve immediate problems, but must be viewed as to how they will play out in the long run. Our advice is almost always tinged with varying amounts of uncertainty, both in the science and in the decisions that will be made based upon that science.

We need a template for what to do. Our competence can take us in the right direction. It is necessary that we base our science on sound principles, many of which have been addressed in this text. Sound science is certainly necessary, but is insufficient in itself to ensure proper and sustainable environmental decisions and actions. A good place to start is adherence to a commonly accepted "code." Environmental science has no universally accepted code. However, much of the environmental engineering mandate is encompassed under engineering professional codes in general, and more specifically in the Code of Ethics of the American Society of Civil Engineers (ASCE) in particular.[1] As evidence, in its most recent amendment on November 10, 1996, the code incorporated the principle of sustainable development.[2]

The Code mandates four principles that engineers abide by to uphold and to advance the "integrity, honor, and dignity of the engineering profession":

1. Using their knowledge and skill for the enhancement of human welfare and the environment;
2. Being honest and impartial and serving with fidelity the public, their employers, and clients;
3. Striving to increase the competence and prestige of the engineering profession; and
4. Supporting the professional and technical societies of their disciplines.

The code further articulates seven fundamental canons:

1. Engineers shall hold paramount the safety, health, and welfare of the public and shall strive to comply with the principles of sustainable development in the performance of their professional duties.
2. Engineers shall perform services only in areas of their competence.
3. Engineers shall issue public statements only in an objective and truthful manner.
4. Engineers shall act in professional matters for each employer or client as faithful agents or trustees, and shall avoid conflicts of interest.
5. Engineers shall build their professional reputation on the merit of their services and shall not compete unfairly with others.
6. Engineers shall act in such a manner as to uphold and enhance the honor, integrity, and dignity of the engineering profession.
7. Engineers shall continue their professional development throughout their careers, and shall provide opportunities for the professional development of those engineers under their supervision.

The first canon is a direct charge to deal properly with environmental contamination. This book is one of many resources available to meet the second canon, or competence. The remaining canons prescribe and proscribe activities to ensure trust. It is important to note that the Code applies to all civil engineers, not just environmental engineers. Thus, even a structural engineer must "hold paramount" the public health and environmental aspects of any project, and must seek ways to ensure that the structure is part of an environmentally sustainable approach. This is an important aspect of sustainability, in that it is certainly not deferred to the so-called environmental professions but is truly an overarching mandate for all professions (including medical, legal, and business-related professionals). That is why environmental decisions must incorporate a wide array of perspectives, while being based in sound science. The first step in this inclusive decision-making process, then, is to ensure that every stakeholder sufficiently understands the data and information gathered when assessing environmental contamination.

Communicating Scientific Information

Risk assessment is a process where information is analyzed to determine the extent that an environmental hazard might cause harm to exposed persons and ecosystems.[3] It requires an understanding of hazards, the adverse effects, the possible exposures of people, ecosystems, and materials, and a means of pulling all of this information together to determine the overall risks of a chemical in the environment. Risk management lays out the approaches needed to address what we are finding out in the risk assessment.[4] Managing risks involves not only the physical and biological scientific results, but also socioeconomic, political, legal, spiritual, ethical, and other human values that frame the risk decision. This is complicated stuff, so we must be faithful to ensure that what we say to our clients, our neighbors, and the public is communicated effectively.

Effective communication begins with the competence, honesty, and integrity of the environmental professionals. There is no place for "spin" or hedging. However, just as one does not expect her seven-year old son to understand the joy of differential equations or quantum mechanics, one must empathize with those who are receiving your reports and briefings. They are usually very intelligent and highly interested people, but they have not "lived" the experience that the engineer or scientist has in the months designing and collecting information, considering possible sources, and calculating risks. They have actually "lived" it in a very different, yet real, sense. The contaminants are in their city, their homes, and maybe in their bodies, so it is unrealistic to expect them to consider all they are hearing with objective and dispassionate eyes and ears.

Psychologists and other social scientists[5] have investigated the reasons for widely differing perceptions of seemingly objective and concrete information. Some of these factors provide insights for sharing complex and complicated environmental information so that those who care to listen clearly understand what is being said.

Factor 1: What Is the Possibility of a Severely Negative or Catastrophic Outcome?

The public is less likely to trust even in a well-designed remediation effort, if the possible negative outcomes are centralized in time and space, i.e., if its *your* specific neighborhood that is being contaminated *right now*, compared to those that are more scattered and random. This is a common challenge since, ordinarily, the scientific folks have not arrived until after contamination has been observed in some manner at a specific site and at a certain time. Ironically, an environmental assessment may very well increase the public's concern by properly investigating and characterizing the problem by collectivizing or "grouping" the negative outcomes in space (e.g., the site's location and the extent of contamination of soil, water, and air have been characterized) and time (e.g., the source has been documented and the movement of the contaminants has been modeled retrospectively and prospectively). For instance, Figure 12.15 presents an example of a spatially and temporally defined problem (cyanide migration).

The engineer should be clear, careful, and sensitive when describing the site and possible remedies. Even when the potential prognosis for site remediation is good, exposures to possible contaminants can be effectively eliminated, and existing technologies have worked well in other similar situations, the public will not automatically be reassured. When the professionals describe what is to be done, the community members may perceive something very different from what the engineers and scientists are trying to convey. The community perception may be that they may be living near another "Love Canal" or "Times Beach."

Factor 2: How Familiar Are the Situation and the Potential Risks?

People fear what they do not understand, which in part explains why so large a segment of the population is quite comfortable with cigarette smoking, but many are very terrified of the storage of spent fuel from nuclear power plants, even when the former hazard accounts for much disease and death. The very nature of nuclear science and information is mysterious to many people.

One major problem with environmental contamination involves the associated nomenclature and vernacular. Although a few contaminants can be well understood by a broad audience, such as the leaking of leaded gasoline from an underground storage tank, most environmental contamination are mixtures of ominous-sounding compounds that challenge the professional to describe the compounds sufficiently so that all parties understand what is at risk.

Another problem is the complex, or at least complex sounding, methods used to test, to model, and to characterize the actual and predicted movement and change of these compounds under different remediation scenarios. For example, the public must understand the difference between "no action" and the other remediation alternatives. In this instance, the engineer must explain that without intervention, the plume of X, Y, and Z contaminants will move 10 meters per year vertically and 100 meters per year horizontally. The engineer will also need to explain how this relates to sensitive receptor sites, such as drinking well intakes and stream inputs. In addition, the chemical, physical, and biological transformation processes must also be explained, so that what may have been released has changed, in part, to other compounds. Thus, in addition to X, Y, and Z, other degradation products X′, X″, and Z′ in the water, soil, and air at various times must be measured. Such equilibrium chemistry is complicated for engineers and scientists (although less so now that you have read this book, one hopes) let alone those members of the community who do not confront it frequently (if at all).

Likewise, all alternative approaches to remedy the situation must explain these same processes and models, including the uncertainties involved in predicting success. These descriptions are further complicated as a function of available engineering controls and remediation steps, each of which must also be explained to the satisfaction of the community. Thus, if a pump-and-treat alternative is being proposed, then all of the chemistry, physics, and biology associated with this technique must be explained. In addition, the public must completely understand how the approach will be evaluated in terms of success. The success is not only to be explained in terms of engineering performance standards like the total volume of water treated and the target level of contaminant removal (e.g., 99.99% removal efficiency), but the quality of the environment following the removal must also be described (e.g., the aquifer's water will contain x ng L^{-1} X, Y, and Z, the soil following treatment will contain y ng kg^{-1} X, X′, Y, and Z).

Factor 3: Can the Processes and Mechanisms Being Proposed or Undertaken Be Explained?

The likelihood of public trust decreases in relation to the complexity of the processes and mechanisms of exposure and risk. When such systems are not well understood, the community is more likely to be concerned about the problem and the proposed remedies. As mentioned in the discussion of Factor 2, contamination and cleanup processes can be highly complex and involve numerous variables. The key is to explain and to provide ample information without overwhelming the audience.

Factor 4: How Certain Is the Science and Engineering?

People tend to lose confidence in science and engineering when, in their view, there is too much uncertainty in outcomes and risks associated with a remedial action or any other important public health or environmental endeavor. Uncertainty in science arises from several sources. Even the most carefully conducted test of a chemical has some degree of variability in the data it provides. If the data are produced from different studies conducted by different laboratories, this will add uncertainty to the data.

Environmental measurement and other technologies continue to change and improve, so comparing historical data (e.g., where detection limits continue to fall) may add uncertainty, at least as perceived by the public. For example, if a table of findings shows that a certain pesticide's concentration in soil was not detected in the 1980s, but was increasingly found in the soil at about 10 micrograms per kilogram ($\mu g\,kg^{-1}$) in the 1990s, the first question the engineer should ask is, what were the detection limits for the pesticide in soil, and how had these limits changed over time? The concentration of the pesticide may not have changed, or even fallen, over the two decades, but the retrospective data neither confirm nor reject this finding.

Multiple measurements by different laboratories will give varying results. The quality assurance/quality control (QA/QC) plan will define data quality objectives (See Chapter 11) that must be met for any study. In dealing with the public, uncertainties in the data and in the information from which decisions will be made must be fully disclosed.

Factor 5: How Much Personal Control Do They Have Over Exposure and Risk?

People are generally more comfortable when they have a modicum of control. Unfortunately, when dealing with hazardous wastes and toxic chemicals, the public may feel alienated by the sophistication of the physics of remedies being conducted by a cadre of outsiders. The public's input must be sought and incorporated into all remediation efforts.

Factor 6: Is the Exposure Voluntary or Involuntary?

Cigarette smoking has shown us that scientific research can provide important, even sound, advice, but predicting how the public will incorporate this advice into their daily lives is difficult. Surely, one important factor in the public's acceptance or rejection of even the most sound scientific advice is whether it interferes with their choices in the matter. People who consider exposure to even dangerous chemicals to be their choice are more likely to accept the risks associated with that exposure. Conversely, the public may reject sound scientific and engineering advice that detracts from their perceived freedom to select the "best" option (e.g. if citizens believe that the environmental scientist or engineer has made up her mind regarding the best approach even before the meeting to discuss options and alternatives).

Factor 7: Are Children or Other Sensitive Subpopulations at Risk?

Children are particularly sensitive to many environmental pollutants. They are growing, so tissue development is in its most prolific stages. In addition, society has stressed—and certainly should!—special levels of protection for infants and children. For example, regulations under the Federal Food Quality Protection Act[6] mandate special treatment of children, evidenced by the so-called "10 X Rule." This rule recommends that, after all other considerations, the exposure calculated for children include ten times more protection; thus, the exposure is multiplied by 10 when children are exposed to toxic substances. Frequently, the prenatal and postnatal toxicities are included when calculating a Reference Dose (RfD). However, as discussed in Chapter 10, uncertainties or an elevated concern for children is not always sufficiently addressed using uncertainty factors in the RfD. Thus, the FQPA requires an additional evaluation of the weight of all relevant evidence. This involves examining the level of concern for how children are particularly sensitive and susceptible to the effects of a chemical, and determining whether traditional uncertainty factors already incorporated into the risk assessment adequately protect infants and children. This is accomplished mathematically in the exposure assessment. The FQPA safety factor for a particular chemical must yield the level of confidence in the hazard and exposure assessments and provide an explicit judgment of the possibility of other residual uncertainties in characterizing the risk to children.

By extension, other sensitive strata of the population also need protection beyond those of the general population.[7] The elderly and asthmatic members of society are more sensitive to airborne particles. Pregnant women are at greater risk from exposure to hormonally active agents, such as phthalates and a number of pesticides. Pubescent females undergo dramatic changes in their endocrine systems and, consequently, are sensitive to certain chemical exposures during this time.

Factor 8: When Are the Effects Likely to Occur?

People may not like acute effects, but they are more likely to accept them than those that manifest themselves only after a protracted latency. For this reason, people will endure some short-term risks to prevent future problems. The engineer should clearly state the acute and chronic outcomes that may result from all phases of remediation.

Factor 9: Are Future Generations at Risk?

If there is any risk to future generations, the public will be concerned. This in part explains many people's discomfort with nuclear power generation and nuclear wastes (e.g., half-lives of hundreds of thousands of years) that will leave a dangerous legacy. Many people perceive the so-called PBTs (persistent, bioaccumulating toxic substances, like dioxins) as something that they do not want to pass along to future generations.

Factor 10: Are Potential Victims Readily Identifiable?

The public may be more concerned about "real" victims than about "statistical" victims. However, it is very difficult to explain what a one-in-a-million risk means, and even more difficult if this risk is described using engineering notation, or risk = 10^{-6}. Vivid examples of Chernobyl and Hiroshima have provided graphic images of real victims of radiation exposure. Love Canal and Bhopal, India have done the same for toxic contaminants. Thus, characterizing the risks "by the numbers" can be unsatisfying for those wanting to know what those numbers mean.

Factor 11: How Much Do People Dread the Outcome?

The psychological concept of *dread* is very important in risk perception and communication. The greater the amount of dread associated with a contaminant, the more concern the public will have about dealing with it. The health effects associated with toxic substances dictate the public's concern. Probably the most dreaded effect is cancer, so professionals must be prepared to address people's concerns about these carcinogens (and avoid expectations that the participants will be coldly objective about the various remediation efforts). Carcinogens are not the only chemicals associated with large dread factors. For example, witness the mothers who have expressed at public hearings their dread of the possible learning disabilities and central nervous system problems in their children when they find out that their drinking water or air had been contaminated by lead from a nearby smelter. This extended dread is particularly important when it also includes risks to children and future generations. Again, telling people that the success of a remedy is to reduce the cancer risk to $<10^{-6}$ or that Pb soil concentrations are falling by X $mg\,kg^{-1}\,year^{-1}$ is unlikely to completely allay their dreads and fears.

Factor 12: Do People Trust the Institution Responsible for Assessing the Risk and Managing the Cleanup?

All institutions have "baggage." No matter how good the environmental professional's reputation, the association with the government agencies and firms that are involved in the project will influence the acceptance by the public of remediation plans. Individual medical doctors and scientists have high trust levels with the public, but the public's distrust of government agencies and corporations has been growing.[8] "Guilt by association" is common. It is realistic to expect that contaminant removal and remediation proposals will be met with a certain amount of skepticism and resistance from the public.

Factor 13: What Is the Media Saying?

If the reports in the newspapers and other parts of the news media have documented a history of problems at a site, and if there is much media attention, the public's concern will be high. The professional must be accessible to the press (following the communication strategy developed by the government agency and other parties), and deal openly and honestly with all inquiries.

Factor 14: What Is the Accident History of This Site or Facility, or of Similar Sites or Facilities?

If the company responsible for cleanup has a poor history of accidents or a track record of incidents related to hazardous chemicals, the public concern can be expected to be heightened. If the types of corrective and remedial actions being proposed have a checkered past, this will also carry over to the plans proposed for a specific site or facilities. This does not mean that actions that have not worked elsewhere should be dismissed out of hand. It does require, however, an accounting of why an action failed and why one does not expect similar failures at this particular site. After satisfying that the conditions are sufficiently different to warrant recommending an action, the professional must provide a strong justification laying out the reasons for expected success. This should also include truthful assurances of redundancies and contingency plans in the event of accidents or failures, however seemingly remote the possibility.

Factor 15: Is the Risk Distributed Equitably?

The history of environmental contamination has numerous examples where certain segments of society are exposed inordinately to chemical hazards. This has been particularly problematic for communities of low socioeconomic status. A landmark study[9] showed that the likelihood of landfill siting and the presence of hazardous waste sites in a community has been disproportionately higher in African American communities. Migrant farm workers, many who are Latinos, can be exposed to higher concentrations of hazardous chemicals where they live and work, in large part due to the nature of their work (e.g., agricultural chemical exposures can be very high shortly after fields are sprayed). In fact, cultural factors can play a large role in how and to what extent persons are exposed to containments. For example, the family work ethic of many Latinos may mean that children may be exposed to extraordinarily high concentrations of pesticides when they join their families to help in farm work. Even a scientifically sound remedial action will be resisted in neighborhoods that have had to deal with injustices in the past. Sensitivity to these experiences should be part of any risk communication plan.

Factor 16: Are the Benefits Clear?

The design team may be well aware of why the remediation is being undertaken. In fact, the benefits of risk and exposure reductions may be so "obvious" that the engineer is tempted to give merely a short consideration and attention to this in meetings with the public. This would be a mistake. For the public to comprehend the plan of action fully, the expected benefits must be clearly articulated. This includes the improvements resulting from hazard reduction, exposure reduction, and prevention of health and environmental effects. If this is not the first meeting, it is still risky to assume that the participants are "up to speed." When in doubt, ask! At the top of the agenda may be: "Here's what we all agreed upon last time."

Factor 17: If There Is Any Failure, Will It Be Reversible?

The potential irreversibility of damages is akin to other public concerns about future generations and controllability. However, in addition, the public is looking to the experts to provide reassurance that the site will be monitored during and long after remediation to prevent catastrophes or at least to catch problems before they become large and irreversible. Consider the interventions described in Chapter 12. The monitoring component of the plan, for example, should stress why measurements are being taken before and after completion of the remedies.

Factor 18: What Is the Stake of Each Person?

Each person's interest in and concern about the project is unique. For example, a person living adjacent to the site may have a greater personal stake in the health issues than one living a mile away. However, the person living a mile away may own property that could become more or less valuable, depending upon the remedial actions selected.

Factor 19: What Is the Origin of the Problem?

Members of the public are generally more tolerant of and patient with remedies to address "natural" disasters than those to address problems caused or exacerbated by humans. Environmental contamination should be considered to be human-derived, even if they are worsened by natural causes. For example, if a tank is ruptured during an earthquake and hazardous chemicals contaminate an aquifer, it should be treated as a human-caused problem, because humans built and installed the tank in the first place. Also, just because something is "natural" does not necessarily mean it is acceptable. For example, trees emit hydrocarbons that exacerbate the tropospheric ozone levels in urban areas. The ozone formation chemistry does not distinguish between anthropogenic and biogenic hydrocarbons, so actions must be taken to protect health irrespective of whether the cause is human-induced, natural or a combination of the two.

Risk perceptions are highly variable, even unpredictable. Openness and full disclosure about the pros and cons of any action is the order of the day. Great care should be taken when sharing information and ideas with the public. A word or phrase may be perfectly clear to the person using it, but might completely unsettle the already nervous and skeptical neighbors of a hazardous waste facility. Recall from Chapter 1 that even within the environmental science community, we have no unanimity on the word *particle*. Like the medical doctor writing a prescription for medication, any ambiguity in technical communication can be dangerous.

Environmental Information Management

The foregoing discussions point to the challenge of conveying meaningful information that truthfully adheres to the data, is underpinned by sound science, and is presented in such a way that all stakeholders can understand it in the process of deciding what actions must be taken. Environmental information can be highly complex and even obtuse. Balancing data and information needs is demanding for the information technologists but, even before the data are generated, environmental professionals have to ensure high-quality and meaningful data and information at every step, from planning the field and lab work, to implementation, to data reduction, and to sharing results in scientific and public forums.

Simply asking upfront what the neighbors and others who will be affected by decisions consider to be important is a good first step. A structured form of asking is helpful, for example a table similar to Table 13.1

TABLE 13.1
Preliminary Results Based upon Literature Reviews and Professional Judgment Following Meetings with Stakeholders (Composite of recommendations by D.A. Vallero, 2003)

Source Type Identified by Community	*Common Names[10] of Analytes to Be Measured*	*Reason for Possible Concern*
Chemical handling and wholesale distributor	Plastics, softening ingredient in personal products like makeup	Hormonally active, may be linked to certain cancers. Found all over ("ubiquitous").
	Ammonia	Irritant, life-threatening at 2400 ppm. Toxic air pollutant. The major problem here is leaks and spills from containers. This also applies to "parked" tanker rail cars and trucks that carry ammonium hydroxide or ammonia gas.
	Products used in cutting oils, soaps, shampoos, cleaners,	Possible links to sex hormone dysfunction (testicular) from animal studies with phthalates.

TABLE 13.1 *(continued)*

Source Type Identified by Community	*Common Names[10] of Analytes to Be Measured*	*Reason for Possible Concern*
	polishers, cosmetics, and drugs, and in making plastics	Can contaminate groundwater and soil.
	Methanol is wood alcohol. MEK, benzene, and toluene are paint stripping ingredients	Several have been suspected to cause cancer. Some, like MEK, may harm the nervous system. Can contaminate groundwater and soil.
	Dry cleaning fluids, degreasers, cleaners	A number of these chlorinated compounds have been linked to cancer.
	Hydrochloric acid, formic acid	Hydrochloric acid is a strong acid and the third leading source of death and injuries from major industrial accidents. When released to the air and contacted with skin, eyes, and airways, it can cause severe irritation, inflammation, and tissue damage (highly corrosive). It is a major air pollutant. Formic acid is a weak acid used in the dye industry. It is somewhat corrosive and will produce carbon monoxide if it breaks down.
Automobile junkyards and abandoned vehicles	Antifreeze	Ethylene glycol is poisonous to pets who drink from green puddles. Can contaminate groundwater and soil. OSHA recommends that air values stay below 50 ppm. This substance mixes with water, so it may easily flow into nearby streams and underlying groundwater and soil.
	Heavy metals	Lead harms the brain and nerves (lowers IQ, associated with diseases like Parkinson's and palsy). Cadmium has been associated with lung cancer and other lung diseases

TABLE 13.1 *(continued)*

Source Type Identified by Community	*Common Names[10] of Analytes to Be Measured*	*Reason for Possible Concern*
		(like emphysema) in humans, but would have to be breathed in (for this type of exposure from junkyards, it would have to be blown off the yards as dust). Other diseases associated with cadmium include those of the heart, bones, kidney, liver, and blood. Chances are, if you lived in the 1950s and 1960s, you were exposed to cadmium as a pretty metallic paint on your popgun or ice tray.
	Chromium—Chrome	Chromium is an essential metal for metabolism, but only in a certain chemical form ("trivalent") and only in small dosages. Other forms, particularly "hexavalent," forms of chromium are toxic and have been associated with cancer. It has been widely used to plate metal, so old cars are likely to be a source as their bumpers and detailing corrode in junkyards. Analyzing for this metal would begin by looking for total chromium. If such a "screening" study shows that the metal exists in the water or soil, much more intricate and expensive tests are needed to see whether it is in the good or bad form.
	Zinc—"Chinese White," "Philosopher's Wool"	Zinc is fine in small amounts (even essential), but at high doses over time it can increase bad cholesterol and decrease good cholesterol. Metals above certain levels are very harmful to fish and wildlife. This heavy metal's harmful effects are worsened in soft water.
		The movement and potential exposure to these metals depends on the chemical characteristics of

TABLE 13.1 *(continued)*

Source Type Identified by Community	*Common Names[10] of Analytes to Be Measured*	*Reason for Possible Concern*
		the soil, especially things like pH and the existence of other substances, especially ions (charged particles).
	Asbestos	Used in car parts like brake liners. Causes lung cancer and asbestosis (a debilitating lung disease). These are significantly worsened in smokers.
	CFCs, "Freon" (This is actually the trade name of one particular type of CFC. "Halon" is another CFC used in some fire extinguishers where water cannot be used, like computer rooms.)	This is really a global pollutant. It is not very toxic, but when released it finds its way to the stratosphere and destroys the ozone layer. The ozone layer protects us from harmful wavelengths of sunlight, so-called ultraviolet or "UV." Exposure to this form of light has caused cancers, including the very virulent form, melanoma.
Aboveground tanks	Volatile organic compounds, hydrocarbons	Leakage from these tanks can present immediate hazard to children and others who are exposed, as well as a long-term hazard if the contaminants leach into the groundwater or run off into creeks.
Open dumps	Highly variable	Aesthetic pollutant. Attraction of disease vectors (rodents, insects). Possible exposures to chemical hazards (similar to "midnight dumping" discussed below).
Abandoned houses and other disintegrating buildings	Physical contaminants (See Appendix 9)	Asbestos, fiberglass and other materials in plasters, siding, pipe covers, and roofing materials. When a house falls down, there is a chance that the asbestos that was once held fast is now released into the air and water. Causes lung cancer and asbestosis (a debilitating lung disease). These

TABLE 13.1 *(continued)*

Source Type Identified by Community	*Common Names[10] of Analytes to Be Measured*	*Reason for Possible Concern*
		are significantly worsened in smokers.
	Tars and combustion products	If the structure catches fire, many chemicals are released to the air. These are suspected of causing cancer.
	Ether, cocaine ingredients, methamphetamine ingredients, narcotic ingredients	So-called "meth labs" and other buildings in which illegal drugs are made and handled present a risk to surrounding populations from the explosive release of chemicals and fires.
Obsolete and underused manufacturing buildings: hosiery mills, tobacco buildings	Nicotine, tars, "brownfield" contaminants, illegal dumping, "midnight dumping"	After manufacturing ends, the structures and soil and water around and under the buildings will be contaminated with chemicals for a long time. Some chemicals are more "persistent" (last longer without breaking down) than others. Tars and organic compounds may remain unaltered for years and are potentially harmful to those who come into contact with them. Other chemicals, like nicotine, that under ordinary conditions would breakdown readily, may remain unchanged for a much larger time if they are not exposed to high temperatures, light, air, or moisture, because they are trapped in floor crevices, under structures, and in dirt. At high enough doses, nicotine is fatal (especially to children). Another possibility of contamination is that since the buildings were abandoned, people have illegally used these areas as "dumps." This can range from dumping construction debris on the site

TABLE 13.1 *(continued)*

Source Type Identified by Community	*Common Names[10] of Analytes to Be Measured*	*Reason for Possible Concern*
		(possible sources of asbestos, heavy metals, and other wastes) to large and small releases of chemicals from tankers (midnight dumping) to hiding chemicals that should be properly disposed of, such as pesticides. To have an idea of what the chemicals are at these sites, careful measurements will have to be made and scientific studies performed. From these studies, we can determine the type and extent of contamination from heavy metals to exotic compounds, depending upon the type of business that was conducted at the site. This may even lead to identifying the culprits (or at least finding out how the chemicals found their way to these sites).
Creek odors	Hydrogen sulfide, rotten egg smell, dead fish smell (amines)	Odors themselves are a nuisance and threat to the quality of life in neighborhoods. They can also be indicators of chemical and biological contamination, such as sewer overflows, chemical dumping into the creek, and residential treatment system failures.
Cement plants	Dust, inks, solvents	At least two types of cement plants exist: (1) mixing operations and (2) rotary kilns. Mixing plants load material into trucks and process the cement for customers, so a lot of dust can be stirred up and find its way to the neighborhoods. Rotary kilns actually manufacture the cement at high temperatures, so they need fuels and sometimes

TABLE 13.1 *(continued)*

Source Type Identified by Community	*Common Names[10] of Analytes to Be Measured*	*Reason for Possible Concern*
		burn toxic wastes that have high caloric values (i.e., the wastes are used as fuel). The release of pollutants can be from spillage as well as from incomplete combustion. In fact, if combustion does not take place properly, the hazardous fuels can become much more hazardous; such as when dioxins and furans are released.

The table likely will subsequently be expanded with columns for actual analytes (e.g., rather than plasticizers, a specific set of phthalates will be targeted; and instead of metals, species of concern, such as hexavalent Cr or methylmercury will be mentioned).

can be shared with all parties following preliminary meetings or interviews. It can be developed with an eye toward actual sampling and analyses to be performed in the future. In the meantime, environmental professionals can be reviewing literature, conducting intensive laboratory studies, or engaging in other activities to help to anticipate the technical needs and objectives of the environmental assessment. These two tasks must intersect at some point, but long before any reports are presented to the clients, and certainly before public pronouncements of "bad" or "good" environmental quality. A possible flow of activities following an inclusive assessment process is shown in Figure 13.1.

The track record of environmental professionals in assessing and communicating risks is spotty, but arguably improving. Unfortunately, one size does not fit all. Certain subpopulations have experienced inordinate disenfranchisement, hostility, and exclusion by the classic "complaint-based" system for environmental assessment and response.[11] These groups are unlikely to place much trust in government agencies, businesses, and professionals, unless hard work is done up front to earn that trust.

One important element of trust is that of the need for informed consent for anyone participating in an environmental study or assessment. While the use of human subjects is more rigorously regulated within the research community (see Figure 13.2 and Table 13.2), the lessons apply to any scientific endeavor. People need to know what they are being asked to do in the assessment. Full disclosure and honesty are always key ingredients to any worthy study.

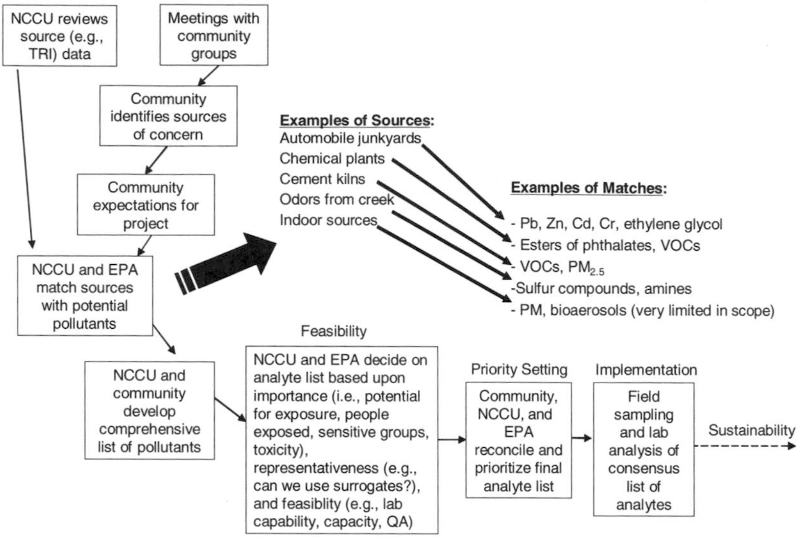

FIGURE 13.1. Recommended process for incorporating community input into an environmental assessment process, North Carolina Central University (Recommended by D.A. Vallero, 2003).

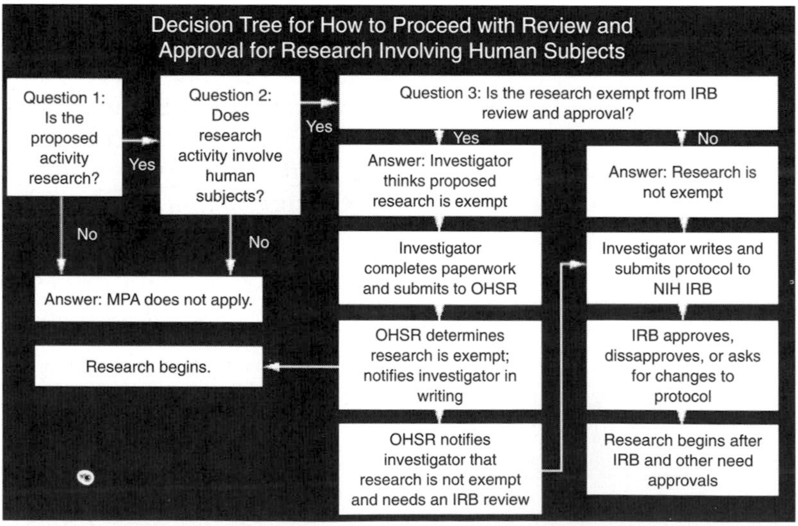

FIGURE 13.2. Steps to be taken in any study involving human subjects.
Note: MPA is a Multiple Project Assurance, or an institution's systematic investigation designed to develop or contribute to generalizable knowledge. (Source: National Institutes of Health System for Human Subjects and the Need for an Institutional Review Board [IRB] Review; and Office of Human Subjects Research [OHSR], U.S. Department of Health and Human Services, 2003, http://ohsr.od.nih.gov/info.)

TABLE 13.2
Institution Review Board (IRB) Protocol Reviews Standards

Regulatory Review Requirement	*Suggested Questions for IRB Discussion*
1. The proposed research design is scientifically sound and will not unnecessarily expose subjects to risk.	(a) Is the hypothesis clear? Is it clearly stated? (b) Is the study design appropriate to prove the hypothesis? (c) Will the research contribute to generalizable knowledge, and is it worth exposing subjects to risk?
2. Risks to subjects are **reasonable** in relation to anticipated benefits, if any, to subjects, **and** the importance of knowledge that may reasonably be expected to result.	(a) What does the IRB consider the level of risk to be? (See risk assessment guide on back of form.) (b) What does the PI consider the level of risk/discomfort/inconvenience to be? (c) Is there prospect of direct benefit to subjects? (See benefit assessment guide on back of form.)
3. Subject selection is equitable.	(a) Who is to be enrolled? Men? Women? Ethnic minorities? Children (rationale for inclusion/exclusion addressed)? Seriously ill persons? Healthy volunteers? (b) Are these subjects appropriate for the protocol?
4. Additional safeguards required for subjects likely to be vulnerable to coercion or undue influence.	(a) Are appropriate protections in place for vulnerable subjects, e.g., pregnant women, fetuses, socially- or economically-disadvantaged, decisionally impaired?
5. Informed consent is obtained from research subjects or their legally authorized representative(s).	(a) Does the informed consent document include the eight required elements? (b) Is the consent document understandable to subjects? (c) Who will obtain informed consent (PI, nurse, other?) and in what setting? (d) If appropriate, is there a children's assent? (e) Is the IRB requested to waive or alter any informed consent requirement?
6. Subject safety is maximized.	(a) Does the research design minimize risks to subjects? (b) Would use of a data and safety monitoring board or other research oversight process enhance subject safety?
7. Subject privacy and confidentiality are maximized.	(a) Will personally identifiable research data be protected to the extent possible from access or use? (b) Are any special privacy and confidentiality issues properly addressed, e.g., use of genetic information?

TABLE 13.2 *(continued)*

Additional Considerations	
1. Ionizing radiation.	If ionizing radiation is used in this protocol, is it medically indicated or for research use only?
2. Collaborative research.	Is this domestic/international collaborative research? If so, are single project assurances (SPAs) or other assurances required for the sites involved?
3. FDA-regulated research	Is an investigational new drug (IND) or investigational device exemption (IDE) involved in this protocol?
4. Other	

See the Code of Federal Regulations (CFR), Title 45, Public Welfare, Part 46, Protection of Human Subjects, revised November 13, 2001, effective December 13, 2001 at http://ohrp.osophs.dhhs.gov/humansubjects/guidance/45cfr46.htm.

The table lists the eight elements from §46.116 required in each consent document. Element 6 is only needed if the research is determined to be greater than minimal risk. The table also includes six elements that must be provided to each human subject when appropriate.

Element	**45 CFR 46.116(a)**
1	A. a statement that the study involves research B. an explanation of the purposes of the research C. the expected duration of the subject's participation D. a description of the procedures to be followed E. identification of any procedures which are experimental
2	a description of any reasonably foreseeable risks or discomforts to the subject
3	a description of any benefits to the subject or to others which may reasonably be expected from the research
4	a disclosure of appropriate alternative procedures or courses of treatment, if any, that might be advantageous to the subject
5	a statement describing the extent, if any, to which confidentiality of records identifying the subject will be maintained
6	A. an explanation as to whether any compensation is available if injury occurs B. an explanation as to whether any medical treatments are available if injury occurs, and, if so C. what they consist of or where further information may be obtained
7	A. an explanation of whom to contact for answers to pertinent questions about the research B. an explanation of whom to contact for answers to pertinent questions about the research subjects' rights C. whom to contact in the event of a research-related injury to the subject
8	A. a statement that participation is voluntary B. refusal to participate will involve no penalty or loss of benefits to which the subject is otherwise entitled C. the subject may discontinue participation at any time without penalty or loss of benefits to which the subject is otherwise entitled

Additional elements (45 CFR 46.116(b) of informed consent (when appropriate, one or more of the following elements of information shall also be provided to each subject):

1	a statement that the particular treatment or procedure may involve risks to the subject (or to the embryo or fetus, if the subject is or may become pregnant) which are currently unforeseeable
2	anticipated circumstances under which the subject's participation may be terminated by the investigator without regard to the subject's consent
3	any additional costs to the subject that may result from participation in the research

Table 13.2 Notes continued

4	A. the consequences of a subject's decision to withdraw from the research B. procedures for orderly termination of participation by the subject
5	a statement that significant new findings developed during the course of the research which may relate to the subject's willingness to continue participation will be provided to the subject
6	the approximated number of subjects involved in the study

These are the minimal regulatory requirements for IRB review, discussion, and documentation in the meeting minutes.

Source: Office of Human Subjects Research [OHSR], U.S. Department of Health and Human Services, 2003, http://ohsr.od.nih.gov/info.

Professional trust can only be gained and cultivated through mutual respect among all parties. This goes beyond the avoidance of condescension and requires genuine inclusiveness of many and varied perspectives.

Notes and Commentary

1. Environmental engineering is a subdiscipline of civil engineering in the United States.
2. American Society of Civil Engineers, 1996, Code of Ethics, Adopted 1914 and most recently amended November 10, 1996, Washington, D.C.
3. National Research Council, 1983, *Risk Assessment in the Federal Government,* National Academy of Sciences, Washington D.C.
4. In fact, the touchstone of William Ruckelshaus's return as Administrator of the U.S. EPA was to disentangle risk assessment from risk management. He saw that the simultaneous attention to assessing and managing risks put science and engineering at a disadvantage to the political, economic, and other social factors embodied in risk management. Thus, upon his return, he declared that the science (risk assessment) would be done as independently as possible. This science, in turn, would be one of the crucial components of managing risks, but not the only one.
5. Vincent Covello, 1992, "Risk Comparisons and Risk Communications," in *Communicating Risk to the Public,* edited by Roger E. Kasperson and P. Stallen, Kluwer, New York, N.Y.
6. The FQPA was enacted on August 3, 1996, to amend the Federal Insecticide, Fungicide, and Rodenticide Act (FIFRA) and the Federal Food, Drug, and Cosmetics Act (FFDCA). Especially important to risk assessment, the FQPA established a health-based standard to provide for a reasonable certainty of no harm for pesticide residues in foods. This new provision was enacted to assure protection from unacceptable pesticide exposure and to strengthen the health protection measures for infants and children from pesticide risks.
7. A very interesting development over the past decade has been the increasing awareness that health research has often ignored a number of polymorphs or subpopulations, such as women and children, and is plagued by the so-called

"healthy worker" effect. Much occupational epidemiology has been based upon a tightly defined population of relatively young and healthy, adult, white males who had already been screened and selected by management and economic systems in place during the twentieth century. Also, health studies have tended to be biased toward adult, white males even when the contaminant or disease of concern was distributed throughout the general U.S. population. For example, much of the cardiac and cancer risk factors for women and children have been extrapolated from studies of adult, white males. Pharmaceutical efficacy studies had also been targeted more frequently toward adult males. This has been changing recently, but the residual uncertainties are still problematic.

8. For an excellent and thorough introduction to emerging paradigms for dealing with environmental problems in disadvantaged communities, see Chapter 6, "Communities of Color Respond to Environmental Threats to Health: The Environmental Justice Framework," in R. Braithwaite, S. Taylor, and J. Austin, 1995, *Building Health Coalitions in the Black Community*, Sage Publications, London, U.K.
9. Commission for Racial Justice, United Church of Christ, 1987, *Toxic Wastes and Race in the United States.*
10. Note: Mention of any compound in this table does not mean that it has been found. The compounds are simply those that have been found in similar source types to those listed here. Every source is unique, so a listed chemical may not be present, and even if it is, it may not have been or ever will be released into the environment.
11. Two examples come to mind. The historical disenfranchisement of African-American communities has been reflected in many examples of environmental racism, whether intended or ancillary to some other priority, such as the PCB landfill siting decisions in Warren County, North Carolina in the 1980s. Also, migrant workers, frequently Latinos and Latinas, experience inordinate exposures to contaminants during agricultural and husbandry operations, such as combined animal feeding operations (CAFOs). In addition to exposures in the upper deciles, fears of reprisals or immigration concerns may take precedence over environmental and public health initiatives. Thus, problems will be underreported or unreported, and such highly exposed subpopulations do not want to risk dealing with the government when their citizenship and work permit status is vulnerable.

CHAPTER 14

Epilogue: Benzene Metabolism Revisited

In the Prologue, we parsed a seemingly simple question. To paraphrase: Is there a factor that accounts for the physical, chemical, and biological reasons that in the field we never experience what has been found in the lab for the degradation of the simplest aromatic compound, benzene? What is it that makes contaminant control so different from the theoretical reaction, at least in terms of stoichiometry? Recall that the reaction is:

$$C_6H_6 + 7.5O_2 \rightarrow 6\,H_20 + 6\,CO_2 + \text{microbial biomass}$$

Reaction 14–1

Although I have endeavored to introduce the many factors that account for the nonlinear nature of environmental contamination and remediation, I must admit that this question has not been answered, at least not in a fully satisfactory manner. Have we failed? Emphatically, no! This simply demonstrates the iterative and step-wise manner of science and inquiry in general. It also particularly highlights the complexity of environmental science.

Certainly, we are better equipped to explain this "simple" chemical reaction. It would be unacceptable to expect the microbes to grow as if they were all exposed to ideal amounts of nutrients and oxygen. Our understanding of gas transfer, especially that of oxygen, helps to explain why only a portion finds its way to the microbes.

Partitioning equations, especially sorption, octanol-water, and Henry's Law constants, explain the reasons that the particles, in this case both the soil and the microbe, vary in their affinity to aqueous and organic phases in water, as well as those of other fluids, especially air versus water (air-water partitioning and Henry's Law). We have also seen that the fundamental chemical reactions themselves are not all that simple. The energy transfer, or bioenergetics, of the microbes highly depends—but not exclusively—on oxidation-reduction (electron donations and acceptances). However, they also are affected by ionization and acid-base relationships, which in turn affect solubility and precipitation reactions.

The biological reactions of growth, metabolism, and reproduction are even more complicated, or at least less certain than purely physicochemical relationships, since microbiological reactions are propagated from the physical and chemical principles. Again, the physics and chemistry are necessary but not sufficient to explain the biology. Thus, the benzene biodegradation is mediated by cellular responses, such as ligand-receptor relations (e.g., the Ah receptor on the microbial cell wall).

The Sensitivity Analysis: An Important Step Beyond Stoichiometry

Why are environmental clean up actions so often more complicated and unpredictable than laboratory studies might indicate they should be? Most scientists have learned to rely on stoichiometry, i.e., the quantities of substances entering into and produced by chemical reactions. We know that when methane combines with oxygen in complete combustion, 16g of methane require 64g of oxygen, and simultaneously 44g of carbon dioxide and 36g of water are produced by this reaction. We know that every chemical reaction requires that all elements in the reaction must be in specific proportions to one another. Certainly, environmental problems cannot ignore the stoichiometry. As Aristotle might say, stoichiometry is a necessary but an insufficient explanation of environmental contamination and remediation. We can be quite certain that when benzene reacts, that it will abide by the stoichiometry of Reaction 14–1. Yet, we know that in the real world that not all of the benzene reacts, even when it seems that there is plenty of oxygen. How then, do we begin to understand some of the other factors controlling the rate and extent of abiotic chemical and biological degradation processes?

One way is to look to the models used by engineers and scientists to predict the fate of a contaminant in the environment. Such models are used widely to predict the spatial and temporal extent of pollution. As can be seen from our benzene discussions in the Prologue, scientists are always concerned with just how certain they need to be about data and information. An important means used by modelers to determine, even quantify, the certainty of data and information needed to make a decision is the sensitivity analysis.

Every reaction in this book is, in fact, a model. A model is simply a system that represents another system, with the aim of helping to explain that target system. Reaction 14–1 explains just how many moles of oxygen are needed to react with benzene to form carbon dioxide and water. It is very robust and sensitive. Any change to the left-hand side will result in concomitant changes in the right-hand side. In fact, the way it is written, the only limits on how much CO_2, water, and microbes that will be generated is the amount of oxygen and hydrocarbons (benzene) that are available.

Of course, the model (reaction) does not show every variable influencing this reaction. Even if one were to pump large volumes of O_2 into an aquifer, it will speed up the degradation of the hydrocarbons but it will still not be immediate. Such a system has a surplus of oxygen, i.e., it is not oxygen limited. Neither is it hydrocarbon limited. But, since these are the only two reactants, how can that be?

Obviously, other factors come into play. For example, the "+" indicates that the two reactants must be in contact with one another, but does not show how abruptly or slowly this contact occurs. Scale and heterogeneity must be factors. So, even if the macro-environment is at an oxygen surplus, the place where the microbes live (e.g., the film around particles) may be oxygen deficient. Or, there may be discontinuities between individual particles, so that some pockets may have very efficient biodegradation, but others are isolated from water, oxygen, and substrate (including the benzene) and the microbes are not happy! A simple stoichiometric model does not describe the conditions away from the tightly controlled laboratory, nor does it give a rate at which the reaction occurs. The stoichiometric model does tell us that biomass will also be produced, but is less specific about them than the abiotic parts of the model. The actual number and species of microbes will vary considerably from place to place. So, our reaction model is very good at expressing exactly how many moles will react and how many moles will be produced, but does not indicate many important conditions and variables outside of a strictly controlled laboratory.

It stands to reason, then, that we might learn something about the complexities and uncertainties in actual environmental problems by deconstructing some of the more complex models in use today. For example, if a model is being used to estimate the size of a plume of a contaminant in groundwater, a number of physical, chemical, and biological variables must be considered. Modelers refer to such variables as model parameters. So, an engineer or hydrologist interested in how far a plume extends and the concentrations of a contaminant within the plume must first identify hydrogeological parameters like aquifer thickness, porosity, transverse and longitudinal dispersivity,[1] source strength and type, recharge of the aquifer, as well as chemical parameters like sorption and degradation rates.

But not all parameters are created equally. Some have major influence on the result with even a slight change while others can change significantly with only a slight change in the result. In the former situation, the result is said to be highly sensitive to the parameter. In the latter, the result is considered to be nearly insensitive. If a result is completely insensitive, the parameter does not predict the outcome at all. This occurs when a parameter may be important for one set of microbes or one class of chemicals (i.e., sensitive), but when the model is used for another set of microbes or chemicals it is completely insensitive. For example, aerobic bacteria may grow according to predictions of an oxygenation parameter in a model, but the same parameter is unimportant in predicting the growth of anaerobic bacteria. What the engineer and scientist want to find out is how much change is induced in a parameter per unit of perturbation. In other words, if the modeled results change 50% with a unit change to parameter A, but change only 5% with the same unit change to parameter B, one could characterize the model as being 10 times more sensitive to parameter A than to parameter B. Modelers are interested in the very same things that we discussed at the beginning of this book. Which variables and parameters limit the change of contaminant concentrations? If we know this, then we can optimize environmental cleanup.

To illustrate environmental parameter sensitivity let us consider a model that has been used by environmental scientists to predict contamination under various scenarios; i.e., the natural attenuation prediction model, Bioplume III. This U.S. EPA model has been subjected to a sensitivity analysis for hydrogeological, physicochemical, and biological parameters. And, pertinent to this discussion, it has been subjected to tests to see just how sensitive benzene contamination is to changes in these parameters.[2]

The Bioplume III sensitivity analysis evaluated five hydrogeological parameters and two chemical parameters:

1. Porosity of the soil or other media;
2. Thickness of the aquifer;
3. Transmissivity[3] of the aquifer;
4. Longitudinal dispersivity;
5. Horizontal dispersivity;
6. Sorption (indirectly indicated by a retardation factor, Rf);[4] and
7. Radioactive decay (as an analog to abiotic chemical half-life).

To test these parameters, the Bioplume model hypothesizes a base case with the characteristics shown in Table 14.1.

The parameters were manipulated to determine the difference in results (i.e., benzene concentrations in the plume) between the base case and other scenarios. These results are shown in Table 14.2. The two most influential hydrogeological parameters on benzene concentrations in the plume appear to be thickness of the aquifer and transmissivity. Benzene concentrations appear to be sensitive to both of the chemical parameters.

In addition the model predicts biodegradation by applying two methods:

1. An overall first-order decay rate that simulates aerobic and anaerobic degradation processes (see Chapter 12) and
2. Specification of background electron acceptor concentrations in the groundwater, matched to a kinetic model (see *Biological Redox Reactions* section in Chapter 7), and a kinetic model is matched to these electron acceptor levels.

TABLE 14.1
Base Case Conditions for Bioplume III Natural Attenuation Model

Characteristic	*Value*
Grid Size	9×10
Cell Size	900 ft × 900 ft
Aquifer Thickness	20 ft
Transmissivity	$0.1\ ft^2 s^{-1}$
Porosity	30%
Longitudinal Dispersivity	100 ft
Transverse Dispersivity	30 ft
Maximum Cell Distance per Movement	0.5
Simulation Time	2.5 yrs
Source and Loading of Contamination	1 injection well @ 0.1 cfs
Contaminant Concentration at Release	$100\ mg\ L^{-1}$
Recharge	0 cfs
Boundary Conditions	Constant head, upgradient and downgradient
Chemical Reactions	None
Biodegradation Reactions	None

Source: U.S. Environmental Protection Agency, 2003, *Bioplume III Natural Attenuation Decision Support System, Users Manual*, Version 1.0, Washington, D.C.

The modeled benzene concentrations in the plume were found to be quite sensitive to biodegradation. Interestingly, however, the benzene concentrations were relatively insensitive to changes in molecular oxygen and only slightly sensitive to the electron acceptor concentrations (i.e., in addition to O_2 in aerobic systems, the model evalutates the anaerobic electron acceptors NO_3, Fe, SO_4, and CO_2). All other things being equal, microbes with the most efficient metabolic mechanisms grow at the fastest rate, so these organisms will overwhelm the growth of microbes with less efficient redox systems. Thus, if O_2 is available in surplus, this will be the preferred reaction in the model. Once a system becomes anaerobic, nitrate is the most preferred redox reaction, followed by solid phase ferric iron, sulfate, and carbon dioxide (the least preferred redox reaction).

A thermodynamically dicated system would give preference, even exclusivity, to the reaction that provides the most energy, so the model uses a sequential process that does not allow the microbes to use any other less preferred electron acceptor until the more preferred acceptor is depleted. However, in reality, when monitoring wells are analyzed near plumes undergoing natural attenuation (i.e., active biodegration), they are seldom entirely depleted in one or more of these electron acceptors. There are seldom such "bright lines" in the field. For example, facultative aerobes, those that can shift from oxygen to anaerobic electron acceptors (especially nitrate), can change electron acceptors even when molecular oxygen is not completely depleted. This can be attributed to the fact that redox potentials

TABLE 14.2
Sensitivity of Benzene Concentrations of Plume to Hydrogeological and Chemical Parameters in the Bioplume III Model

Parameter	*Value (* base case)*	*Maximum Benzene Concentration in Plume ($mg L^{-1}$)*	*Plume Length (number of cells)*	*Plume Width (number of cells)*
Porosity	15%	75	6	5
	30%*	67	4	3
	45%	80	4	3
Aquifer thickness (ft)	10	75	6	5
	20*	67	4	3
	40	47	2	2
Transmissivity ($ft^2 s^{-1}$)	0.01	90	3	3
	0.1*	67	4	3
	0.2	57	5	3
Longitudinal dispersivity (ft)	10	70	3	3
	50	69	4	3
	100*	67	4	3
Transverse dispersivity (ft)	10	68	4	3
	30*	67	4	3
	60	66	4	3
Retardation factor	1*	67	4	3
	2	49	3	2
	5	28	2	1
Abiotic chemical half-life (s)	0*	67	4	3
	1×10^7	20	2	2
	2×10^7	33	2	3

Source: U.S. Environmental Protection Agency, 2003, *Bioplume III Natural Attenuation Decision Support System, Users Manual*, Version 1.0, Washington, D.C.

for oxygen and nitrate are not substantially different (at pH 7, $O_2 = +820$ volts and $NO_3 = +740$ volts, compared to $CO_2 = -240$ volts). Also, the apparent divergence from pure thermodynamics in the field may simply be a sampling artifact, which can be attributed to the way monitoring is conducted. For example, monitoring wells do not collect water from a "point." Rather, the screens (the perforated regions of underground piping where water enters) are set at 1.5 to 3 m intervals, so waters will mix from different vertical horizons. Thus, if different reactions are occuring with depth, these are actually aggregated into a single water sample.

The sequencing of electron acceptors is akin to the rate limitation phenomena discussed in Chapter 4. That is, when a contaminant degrades sequentially, the slowest degradation step has the greatest influence on the time it takes the chemical to break down. If this most sensitive step can be sped up, the whole process can be sped up. Conversely, if an engineer or sci-

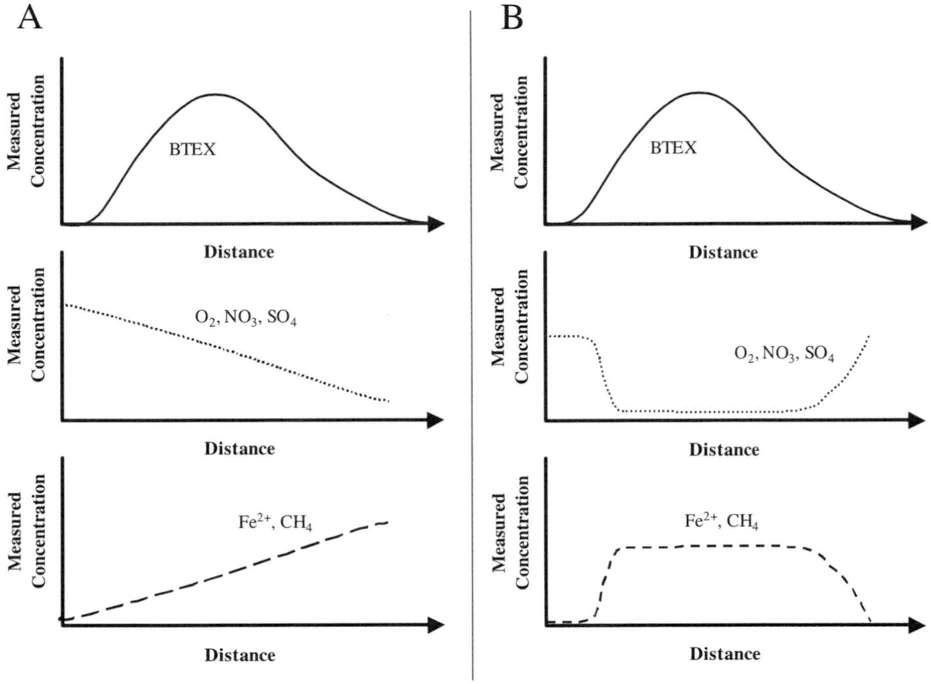

FIGURE 14.1. Two possible hypotheses for how microbes degrade benzene, toluene, ethyl benzene, and xylenes (BTEX): A. Rate of biodegradation is limited by microbial kinetics: Concentrations of anaerobic electron acceptors (nitrate and sulfate) decrease at a constant rate downgradient from the pollutant source, with a concominant increase in the concentrations of the by-products of these anaerobic reactions (ferrous iron and methane); and, B. Rate of biodegradation is relatively fast (days, not years, so compared to many groundwater replenishment rates, this can be characterized as instantaneous): Virtually all of the nitrate and sulfate anaerobic electron acceptors are depleted, while the iron and methane by-products of these anaerobic reactions show the highest concentrations near the contaminant source. In both A and B the total concentrations of the by-products are inversely related to the total concentrations of the principal electron acceptors in the anaerobic reactions overall. (Adapted from U.S. Environmental Protection Agency, 2003, *Bioplume III Natural Attenuation Decision Support System, Users Manual,* Version 1.0, Washington, D.C.)

entist devotes much time and effort to one of the faster steps in the degradation sequence, little or no enhancement to the degradation process may occur. Thus, the model seems to point to the need to take care to avoid, or at least not to overgeneralize, the common assumption that a contamination plume is limited by oxygen or even other redox conditions. Adding iron to an anaerobic system or pumping air into an aerobic stratum of an aquifer will help, but only so much. Figure 14.1 demonstrates a way to apply microbial kinetics limits to redox.

Another difference between the lab and the field is the presence of confounding chemical mixtures in real contamination scenarios. For example, leaking underground storage tanks (LUSTs) are a widespread problem. It is tempting to think that since these tanks contain refined fuels, that most spills will be similar. However, as we have discussed throughout this text, each compound has specific physicochemical properties that will affect its reactivity and movement in the environment. As evidence, benzene, toluene, ethyl benzene, and xylenes (so called BTEX) usually comprise only a small amount (ranging from about 15 to 26%) of the mole fraction of gasoline or jet fuel.[5] However, largely because the BTEX compounds have high aqueous solubilities (152 to 1780 $mg L^{-1}$) compared to the other organic constituents (0.004 to 1230 $mg L^{-1}$) in these fuels, they often account for more than two-thirds of the amount of the contaminants that migrate away from the LUST. Also, soils are seldom homogeneous, so even if the contaminant is well characterized, how it will react and move are largely affected by the media's characteristics. For example, each contaminated soil's sorption partitioning should be documented (e.g., Freundlich isotherms determined) with respect to the specific contaminants.

Interdependencies between a Contaminant and a Substrate

Another precaution in ascertaining why benzene or any other contaminant may behave differently in the environment than is theoretically expected is that complexities and scale effects often come into play. For example, sorption and bioconcentration are two equilibrium phenomena that are not explicitly inherent to the contaminant, but are, in fact, functions of the physicochemical properties of the contaminant and those of the substrate, i.e. the particle and organism, respectively. Recall that the soil partition coefficient (K_d) is the experimentally derived ratio of a contaminant's concentration in the solid matrix to the contaminant concentration in the liquid phase at chemical equilibrium. The K_d is also called the distribution coefficient because of the relative affinity to sorb to solid particles is an important factor in its potential to leach, especially through soils. The other frequently reported liquid to solid phase partitioning coefficient is the organic carbon partitioning coefficient (K_{oc}), which is the ratio of the contaminant concentration sorbed to organic matter in the matrix (soil or sediment) to the concentration in the aqueous phase. Thus, the K_{oc} is derived from the quotient of a contaminant's K_d and the fraction of organic matter (OM) in the matrix:

$$K_{oc} = \frac{K_d}{OM} \qquad \text{Equation 14–1}$$

So, the K_{oc} will vary from soil to soil. For example, TCDD has been shown to have a mean log K_{oc} of 7.39 for ten contaminated soils collected from New Jersey and Missouri, but its partitioning in other soils must also be experimentally determined. To illustrate the importance of how the substrate or medium can strongly influence the sorption of a contaminant, Figure 14–2 shows the log K_{oc} values for the PAH, phenanthrene can range nearly three orders of magnitude, depending on the matrix. The figure also points to some promising methods to collect and treat persistent contaminants, since granulated activated carbon has such a strong affinity to PAHs.

The variability and interdependencies are also demonstrated in the differences in theoretically expected sorption based on chemical structures and inherent molecular properties from those observed in the field. These values from Table 5.1, shown graphically in Figure 14.3, demonstrate that while the expected affinity for organic carbon is near that of the calculated K_{oc} values, they are certainly not identical ($R^2 = 0.97$). The outliers in the figure indicate the importance of soil type, e.g. hexachlorobenzene's calculated K_{oc} is about 55,000 but its measured value was 80,000, and pentachlorobenzene's calculated K_{oc} is about 17,000, but its measured K_{oc} was about 32,000. Such variability may help to explain variations in the field.

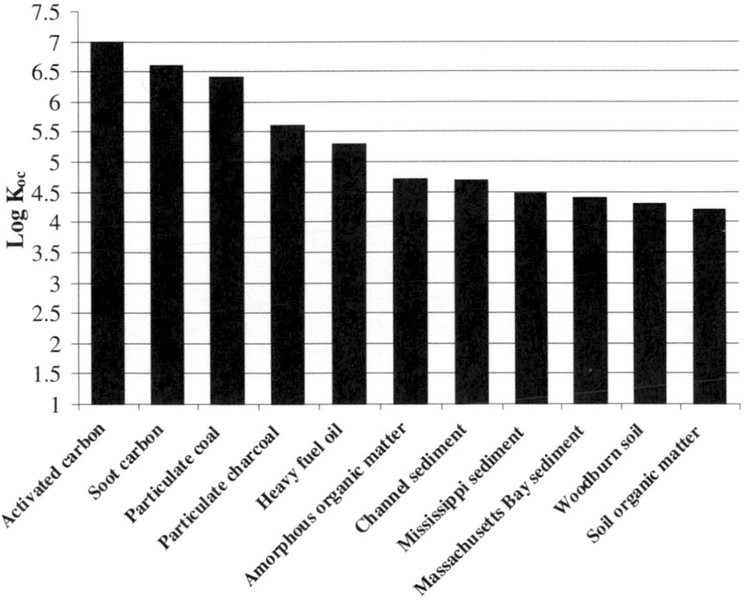

FIGURE 14.2. Comparison of reported values of log K_{oc} for phenanthrene sorption on different types of organic matter found in soils and sediments. (Source: S.W. McNamara, R.G. Luthy, and D.A. Dzombak, 2002, Bioavailability and Biostabilization of PCBs in Soils, NCER Assistance Agreement Final Report, U.S. Environmental Protection Agency, Grant No. R825365-01-0.)

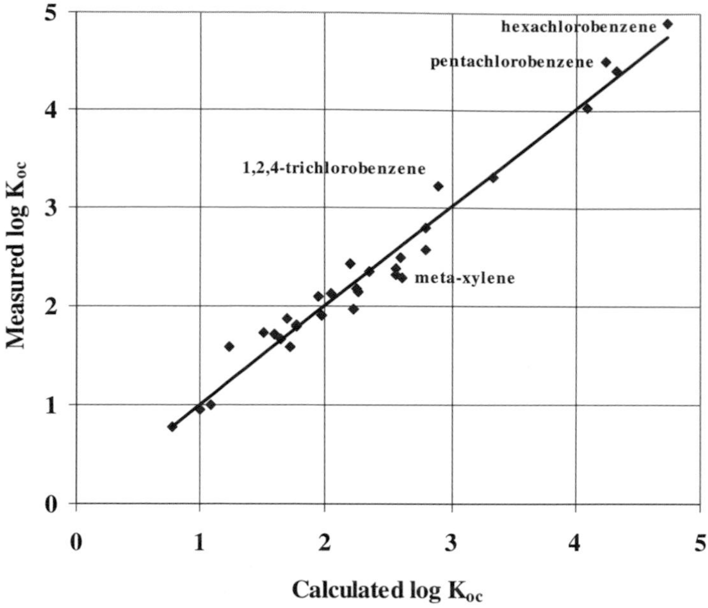

FIGURE 14.3. Comparison of calculated and experimental organic carbon partitioning coefficients for contaminants listed in Table 5.1.

The bioaccumulation and bioconcentration of a contaminant is also highly dependent upon the genus, age and other characteristics of the organism that has been exposed to the contaminant. Usually, the BCF reflects uptake and depuration results from water in laboratory experiments. For example, BCFs for mercury can vary significantly, with the highest factors determined for methyl mercury. BCFs for methyl mercury in brook trout range from 69,000 to 630,000, depending on the tissue analyzed, and BCFs for inorganic mercury (mercuric chloride) in saltwater species range from 129 in adult lobsters (*Homarus americanus*) to 10,000 in oysters (*Crassostrea virginica*).[6] Therefore, expectations of kinetics and equilibrium cannot always reliably be predicted from published partitioning coefficients unless the substrates are also well characterized.

These contaminant-substrate interdependencies are examples of how complexities in reactions and environmental systems can dictate the fate of a contaminant. For example, knowing the molecular structure of the specific chemical species and its phases tells much about the complexity of the reaction, but to understand the whole environmental system, one must determine the spatial and temporal scales, along with the initial and boundary conditions of the system. So, the phenomena that are occurring at the microscopic scale may be different from those at a larger scale, e.g. conditions around a clay particle may differ in degree and kind from what is occurring in the entire soil column. And, the specific soil system being monitored may not reflect much at all about all the soil that has been contaminated, due to the normal (or abnormal) heterogeneity of soils. So, the chemical reactions increase in complexity from those involving elementary reactions to highly complex, multiple pathway reactions, such as those involving biodegradation. Environmental complexities, then, can be driven by the components of a reaction and by scale. Components, including contaminant concentrations and substrate properties, help to describe the relative reaction complexities, while scale helps to define the complexity of the whole system. Therefore, if what we know about benzene degradation is predominantly learned from laboratory experiments at the micro-scale,[7] and from empirical and experiential knowledge gained from mega-scale black boxes, it may not be so surprising that we do not know more about the midrange and macro-scale phenomena.

Ease of implementation and sensitivity are both important considerations when deciding how to address environmental contamination. Unfortunately, steps that are readily available may be relatively insensitive to the intended outcome. Sometimes, however, immediate and relatively inexpensive measures can be taken that are sensitive, such as pumping air and water to speed up biodegradation in an aquifer that has already shown natural attenuation. This is analogous to the business world concept of "low hanging fruit." Managers are encouraged to make improvements that are relatively easy and that pay immediate dividends, before moving on to the more intractable problems. For example, if a survey shows that employees are unhappy and have low morale because a current policy does not allow them to eat their lunches at their desks, and no good reason can be found for this policy, a manager can simply change the policy at no cost to the company and reap immediate results. However, if the same survey showed that everyone in the organization needs to be retrained at considerable costs to the company, this would call for a more thoughtful and laborious correction pathway. The former improvement (i.e., eating at one's desk) may not greatly affect the bottom line, but it is easy to implement. The latter improvement (training) may greatly influence the bottom line (i.e., profit is more sensitive to a well-trained workforce), but it is difficult to implement. So, it may be that benzene degradation is highly sensitive to soil type, but there may be little that the engineer can do about it, i.e., soil type is a sensitive parameter, but almost impossible to change (outside of complete soil removal, which has been done for severely contaminated sites).

Thus, the sound wisdom displayed by my three colleagues expressed in the Prologue now rings true. Their advice was:

1. Start with first principles. Remember that oxygen is not very soluble in water, so base any estimation on how much finds its way to the

microbe on scientifically sound information. Also, keep in mind that in the field, critically important assumptions in most laboratory studies, like complete mixing, are not available in the field.

2. Admit what you do not fully understand. However, do not get "bogged down" in seeking perfect understanding, or succumb prematurely to the law of diminishing returns. That is why the age-old practice of the "black box" is so vital to many remediation efforts, as it is to other heuristic endeavors, such as recycling. In other words, if you explain why 90% of the microbes or alligators or people behave the way they do, it may be okay for now, but ensure that the research community is working on explanations for the unexplained 10%.
3. Finally, the universally accepted "right answer" to environmental questions is "it depends." Field situations include so many possible uncertainties and variabilities that it would be unwise to expect a remediation project, such as benzene cleanup in soils, to behave completely as the lab studies predict. However, if the process is so different (e.g., the amount of benzene is not decreasing to any measurable extent), such inconsistency is reason to investigate first principles. For example, if the pumps are not delivering any water to the microbes, there is no possibility of their growth, and hence no biodegradation. Thus, if the trend is in the right direction, but the rates are lower (and sometimes higher if we are fortunate), then we may have a sufficient understanding of the processes and mechanisms and do not need to worry so much about the "noise" (i.e., slight differences between expected versus observed degradation rates). It should be some consolation to engineers, for example, that even highly controlled laboratory studies of partitioning coefficients differ from theoretically predicted coefficients based upon chemical structures of the compound (e.g., the differences in some of this and other books' table columns for calculated versus empirical values for organic carbon and Henry's Law coefficients).

Thus, we should not be dismayed by the inability to explain fully all of the processes that make for contaminant transport, transformation, and fate. But, as scientists, we should continue to work toward decreasing uncertainty in environmental decisions. This book has addressed the complicated and interdisciplinary nature of environmental contamination. Good data make for reliable information. Reliable information adds to scientific and societal knowledge. Knowledge, with time and experience, leads to wisdom (Figure 14.4). Environmental assessment and protection need to include every step in the "wisdom cascade"[8]:

The means for addressing contamination of the environment have changed significantly in the past 50 years. One of the keys to the many successes has been our ability to work with natural processes. In fact, many of

Concerns and Interests

↓

Data

↓

Information

↓

Knowledge

↓

Wisdom

FIGURE 14.4. The Wisdom Cascade.

the so-called breakthroughs have simply been accelerated or expanded versions of natural systems that have been very effective in dealing with wastes. Such natural processes can be seen in energy transfer systems between trophic levels in ecosystems, including the oxidation-reduction systems of decomposers. This in no way diminishes the absolutely astounding successes gained by environmental scientists and engineers. In fact, the successes have been in no small part because of the observation skills of these pioneers. To paraphrase one of these innovators, if one is looking for a solution to an environmental problem, one only need look under one's feet.[9] He meant that literally. If we need to degrade benzene, dioxins, or whatever, there is a good chance that the bacteria, fungi, plankton, or other "chemical degradation factories" are ready to help.

That is not to say that it is easy. For example, getting just the right recipe for success requires a systematic approach. The lab work must complement the field studies. Each of these must be used to improve models. The bottom line is that we are in the business of closing gaps; gaps in data, gaps in knowledge, and gaps in science. Engineers must stay abreast of these developments. Like most professions, environmental science and engineering is a life-long endeavor. Receiving the degree is the gateway, but the career requires us to be continuously honing our skills (in fact, it is one of the canons of the engineering profession). We must be prepared for the new challenges that await us, not forgetting what we learned from those before us, yet being willing to think anew about how to address environmental contaminants.

The bad news is that everyday we are challenged with new and previously unexpected sources and risks from contaminants. The good news is that we are ever growing in wisdom in how to address these risks. Let us continue in our noble pursuit!

Notes and Commentary

1. Dispersivity (D) is defined as the ratio of the hydrodynamic dispersion coefficient (d) to the pore water velocity (v); thus $D = \frac{d}{v}$.
2. For another excellent sensitivity analysis that illustrates the importance of numerous parameters, see J.E. Odencrantz, J.M. Farr, and C.E. Robinson, 1992, "Transport model parameter sensitivity for soil cleanup level determinations using SESOIL and AT123D in the context of the California Leaking Underground Fuel Tank Field Manual," *Journal of Soil Contamination*, 1(2): 159–182. The study found that benzene concentrations are most sensitive to biodegradation rate, climate, effective solubility, and soil organic carbon content.
3. Transmissivity is the rate at which water passes through a unit width of the aquifer under a unit hydraulic gradient. It is equal to the hydraulic conductivity multiplied by the thickness of the zone of saturation. It is expressed as volume per time per length such as gallons per day per foot ($gal\,d^{-1}\,ft^{-1}$) or liters per day per meter ($L\,d^{-1}\,m^{-1}$).
4. Retardation represents the extent to which a contaminant is slowed down compared to if it were transferred entirely with the advective movement of the fluid (usually water). For example, if the water in an aquifer or vadose zone is moving at $1 \times 10^{-5}\,cm\,s^{-1}$, but due to sorption and other partitioning mechanisms the contaminant is only moving at $1 \times 10^{-6}\,cm\,s^{-1}$, the retardation factor (Rf) = 10, so an Rf of 10 means that the contaminant is moving at 1/10 the velocity of the water. Rf is a correction factor that accounts for the degree to which a contaminants' velocity is affected by sorption in the ground water system. An Rf calculation must consider the bulk density of the media, porosity, and the distribution coefficient (K_d).
5. See for example: P.C. Johnson, M.W. Kemblowski, and J.D. Colthart, 1990a. "Quantitative Analysis of Cleanup of Hydrocarbon-Contaminated Soils by In-Situ Soil Venting," *Ground Water*, Vol. 28, no. 3, May–June 1990, pp. 413–429; P.C. Johnson, C.C. Stanley, M.W. Kemblowski, D.L. Byers, and J. D. Colthart, 1990b. "A Practical Approach to the Design, Operation, and Monitoring of In Site Soil-Venting Systems," *Ground Water Monitoring and Remediation*, Spring 1990, pp. 159–178; and M.E. Stelljes, and G.E. Watkin, 1993. "Comparison of Environmental Impacts Posed by Different Hydrocarbon Mixtures: A Need for Site Specific Composition Analysis," in *Hydrocarbon Contaminated Soils and Groundwater*, Vol. 3, P.T. Kostecki and E.J. Calabrese, Eds., Lewis Publishers, Boca Rotan, Fla., p. 554.
6. National Oceanic and Atmospheric Administration, 2004, Toxic Chemicals in Coastal Environments Mercury in Aquatic Habitats, http://response.restoration.noaa.gov/cpr/sediment/mercury.html
7. A demonstration of the importance of scale and complexity in environmental systems was brought home during a recent seminar held by a Duke graduate student regarding research being directed by my colleague, Zbigniew Kabala. The gist of the discussion was how at the small or micro-scale, the geometry

of a conduit can have profound effects on whether flow is laminar or turbulent. In fact, engineers generally expect the flow between laminar and turbulent conditions, i.e. the critical flow region to have Reynolds Numbers >2000 and <4000 (See Chapter 3). Critical flow may also be defined as a flow with velocity = 0 at the walls and twice the average velocity at the center of the conduit (laminar) and a flow with no relationship to the proximity of the wall due to mixing (turbulent). The seminar pointed out that at very low Reynolds number, in small conduits, flows behaved more turbulently than would be expected in larger systems. In fact, the visual demonstration of the flow using dye showed that the size, and especially the shape of the pockets lateral to the flow, changed the critical range substantially, even to the point where a finite amount of the fluid remained in the pockets (adhering to the walls) well after the rest of the flow had moved downstream. In other words, when clear water was sent through the conduit, some of the blue dye remained out of the streamlines. This may help to explain why, even though at the meso- or macro-scale, large amounts of water are used to flush an aquifer, core samples of soil or other unconsolidated material may still contain measurable concentrations of contaminants, even if they have relatively high aqueous solubility, like benzene.

8. Although this cascade is depicted to be entirely sequential, it is also actually a parallel process. New information is gained continuously, as new data are collected. Feedback loops, blind alleys, and "cul-de-sacs" are constantly encountered in any process leading toward new knowledge and, hopefully, greater wisdom. This process has obviously been taking place in environmental science and engineering. Simply compare what is known about wastewater treatment, hazardous waste remediation, air quality control, and drinking water technology in the middle of the 20th century to the present. New challenges must continue to be met by employing what we have been given by our predecessors. However, even with such a solidly proven track record, we must sometimes think anew and come up with solutions that are "outside the box." That is, we must be mindful of what is working and what is falling short. If the "standard" ways of dealing with a particular pollutant are not delivering the necessary results, perhaps it is time to consider a bold and different approach.
9. This is paraphrased from a lecture given by Ross McKinney two decades ago. McKinney, in my opinion, is a technological optimist. At the time, some recalcitrant chemical compounds were not amenable to biological processing. I remember asking McKinney about dioxins and whether they would be able to be broken down on a large scale using microbes. His answer was basically that he could see no reason that they would not, given the right combination of conditions needed to acclimate them to the compounds as an energy source. Time has often proven McKinney correct, as I expect will be the case for dioxins.

Glossary of Environmental Sciences and Engineering Terminology[1]

A horizon The uppermost layer of a mineral soil containing organic matter. This layer has the soil's largest amount of biological activity and removal of soil material by chemical suspension and solution.

Abandoned waste site A hazardous waste site that has been closed and is no longer in operation, and the original owner/operator is no longer in business.

Abandoned well A permanently discontinued well or one in a state of such disrepair that it cannot be used for its intended purpose.

Abatement Amelioration or reduction of degree or intensity of, or eliminating, pollution.

Abatement debris Waste from remediation activities.

Abiotic Not relating to living things, not alive. Opposite of *biotic*.

Abiotic factors Nonliving influences on an organism's functions.

Absolute error In statistics and quality assurance, the difference between the measured value and the true value.

Absorbed dose In exposure assessment, the amount of a substance that penetrates and passes through an exposed organism's absorption barriers (e.g., skin, lung tissue, gastrointestinal tract) through physical or biological processes. The term is synonymous with *internal dose*.

Absorption (1) Penetration and collection of a chemical within a surface of a body (contrast with *adsorption*). A form of sorption. (2) The uptake of chemicals by an organism, making the chemical available to metabolic processes. (3) In soil science, the movement of ions and water into plants. Active soil absorption uses the plant's metabolic processes to remove chemicals, while passive absorption depends upon chemical diffusion.

Absorption barrier Any of the exchange sites of the body that permit uptake of various substances at different rates (e.g., skin, lung tissue, and gastrointestinal tract wall).

Absorption factor The fraction of a chemical that reaches the cells and tissues of an organism. This is one of the factors of the exposure calculation.

Accelerated flow Form of a varied flow where velocity is increasing while the depth is decreasing.

Acceleration Rate of change in velocity with respect to time (m sec^{-2}).

Acceptable daily intake (ADI) The daily dose of a chemical that has been determined by research and scientific investigation to be free from adverse effects in the general human population after a lifetime of exposure. Amount of a substance representing a daily dose that is highly likely to be safe for noncancer effects over an extended time period. Lest strictly defined term than *reference dose*.

Accident site Location of an unexpected occurrence resulting in a release of hazardous materials.

Acclimatization An organism's physiological and behavioral adjustments to changes in its environment.

Accuracy The degree to which a measurement or statistic reflects the true value. More accurate measurements have lower absolute error.

Acid Corrosive compound with the following characteristics: (1) reacts with metals, yielding hydrogen; (2) reacts with a base, forming a salt; (3) dissociates in water, yielding hydrogen or hydronium ions; (4) pH < 7.0; and (5) neutralizes bases or alkaline media.

Acid-catalyzed hydrolysis Enhanced hydrolysis resulting from protonation.

Acid rain Precipitation with pH values below 5.6, usually due to higher than average concentrations of nitric and sulfuric acids formed primarily by nitrogen oxides and sulfur oxides released into the troposphere from fossil fuel combustion. Acid rain may be in the form of wet precipitation (rain, snow, or fog) or dry precipitation (when acidic gases sorb to particulate matter). The pH of normal rain is about 5.6, due to the dissolution of carbon dioxide and the formation of carbonic acid. Also called acid precipitation or acid deposition.

Action level (AL) (1) Concentration of an airborne chemical determined by the Occupational Safety and Health Administration at which any workers exposed to this level must be tested. The AL is frequently calculated as one-half the permitted level of exposure. [See 29 Code of Federal

Regulations (CFR) 1910.1001-1047.] Action levels are thresholds for contamination at which the need for decontamination is required. (2) In the Superfund program, existence of a contaminant concentration in the environment sufficiently high to warrant action or trigger a response.

Activated carbon (AC) Highly adsorbent form of carbon (due to the great amount of particle surface area) used to trap odors and toxic substances from liquid or gaseous emissions. AC is used to remove dissolved organic matter from wastewater and halogenated hydrocarbons and other toxic compounds from drinking water. It is also used in evaporative control systems of motor vehicles.

Activated sludge The suspended solids, predominantly composed of the living biomass of microbes, found in the wastewater treatment plant's aeration tanks. Activated sludges are rich in bacteria.

Activation Toxicological term for rendering a substance more toxic after being transformed biochemically after entering an organism. The transformations are mediated by biological catalysts (enzymes). An example of activation is the metabolism of the polcyclic aromatic hydrocarbon, benzo(a)pyrene. The compound becomes more toxic and carcinogenic when it is metabolized to an epoxide form, wherein an oxygen atom joins with two of benzo(a)pyrene's carbon atoms.

Active chemical A chemical that readily combines with other chemicals.

Active ingredients The chemicals in a pesticide or pharmaceutical formulation that are responsible for the target effect. The other ingredients are inert.

Activity Thermodynamic function applied instead of concentration in equilibrium constants for reactions for gases and solutions other than ideal gases and solutions.

Activity coefficient Factor that relates activity to the concentration of a substance in solution. The activity coefficient is dependent upon the ionic strength of the solution. Ions of similar size and with the same charge have similar activity coefficients.

Activity pattern What people are doing when they are being studied, including the actual physical activity (including exertion levels), physical location, and time of day the activity takes place. Example activities include watching TV at home from 6:30 to 7:30 p.m., sleeping at home from 11:00 p.m. to 7:00 a.m., jogging in the park from 6:00 a.m. to 7:00 a.m., or driving in a car from 8:00 to 8:30 a.m. The HAPEM4 model extracts activity pattern data from the EPA's Comprehensive Human Activity Database (CHAD).

Acute effect A disease or other adverse health outcome wherein symptoms occur shortly after an exposure to a chemical.

Acute exposure Usually, the amount of an exposure received in one day or less.

Acute toxicity The ability of a chemical to cause adverse effects from an acute exposure.

Addition reaction Chemical reaction, in which a molecule is added to another molecule that contains a double bond, and converts the double bond into a single bond.

Adenosine triphosphate (ATP) Macromolecule that carries energy to a cell. Energy is first stored in a high-energy bond (p_i) between the second and third phosphate groups on the ATP molecule.

Adiabatic expansion Expansion of air when the air mass rises without heat exchange with the surroundings.

Adiabatic process A change where no loss or gain in heat is involved.

Administered dose The amount of a chemical given to an organism.

Adsorption Collection of a chemical on a surface of a body. A form of sorption of physical bonding or adherence of ions or molecules onto a solid surface.

Advection Transport of solutes by mass flow of a medium; for example, the transport of a contaminant in water along stream lines at the average linear seepage flow velocity.

Aeration Injecting air or molecular oxygen into a medium to promote biological degradation of organic matter in water. The process may be passive (allowing the wastewater to be exposed to air), or active (mechanically mixing or using a bubbling device to introduce the air).

Aeration tank Chamber used to inject air into water.

Aerobe Microbe that requires free, molecular oxygen.

Aerobic Requiring the presence of molecular oxygen (O_2). Term is applied to microbial processes that decompose organic wastes and, if complete, yield carbon dioxide (CO_2) and water.

Aerosol (1) Synonym for particle (air pollutant). (2) Colloidal dispersion of liquid or solid particles in a gas, as in a mist or smoke; e.g., aerosol sprays use a propellant liquefied under pressure.

Affinity (1) Chemical attraction or force that causes the atoms of certain elements or compounds to combine with atoms of another element or compound and remain in the combined state. (2) Attractiveness of an environmental compartment to a contaminant. For example, lipophilic compounds have a greater affinity for the organic phase of sediments than to the aqueous phase, owing to their low water solubility and high octanol-water partition coefficient.

Aggregate exposure Sum total of all exposure to pesticides through inhalation, or dermal, oral, or other routes.

Air change (or air flush) Complete replacement of air from a defined spaced. Expressed as "air exchange rate."

Air column A volume of air extending from the surface upwardly. Analogous to *water column*.

Air sparging Groundwater treatment technology that introduces air or other gases beneath the water table to remove volatile organic compounds. Sparging combines volatilization (see *air stripping*) and bioremediation processes. Examples include in-well aeration and aquifer air injection. Sparging's bubbling action strips volatile compounds from the groundwater, forcing them to move upwardly into the overlying, unsaturated soils, where they are trapped and treated.

Air stripping Removal of volatile organic compounds from groundwater using a pump to create an air stream that picks up the volatile compounds. Typically, stripping passes the air across the media of a high surface area, which increases the available surface area. At the same time a countercurrent of air is provided by a blower. Contaminants are stripped from water by mass transfer to the air, from which the contaminants are then either discharged to the atmosphere (if the concentrations are below regulatory standards), or treated with additional vapor phase treatment or capture systems (i.e., a pump-and-treat system).

Air toxic Air pollutant known to cause or suspected of causing cancer or other serious health problems. The Clean Air Act Amendments of 1990 enumerated 184 compounds as "air toxics." Health concerns may be associated with both short- and long-term exposures to these pollutants. Many air toxics are known to have respiratory, neurological, immune, or reproductive effects, particularly for more susceptible sensitive populations such as children.

Algae Simple plants without roots, stems, or leaves, but that photosynthesize.

Aliphatic compounds Organic compounds with carbon atoms linked in a chainlike formation; including alkanes and alkenes.

Aliquot Discrete sample used for analysis.

Alkalinity Capacity of water to neutralize acids without a large pH change. Alkalinity is attributable to the water's ionic strength from carbonates, bicarbonates, hydroxides, borates, silicates, and phosphates. Units are equivalents (eq) or microequivalents (μeq).

Alkanes Hydrocarbon compounds (e.g., CH_3-CH_2-CH_3) that do not contain double or triple bonds between carbons. Alkanes form straight chains, e.g., the six-carbon hexane, or cyclic structures, such as the six-carbon cyclo-hexane.

Alkenes Straight chain hydrocarbon compounds that contain one or more double bonds between carbons.

Alkynes Straight chain hydrocarbon compounds that contain on or more triple bonds between carbons.

Allergen A substance that elicits an antibody response and is responsible for producing allergic reactions by inducing formation of IgE. IgE is one of a group of immune system mediators. IgE antibodies, when bound to basophiles in circulation or mast cells in tissue, cause these cells to release chemicals when they come into contact with an allergen. These chemicals can cause injury to surrounding tissue—the visible signs of an allergy. Fungal allergens are proteins found in either the mycelium or spores. Only a few fungal allergens have been characterized, but all fungi are thought to be potentially allergenic.

Alluvium Material deposited by water, such as soil, silt, or clay.

Ambient Surrounding, as in the surrounding environment. For example, ambient air refers to the air surrounding a person through which pollutants can be carried. Ambient air or water is usually distinguished from emissions or effluents. Likewise, ambient exposure is distinct from indoor or personal exposures.

Amphiphilicity The property of a substance indicating that it is soluble in both polar (i.e., hydrophilic) and non-polar (i.e., hydrophobic) substances. Examples include detergents and several alcohols.

Amphoteric behavior Aqueous complex's ability to have a positive, negative, or neutral charge.

Anabolism Process in which an organism uses energy to build organic compounds, such as enzymes and nucleic acids, for its biological functions.

Anaerobe Microbe that grows only or thrives in the absence of free, molecular oxygen.

Anaerobic Microbial processes that do not use molecular oxygen (O_2). Anaerobic microbes usually cannot live in the presence of O_2.

Anaerobic decomposition Reduction of the net energy level and altering the chemical composition of organic matter caused by microorganisms in an oxygen-free environment.

Analyte The target chemical to be analyzed. For example, if a heavy metals study is being conducted in a stream, the analytes may be cadmium, zinc, chromium, and lead.

Anion Negatively charged ion.

Anion Exchange Capacity (ANC) Measure of surface charge of an anion reported in equivalents of exchangeable ions per unit weight of the solid.

Anisotropic/Anisotropy Possessing different properties or behaving differently in different directions.

Antagonism (1) Interference or inhibition of an effect of one agent by the action of another agent. (2) Interaction of two or more chemicals that results in an effect that is less than the sum of what their effects would be if acting independently.

Anthropogenic Of or relating to humans; human-made or human-caused. Compare to *biogenic*.

Applied dose Amount of a chemical given to an organism to determine dose-response relationships. Applied dose does not differentiate the amount of the chemical administered versus the amount absorbed by the organism.

Aquiclude Confining underground layer with sufficiently low porosity and hydraulic conductivity so as to be excluded from the regional groundwater flow system. The difference between an aquiclude and an aquifuge is that the former is a saturated layer, but with extremely low porosity and/or limited pore interconnections, while the latter has either zero porosity or the pore spaces do not interconnect.

Aquifer Underground layer of the earth that can transmit a sufficient amount of water as a source for a water supply.

Aquitard Saturated underground layer that contributes to the regional groundwater flow regime, but lacks the hydraulic conductivity to be used as a source of water supply.

Aromatic hydrocarbons Organic compounds with one or more benzene rings (a six-carbon ring structure with alternating double bonds between carbons).

Artesian formation A confined groundwater system under hydrostatic pressure. An artesian aquifer's piezometric height is at a higher elevation than the aquifer's height.

ASPEN (Assessment System for Population Exposure Nationwide) model A computer simulation air dispersion model used to estimate toxic air pollutant concentrations, used in the National Air Toxics Assessment. The ASPEN model takes accounts for the rate of release, location of release, the height from which the pollutants are released, wind speeds and directions from the meteorological stations nearest to the release, chemical degradation of the pollutants in the atmosphere after emission, deposition of particles, and transformation of one pollutant into another compound via secondary formation. The model estimates toxic air pollutant concentrations for every census tract in the continental United States, Puerto Rico, and the Virgin Islands.

Autotroph Organism that can make all of its components from inorganic substances.

Background concentrations Contributions to contaminant concentrations from natural sources, previous emissions that have persisted and remain in the environment, and long-range transport of pollutants from distant sources.

Base Substance usually capable of freeing OH^- anions when dissolved in water. Can weaken a strong acid, reacts with an acid, forming a salt, and has a $pH > 7$.

Base-catalyzed hydrolysis Enhanced hydrolysis resulting from the attacks of hydroxyl ions.

Base flow Amount of a stream's discharge coming from groundwater flow.

Benchmark Chemical concentration used to calculate the hazard quotient, or the likelihood that a chemical will be associated with an adverse health or environmental outcome. Examples include quality criteria for water, sediment and air, and thresholds, such as the no observable adverse effect level (NOAEL).

Bernoulli equation Fluid mechanics expression of the conservation of energy that states that for a non-viscous, incompressible fluid in steady flow, the sum of pressure, potential and kinetic energies per unit volume is constant at any point.

Bioaccumulation Rate of increase of a chemical in an organism resulting from an excess of chemical intake versus the organisms ability to detoxify and eliminate the chemical.

Bioassay Method of testing a contaminant's effects on living organisms.

Bioavailability The ease with which a substance (contaminant or nutrient) can be taken up and distributed within an organism. Usually describes the extent and rate of absorption by a parent compound within the systemic circulation following exposure.

Biochemical oxygen demand (BOD) Amount of oxygen consumed during the biochemical oxidation of matter over a specified period of time. For example, the BOD_5 measures the quantity of oxygen consumed in five days.

Biocide Chemical that limits the growth of or kills living things, especially microorganisms such as bacteria or fungi.

Bioconcentration Buildup of a chemical in organisms, reaching concentrations above that found in the surrounding environment. Accumulation of a contaminant in tissues of an organism to levels greater than in the surrounding medium in which the organism lives.

Bioconcentration factor (BCF) Expression of the extent of chemical partitioning at equilibrium between a biological medium (e.g., fish or plant tissue) and an external medium (e.g., surface water). The higher the BCF, the greater the likelihood that a chemical will accumulate in living tissue.

Biodegradable A characteristic of a substance indicating its ability to break down under natural conditions.

Biodegradation (1) Decrease in concentration of a substance via naturally occurring microbial activity. (2) Process by which an organic molecule becomes transformed by biological means.

Biodegradation rate Mass of contaminant metabolized by microorganisms per unit time. For example, the biodegradation rate in soil can be normalized to the mass of soil and expressed as mg contaminant degraded per kg soil per day (mg kg^{-1} day^{-1}).

Biodiversity Expression of the variety and variability among living organisms in an ecosystem.

Biogenic Of natural origin (rather than of human origin, i.e., anthropogenic).

Biological treatment Engineering and technologies that employ bacteria and other microbes to consume waste.

Biologically effective dose Amount of chemical taken up by a person to cause an adverse effect. This is usually organ-specific (e.g., dose leading to a specific type of liver damage).

Biomarker Measurement of an interaction between a biological system and a chemical. Exposure biomarkers of a chemical are measurements of the chemical itself or its metabolite, indicating that the organism has been exposed to a chemical (e.g., the concentration of cotinine in a person's blood as an indication of nicotine exposure). Effect biomarkers are measurements of biochemical, behavioral, or physiological changes in an organism resulting from an exposure to a chemical (e.g., the reduction in sperm count and lower testosterone blood levels following an exposure to an antiandrogen, such as DDE).

Biomass The collective of all the living material in a given area.

Biome The whole community of living organisms in a single major ecological area. See *Biotic community*.

Bioremediation Use of enzymatic and metabolic actions of microbes to degrade and to detoxify contaminants. Contrast with phytoremediation.

Biosolids Sludge.

Biota Animal, plant, and other living things of a given region.

Biotechnology Techniques making use of organisms or parts of organisms to produce products, or to develop microorganisms to remove toxic compounds from bodies of water, or act as pesticides.

Biotic community Naturally-occurring assemblage of mutually sustaining and interdependent plants and animals that live in the same environment.

Biotransformation (1) Organisms' conversion of a substance into other compounds. Biodegradation is a type of biotransformation. (2) Conversion of a chemical within an organism after absorption. This usually leads to a less toxic compound compared to the compound to which the organism was exposed.

Bioventing Treatment process that aerates the vadose zone soils using installed vents to stimulate *in situ* biological activity and to optimize biodegradation of organic compounds with some volatilization occurring.

"Black mold" This poorly defined term, which has no scientific meaning (also called "toxic black mold"), has been associated with *Stachybotrys chartarum*. While only a few molds are truly black, many appear black. Not all molds that appear to be black are Stachybotrys.

Blank Artificial sample designed to monitor the introduction of artifacts into the sampling and analytical process.

Body burden Total amount of a chemical to which a person has been exposed from all sources over time.

BTEX Standard abbreviation for four common aromatic constituents of gasoline: benzene; toluene; ethyl-benzene; and total xylenes. BTEX is a common indicator when investigating fuel releases and spills. BTEX releases are volatile, mobile, and leachable; are sources of highly flammable and explosive vapors; pose a serious threat to human health (i.e., they are known or suspected human carcinogens or neurotoxins); have relatively high water solubilities; are readily adaptable to gas chromatography detection; and have associated maximum contaminant levels (MCLs) for drinking water.

Buffer Substance that resists the change in pH that would otherwise be produced by adding acids or bases to a solution.

Bulk density Density/volume ratio for a solid, especially a soil, not corrected for the voids contained in the bulk of material; units are kg m^{-3}.

Bulk transport Movement of a contaminant along with the general movement of a fluid, such as in an air mass or in flowing water. (See *Advection*).

Cancer New, malignant growth. Carcinoma is a malignant epithelial tumor that affects surrounding tissue and can lead to metastases (movement of the cancer cells to other parts of the body). Sarcoma is a malignant, connective tissue tumor of anaplastic cells that resembles the supporting tissues.

Cancer Risk Evaluation Guides (CREGs) Estimated contaminant concentrations in air, soil, or water expected to cause no greater than one excess cancer in a million (risk $\leq 10^{-6}$) persons exposed over a lifetime. CREGs are calculated from the EPA's cancer slope factors.

Cap The final, permanent layer of impermeable material, such as compacted clay or synthetics, on top of a landfill. Part of a closure.

Capillary action Means by which a liquid moves through the porous spaces in a matrix, such as soil, plant roots, and the capillary blood vessels in animals, due to the forces of adhesion, cohesion, and surface tension. Capillary action is essential in carrying substances and nutrients from one place to another in plants and animals.

Carcinogen Substance that causes, or is suspected of causing, cancer.

Cation A positively charged ion.

Census tracts Land areas defined by the U.S. Bureau of the Census that vary in size but typically contain about 4000 residents each. Census tracts are usually smaller than two square miles in size in cities, but much larger in rural areas.

Chain-of-custody Procedures used to minimize the possibility of tampering with environmental samples. Includes every step, from sample preparation to analysis and sample storage and archiving.

Characteristic waste Substances defined as hazardous wastes owing to their ignitability, corrosivity, reactivity, or toxicity [from 40 Code of Federal Regulations (CFR), Part 261, Subpart C].

Charge separation Electrical charge moved from one conductor to another while maintaining no net charge.

Chelating agents Organic compounds that are able to withdraw ions from their water solutions into soluble complexes.

Chemical bond Force holding atoms together. Ionic bonds transfer electrons from one atom to another. Covalent bonds share electrons among atoms. The covalent bond is stronger than the ionic bond, so covalently bonded compounds are more difficult to degrade.

Chemical equilibrium Equality of chemical reactions in forward and reverse directions.

Chemical oxygen demand (COD) Measure of all the oxidizable matter found in a sample, a part of which is biochemical oxygen demand, but COD also includes abiotic chemistry.

Chemical treatment Technology using abiotic, chemical processes to treat waste.

Chemicals of potential concern Chemicals at a site about which available data indicate the need for conducting a quantitative risk assessment.

Chiral Compound that has different left-handed and right-handed stereoisomeric forms; i.e., mirror images. The left versus right chirality may be the difference between efficaciousness and toxicity in a pharmaceutical or pesticide. The chirals often also have very different half-lives in the environment (likely owing to the greater similarity of one chiral to naturally occurring substances).

Chlorinated hydrocarbon Class of persistent, pesticides, notably DDT, wherein Cl atoms substitute for H atoms, that linger in the environment and have a strong ability to bioaccumulate. Examples include DDT, aldrin, dieldrin, heptachlor, chlordane, lindane, endrin, mirex, benzene, hexachloride, and toxaphene.

Chlorofluorocarbons (CFCs) Family of chemicals commonly used in air conditioning and refrigeration systems, but also as solvents and aerosol propellants. CFCs have been known to drift into the upper atmosphere where their chlorine components destroy ozone, and as such have been implicated as a major contributor to stratospheric ozone depletion.

Chromophore Group of atoms in a molecule responsible for absorbing the UV, visible light, or other electromagnetic radiation.

Chronic exposure Exposures lasting more than six months.

Chronic reference dose (RfD) Estimated lifetime daily exposure level for the human population, likely to be without an appreciable risk of deleterious effects, protecting the population from long-term exposure to a chemical (>7 years).

Clay Soil particles with grain sizes less than 0.002 mm in diameter.

Cleanup Set of steps taken to address a release or threat of release of a hazardous substance. Synonyms include *remedial action, removal action, response action*, or *corrective action*.

Closed system Processes designed and used so that chemicals are not released. Closed systems are measures to control exposures to hazardous materials in industrial operations.

Cohort Group of people within a population who are assumed to have identical exposures during a specified exposure period. The use of cohorts is a necessary simplifying assumption for modeling exposures of a large population. For the exposure assessment, the population is divided into a set of cohorts such that (1) each person is assigned to one and only one cohort, and (2) all the cohorts combined encompass the entire population.

Colloid Fine particle with a diameter range of 1 to 500 μm. Due to small mass and charge, colloids do not easily settle in a water column without chemical treatment.

Comparative biology Use of animal testing to determine toxicity of a chemical and to develop models of how the chemical may behave in humans.

Comparison values Screening values in the preliminary identification of "contaminants of concern" at a hazardous waste site.

Competence Ability of a fluid to hold suspended matter, directly proportional to the velocity of the fluid.

Complexation Process whereby two or more solutes join chemically to form a new chemical complex, such as a cation joining a ligand to form a metal chelate.

Composite sample Sample taken to represent the loadings or concentrations of pollutants over a designated time periods, collected at regular time intervals, and pooled into one large sample. Samples can be composited according to a specified period of time (e.g., a 12-hour composite indoor air sample) or flow rate (e.g., every 1000 L of air flowing through a filter).

Composting Engineered degradation of organic wastes using "aerobic" microbes to generate a high nutrient substance (i.e., compost).

Compressibility Property of a fluid wherein density ρ changes appreciably within a domain of interest. Typically, compression occurs at high velocities, so engineers often assume that most environmental gases and liquids behave as compressible fluids under general environmental conditions.

Concentration gradient Change in concentration of a chemical [C] over a unit length. Diffusion occurs from higher to lower [C] (law of potentialities). Rates of diffusion increase with increasing [C] in a medium.

Congener Compounds possessing similar structures to another compound.

Consent decree Binding agreement by both parties that settles a lawsuit (this can be between commercial parties, between commercial and governmental parties, or between private citizens and governmental and/or commercial parties).

Contaminant Substance that exists in an environment where it does not belong or is found in concentrations that can elicit adverse effects. Substance that harms health, ecosystem integrity, or material properties.

Contamination Process by which harmful substances or microorganisms reach environmental media, including air, water, soil, and food.

Contaminants of concern Chemical found at the site that the health professionals select to be analyzed for potential human health effects.

Contingency plan Document laying out the organized, planned, and coordinated courses of action to be taken in the event of an accident where toxic chemicals, hazardous wastes, or radioactive materials are released.

Convection Transfer by a moving fluid, such as air or water. In soils, convection is the bulk water transport (flux) through porous media.

Corrective action Required cleanup of hazardous substance releases. Often refers to releases prior to the passage of the Resource Conservation and Recovery Act of 1976 (42 U.S.C. s/s 6901 et seq.).

Corrosivity Hazardous waste characteristic for any waste with pH $\leq$ 2.0 or $\geq$12.5.

Co-Solvation Process where a substance (the cosolute) is first dissolved in one solvent, and then the new solution is mixed with another solvent (the cosolvent). This can be an important contaminant transport phenomenon. For example, a hydrophobic compound like a polychlorinated biphenyl may have resisted being dissolved in water, but it may be able to migrate into and within groundwater if it is first dissolved in an organic solvent (like chlorobenzene) that migrates downward because its density is less than that of water (i.e., the PCB is transported in the chlorobenzene, which then undergoes cosolvation with the water).

Cost-benefit analysis Approach for determining whether a project is worth pursuing, taking into consideration expected costs and expected benefits. The costs and benefits can be either monetized (based upon dollar value), or nonmonetized (some other system of value). The analysis should yield a benefit/cost ratio. If the ratio is greater than 1, the project may be worth pursuing, but engineers are usually seeking projects where the ratio is much larger than 1.

Cradle-to-grave Requirement of the Resource Conservation and Recovery Act that a substance be accounted for from generation, through transport, to its ultimate treatment and disposal.

Critical effect First adverse effect, or its known precursor, occurring in the most sensitive species as the dose rate of an agent increases.

Cumulative risk Risk of a common toxic effect associated with concurrent exposure by all pathways and routes from a group of contaminants that share a common mechanism of toxicity.

Darcy's Law Expression of the laminar flow of water through porous media. Darcy's Law states that the velocity of the water through porous media is proportional to the hydraulic gradient.

Data quality objective (DQO) Quantitative and qualitative statements that specify the appropriate quantity and quality of data to be collected during field activities to test hypotheses, or to support specific decisions or regulatory actions. Part of a quality assurance/quality control plan.

Daughter product Compound formed from the biodegradation of another. For example, cis 1,2-Dichloroethene (cis 1,2-DCE) is commonly a daughter product of trichloroethene (TCE).

***De minimus* risk** A risk not deemed to be important or of no public health concern. For carcinogens, this has recently been deemed by some to be less than "one in a million" risk (1×10^{-6}).

Debris Grass cuttings, tree trimmings, stumps, street sweepings, roofing and construction wastes, and similar waste material.

Decay constant Constant expressing the probability that a chemical compound will decay in a given time interval. An expression of the persistence or half-life of a substance.

Dehalogenation Removal of chlorine, bromine, or other halogen atoms from a molecule. This process usually reduces the toxicity and persistence of chemical contaminants, so it can be an important part of cleanup efforts at sites contaminated with halogenated wastes. The most common type of dehalogenation is **dechlorination**, such as the removal of chlorine atoms from PCBs using anaerobic biodegradation.

Delisting Formally removing a substance from the U.S. EPA's listing of regulated materials, as new data are made available.

Denitrifying bacteria Soil bacteria that can convert nitrogenous compounds to gaseous nitrogen (N_2).

Dense nonaqueous phase liquid (DNAPL) Organic compound that sinks in groundwater.

Deposition Process that transports a substance from the atmosphere to the earth's surface by either impingement (dry deposition) or precipitation (wet deposition).

Depuration Any process that results in reducing concentrations or completely eliminating a contaminant from an organism. Antonym of *bioconcentration*.

Dermal Relating to the skin.

Designated facility Entity that treats, stores, or disposes of hazardous wastes.

Detection limit Minimum amount of a chemical that can be measured (specific to instruments and laboratories).

Developmental toxicity Impairment of the developing organism. Often linked in risk assessment to "reproductive toxicity" and, recently, to "endocrine disruption."

Diatomic Molecule containing two atoms. Diatomic elements are those seven elements that react with themselves to form diatomic molecules, i.e., H_2, N_2, O_2, F_2, Cl_2, Br_2, and I_2.

Diffusion Transport of a chemical through an environmental medium based upon a concentration gradient of the chemical. Diffusion is the net transport of solutes within the liquid, solid, or gas phase resulting from random (Brownian) motion of individual molecules in response to a concentration or other gradient. See *Fick's Law*.

Diffusion coefficient (K_d) Measure of the extent of chemical partitioning between soil or sediment and water, not adjusted for dependency upon organic carbon. The higher the K_d, the more likely a chemical is to bind to soil or sediment than to remain in water, which affects the efficiency of water-based remediation.

Diffusivity Movement of a molecule in a liquid or gas medium due to differences in concentration. Diffusivity is used to estimate the rate of volatilization of a pure substance from a surface or to estimate a Henry's Law constant for chemicals with low water solubility. The higher the diffusivity, the more likely a chemical will move in response to concentration gradients.

Digestion Biochemical decomposition of organic matter, due to partial gasification, liquefaction, and mineralization of pollutants.

Dilution ratio Relationship between the volume of water in a water body and the volume of incoming water. The dilution ratio is a factor in a stream's ability to assimilate waste.

Dioxin Any isomer or congener of the class of compounds dibenzo-para-dioxins.

Dipole Intermolecular force, wherein a molecule has two opposite electrical poles, or regions, separated by a finite distance. Attraction between the oppositely charged poles of adjacent molecules.

Dipole moment Expression of the degree of polarization of a molecule, i.e., the size of the dipole.

Direct filtration Water treatment method employing the addition of coagulant chemicals, flash mixing, coagulation, minimal flocculation, and filtration. Sedimentation is not part of the direct filtration process.

Dirty Dozen Nickname for the twelve toxic substances targeted by the United Nations Environment Programme (UNEP) for early action under a global convention. The compounds are dioxins, furans, polychlorinated biphenyls (PCBs), DDT, chlordane, heptachlor, hexachlorobenzene, toxaphene, aldrin, dieldrin, endrin, and mirex.

Dirty 33 Nickname for urban air toxics that are deemed to pose the greatest potential health threat in urban areas.

Discharge (1) Volume of water passing a given location within a given period of time ($m^3 sec^{-1}$). (2) Discharge of effluent from a facility or of chemical emissions into the air through designated venting mechanisms or stacks.

Dispersant Chemical agent used to break up concentrations of organic material such as spilled oil.

Dispersion Net transport or mixing resulting from differential advection in environmental media, such as through soil pores of varying diameters or in an air mass. In groundwater, hydrodynamic dispersion is the process whereby a contaminant dissolved in groundwater spreads out in the direction coincident to and perpendicular to groundwater flow. This dilutes the concentration of the contaminant. Dispersion is the net effect of mechanical mixing and molecular diffusion on a dissolved contaminant that results in dilution of the contaminant. The mixing results from differences in flow path length and the varying velocity of different molecules.

Dispersion model A mathematical approach to predict the transport of chemicals from one site to another. Models may be stochastic (statistically-based) or deterministic (including the scientific attributes and parameters expected to drive the movement of chemicals).

Disposal Discharge, deposition, injection, dumping, spilling, leaking, or placing of any solid waste or hazardous waste into or on any land or water with the potential that the waste or any of its constituents may enter the environment.

Dissociation Break up of a chemical into components (e.g., ions) in a solution.

Dose-response Relationship of adverse effects in an organism to the amount of chemical to which the organism is exposed.

Dosimetry Measurement of amount of a toxic substance.

Downgradient Direction in which groundwater flows; analogous to "downstream" for surface water.

Downgradient well Well used to sample groundwater that has passed beneath a facility that may release contaminants, such as a landfill.

Dredging Removal of sediments and accumulated material from surface waters.

Dry weight The weight of a sample based on percent solids. The weight after drying in an oven. Commonly used for soil and sediment sampling.

Dumpster Large container for solid waste materials.

Duplicate sample Identical split samples of individual samples to be analyzed by the laboratory to provide a measure of method reproducibility.

Ecological risk assessment Estimating the contributing factors and the effects of human activities on ecosystems.

Ecology Scientific discipline concerned with the relationships between all organisms and the environment.

Ecosystem A system of living organisms in a particular environment of other living things (biotic environment) and nonliving things (abiotic environment).

Educt Substance extracted from a mixture.

Effluent Discharge flowing from a conveyance. Analogous to emissions to the air.

Effluent guidelines Technical documents that set discharge limits for particular types of industries for specific contaminants.

Effluent limitations Limits and ranges (e.g., pH) on contaminants that may be discharged by a facility; calculated so that water quality standards will not be violated even at low stream flows.

Electrical resistivity Noninvasive, geophysical measurement method for determining types of underlying strata and their characteristics. Resistivity measurements are made by placing electrodes on the ground surface, sending through an electric current, and determining the decrease in potential between the electrode locations. Resistance is the reciprocal of conductivity. In general, dry, coarse-grained matrices have higher resistance than finer-grained, moisture-laden matrices. Such relationships can be quantified and used as input for inverse models to estimate permeability, porosity and chemical concentrations in groundwater. Units of resistance are *ohms* and units of conductance are *mhos*.

Electrodialysis Process for extracting salts from water that uses a membrane with an electrical current to separate the ions. Positive ions go through one membrane, while the negative ions flow through a different membrane, leaving the end product of water with lower salt concentrations.

Electronegativity Relative ability of an atom or molecule to attract electrons to itself.

Emergency response Condition that occurs unexpectedly and relatively quickly that must be addressed, such as a spill or other contaminant release, a terrorist act, or a natural disaster. The emergency response plan is document that anticipates various emergency situations and provides contingencies to address them.

Emergency response team (ERT) Group of experts and responders who have been trained to assist following a spill or release of contaminants. Tasks include measuring the extent of contamination and the implementation of decontamination efforts.

Emission Release or discharge of a substance into the atmosphere. Analogous to discharges in water.

Enantiomers One in a pair of optical isomers.

Encapsulation Method for disposing of hazardous substance that employs an impervious container made of materials highly resistant to physical, chemical, and biological degradation. The container is also sealed within a more durable container (e.g., steel, plastic, and/or concrete) of sufficient thickness and strength to resist physical damage during storage. Similarly, asbestos encapsulation is a process of coating asbestos-containing material to keep the fibers in place and to prevent contact with the friable asbestos fibers.

Endocrine disruption Dysfunction of normal hormonal processes in human and wildlife. Three types of endocrine disruption can occur: agonism, antagonism, and indirect. Endocrine agonists are chemicals that mimic estrogen, testosterone, and other hormones because the chemicals are in some way similar in chemical structure (e.g., functional groups) to that of the natural hormone, and are able to bind with the receptors of a cell. Antagonists are chemicals that interfere or block normal hormone receptors, so that the cell cannot produce sufficient amounts of a hormone (e.g., antiandrogens shut off testosterone receptors, causing a net increase in estrogens, and feminizing the organism). Indirect disruptors may be agonists or antagonists for nonendocrine systems, such as the neurological or immune systems, but the changes to these other systems leads to endogenous changes that ultimately affect the normal hormonal functions, i.e., reproductive, developmental and physiological homeostasis.

Endpoint Disease or harmful outcome associated with exposures to a chemical that is the focus of a study.

Enthalpy In thermodynamics, the sum of the internal energy plus the product of the pressure times the volume of the gas in a system.

Entropy In thermodynamics, the expression of the amount of disorder in a system. Measure of the amount of energy in a system available to do work.

Environmental audit Investigation of a company's compliance with a range of environmental regulations, or an assessment of potential environmental liabilities, such as a condition of a real estate transaction.

Environmental engineering Application of the principals of physics, chemistry, and biology to address and to prevent environmental and public health problems. The profession has evolved from the more confined field of "sanitary engineering," which was concerned with the design and operation of environmental facilities, e.g., drinking water plants, wastewater treatment facilities, and air pollution abatement equipment. The field presently addresses hazardous waste management, risk assessment, and ecosystem protection.

Environmental Impact Statement (EIS) A document required by the National Environmental Policy Act of 1969 for any proposed federal action that may significantly affect environmental quality. At a minimum, the EIS includes analyses of the (1) environmental impact of the proposed action, (2) alternatives to the proposed action, and (3) irreversible or irretrievable resource commitments that can result from the action.

Environmental justice Inclusion of race and social issues in environmental decisions. Combination of environmental and social justice.

Environmental Media Evaluation Guides (EMEGs) Concentrations of a chemical in the air, soil, or water below which noncancer effects are not expected to be associated with exposures over a specified duration of exposure.

Epidemiology Study of the occurrence and distribution of diseases in humans or adverse effects in ecosystems. Epidemiology considers the factors that influence the distribution of these effects. Descriptive epidemiology is concerned with the delineation of diseases and adverse effects in populations and subpopulations (known as "polymorphs"). Analytical epidemiology delves further into the potential reasons for such occurrences.

Equilibrium Steady-state condition where no net gain or net loss occurs, i.e., inflow equals outflow.

Equilibrium constant Constant representative of the ratio of concentrations of both sides of a reversible reaction, where the forward and reverse rates are equal. Equilibrium constants are controlled by thermodynamics, and can include partition coefficients.

Estuary Complex ecosystem of the mixing zone of fresh and saline waters between a river and near-shore ocean waters.

Exogenous Taking place outside of an organism.

Exposure Contact of a person (or organism) with a chemical agent, quantified as the mass of the chemical available at the exchange boundaries (e.g., lungs, skin, digestive tract) and available to be absorbed. If the organism is exposed in the medium of release (e.g., a person breaths air that contains a chemical released from a stack), such exposure is considered "direct." If the exposure occurs only after the chemical has moved through various media (e.g., the chemical is deposited onto a water body and is taken up by a fish eaten by a person), then the exposure is "indirect." Units are in mass of chemical per mass of body weight per time ($mg\,kg^{-1}\,d^{-1}$).

Exposure pathway Physical course taken by a chemical to reach the exposed organism.

Extraction Procedure (EP) Test Leach test required by the U.S. Environmental Protection Agency to estimate the likelihood that a waste would transport 15 toxic metals and organic compounds into groundwater. Modified and expanded to address 40 chemicals in 1990 (renamed the Toxicity Characteristic Leaching Procedure).

False negative (1) Test result that states that a contaminant is absent in a sample when, in fact, the contaminant is present; e.g., a lab shows that dioxin is not detected in soil, but in fact, due to instrument error, the soil contains concentrations that should have been detected. (2) Finding that an agent is not the cause of disease or adverse effects, but the agent does in fact cause the disease of effect.

False positive (1) Test result that states that a contaminant is present in a sample when, in fact, the contaminant is not present or is at a significantly lower concentration than what is shown by the test; e.g., a test shows a compound in water above the water quality standard, but the compound is really present at levels below the standard. (2) Finding that an agent is a cause of disease or adverse effect, but in fact the association is erroneous.

Fate The ultimate site of a pollutant following its release. The pollutant will undergo numerous stages before reaching its fate. These include physical transport, chemical (including photochemical and biochemical) and biological transformations, sequestration and storage. Fate is often described according to environmental media or compartments. For example, a chemical may have an affinity for sediments, so its physicochemical properties may make for low residence time in water and air, driving its fate toward the sorption onto sediment particles.

Fick's Law Fick's first law of diffusion states that the rate of diffusion of one material through a different material is proportional to the cross-sectional area of diffusion, the "concentration gradient" of the first material, and a coefficient of diffusion. The law is expressed as: $M/A = -D(dC/dx)$, where M is the mass transfer rate, A is the cross-sectional area, D is the coefficient of diffusion, and dC/dx is the concentration gradient.

Field blank Aliquot of reagent water or other reference substance placed in a sample container in the laboratory or the field and treated in exactly the same way as the actual samples, such as being exposed to sampling site conditions, storage, preservation, and all analytical procedures. The purpose of the field blank is to determine if the field or sample transporting procedures and environments have contaminated the sample.

Fixation Immobilization of wastes by combining them with relatively inert and stable materials, such as fly ash, concrete, or refractory clays.

Flume Open channel flow section shaped to provide a change in the channel area and/or slope. This results in an increased velocity and change in the level of the liquid flowing through the flume. A common flume design has three sections: (1) a converging section; (2) a throat section; and (3) a diverging section. Flow rate through the flume is a function of the liquid level at a specified point in the flume.

Flux Flow rate through a cross-sectional area (mass or volume per unit area per time).

Free radical Molecule with an unpaired electron, making it very reactive.

Fugacity The propensity of a chemical to move out of a compartment, such as from water to air. Expressed by Henry's Law constants.

Fugitive emissions Contaminants released to the air via conveyances other than from stacks or vents.

Fungi Neither animals nor plants, fungi are classified in their own kingdom. The fungi kingdom includes more than 100,000 species, with some estimates as high as 10 million species, including molds and yeasts.

Gas-side impedance The difference in a contaminant's partial pressure in the bulk gas and at the interface between the gas-side film and the liquid-side film.

Generator Producer of hazardous wastes.

Gold standard Accepted reference standard or diagnostic test for a disease or other adverse outcome.

Grab sample Instantaneous sample collected at a single point in time, without regard to flow or time. Compare to *Composite sample.*

Green chemistry Application of a set of principles to reduce or to eliminate the use or generation of hazardous substances in the design, manufacture and application of chemical products. The goal of green chemistry is to prevent a harmful chemical from being produced in the first place, by using computational sciences, informatics, and structure activity relationships to predict toxicity, persistence, bioaccumulation and other characteristics of a compound.

Groundwater or ground water Water that has infiltrated through the soil and is stored underground, usually for long time periods.

Groups Vertical columns in the periodic table.

Half-life ($T^{1/2}$) Time needed to decrease the concentration or mass of a chemical by one-half. Half-lives may be biological (by metabolism and elimination processes), chemical (by transformation and reactions), or radiological (instability of the atom's nucleus).

Hazard A physical agent or chemical substance that can harm human health or environmental resources.

Hazard index (HI) Sum of hazard quotients for substances that affect the same target organ or organ system. Different pollutants may cause similar adverse health effects, so it is often appropriate to combine hazard quotients associated with different substances. Ideally, hazard quotients should be combined for pollutants that cause adverse effects by the same toxic mechanism.

Hazard quotient Ratio of the potential exposure to the substance and the level at which no adverse effects are expected. If the hazard quotient is less than 1, then no adverse health effects are expected as a result of exposure. If the hazard quotient is greater than 1, then adverse health effects are possible. The quotient only expresses the possibility of adverse outcomes, but exceeding 1 does not necessarily mean that adverse effects will occur.

Hazard Ranking System (HRS) Process that screens the threats of each site to determine if the site should be listed on the National Priority Listing (NPL) of most serious sites identified for possible long-term cleanup, and what the rank of a listed site should be.

Hazardous Quality of a substance giving it the potential for an adverse outcome (harm).

Hazardous Air Pollutant Exposure Model, Version 4 (HAPEM4) A computer model that has been designed to estimate inhalation exposure for specified population groups and air toxics. Through a series of calculation routines, the model makes use of census data, human activity patterns, ambient air quality levels, climate data, and indoor/outdoor concentration relationships to estimate an expected range of inhalation exposure concentrations for groups of individuals.

Hazardous substance Chemical that will threaten human health and the environment if released in sufficient quantities. Specifically defined by various agencies, including the Department of Transportation and the Occupational Safety and Health Administration.

Hazardous waste 1. Substance that has been produced as a by-product of human activities with the potential of harming human health or environmental resources. Hazardous wastes must possess at least one of four characteristics—ignitable, corrosive, reactive, or toxic—or they must appear on a U.S. Environmental Protection Agency special list. 2. Solid waste that has been listed as possessing hazardous characteristics under the Resource Conservation and Recovery Act.

Hazardous waste management Comprehensive approach for dealing with hazardous wastes, including pollution prevention, exchanges, and engineering approaches.

Head of liquid Depth of flow.

Headspace Zone above the contents of a closed container. Three-dimensional space in a container above a liquid that receives gases that have partitioned from the liquid.

Heavy metal Metallic elements of atomic weights at or above iron (55.847). Some important environmental heavy metals are cadmium, copper, lead, mercury, manganese, zinc, chromium, tin, thallium, and selenium. Usually, the metalloid arsenic is also included in this listing of environmental heavy metals.

Henry's Law constant (K_H) Ratio of a chemical compound's mass in the gas phase to its mass in the aqueous phase. An expression of fugacity.

Heterogeneous reaction Chemical reaction that takes place in more than one phase.

High-end exposure Denotes that a person living at the centroid (center of the population mass) of a census tract and engaging in a range of activities that tend to produce higher exposures and risks than is typical. The high-end group may be represented as an upper percentile, such as the 10th percentile of a population who engage in activities associated with elevated risk (i.e., more often than 90% of individuals in the population).

Homeostasis The body's ability to maintain a relatively consistent internal environment.

Homogeneous reaction Chemical reaction that takes place in a single phase.

Horizon Horizontal soil layer (from top: A horizon, highest organic material content); B horizon (nutrients from leaching); and C horizon (partially weathered parent rock).

Hydraulic conductivity (K) Coefficient expressing the permeability of an aquifer.

Hydraulic gradient (i) Rate of change in hydraulic head over a unit distance. Also, the head loss over a horizontal distance (dimensionless).

Hydraulic head (i or h) Height of the water column. Elevation of the water surface above a plane of reference (e.g., mean sea level). Expressed in units of length.

Hydraulic head loss (Δh) Decrease in the height of a water column.

Hydrogen bond The bond formed when a hydrogen atom bonded to atom A in a molecule forms an additional bond to atom B in the same or another molecule. The H-bond is strongest when atoms A and B are very electronegative (e.g., fluorine, oxygen, or nitrogen atoms).

Hydrolysis Chemical reaction in which a molecule of water or a hydroxide ion replaces an atom or group of atoms of another molecule, making the transformation products more polar.

Hydrophilic Propensity to dissolve in water.

Hydrophobic Resistance to dissolving in water. Usually synonymous with lipophilic (fat soluble).

Hysteresis A system's dependence on its previous history (often resulting in a lag of an effect behind the cause).

Ignitability Hazardous characteristic of a chemical pertaining to its likelihood to catch fire.

Impermeability Resistant to passage of a liquid.

In situ In place. For example, *in situ* remediation or treatment occurs where the contamination exists, rather than being removed and treated elsewhere.

***In utero* exposure** Contact with a chemical through the placenta, during an organism's gestation period.

In vitro In glass. Experiments that are performed in test tubes and other laboratory apparatus.

In vivo In a living organism. Experiments that are performed on living organisms.

Incineration Combustion of organic materials.

Incomplete combustion Any high temperature oxidation that does not result in the production of carbon dioxide and water. Products of incomplete combustion include the dioxins and polycyclic aromatic hydrocarbons, as well as other substituted aromatic compounds like hexachlorobenzene.

Isomer Chemical compound with the same molecular formula of another, but where the atoms are arranged differently.

Isotope An element's forms with differing atomic weights, e.g., ^{204}Pb, ^{206}Pb, ^{207}Pb, ^{208}Pb or ^{12}C and ^{14}C.

Karst topography Structure of land surface resulting from fractured limestone, dolomite, gypsum beds, and other rocks, and subsequent dissolution. Karst topography is characterized by closed depressions, sinkholes, caves, and underground drainage. One of the few groundwater systems that experiences turbulent flow.

Kernel Combined unit consisting of an atom's nucleus plus the electrons of its filled, inner shells.

Laminar flow Flow having a smooth appearance and lacking the mixing phenomena and eddies of common turbulent flow systems.

Landfill Engineered system for disposal of solid waste on land. Refuse is spread and compacted and a cover of soil applied to minimize leakage and stabilize the wastes.

Leachate Liquid (commonly water) that percolates through a landfill or other system and has collected dissolved and suspended contaminants from the waste and, without intervention, will carry these pollutants off-site via groundwater and surface water transport.

Leachate collection system Arrangement of catchments and piping underlying a landfill or other waste site that is engineered to catch and to remove water that migrates through the site.

Lethal concentration 50 (LC_{50}) Concentration of a contaminant that results in the death of half of the test organisms within a designated period of time; a common measure of acute toxicity. The lower the LC_{50}, the more toxic the compound.

Lethal dose 50 (LD_{50}) Dose of a contaminant that results in the death of half of the test organisms within a designated period of time. The lower the LD_{50}, the more toxic the compound.

Lifetime average daily dose (LADD) Total dose that a person receives over a lifetime. A measure of chronic exposure.

Liquid-side impedance The difference in a contaminant's concentration in the bulk liquid and at the interface between the gas-side film and the liquid-side film.

Lowest observable adverse effects level (LOAEL) Lowest level of exposure to a chemical where an adverse effect in the exposed population increases significantly (statistically and biologically) compared to the unexposed population.

Material safety data sheet (MSDS) Printed material concerning a hazardous substance or extremely hazardous substance, including its physical properties, hazards to personnel, fire and explosion potential, safe handling recommendations, health effects, firefighting techniques, reactivity, and proper disposal.

Maximum Contaminant Level (MCL) Highest concentration of certain contaminants permitted in drinking water supplied by a public water system. MCLs are set as close to MCLGs (see next definition) as possible, considering costs and technology.

Maximum Contaminant Level Goal (MCLG) Highest concentration of a contaminant that is associated with no adverse health effects from drinking water containing that contaminant over a lifetime. For carcinogens, the MCLGs are set at zero.

Maximum daily dose (MDD) Highest dose received in a 24-hour period during an exposure period.

Methemoglobinemia Oxygen deficiency that can be caused by increased levels of nitrates in the blood, particularly to infants and small children. Also known as "blue baby syndrome."

Microenvironment (μE) The finite space where human contact with a pollutant occurs and that can be treated as a well-characterized, relatively homogenous location with respect to pollutant concentrations for a specified time period. On a national scale, the U.S. EPA exposure model HAPEM4 considers cohort activities in 37 μE locations, including: indoors in residences, offices, stores, schools, restaurants, churches, manufacturing facilities, auditoriums, health care facilities, service stations, other public buildings, and garages; locations where people spend time in outdoor parking lots and garages, near roadways, on motorcycles, at service stations, at construction site, on residential grounds, at school, attending events at sports arenas, visiting parks, and on golf courses; and in-vehicle locations (e.g., driving or riding in a car, bus, truck, train/subway, or airplane).

Microenvironmental exposure An estimate of a person's potential contact with a chemical agent measured from the immediate local environment (e.g., indoor air in a home or in a vehicle). Units are in mass of chemical per mass of body weight per time ($mg\,kg^{-1}\,d^{-1}$).

Mineralization Degradation of an organic or organometallic compound that progresses toward the inorganic chemical species.

Minimal risk levels (MRLs) Estimates of daily human exposure to a chemical agent ($mg\,kg^{-1}\,d^{-1}$) that are not expected to be associated with any appreciable risk of noncancer effects over a specified duration of exposure.

Modifying factor Factor that reflects the results of qualitative assessments of the studies used to determine the threshold values (dimensionless, usually factors of 10).

Mole Molecular weight of chemical in grams (gram-mole or gmole). In engineering, sometimes molecular weight in pounds (pound-mole or pmole).

Mycotoxin Compounds produced by "toxigenic fungi" that are toxic to humans or animals.

National Priority Listing (NPL) Annual list compiled by the U.S. Environmental Protection Agency of the hazardous wastes sites in most need of cleanup.

Nephelometric turbidity unit (NTU) Unit of measurement that indicates water turbidity.

Neutralization The chemical process that renders the acidic or basic characteristics of a fluid to those of deionized water ($pH = 7$).

No action alternative Status quo. No additional intervention.

No observable adverse effect level (NOAEL) Highest dose where no adverse effects are seen.

Nonaqueous phase liquid (NAPL) Organic compound that floats within the groundwater.

Nondietary ingestion Exposures through the mouth to the digestive tract from sources other than eating food, including pica (e.g., children eating paint chips) and contaminants transferred by hand to mouth from surfaces.

Nonroad mobile sources Vehicles and equipment not found on roads and highways (e.g., airplanes, trains, lawn mowers, construction vehicles, farm machinery).

Obligate aerobe Microbe that uses oxygen exclusively as its electron acceptor. Thus, the presence of molecular oxygen is a requirement for these microbes.

Obligate anaerobe Microbe that can only grow in the absence of oxygen; the presence of molecular oxygen either inhibits growth or kills the organism.

Occupational exposure (1) Exposure in the working environment (as distinguished from residential or ambient environmental exposure). (2) Dermal, eye, mucous membrane, or parental contact with blood or other potentially infectious materials that results from the performance of an employee's duties.

Octanol-water partition constant (K_{ow}) Expression of a compound's affinity for either an organic medium (less polar) versus its affinity for the aqueous medium (more polar).

Onroad mobile sources Vehicles usually traveling on roads and highways (e.g., cars, trucks, buses).

Organic carbon-normalized sorption coefficients (K_{oc}) Measure of extent of adsorption of an organic compound to soil or sediment particles. Expressed as the ratio of mass of adsorbed carbon per unit mass of total organic carbon.

Overall confidence Statistical confidence level for a pollutant concentration based on the combined uncertainties; e.g., uncertainties associated with emissions estimations ambient concentration modeling, and exposure modeling. This could be a factor in preparing a data quality objective for an environmental study.

Oxidation (1) Loss of electrons from a compound. Oxidation can supply energy that microorganisms use for growth. Frequently, oxidation results in the addition of an oxygen atom and/or the loss of a hydrogen atom. (2) Addition of oxygen to degrade organic waste or chemicals such as benzene, phenols, and organic sulfur compounds, and even inorganic compounds like cyanide, in sewage and wastes using biological and chemical processes.

Oxidation-Reduction Potential (ORP) Electric potential needed to transfer electrons from one compound or element (the oxidant) to another compound or element (the reductant); used as an indirect and qualitative indicator of the state of oxidation in water treatment systems.

Oxygenated solvent Organic solvent that contains oxygen in its molecular structure. Alcohols and ketones are examples.

Ozonation Application of ozone to water for disinfection, as well as for taste and odor control.

Ozone layer Protective layer in the stratosphere that absorbs some of the sun's ultraviolet light, reducing the amount of potentially harmful radiation reaching the earth's surface.

Partial ionic character Strengthening of molecular bonds when molecules are neither symmetrical nor completely electrically neutral. For example, the oxygen atom has a greater attraction for electrons than does the sole proton of the hydrogen atom. Molecules forming these bonds may or may not have partial electrical charges.

Partition coefficient Factor expressing the ratio of a chemical's respective concentration in each of two phases into which the chemical has transferred. Partition coefficients can be equilibrium constants, but can also be expressions of phase transfer under nonequilibrium conditions.

Permeability Ease with which a fluid passes through a substance.

Permissible exposure limit (PEL) Workplace exposure limits for contaminants established by the U.S. Occupational Safety and Health Administration.

Personal exposure Actual contact of a person with a chemical agent, quantified as the mass of the chemical available at the exchange boundaries (e.g., lungs, skin, digestive tract) and available to be absorbed. Personal exposure can be measured directly (see *Personal exposure monitors*) or modeled from chemical measurements in the ambient environment or in microenvironments (e.g., indoor air). Units are in mass of chemical per mass of body weight per time ($mg\,kg^{-1}\,d^{-1}$).

Personal exposure monitors (PEMs) Devices placed on or carried by people to determine actual, personal exposures (as contrasted with "ambient measurements" and "microenvironmental exposures"). These can be active monitors (those that include a pump to gather samples) or passive monitors (those that are based upon diffusion and Fick's Law).

pH Potential hydrogen ion concentration $[H^+]$ of a solution. Calculated as the negative logarithm of $[H^+]$. Thus a pH 5 solution has two orders of magnitude or 100 times the $[H^+]$ of a pH 7 (neutral) solution. The $[H^+]$ and the hydroxide ion concentration, [OH^-], are equal in a pH 7 solution. Measures of pH are often erroneously used synonymously with alkalinity/acidity. However, the latter include numerous other ions, besides H^+ and OH^-.

Photolysis Light-initiated chemical reaction that degrades a chemical. Direct photolytic reactions form new compounds by absorbing the light, while indirect photolysis results from having a molecule first absorb the light energy and then transfer it to another compound that is degraded.

Photon Distinct packet (i.e., "particle") of electromagnetic radiation.

Physiologically based pharmacokinetic (PBPK) modeling Simulation of absorption, distribution, metabolism, and excretion of chemical agents in a biological system over time based upon physiological factors and variables, such as respiration rates, blood flow, and endocrine processes.

Phytoremediation Method of remediating contaminated soil, sludge, sediments, or groundwater by the direct use of living green plants *in situ* or through contaminant removal, degradation, or containment. Growing and, in some instances, harvesting plants on a contaminated site is a passive technique used to clean up sites with shallow (i.e., at root depth), low to moderate levels of contamination. Phytoremediation can be combined with other cleanup techniques (e.g., treating the plants that translocate heavy metals). Phytoremediation has been used at sites contaminated with metals, pesticides, solvents, explosives, crude oil, polyaromatic hydrocarbons, and landfill leachates.

Pica Ingestion of nonfood substances.

Point of Departure (PoD) Dose that can be considered to be in the range of observed responses, without significant extrapolation. For example, modeling from animal or epidemiological data uses the PoD to demark the beginning of extrapolation to determine risk associated with human exposures to the same contaminant.

Point-source pollution Water pollution from a single point, such as sewage outfall.

Pollution Anything that degrades the quality of the environment.

Pollution prevention Substituting, changing, or improving equipment, processes, and activities so that less wastes or toxic chemicals are produced. For example, recycling, "clean" technologies, energy conservation, and waste minimization approaches have been applied to manufacturing, agricultural, and transportation activities. Another type of pollution prevention includes the so-called "green" chemistry and engineering initiatives, where compounds and processes are designed to prevent pollution from the outset, and possible sources of contamination are simply removed during the design phase.

Polychlorinated biphenyls (PCBs) Group of persistent and toxic industrial chemical compounds formerly used in industrial processes and as dielectric fluids in electrical transformers. PCBs were frequently found in industrial wastes, and subsequently found their way into surface and groundwaters. Although PCBs were virtually banned in 1979 with the passage of the Toxic Substances Control Act, they continue to appear in human and other tissues, owing to their persistence and bioaccumulation potential.

Polycyclic aromatic hydrocarbon (PAH) Compound in a group of over 100 different chemicals formed from incomplete combustion. When coal, oil and gas, garbage, or other organic substances like tobacco or charbroiled meat are burned, numerous mixtures of PAHs are released. Since most of these compounds are semivolatile, the phases in the atmosphere may be gaseous, liquid, or solid. The pure forms of PAHs usually are colorless, white, or pale yellow-green solids.

Polycyclic organic matter (POM) Class of compounds, including the polycyclic aromatic hydrocarbon compounds (PAHs), that are formed primarily from combustion and found in the atmosphere in particulate form. However, since many of these compounds are semivolatile, a portion may also be found in the gas phase. Because of limited emissions data, POM pollution is often indicated by either the group of 7 or the group of 16 individual PAH species about which most data are available. These two groups are referred to respectively as "7-PAH" and "16-PAH."

Polymerization Reaction wherein two or more identical molecules and/or ions join chemically to form large molecules or complexes (e.g., polyvinyl chlorides).

Pore water velocity Ratio of soil water flux density to volume water content. Pore water velocity is an expression of the mean velocity of liquid phase transport through soil or sediment pores.

Porosity Measure of water-bearing capacity of a substance.

Porous media Any solid phase with pore space (e.g., soil, aquifer matrix); any substance with a porosity >0.

Postclosure plan Steps to be taken by a hazardous waste facility to protect groundwater and to prevent exposures following cleanup. Requires environmental monitoring, reporting, waste containment, security, and other actions to prevent exposures for 30 years following closure of the site.

Potable water Water of a quality considered suitable for drinking.

Precipitation Reaction resulting in a solute exceeding its solubility limit and changing to the solid phase.

Preliminary assessment/Site inspection (PA/SI) First stage of collecting data and evaluating a site that contains a hazardous waste. Required under the Comprehensive Environmental Response, Compensation, and Liability Act of 1980 (42 U.S.C. s/s 9601 et seq.).

Primary wastewater treatment Initial stage of the wastewater-treatment process that uses mechanical methods, such as filters and scrapers, to remove pollutants. Solid material in sewage settles and is collected during this treatment process.

Products of incomplete combustion Compounds formed from thermal processes of incineration and manufacturing. Includes dioxins, furans, hydrocarbons, and polycyclic aromatic hydrocarbons.

Protonation Attack of chemical compound by hydrogen ions.

Pump and treat system Groundwater remediation system consisting of two stages: (1) physical removal of contaminated water; and (2) chemical and/or biological treatment of the removed water. After treatment, the clean water is returned to the aquifer.

Pyrolysis Decomposition with heat in the absence of oxygen.

Quality assurance/Quality control (QA/QC) Approaches and procedures employed to ensure accurate and reliable results from environmental studies. A QA/QC plan includes field and laboratory protocols for sample collection and handling, and blanks, duplicates, and split samples in the laboratory.

Raoult's Law States that the vapor pressure of a solvent above a solution is equal to the product of the mole fraction of a solvent in the solution and the vapor pressure of the pure solvent.

Rate constant Proportionality constant for the rate of a chemical reaction.

Reactivity Hazardous property of a chemical owing to the chemical's high likelihood to react chemically with other substances in the environment.

Receptor (1) Person, organism, or material that is exposed to a contaminant. (2) Location on a cell wall or within a cell that binds to a chemical.

Receptor model Models used to determine pollutant sources (usually air pollution) that are based upon measurements of tracers taken in the ambient environment. Contrast with dispersion model.

Recharge Water that is added to an aquifer; e.g., when rainfall percolates into the ground.

Reclaimed wastewater Treated wastewater used for beneficial purposes, such as irrigating certain plant life.

Record of Decision (ROD) Document that contains the selected remedial action to be taken at a site, based upon the results of the remedial investigation/feasibility study.

Recycled water Water used more than once before returning to a hydrologic system.

Reference concentration (RfC) Estimate of the daily inhalation exposure to a human population that is likely to be without appreciable risk of adverse effects during a lifetime. Like the reference dose, the RfC uncertainty spans perhaps an order of magnitude and incorporates sensitive subgroups, e.g., children, asthmatics and the elderly. It can be derived from various types of human or animal data, with uncertainty factors generally applied to reflect limitations of the data used. See *Reference dose.*

Reference dose (RfD) Estimate of the daily exposure to a human population that is likely to be without adverse effects during a lifetime. See *Reference concentration (RfC).*

Remedial design/Remedial action (RD/RA) Specification of remedies that will be undertaken at a site and all plans for meeting cleanup standards for all environmental media.

Remedial investigation/Feasibility study (RI/FS) Formal study following the initial investigation of a hazard to assess the nature and the extent of contamination.

Remediation Process of cleaning up a contaminated site.

Reservoir Surface water basin, either natural or artificial, that is stored or released to control the flow of water and to prevent flooding.

Restricted use pesticide Pesticide that can be sold to or used by only certified applicators.

Reverse osmosis (1) Advanced water or wastewater treatment using a semi-permeable membrane to separate waters from pollutants. An external force is used to reverse the normal osmotic process that results in the solvent moving from a solution of higher concentration to one of lower concentration. (2) Membrane process used to remove dissolved salts from water. Water passes through a fine membrane but salt molecules are retained. The brine waste is removed and disposed. This process differs from electrodialysis.

Risk Probability of an adverse outcome resulting from an exposure to a chemical.

Risk assessment paradigm Scientific framework for assessing risks. The paradigm usually includes hazard identification, dose-response determinations, exposure assessments, and risk characterization.

Risk factor Characteristic (e.g., ethnicity, race, sex, age, or obesity) or variable (e.g., smoking, fat intake, alcohol consumption, or level of exposure) associated with increased probability of an adverse effect. Some standard risk factors used in general risk assessment calculations include mean breathing rates, body weight, and human life span.

Risk management Set of engineering and policy approaches that are employed to prevent, remove, treat, exchange, and recycle wastes identified and characterized by a risk assessment.

Rodenticide Type of chemical used to kill rats and other rodents.

Route of exposure How a chemical enters an organism after contact, e.g., by dermal, ingestion, or inhalation.

Sand A soil or detritus particle with a grain size larger than silt and smaller than gravel. Generally, sand grains range between 0.07 to less than 5 mm diameter.

Saturation effect Point on the dose-response curve where the curve's slope is no longer linear and begins to become less steep.

Scientific method Structured approach to science consisting of initial observations and formulation of research or study objectives, forming hypotheses, collecting information (including experimentation), analyzing that information and summarizing the results, reaching a conclusion, and identifying the need for new areas of scientific inquiry and endeavor. Many attribute this process to Robert Boyle and the Royal Society of London in the 17th century.

Screen/screening Process for evaluating substances using tests or assays ("screens") to detect a specific property or behavior in the environment. Results from screening tests and studies may be used to conduct more focus on environmental assessments for those contaminants found with the screens.

Sediment (1) Matter that was once suspended in a liquid and has subsequently settled to the bottom as a multiphase system of organic (including living microbes) and inorganic matter. (2) Material, such as soil, that has eroded and washed into a water body.

Sedimentation Process of the deposition of solids from the water column, e.g., to form sediments at the bottom of a surface water body.

Semivolatile organic compound (SVOC) Compound that may exist in various physical states, depending upon environmental conditions. Generally includes compounds with vapor pressures less than 10^{-2} kilopascals but greater than 10^{-5} kilopascals.

Sensitivity (1) In epidemiology, probability of the test finding disease or other adverse outcome among members of a population who, in fact, have the disease, or the proportion of a portion of a population with the adverse effect who have a positive test result. Sensitivity is calculated as: Sensitivity = true positives / (true positives + false negatives). Compare to *specificity*. (2) In analytical chemistry, extent that a small change in concentration of an analyte will engender a large change in the analytical measurement.

Series Horizontal columns in the periodic table.

7-PAH Chemical group that is frequently used as an indication of polycyclic aromatic hydrocarbon and polycyclic organic matter pollution. The group includes seven compounds: Benz[a]anthracene, Benzo[b]fluoranthene, Benzo[k]fluoranthene, Benzo[a]pyrene, Chrysene, Dibenz[a,h]anthracene, and Indeno[1,2,3-cd]pyrene. The 7-PAH are a subset of another indicator group, 16-PAH. All seven species that comprise the 7-PAH are probably human carcinogens.

Significant figure Number that is considered to be correct within a specified or implied limit of error.

Silt Grains of soil finer than sand and coarser than clay (commonly between 0.07 and 0.002 mm grain diameter).

Sink Process or location in the environment that removes a contaminant from a medium or compartment. For example, trees can be sinks for carbon dioxide, and soils with certain levels of moisture can be sinks for tropospheric sulfur dioxide.

16-PAH Chemical group used as an indication of and surrogate for the large group of polycyclic aromatic hydrocarbon and polycyclic organic matter pollution, comprised of the sum of 16 specific compounds: acenaphthene; acenaphthylene; anthracene; benzo(k)fluoranthene; benz(a)anthracene; benzo(a)pyrene; benzo(b)fluoranthene; benzo(ghi)perylene; chrysene; dibenz(a,h)anthracene; fluoranthene; fluorine; indeno(1,2,3-cd)pyrene; naphthalene; phenanthrene; and pyrene.

slope factor (SF) See *Toxicity slope factor.*

Slurry Mixture of solids and fluid that can be pumped.

Solid waste Solid, semisolid, liquid, or contained gaseous materials discarded from industrial, commercial, mining, or agricultural operations, and from community activities. This includes garbage, construction debris, commercial refuse, sludge from water supply or waste treatment plants, and air pollution control facilities.

Solubility Mass of a substance (i.e., solute) that will dissolve in a unit volume of a solvent. If the solvent is water, this is known as aqueous solubility.

Solvation Process whereby a solute particle is surrounded with solvent particles.

Sorption Solid-liquid distribution of a chemical in a given volume of an aquatic compartment (see *Absorption* and *Adsorption*).

Source apportionment Sorting out the various sources, often by using receptor models.

Source reduction Design, manufacture, purchase, or use of materials to reduce the quantity or toxicity of waste to be generated.

Speciation Determination of the specific chemical form or compound in which an element occurs in a sample, e.g., determination of whether chromium occurs as trivalent or hexavalent ions, or as part of an organometallic molecule. Quantification of the distribution of each chemical form found in the environment. For example, in a study of mercury in a wetland sample, there may be 50 ng L^{-1} zero valence mercury (Hg^0), 5 ng L^{-1} monomethyl mercury, and 2 ng L^{-1} dimethyl mercury. A nonspeciated analysis of the same sample could only show 57 ng L^{-1}.

Species differences in sensitivity Differing responses by different biological species to the actions of a pollutant. For example, a rat's endocrine system may be particularly vulnerable to a compound, while a Rhesus monkey may show no effects whatsoever. Or, a rat may generate tumors of the fore stomach, but humans may develop tumors in the large intestines after exposure to the same chemical. A special form of this difference is "species-specific sensitivity," where a response to a toxic substance is characteristic for particular species of living organism. An example might include the raven family of birds' particularly high mortality rate when exposed to the West Nile virus.

Specific gravity Ratio of the weight of a given volume of a substance to that of a given volume of water.

Specific yield Ratio of water volume that a given mass of saturated soil will yield by gravity.

Specificity (1) Probability of a test finding no adverse effect in members of a population who, in fact, do not have the adverse effect. Specificity is found as: Specificity = true negatives / (true negatives + false positives). Compare to *sensitivity*. (2) In analytic chemistry, extent to which a method provides a response from the detection system that is considered exclusively characteristic of the analyte. For example, in mass spectrometry, the presence of a particular fragment is only found in one class of chemicals, so the fragment is specific to that chemical class. In reality, few fragments are completely specific to any chemical, so it becomes a matter of the level of specificity; e.g., 99.87% of the time in environmental studies, this fragment is associated with this isomer.

Spike Known amounts of specific chemical constituents added by the laboratory to selected samples to determine recovery efficiencies of specific analytical methods within the actual sample matrices. So, if a solvent is spiked to have 10 ng L^{-1} of tetrachloromethane, but the chromatography quantitiation shows 9.8 ng L^{-1}, then this −2% difference may reflect a problem in detection, or at least indicates the confidence of the analysis. Another type of spike or "dosing" is used to determine extraction efficiencies or recoveries from trapped material. For example, if contaminants are collected using polyurethane foam (PUF), it must be determined beforehand how efficiently the contaminant can be removed from the PUF. So, if the PUF is spiked with 100 ng of the compound and the extraction only removes 90 ng of the spiked compound, the recovery is only 90%.

Spore General term for a reproductive structure in fungi, bacteria, and some plants. In fungi, the spore is the structure that may be used for dissemination and may be resistant to adverse environmental conditions.

Stachybotrys Genus that includes approximately 10 species and occurs mainly on dead plant materials. Of these, *Stachybotrys* chartarum is the most common.

Standard Industrial Classification (SCI) code A method of grouping industries with similar products or services and assigning codes to these groups.

Standardized mortality ratio (SMR) The comparison of observed deaths in a population to the expected number of deaths as derived from rates in a standard population with adjustment of age and possibly other factors such as sex or race.

Steady-state inhibition Point in time when continued dosing at the same level ceases to increase the effect. For example, steady-state inhibition is achieved when an animal dosed with a pesticide no longer has an increase in cholinesterase inhibition.

Stoichiometry Relationship between the masses of the reactants and the masses of products of a chemical reaction. In a balanced reactions, the proportionality between the reactant and product masses must be maintained.

Stratigraphy Arrangement of geologic strata.

Stripping Removal of organic compounds (usually volatile organic compounds) from a soil or other contaminated matrix. The compounds are volatilized and transferred into a gas flow. The gas is then collected and treated.

Subatomic particles Particles comprising and less complex than an atom. Generally, in environmental sciences, the most important subatomic particles are electrons, protons, and neutrons. Also known as fundamental particles.

Surface tension Force holding a liquid in pore spaces of a matrix preventing flow due to gravity.

Surface water Water that is naturally open to the atmosphere (i.e., oceans, lakes, streams, and puddles).

Surfactant Substance, like a detergent, that lowers the surface or interfacial tension of the medium in which it is dissolved. Molecule with a charged, hydrophilic head group, and a hydrophobic, hydrocarbon tail.

Tensor Vector whose magnitude depends on direction, e.g., a wind that gusts at 30 knots from the north and 10 knots from the east.

Teratogenesis Nonhereditary birth defects in a developing unborn child by exogenous factors such as physical or chemical agents (i.e., teratogens) acting in the womb to interfere with normal embryonic development.

Texture Grain size of soil particles.

Threshold Lowest dose of a chemical at which a specified effect is observed and below which it is not observed. See *No observable adverse effect level.*

Threshold level Time-weighted average pollutant concentration values, exposure above which is likely to adversely affect human health; used more commonly for occupational than ambient environmental exposures.

Threshold limit value (TLV) Concentration of an airborne substance that a healthy person can be exposed to for a 40-hour work week without adverse effect; a workplace exposure standard. The TLV represents the conditions under which all workers may be exposed day after day without an expected adverse effect. TLVs may be expressed as: (1) TLV-TWA—time-weighted average (TWA) based on an allowable exposure averaged over a normal 8-hour workday or 40-hour workweek; (2) TLV-STEL—short-term exposure limit or maximum concentration for a brief, specified period of time, depending on a specific chemical (TWA must still be met); and (3) TLV-C—ceiling exposure limit or maximum exposure concentration not to be exceeded under any circumstances (TWA must still be met).

Tolerance (1) Ability of an organism to withstand adverse conditions, such as exposure to toxic substances. (2) In adherence to the Federal Insecticide, Fungicide, and Rodenticide Act, the quantity of pesticide that may safely remain in or on raw farm products at the time of sale. When a pesticide is registered for use on a food or feed crop, a tolerance is established by the U.S. EPA and enforced by the Food and Drug Administration and the Department of Agriculture. (3) In engineering, specification of properties needed to consider variability (e.g. a pipe has a tolerance of ± 0.5 mm of interior diameter).

Total dissolved solids (TDS) Amount of dissolved material in a given volume of water.

Total suspended particulates (TSP) Amount of all liquid and solid phase particles, including both the fine and course fractions (aerosols) in a given volume of air (analogous to total suspended solids in water).

Total suspended solids (TSS) Amount of solid phase matter in a given volume of water (analogous to total suspended particulates in air).

Toxic Ability to cause adverse effects.

Toxic equivalency factor (TEF) Aggregate means of estimating the risks associated with exposure to chemical classes of highly toxic groups, such as the chlorinated dioxins and furans, polycyclic aromatic hydrocarbons, and polychlorinated biphenyls. Usually compared to the most toxic isomer (e.g., tetrachorodibenzo-p-dioxin for dioxins and furans).

Toxic Release Inventory (TRI) Database of annual toxic releases reported by industry on about 350 toxic chemicals within 22 chemical categories that may be released directly to air, water, or land, inject underground, or transfer to off-site facilities. The information is made available to the public under the "Community Right-to-Know" portion of the Superfund law.

Toxicity Characteristic of chemical wherein it can cause acute or chronic adverse effects in humans or wildlife.

Toxicity characteristic leaching procedure (TCLP) Test designed to provide replicable results for organic compounds and to yield the same type of results for inorganic substances as those from the original extraction procedure (i.e., the so-called EP Tox Test) test; 25 organic compounds were added to the original EP list.

Toxicity slope factor Dimensionless slope factors used to calculate the estimated probability of increased occurrence of adverse outcome over a person's lifetime (for cancer, this is the so-called "excess lifetime cancer risk" or ELCR). Like the reference doses, slope factors follow exposure pathways.

Toxicity testing Biological testing (e.g., with invertebrate, fish, or small mammal) to determine the existence and severity of adverse effects of a substance. Usually, extrapolated to estimate possible adverse effects in humans using models.

Toxicological profile Document prepared for a specific substance in which scientists interpret available information on the chemical to specify hazardous exposure levels. The profile also identifies knowledge gaps and uncertainties about the chemical.

Transformation Change in the chemical form of a substance after its release. This transformation can take place in the ambient environment (e.g., via hydrolytic processes) or within an organism via metabolism (known as biotransformation).

Transgenerational effects Adverse effects in the progeny or later generations of individuals who were actually exposed to a contaminant. DES is the classic example. Women were prescribed DES with little or no effect on them, but their daughters developed cervical cancers that have been linked to the DES exposures during their mothers' pregnancies.

Transmissivity Rate at which water moves through a unit width of an aquifer under a unit hydraulic gradient.

Transport Means by which a contaminant moves within and among compartments in the environment, especially by advection, dispersion, and diffusion.

Treatability Subjection of a waste to a treatment process to determine whether it is amenable to treatment or to determine the treatment efficiency or optimal process conditions for treatment.

Treatment, Storage, and Disposal (TSD) Facility Place where a hazardous substance is treated, stored, or disposed. TSD facilities are regulated by the U.S. EPA and states under the Resource Conservation and Recovery Act of 1976.

Troposphere Layer of the atmosphere closest to the earth's surface.

Turbidity Capacity of light to pass through water. An indirect measure of suspended solids in water. Indication of how cloudy water is due to suspended solids (often colloids).

Turbulent flow Flow characterized by mixing actions throughout the flow field, with the mixing caused by eddies of varying size within the flow.

Two-film model Theory that molecules move between the gas and liquid phases by first passing through two films, one on the gas side and one on the liquid side. The films join at an interface.

Typical It means a person living at the centroid (center of population mass) of a census tract and engaging in a range of activities (indoors and outdoors) that are representative of those in which individuals might engage. Does not refer to a specific individual or even the average over a group of individuals.

Ultraviolet (UV) light Electromagnetic radiation of wavelengths just shorter than visible light. The radiation from one part of the spectrum (UV-A) enhances plant life and is useful in some medical and dental procedures; UV wavelengths from other parts of the spectrum (UV-B) can cause skin cancer or other tissue damage. The ozone layer in the stratosphere partly shields the inhabitants on the earth's surface from ultraviolet rays reaching the earth's surface.

Uncertainty factor Adjustment to toxicity data to set acceptable human dose levels to protect against noncancer adverse outcomes. Designed to account for the large amounts of uncertainty from animal testing and other health data.

Unconsolidated aquifer Underground-water-bearing stratum composed of loose geologic materials, such as gravel or sand.

Uncontrolled site Abandoned hazardous waste site where wastes have been or are being released or may be released.

Underground storage tank (UST) Tank and associated underground piping that has 10% or more of its volume beneath the ground surface.

Unit risk estimate (URE) The Unit Risk Estimate is the upper-bound excess lifetime cancer risk estimated to result from continuous exposure to an agent at a concentration of 1 $\mu g/m^3$ in air. The interpretation of the Unit Risk Estimate would be as follows: if the Unit Risk Estimate = 1.5×10^{-6} per μg/m3, 1.5 excess tumors are expected to develop per 1,000,000 people if exposed daily for a lifetime to 1 μg of the chemical in 1 cubic meter of air. Unit Risk Estimates are considered upper-bound estimates, meaning they represent a plausible upper limit to the true value. (Note that this is usually not a true statistical confidence limit.) The true risk is likely to be less, but could be greater.

Upper-bound A plausible upper limit to the true value of a quantity, usually not a true statistical confidence limit.

Upper-bound lifetime cancer risk A plausible upper limit to the true probability that an individual will contract cancer over a 70-year lifetime as a result of a given hazard (such as exposure to a toxic chemical). This risk can be measured or estimated in numerical terms (e.g., one chance in a hundred).

Upper Confidence Limit (UCL) The Upper Confidence Limit is the upper bound of a confidence interval around any calculated statistic, most typically an average. For example, the 95% confidence interval for an average is the range of values that will contain the true average (i.e., the average of the full statistical population of all possible data) 95% of the time. In other words, we can say with 95% certainty that the "true" average will exceed the UCL only 2.5% of the time. The EPA has based most Unit Risk Estimates on the Upper Confidence Limit of response data or of fitted curves, to avoid underestimating the true Unit Risk Estimate in the face of uncertainty.

Urban Consistent with the definition the EPA used in its analyses to support the Integrated Urban Air Toxics Strategy, a county was considered "urban" if, based on 1990 census data, it either includes a metropolitan statistical area with a population greater than 250,000 or the U.S. Census Bureau designates more than 50% of the population as "urban." This definition does not necessarily apply for any regulatory or implementation purpose.

Urban air toxics Air pollutants deemed to pose the greatest potential health threat in urban areas. The Clean Air Act required that the U.S. EPA identify a list of at least 30 air toxics that pose the greatest potential health threat in urban areas as a subset of the 188 compounds addressed by the Act. The 33 compounds are:

Urban Air Toxics

1. acetaldehyde
2. acrolein
3. acrylonitrile
4. arsenic compounds
5. benzene
6. beryllium compounds
7. 1,3-butadiene
8. cadmium compounds
9. carbon tetrachloride
10. chloroform
11. chromium compounds
12. coke oven emissions
13. 1,3-dichloropropene
14. diesel particulate matter
15. ethylene dibromide
16. ethylene dichloride
17. ethylene oxide
18. formaldehyde
19. hexachlorobenzene
20. hydrazine
21. lead compounds
22. manganese compounds
23. mercury compounds
24. methylene chloride
25. nickel compounds
26. perchloroethylene
27. polychlorinated biphenyls
28. polycyclic organic matter
29. propylene dichloride
30. quinoline
31. 1,1,2,2-tetrachloroethane
32. trichloroethylene
33. vinyl chloride

Vadose zone Underground strata above the water table (i.e., unsaturated zone). Also called the zone of aeration.

Vapor The gas given off by substances that are solids or liquids at ordinary atmospheric pressure and temperatures.

Vapor density Weight of a pure vapor or gas compared with the weight of an equal volume of dry air at the same temperature and pressure. A vapor density less than one indicates that the material is lighter than air and may rise. A vapor density greater than one means the substance is heavier than air and will stay low to the ground.

Vapor dispersion The movement of vapor clouds or plumes in the air due to wind, gravity, spreading, and mixing.

Vapor pressure (P^0) Pressure exerted by a gaseous substance in equilibrium with its liquid or solid phase. The pressure exerted by a chemical vapor in equilibrium with its solid or liquid form at a given temperature. P^0 is used to calculate the rate of volatilization of a pure substance from a surface, or in estimating a Henry's Law constant for chemicals with low water solubility. The higher the vapor pressure, the more likely a chemical is to volatilize and exist in a gaseous state. Units are Pascals, atmospheres, and other units of pressure.

Vaporization Transfer of a chemical substance from the liquid or solid state to the gaseous state.

Variance Sum of the squares of the difference between the individual values of a set of numbers and the arithmetic mean of the set, divided by one less than the number of values.

Vector Measure that has magnitude and direction, e.g., acceleration of an aerosol moving in an air mass.

Velocity (v) Measure of the direction and rate of movement.

Vent Connection and piping serving as conduits for gases to enter and to exit equipment. One of the sources that must be considered in an air-quality emissions inventory.

Vent well Well designed to facilitate injection or extraction of air to and from a contaminated soil area.

Viscosity (μ) Fluid's resistance to flow. Absolute or dynamic viscosity of liquids increases with decreasing temperature. For gases, absolute viscosity increases with increasing temperature. Expressed as mass per length-time (e.g., $kg\,m^{-1}\,s^{-1}$). A common viscosity unit is the poise. One poise equals $1.0\,g\,m^{-1}\,s^{-1}$. Kinematic viscosity is absolute viscosity of a fluid divided by the density of the fluid.

Vitrification Immobilization of waste by converting it into a high-strength glass or glasslike substance. Vitrification is used to treat excavated waste or soil *in situ*. It is commonly used to treat radioactive material, and soils with high concentrations of volatile organic compounds and metals.

Volatile Any property of a substance indicating that it evaporates quickly (i.e., it has a relatively high vapor pressure).

Volatile organic compound (VOC) Compound with an affinity for the gas phase, usually with vapor pressures greater than 10^{-2} kilopascals.

Waste exchange Practice of matching the chemicals considered wastes from companies, laboratories, government agencies, and other entities with entities where those same chemicals are needed. Active waste exchange makes use of an organization (e.g., a clearinghouse) to arrange the transfer of waste chemicals from a waste generator to an entity needing the chemicals, whereas a passive waste exchange is one where information is made available more generally and the interested parties are responsible for working together (e.g., an "adopt a chemical" program that advertises available chemicals on the Internet).

Water column A hypothetical volume (often cylindrical) extending from the surface to the bottom of a surface water system. This volume is a parcel from which water quality and water resource measurements are taken.

Water-filled pore space (WFPS) An expression of the amount of water that can be stored by an unconsolidated material.

Water quality standard Combination of a designated use and the maximum concentration of a pollutant expected to protect that use for any given body of water. For example, in a trout stream, the standard for iron should not exceed $1\,mg\,L^{-1}$.

Water table Boundary between the saturated and unsaturated zones (i.e., between the vadose zone and the zone of saturation). Generally, the level to which water will rise in an unconfined aquifer. This is also the boundary where the pore water pressure of the aquifer material is equal to atmospheric pressure.

Water treatment Processing of source water to make it potable (consumable).

Weight-of-evidence Extent to which the available data links cause with effect. In toxicology and risk assessment, the extent to which an agent is linked to an adverse outcome, such as whether a substance causes cancer, immune dysfunction, reproductive and developmental disorders, hormonal effects, and neurotoxicity.

Weight-of-evidence for carcinogenicity A system used by the EPA for characterizing the extent to which the available data support the hypothesis that an agent causes cancer in humans. Under the EPA's 1986 risk assessment guidelines, the weight-of-evidence is described by categories "A through E":

- Group A (human carcinogen)
- Group B (probable human carcinogen)
  - Group B1: Compounds for which limited human data suggest a cause-and-effect relationship between exposure and cancer incidence (rate of occurrence) in humans
  - Group B2: Compounds for which animal data are sufficient to demonstrate a cause-and-effect relationship between exposure and cancer incidence (rate of occurrence) in animals, and human data are inadequate or absent
- Group C (possible human carcinogen)
- Group D (not classifiable as to human carcinogenicity)
- Group E (evidence of noncarcinogenicity)

Weir Device installed in a channel to gauge the flow rate of liquid. Essentially, the design of a weir is a dam built across an open channel over which the liquid flows, usually through a notch.

Wellhead protection area Surface and subsurface zone surrounding a well or well field that supplies a public water system that must be protected because contaminants may reach the well water.

Wetland Area that is inundated with or saturated by surface or groundwater frequently enough or for sufficient duration to support plants, birds, animals, and aquatic life. Wetlands generally include swamps, marshes, bogs, estuaries, and other inland and coastal areas, and are federally protected. Wetlands frequently serve as recharge/discharge areas and are known as "nature's kidneys" since they help purify water. Wetlands also have been referred to as natural sponges that absorb flood waters, functioning like natural tubs to collect overflow. Wetlands are important wildlife habitats, breeding grounds, and nurseries because of their biodiversity. Many endangered species as well as countless estuarine and marine fish and shellfish, mammals, waterfowl, and other migratory birds use wetland habitat for growth, reproduction, food, and shelter. Wetlands are among the most fertile, natural ecosystems in the world since they produce great volumes of food (plant material).

Wood preservatives Active ingredients used to protect wood from insects, fungi, and other types of chemical and biological degradation. These include inorganics (e.g., arsenic, chromium, and copper compounds), organics (e.g., pentachlorophenol, creosote, creosote-coal tar, and creosote petroleum), and organometallics (e.g., copper naphthenate).

Worker protection standards Limits on contaminant concentrations designed to reduce risks of illness or injury from occupational exposures.

Yield Quantity of water, expressed as a total quantity per year or as a rate of flows, that can be collected for a given use from surface or groundwater sources.

Z-list Toxic and Hazardous Substances Tables (Z-1, Z-2, and Z-3) of air contaminants issued by the Occupational Safety and Health Administration. OSHA considers any material found in these tables to be hazardous. http://www.epa.gov/pesticides/glossary/index.html-a.

Zone of aeration See *Vadose zone.*

Zone of saturation Underground stratum or strata with all pore spaces (i.e., interstices) filled with water that is under higher pressure than that of the atmosphere. The saturation zone is below the vadose zone.

Notes and Commentary

1. In addition to my own working definitions, the sources for the terms in this glossary include the following texts, as well as others cited in earlier chapters:

American Industrial Hygiene Association Glossaries.

V. Covello, 1992, "Risk Comparisons and Risk Communications," in *Communicating Risk to the Public*, ed. by Roger E. Kasperson and P. Stallen, Kluwer, New York, N.Y.

M. Derelanko, 1999, *CRC Handbook of Toxicology*, "Risk Assessment," M. J. Derelanko and M.A. Hollinger, eds., CRC Press, Boca Raton, Fla.

C. Klaasssen, 1996, *Casarett and Doull's Toxicology: The Basic Science of Poisons*, Fifth Edition, McGraw-Hill, New York, N.Y.

G. Rand, ed., 1995, *Fundamentals of Aquatic Toxicology*, Taylor & Francis, Washington, D.C.

U.S. Environmental Protection Agency, 1997, *Exposure Factors Handbook, Volume 1*; EPA/600/P-95/002FA; Washington, D.C.

U.S. Environmental Protection Agency, 2002, *Terms of Environment*, http://www.epa.gov/OCEPAterms/intro.htm.

U.S. Environmental Protection Agency, 2004, National Air Toxics Assessment; U.S. EPA, *Pesticides Glossary.*

D. Vallero, 2003, *Engineering the Risks of Hazardous Wastes*, Butterworth-Heinemann, Boston, Mass.

B. Wyman and L. Stevenson, 2001, *The Facts on File Dictionary of Environmental Science*, Checkmark Books, New York, N.Y.

APPENDIX 1

Information Needed to Prepare Environmental Impact Statements

National Environmental Policy Act of 1969

Title of Guidance	*Summary of Guidance*	*Citation*	*Relevant Regulation/Documentation*
Forty Most-Often-Asked Questions Concerning CEQ's National Environmental Policy Act Regulations	Provides answers to 40 questions most frequently asked concerning implementation of NEPA.	46 FR 18026, dated March 23, 1981	40 CFR Parts 1500–1508
Implementing and Explanatory Documents for Executive Order 12114, Environmental Effects Abroad of Major Federal Actions	Provides implementing and explanatory information for EO 12114. Establishes categories of Federal activities or programs as those that significantly harm the natural and physical environment. Defines which actions are excluded from the order and those that are not.	44 FR 18672, dated March 29, 1979	EO 12114, Environmental Effects Abroad of Major Federal Actions
Publishing of Three Memoranda for Heads of Agencies on: —Analysis of Impacts on	1/2 Discusses the irreversible conversion of unique agricultural lands by Federal Agency action (e.g., construction activities, developmental grants, and federal land management). Requires identification of	45 FR 59189, dated September 8, 1980	1/2 Farmland Protection Policy Act (7 U.S.C. §4201 et seq.) 3 The Wild and Scenic Rivers Act of 1965 (16 U.S.C. §1271 et seq.)

Prime or Unique Agricultural Lands (Memoranda 1 and 2)	and cooperation in retention of important agricultural lands in areas of impact of a proposed agency action. The agency must identify and summarize existing or proposed agency policies, to preserve or mitigate the effects of agency action on agricultural lands.
—Interagency Consultation to Avoid or Mitigate Adverse Effects on Rivers in the Nationwide Inventory (Memorandum 3)	3 "Each Federal agency shall, as part of its normal planning and environmental review process, take care to avoid or mitigate adverse effects on rivers identified in the Nationwide Inventory prepared by the Heritage Conservation and Recreation Service in the Department of the Interior." Implementing regulations includes determining whether the proposed action: affects an Inventory river; adversely affects the natural, cultural, and recreation values of the Inventory river segment; forecloses options to

Title of Guidance	*Summary of Guidance*	*Citation*	*Relevant Regulation/Documentation*
	classify any portion of the Inventory segment as a wild, scenic, or recreational river area, and incorporates avoidance/mitigation measures into the proposed action to maximum extent feasible within the agency's authority.		
Memorandum for Heads of Agencies for Guidance on Applying Section 404(r) of the Clean Water Act at Federal Projects That Involve the Discharge of Dredged or Fill Materials into Waters of the U.S., Including Wetlands	Requires timely agency consultation with U.S. Army Corps of Engineers (COE) and the U.S. Environmental Protection Agency (EPA) before a Federal project involves the discharge of dredged or fill material into U.S. waters, including wetlands. Proposing agency must ensure, when required, that the EIS includes written conclusions of the EPA and COE (generally found in Appendix).	Council on Environmental Quality, dated November 17, 1980	Clean Water Act (33 U.S.C. §1251 et seq.) EO 12088, Federal Compliance with Pollution Control Standards
Scoping Guidance	Provides a series of recommendations distilled from agency research	46 FR 25461, dated May 7, 1981	40 CFR Parts 1500–1508

	regarding the scoping process. Requires public notice; identification of significant and insignificant issues; allocation of EIS preparation assignments; identification of related analysis requirements in order to avoid duplication of work; and the planning of a schedule for EIS preparation that meshes with the agency's decision-making schedule.		
Guidance Regarding NEPA Regulations	Provides written guidance on scoping, categorical exclusions adoption regulations, contracting provisions, selecting alternatives in licensing and permitting situations, and tiering.	48 FR 34263, dated July 28, 1983	40 CFR Parts 1501, 1502, and 1508
National Environmental Policy Act (NEPA) Implementation Regulations, Appendices I, II, and III	Provides guidance on improving public participation, facilitating agency compliance with NEPA and CEQ implementing regulations. Appendix I updates required NEPA contacts; Appendix II compiles a list of Federal and Federal-State Agency Offices with jurisdiction by	49 FR 49750, dated December 21, 1984	40 CFR Part 1500

Title of Guidance	*Summary of Guidance*	*Citation*	*Relevant Regulation/Documentation*
	law or special expertise in environmental quality issues; and Appendix III lists the Federal and Federal-State Offices for receiving and commenting on other agencies' environmental documents.		
Incorporating Biodiversity Considerations into Environmental Impact Analysis under the National Environmental Policy Act	Provides for "acknowledging the conservation of biodiversity as national policy and incorporates its consideration in the NEPA process"; encourages seeking out opportunities to participate in efforts to develop regional ecosystem plans; actively seeks relevant information from sources both within and outside government agencies; encourages participating in efforts to improve communication, cooperation, and collaboration between and among governmental and nongovernmental entities; improves the availability of	Council on Environmental Quality, Washington, D.C., dated January 1993	Not applicable

	information on the status and distribution of biodiversity, and on techniques for managing and restoring it; and expands the information base on which biodiversity analyses and management decisions are based.		
Pollution Prevention and the National Environmental Policy Act	Pollution-prevention techniques seek to reduce the amount and/or toxicity of pollutants being generated, promote increased efficiency of raw materials and conservation of natural resources and can be cost-effective. Directs Federal agencies that to the extent practicable, pollution prevention considerations should be included in the proposed action and in the reasonable alternatives to the proposal, and to address these considerations in the environmental consequences section of an EIS and EA (when appropriate).	58 FR 6478, dated January 29, 1993	EO 12088, Federal Compliance with Pollution Control Standards
Considering Cumulative Effects under the	Provides a "framework for advancing environmental cumulative impacts analysis	January 1997	40 CFR §1508.7

Title of Guidance	*Summary of Guidance*	*Citation*	*Relevant Regulation/Documentation*
National Environmental Policy Act	by addressing cumulative effects in either an environmental assessment (EA) or an environmental impact statement." Also provides practical methods for addressing coincident effects (adverse or beneficial) on specific resources, ecosystems, and human communities of all related activities, not just the proposed project or alternatives that initiate the assessment process.		
Environmental Justice Guidance Under the National Environmental Policy Act	Provides guidance and general direction on Executive Order 12898, which requires each agency to identify and address, as appropriate, "disproportionately high and adverse human health or environmental effects of its programs, policies, and activities on minority populations and low-income populations."	Council on Environmental Quality, Washington, D.C., dated December 10, 1997	EO 12898, Federal Actions to Address Environmental Justice in Minority Populations and Low-Income Populations

Source: National Aeronautics and Space Administration, 2001, *Implementing the National Environmental Policy Act and Executive Order 12114*, Chapter 2.

Format of an Environmental Impact Statement

Cover Sheet (See next table for information to be included) Appendices

EXECUTIVE SUMMARY

TABLE OF CONTENTS

LIST OF ABBREVIATIONS AND ACRONYMS

MEASUREMENT CONVERSION TABLES

CHAPTERS:

1. PURPOSE AND NEED FOR THE ACTION
2. DESCRIPTION AND COMPARISON OF ALTERNATIVES
   - Description of proposed action and each reasonable alternative, including No-Action
   - Brief description of alternatives not considered in detail; explain why
   - Summary of environmental impacts of proposed action and reasonable alternatives, including No-Action
3. DESCRIPTION OF THE AFFECTED ENVIRONMENT
   - Appropriate-level descriptions of the physical, natural, and socioeconomic aspects of the environment that will be impacted, including, but not limited to, air quality, historical/cultural resources, threatened or endangered species and habitats, wetlands, floodplains, and other sensitive/protected resources
4. ENVIRONMENTAL CONSEQUENCES
   - Impact analyses for the proposed action and reasonable alternatives, including No-Action
   - Mandatory subsections
   - Relationship between Short-Term Use of the Human Environment and the Maintenance and Enhancement of Long-Term Productivity
   - Irreversible and Irretrievable Commitments of Resources
5. MITIGATION AND MONITORING (optional; can be incorporated into Chapter 4 if appropriate)
6. REFERENCES
7. LIST OF PREPARERS
8. AGENCIES, ORGANIZATIONS, AND INDIVIDUALS CONSULTED
   - Consulting Agencies
   - Distribution List
9. INDEX

Appendices (Final EIS must have a "Response to Comments" chapter; as either an appendix or in a separate volume.)

Source: National Aeronautics and Space Administration, 2001, *Implementing the National Environmental Policy Act and Executive Order 12114*, Chapter 6.

Required Cover Sheet for an Environmental Impact Statement

POPULAR NAME of PROPOSAL	
INCLUDES TYPE (e.g., DRAFT or FINAL)	
Lead Agency:	NASA, State name of Sponsoring Entity; name(s) of cooperating agency(ies) if appropriate
Point of Contact for Information:	Name, title, address, and phone number of NASA Point of Contact
Date:	Date of Issuance (recommend using month and year)
Abstract:	Succinct statement of proposed action; brief abstract of the EIS, stating proposed action, alternatives examined, and summary of key findings (the abstract may be printed on a separate page, if necessary).

Source: National Aeronautics and Space Administration, 2001, *Implementing the National Environmental Policy Act And Executive Order 12114*, Chapter 6.

APPENDIX 2

Safe Drinking Water Act Contaminants and Maximum Contaminant Levels

TABLE A.2.1
Microorganisms

Contaminant	*MCLG** *(mg L^{-1})*†	*MCL or TT** *(mg L^{-1})*†	*Potential Health Effects from Ingestion of Water*	*Sources of Contaminant in Drinking Water*
Cryptosporidium	zero	TT‡	Gastrointestinal illness (e.g., diarrhea, vomiting, cramps)	Human and fecal animal waste
Giardia lamblia	zero	TT‡	Gastrointestinal illness (e.g., diarrhea, vomiting, cramps)	Human and animal fecal waste
Heterotrophic plate count	n/a	TT‡	HPC has no health effects; it is an analytic method used to measure the variety of bacteria that are common in water. The lower the concentration of bacteria in drinking water, the better maintained the water system is.	HPC measures a range of bacteria that are naturally present in the environment
Legionella	zero	TT‡	Legionnaire's Disease, a type of pneumonia	Found naturally in water; multiplies in heating systems

Total Coliforms (including fecal coliform and *E. Coli*)	zero	5.0%¶	Not a health threat in itself; it is used to indicate whether other potentially harmful bacteria may be present§	Coliforms are naturally present in the environment; as well as feces; fecal coliforms and *E. coli* only come from human and animal fecal waste.
Turbidity	n/a	TT‡	Turbidity is a measure of the cloudiness of water. It is used to indicate water quality and filtration effectiveness (e.g., whether disease-causing organisms are present). Higher turbidity levels are often associated with higher levels of disease-causing microorganisms such as viruses, parasites and some bacteria. These organisms can cause symptoms such as nausea, cramps, diarrhea, and associated headaches.	Soil runoff
Viruses (enteric)	zero	TT‡	Gastrointestinal illness (e.g., diarrhea, vomiting, cramps)	Human and animal fecal waste

Source: U.S. Environmental Protection Agency, 2002, Report EPA 816-F-02-013.

TABLE A.2.2
Disinfection Byproducts

Contaminant	*MCLG** $(mg\,L^{-1})^{\dagger}$	*MCL or TT** $(mg\,L^{-1})^{\dagger}$	*Potential Health Effects from Ingestion of Water*	*Sources of Contaminant in Drinking Water*
Bromate	zero	0.010	Increased risk of cancer	By-product of drinking water disinfection
Chlorite	0.8	1.0	Anemia; infants and young children: nervous system effects	By-product of drinking water disinfection
Haloacetic acids (HAAs)	n/a[‖]	0.060	Increased risk of cancer	By-product of drinking water disinfection
Total Trihalomethanes (TTHMs)	none[#] — n/a[‖]	0.10 — 0.080	Liver, kidney, or central nervous system problems; increased risk of cancer	By-product of drinking water disinfection

TABLE A.2.3
Disinfectants

Contaminant	*MCLG** $(mg\,L^{-1})^{\dagger}$	*MRDL** $(mg\,L^{-1})^{\dagger}$	*Potential Health Effects from Ingestion of Water*	*Sources of Contaminant in Drinking Water*
Chloramines (as Cl_2)	MRDLG = 4*	MRDL = 4.0*	Eye/nose irritation; stomach discomfort, anemia	Water additive used to control microbes
Chlorine (as Cl_2)	MRDLG = 4*	MRDL = 4.0*	Eye/nose irritation; stomach discomfort	Water additive used to control microbes
Chlorine dioxide (as ClO_2)	MRDLG = 0.8*	MRDL = 0.8*	Anemia; infants and young children: nervous system effects	Water additive used to control microbes

TABLE A.2.4
Inorganic Chemicals

Contaminant	*MCLG** *(mg L^{-1})†*	*MCL or TT** *(mg L^{-1})†*	*Potential Health Effects from Ingestion of Water*	*Sources of Contaminant in Drinking Water*
Antimony	0.006	0.006	Increase in blood cholesterol; decrease in blood sugar	Discharge from petroleum refineries; fire retardants; ceramics; electronics; solder
Arsenic	0#	0.010 as of 01/23/06	Skin damage or problems with circulatory systems, and may have increased risk of getting cancer	Erosion of natural deposits; runoff from orchards, runoff from glass and electronics production wastes
Asbestos (fiber >10 micrometers)	7 million fibers per liter	7 MFL	Increased risk of developing benign intestinal polyps	Decay of asbestos cement in water mains; erosion of natural deposits

Barium	2	2	Increase in blood pressure	Discharge of drilling wastes; discharge from metal refineries; erosion of natural deposits
Beryllium	0.004	0.004	Intestinal lesions	Discharge from metal refineries and coal-burning factories; discharge from electrical, aerospace, and defense industries
Cadmium	0.005	0.005	Kidney damage	Corrosion of galvanized pipes; erosion of natural deposits; discharge from metal refineries; runoff from waste batteries and paints
Chromium (total)	0.1	0.1	Allergic dermatitis	Discharge from steel and pulp mills; erosion of natural deposits
Copper	1.3	TT**; Action Level = 1.3	Short-term exposure: gastrointestinal distress	Corrosion of household plumbing systems; erosion of natural deposits

TABLE A.2.4 *(continued)*

Contaminant	*MCLG* (mg L^{-1})†*	*MCL or TT* (mg L^{-1})†*	*Potential Health Effects from Ingestion of Water*	*Sources of Contaminant in Drinking Water*
			Long-term exposure: liver or kidney damage	
			People with Wilson's Disease should consult their personal doctor if the amount of copper in their water exceeds the action level	
Cyanide (as free cyanide)	0.2	0.2	Nerve damage or thyroid problems	Discharge from steel/metal factories; discharge from plastic and fertilizer factories
Fluoride	4.0	4.0	Bone disease (pain and tenderness of the bones); children may get mottled teeth	Water additive that promotes strong teeth; erosion of natural deposits; discharge from fertilizer and aluminum factories

Lead	zero	TT**; Action Level = 0.015	Infants and children: delays in physical or mental development; children could show slight deficits in attention span and learning abilities Adults: kidney problems; high blood pressure	Corrosion of household plumbing systems; erosion of natural deposits
Mercury (inorganic)	0.002	0.002	Kidney damage	Erosion of natural deposits; discharge from refineries and factories; runoff from landfills and croplands
Nitrate (measured as Nitrogen)	10	10	Infants below the age of six months who drink water containing nitrate in excess of the MCL could become seriously ill and, if untreated, may die. Symptoms include shortness of breath and blue-baby syndrome.	Runoff from fertilizer use; leaching from septic tanks, sewage; erosion of natural deposits

TABLE A.2.4 *(continued)*

Contaminant	*MCLG* $(mg\,L^{-1})^{\dagger}$*	*MCL or TT* $(mg\,L^{-1})^{\dagger}$*	*Potential Health Effects from Ingestion of Water*	*Sources of Contaminant in Drinking Water*
Nitrite (measured as Nitrogen)	1	1	Infants below the age of six months who drink water containing nitrite in excess of the MCL could become seriously ill and, if untreated, may die. Symptoms include shortness of breath and blue-baby syndrome.	Runoff from fertilizer use; leaching from septic tanks, sewage; erosion of natural deposits
Selenium	0.05	0.05	Hair or fingernail loss; numbness in fingers or toes; circulatory problems	Discharge from petroleum refineries; erosion of natural deposits; discharge from mines
Thallium	0.0005	0.002	Hair loss; changes in blood; kidney, intestine, or liver problems	Leaching from ore-processing sites; discharge from electronics, glass, and drug factories

TABLE A.2.5
Organic Chemicals

Contaminant	*MCLG* ($mg\,L^{-1}$)†*	*MCL or TT* ($mg\,L^{-1}$)†*	*Potential Health Effects from Ingestion of Water*	*Sources of Contaminant in Drinking Water*
Acrylamide	zero	TT***	Nervous system or blood problems; increased risk of cancer	Added to water during sewage/wastewater treatment
Alachlor	zero	0.002	Eye, liver, kidney or spleen problems; anemia; increased risk of cancer	Runoff from herbicide used on row crops
Atrazine	0.003	0.003	Cardiovascular system or reproductive problems	Runoff from herbicide used on row crops
Benzene	zero	0.005	Anemia; decrease in blood platelets; increased risk of cancer	Discharge from factories; leaching from gas storage tanks and landfills
Benzo(a)pyrene (PAHs)	zero	0.0002	Reproductive difficulties; increased risk of cancer	Leaching from linings of water storage tanks and distribution lines
Carbofuran	0.04	0.04	Problems with blood, nervous system, or reproductive system	Leaching of soil fumigant used on rice and alfalfa
Carbon tetrachloride	zero	0.005	Liver problems; increased risk of cancer	Discharge from chemical plants and other industrial activities

TABLE A.2.5 *(continued)*

Contaminant	*MCLG* (mg L^{-1})†*	*MCL or TT* (mg L^{-1})†*	*Potential Health Effects from Ingestion of Water*	*Sources of Contaminant in Drinking Water*
Chlordane	zero	0.002	Liver or nervous system problems; increased risk of cancer	Residue of banned termiticide
Chlorobenzene	0.1	0.1	Liver or kidney problems	Discharge from chemical and agricultural chemical factories
2,4-D	0.07	0.07	Kidney, liver, or adrenal gland problems	Runoff from herbicide used on row crops
Dalapon	0.2	0.2	Minor kidney changes	Runoff from herbicide used on rights of way
1,2-Dibromo-3-chloropropane (DBCP)	zero	0.0002	Reproductive difficulties; increased risk of cancer	Runoff/leaching from soil fumigant used on soybeans, cotton, pineapples, and orchards
o-Dichlorobenzene	0.6	0.6	Liver, kidney, or circulatory system problems	Discharge from industrial chemical factories
p-Dichlorobenzene	0.075	0.075	Anemia; liver, kidney or spleen damage; changes in blood	Discharge from industrial chemical factories
1,2-Dichloroethane	zero	0.005	Increased risk of cancer	Discharge from industrial chemical factories
1,1-Dichloroethylene	0.007	0.007	Liver problems	Discharge from industrial chemical factories

cis-1,2-Dichloroethylene	0.07	0.07	Liver problems	Discharge from industrial chemical factories
trans-1,2-Dichloroethylene	0.1	0.1	Liver problems	Discharge from industrial chemical factories
Dichloromethane	zero	0.005	Liver problems; increased risk of cancer	Discharge from drug and chemical factories
1,2-Dichloropropane	zero	0.005	Increased risk of cancer	Discharge from industrial chemical factories
Di(2-ethylhexyl) adipate	0.4	0.4	Weight loss, liver problems, or possible reproductive difficulties	Discharge from chemical factories
Di(2-ethylhexyl) phthalate	zero	0.006	Reproductive difficulties; liver problems; increased risk of cancer	Discharge from rubber and chemical factories
Dinoseb	0.007	0.007	Reproductive difficulties	Runoff from herbicide used on soybeans and vegetables
Dioxin (2,3,7,8-TCDD)	zero	0.00000003	Reproductive difficulties; increased risk of cancer	Emissions from waste incineration and other combustion; discharge from chemical factories
Diquat	0.02	0.02	Cataracts	Runoff from herbicide use

TABLE A.2.5 *(continued)*

Contaminant	*MCLG** $(mg\,L^{-1})^{\dagger}$	*MCL or TT** $(mg\,L^{-1})^{\dagger}$	*Potential Health Effects from Ingestion of Water*	*Sources of Contaminant in Drinking Water*
Endothall	0.1	0.1	Stomach and intestinal problems	Runoff from herbicide use
Endrin	0.002	0.002	Liver problems	Residue of banned insecticide
Epichlorohydrin	zero	TT***	Increased cancer risk, and over a long period of time, stomach problems	Discharge from industrial chemical factories; an impurity of some water treatment chemicals
Ethylbenzene	0.7	0.7	Liver or kidneys problems	Discharge from petroleum refineries
Ethylene dibromide	zero	0.00005	Problems with liver, stomach, reproductive system, or kidneys; increased risk of cancer	Discharge from petroleum refineries
Glyphosate	0.7	0.7	Kidney problems; reproductive difficulties	Runoff from herbicide use
Heptachlor	zero	0.0004	Liver damage; increased risk of cancer	Residue of banned termiticide
Heptachlor epoxide	zero	0.0002	Liver damage; increased risk of cancer	Breakdown of heptachlor
Hexachlorobenzene	zero	0.001	Liver or kidney problems; reproductive difficulties; increased risk of cancer	Discharge from metal refineries and agricultural chemical factories

Hexachlorocyclopentadiene	0.05	0.05	Kidney or stomach problems	Discharge from chemical factories
Lindane	0.0002	0.0002	Liver or kidney problems	Runoff/leaching from insecticide used on cattle, lumber, gardens
Methoxychlor	0.04	0.04	Reproductive difficulties	Runoff/leaching from insecticide used on fruits, vegetables, alfalfa, livestock
Oxamyl (Vydate)	0.2	0.2	Slight nervous system effects	Runoff/leaching from insecticide used on apples, potatoes, and tomatoes
Polychlorinated biphenyls (PCBs)	zero	0.0005	Skin changes; thymus gland problems; immune deficiencies; reproductive or nervous system difficulties; increased risk of cancer	Runoff from landfills; discharge of waste chemicals
Pentachlorophenol	zero	0.001	Liver or kidney problems; increased cancer risk	Discharge from wood preserving factories
Picloram	0.5	0.5	Liver problems	Herbicide runoff
Simazine	0.004	0.004	Problems with blood	Herbicide runoff
Styrene	0.1	0.1	Liver, kidney, or circulatory system problems	Discharge from rubber and plastic factories; leaching from landfills

TABLE A.2.5 (*continued*)

Contaminant	*MCLG* (mg L^{-1})†*	*MCL or TT* (mg L^{-1})†*	*Potential Health Effects from Ingestion of Water*	*Sources of Contaminant in Drinking Water*
Tetrachloroethylene	zero	0.005	Liver problems; increased risk of cancer	Discharge from factories and dry cleaners
Toluene	1	1	Nervous system, kidney, or liver problems	Discharge from petroleum factories
Toxaphene	zero	0.003	Kidney, liver, or thyroid problems; increased risk of cancer	Runoff/leaching from insecticide used on cotton and cattle
2,4,5-TP (Silvex)	0.05	0.05	Liver problems	Residue of banned herbicide
1,2,4-Trichlorobenzene	0.07	0.07	Changes in adrenal glands	Discharge from textile finishing factories
1,1,1-Trichloroethane	0.20	0.2	Liver, nervous system, or circulatory problems	Discharge from metal degreasing sites and other factories
1,1,2-Trichloroethane	0.003	0.005	Liver, kidney, or immune system problems	Discharge from industrial chemical factories
Trichloroethylene	zero	0.005	Liver problems; increased risk of cancer	Discharge from metal degreasing sites and other factories
Vinyl chloride	zero	0.002	Increased risk of cancer	Leaching from PVC pipes; discharge from plastic factories
Xylenes (total)	10	10	Nervous system damage	Discharge from petroleum factories; discharge from chemical factories

TABLE A.2.6
Radionuclides

Contaminant	*MCLG* ($mg\,L^{-1}$)†*	*MCL or TT* ($mg\,L^{-1}$)†*	*Potential Health Effects from Ingestion of Water*	*Sources of Contaminant in Drinking Water*
Alpha particles	none*** ——— zero	15 picocuries per liter ($pCi\,L^{-1}$)	Increased risk of cancer	Erosion of natural deposits of certain minerals that are radioactive and may emit a form of radiation known as alpha radiation
Beta particles and photon emitters	none[7] ——— zero	4 millirems per year	Increased risk of cancer	Decay of natural and man-made deposits of certain minerals that are radioactive and may emit forms of radiation known as photons and beta radiation
Radium 226 and Radium 228 (combined)	none[7] ——— — zero	$5\,pCi\,L^{-1}$	Increased risk of cancer	Erosion of natural deposits
Uranium	zero	$30\,ug\,L^{-1}$ as of 12/08/03	Increased risk of cancer, kidney toxicity	Erosion of natural deposits

Notes
* Table Legend:

MCL = Maximum Contaminant Level: The highest level of a contaminant that is allowed in drinking water. MCLs are set as close to MCLGs as feasible using the best available treatment technology and taking cost into consideration. MCLs are enforceable standards.

MCLG = Maximum Contaminant Level Goal: The level of a contaminant in drinking water below which there is no known or expected risk to health. MCLGs allow for a margin of safety and are nonenforceable public health goals.

MRDL = Maximum Residual Disinfectant Level: The highest level of a disinfectant allowed in drinking water. There is convincing evidence that addition of a disinfectant is necessary for control of microbial contaminants.

MRDLG = Maximum Residual Disinfectant Level Goal: The level of a drinking water disinfectant below which there is no known or expected risk to health. MRDLGs do not reflect the benefits of the use of disinfectants to control microbial contaminants.

TT = Treatment Technique: A required process intended to reduce the level of a contaminant in drinking water.

† Units are in milligrams per liter ($mg\,L^{-1}$) unless otherwise noted. Milligrams per liter are equivalent to parts per million under standard environmental conditions.

‡The EPA's surface water treatment rules require systems using surface water or groundwater under the direct influence of surface water to (1) disinfect their water, and (2) filter their water or meet criteria for avoiding filtration so that the following contaminants are controlled at the following levels:

- Cryptosporidium (as of 1/1/02 for systems serving >10,000 and 1/14/05 for systems serving <10,000) 99% removal.
- *Giardia lamblia:* 99.9% removal/inactivation
- Viruses: 99.99% removal/inactivation
- *Legionella:* No limit, but the EPA believes that if *Giardia* and viruses are removed/inactivated, *Legionella* will also be controlled.
- Turbidity: At no time can turbidity (cloudiness of water) go above 5 nephelolometric turbidity units (NTU); systems that filter must ensure that the turbidity go no higher than 1 NTU (0.5 NTU for conventional or direct filtration) in at least 95% of the daily samples in any month. As of January 1, 2002, turbidity may never exceed 1 NTU, and must not exceed 0.3 NTU in 95% of daily samples in any month.
- HPC: No more than 500 bacterial colonies per milliliter.
- Long-Term 1 Enhanced Surface Water Treatment (Effective Date: January 14, 2005); Surface water systems or (GWUDI) systems serving fewer than 10,000 people must comply with the applicable Long-Term 1 Enhanced Surface Water Treatment Rule provisions (e.g., turbidity standards, individual filter monitoring, Cryptosporidium removal requirements, and updated watershed control requirements for unfiltered systems).
- Filter Backwash Recycling; the Filter Backwash Recycling Rule requires systems that recycle to return specific recycle flows through all processes of the system's existing conventional or direct filtration system or at an alternate location approved by the state.

¶ More than 5.0% samples total coliform-positive in a month. (For water systems that collect fewer than 40 routine samples per month, no more than one sample can be total coliform-positive per month.) The U.S. EPA requires that total coliform samples be collected at sites which are representative of water quality throughout the distribution system according to a written sample siting plan subject to state review and revision. Samples must be collected at regular time intervals throughout the month except groundwater systems serving 4900 persons or fewer may collect them on the same day. Monthly sampling requirements are based on population served (see Table A.2.7 for the minimum sampling frequency). A reduced monitoring frequency may be available for systems serving 1000 persons or fewer and using only groundwater if a sanitary survey within the past 5 years shows the system is free of sanitary defects (the frequency may be no less than 1 sample per quarter for community and 1 sample per year for noncommunity systems). Each total coliform-positive routine sample must be tested for the presence of fecal coliforms or *E. coli*. If any routine sample is total coliform-positive, repeat samples are required.

Repeat Sampling Requirement:

Within 24 hours of learning of a total coliform-positive routine sample result, at least 3 repeat samples must be collected and analyzed for total coliforms:

- One repeat sample must be collected from the same tap as the original sample.
- One repeat sample must be collected within five service connections upstream.
- One repeat sample must be collected within five service connections downstream.
- Systems that collect 1 repeat sample per month or fewer must collect a 4th REPEAT sample.

If any REPEAT sample is total coliform-positive:

- The system must analyze that total coliform-positive culture for fecal coliforms or *E.coli*.
- The system must collect another set of REPEAT samples, as before, unless the MCL has been violated and the system has notified the state.

In addition, a positive routine or repeat total coliform result requires a minimum of five routine samples be collected the following month the system provides water to the public unless waived by the state.

§Fecal coliform and *E. coli* are bacteria whose presence indicates that the water may be contaminated with human or animal wastes. Disease-causing microbes (pathogens) in these wastes can cause diarrhea, cramps, nausea, headaches, or other symptoms. These pathogens may pose a special health risk for infants, young children, and people with severely compromised immune systems.
||Although there is no collective MCLG for this contaminant group, there are individual MCLGs for some of the individual contaminants:

- Trihalomethanes: bromodichloromethane (zero); bromoform (zero); dibromochloromethane ($0.06\,mg\,L^{-1}$). Chloroform is regulated with this group but has no MCLG.
- Haloacetic acids: dichloroacetic acid (zero); trichloroacetic acid ($0.3\,mg\,L^{-1}$). Monochloroacetic acid, bromoacetic acid, and dibromoacetic acid are regulated with this group but have no MCLGs.

#MCLGs were not established before the 1986 Amendments to the Safe Drinking Water Act. Therefore, there is no MCLG for this contaminant.
**Lead and copper are regulated by a Treatment Technique that requires systems to control the corrosiveness of their water. If more than 10% of tap water samples exceed the action level, water systems must take additional steps. For copper, the action level is $1.3\,mg\,L^{-1}$, and for lead it is $0.015\,mg\,L^{-1}$.
***Each water system must certify, in writing, to the state (using third-party or manufacturer's certification) that when acrylamide and epichlorohydrin are used in drinking water systems, the combination (or product) of dose and monomer level does not exceed the levels specified, as follows:

- Acrylamide = 0.05% dosed at $1\,mg\,L^{-1}$ (or equivalent)
- Epichlorohydrin = 0.01% dosed at $20\,mg\,L^{-1}$ (or equivalent)

National Secondary Drinking Water Regulations

National Secondary Drinking Water Regulations (NSDWRs or "secondary standards") are nonenforceable guidelines regulating contaminants that may cause cosmetic effects (such as skin or tooth discoloration) or aesthetic effects (such as taste, odor, or color) in drinking water. The EPA recommends secondary standards to water systems but does not require systems to comply. However, states may choose to adopt them as enforceable standards.

TABLE A.2.7
Public Water System Routine Monitoring Frequencies (Source: U.S. EPA)

Minimum Population	*Required Samples per Month*
25–1,000	1
1,000–2,500	2
2,501–3,300	3
3,301–4,100	4
4,101–4,900	5
4,901–5,800	6
5,801–6,700	7
6,701–7,600	8
7,601–8,500	9
8,501–12,900	10
12,901–17,200	15
17,201–21,500	20
21,501–25,000	25
25,001–33,000	30
33,001–41,000	40
41,001–50,000	50
50,001–59,000	60
59,001–70,000	70
70,001–83,000	80
83,001–96,000	90
96,001–130,000	100
130,001–220,000	120
220,001–320,000	150
320,001–450,000	180
450,001–600,000	210
600,001–780,000	240
780,001–970,000	270
970,001–1,230,000	300
1,230,001–1,520,000	330
1,520,001–1,850,000	360
1,850,001–2,270,000	390
2,270,001–3,020,000	420
3,020,001–3,960,000	450
≥3,960,001	480

TABLE A.2.8
Secondary Drinking Water Standards

Contaminant	*Secondary Standard*
Aluminum	0.05 to 0.2 $mg L^{-1}$
Chloride	250 $mg L^{-1}$
Color	15 (color units)
Copper	1.0 $mg L^{-1}$
Corrosivity	noncorrosive
Fluoride	2.0 $mg L^{-1}$
Foaming Agents	0.5 $mg L^{-1}$
Iron	0.3 $mg L^{-1}$
Manganese	0.05 $mg L^{-1}$
Odor	3 threshold odor number
pH	6.5–8.5
Silver	0.10 $mg L^{-1}$
Sulfate	250 $mg L^{-1}$
Total Dissolved Solids	500 $mg L^{-1}$
Zinc	5 $mg L^{-1}$

APPENDIX 3

Toxic Compounds Listed in the 1990 Clean Air Act Amendments

CAS #	*Chemical or Class*
75070	Acetaldehyde
60355	Acetamide
75058	Acetonitrile
98862	Acetophenone
53963	2-Acetylaminofluorene
107028	Acrolein
79061	Acrylamide
79107	Acrylic acid
107131	Acrylonitrile
8107051	Allyl chloride
92671	4-Aminobiphenyl
62533	Aniline
90040	o-Anisidine
1332214	Asbestos
71432	Benzene (including from gasoline)
92875	Benzidine
98077	Benzotrichloride
100447	Benzyl chloride
92524	Biphenyl
117817	Bis (2-ethylhexyl) phthalate (DEHP)
542881	Bis(chloromethyl) ether

CAS #	*Chemical or Class*
75252	Bromoform
106990	1,3-Butadiene
156627	Calcium cyanamide
105602	Caprolactam
133062	Captan
63252	Carbaryl
75150	Carbon disulfide
56235	Carbon tetrachloride
463581	Carbonyl sulfide
120809	Catechol
133904	Chloramben
57749	Chlordane
7782505	Chlorine
79118	Chloroacetic acid
532274	2-Chloroacetophenone
108907	Chlorobenzene
510156	Chlorobenzilate
67663	Chloroform
107302	Chloromethyl methyl ether
126998	Chloroprene
19773	Cresols/Cresylic acid (isomers and mixture)
95487	0-Cresol
108394	m-Cresol
106445	p-Cresol
98828	Cumene
94757	2,4-D, salts and esters
3547044	DDE
334883	Diazomethane
132649	Dibenzofurans
96128	1,2-Dibromo-3-chloropropane
84742	Dibutylphthalate
106467	1,4-Dichlorobenzene(p)
91941	3,3′-Dichlorobenzidene
111444	Dichloroethyl ether (Bis(2chloroethyl)ether)
542756	1,3-Dichloropropene
62737	Dichlorvos
111422	Diethanolamine
121697	N,N-Diethyl aniline (N,N-Dimethylaniline)
64675	Diethyl sulfate
119904	3,3-Dimethoxybenzidine
60117	Dimethyl aminoazobenzene
119937	3,3-Dimethylbenzidine
79447	Dimethyl carbamoyl chloride
68122	Dimethyl formamide
57147	1,1 Dimethylhydrazine
131113	Dimethyl phthalate
77781	Dimethyl sulfate

CAS #	Chemical or Class
534521	4,6-Dinitro-o-cresol, and salts
51285	2,4-Dinitrophenol
121142	2,4-Dinitrotoluene
123911	1,4-Dioxane (1,4-Diethyleneoxide)
122667	1,2-Diphenylhydrazine
106898	Epichlorohydrin (l-Chloro-2,3-epoxypropane)
106887	1,2-Epoxybutane
140885	Ethyl acrylate
100414	Ethyl benzene
51796	Ethyl carbamate (Urethane)
75003	Ethyl chloride (Chloroethane)
106934	Ethyl enedibromide (Dibromoethane)
107062	Ethyl enedichloride (1,2-Dichloroethane)
107211	Ethylene glycol
151564	Ethyleneimine (Aziridine)
75218	Ethylene oxide
96457	Ethylene thiourea
75343	Ethylidene dichloride (1,1-Dichloroethane)
50000	Formaldehyde
76448	Heptachlor
118741	Hexachlorobenzene
87683	Hexachlorobutadiene
77474	Hexachlorocyclopentadiene
67721	Hexachloroethane
822060	Hexamethylene-1,6-diisocyanate
680319	Hexamethylphosphoramide
110543	Hexane
302012	Hydrazine
7647010	Hydrochloric acid
7664393	Hydrogen fluoride (Hydrofluoric acid)
123319	Hydroquinone
78591	Isophorone
58899	Lindane (all isomers)
108316	Maleic anhydride
67561	Methanol
72435	Methoxychlor
74839	Methyl bromide (Bromomethane)
74873	Methyl chloride (Chloromethane)
71556	Methyl chloroform (1,1,1-Trichloroethane)
78933	Methyl ethyl ketone (2-Butanone)
60344	Methyl hydrazine
74884	Methyl iodide (Iodomethane)
108101	Methyl isobutyl ketone (Hexone)
624839	Methyl isocyanate
80626	Methyl methacrylate
1634044	Methyl tert butyl ether
101144	4,4-Methylene bis (2-chloroaniline)

CAS #	*Chemical or Class*
75092	Methylene chloride (Dichloromethane)
101688	Methylene diphenyl diisocyanate (MDI)
101779	4,4′-Methylenedianiline
91203	Naphthalene
98953	Nitrobenzene
92933	4-Nitrobiphenyl
100027	4-Nitrophenol
79469	2-Nitropropane
684935	N-Nitroso-N-methylurea
62759	N-Nitrosodimethylamine
59892	N-Nitrosomorpholine
56382	Parathion
82688	Pentachloronitrobenzene (Quintobenzene)
87865	Pentachlorophenol
108952	Phenol
106503	p-Phenylenediamine
75445	Phosgene
7803512	Phosphine
7723140	Phosphorus
85449	Phthalic anhydride
1336363	Polychlorinated biphenyls (Aroclors)
1120714	1,3-Propane sultone
57578	beta-Propiolactone
123386	Propionaldehyde
114261	Propoxur (Baygon)
78875	Propylene dichloride (1,2-Dichloropropane)
75569	Propylene oxide
75558	1,2-Propylenimine (2-Methyl aziridine)
91225	Quinoline
106514	Quinone
100425	Styrene
96093	Styrene oxide
1746016	2,3,7,8-Tetrachlorodibenzo-p-dioxin
79345	1,1,2,2-Tetrachloroethane
127184	Tetrachloroethylene (Perchloroethylene)
7550450	Titanium tetrachloride
108883	Toluene
95807	2,4-Toluene diamine
584849	2,4-Toluene diisocyanate
95534	o-Toluidine
8001352	Toxaphene (chlorinated camphene)
120821	1,2,4-Trichlorobenzene
79005	1,1,2-Trichloroethane
79016	Trichloroethylene
95954	2,4,5-Trichlorophenol
88062	2,4,6-Trichlorophenol
121448	Triethylamine

CAS #	Chemical or Class
1582098	Trifluralin
540841	2,2,4-Trimethylpentane
108054	Vinyl acetate
593602	Vinyl bromide
75014	Vinyl chloride
75354	Vinylidene chloride (1,1-Dichloroethylene)
1330207	Xylenes (isomers and mixture)
95476	o-Xylenes
108383	m-Xylenes
106423	p-Xylenes
NA	Antimony compounds
NA	Arsenic compounds (inorganic, including arsine)
NA	Beryllium compounds
NA	Cadmium compounds
NA	Chromium compounds
NA	Cobalt compounds
NA	Coke oven emissions
NA	Cyanide compounds[1]
NA	Glycol ethers[2]
NA	Lead compounds
NA	Manganese compounds
NA	Mercury compounds
NA	Mineral fibers[3]
NA	Nickel compounds
NA	Polycyclic organic matter[4]
NA	Radionuclides (including radon)[5]
NA	Selenium compounds

Note: For all listings above that contain the word "compounds" and for glycol ethers, the following applies: unless otherwise specified, these listings are defined as including any unique chemical substance that contains the named chemical (i.e., antimony, arsenic, etc.) as part of that chemical's infrastructure.

[1] X′CN where X = H′ or any other group where a formal dissociation may occur. For example, KCN or $Ca(CN)_2$.

[2] Includes mono- and di- ethers of ethylene glycol, diethylene glycol, and triethylene glycol $R\text{-}(OCH_2CH_2)_n\text{-}OR'$ where n = 1, 2, or 3: R = alkyl or aryl groups; R′ = R, H, or groups which, when removed, yield glycol ethers with the structure: $R\text{-}(OCH_2CH)_n\text{-}OH$. Polymers are excluded from the glycol category.

[3] Includes glass, rock, or slag fibers (or other mineral derived fibers) of average diameter 1 micrometer or less.

[4] Includes organic compounds with more than one benzene ring, and which have a boiling point greater than or equal to 100°C.

[5] A type of atom that spontaneously undergoes radioactive decay.

APPENDIX 4

Physical Constants

Quantity	*Symbol*	*Value*	*Unit*	*Relative Std. Uncert.* u_r
Avogadro constant	N_A, L	$6.02214199\ (47) \times 10^{23}$	mol^{-1}	7.9×10^{-8}
atomic mass constant				
$m_u = {}^1/_{12}\ m(^{12}C) = 1\,u$	m_u	$1.66053873\ (13) \times 10^{-27}$	kg	7.9×10^{-8}
$= 10^{-3}\,kg\,mol^{-1}/N_A$				
energy equivalent	$m_u c^2$	$1.49241778\ (12) \times 10^{-10}$	J	7.9×10^{-8}
in MeV		931.494013 (37)	MeV	4.0×10^{-8}
Faraday constant[a] $N_A e$	F	96485.3415 (39)	$C\,mol^{-1}$	4.0×10^{-8}
molar Planck constant	$N_A h$	$3.990312689\ (30) \times 10^{-10}$	$J\,s\,mol^{-1}$	7.6×10^{-9}
	$N_A hc$	0.11962656492 (91)	$J\,m\,Mol^{-1}$	7.6×10^{-9}
molar gas constant	R	8.314472 (15)	$J\,mol^{-1}\,K^{-1}$	1.7×10^{-6}
Boltzmann constant R/N_A	k	$1.3806503\ (24) \times 10^{-23}$	$J\,K^{-1}$	1.7×10^{-6}
in $eV\,K^{-1}$		$8.617342\ (15) \times 10^{-5}$	$eV\,K^{-1}$	1.7×10^{-6}
	k/h	$2.0836644\ (36) \times 10^{10}$	$Hz\,K^{-1}$	1.7×10^{-6}
	k/hc	69.50356 (12)	$m^{-1}\,K^{-1}$	1.7×10^{-6}
molar volume of ideal gas RT/p				
$T = 273.15\,K$, $p = 101.325\,kPa$	V_m	$22.413996\ (39) \times 10^{-3}$	$m^3\,mol^{-1}$	1.7×10^{-6}
Loschmidt constant N_A/V_m	n_0	$2.6867775\ (47) \times 10^{25}$	m^{-3}	1.7×10^{-6}
$T = 273.15\,K$, $p = 100\,kPa$	V_m	$22.710981\ (40) \times 10^{-3}$	$m^3\,mol^{-1}$	1.7×10^{-6}

Sackur-Tetrode constant				
(absolute entropy constant)[b]				
$^5/_2 + \ln[(2\pi m_u kT_1/h^2)^{3/2} kT_1/p_0]$				
$T_1 = 1\,K$, $p_0 = 100\,kPa$	S_0/R	−1.1517048 (44)		3.8×10^{-6}
$T_1 = 1\,K$, $p_0 = 101.325\,kPa$		−1.1648678 (44)		3.7×10^{-6}
Stefan-Boltzmann constant				
$(\pi^2/60)k^4/h^3c^2$	σ	$5.670400\ (40) \times 10^{-8}$	$W\,m^{-2}\,K^{-4}$	7.0×10^{-6}
first radiation constant $2\pi hc^2$	c_1	$3.74177107\ (29) \times 10^{-16}$	$W\,m^2$	7.8×10^{-8}
first radiation constant for spectral radiance $2hc^2$	c_u	$1.191042722\ (93) \times 10^{-16}$	$W\,m^2\,sr^{-1}$	7.8×10^{-8}
second radiation constant hc/k	c_2	$1.4387752\ (25) \times 10^{-2}$	m K	1.7×10^{-6}
Wien displacement law constant				
$b = \lambda_{max}T = c_2/4.965114231\ \ldots$	b	$2.8977686\ (51) \times 10^{-3}$	m K	1.7×10^{-6}

[a] The numerical value of F to be used in coulometric chemical measurements is 96485.3432 (76) [7.9×10^{-5}] when the relevant current is measured in terms of representations of the volt and ohm based on the Josephson and quantum Hall effects and the internationally adopted conventional values of the Josephson and von Klitzing constants K_{1-90} and R_{1-90} given in the "Adopted values" table.

[b] The entropy of an ideal monoatomic gas of relative atomic mass A_r is given by $S - S_0 + {}^3/_2R$ in $A_r - R\ln(p/p_0) + {}^5/_2R\ln(T/K)$.

These are published by the federal government, NIST, so I assume permission is not needed so long as we give the citation: Sources: P.J. Mohr and B.N. Taylor, 1998, CODATA Recommended Values of the Fundamental Physical Constants; *Journal of Physical and Chemical Reference Data*, 1999, Vol. 28, no. 6; and *Reviews of Modern Physics*, 2000, Vol. 72, no. 2 (via physics.nist.gov/constants).

APPENDIX 5

Universal Constants

Quantity	*Symbol*	*Value*	*Unit*	*Relative Std. Uncert.* u_r
Speed of light in vacuum	c, c_0	299792458	$\mathrm{m\,s^{-1}}$	(exact)
magnetic constant	μ_0	$4\pi \times 10^{-7}$	$\mathrm{N\,A^{-2}}$	
		$= 12.566370614 \ldots \times 10^{-7}$	$\mathrm{N\,A^{-2}}$	(exact)
Electric constant $1/\mu_0 c^2$	ε_0	$8.854187817 \ldots \times 10^{-12}$	$\mathrm{F\,m^{-1}}$	(exact)
characteristic impedance				
of vacuum $\sqrt{\mu_0/\varepsilon_0} = \mu_0 c$	Z_0	375.730313461 ...	Ω	(exact)
Newtonian constant				
of gravitation	G	$6.673\,(10) \times 10^{-11}$	$\mathrm{m^3\,kg^{-1}\,s^{-2}}$	1.5×10^{-3}
	G/hc	$6.707\,(10) \times 10^{-39}$	$(\mathrm{GeV}/c^2)^{-2}$	1.5×10^{-3}
Planck constant	h	$6.62606876\,(52) \times 10^{-34}$	J s	7.8×10^{-8}
in eV s		$4.13566727\,(16) \times 10^{-15}$	eV s	3.9×10^{-8}
$h/2\pi$	h	$1.054571596\,(82) \times 10^{-34}$	J s	7.8×10^{-8}
in eV s		$6.58211889\,(26) \times 10^{-16}$	eV s	$3.9 - 10^{-8}$
Planck mass $(hc/G)^{1/2}$	m_p	$2.1767\,(16) \times 10^{-8}$	kg	7.5×10^{-4}
Planck length $h/m_p c = (hG/c^3)^{1/2}$	l_p	$1.6160\,(12) \times 10^{-35}$	m	7.5×10^{-4}
Planck time $l_p/c = (hG/c^5)^{1/2}$	t_p	$5.3905\,(40) \times 10^{-44}$	S	7.5×10^{-4}

These are published by the federal government, NIST, so I assume permission is not needed so long as we give the citation: Sources: P.J. Mohr and B. N. Taylor, 1998, CODATA Recommended Values of the Fundamental Physical Constants; *Journal of Physical and Chemical Reference Data*, 1999, Vol. 28, no. 6; and *Reviews of Modern Physics*, 2000, Vol. 72, no. 2 (via physics.nist.gov/constants).

APPENDIX 6

Constants Frequently Applied in the Physical Sciences

Frequently Used Constants

Quantity	*Symbol*	*Value*	*Unit*	*Relative Std. Uncert.* u_r
speed of light in vacuum	c, c_0	299792458	$m\,s^{-1}$	(exact)
magnetic constant	μ_0	$4\pi \times 10^{-7}$	$N\,A^{-2}$	
		$= 12.566370614 \ldots \times 10^{-7}$	$N\,A^{-2}$	(exact)
electric constant $1/\mu_0 c^2$	ε_0	$8.854187817 \ldots \times 10^{-12}$	$F\,m^{-1}$	(exact)
Newtonian constant of gravitation	G	$6.673\ (10) \times 10^{-11}$	$m^3\,kg^{-1}\,s^{-2}$	1.5×10^{-3}
Planck constant	h	$6.62606876\ (52) \times 10^{-34}$	$J\,s$	7.8×10^{-8}
$h/2\pi$	h	$1.054571596\ (82) \times 10^{-34}$	$J\,s$	7.8×10^{-8}
elementary charge	e	$1.602176462\ (63) \times 10^{-19}$	C	3.9×10^{-8}
magnetic flux quantum $h/2e$	Φ_0	$2.067833636\ (81) \times 10^{-15}$	Wb	3.9×10^{-8}
conductance quantum $2e^2/h$	G_0	$7.748091696\ (28) \times 10^{-5}$	S	3.7×10^{-9}
electron mass	m_e	$9.10938188\ (72) \times 10^{-31}$	kg	7.9×10^{-8}
proton mass	m_p	$1.67262158\ (13) \times 10^{-27}$	kg	7.9×10^{-8}
proton-electron mass ratio	m_p/m_e	1836.1526675 (39)		2.1×10^{-9}
fine-structure constant $e^2/4\pi\varepsilon_0 hc$	α	$7.297352533\ (27) \times 10^{-3}$		3.7×10^{-9}
inverse fine-structure constant	α^{-1}	137.03599976 (50)		3.7×10^{-9}
Rydberg constant $\alpha^2 m_e c/2h$	R_∞	10973731.568549 (83)	m^{-1}	7.6×10^{-12}
Avogadro constant	N_A, L	$6.02214199\ (47) \times 10^{23}$	mol^{-1}	7.9×10^{-8}
Faraday constant $N_A e$	F	96485.3415 (39)	$C\,mol^{-1}$	4.0×10^{-8}
molar gas constant	R	8.314472 (15)	$J\,mol^{-1}\,K^{-1}$	1.7×10^{-6}
Boltzmann constant R/N_A	k	$1.3806503\ (24) \times 10^{-23}$	$J\,K^{-1}$	1.7×10^{-6}
Stefan-Boltzmann constant $(\pi^2/60)k^4/h^3c^2$	σ	$5.670400\ (40) \times 10^{-8}$	$W\,m^{-2}\,K^{-4}$	7.0×10^{-6}
Non-SI units accepted for use with the SI				
electron volt: (e/C) J	eV	$1.602176462\ (63) \times 10^{-19}$	J	3.9×10^{-8}
(unified) atomic mass unit $1\,u = m_u = \frac{1}{12}m(^{12}C) = 10^{-3}\,kg\,mol^{-1}/N_A$	u	$1.66053873\ (13) \times 10^{-27}$	kg	7.9×10^{-8}

These are published by the federal government, NIST, so I assume permission is not needed so long as we give the citation: Sources: P.J. Mohr and B.N. Taylor, 1998, CODATA Recommended Values of the Fundamental Physical Constants; *Journal of Physical and Chemical Reference Data*, 1999, Vol. 28, no. 6; and *Reviews of Modern Physics*, 2000, Vol. 72, no. 2 (via physics.nist.gov/constants).

APPENDIX 7

Periodic Table of Elements

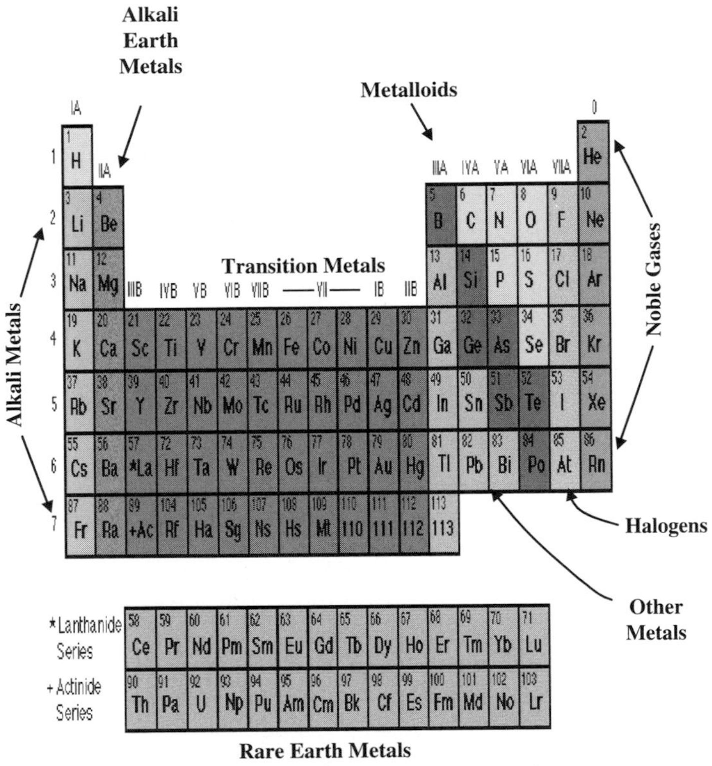

APPENDIX 8

Minimum Risk Levels for Chemicals

Chemical	*Route*	*Duration*	*MRL*[1]	*Factors*	*Endpoint*
ACENAPHTHENE	Oral	Intermittent	$0.6\,mg\,kg^{-1}\,d^{-1}$	300	Liver
ACETONE	Inhalation	Acute	26 ppm	9	Nervous system
		Intermittent	13 ppm	100	Nervous system
		Chronic	13 ppm	100	Nervous system
	Oral	Intermittent	$2\,mg\,kg^{-1}\,d^{-1}$	100	Blood
ACROLEIN	Inhalation	Acute	0.00005 ppm	100	Eye
		Intermittent	0.000009 ppm	1000	Lung
	Oral	Chronic	$0.0005\,mg\,kg^{-1}\,d^{-1}$	100	Blood
ACRYLONITRILE	Inhalation	Acute	0.1 ppm	10	Nervous system
	Oral	Acute	$0.1\,mg\,kg^{-1}\,d^{-1}$	100	Developmental
		Intermittent	$0.01\,mg\,kg^{-1}\,d^{-1}$	1000	Reproductive
		Chronic	$0.04\,mg\,kg^{-1}\,d^{-1}$	100	Blood
ALDRIN	Oral[2]	Acute	$0.002\,mg\,kg^{-1}\,d^{-1}$	1000	Developmental
		Chronic	$0.00003\,mg\,kg^{-1}\,d^{-1}$	1000	Liver
ALUMINUM	Oral	Intermittent	$2.0\,mg\,kg^{-1}\,d^{-1}$	30	Nervous system
AMMONIA	Inhalation	Acute	0.5 ppm	100	Lung
		Chronic	0.3 ppm	10	Lung
	Oral	Intermittent	$0.3\,mg\,kg^{-1}\,d^{-1}$	100	Other
ANTHRACENE	Oral	Intermittent	$10\,mg\,kg^{-1}\,d^{-1}$	100	Liver
ARSENIC	Oral[3]	Acute	$0.005\,mg\,kg^{-1}\,d^{-1}$	10	Gastrointestinal
		Chronic	$0.0003\,mg\,kg^{-1}\,d^{-1}$	3	Dermal
ATRAZINE	Oral[4]	Acute	$0.01\,mg\,kg^{-1}\,d^{-1}$	100	Body weight
BENZENE	Inhalation	Acute	0.05 ppm	300	Immune system
		Intermittent	0.004 ppm	90	Nervous system
BERYLLIUM	Oral[5]	Chronic	$0.001\,mg\,kg^{-1}\,d^{-1}$	100	Gastrointestinal
BIOALLETHRIN	Oral[6]	Acute	$0.0007\,mg\,kg^{-1}\,d^{-1}$	300	Developmental
BIS(CHLOROMETHYL) ETHER	Inhalation	Intermittent	0.0003 ppm	100	Lung
BIS(2-CHLOROETHYL) ETHER	Inhalation	Intermittent	0.02 ppm	1000	Body weight
BORON	Oral	Intermittent	$0.01\,mg\,kg^{-1}\,d^{-1}$	1000	Developmental

BROMODICHLOROMETHANE	Oral	Acute	0.04 mg kg^{-1} d^{-1}	1000	Liver
		Chronic	0.02 mg kg^{-1} d^{-1}	1000	Kidney
BROMOFORM	Oral	Acute	0.6 mg kg^{-1} d^{-1}	100	Nervous system
		Chronic	0.2 mg kg^{-1} d^{-1}	100	Liver
BROMOMETHANE	Inhalation	Acute	0.05 ppm	100	Nervous system
		Intermittent	0.05 ppm	100	Nervous system
		Chronic	0.005 ppm	100	Nervous system
	Oral	Intermittent	0.003 mg kg^{-1} d^{-1}	100	Gastrointestinal
CADMIUM	Oral	Chronic	0.0002 mg kg^{-1} d^{-1}	10	Kidney
CARBON DISULFIDE	Inhalation	Chronic	0.3 ppm	30	Nervous system
	Oral	Acute	0.01 mg kg^{-1} d^{-1}	300	Liver
CARBON TETRACHLORIDE	Inhalation	Acute	0.2 ppm	300	Liver
		Intermittent	0.05 ppm	100	Liver
	Oral	Acute	0.02 mg kg^{-1} d^{-1}	300	Liver
		Intermittent	0.007 mg kg^{-1} d^{-1}	100	Liver
CESIUM	Radiation	Acute	4 mSv	3	Developmental
		Chronic	1 mSv/yr	3	Other
CHLORDANE	Inhalation	Intermittent	0.0002 mg m^{-3}	100	Liver
		Chronic	0.00002 mg m^{-3}	1000	Liver
	Oral	Acute	0.001 mg kg^{-1} d^{-1}	1000	Liver
		Intermittent	0.0006 mg kg^{-1} d^{-1}	100	Liver
		Chronic	0.0006 mg kg^{-1} d^{-1}	100	Liver
CHLORDECONE	Oral	Acute	0.01 mg kg^{-1} d^{-1}	100	Nervous system
		Intermittent	0.0005 mg kg^{-1} d^{-1}	100	Kidney
		Chronic	0.0005 mg kg^{-1} d^{-1}	100	Kidney
CHLORFENVINPHOS	Oral	Acute	0.002 mg kg^{-1} d^{-1}	1000	Nervous system
		Intermittent	0.002 mg kg^{-1} d^{-1}	1000	Immune system
		Chronic	0.0007 mg kg^{-1} d^{-1}	1000	Nervous system
CHLOROBENZENE	Oral	Intermittent	0.4 mg kg^{-1} d^{-1}	100	Liver
CHLORODIBROMOMETHANE	Oral	Acute	0.04 mg kg^{-1} d^{-1}	1000	Kidney
		Chronic	0.03 mg kg^{-1} d^{-1}	1000	Liver

Chemical	*Route*	*Duration*	*MRL*[1]	*Factors*	*Endpoint*
CHLOROETHANE	Inhalation	Acute	15 ppm	100	Developmental
CHLOROFORM	Inhalation	Acute	0.1 ppm	30	Liver
		Intermittent	0.05 ppm	100	Liver
		Chronic	0.02 ppm	100	Liver
	Oral	Acute	0.3 $mg\,kg^{-1}\,d^{-1}$	100	Liver
		Intermittent	0.1 $mg\,kg^{-1}\,d^{-1}$	100	Liver
		Chronic	0.01 $mg\,kg^{-1}\,d^{-1}$	1000	Liver
CHLOROMETHANE	Inhalation	Acute	0.5 ppm	100	Nervous system
		Intermittent	0.2 ppm	300	Liver
		Chronic	0.05 ppm	1000	Nervous system
CHLORPYRIFOS	Oral	Acute	0.003 $mg\,kg^{-1}\,d^{-1}$	10	Nervous system
		Intermittent	0.003 $mg\,kg^{-1}\,d^{-1}$	10	Nervous system
		Chronic	0.001 $mg\,kg^{-1}\,d^{-1}$	100	Nervous system
CHROMIUM(VI), AEROSOL MISTS	Inhalation	Intermittent	0.000005 $mg\,m^{-3}$	100	Lung
CHROMIUM(VI), PARTICULATES	Inhalation	Intermittent	0.001 $mg\,m^{-3}$	30	Lung
COBALT	Inhalation[14]	Chronic	0.0001 $mg\,m^{-3}$	10	Lung
	Oral	Intermittent	0.01 $mg\,kg^{-1}\,d^{-1}$	100	Blood
	Radiation	Acute	4 mSv	3	Developmental
		Chronic	1 $mSv\,yr^{-1}$	3	Other
CRESOL, META-	Oral	Acute	0.05 $mg\,kg^{-1}\,d^{-1}$	100	Lung
CRESOL, ORTHO-	Oral	Acute	0.05 $mg\,kg^{-1}\,d^{-1}$	100	Nervous system
CRESOL, PARA-	Oral	Acute	0.05 $mg\,kg^{-1}\,d^{-1}$	100	Nervous system
CYANIDE, SODIUM	Oral	Intermittent	0.05 $mg\,kg^{-1}\,d^{-1}$	100	Reproductive
CYCLOTETRAMETHYLENE TETRANITRAMINE (HMX)	Oral	Acute	0.1 $mg\,kg^{-1}\,d^{-1}$	1000	Nervous system
		Intermittent	0.05 $mg\,kg^{-1}\,d^{-1}$	1000	Liver

CYCLOTRIMETHYLENE TRINITRAMINE (RDX)	Oral	Acute	$0.06\,mg\,kg^{-1}\,d^{-1}$	100	Nervous system
		Intermittent	$0.03\,mg\,kg^{-1}\,d^{-1}$	300	Reproductive
DDT, P,P′-	Oral[8]	Acute	$0.0005\,mg\,kg^{-1}\,d^{-1}$	1000	Developmental
		Intermittent	$0.0005\,mg\,kg^{-1}\,d^{-1}$	100	Liver
DELTAMETHRIN	Oral[9]	Acute	$0.002\,mg\,kg^{-1}\,d^{-1}$	300	Developmental
DI(2-ETHYLHEXYL)PHTHALATE	Oral[10]	Intermittent	$0.01\,mg\,kg^{-1}\,d^{-1}$	300	Developmental
DI-N-BUTYL PHTHALATE	Oral[11]	Acute	$0.5\,mg\,kg^{-1}\,d^{-1}$	100	Developmental
DI-N-OCTYL PHTHALATE	Oral	Acute	$3\,mg\,kg^{-1}\,d^{-1}$	300	Liver
		Intermittent	$0.4\,mg\,kg^{-1}\,d^{-1}$	100	Liver
DIAZINON	Inhalation	Intermittent	$0.009\,mg\,m^{-3}$	30	Nervous system
	Oral	Intermittent	$0.0002\,mg\,kg^{-1}\,d^{-1}$	100	Nervous system
DICHLORVOS	Inhalation	Acute	0.002 ppm	100	Nervous system
		Intermittent	0.0003 ppm	100	Nervous system
		Chronic	0.00006 ppm	100	Nervous system
	Oral	Acute	$0.004\,mg\,kg^{-1}\,d^{-1}$	1000	Nervous system
		Intermittent	$0.003\,mg\,kg^{-1}\,d^{-1}$	10	Nervous system
		Chronic	$0.0005\,mg\,kg^{-1}\,d^{-1}$	100	Nervous system
DIELDRIN	Oral[12]	Intermittent	$0.0001\,mg\,kg^{-1}\,d^{-1}$	100	Nervous system
		Chronic	$0.00005\,mg\,kg^{-1}\,d^{-1}$	100	Liver
DIETHYL PHTHALATE	Oral	Acute	$7\,mg\,kg^{-1}\,d^{-1}$	300	Reproductive
		Intermittent	$6\,mg\,kg^{-1}\,d^{-1}$	300	Liver
DIISOPROPYL METHYLPHOSPHONATE (DIMP)	Oral	Intermittent	$0.8\,mg\,kg^{-1}\,d^{-1}$	100	Blood
		Chronic	$0.6\,mg\,kg^{-1}\,d^{-1}$	100	Blood
DISULFOTON	Inhalation	Acute	$0.006\,mg\,m^{-3}$	30	Nervous system
		Intermittent	$0.0002\,mg\,m^{-3}$	30	Nervous system
	Oral	Acute	$0.001\,mg\,kg^{-1}\,d^{-1}$	100	Nervous system
		Intermittent	$0.00009\,mg\,kg^{-1}\,d^{-1}$	100	Developmental
		Chronic	$0.00006\,mg\,kg^{-1}\,d^{-1}$	1000	Nervous system

Chemical	*Route*	*Duration*	*MRL*[1]	*Factors*	*Endpoint*
ENDOSULFAN	Oral	Intermittent	$0.005\,mg\,kg^{-1}\,d^{-1}$	100	Immune system
		Chronic	$0.002\,mg\,kg^{-1}\,d^{-1}$	100	Liver
ENDRIN	Oral	Intermittent	$0.002\,mg\,kg^{-1}\,d^{-1}$	100	Nervous system
		Chronic	$0.0003\,mg\,kg^{-1}\,d^{-1}$	100	Nervous system
ETHION	Oral	Acute	$0.002\,mg\,kg^{-1}\,d^{-1}$	30	Nervous system
		Intermittent	$0.002\,mg\,kg^{-1}\,d^{-1}$	30	Nervous system
		Chronic	$0.0004\,mg\,kg^{-1}\,d^{-1}$	150	Nervous system
ETHYLBENZENE	Inhalation	Intermittent	1.0 ppm	100	Developmental
ETHYLENE GLYCOL	Inhalation	Acute	0.5 ppm	100	Kidney
	Oral	Acute	$2.0\,mg\,kg^{-1}\,d^{-1}$	100	Developmental
		Chronic	$2.0\,mg\,kg^{-1}\,d^{-1}$	100	Kidney
ETHYLENE OXIDE	Inhalation	Intermittent	0.09 ppm	100	Kidney
FLUORANTHENE	Oral	Intermittent	$0.4\,mg\,kg^{-1}\,d^{-1}$	300	Liver
FLUORENE	Oral	Intermittent	$0.4\,mg\,kg^{-1}\,d^{-1}$	300	Liver
FLUORIDE, SODIUM	Oral[13]	Chronic	$0.06\,mg\,kg^{-1}\,d^{-1}$	10	Musculoskeletal
FLUORINE	Inhalation[14]	Acute	0.01 ppm	10	Lung
FORMALDEHYDE	Inhalation	Acute	0.04 ppm	9	Lung
		Intermittent	0.03 ppm	30	Lung
		Chronic	0.008 ppm	30	Lung
	Oral	Intermittent	$0.3\,mg\,kg^{-1}\,d^{-1}$	100	Gastrointestinal
		Chronic	$0.2\,mg\,kg^{-1}\,d^{-1}$	100	Gastrointestinal
FUEL OIL NO. 2	Inhalation	Acute	$0.02\,mg\,m^{-3}$	1000	Nervous system
HEXACHLOROBENZENE	Oral[15]	Acute	$0.008\,mg\,kg^{-1}\,d^{-1}$	300	Developmental
		Intermittent	$0.0001\,mg\,kg^{-1}\,d^{-1}$	90	Reproductive
		Chronic	$0.00002\,mg\,kg^{-1}\,d^{-1}$	1000	Developmental
HEXACHLOROBUTADIENE	Oral	Intermittent	$0.0002\,mg\,kg^{-1}\,d^{-1}$	1000	Kidney
HEXACHLOROCYCLOHEXANE, α-	Oral	Chronic	$0.008\,mg\,kg^{-1}\,d^{-1}$	100	Kidney
HEXACHLOROCYCLOHEXANE, β-	Oral	Acute	$0.2\,mg\,kg^{-1}\,d^{-1}$	100	Nervous system
		Intermittent	$0.0006\,mg\,kg^{-1}\,d^{-1}$	300	Liver
HEXACHLOROCYCLOHEXANE, γ-[16]	Oral	Acute	$0.01\,mg\,kg^{-1}\,d^{-1}$	100	Nervous system
		Intermittent	$0.00001\,mg\,kg^{-1}\,d^{-1}$	1000	Immune system

HEXACHLOROCYCLOPENTADIENE	Inhalation	Intermittent	0.01 ppm	30	Lung
		Chronic	0.0002 ppm	90	Lung
	Oral	Intermittent	$0.1\ mg\ kg^{-1}\ d^{-1}$	100	Kidney
HEXACHLOROETHANE	Inhalation	Acute	6 ppm	30	Nervous system
		Intermittent	6 ppm	30	Nervous system
	Oral	Acute	$1\ mg\ kg^{-1}\ d^{-1}$	100	Liver
		Intermittent	$0.01\ mg\ kg^{-1}\ d^{-1}$	100	Liver
HEXAMETHYLENE DIISOCYANATE	Inhalation	Intermittent	0.00003 ppm	30	Lung
		Chronic	0.00001 ppm	90	Lung
HEXANE, N-	Inhalation	Chronic	0.6 ppm	100	Nervous system
HYDRAZINE	Inhalation	Intermittent	0.004 ppm	300	Liver
HYDROGEN FLUORIDE	Inhalation[17]	Acute	0.03 ppm	30	Lung
		Intermittent	0.02 ppm	30	Lung
HYDROGEN SULFIDE	Inhalation	Acute	0.07 ppm	30	Lung
		Intermittent	0.03 ppm	30	Lung
IODIDE	Oral[18]	Acute	$0.01\ mg\ kg^{-1}\ d^{-1}$	1	Endocrine system
		Chronic	$0.01\ mg\ kg^{-1}\ d^{-1}$	1	Endocrine system
ISOPHORONE	Oral	Intermittent	$3\ mg\ kg^{-1}\ d^{-1}$	100	Other
		Chronic	$0.2\ mg\ kg^{-1}\ d^{-1}$	1000	Liver
JP-4 (Jet Fuel)	Inhalation	Intermittent	$9\ mg\ m^{-3}$	300	Liver
JP-5/JP-8 (Jet Fuel)	Inhalation	Intermittent	$3\ mg\ m^{-3}$	300	Liver
JP-7 (Jet Fuel)	Inhalation	Chronic	$3\ mg\ m^{-3}$	300	Liver
KEROSENE	Inhalation	Intermittent	$0.01\ mg\ m^{-3}$	1000	Liver
MALATHION	Inhalation[19]	Acute	$0.2\ mg\ m^{-3}$	100	Nervous system
		Intermittent	$0.02\ mg\ m^{-3}$	1000	Lung
	Oral	Intermittent	$0.02\ mg\ kg^{-1}\ d^{-1}$	10	Nervous system
		Chronic	$0.02\ mg\ kg^{-1}\ d^{-1}$	100	Nervous system
MANGANESE	Inhalation	Chronic	$0.00004\ mg\ m^{-3}$	500	Nervous system
MERCURIC CHLORIDE	Oral	Acute	$0.007\ mg\ kg^{-1}\ d^{-1}$	100	Kidney
		Intermittent	$0.002\ mg\ kg^{-1}\ d^{-1}$	100	Kidney

Chemical	*Route*	*Duration*	*MRL*[1]	*Factors*	*Endpoint*
MERCURY	Inhalation	Chronic	0.0002 mg m^{-3}	30	Nervous system
METHOXYCHLOR	Oral[20]	Intermittent	0.005 mg kg^{-1} d^{-1}	1000	Reproductive
METHYL PARATHION	Oral	Intermittent	0.0007 mg kg^{-1} d^{-1}	300	Nervous system
		Chronic	0.0003 mg kg^{-1} d^{-1}	100	Blood
METHYL-Tertiary-BUTYL ETHER	Inhalation	Acute	2 ppm	100	Nervous system
		Intermittent	0.7 ppm	100	Nervous system
		Chronic	0.7 ppm	100	Kidney
	Oral	Acute	0.4 mg kg^{-1} d^{-1}	100	Nervous system
		Intermittent	0.3 mg kg^{-1} d^{-1}	300	Liver
METHYLENE CHLORIDE	Inhalation	Acute	0.6 ppm	100	Nervous system
		Intermittent	0.3 ppm	90	Liver
		Chronic	0.3 ppm	30	Liver
	Oral	Acute	0.2 mg kg^{-1} d^{-1}	100	Nervous system
		Chronic	0.06 mg kg^{-1} d^{-1}	100	Liver
METHYLMERCURY	Oral	Chronic	0.0003 mg kg^{-1} d^{-1}	4	Developmental
MIREX	Oral	Chronic	0.0008 mg kg^{-1} d^{-1}	100	Liver
MUSTARD GAS	Inhalation[21]	Acute	0.0002 mg m^{-3}	900	Lung
	Oral	Acute	0.5 μg kg^{-1} d^{-1}	1000	Developmental
		Intermittent	0.02 μg kg^{-1} d^{-1}	1000	Gastrointestinal
N-NITROSODI-N-PROPYLAMINE	Oral	Acute	0.095 mg kg^{-1} d^{-1}	100	Liver
NAPHTHALENE	Inhalation	Chronic	0.002 ppm	1000	Lung
	Oral	Acute	0.05 mg kg^{-1} d^{-1}	1000	Nervous system
		Intermittent	0.02 mg kg^{-1} d^{-1}	300	Liver
NICKEL	Inhalation	Chronic	0.0002 mg m^{-3}	30	Lung
PENTACHLOROPHENOL	Oral	Acute	0.005 mg kg^{-1} d^{-1}	1000	Developmental
		Intermittent	0.001 mg kg^{-1} d^{-1}	1000	Reproductive
		Chronic	0.001 mg kg^{-1} d^{-1}	1000	Endocrine system
PHOSPHORUS, WHITE	Inhalation	Acute	0.02 mg m^{-3}	30	Lung
	Oral	Intermittent	0.0002 mg kg^{-1} d^{-1}	100	Reproductive

POLYBROMINATED BIPHENYLS (PBBs)	Oral	Acute	0.01 mg kg^{-1} d^{-1}	100	Endocrine system
POLYCHLORINATED BIPHENYLS (PCBs) Aroclor 1254	Oral	Intermittent	0.03 μg kg^{-1} d^{-1}	300	Nervous system
		Chronic	0.02 μg kg^{-1} d^{-1}	300	Immune system
PROPYLENE GLYCOL DINITRATE	Inhalation	Acute	0.003 ppm	10	Nervous system
		Intermittent	0.00004 ppm	1000	Blood
		Chronic	0.00004 ppm	1000	Blood
PROPYLENE GLYCOL	Inhalation	Intermittent	0.009 ppm	1000	Lung
SELENIUM	Oral[22]	Chronic	0.005 mg kg^{-1} d^{-1}	3	Skin
STRONTIUM	Oral[23]	Intermittent	2 mg kg^{-1} d^{-1}	30	Musculoskeletal
STYRENE	Inhalation	Chronic	0.06 ppm	100	Nervous
	Oral	Intermittent	0.2 mg kg^{-1} d^{-1}	1000	Liver
SULFUR DIOXIDE	Inhalation	Acute	0.01 ppm	9	Lung
TETRACHLOROETHYLENE	Inhalation	Acute	0.2 ppm	10	Nervous system
		Chronic	0.04 ppm	100	Nervous system
	Oral	Acute	0.05 mg kg^{-1} d^{-1}	100	Developmental
TITANIUM TETRACHLORIDE	Inhalation	Intermittent	0.01 mg m^{-3}	90	Lung
		Chronic	0.0001 mg m^{-3}	90	Lung
TOLUENE	Inhalation	Acute	1 ppm	10	Nervous system
		Chronic	0.08 ppm	100	Nervous system
	Oral	Acute	0.8 mg kg^{-1} d^{-1}	300	Nervous system
		Intermittent	0.02 mg kg^{-1} d^{-1}	300	Nervous system
TOXAPHENE	Oral	Acute	0.005 mg kg^{-1} d^{-1}	1000	Liver
		Intermittent	0.001 mg kg^{-1} d^{-1}	300	Liver
TRICHLOROETHYLENE	Inhalation	Acute	2 ppm	30	Nervous system
		Intermittent	0.1 ppm	300	Nervous system
	Oral	Acute	0.2 mg kg^{-1} d^{-1}	300	Developmental
URANIUM, HIGHLY SOLUBLE SALTS	Inhalation	Intermittent	0.0004 mg m^{-3}	90	Kidney
		Chronic	0.0003 mg m^{-3}	30	Kidney
	Oral	Intermittent	0.002 mg kg^{-1} d^{-1}	30	Kidney

Chemical	*Route*	*Duration*	*MRL*[1]	*Factors*	*Endpoint*
URANIUM, INSOLUBLE COMPOUNDS	Inhalation	Intermittent	$0.008\,mg\,m^{-3}$	30	Kidney
VANADIUM	Inhalation	Acute	$0.0002\,mg\,m^{-3}$	100	Lung
	Oral	Intermittent	$0.003\,mg\,kg^{-1}\,d^{-1}$	100	Kidney
VINYL ACETATE	Inhalation	Intermittent	0.01 ppm	100	Lung
VINYL CHLORIDE	Inhalation	Acute	0.5 ppm	100	Developmental
		Intermittent	0.03 ppm	300	Liver
	Oral	Chronic	$0.00002\,mg\,kg^{-1}\,d^{-1}$	1000	Liver
XYLENE, meta-	Oral	Intermittent	$0.6\,mg\,kg^{-1}\,d^{-1}$	1000	Liver
XYLENE, para-	Oral	Acute	$1\,mg\,kg^{-1}\,d^{-1}$	100	Nervous system
XYLENES, total	Inhalation	Acute	1 ppm	100	Nervous system
		Intermittent	0.7 ppm	300	Developmental
		Chronic	0.1 ppm	100	Nervous system
	Oral	Intermittent	$0.2\,mg\,kg^{-1}\,d^{-1}$	1000	Kidney
ZINC	Oral	Intermittent	$0.3\,mg\,kg^{-1}\,d^{-1}$	3	Blood
		Chronic	$0.3\,mg\,kg^{-1}\,d^{-1}$	3	Blood
1-METHYLNAPHTHALENE	Oral	Chronic	$0.07\,mg\,kg^{-1}\,d^{-1}$	1000	Lung
1,1-DICHLOROETHENE	Inhalation	Intermittent	0.02 ppm	100	Liver
	Oral	Chronic	$0.009\,mg\,kg^{-1}\,d^{-1}$	1000	Liver
1,1-DIMETHYLHYDRAZINE	Inhalation	Intermittent	0.0002 ppm	300	Liver
1,1,1-TRICHLOROETHANE	Inhalation	Acute	2 ppm	100	Nervous system
		Intermittent	0.7 ppm	100	Nervous system
1,1,2-TRICHLOROETHANE	Oral	Acute	$0.3\,mg\,kg^{-1}\,d^{-1}$	100	Nervous system
		Intermittent	$0.04\,mg\,kg^{-1}\,d^{-1}$	100	Liver
1,1,2,2-TETRACHLOROETHANE	Inhalation	Intermittent	0.4 ppm	300	Liver
	Oral	Intermittent	$0.6\,mg\,kg^{-1}\,d^{-1}$	100	Body weight
		Chronic	$0.04\,mg\,kg^{-1}\,d^{-1}$	1000	Lung
1,2-DIBROMO-3-CHLOROPROPANE	Inhalation	Intermittent	0.0002 ppm	100	Reproductive
	Oral	Intermittent	$0.002\,mg\,kg^{-1}\,d^{-1}$	1000	Reproductive

1,2-DICHLOROETHENE, cis-	Oral	Acute	1 $mg\,kg^{-1}\,d^{-1}$	100	Blood
		Intermittent	0.3 $mg\,kg^{-1}\,d^{-1}$	100	Blood
1,2-DICHLOROETHANE	Inhalation	Chronic	0.6 ppm	90	Liver
	Oral	Intermittent	0.2 $mg\,kg^{-1}\,d^{-1}$	300	Kidney
1,2-DICHLOROPROPANE	Inhalation	Acute	0.05 ppm	1000	Lung
		Intermittent	0.007 ppm	1000	Lung
	Oral	Acute	0.1 $mg\,kg^{-1}\,d^{-1}$	1000	Nervous system
		Intermittent	0.07 $mg\,kg^{-1}\,d^{-1}$	1000	Blood
		Chronic	0.09 $mg\,kg^{-1}\,d^{-1}$	1000	Blood
1,2-DICHLOROETHENE, trans-	Inhalation	Acute	0.2 ppm	1000	Liver
		Intermittent	0.2 ppm	1000	Liver
	Oral	Intermittent	0.2 $mg\,kg^{-1}\,d^{-1}$	100	Liver
1,2-DIMETHYLHYDRAZINE	Oral	Intermittent	0.0008 $mg\,kg^{-1}\,d^{-1}$	1000	Liver
1,2,3-TRICHLOROPROPANE	Inhalation	Acute	0.0003 ppm	100	Lung
	Oral	Intermittent	0.06 $mg\,kg^{-1}\,d^{-1}$	100	Liver
1,3-DICHLOROPROPENE	Inhalation	Intermittent	0.003 ppm	100	Lung
		Chronic	0.002 ppm	100	Lung
1,3-DINITROBENZENE	Oral	Acute	0.008 $mg\,kg^{-1}\,d^{-1}$	100	Reproductive
		Intermittent	0.0005 $mg\,kg^{-1}\,d^{-1}$	1000	Blood
1,4-DICHLOROBENZENE	Inhalation	Acute	0.8 ppm	100	Developmental
		Intermittent	0.2 ppm	100	Blood
		Chronic	0.1 ppm	100	Blood
	Oral	Intermittent	0.4 $mg\,kg^{-1}\,d^{-1}$	300	Liver
2-BUTOXYETHANOL	Inhalation	Acute	6 ppm	9	Blood
		Intermittent	3 ppm	9	Blood
		Chronic	0.2 ppm	3	Blood
	Oral	Acute	0.4 $mg\,kg^{-1}\,d^{-1}$	90	Blood
		Intermittent	0.07 $mg\,kg^{-1}\,d^{-1}$	1000	Blood

Chemical	*Route*	*Duration*	*MRL*[1]	*Factors*	*Endpoint*
2,3,4,7,8-PENTACHLORODIBENZOFURAN	Oral	Acute	0.001 $\mu g\,kg^{-1}\,d^{-1}$	3000	Immune system
		Intermittent	0.00003 $\mu g\,kg^{-1}\,d^{-1}$	3000	Liver
2,3,7,8-TETRACHLORODIBENZO-para-DIOXIN	Oral	Acute	0.0002 $\mu g\,kg^{-1}\,d^{-1}$	21	Immune system
		Intermittent	0.00002 $\mu g\,kg^{-1}\,d^{-1}$	30	Lymph Nodes
		Chronic	0.000001 $\mu g\,kg^{-1}\,d^{-1}$	90	Developmental
2,4-DICHLOROPHENOL	Oral	Intermittent	0.003 $mg\,kg^{-1}\,d^{-1}$	100	Immune system
2,4-DINITROPHENOL	Oral	Acute	0.01 $mg\,kg^{-1}\,d^{-1}$	100	Body weight
2,4-DINITROTOLUENE	Oral	Acute	0.05 $mg\,kg^{-1}\,d^{-1}$	100	Nervous system
		Chronic	0.002 $mg\,kg^{-1}\,d^{-1}$	100	Blood
2,4,6-TRINITROTOLUENE	Oral	Intermittent	0.0005 $mg\,kg^{-1}\,d^{-1}$	1000	Liver
2,6-DINITROTOLUENE	Oral	Intermittent	0.004 $mg\,kg^{-1}\,d^{-1}$	1000	Blood
4-CHLOROPHENOL	Oral	Acute	0.01 $mg\,kg^{-1}\,d^{-1}$	100	Liver
4,4′-METHYLENEBIS(2-CHLOROANILINE)	Oral	Chronic	0.003 $mg\,kg^{-1}\,d^{-1}$	3000	Liver
4,4′-METHYLENEDIANILINE	Oral	Acute	0.2 $mg\,kg^{-1}\,d^{-1}$	300	Liver
		Intermittent	0.08 $mg\,kg^{-1}\,d^{-1}$	100	Liver
4,6-DINITRO-O-CRESOL	Oral	Acute	0.004 $mg\,kg^{-1}\,d^{-1}$	100	Nervous system
		Intermittent	0.004 $mg\,kg^{-1}\,d^{-1}$	100	Nervous system

Source: ATSDR, 2003, Minimum Risk Levels for Hazardous Substances, http://www.atsdr.cdc.gov/mrls.html.
[1] All MRLs are established as final by the Agency for Toxic Substance and Disease Registry, except where otherwise noted.
[2] Draft MRL.
[3] Provisional oral MRL.
[4] Draft MRL.
[5] Draft MRL.
[6] Draft MRL.
[7] Draft MRL.
[8] Draft MRL.
[9] Draft MRL.
[10] Provisional oral MRL.
[11] Draft MRL.
[12] Draft MRL.
[23] Draft MRL.
[24] Draft MRL.
[25] Draft MRL.
[26] Known commercially as "lindane."
[27] Draft MRL.
[28] Draft MRL.
[29] Draft MRL.
[20] Draft MRL.
[21] Draft MRL.
[22] Draft MRL.
[23] Draft MRL.

APPENDIX 9

Physical Contaminants

This book focuses primarily on chemical contaminants, but some contaminants are better classified according to their physical rather than chemical properties. In particular, environmental scientists are concerned with particles and fibers, not so much because of their chemical composition, but more because of their the irritations and other effects elicited when particles and fibers come into contact with biotic tissues.

Particulate Matter

Particulate matter (PM) includes particles found in the air, such as dust, dirt, soot, smoke, and liquid droplets.[1] Unlike other U.S. criteria pollutants subject to the National Ambient Air Quality Standards (O_3, CO, SO_2, NO_2, and Pb), PM is not a specific chemical entity but is a mixture of particles from different sources and of different sizes, compositions, and properties. The chemical composition of tropospheric particles includes inorganic ions, metallic compounds, elemental carbon, organic compounds, and crustal (e.g. carbonates and compounds, alkali and rare earth elementals) substances. For example, the mean 24-hour $PM_{2.5}$ concentration measured near Baltimore, Maryland in 1999 was composed of 38% sulfate, 13% ammonium, 2% nitrate, 36% organic carbon, 7% elemental carbon, and 4% crustal matter.[2] In addition, some atmospheric particles can be hygroscopic, i.e., they contain particle-bound water. The organic fraction can be particularly difficult to characterize, since it often contains thousands of organic compounds.

Particle size is determined by how the particle is formed, for example, combustion can generate very small particle, while coarse particle are often formed by mechanical processes (See Figure A.9.1). Particles, if they are small and have low mass, can be suspended in the air for long periods of time. Particles may be sufficiently large (e.g. $>10\,\mu m$ aerodynamic diameter) as to be seen as smoke or soot (See Figure A.9.2), while others are very small ($<2.5\,\mu m$). Sources of particles are highly variable; e.g., emitted

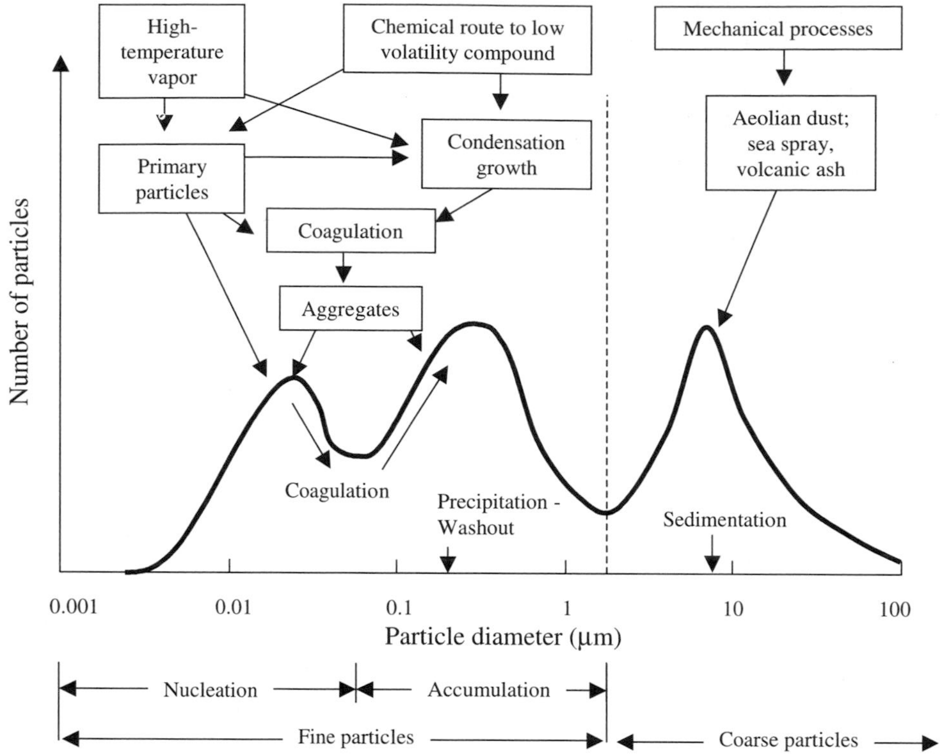

FIGURE A.9.1. Prototypical size distribution of tropospheric particles with selected sources and pathways of how the particles are formed. Dashed line is approximately 2.5 μm diameter. Adapted from: United Kingdom Department of Environment, Food, and Rural Affairs, Expert Panel on Air Quality Standards, 2004, *Airborne Particles: What Is the Appropriate Measurement on which to Base a Standard? A Discussion Document.*

directly to the air from stationary sources, such as factories, power plants, and open burning, and from moving vehicles (known as "mobile sources"), especially those with internal combustion engines. Area or non-point sources of particles include construction, agricultural activities such as plowing and tilling, mining, and forest fires.

Particles may also form from gases that have been previously emitted, e.g., when gases from burning fuels react with sunlight and water vapor. A common production of such "secondary particles" occurs when gases undergo chemical reactions in the atmosphere involving O_2 and water vapor (H_2O). Photochemistry can be an important step in secondary particle formation, resulting when chemical species like ozone (O_3) are involved in step reactions with radicals, e.g., the hydroxyl ($\cdot OH$) and nitrate ($\cdot NO_3$) radicals. Photochemistry also occurs in the presence of air pollutant gases like sulfur

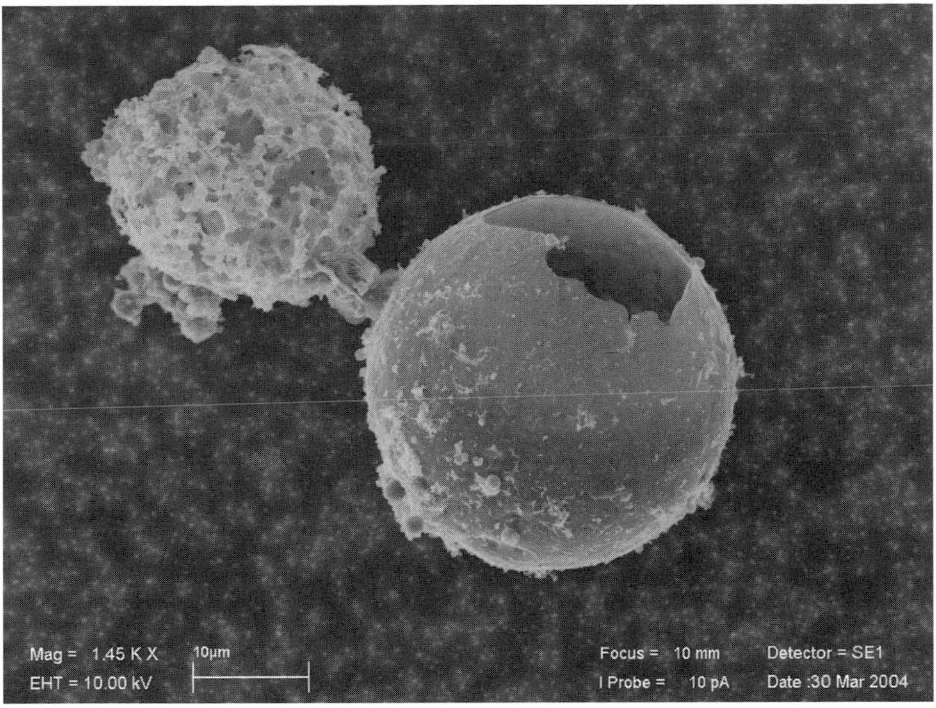

FIGURE A.9.2. Scanning electron micrograph of coarse particles emitted from an oil-fired power plant. Diameters of the particles are greater than 20 μm optical diameter. Both particles are hollow, so their aerodynamic diameter is significantly smaller than if they were solid. (Source: Source characterization study by R. Stevens, M. Lynam, and D. Proffitt, 2004. Photo courtesy of R. Willis, Man Tech Environmental Technology, Inc., 2004; used with permission.)

dioxide (SO_2), nitrogen oxides (NO_X), and organic gases emitted by anthropogenic and natural sources. In addition, nucleation of particles from low-vapor pressure gases emitted from sources or formed in the atmosphere, condensation of low-vapor pressure gases on aerosols already present in the atmosphere, and coagulation of aerosols can contribute to the formation of particles. The chemical composition, transport, and fate of particles are directly associated with the characteristics of the surrounding gas. The term "aerosol" is often used synonymously with PM. An aerosol can be a suspension of solid or liquid particles in air, and an aerosol includes both the particles and all vapor or gas phase components of air.

Smaller particles are particularly problematic because they can travel longer distances and are associated with numerous health effects. Generally, the mass of PM falling in two size categories is measured, i.e., $\leq$2.5 micron diameter, and $\geq$2.5 micron $\leq$10 micron diameter. These measurements are taken by instruments (See Figure A.9.3) with inlets using size exclusion mechanisms to segregate the mass of each size fraction (i.e., "dichotomous" samplers). Particles with diameters ≥10 microns are generally of less concern, however, they are occasionally measured if a large particulate emitting source (e.g., a coal mine) is nearby, since these particles rarely travel long distances.

Mass can be determined for a predominantly spherical particle by microscopy, either optical or electron, by light scattering and Mie theory, by the particle's electrical mobility, or by its aerodynamic behavior. However, since most particles are not spherical, PM diameters are often described using an equivalent diameter, i.e. the diameter of a sphere that would have the same fluid properties. Another term, optical diameter, is the diameter of a spherical particle that has an identical refractive index as the particle. Optical diameters are used to calibrate the optical particle sizing instruments, which scatter the same amount of light into the solid angle measured. Diffusion and gravitational settling are also fundamental fluid phenomena used to estimate the efficiencies of PM transport, collection, and removal processes, such as in designing PM monitoring equipment and ascertaining the rates and mechanisms of how particles infiltrate and deposit in the respiratory tract.

Particulate Matter Example

A $PM_{2.5}$ particle sampler was operated for a period of 24 hours at a flow rate of 16.67 liters per minute. The filter was weighed before and after sampling and had masses of 0.130058 gram and 0.130627 gram, respectively. What was the average mass concentration in the atmosphere that day (in units of $\mu g\ m^{-3}$)?

Solution

Mass difference: 0.130627 gram − 0.130058 gram = 0.000569 gram (or 569 micrograms)

16.67 liters/minute = $0.01667\,m^3\,min^{-1}$

Volume of air sampled: 24 hours × $60\,min\,hr^{-1} \times 0.01667\,m^3\,min^{-1} = 24\,m^3$

Concentration: 569 micrograms ÷ $24\,m^3 = 23.7\,\mu g\,m^{-3}$

Only for very small diameter particles is diffusion sufficiently important that the Stokes diameter is often used. The Stokes diameter for a particle is the diameter of a sphere with the same density and settling velocity as the particle. The Stokes diameter is derived from the aerodynamic drag force caused by the difference in velocity of the particle and the surround-

FIGURE A.9.3. Sampler for measuring particles with aerodynamic diameters ≤2.5 microns. Each sampler has an inlet (top) that takes in particles ≤10 microns. An impacter downstream in the instrument cuts the size fraction to ≤2.5 microns, which is collected on a Teflon filter. The filter is weighed before and after collection. The Teflon construction allows for other analyses, e.g., X-ray fluorescence to determine inorganic composition of the particles. Quartz filters would be used if any subsequent carbon analyses are needed. (Photo courtesy of U.S. EPA.)

ing fluid. Thus, for smooth, spherical particles, the Stokes diameter is identical to the physical or actual diameter. The aerodynamic diameter (D_{pa}) for all particles greater than 0.5 micrometer can be approximated[34] as the product of the Stokes particle diameter (D_{ps}) and the square root of the particle density (ρ_p):

$$D_{pa} = D_{ps}\sqrt{\rho_p} \qquad \text{Equation A.9.1}$$

If the units of the diameters are in μm, the units of density are g cm^{-3}.

Fine particles (<2.5 μm) generally come from industrial combustion processes (See Figure A.9.4) and from vehicle exhaust. This smaller sized fraction has been closely associated with increased respiratory disease, decreased lung functioning, and even premature death, probably due to their ability to bypass the body's trapping mechanisms, such as cilia in the lungs, and nasal hair filtering. Some of the diseases linked to PM exposure include aggravation of asthma, chronic bronchitis, and decreased lung function.

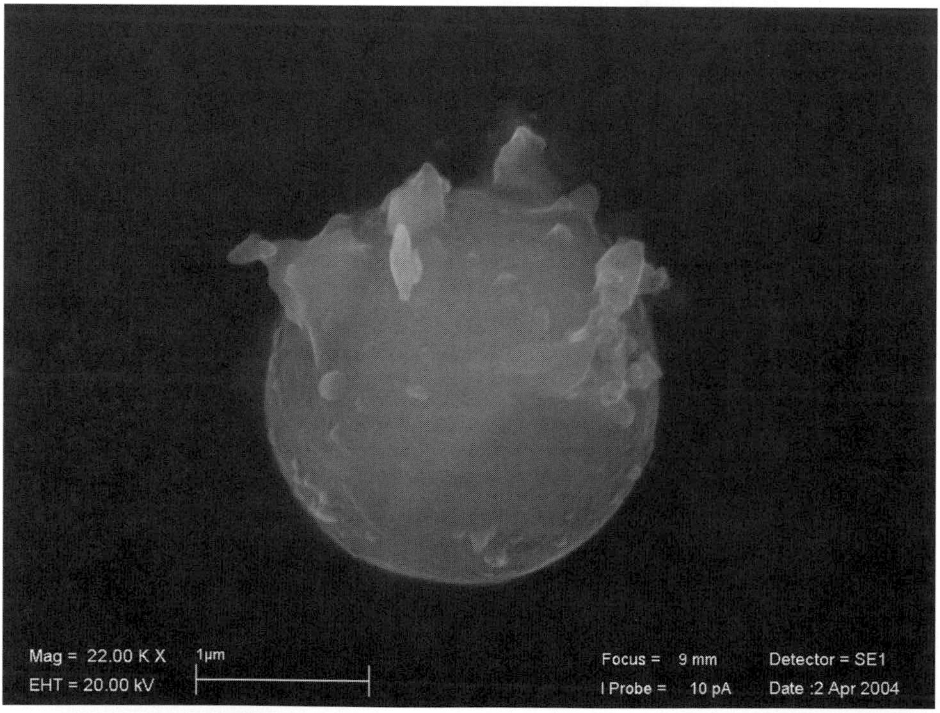

FIGURE A.9.4. Scanning electron micrograph of spherical aluminosilicate fly ash particle emitted from an oil-fired power plant. Diameter of the particle is approximately 2.5 μm. (Photo courtesy of R. Willis, Man Tech Environmental Technology, Inc., 2004; used with permission.)

In addition to health impacts, PM is also a major contributor to reduced visibility, including near national parks and monuments. Also, particles can be transported long distances and serve as vehicles on which contaminants are able to reach water bodies and soils. Acid deposition, for example, can be as dry or wet. Either way, particles play a part in acid rain. In the first, the dry particles enter ecosystems and potentially reduce the pH of receiving waters. In the latter, particles are washed out of the atmosphere and, in the process; lower the pH of the rain. The same transport and deposition mechanisms can also lead to exposures to persistent organic contaminants like dioxins and organochlorine pesticides, and heavy metals like mercury that have sorbed in or on particles.

Routes of exposure for PM are similar to those of chemical contaminants. In fact, the lifetime average daily dose (LADD) equations include provisions for particle exposure (See Chapter 9). Also, particles can serve as vehicles for carrying chemical contaminants. For example, compounds that are highly sorptive (e.g. those with large K_{oc} partitioning coefficients) can use particles as a means for long-range transport. Also, charge differences between the particle and ions (particularly metal cations) will also make particles a means by which contaminants are transported.

Fibers

Generally, when environmental scientists discuss particles, they mean those that are somewhat spherical or angular like soil particles. Particles that are highly elongated are usually differentiated as "fibers." Environmentally important fibers include fiberglass, fabrics, and minerals (See Figures A.9.5 and A.9.6). Exposure to fiberglass and textile fibers are most commonly found in industrial settings, such as has been associated with the health problems of textile workers exposed to fibrous matter in high doses for many years. For example chronic exposure to cotton fibers has led the ailment, byssinosis, also referred to as "brown lung disease," which is characterized by the narrowing of the lung's airways. However, when discussing fibers, it is highly likely that first contaminant to come to mind is asbestos, a group of highly fibrous minerals with separable, long, and thin fibers. Separated asbestos fibers are strong enough and flexible enough to be spun and woven. Asbestos fibers are heat resistant, making them useful for many industrial purposes. Because of their durability, asbestos fibers that get into lung tissue will remain for long periods of time.

There are two general types of asbestos, *amphibole* and *chrysotile*. Some studies show that amphibole fibers stay in the lungs longer than chrysotile, and this tendency may account for their increased toxicity.

Generally, health regulations classify asbestos into six mineral types: chrysotile, a serpentine mineral with long and flexible fibers; and five amphiboles, which have brittle crystalline fibers. The amphiboles include

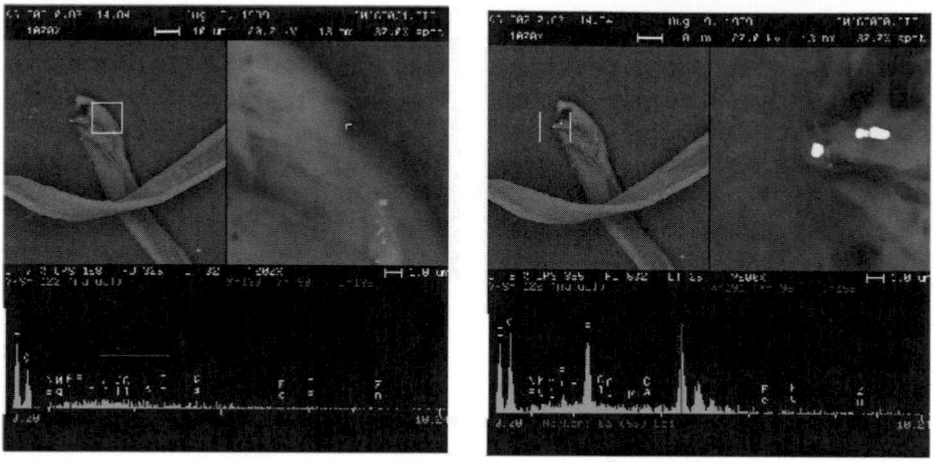

FIGURE A.9.5. Scanning electron micrograph of cotton fibers. Acquired using an Aspex Instruments, Ltd. Scanning electron microscope. Source: U.S. Environmental Protection Agency, 2004. (Photo courtesy of T. Conner, used with permission.)

actinolite asbestos, tremolite asbestos, anthophyllite asbestos, crocidolite asbestos, and amosite asbestos (See Figure A.9.7).

Asbestos Routes of Exposure

Ambient air concentrations of asbestos fibers are about 10^{-5} to 10^{-4} fibers per milliliter (fibers mL^{-1}), depending on location. Human exposure to concentrations much higher than 10^{-4} fibers mL^{-1} is suspected of causing health effects.[5] Asbestos fibers are very persistent and resist chemical degradation (i.e., they are inert under most environmental conditions) so their vapor pressure is nearly zero meaning they do not evaporate, nor do they dissolve in water. However, segments of fibers do enter the air and water as asbestos-containing rocks and minerals are weathered naturally or when extracted during mining operations. One of the most important exposures is when manufactured products (e.g., pipe wrapping and fire-resistant materials) begin to wear down. Small diameter asbestos fibers may remain suspended in the air for a long time and be transported advectively by wind or water before sedimentation. Like particles, heavier fibers settle more quickly. Asbestos seldom moves substantially via soil. They are generally not broken down to other compounds in the environment and will remain virtually unchanged over long periods. Although most asbestos is highly persistent, chrysotile, the most commonly encountered form, may break down slowly in acidic environments. Asbestos fibers may break into shorter strands and, therefore, increased number of fibers, by mechanical processes, e.g. grinding and pulverization. Inhaled fibers may get trapped in the lungs and with

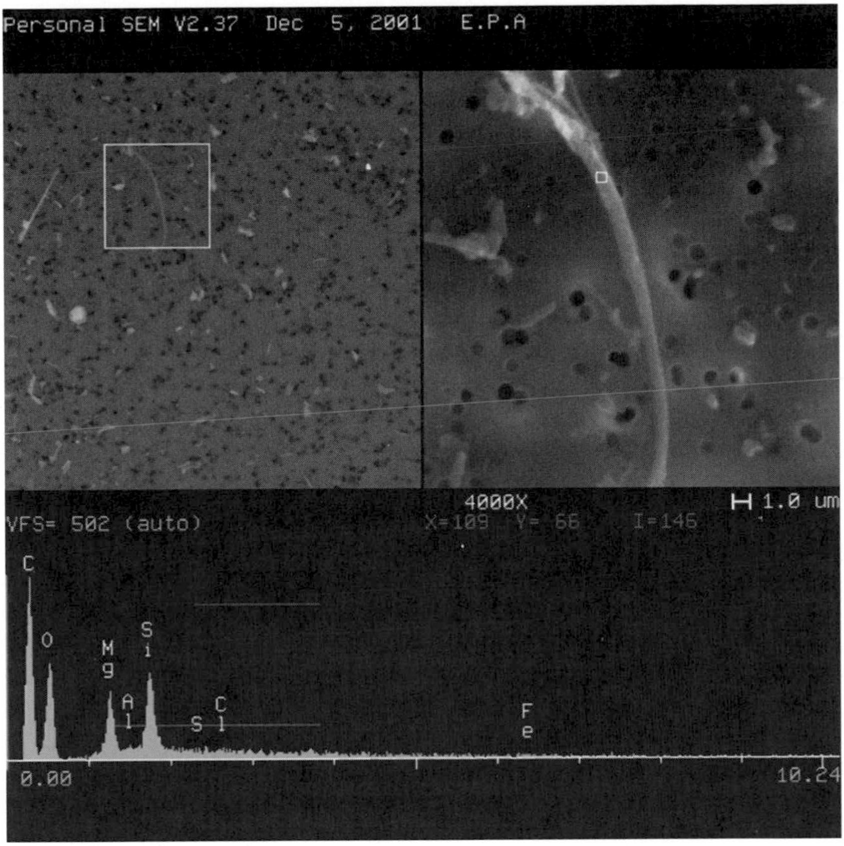

FIGURE A.9.6. Scanning electron micrograph of fibers in dust collected near the World Trade Center, Manhattan, NY in September 2001. Acquired using an Aspex Instruments, Ltd. Scanning electron microscope. The bottom of the micrograph represents the elemental composition of the highlighted 15 μm long fiber by energy dispersive spectroscopy (EDS). This composition (i.e., O, Si, Al, and Mg) and the morphology of the fibers indicate they are probably asbestos. The EDS carbon peak results from the dust being scanned on a polycarbonate filter. (Source: U.S. Environmental Protection Agency, 2004. Photo courtesy of T. Conner, used with permission.)

chronic exposures build up over time. Some fibers, especially chrysotile, can be removed from or degraded in the lung with time.

In 1989, the U.S. EPA established a ban on new uses of asbestos. Other EPA actions include:

- Regulations that require school systems to inspect for asbestos and, if damaged asbestos is found, to eliminate or reduce the exposure,

FIGURE A.9.7. Scanning electron micrograph of asbestos fibers (amphibole) from a former vermiculite-mining site near Libby, Montana. (Source: U.S. Geological Survey and U.S. Environmental Protection Agency, Region 8, Denver, Colorado.)

either by removing the asbestos or by encapsulating it to prevent its migration into the air

- Guidance and support for reducing asbestos exposure in other public buildings
- Regulation of the release of asbestos from factories and during building demolition or renovation to prevent asbestos from getting into the environment
- Rules for the disposal of waste asbestos materials or products, requiring these to be placed only in approved locations.
- Proposal of a concentration limit of 7000 fibers mL^{-1} for long fibers (length $\geq 5\,\mu m$) in drinking water.

In addition, the U.S. Food and Drug Administration (FDA) regulates asbestos in the preparation of drugs and restricts the use of asbestos in food-packaging materials, and the National Institute for Occupational Safety and

Health (NIOSH) recommends that inhalation exposures not exceed 100,000 fibers with lengths greater than or equal to 5 μm per m^3 of air (0.1 fibers mL^{-1}). The Occupational Safety and Health Administration (OSHA) has established an enforceable limit on the average 8-hour daily concentration of asbestos allowed in air in the workplace to be 100,000 fibers with lengths greater than or equal to 5 μm per m^3 of air (0.1 fibers mL^{-1}).

Most asbestos health effects information is derived from epidemiological studies of highly exposed persons, including those exposed to fibers $\geq$5 μm long in occupational settings with air concentrations as high as 5 fibers mL^{-1}. Ongoing inhalation of fibers $\geq$5 μm fibers can lead to the slow buildup of scar tissue in the lungs and in the pleural membranes (i.e., pleural plagues) surrounding the lungs, known as asbestosis. This tissue does not expand and contract like normal lung tissue, diminishing air-exchange efficiencies and labored breathing and coughing. Blood flow to the lung may also be decreased, enlarging the heart. Pleural plaques are quite common in people occupationally exposed to asbestos, but are occasionally seen in persons residing in areas with high environmental levels of asbestos. Effects on breathing from pleural plaques alone are usually not serious.

Asbestos workers have increased chances of getting two principal types of cancer: cancer of the lung tissue itself and mesothelioma, a cancer of the thin membrane that surrounds the lung and other internal organs. These diseases do not develop immediately following exposure to asbestos, but appear only after a number of years. There is also some evidence from studies of workers that breathing asbestos can increase the chances of getting cancer in other locations (for example, the stomach, intestines, esophagus, pancreas, and kidneys), but this is less certain. Members of the public who are exposed to lower levels of asbestos may also have increased chances of getting cancer, but the risks are usually small and are difficult to measure directly. Lung cancer is usually fatal, while mesothelioma is almost always fatal, often within a few months of diagnosis. Some scientists believe that early identification and intervention of mesothelioma may increase survival.

The most important asbestos risk factors for asbestos-related diseases are length of exposure, air concentration of asbestos during the exposure, and smoking. Cigarette smoking and asbestos exposure are synergistic (i.e., the risk of disease is multiplied if a person is both exposed to asbestos and smokes; See Table A.9.1). There is an ongoing scientific debate about the differences in the extent of disease caused by different fiber types and sizes. Some of these differences may be due to the physical and chemical properties of the different fiber types. For example, several studies suggest that amphibole asbestos types (tremolite, amosite, and especially crocidolite) may be more harmful than chrysotile, particularly for mesothelioma. Other data indicate that fiber size dimensions (length and diameter) are important factors for cancer-causing potential. Some data indicate that fibers with lengths greater than 5.0 μm are more likely to cause injury than fibers with

TABLE A.9.1
Estimated lifetime excess lung cancer risks due to continuous exposure to asbestos (cases per million population), calculated with a confidence interval = 0.01. Numbers in parentheses are the estimated ranges with a lower limit = 0 and upper limit calculated from a confidence interval = 0.1. Source: California Air Resources Board, 1986, Staff Report: Initial Statement of Reasons for Rulemaking. The report also found similar synergistic effects for mesothelioma.

	Excess Lung Cancer Risk				
Exposure Group by Asbestos Dose (fibers m^{-3})	*8*	*50*	*80*	*500*	*2000*
Male Smokers	1 (0–9)**	6 (0–55)	9 (0–88)	55 (0–550)	221 (0–2210)
Female Smokers	1 (0–5)	2 (0–25)	5 (0–41)	25 (0–250)	101 (0–1010)
Male Nonsmokers	1 (0–1)	1 (0–8)	1 (0–11)	8 (0–75)	29 (0–290)
Female Nonsmokers	1 (0–1)	1 (0–3)	1 (0–5)	3 (0–28)	11 (0–110)

Source: California Air Resources Board, 1986, Staff Report: Initial Statement of Reasons for Rulemaking.

lengths less than 2.5 μm. Additional data indicate that short fibers can contribute to injury. This appears to be true for mesothelioma, lung cancer, and asbestosis. However, fibers thicker than 3.0 μm are of lesser concern, because they have little chance of penetrating to the lower regions of the lung.

Some groups of people who have been exposed to asbestos fibers in drinking water have higher-than-average death rates from cancer of the esophagus, stomach, and intestines. However, it is very difficult to tell whether this is caused by asbestos or by other causes. High-dose animal testing of asbestos in food have not shown elevated fatal cancers, however, an increase in the number nonfatal tumors did occur in the intestines of rats in one study.

Asbestos exposure can occur in the workplace. Asbestos exposure may occur in a number of occupational settings, including: pipe and steamfitting, plumbing, auto mechanic (brake repair), dry wall finishing, carpentry, roofing, electrical, welding, mining, boilermaking, and shipyards. The best means for determining possible asbestos exposures is a thorough examination by a physician.

Asbestos measurements are sampled at an airflow rater of $\geq$0.5 L min^{-1}. For sufficient precision and accuracy, flow and time are adjusted to give a fiber density ranging from 100 to 1300 fibers per mm^{-1}. Collection efficiency for fibers does not appear to be a function of flow rates between 0.5 and 16 L min^{-1}, and longer fibers (>3 μm) may be lost due to aspiration and deposition onto the inlet. Also, sampling at a flow rate of 1 to 4 L min^{-1} for 8 hours is desirable when asbestos concentrations are about 0.1 fibers cm^{-3}, but in dusty environments short, consecutive ($\leq$400 L) samples should be taken so that the filter does not become overloaded with dust.[6] The filters are then analyzed using phase contrast microscopy (PCM) or transmission electron microscopy (TEM). Generally, PCM counts fibers ≥5 microns long with a minimum aspect (length to width ratio) of 3 : 1, while TEM counts fibers of any length, and asbestos structures too thin to be analyzed by PCM. Thus PCM may underestimate airborne asbestos because shorter fibers are missed; meaning that TEM is a more definitive asbestos approach and SCM is a screening approach that, if positive, may call for more reliable and precise measurements.[7]

Notes and Commentary

1. United Kingdom Department of Environment, Food, and Rural Affairs, Expert Panel on Air Quality Standards, 2004, *Airborne Particles: What Is the Appropriate Measurement on which to Base a Standard? A Discussion Document.*
2. G. Bonne, P. Mueller, L.W. Chen, B.G. Doddridge, W.A. Butler, P.A. Zawadzki, J.C. Chow, R.J. Tropp, and S. Kohl, *Proceeding of the PM2000: Particulate Matter and Health Conference,* "Composition of PM2.5 in the Baltimore-Washington Corridor," Air & Waste Management Association, Washington, DC, pp. W17–18, Jan 2000.
3. Aerosol textbooks provide methods to determine the aerodynamic diameter of particles less than 0.5 micrometer. For larger particles gravitational settling is more important and the aerodynamic diameter is often used.
5. For more information on asbestos exposure, see the Public Health Statement on Asbestos, see ATSDR, 2001, Public Health Statement for Asbestos, http://www.atsdr.cdc.gov/toxprofiles/phs61.html.
6. Centers for Disease Control and Prevention, 1994, Method 7400: Asbestos and Other Fibers by PCM, Issue 2.
7. J. Wellings, 1999, "PCM versus TEM for Analysis of Airborne Asbestos," *The Cohen Group Newsletter*, Vol. 1, Issue 3.

Index